WILEY SERIES IN NUMERICAL METHODS IN ENGINEERING

Consulting Editors
R. H. Gallagher, *College of Engineering, University of Arizona*
and
O. C. Zienkiewicz, *Department of Civil Engineering, University College Swansea*

Rock Mechanics in Engineering Practice
Edited by K. G. Stagg and O. C. Zienkiewicz

Optimum Structural Design: Theory and Applications
Edited by R. H. Gallagher and O. C. Zienkiewicz

Finite Elements in Fluids
Vol. 1 Viscous Flow and Hydrodynamics
Vol. 2 Mathematical Foundations, Aerodynamics and Lubrication
Edited by R. H. Gallagher, J. T. Oden, C. Taylor and O. C. Zienkiewicz

Finite Elements in Geomechanics
Edited by G. Gudehus

Numerical Methods in Offshore Engineering
Edited by O. C. Zienkiewicz, R. W. Lewis and K. G. Stagg

Finite Elements in Fluids—Vol. 3
Edited by R. H. Gallagher, O. C. Zienkiewicz, J. T. Oden, M. Morandi Cecchi and C. Taylor

Energy Methods in Finite Element Analysis
Edited by R. Glowinski, E. Rodin and O. C. Zienkiewicz

Finite Elements in Electrical and Magnetic Field Problems
Edited by M. V. K. Chari and P. Silvester

Numerical Methods in Heat Transfer
Edited by R. W. Lewis, K. Morgan and O. C. Zienkiewicz

Finite Elements in Biomechanics
Edited by R. H. Gallagher, B. R. Simon, P. C. Johnson and J. F. Gross

Soil Mechanics—Transient and Cyclic Loads
Edited by G. N. Pande and O. C. Zienkiewicz

Finite Elements in Fluids—Vol. 4
Edited by R. H. Gallagher, D. Norrie, J. T. Oden and O. C. Zienkiewicz

Foundations of Structural Optimization: A Unified Approach
Edited by A. J. Morris

Foundations of Structural Optimization: A Unified Approach

Edited by
A. J. Morris
Reader in Aircraft Structures, Cranfield Institute of Technology

A Wiley–Interscience Publication

JOHN WILEY & SONS
Chichester · New York · Brisbane · Toronto · Singapore

M. C. Petiau	*Avions Marcel Dassault–Breguet, 78 Quai Carnot, 92—Saint Cloud, BP 300 – 92214 Saint Cloud Cedex, France*
J. Sobieski	*National Aeronautics & Space Administration, Langley Research Center, Hampton, Virginia 23665, USA*
K. Svanberg	*Contract Research Group for Applied Mathematics, Royal Institute of Technology, S-100 44 Stockholm, Sweden*
R. Thom	*Advanced Mobility Systems, Bell Aerospace Company, P.O. Box 4, Buffalo, New York 14240, USA*

Preface

During the course of the last decade structural optimization has passed through a period of rapid development during which it has moved from an assorted collection of interesting techniques to a mature discipline routinely used in engineering design. A variety of approaches have emerged which are used in a wide range of industries and have found a permanent place in several of the computer-aided design systems which are becoming commercially available.

This book gives a comprehensive treatment of the subject by a group of leading experts in the field and consists of lectures presented at a NATO Advanced Study Institute on Modern Structural Optimization held at the University of Liège, Belgium, August 1980. Although the book has multiple authorship the first 12 chapters form a complete text which provides a unified description of the subject. In these chapters all the main techniques for the solution of minimum weight structural design problems are given in detail together with the necessary mathematical background and an outline of basic structural analysis. These chapters are, therefore, suitable for an advanced undergraduate or postgraduate course designed to present a modern and comprehensive approach to structural optimization. To assist the teaching process certain of the earlier chapters are provided with problem sets which allow the student an opportunity to consolidate the knowledge gained during the reading of the chapter itself. Material from the book has been used in a postgraduate course at Imperial College, London, and in short courses for industrial design engineers.

Care has also been taken to provide a practical text suitable for day-to-day use in the design office or in consultancy. Algorithms are described in sufficient detail to allow the reader to quickly program and implement solution procedures and guidance is given on how techniques may be 'tuned' and performance improved. Examples are provided of the solution of actual design problems which exploit the methods described in the more theoretical sections. Although these examples are taken primarily from the aircraft industry this simply reflects the industrial background of the authors and does not imply any restriction on the range of application of these procedures. Indeed, examples can be found from applications across the entire engineering spectrum in which structural optimization methods are used either directly to seek solutions to optimal design problems or indirectly as the basic module in a computer-aided design system.

The final part of the book consists of special applications of the structural optimization theory and techniques described in the first part, together with an outline of the use of geometric programming. Much of the work presented in these chapters represents current research and provides aspects of the subject which could be used by new workers wishing to start their own lines of research.

Finally, I would like to thank the NATO Scientific Affairs Division under Dr Mario di Lullo for providing both finance and guidance for the Advanced Study Institute which brought the contributors to this volume together. I am also indebted to Dr Claude Fleury and Professor Michele Geradin of the Université de Liège and my wife Lilian for considerable administrative help in organizing the Advanced Study Institute at the Sart Tilman Campus of Liège University. It is a pleasure to acknowledge the considerable assistance of Mr 'Roly' Piggot and Dr Peter Bartholomew in preparing Chapters 2 and 3 respectively.

September, 1981   A. J. MORRIS

Contents

Preface vii
1 Introduction 1
   1.1 Rationale 1
   1.2 Definition of terms 2
   1.3 Short history of structural optimization 4

2 Basic Mathematical Concepts 11
   2.1 Introduction 11
   2.2 Sets 11
   2.3 n-dimensional space 13
   2.4 Functions in n-dimensional space 16
   2.5 Open and closed sets 18
   2.6 Sequences 19
   2.7 Convexity 20
   2.8 Conditions for minima 23
   2.9 Conditions for constrained minima 26
   2.10 Duality 31
   2.11 Algorithms 36
   2.12 Efficiency of algorithms 39

3 Elementary Concepts in Structural Optimization 43
   3.1 Introduction 43
   3.2 Statically determinate structures 43
        Analysis 44
        Optimum design of strength-critical frameworks 46
        Optimum design of stiffness-critical frameworks 49
        Convexity 49
        The single displacement-constrained problem 52
        Multiple displacement-constrained problem 57

3.3	Statically indeterminate structures.	64
	Analysis	64
	Optimum design of strength-critical frameworks	66
	Displacement-constrained statically indeterminate structures.	71
3.4	General structures.	72
	Finite element analysis and derivatives	73
	Stress- and displacement-constrained finite element structural models	75
	Minimum-weight design subject to buckling constraints	78
Problems		80

4 Finite Elements and Optimization — 83

4.1	Introduction	83
4.2	Calculation of sensitivities for static loads	85
4.3	Example—cantilever beam subject to a displacement constraint.	86
4.4	Sensitivity calculation of vibration frequencies and modes	89
4.5	Example—a beam subject to a frequency constraint	90
4.6	Derivatives of transient response	92

5 Optimality Criterion Methods in Structural Optimization — 99

5.1	Introduction	99
5.2	Basic equations of analysis.	101
	Displacement method	101
	Scaling of the design.	103
5.3	Displacement constraints.	104
	Single displacement constraint	105
	Multiple displacement constraints.	109
5.4	Stress constraints	118
	Fully stressed design method	119
	Stress constraint with equivalent displacement constraint.	120
5.5	Algorithm with the reciprocal design variable	124
	Optimality criterion	124
	Recurrence relations.	125
	Evaluation of the Lagrange multiplier.	126
5.6	System stability constraint	126
5.7	Effect of structural symmetry on the algorithm.	128
5.8	Optimization procedure	129
5.9	Illustrative problems.	133

	5.10 Summary and conclusions	160
	Appendix Detailed iteration histories.	164

6 Optimality Criteria using a Force Method Analysis Approach 237
6.1 Introduction 237
6.2 Technical discussion 239
    Optimality criteria 239
    Fundamentals of the force method 240
    Role of the force method in structural optimization 244
    Equation solution and constraint detection 246
    Linear programming 247
    Constraint discard and the generalized
    Newton–Raphson technique 258
    Displacement constraints 263
6.3 Conclusions 269

7 Introduction to Mathematical Programming Methods 273
7.1 Introduction 273
    Classification of mathematical programming problems 274
7.2 Primal and dual statement 276
7.3 Descent methods for minimization 279
7.4 The method of steepest descent 282
7.5 Line-search techniques 286
    One-point pattern: Newton–Raphson iteration 286
    Two-point pattern 287
    Three-point pattern: quadratic interpolation 293
Problems 296

8 Unconstrained and Linearly Constrained Minimization 299
8.1 Introduction 299
8.2 Newton-type methods for unconstrained optimization 300
    The quadratic approximation 300
    The standard Newton algorithm 301
    Order of convergence of the Newton method 301
    Safeguarding the Newton method 302
    Remedies against failure 303
8.3 Conjugate direction methods 305
    Difinition of conjugate directions 305
    Expansion of an arbitrary vector 306
    Minimization of a quadratic function 306

The general conjugate direction algorithm 307
The expanding-subspace theorem 308
The conjugate gradient algorithm 308
Application to non-quadratic objective functions 310
8.4 Quasi-Newton methods 311
Principle of quasi-Newton methods 311
The quasi-Newton equation 311
Quasi-Newton algorithms 313
Rank-one updates 313
Application to the quadratic case 313
Rank-two update: the Davidson–Fletcher–Powell
method 315
8.5 A global view of unconstrained minimization 318
8.6 The gradient projection method 319
8.7 First- and second-order projections methods 325
8.8 The active-set strategy 328
8.9 Special treatment of side constraints 330
Problems 332

9 General Nonlinear Programming Methods 335
9.1 Introduction 335
9.2 Barrier function methods 336
9.3 Penalty function methods 341
9.4 The multiplier methods 344
9.5 Primal methods 350
Gradient projection for nonlinear constraints 350
Methods of feasible directions 352
9.6 Linearization methods 354
Recursive linear programming 354
Kelley's cutting plane method 355
Method of approximation programming 356
Recursive quadratic programming methods 357
Problems 358

10 Reconcilation of Mathematical Programming and
Optimality Criteria Methods 363
10.1 Introduction 363
The structural optimization problem 364
10.2 The generalized optimality criterion (dual solution
scheme) 365
10.3 The mixed method (primal solution scheme) 370
10.4 Relations between OC and MP approaches 376
The generalized OC as a linearization method 376

	The approximation concepts approach	379
	the constraint gradients	380
	Stress-ratioing	380
10.5	Optimization algorithms	383
	Second-order primal algorithm—PRIMAL-2	384
	Second-order dual algorithm—DUAL-2	385
10.6	Numerical examples using primal and dual algorithms	386
	72-bar truss	386
	63-bar truss	388
	I-beam	390
	Delta wing	395
	Aircraft spoiler	398
10.7	Concluding remarks	402
Problems		402

11 From a 'Black Box' to a Programming System 405

11.1	Introduction	405
11.2	Basic implementation schemes	407
	Special-purpose 'black box'	407
	Optimization algorithm as a subroutine	409
	Programming system	413
11.3	Components of the programming system and their relationship	414
	Analyser	415
	Optimizer	418
	Interface processors	418
	Terminator	419
	Connecting network	420
11.4	Organization of the execution flow	421
	Matrix of options	421
	Procedure organization for each option	422
	Generation of initial cross-sectional dimensions by a fully stressed design	424
	Cost of the data manipulations	424
	Adapting a programming system to a particular application	425
11.5	Application examples for programming systems	426
	Example of wing flutter resizing	427
	Miscellaneous examples	428
	Summary of applications	437
11.6	Concluding remarks	437
Appendix A	Selected techniques for the analyser approximate analysis	438

Appendix B   Selected techniques for a post-processor. 441
Appendix C   Two-level optimization method. 445

12 Optimization of Aircraft Structures 451
12.1   Analysis methods. 451
12.2   Structural optimization method 457
      Design variables 458
      Optimization constraints 458
      Optimization algorithm 458
      Final touches 458
12.3   Computation of constraints and constraint derivatives 460
      Derivatives in static optimization 460
      Derivatives in dynamic optimization 461
12.4   Explicit optimization and sub-iteration 462
      Explicit optimization 462
      Sub-iteration process 463
12.5   Examples and comments 463
      Optimization of a metallic delta wing 463
      Optimization of a carbon epoxy empennage under stress
      and flutter speed constraints. 467
      Optimization of a carbon epoxy wing with a vertical fin
      on a fighter aircraft. 469
12.6   Development. 478
      Maximization of margins 478
      'Sparsity' of the partial derivatives matrix 479
12.7   Conclusions 481
Appendix 1   Static aeroelastic coefficients analysis and
              derivatives 482
Appendix 2   Derivation of flutter speed 483

SPECIAL APPLICATIONS AND TECHNIQUES   485

13 Structural Optimization with Material Selection 487
13.1   Introduction 487
13.2   Problem formulation 488
13.3   Approximate problem 491
13.4   The dual method 493
13.5   Numerical examples 496
13.6   Conclusions 501

14 Optimal Geometry in Truss Design 513
14.1   Introduction 513
14.2   Formulation of the optimization problem 515

14.3	Gradients of the constraint functions.	516
	Displacement constraints	517
	Stress constraints.	519
	Buckling constraints	520
	Consideration of the gravity loads	520
14.4	Independent variables.	521
14.5	Method for solving the optimization problem.	523
	General approach, flow chart	523
	How to choose $w^{(k)}$ and $g_j^{(k)}$.	524
	How to obtain perfectly feasible solutions	527
14.6	Using duality to solve the subproblems.	529
14.7	The influence of the loading magnitude on the optimal solution	531
	Displacement constraints	532
	Stress constraints.	532
	Buckling constraints	533
	Mixed constraints	534
14.8	Test problems	536
	3-bar pyramid	536
	39-bar tower.	538
Appendix	Proof of some lemmas.	541

15 Shape Optimal Design of Elastic Structural Elements 545

15.1	The shape optimization problem.	545
15.2	Shape optimal design.	545
	Numerical methods for shape optimization.	545
	Shape of cross-section of shafts in torsion	547
	Shapes of holes in planar solids	547
	Miscellaneous shape optimal design problems.	548
	Related literature on domain optimization	548
15.3	Analysis of the effect of shape variation	548
15.4	Maximum torsional stiffness of a shaft.	551
15.5	Iterative shape optimal design of a shaft	553

16 Optimization of Structures in which Repeated Eigenvalues Occur. 559

16.1	Introduction.	559
16.2	Examples of structural optimization problems with multiple eigenvalues	559
	A simple spring–mass optimal problem with repeated natural frequencies.	560
	A column problem with repeated eigenvalues.	561

- 16.3 Formal Langrange multiplier analysis of the two degree of freedom spring–mass example 565
- 16.4 Directional derivatives of eigenvalues 568
- 16.5 Rigorous optimality criteria 571

17 Structural Optimization by Geometric Programming 573
- 17.1 Introduction to geometric programming 573
- 17.2 Lagrangian interpretation 579
- 17.3 Reduction to independent variables 585
- 17.4 The generation of minimum-weight designs for statically determinate pin-jointed frameworks 587
- 17.5 Geometric programming and general problems 592
  - Application to a design problem 595
- 17.6 Condensed geometric programming 600
  - Some properties of condensed structural optimization problems 604
  - Conclusions 609

Index 611

Foundations of Structural Optimization: A Unified Approach
Edited by A. J. Morris
© 1982 John Wiley & Sons Ltd.

Chapter 1
Introduction

1.1 RATIONALE

Design in any engineering discipline is a complex process by means of which a product is generated to satisfy a perceived market requirement. In the commercial environment there will normally be more than one producer and emphasis has to be placed on matching the requirement in the most efficient and effective manner. This is true even in the most sheltered industries, such as the production of defence equipment, where the effects of competition may not always be apparent. Protection from such forces cannot be complete and a given industry cannot shut itself off indefinitely from outside influence. Past experience indicates that once the presence of market forces is observed the effect can be dramatic and many large and seemingly healthy organizations have, almost overnight, found themselves swept away. Efficiency is, therefore, essential in the design process leading to the creation of any specific product even when a government is present to act as a protective buffer.

Although efficiency in design implies the generation of products which are cost-effective, it also means that the designs must be in advance of the competition. Advances are achieved through a mixture of innovation and evolution, with the blend between these two aspects depending on the time available. For example, during the Second World War the US and other aircraft industries produced a plethora of new designs each requiring a short design cycle time. Thus, rapid progress was made in aircraft technology through a sequence of designs each of which relied on its predecessors and emphasis was placed on the evolution of designs rather than innovative leaps. The modern situation for most major industries, including the aircraft industry, is the complete inverse of the wartime experience. Most designs now require a long gestation period with much more emphasis being placed on innovation to compensate for long design cycle times. In order to introduce a large measure of innovation with the required degree of confidence and safety, a heavy reliance must be placed on the rôle of theory in design. Since our aim is to produce efficient designs an important contribution in this area is made by optimization theory and the computer-based methods which result from the practical application of this theory.

The design process consists in both selecting a suitable concept and, in the final stage, completing a detailed design study which fixes the size, materials, layout, etc., of the actual product. The selection of an appropriate concept necessitates the examination of a range of models at the project stage with judgements made concerning the likely contenders. At the detailed design stage we are attempting to hone down the selected model to its most efficient and cost-effective form. Although the optimization methods described here are primarily concerned with this 'honing' process, they nevertheless have an important contribution to make at the project stage. In the case of a specific concept, optimization provides a logical method for generating a 'best' design. Where a variety of concepts are in competition it provides an attractive means for aiding technical judgement by generating the 'best' design within each model. Thus, comparisons can be made on the basis of the same datum point, i.e. we can compare a range of optimum designs.

The motivation for the present work is to provide the structural design engineer with the tools of optimization theory and practice in order to exploit the benefits outlined above. A unified approach is presented which provides a complete understanding of modern, fast, computer algorithms and indicates the interrelation between the various techniques. The notation and basic theory are described permitting not only an understanding of the advanced theories of later chapters but also equipping the reader with the knowledge to read the most modern contributions in the research journals. Some of this work has been developed by looking at the problem from a structural engineering point of view; other aspects derive from the allied discipline of mathematical programming. This latter is an attempt to place the most general type of optimization solution methods on a firm mathematical basis with particular attention given to solution speed and guaranteeing convergence and stability.

The impact of structural optimization is growing due to the economic pressures experienced by the design community. Nevertheless, it is possible to survive without the application of these techniques, though this is becoming increasingly difficult in advanced industries such as aeronautical engineering. However, once the new, class 6, supercomputers become routinely available then optimization will be an essential part of the design process.

1.2 DEFINITION OF TERMS

Turning to structural design the theory we referred to in the previous section has at least two aspects. One of these, analysis, allows for an accurate description of the behaviour of the structure under the action of the applied loads and for this process the finite element technique is used in most industries. This is also the method assumed in the remaining chapters where computer-based analysis methods are referred to or employed. The second aspect is, of course, the creation of designs with high efficiency which are in some ways better than possible alternatives. In order to achieve an optimum design it is necessary to have some function for which a minimum value is sought. This is known as the *objective function* or *cost function*. It is tempting to equate the objective function with some form of financial outlay, perhaps cost of manufacture or direct operating cost, etc. However, this often turns out to be inappropriate due to both the difficulty of giving a précise definition to this type of cost and the dependence of these functions on highly variable factors such as wage or interest rates, new manufacturing processes, etc. The more usual practice in engineering design is to seek a more direct function which can stand in place of cost, reflecting the influence of cost variations, yet possessing an intrinsic merit of its own. In many industries and particularly in the aircraft industry the mass or weight of the structure is taken as the objective function on the basis of this argument. Thus, much of the theory developed in later chapters assumes weight is the objective function, though many of the general techniques associated with mathematical programming methods can treat alternative objective functions.

Once an objective function has been selected it becomes clear that an additional choice must be made concerning the quantities making up this function and which are varied to achieve the desired optimum. These are called *design variables* and their number and type will depend upon the problem. For example, in the case of minimum-weight design of a pin-jointed framework consisting simply of bars, the variables are normally the bar cross-sectional areas; whilst in the case of a rigid-jointed framework these might change to second moments of area. Alternatively, a design variable may link several structural parameters, thus, a number of bar cross-sectional areas in a framework may be represented by a single design variable. Clearly this type of linking implies that there is a defined relationship between parameters linked to a particular design variable. Taking a simple example, three of the cross-sectional areas $\{x_1, x_2, x_3\}$ of a given framework may be linked to the single design variable y through the relationship $\alpha_1 x_1 = \alpha_2 x_2 = \alpha_3 x_3 = y$ where $\alpha_1, \alpha_2, \alpha_3$ are predetermined constants. This type of variable linking is used both to control the size of the optimization problem and to remove certain instabilities which can arise in the solution process. In all the cases treated in this book the number of design variables are finite and often linked to individual finite elements. Thus, we are dealing with the solution of problems in a finite-dimensional space.

The selection of the objective function and design variables partially describes the optimization problem which is completed by defining the constraints which limit the range of the design variables. These require that, for example, the stresses within the structural members or displacements at prescribed positions on the structure should be limited by prescribed values. In general these are defined by a finite number of equations in the form of equalities or inequalities,

$$g_j(x) \geq 0 \quad j = 1, \ldots, m$$
$$h_k(x) = 0 \quad k = 1, \ldots, p$$

i.e. there are m inequality and p equality constraints. Although equality constraints can arise they are not usually found in structural design problems. However, additional inequality constraints which directly limit the size of the design variables rather than functions of the design variables are common. These are known as gauge constraints and occur because there are normally practical engineering reasons for restricting the size of structural members. Gauge constraints often refer to lower bounds on variables of the form $x_i \geq \bar{x}_i$ ($i = 1, 2, \ldots, n$) though upper bound also occur in some practical problem $\bar{x}_i \geq x_i (i = 1, 2, \ldots, n)$.

The design variables describe a finite-dimensional space whose boundaries are defined by a set of surfaces where the inequality and equality constraints are satisfied. The interior of this region is known as the *feasible region* and represents design variables which satisfy or do not violate the constraints. For most design problems the optimizing point is located on the boundary of the feasible region and in such circumstances the optimum can be visualized as resting against a subset of the total number of constraints. These are known as the *active constraints*. A schematic diagram outlining some of these definitions is shown in Fig. 1.1 where $g_1(x)$ is the only active constraint at the optimum whilst the remaining three constraints are passive.

1.3 SHORT HISTORY OF STRUCTURAL OPTIMIZATION

Structural analysis and structural optimization are, in many ways, two aspects of the same subject. As a result they have been developed together, though at somewhat different rates, with the theory of analysis usually receiving the most attention. Thus, many of the major contributors to structural analysis theory and practice have also made important discoveries in structural optimization.

In reviewing the history of our subject it is, therefore, convenient to start with the work of Galileo [1] who represents one of the earliest workers to attempt to describe the state of stress in beams under the action of a bending load. Although his assumption regarding the stress distribution in this case was incorrect he, nevertheless, attacked the problem of finding the optimum shape of a variable-depth beam. This initial research was extended in 1687 by Johan Bernoulli [2] who took an interest in the subject of beam design. Bernoulli made the assumption that plane sections remain plane and also invoked Hooke's law. This combination leads to a linearly varying stress distribution through the beam section which gives rise to an axial stress resultant similar to the Galileo theory. In the course of this work he was eager to apply the 'new' differential calculus and on the basis of his theory examined the problem of designing a beam of uniform

Introduction

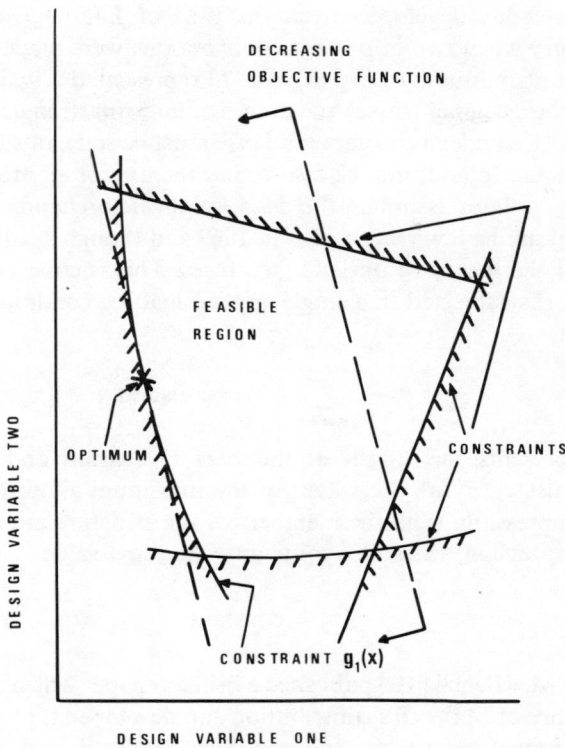

Fig. 1.1 Generalized optimization problem

strength. This same year saw Newton solving the problem of the catenary which is an unattractive structural form.

From the point of view of beam design the major breakthrough came with the publications of Parent [3,4] in 1708 and 1710. He realized that the basic assumptions on the stress distribution must give rise to no axial stress resultant and introduced the concept of the neutral axis. Using his new theory he advanced optimum structural design by solving the problem of the beam of uniform strength under the action of moving loads. Lagrange [5,6], who was ignorant of Parent's earlier work, rediscovered the neutral axis concept and went on in 1770–73 to examine the design of columns under axial loads and also looked at the shape of the elastic curve. This study examined the minimum-weight design of a column with an axisymmetric shape. His solution indicated that the optimum design took a circular cylindrical form of constant section which is incorrect when self-weight is taken into account. The problem was, however, correctly solved by Clausen [7] in 1851.

The subject steadily developed from the time of Lagrange until the late nineteenth century when two important contributions were made by Lévy and Maxwell. The contribution by Lévy [8] in 1873 represented a major study into the problem of the design of trusses and arches of uniform strength. He was able to show that a lattice under a constant load case must be statically determinate in order to be optimal. In addition, he also found the axis of an arch of uniform strength directly without assuming the dead weight beforehand. The paper by Maxwell [9] appeared a few years earlier in 1869 and though small, represents a turning point in the theory of optimal structures. This is concerned with pin-jointed frameworks subjected to a single applied loading condition and proves the theorem that

$$\sum_{i=1}^{m} l_i^t \sigma_i^t - \sum_{j=m+1}^{n} l_j^c \sigma_j^c = \text{constant}$$

where l_i^t, l_j^c represents the length of the bars in tension and compression respectively, whilst σ_i^t, σ_j^c are the values of the maximum allowable stresses in tension and compression. If all the members of the structure are in tension (or similarly in compression) then all layouts give the same value

$$\sum_{i=1}^{n+m} l_i^t \sigma_i^t = \text{constant}$$

In 1904, A. G. M. Michell [10] published a brilliant paper which demonstrated the full significance of Maxwell's contribution and developed it to provide a tool for designing optimal layouts for the minimum-weight design of pin-jointed frameworks subject to stress constraints. The theory only admits of problems where a single load set is applied and thus, from Lévy's theorem, the resulting structure is statically determinate and potentially unstable if additional loads are applied. Nevertheless, Michell has provided a technique for generating an optimal orthogonal layout given the positions, in a two-dimensional space, of the points of application of the applied load and the positions of the supports. The direct application of the technique is difficult for all but the simplest cases and a full exploitation of the basic principles had to await the advent of the computer. Unfortunately, the solutions obtained are usually too impractical but they do provide a basis for comparison against which practical designs may be measured. In a sense the theory plays a similar rôle in structural optimization as that provided by the Carnot cycle in the design of realistic engine thermodynamic cycles.

Although minimum-weight design was important for many structures the real drive to generate this type of optimum came as a result of competitive aircraft design. Initially aircraft structural design was subordinate to the aerodynamic and stability requirements but the introduction of aluminium alloy monocoque construction and the pressures of the Second World War radically changed the situation. The primary constraints on this type of structure were generated by the need to avoid buckling in the skins and spar/webs of the emerging designs. Although Wagner [11] examined the design of simple elements subject to buckling, the first attack on the principles of minimum weight designs of compression structures was made by Smith and Cox [12] in 1943. Attention was then directed at stiffened panels and in 1944 Zahorski [13] considered the problem of the minimum-weight design of stringer panels. During this same period Leggett and Hopkins [14] and subsequently Wittrick [15] looked at the design of sandwich panels subjected to buckling constraints. Following on from this early work a body of knowledge has grown up on the minimum-weight design of structures subject to buckling constraints. In essence, the approach adopted is that used in earlier work which concentrated on creating structures of uniform strength where a specified number of constraints are regarded as active. The solution thus necessitates that there are as many active constraints as design variables which implies that several buckling modes occur simultaneously. For example, in the case of the minimum-weight design of a circular cylindrical thin shell under end load with the wall thickness x_1 and the mean diameter x_2 as the design variables, two constraints are required to obtain a solution. These are taken as the overall 'column' buckling mode and the local buckling mode, both in the elastic range. The optimum, therefore, is found by equating the buckling loads for these two modes as illustrated in Fig. 1.2. The concept of seeking optimum designs by coalescing buckling modes has been extensively used within the aircraft industry, but recent studies have shown that this can be a dangerous procedure.

With the advent of the digital computer the philosophy associated with seeking designs of uniform strength or optimum derived by coalescing of buckling modes was further developed. For the most part early work concentrated on generating optimization algorithms for structural designs which are strength critical. The constraints relate to stress levels in the various members; thus for plane trusses these are the direct stresses in the bars whilst for more complex structures one of the standard yield criteria of Tresca or Von Mise might be employed. Because it is assumed that each design variable at the solution has its value limited by one of the stress constraints, the resulting structure is called a fully stressed design. In the case of statically determinate structures subject to a single load a fully stressed design is also an optimum design. For more general structures with multiple loading cases there is no guarantee that such designs are in fact optimal.

Computationally the generation of a simple algorithm is straightforward and the earliest technique which is still the most popular, is the stress-ratioing method. In this method the fully stressed design is sought by multiplying the design variables by the ratio of the current value of the constraint to the constraint limit.

For example, if we take a pin-jointed framework where the design variables are the bar cross-sectional areas $x_i(i = 1, \ldots, n)$ subject to the constraints that direct stresses in the bars $\sigma_i(i = 1, \ldots, n)$ do not exceed prescribed values

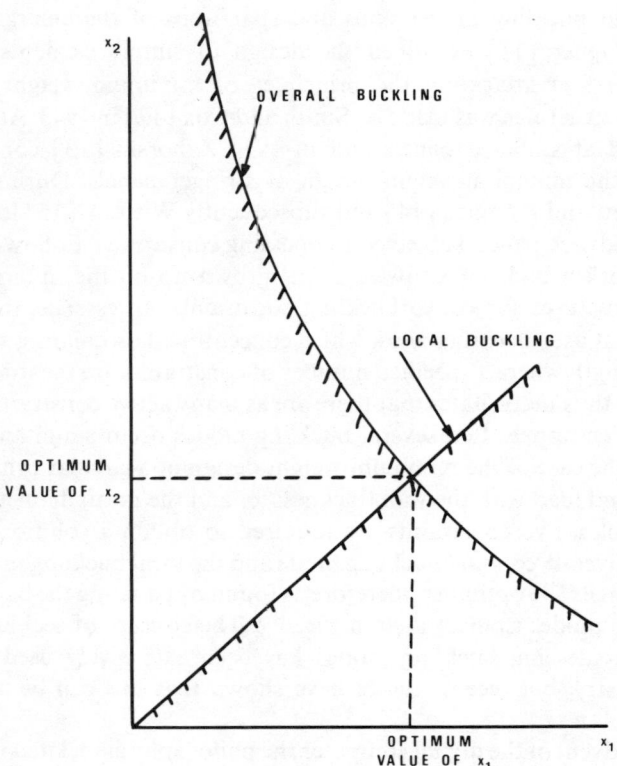

Fig. 1.2 Optimum design by coalescing of buckling modes

$\bar{\sigma}_i (i = 1, \ldots, n)$, then the ratioing formula is given by

$$x_i^* = x_i \left(\frac{\sigma_i}{\bar{\sigma}_i} \right)$$

In more complex structures, modelled by finite elements, the design variables may be more general than simple cross-sectional areas with constraint functions defined by a plastic yield criterion. Nevertheless, the basic ratioing formula remains the same and the simple nature of the computer algorithm is preserved. Because of this the algorithm is an essential part of any general structural optimization system which allows the user access to more than one optimization method. It is an effective technique for pruning down the number of active stress constraints and provides a simple method for making rapid, yet stable, progress in the initial stages of an optimizing sequence.

A. J. M.

REFERENCES

1. G. L. Galileo, *Discorsi e dimonstrazioni matematiche intomo a due nuove scienze attenenti alla mecanica et i movimenti locali*, Leida, 1638.
2. J. Bernoulli, Letter to Leibnitz on beams of uniform strength, 1687.
3. A. Parent, Des resistances des poutres, et des poutres de plus grande resistance, independamment de tout systeme physique, *Mem. Acad. Roy. Sci. Paris*, 1708.
4. A. Parent, Des points de la rupture des figures. D'en deduire celles qui sont partout d'une resistance egale, *Mem. Acad. Roy. Sci. Paris*, 1710.
5. J. L. Lagrange, Sur la figure des collones. *Miscellanes Taurinensia*, Vol. 5 1770–3; also *Oeuvres de Lagrange*, Vol. 2, pp. 125–70 Gauthier-Villars, Paris, 1868.
6. J. L. Lagrange, Sur la force des ressorts plies, *Mem. Acad. Berlin*, 1972.
7. T. Clausen, Column of minimum weight, *Bull. Phys-Math. Acad. St Petersburg*, Vol. 9, 1851.
8. M. Lévy, La statique graphie et ses applications aux constructions, *Comptes Rendu*, 1875.
9. C. Maxwell, *Scientific Papers*, 11, p. 175, Cambridge University Press, 1890.
10. A. G. M. Michell, The limit of economy of material in frames structures. *Phil. Mag*, **8**, 589–97, 1904.
11. H. Wagner, Remarks on airplane struts and girders under compressive and bending stresses, index values, *N.A.C.A. Tech. Mem. No. 500*, 1929.
12. H. L. Cox and H. E. Smith, Structures of minimum weight, Aeronautical Research Council, *Reports & Memoranda No. 1923*, Nov. 1945.
13. A. Zahorski, Effects of material distribution on strength of panels, *Journ. of Aero. Sci.*, **11**(3), 247–53, 1944.
14. D. M. A. Leggett and H. G. Hopkins, Sandwich panels and cylinders under compressive end loads, Aeronautical Research Council, *Reports and Memoranda No. 2262*, Aug. 1949.
15. W. H. Wittrick, A theoretical analysis of the efficiency of sandwich construction under compressive end load, Aeronautical Research Council, *Reports and Memoranda No. 2016*, April 1945.

Edited by A. J. Morris
Foundations of Structural Optimization: A Unified Approach
© 1982 John Wiley & Sons Ltd.

Chapter 2
Basic Mathematical Concepts

2.1 INTRODUCTION

In Chapter 3, we shall discuss some of the more important classical methods of structural optimization. For all their limitations, these methods are very effective in special circumstances, but the engineer is often faced with the need to find optimum designs for problems to which they cannot be applied. In this type of situation, general mathematical methods of optimization can be used to characterize the optimum solution. Many problems are so complex that their solutions cannot be located by closed form mathematical methods, and systematic search techniques have been developed for use in such cases. The study of these mathematical methods and search techniques is the concern of a branch of numerical analysis known as mathematical programming.

Mathematical analysis requires a framework within which ideas can be concisely and rigorously expressed. The purpose of this chapter is to outline some of the mathematical concepts which form the framework for mathematical programming methods. Because the emphasis of this book is on applications of these ideas, rather than on the underlying theory, most results are justified by appeal to illustrative examples, and proofs are given only where they serve as an aid to understanding. At appropriate points, reference is made to texts which give a more complete mathematical treatment.

The reader is assumed to be familiar with the elementary properties of real numbers and matrices, and with the ideas of continuity and differentiability of functions of several variables. It is hoped that this chapter will lay the necessary foundation for the study, not only of the later chapters of this book, but also of the relevant parts of the extensive literature on mathematical programming.

2.2 SETS

Since its development at the end of the nineteenth century, set theory has become one of the cornerstones of modern mathematics and has exerted a great influence on mathematical thought as a whole. Any study of structural optimization which considers mathematical programming methods must employ some aspects of this theory. In this section we introduce the basic notation of set theory and some of the elementary concepts required for later use. More detailed introductions to the subject can be found in texts concerned with mathematical analysis[1, 2, 3].

For our purposes, a *set* may be defined by saying simply that it is a collection of objects; thus the collection of books in a library would constitute a set. The individual objects which together make up a set are called the *elements* or *members* of the set. If A denotes a set (e.g. the books in a certain library) and b denotes one of its elements (a particular book), then we write

$$b \in A$$

which may be read as 'b belongs to A'; we also say 'A contains b'. The contrary (i.e. b is not a book or is a book not in the library) is denoted by

$$b \notin A.$$

A particular set is defined by specifying its elements; this may be done by listing them or by giving a rule for identifying them. Thus we write

$$A = \{a, b, c, \dots\}$$

which is read 'A is the set whose elements are $a, b, c, \dots$' (e.g. the library catalogue) or

$$B = \{x : x \text{ has properties } P, Q, R, \dots\}$$

which is read 'B is the set of all objects x possessing properties $P, Q, R, \dots$' (e.g. the properties of being a book and of being in our library). If in the definition of the set B there is no object which possesses the stated properties, then B is the set with no elements, called the *empty* set or *null* set and denoted by the symbol ϕ. For example,

$$\phi = \{x : x \text{ is an integer, } 1 < x < 2\}$$

If two sets A and B are related in such a way that every element of B is also an element of A, then we say that B is a *subset* of A and write

$$B \subset A \text{ or } A \supset B$$

which may be read 'B is included in A' and 'A includes B' respectively (e.g. B is the set of fiction books in our library). It is often helpful to represent sets pictorially as in Fig. 2.1. Here points inside the left-hand circle represent elements of set A and those inside the right-hand circle represent elements of set B; this representation is called a Venn diagram. The reader will observe that in this case it is neither true that A includes B nor that B includes A; however, the set represented by the shaded area in the diagram is included both in A and in B. This set, comprising all the elements common to both A and B is called their *intersection* and written

$$A \cap B = \{x : x \in A \text{ and } x \in B\}$$

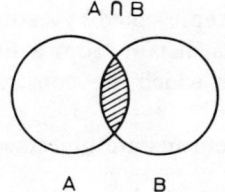

Fig. 2.1 The intersection of two sets A and B

We say A and B are *disjoint* if they have no element in common, i.e. if

$$A \cap B = \phi$$

Another set which can be derived from two given sets A and B is their *union*, consisting of all the elements belonging to A or B (or both); it is written

$$A \cup B = \{x : x \in A \text{ or } x \in B\}$$

Two examples of the union of two sets are represented by the shaded areas in Fig. 2.2; in the right-hand example the sets A and B are disjoint.

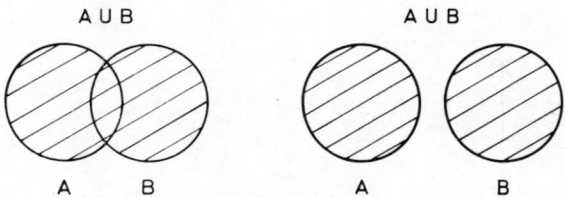

Fig. 2.2 The union of two sets A and B

The *direct product* (or *cartesian product*) of two sets A and B is the set of all ordered pairs which can be made up by taking first an element of A and then an element of B. It is written

$$A \times B = \{(x, y) : x \in A, y \in B\}$$

For example, if A is a set of routes from London to Paris and B a set of routes from Paris to Rome, then $A \times B$ consists of routes from London to Rome via Paris. The direct product of n sets A_i ($i = 1, 2, \ldots, n$) is

$$A_1 \times A_2 \times \ldots \times A_n = \{(x_1, x_2, \ldots, x_n) : x_i \in A_i \text{ for all } i\}$$

an example of this type of product set is given in the next section.

2.3 n-dimensional space

In this section, we introduce special sets with which we shall be mainly concerned in this book. We begin with the set R of all real numbers, often called the *real line* because it can conveniently be represented by a straight line, the number r being represented by a point r units distant from a fixed point, the *origin*, which represents the number 0. Sets which are constructed in this way from real numbers are called *point sets*.

Consider next sets whose elements are objects of the form

$$\mathbf{x} = (x_1, x_2, \ldots, x_n)$$

where n is a positive integer, each x_i is a real number and $\mathbf{x}$ represents a vector. If we have two such vectors $\mathbf{x}$ and $\mathbf{y}$ we say they are equal if and only if $x_i = y_i$ for $i = 1, 2, \ldots, n$. These objects are *real n-vectors* and the set of all real n-vectors is called *real n-space* and denoted by R^n; the integer n is the *dimension* of the space. We can identify R^1 with the real line R, defined above, and it is only a short step to regard R^2 as the *real plane*, the 2-vector $\mathbf{x}$ being represented by a point whose coordinates with respect to a fixed set of axes are (x_1, x_2). An example of a two-dimensional point set lying in the real plane is shown in Fig. 2.3. The geometrical representation can clearly be extended to R^3 and it continues to provide intuitive assistance even for spaces of higher dimension. The reader will observe that R^n is the direct product of n real lines.

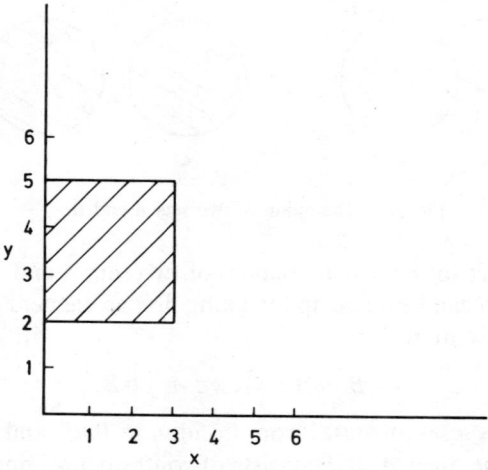

Fig. 2.3 A two-dimensional point set lying in the real plane

R^n is an example of a *real vector space* (or *linear space*), which is a set V whose elements, called *vectors*, can be multiplied by real numbers and added together in a natural way. In R^n these operations are carried out as follows:

For $a \in R$ and $\mathbf{x} \in R^n$, $a\mathbf{x} = (ax_1, ax_2, \ldots, ax_n)$
For $\mathbf{x}$ and $\mathbf{y} \in R^n$, $\mathbf{x} + \mathbf{y} = (x_1 + y_1, x_2 + y_2, \ldots, x_n + y_n)$

A fuller treatment of vector space theory can be found in Ref.[4]. Here we simply give a brief introduction to two concepts which will be of great importance later in the book: linear independence and orthogonality.

Suppose that a vector $\mathbf{x} \in R^n$ can be written as a linear combination of k other vectors $\mathbf{x}^{(1)}, \mathbf{x}^{(2)}, \ldots, \mathbf{x}^{(k)}$:

$$\mathbf{x} = a_1 \mathbf{x}^{(1)} + a_2 \mathbf{x}^{(2)} + \ldots + a_k \mathbf{x}^{(k)} = \sum_{i=1}^{k} a_i \mathbf{x}^{(i)}$$

where $a_i \in R$ and $\mathbf{x}^{(i)} \in R^n$ for $i = 1, 2, \ldots, k$; then we say that $\mathbf{x}$ is *linearly dependent* on the set $K = \{\mathbf{x}^{(1)}, \mathbf{x}^{(2)}, \ldots, \mathbf{x}^{(k)}\}$. As an example, consider the three vectors $\mathbf{x}^{(1)} = (-1, 2, 2)$, $\mathbf{x}^{(2)} = (1, 2, -1)$, $\mathbf{x}^{(3)} = (1, 12, 3/2)$; these are linearly dependent since

$$5\mathbf{x}^{(1)} + 7\mathbf{x}^{(2)} - 2\mathbf{x}^{(3)} = 0$$

and by using the modified notation $\mathbf{x} = \mathbf{x}^{(3)}$ we have,

$$\mathbf{x} = \frac{5}{2}\mathbf{x}^{(1)} + \frac{7}{2}\mathbf{x}^{(2)}$$

For this particular example the set K is clearly equal to $\{\mathbf{x}^{(1)}, \mathbf{x}^{(2)}\}$. More generally the set of all vectors linearly dependent on K is itself a vector space and is called the space *spanned* by K; it is a *subspace* of R^n. A special example of a spanning vector in a particular subspace of R^n, say R^2, would be the conventional cartesian coordinate system. Any set of vectors in R^n, one member of which is linearly dependent on the remainder, is a *linearly dependent set*. Thus K is a linearly dependent set if and only if there are real numbers $a_1, a_2, \ldots, a_k$ (at least one of which is not zero) such that

$$\sum_{i=1}^{k} a_i \mathbf{x}^{(i)} = 0$$

If this equation only holds when $a_i = 0$ for $i = 1, 2, \ldots, k$ then K is a *linearly independent set*. It can be shown that no linearly independent set in R^n can contain more than n vectors and it follows that any linearly independent set of n vectors in R^n spans the whole space; no smaller set can do this.

To illustrate these ideas, consider first a single vector $\mathbf{x}^{(1)}$ in R^3, say $(1, 0, 0)$. The subspace spanned by $\{\mathbf{x}^{(1)}\}$ is the set

$$\{(a, 0, 0) : a \in R\}$$

i.e. the x_1-axis. The subspace spanned by any single non-zero vector $\mathbf{x} \in R^n$ consists of the straight line through x and the origin 0. Now let $\mathbf{x}^{(2)} = (0, 1, 0)$; the set $\{\mathbf{x}^{(1)}, \mathbf{x}^{(2)}\}$ is linearly independent and spans the plane containing the x_1- and x_2-axes. Any two linearly independent vectors $\mathbf{x}$ and $\mathbf{y}$ in R^n span the plane containing the lines $0x$ and $0y$. Finally, if $\mathbf{x}^{(3)} = (0, 0, 1)$, then $\{\mathbf{x}^{(1)}, \mathbf{x}^{(2)}, \mathbf{x}^{(3)}\}$ is linearly independent and spans the whole of R^3. A second example of a set

spanning R^3 is defined by
$$K = \{(3, 0, 0), \quad (4, 0, 0), \quad (0, 1, 0), \quad (0, 0, 1)\}$$
since any vector $(y^{(1)}, y^{(2)}, y^{(3)})$ can be expressed as linear combination of K:
$$(y^{(1)}, y^{(2)}, y^{(3)}) = a_1 x^{(1)} + a_2 x^{(2)} + a_3 x^{(3)} + a_4 x^{(4)}$$
Clearly the component $y^{(1)}$ is given by $3a_1 + 4a_2$ where the coefficients a_1, a_2 cannot be unique.

The first of the two examples given above is special in that unique values for the coefficients of $x^{(1)}, x^{(2)}, x^{(3)}$ yields a unique value in R^3 and is called a *basis*.

Two vectors x and y are *orthogonal* if
$$\sum_{i=1}^{n} x_i y_i = 0$$

A set of vectors is orthogonal if every pair of its members are orthogonal. The set $\{(1, 0, 0), (0, 1, 0), (0, 0, 1)\}$ described above is orthogonal. An orthogonal set is necessarily linearly independent, but the converse is not true; for example, the set $\{(1, 0, 0), (1, 1, 0), (1, 1, 1)\}$ is linearly independent and spans R^3 but is not orthogonal.

We define the distance between two vectors x and y in R^n by the familiar formula
$$d(x, y) = \left[\sum_{i=1}^{n} (x_i - y_i)^2 \right]^{1/2}$$

This agrees in three-dimensional space with our usual idea of distance; the space R^n with this distance function is *Euclidean n-space E^n*. The *length* or *norm* of a vector x in E^n is $d(0, x)$ and is denoted by $|x|$.

An n-vector may also be thought of as a matrix with n rows and one column (a *column vector*) or with one row and n columns (a *row vector*); the vector space operations defined above then agree with the usual matrix operations. In this book, we shall always regard vectors as column vectors; thus the orthogonality condition given above will be written
$$x^T y = 0$$
where the superscript T denotes matrix transposition.

2.4 FUNCTIONS IN n-DIMENSIONAL SPACE

The concept of *function* will be familiar to the reader; in this section, we review briefly those aspects which we shall need later. Thus in the context of set theory, a function on a set A into a set B may be defined as a subset f of the direct product $A \times B$, with the following properties:
(a) If $a \in A$, there is an element $b \in B$ such that $(a, b) \in f$; we then write $b = f(a)$.

(b) If $(a, b_1) \in f$ and $(a, b_2) \in f$, then $b_1 = b_2$. This simply means that $f(a)$ is uniquely determined by a and is called the *image* of a under f. We have thus restricted ourselves to single-valued functions $f: A \to B$ where f associates elements in A with those of B uniquely.

The set A is called the *domain* of definition of f and the set B is the *range* of f. If A is a subset of E^n and $f(A)$ is a set of real numbers, then f is a *real function of n variables*. If $f(B)$ is a subset of E^m, then f is a *vector function*. In the general case where A and B are two arbitrary sets then f is usually called a *mapping*.

Let $f(x)$ denote a real function of the n real variables $x_1, x_2, \ldots, x_n$. The n-vector whose ith component is the partial derivative $\partial f/\partial x_i$ will assume great importance in later sections of this book; it is called the *gradient* of f and denoted by grad f or ∇f or $\nabla_x f$ or f_x or G. Of course, many functions do not have derivatives with respect to all their variables; if ∇f exists for all x in a subset D of E^n then f is said to be *differentiable* in D. If ∇f exists in D, it is a vector function of x in D; if all its components can be differentiated with respect to all the variables, then f_2 is *twice differentiable* in D; the symmetric matrix H such that $g_{ij} = \partial^2 f/\partial x_i \partial x_j$ is called the *Hessian matrix* of f.

If $f(x)$ is an m-vector function of n variables, and all of the components of $f(x)$ are differentiable in $D \subset E^n$, then f is differentiable in D and the $m \times n$ matrix whose (i, j)th component is $\partial f_j/\partial x_i$ is the *Jacobian matrix* f_x or $\nabla_x f$ of f; each column of f_x is the gradient of a component of f.

It is useful to think of a real-valued function $f(x)$ in terms of its *graph*, which is the set $\{(x, x_{n+1}): x \in D, x_{n+1} = f(x)\}$ in E^{n+1}; according to our set theory definition, a function is the same as its graph! When $n = 1$, this graph is just what we normally mean by a graph (Fig. 2.8 for example). When $n = 2$, the graph is a surface in three-dimensional space, and for higher dimensions we continue to think of the graph as a surface in $(n + 1)$-dimensional space.

If the function f is linear, we can write

$$f(x) = f(x^0) + (x - x^0)^T \nabla f(x^0)$$

where x and x^0 are any two points of E^n. The graph of a linear function is a line ($n = 1$) or a plane ($n = 2$) or in general a *hyperplane*. The gradient $\nabla f(x)$ of a linear function is independent of x.

If f is quadratic, we can write

$$f(x) = f(x^0) + (x - x^0)^T \nabla f(x^0) + \tfrac{1}{2}(x - x^0)^T H(x^0)(x - x^0)$$

and in this case the Hessian matrix $H(x)$ is independent of x, while the gradient is given by

$$\nabla f(x) = \nabla f(x^0) + H(x^0)(x - x^0)$$

If f is a differentiable nonlinear function, the hyperplane given by

$$x_{n+1} = f(x^0) + (x - x^0)^T \nabla f(x^0)$$

is *tangent* to $f(x)$ at $\mathbf{x} = \mathbf{x}^0$; the reader will readily verify that this agrees with the usual concepts of tangency in two and three dimensions.

2.5 OPEN AND CLOSED SETS

In this section and the next we consider some concepts, mainly from the theory of metric spaces and topology, which we shall need when discussing the convergence of optimization algorithms.

We begin by describing open and closed subsets of the real line. We define the *open interval* with *end points* a and b as

$$(a, b) = \{x : a < x < b\}$$

It is noted that in an open interval x can never attain its end points a, b, though it can get arbitrarily close. The *closed* interval with end points a and b is

$$[a, b] = \{x : a \leqslant x \leqslant b\}$$

A *half-open* interval is one which is open at one end and closed at the other, for example

$$[a, b] = \{a : a < x \leqslant b\}$$

An open ε-*neighbourhood* of a point $a \in R$ is an open interval with end points $a - \varepsilon$ and $a + \varepsilon$, where $\varepsilon > 0$; any set which includes an ε-neighbourhood of a is called a *neighbourhood* of a.

If A is a subset of R and x is a point such that, for some $\varepsilon > 0$, the ε-neighbourhood of x is a subset of A; that is the neighbourhood is entirely contained in the interval A, then x is an *interior point* of A. The set of all interior points of A is called the *interior* of A. If every ε-neighbourhood of x contains both points in A and points not in A, then x is a *boundary point* of A; the set of all such points is the *boundary* ∂A of A. Note that a boundary point of A may or may not belong to A. The end points a and b are the boundary points of all the intervals defined above, and the open interval (a, b) is the interior of all these intervals.

A subset of R is *open* if all its points are interior points, and *closed* if it contains all its boundary points; thus (a, b) is open, $[a, b]$ is closed and $(a, b]$ is neither open nor closed. The *closure* $\bar{A}$ of a set A is defined as $A \cup \partial A$, which is the smallest closed set containing A. Thus the closure of the interval $(a, b) = x : a < x < b$ is $[a, b] = [x : a \leqslant x \leqslant b]$.

These ideas, except that of 'interval', extend easily into n-dimensional space. An ε-neighbourhood of a point $a \in E^n$ is a set of the form

$$N_\varepsilon(a) = \{x : d(x, a) < \varepsilon\}$$

where $\varepsilon > 0$ and d is the distance function defined in Section 2.3. The remaining definitions are exactly as in the preceding paragraphs.

2.6 SEQUENCES

Consider a set A of points in E^n and suppose that we have a rule for determining a point $a^{(i)} \in A$ corresponding to every positive integer i. The points $a^{(1)}$, $a^{(2)}, \ldots$ form a *sequence*, denoted by $\{a^{(i)}\}$; the member $a^{(i)}$ is called the ith term of the sequence. This is a situation which frequently arises when we are searching for the solution to an optimization problem and have an initial estimate $a^{(1)}$ for the optimizing point and a rule for calculating, from any estimate $a^{(i)}$, a better estimate $a^{(i+1)}$. Thus we must study the properties of such sequences [5,6].

As examples, we consider the following three sequences of real numbers:

$$S_1 = \{1 - 1/(2i)\} \qquad = \tfrac{1}{2}, \tfrac{3}{4}, 5/6, \ldots$$
$$S_2 = \{(-1)^i(1 - 1/i)\} \qquad = 0, \tfrac{1}{2}, -\tfrac{2}{3}, \ldots$$
$$S_3 = \{2^i\} \qquad = 2, 4, 8, \ldots$$

Sequence S_1 is a *subsequence* of S_2, i.e. the members of S_1 are a subset of those of S_2 and occur in the same order. Had we wished to explicitly state the range of values taken by the sequence this may be done by placing the upper and lower values outside of the brackets denoting the sequence. For example, $\{\tfrac{1}{i}\}_1^\infty$, means that the sequence $1, \tfrac{1}{2}, \tfrac{1}{3}, \ldots$ takes an infinite number of terms.

In order to keep the notation compact when the function inside the brackets is complicated it is customary to write the sequence as $\{x^i\}_1$ and define x^i separately. Thus the sequence S_1 could be written as

$$\{x^i\}_1^\infty, \quad x^i = 1 - 1/(2i)$$

Our main interest lies in the terms of the sequence corresponding to large values of i. Consider first S_1. As i increases, the terms of this sequence become closer and closer to 1; indeed, for any ε-neighbourhood of 1, however small, we can choose i in such a way that every term after the ith lies within $N_\varepsilon(1)$. We say that the sequence *converges* to 1 or has 1 as its *limit*; a sequence which has a limit is *convergent*. The limit of the sequence $a^{(i)}$ is denoted by $\lim_{i \to \infty} a^{(i)}$ or sometimes by $a^{(\infty)}$.

Consider next the sequence S_2. In this case there is no unique point to which the sequence converges since, for large values of i, it oscillates between the two points $1, -1$. Another way of describing this phenomenon would be to say that for $i \to \infty$ the terms of S_2 cluster about the points $1, -1$ and thus these points are called *cluster points* (or *accumulation points* or *limiting points*). We have observed that S_2 has a subsequence, i.e. S_1, which converges to 1. This is a general property in that any sequence containing n cluster points can be broken down into n subsequences each converging to a single limit. The sequence S_3 also fails to converge, though in a different manner: it has no cluster points. Sequences such as S_2 and S_3, which do not converge, are *divergent*.

We defined our examples as subsets of the set of all real numbers. But suppose we now regard S_1 as a sequence of points in the open interval $(-1, +1)$. This interval certainly contains every term of S_1, but it does not contain the limit of the sequence. This leads us to ask the following question: can a set have the property that every infinite subset has a cluster point in the set? Before answering this question, we introduce the concept of boundedness.

A bounded set in E^n is one whose members are all within a given distance of the origin; precisely, $A \subset E^n$ is *bounded* if there is a real number M such that, for every $x \in A$, $d(0, x) \leq M$. For sets of real numbers, we can take this idea a little further. Suppose that A is a non-empty bounded set of real numbers. M is an *upper bound* of A if $x \leq M$ for all $x \in A$, and m is a *lower bound* of A if $x \geq m$ for all $x \in A$. Consider, for example, the half-open interval $(-1, +1]$; any number less than or equal to -1 is a lower bound of this set, and any number greater than or equal to $+1$ is an upper bound. The smallest number which is an upper bound of A is the *supremum* or *least upper bound* of A, written $\sup A$ or $\sup_{x \in A} x$; the largest number which is lower bound of A is the *infimum* or *greatest lower bound* of A, written $\inf A$ or $\inf_{x \in A} x$. Every non-empty bounded set of real numbers has a supremum and an infimum and these may or may not themselves be members of the set. Thus $(-1, +1]$ has $+1$ as both supremum and maximum element, and -1 as infimum; it has no minimum element since we get arbitrarily close to -1 but never achieve it.

We now return to our question about cluster points. Let A be an arbitrary set in E^n; it can be shown that every infinite subset of A has at least one cluster point if and only if A is closed and bounded. Sets with this property are *compact*; thus the open interval $(0, +1)$, in which the sequence $\{1/i\}_1^\infty$ has no cluster point, which implies that the sequence is unbounded since for all $x_i \in A$, $A = \{x | x \in R, 0 < x < 1\}$ there exists an arbitrarily small number ε such that $\varepsilon > 0, \varepsilon > x^i > 0$. In other words, we can never find the smallest value for $1/x^i$ since for any given i there are always infinitely more smaller values existing in the interval $(0, +1/x^i)$.

A similar situation arises in the case of sequence S_2 if we use the half-open interval $(-1, +1]$. Although the cluster point at $x = +1$ has been included the other cluster point at $x = -1$ is outside the interval. Thus once again it is possible to get arbitrarily close to $x = -1$ but we can never achieve this point and an iterative procedure designed to reach $x = -1$ would give rise to problems. The pattern evolving is that if any cluster point is omitted from the range of a sequence this omission will lead to ill behaviour.

Had the closed interval $[0, +1]$ been selected for the sequence $\{1/x\}_1^\infty$ and $[-1, +1]$ for S_2 all the cluster points would have been contained within the respective intervals which would then be compact.

2.7 CONVEXITY

When we come to discuss conditions for the existance of solutions to optimization problems and to examine the performance of algorithms for solving them, we shall find that the geometrical shape of certain sets of points in E^n is of great importance. In this section, we discuss the principal geometric concept which we shall need, convexity [5, 6, 7].

Let a and b be two points in a set $A \subset E^n$. The *line segment ab* joining them is the set
$$\{x : x = (1-t)a + tb, 0 \leqslant t \leqslant 1\}$$
and each member of this set is a *convex combination* of a and b. The set consists of a (corresponding to $t = 0$), every point of the straight line joining a to b and lying between them ($0 < t < 1$) and b ($t = 1$). A convex combination of N points $a_1, a_2, \ldots, a_N$ is any point
$$x = \sum_{i=1}^{N} t_i a_i \quad \text{where all } t_i > 0 \text{ and} \quad \sum_{i=1}^{N} t_i = 1.$$

A set A is *convex* if the line segment joining a and b is included in A whenever a and b are points of A. Figure 2.4 illustrates the following two sets in E^2:
$$A = \{x = (x_1, x_2) : x_1 \geqslant 0, x_2 \geqslant 0, x_1 - x_2^2 \geqslant 4\}$$
$$B = \{x = (x_1, x_2) : x_1 \geqslant 0, x_2 \geqslant 0, x_1 - x_2^2 \leqslant 4\}.$$

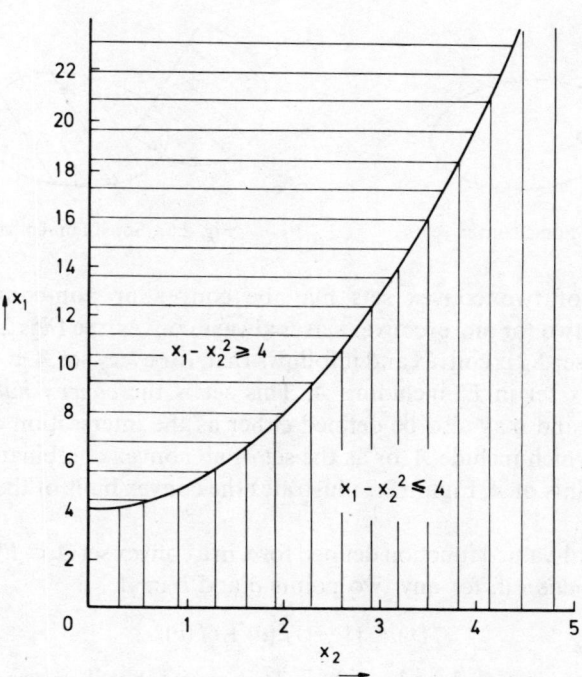

Fig. 2.4 An example of convexity and non-convexity

Let $a = (a_1, a_2)$ and $b = (b_1, b_2)$ be any two points of A and let x be a convex combination of a and b. Then

$$x_1 - x_2^2 = (1-t)a_1 + tb_1 - [(1-t)a_2 + tb_2]^2 \text{ for some } t \in [0, 1]$$
$$= (1-t)(a_1 - a_2^2) + t(b_1 - b_2^2) + (1-t)t(a_2 - b_2)^2$$
$$> 4.$$

Thus $x \in A$ and we have shown A to be convex. Conversely, by considering any two points a and b on the boundary between A and B and showing that the interior of the line segment ab (which we know is in A) is outside B, we can establish that B is not convex. Further examples are given in the next two figures; Fig. 2.5 illustrates convex sets and Fig. 2.6 non-convex sets.

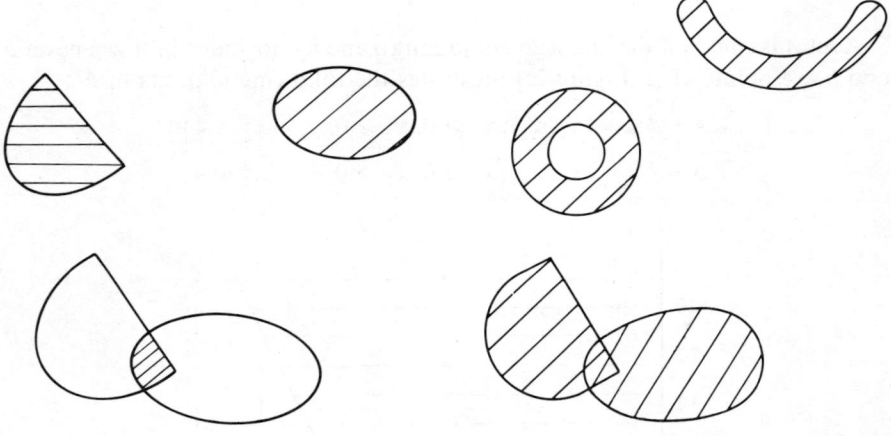

Fig. 2.5 Some convex sets   Fig. 2.6 Some non-convex sets

The union of two convex sets may be convex or non-convex, but the intersection of two (or more) convex sets is always convex; see Figs 2.5 and 2.6 for examples. The set E^n is convex and it follows that, for every set $A \subset E^n$, there is a smallest convex set in E^n including A. This set is the *convex hull* (or *convex envelope*) of A and may also be defined either as the intersection of all convex subsets of E^n which include A, or as the set of all convex combinations of finite numbers of points of A. Figure 2.7 illustrates the convex hulls of the non-convex sets of Fig. 2.6.

If $f(x)$ is a real-valued function defined for $\mathbf{x}$ in a convex set $A \subset E^n$, we say $f(x)$ is a *convex function* if, for any two points a and b in A,

$$f(x) \leq (1-t)f(a) + tf(b)$$

whenever $\mathbf{x} = (1-t)\mathbf{a} + t\mathbf{b}$ and $t \in [0, 1]$. This means that if, over a line segment in A, we approximate $f(x)$ by a linear function equal to $f(x)$ at the two end points, then at every intermediate point the value of $f(x)$ is smaller than (or equal to) our

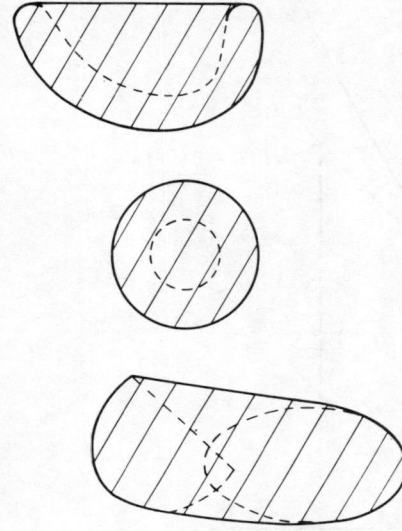

Fig. 2.7 Examples of convex hulls for the non-convex sets illustrated in Fig. 2.6

approximation. We say $f(x)$ is a *concave function* if $-f(x)$ is a convex function. Note that a linear function is both convex and concave, since linearity implies that

$$f(x) = (1-t)f(a) + tf(b)$$

whenever $\mathbf{x} = (1-t)\mathbf{a} + t\mathbf{b}$.

If $f(x)$ is differentiable, then it is convex on the convex set A if, and only if,

$$f(y) \geq f(x) + (\mathbf{y} - \mathbf{x})^T \nabla f(x)$$

for all $\mathbf{x}$ and $\mathbf{y}$ in A. We can express this geometrically as follows: a tangent to a convex function always lies below the graph of the function, as shown in Fig. 2.8.

If the second derivatives of $f(x)$ exist then the function is convex on the convex set A if, and only if, the Hessian $H(x)$ is positive semi-definite. We recall that a matrix H is positive semi-definite if, for any vector $\mathbf{x}$, $\mathbf{x}^t H \mathbf{x} > 0$, (example $1 + x_1^4 + x_2^4$).

2.8 CONDITIONS FOR MINIMA

In the next two sections, we consider the conditions which characterize the solutions to various types of minimization problem. [7,8] Although we begin with some general definitions the present section is concerned primarily with unconstrained optimization, leaving constrained problems to Section 2.9.

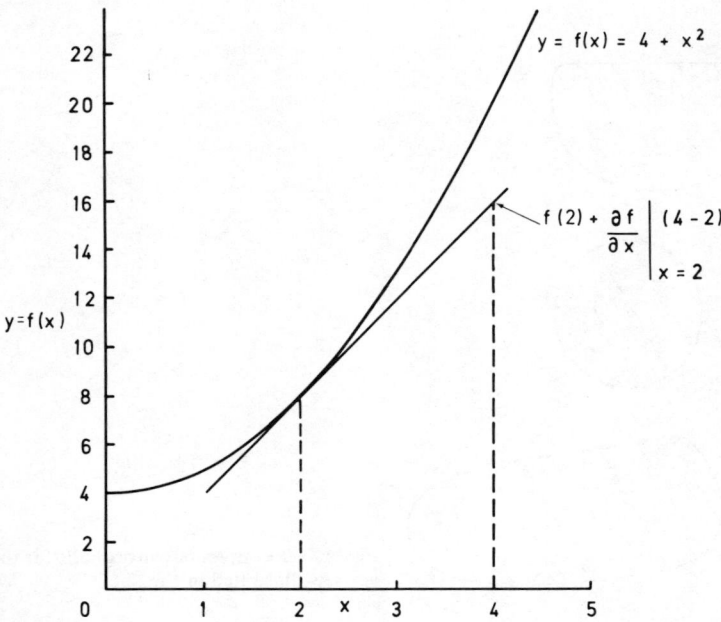

Fig. 2.8 A tangent to a convex function always lies below the graph of the function

The most general *constrained optimization problem* which we shall discuss is the following:

$$\begin{aligned}
\text{minimize} \quad & f(x) \\
\text{subject to} \quad & g_j(x) \geqslant 0 \quad \text{for } j = 1, 2, \ldots, m \\
\text{and} \quad & h_k(x) = 0 \quad \text{for } k = 1, 2, \ldots, p
\end{aligned} \qquad \text{(CP1)}$$

where $f(x)$, $g_j(x)$ and $h_k(x)$ are real-valued functions of the vector $\mathbf{x} = (x_1, x_2, \ldots, x_n) \in E^n$. More concisely,

$$\min f(x) \quad \text{for } x \in P \subset E^n$$

where $P = \{x : x \in E^n, g(x) \geqslant 0 \in E^m, h(x) = 0 \in E^p\}$. The set P is called the *feasible region* and its elements are the *feasible points* of the problem, the remaining points of E^n being *infeasible*. The relations $g(x) \geqslant 0$ and $h(x) = 0$ are called respectively *inequality constraints* and *equality constraints*; feasible points are said to *satisfy* the constraints. The components x_i of $\mathbf{x}$ are the *variables* of the problem, and $f(x)$ is the *objective function*. We shall assume that the objective and constraint functions are twice differentiable throughout E^n; any case where this does not hold will require special treatment, usually aimed at showing that the derivatives do exist wherever they are needed.

A point $\mathbf{x}^*$ is called a *local solution* of (CP1), or a *local minimum* of $f(x)$ subject to the constraints, if there is a neighbourhood N of $\mathbf{x}^*$ such that $f(x) \geqslant f(\mathbf{x}^*)$

whenever $\mathbf{x} \in N \cap P$; that is, we cannot move away from $\mathbf{x}^*$, in a direction in which $f(x)$ decreases, unless we leave the feasible region. If $f(x) \geq f(x^*)$ for all $\mathbf{x} \in P$, then $\mathbf{x}^*$ is the *global solution* to (CP1) and the *global minimum* of $f(x)$ subject to the constraints. Clearly, a global solution must be a local solution, but the reverse is not true in general. Figure 2.9 shows a function of one variable which, in the absence of any constraints, has two local minima in addition to its global minimum; this function also has two local maxima and, if we assume $f(x)$ to grow without limit for large positive and negative values of $\mathbf{x}$, it has no global maximum.

In this section, we discuss the *unconstrained minimization problem*

$$\min f(x) \quad \text{for } \mathbf{x} \in E^n$$

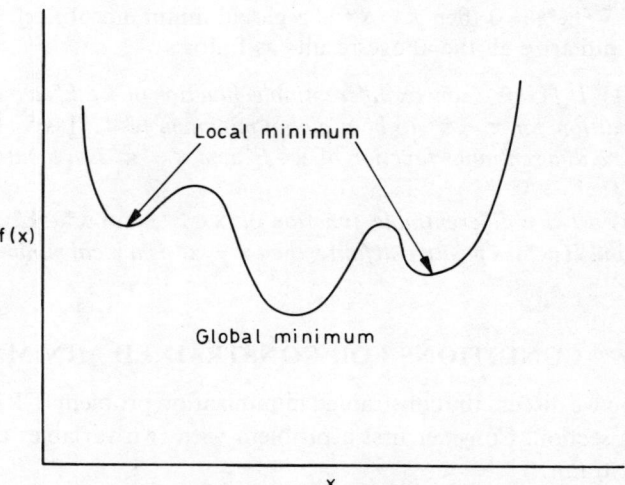

Fig. 2.9 An example of a function with both a global minimum and two local minima

Although engineering problems rarely give rise to problems without constraints, examination of the unconstrained case provides the basic theory which we can develop to deal with constrained problems. We shall see in Chapters 7 and 8 that some of the most effective techniques for solving constrained problems involve solving sequences of related unconstrained problems.

If $\phi(\theta)$ is a differentiable function of a single real variable θ, we know that a *necessary condition* for $\theta = \theta^*$ to be a local minimum of $\phi(\theta)$ is $\phi'(\theta^*) = 0$, and that *sufficient condition* is $\phi'(\theta^*) = 0$ together with $\phi''(\theta^*) > 0$. Suppose now that $f(x)$ is a real function defined on E^n and consider the values taken by $f(x)$ along the line $\mathbf{x} = \mathbf{x}^* + \theta \mathbf{u}$ where $\mathbf{x}^*$ and $\mathbf{u}$ are given n-vectors and θ is a real variable. Let

$\phi(\theta) = f(x^* + \theta u)$; then

$$\phi'(0) = \mathbf{u}^T \nabla f(x^*)$$

and

$$\phi''(0) = \mathbf{u}^T H(x^*) \mathbf{u}$$

where $H(x)$ is the Hessian matrix of f. If $\mathbf{x} = \mathbf{x}^*$ is a local minimum of $f(x)$, then $\theta = 0$ must minimize $\phi(\theta)$ for any given $\mathbf{u}$, i.e. $\phi'(0) = 0$ for any $\mathbf{u}$, and this can only be the same if $\nabla f(x^*) = 0$. Conversely, if $\nabla f(x^*) = 0$ and $\mathbf{u}^T H(x^*) \mathbf{u} > 0$ for all non-zero $u \in E^n$, then in whatever direction we move away from $\mathbf{x}^*$ the value of $f(x)$ increases, showing that $\mathbf{x} = \mathbf{x}^*$ is a local minimum of $f(x)$. If $f(x)$ is convex, we can obtain a stronger result, since then

$$f(x) \geqslant f(x^*) + (\mathbf{x} - \mathbf{x}^*)^T \nabla f(x^*) \quad \text{for all } x \in E^n$$

and hence if $\nabla f(x^*) = 0$ then $\mathbf{x} = \mathbf{x}^*$ is a global minimum of $f(x)$.

We can summarize all the above results as follows.

Theorem (1) If $f(x)$ is a convex differentiable function of $\mathbf{x} \in E^n$, a necessary and sufficient condition for $\mathbf{x} = \mathbf{x}^*$ to be a global minimum of $f(x)$ is $\nabla f(x^*) = 0$.
(2) If $f(x)$ is a differentiable function of $\mathbf{x} \in E^n$ and $\mathbf{x} = \mathbf{x}^*$ is a local minimum of $f(x)$, then $\nabla f(x^*) = 0$.
(3) If $f(x)$ is a twice differentiable function of $\mathbf{x} \in E^n$, and $\mathbf{x}^* \in E^n$ is such that $\nabla f(x^*) = 0$ and $H(x^*)$ is positive-definite, then $\mathbf{x} = \mathbf{x}^*$ is a local minimum of $f(x)$.

2.9 CONDITIONS FOR CONSTRAINED MINIMA

In this section we discuss the constrained minimization problem (CP1) defined in the previous section. Consider first a problem with two variables and a single equality constraint:

$$\begin{aligned} &\min f(x) \quad \text{for } x = (x_1, x_2) \in E^2 \\ &\text{subject to} \quad c(x_1, x_2) = 0 \end{aligned}$$

The feasible region for this problem consists of those points of E^2 for which the equation $c(x) = 0$ holds; suppose that we can solve this equation for x_2 in terms of x_1, say

$$x_2 = y(x_1)$$

Then, for any selected value of x_1 we can calculate the unique value of x_2 which makes (x_1, x_2) feasible; we may thus regard x_1 as an independent variable (or *decision variable*) and x_2 as a dependent variable (or *state variable*). Writing

$$F(x_1) = f\{x_1, y(x_1)\}$$

we see that our constrained problem is equivalent to the unconstrained problem

$$\min F(x_1) \quad \text{for } x_1 \in E^1$$

If x_1^* is a solution to this problem, then $F'(x_1^*) = 0$. But

$$\frac{dF}{dx_1} = \frac{\partial f}{\partial x_1} + \frac{\partial f}{\partial x_2}\frac{dy}{dx_1}$$

We can gain some information on $y'(x_1)$ by observing that

$$C(x_1) \equiv c(x_1, y(x_1)) = 0 \quad \text{for all } x_1$$

and hence that

$$\frac{dC}{dx_1} = \frac{\partial c}{\partial x_1} + \frac{\partial c}{\partial x_2}\frac{dy}{dx_1} = 0$$

or

$$\frac{dy}{dx_1} = -\frac{\partial c}{\partial x_1}\bigg/\frac{\partial c}{\partial x_2}$$

The condition $F'(x_1^*) = 0$ can be written

$$\frac{\partial f}{\partial x_1} - \left(\frac{\partial f}{\partial x_2}\bigg/\frac{\partial c}{\partial x_2}\right)\frac{\partial c}{\partial x_1} = 0$$

and if we add the identity

$$\frac{\partial f}{\partial x_2} - \left(\frac{\partial f}{\partial x_2}\bigg/\frac{\partial c}{\partial x_2}\right)\frac{\partial c}{\partial x_2} = 0$$

we arrive at the relation

$$\nabla f - \lambda \nabla c = 0$$

where $\lambda = \dfrac{\partial f}{\partial x_2}\bigg/\dfrac{\partial c}{\partial x_2}$ is the *Lagrange multiplier* corresponding to the equation constraint and the left-hand side is known as the *constrained derivative*. For the more general case where x is an n-vector and c an m-vector, the necessary condition for $\mathbf{x}^*$ to be a solution is that there should be an m-vector λ of Lagrange multipliers such that

$$\nabla f(\mathbf{x}^*) - \lambda^T \nabla c(\mathbf{x}^*) = 0$$

We next introduce an interpretation of the Lagrange multipliers which will be of use when we consider inequality constraints. Returning to our simple problem with two variables and one constraint, we generalize the constraint slightly to

$$c(x) = b$$

and observe that the solution $\mathbf{x}^*$ must be a function of the quantity b which occurs in the constraint; let us write

$$\mathbf{x}^* = X(b)$$

For all b, $X(b)$ must satisfy

$$c\{X(b)\} = b$$

and
$$\nabla f\{X(b)\} - \lambda \nabla c\{X(b)\} = 0$$
$$\frac{df\{X(b)\}}{db} = \frac{\partial f}{\partial x_1} X'_1(b) + \frac{\partial f}{\partial x_2} X'_2(b)$$
$$= \nabla f\{X(b)\}^T \frac{dX}{db}$$

and
$$1 = \frac{dc\{X(b)\}}{db} = \nabla c\{X(b)\}^T \frac{dX}{db}$$

By employing these last two equations and noting that
$$\nabla f\{X(b)\} \frac{dX}{db} - \lambda \nabla c\{X(b)\} \frac{dX}{db} = 0$$

we obtain
$$\lambda = \frac{df\{X(b)\}}{db}$$

In the more general case when there are m constraints, we have
$$\lambda_i = \frac{\partial f\{X(b)\}}{\partial b_i} \qquad \text{for } i = 1, 2, \ldots, m$$

We illustrate these ideas with a simple example:
$$\min f(x) = x_1^2 + x_2^2$$
$$\text{subject to } c(x) = x_1 + x_2 = b$$
Here
$$\nabla f = (2x_1, 2x_2)$$
and
$$\nabla c = (1, 1)$$
The necessary conditions are
$$2x_i^* - \lambda = 0 \qquad \text{for } i = 1 \text{ and } 2$$
and so to satisfy the constraint we must have
$$b = x_1^* + x_2^* = \lambda$$
and hence
$$x_1^* = x_2^* = \tfrac{1}{2}b$$
This gives
$$f(x^*) = \tfrac{1}{2}b^2$$
notice that
$$\frac{df}{db} = b = \lambda$$

Now we can discuss problems with inequality constraints, beginning with another small modification to our two-variable problem, making the constraint

$$c(x) \geq 0$$

We can rewrite this problem as

$$\min \phi(b) \quad \text{subject to } b > 0$$

where b is a new variable and $\phi(b)$ is the solution to the previous problem, i.e.

$$\phi(b) = \min f(x) \quad \text{subject to } c(x) = b$$

Thus, if x^* is a solution to our new problem, there must be a multiplier λ such that

$$\nabla f(x^*) - \lambda \nabla c(x^*) = 0$$

and

$$\frac{d\phi}{db} = \lambda$$

The condition for $b = b^*$ to minimize $\phi(b)$ there exist two possibilities, either the constraint is active or inactive. When the constraint is not active, i.e. $b^* > 0$ then $c(x^*) > 0$ in which case the minimum is unconstrained, hence $\nabla f(x^*) = 0$ requiring $\lambda = 0$; if the constraint is active the $b^* = 0$ i.e. $c(x^*) = 0$ and the constrained derivative must vanish, thus

$$\nabla f(x^*) - \lambda \nabla c(x^*) = 0$$

and this requires

$$\lambda = d\phi/db > \theta$$

These can be written in the condensed form

$$\lambda c(x^*) = 0, \ \lambda \geq 0, \ c(x^*) \geq 0$$

Thus, to sum up, our condition for x^* to solve the problem is the existence of λ such that

$$\nabla f(x^*) - \lambda \nabla c(x^*) = 0$$
$$\lambda c(x^*) = 0$$
$$\lambda \geq 0 \quad \text{and} \quad c(x^*) \geq 0$$

If $c(x^*) = 0$ the constraint is said to be *active* at the solution, otherwise it is *inactive*.

The examples discussed above illustrate the main ideas of the general theorem which we are about to state. However, before stating this theorem, we must impose a restriction on our general optimization problem in order to exclude some exceptional cases to which the theorem does not apply. For any point $x^* \in E^n$, suppose the inequality constraints are ordered in such a way that

$$g_i(x^*) = 0 \quad \text{for } i = 1, 2, \ldots, m'$$
$$g_i(x^*) > 0 \quad \text{for } i = m'+1, \ldots, m$$

Then $\mathbf{x}^*$ is said to satisfy the *constraint qualification* if, for any $\mathbf{u} \in E^n$ satisfying
$$\mathbf{u}^T \nabla g_i(x^*) \geq 0, \qquad \text{for } i = 1, 2, \ldots, m'$$
there is a curve
$$\mathbf{x} = y(t), \qquad t \geq 0$$
which is tangent at $\mathbf{x} = \mathbf{x}^*$ to the line
$$\mathbf{x} = \mathbf{x}^* + t\mathbf{u}$$
and which lies in the feasible region in some ε-neighbourhood of $\mathbf{x}^*$, i.e. there is an $\varepsilon > 0$ and $t_0 > 0$ such that if
$$0 \leq t < t_0$$
then
$$y(t) \in N_\varepsilon(x^*) \cap \{x : g(x) \geq 0, h(x) = 0\}$$

If $\mathbf{x}^*$ satisfies the constraint qualification, it is a *regular* point for the optimization problem (CP1); regularity is a more general property which we shall not define here. A useful condition which ensures that $\mathbf{x}^*$ satisfies the constraint qualification is that the gradient vectors $\nabla g_i(x^*)$ and $\nabla h_j(x^*)$ of the active constraints be linearly independent.

Theorem *If $\mathbf{x}^*$ is a regular point of the general constrained optimization problem (CP1) and is also a minimizing point, then there exist $\lambda \in E^m$ and $\mu \in E^p$ such that*

$$\nabla f(x^*) - \lambda^T \nabla g(x^*) - \mathbf{u}^T \nabla h(x^*) = 0$$
$$\lambda^T g(x^*) = 0$$
$$\lambda \geq 0 \qquad g(x^*) \geq 0 \qquad h(x^*) = 0$$

These are known as the Kuhn–Tucker conditions.

If f is convex, the g_i are concave, and the h_i are linear (such a problem is a *convex programming problem*) and the Kuhn–Tucker conditions hold, then $\mathbf{x}^*$ is a solution to the problem.

If, in some ε-neighbourhood of $\mathbf{x}^*$, f is convex, the g_i concave and the h_i linear, then the Kuhn–Tucker conditions are sufficient to prove $\mathbf{x}^*$ a local solution to (CP1).

As an example of the Kuhn–Tucker conditions consider the problem

$$\min f(x) = x_1^2 + x_2^2$$

subject to
$$g_1(x) = x_2 - x_1^2 \geq 0$$
$$g_2(x) = x_2 + x_1 - 1 \geq 0$$

which is shown in Fig. 2.10. By applying the equations of Section 2.7 it can be established that $f(x)$ is convex whilst $g_1(x)$ and $g_2(x)$ are concave. Thus, in order to satisfy the Kuhn-Tucker conditions non-negative multipliers λ_1, λ_2 are

Basic Mathematical Concepts

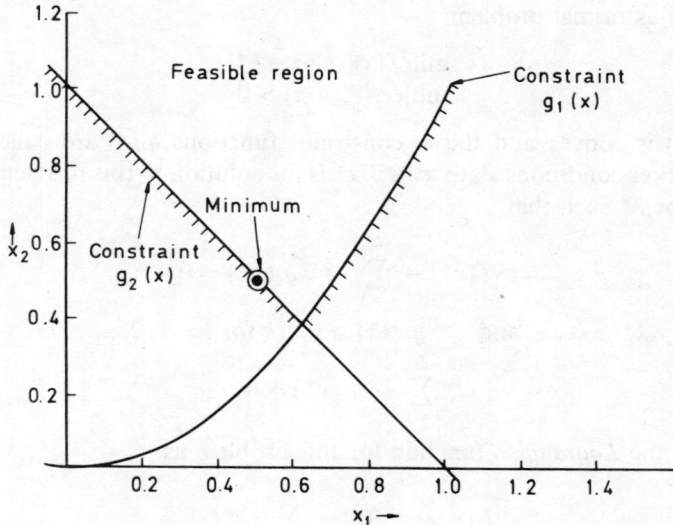

Fig. 2.10 Graphical illustration of the problem
minimize $f(x) = x_1^2 + x_2^2$
subject to $g_1(x) = x_2 - x_1^2 \geq 0$
$g_2(x) = x_2 + x_1 - 1 \geq 0$

required such that,

$$\frac{\partial f}{\partial x_1} - \lambda_1 \frac{\partial g_1}{\partial x_1} - \lambda_2 \frac{\partial g_2}{\partial x_1} = 0$$

$$\frac{\partial f}{\partial x_2} - \lambda_1 \frac{\partial g_1}{\partial x_2} - \lambda_2 \frac{\partial g_2}{\partial x_2} = 0$$

$$\lambda_1 (x_2 - x_1^2) = 0$$

$$\lambda_2 (x_2 + x_1 - 1) = 0$$

By choosing $\lambda_1 = 0$, $\lambda_2 = 1$ then these equations are satisfied by $x_1 = x_2 = \frac{1}{2}$ and since these values are feasible they constitute the minimizing point for the problem.

2.10 DUALITY

In this section, we explore the relations which exist between any given convex nonlinear programming problem, here called the *primal* problem, and a second nonlinear programming problem, called the *dual* problem. [5, 8] Besides giving us useful theoretical insights, this relationship (or duality) between the two problems is put to practical use in some algorithms.

We take as primal problem:

$$\min f(x) \quad x \in E^n$$
$$\text{subject to } g(x) > 0$$

where $f(x)$ is convex and the m constraint functions $g_i(x)$ are concave. The Kuhn–Tucker conditions state that, if x^* is the solution to this problem, there is an m-vector λ^* such that

$$\nabla f(x^*) - \sum_{i=1}^{m} \lambda_i^* \nabla g_i(x^*) = 0$$

$$\lambda_i^* \geq 0 \quad \text{and} \quad g_i(x^*) \geq 0 \quad \text{for } i = 1, 2, \ldots, m$$

$$\sum_{i=1}^{m} \lambda_i^* g_i(x^*) = 0$$

We define the *Lagrangian* function for this problem as

$$L(x, \lambda) = f(x) - \sum_{i=1}^{m} \lambda_i g_i(x)$$

and observe that the first of the Kuhn–Tucker conditions may be written

$$\nabla_x L(x^*, \lambda^*) = 0$$

which is also a necessary condition for $x = x^*$ to minimize $L(x, \lambda^*)$. Note that, in view of the final Kuhn–Tucker condition, $L(x^*, \lambda^*) = f(x^*)$.

In fact, for any x, since

$$f(x) \geq f(x^*) + (x - x^*)^T \nabla f(x^*)$$
$$-g(x) \geq -g(x^*) - (x - x^*)^T \nabla g(x^*)$$

and $\lambda_i^* \geq 0$ for $i = 1, 2, \ldots, m$, we have

$$L(x, \lambda^*) = f(x) - \sum_{i=1}^{m} \lambda_i^* g_i(x)$$
$$\geq L(x^*, \lambda^*) + (x - x^*)^T L_x(x^*, \lambda^*)$$
$$= L(x^*, \lambda^*)$$

Next, we examine the dependence of $L(x^*, \lambda)$ on λ. For any $\lambda \geq 0$,

$$L(x^*, \lambda) = L(x^*, \lambda^*) - \sum_{i=1}^{m} (\lambda_i - \lambda_i^*) g_i(x^*)$$
$$\leq L(x^*, \lambda^*)$$

by the second and third Kuhn–Tucker conditions. Thus, for any x and any $\lambda \geq 0$,

$$L(x, \lambda^*) \geq L(x^*, \lambda^*) \geq L(x^*, \lambda)$$

Any point $(x^*, \lambda^*) \in E^{n+m}$ which satisfies these conditions is a *saddle point* of the function $L(x, \lambda)$.

The saddle point conditions suggest that to problems associated with the minimization problem whose solution is $\mathbf{x} = \mathbf{x}^*$, there is a maximization problem with the solution $\lambda = \lambda^*$. Of course, when solving our problem we do not know either $\mathbf{x}^*$ or λ^*, so we cannot directly use the functions $L(x, \lambda^*)$ and $L(x^*, \lambda)$. Suppose, however, that for each $\mathbf{x}$ we can find $\lambda = \lambda(x)$ so as to maximize $L(x, \lambda)$ subject to $\lambda \geqslant 0$ and likewise that for each $\lambda \geqslant 0$ we can find $\mathbf{x} = \mathbf{x}(\lambda)$ to minimize $L(x, \lambda)$ unconstrained. Then it follows from the saddle point conditions that

$$L\{x, \lambda(x)\} \geqslant L(x^*, \lambda^*) \geqslant L\{x(\lambda), \lambda\}$$

We can make two useful deductions from this:
(1) For given $\mathbf{x}$ and $\lambda \geqslant 0$, we can use $L\{x, \lambda(x)\}$ and $L\{x(\lambda), \lambda\}$ as upper and lower bounds for $f(x^*)$.
(2) The problem: choose $\lambda \geqslant 0$ to *maximize* $L\{x(\lambda), \lambda\}$ has a solution $\lambda = \lambda^*$, with $\mathbf{x}^* = \mathbf{x}(\lambda^*)$. This is, in essence, the dual problem referred to at the start of the section; however, in the statement of the theorem which follows, we avoid reference to the function $\mathbf{x}(\lambda)$, which may not always exist, and substitute the condition $L_x(x, \lambda) = 0$.

Theorem of duality *Let the primal problem be*

$$\min f(x), \; \mathbf{x} \in E^n, \quad \text{subject to } g(x) \geqslant 0$$

where $f(x)$ *is a convex function and* $g(x)$ *is a vector of* m *concave functions, and define the dual problem as*

$$\max L(x, \lambda), \; (\mathbf{x}, \lambda) \in E^{n+m}$$
$$\text{subject to } \lambda \geqslant 0 \text{ and } \nabla_x L(x, \lambda) = 0$$

where
$$L(x, \lambda) = f(x) - \lambda^T g(x).$$

Then, if $\mathbf{x}^*$ *is a solution to the primal problem, there is a* λ^* *such that* $(\mathbf{x}^*, \lambda^*)$ *solves the dual problem and*

$$f(x^*) = L(x^*, \lambda^*).$$

We illustrate the above ideas with a simple example in which there is only one variable. The primal problem is

$$\min f(x) \equiv \tfrac{1}{2}x^2$$
$$\text{subject to } g(x) \equiv x - 1 \geqslant 0$$

The Lagrangian function is

$$L(x, \lambda) = \tfrac{1}{2}x2 - \lambda(x - 1)$$

and the Kuhn–Tucker conditions are

$$x^* - \lambda^* = 0$$
$$\lambda^* \geqslant 0 \quad \text{and} \quad x^* - 1 \geqslant 0$$
$$\lambda^*(x^* - 1) = 0$$

i.e. either $\lambda^* = 0$ or $x^* = 1$. The solution is

$$x^* = 1, \lambda^* = 1$$

We examine the functions $\lambda(x)$ and $x(\lambda)$ defined above. $\lambda = \lambda(x)$ maximizes $L(x, \lambda)$ for $\lambda \geqslant 0$:

if $x < 1$, $\lambda(x) = +\infty$ and $L\{x, \lambda(x)\} = +\infty$
if $x = 1$, $\lambda(x)$ is indeterminate and $L\{x, \lambda(x)\} = \frac{1}{2}$
if $x > 1$, $\lambda(x) = 0$ and $L\{x, \lambda(x)\} = \frac{1}{2}x^2$

Figure 2.11(a) shows the graph of $L\{x, \lambda(x)\}$ and emphasizes that the problem of choosing x to minimize $L\{x, \lambda(x)\}$ is simply the primal problem. Turning to $x(\lambda)$, we observe that

$$L(x, \lambda) = \tfrac{1}{2}(x - \lambda)^2 + \lambda - \tfrac{1}{2}\lambda^2$$

Thus

$$x(\lambda) = \lambda \quad \text{and} \quad L\{x(\lambda), \lambda\} = \lambda - \tfrac{1}{2}\lambda^2.$$

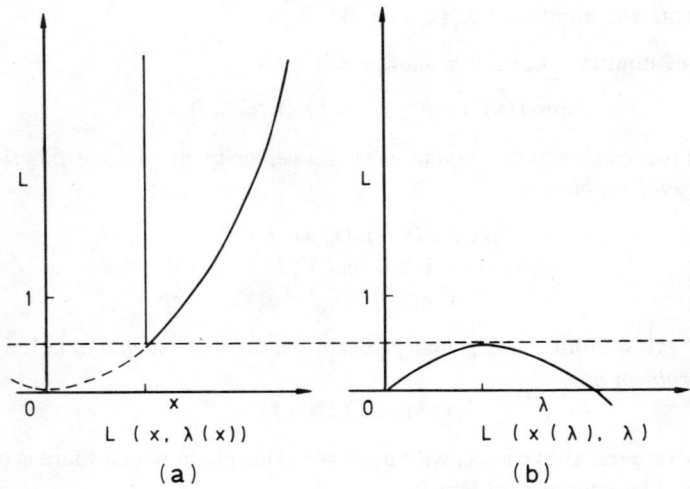

Fig. 2.11 Primal and dual problems for the Lagrangian $L(x, \lambda) \geqslant \tfrac{1}{2}x^2 - \lambda(x - 1)$

Figure 2.11(b) shows the graph of $L\{x(\lambda), \lambda\}$. Maximizing this function gives $\lambda^* = 1$ and $x^* = x(\lambda^*) = 1$, as before; notice that we have been able to calculate λ^*, even though $\lambda(x^*)$ is indeterminate.

As a further illustration, we consider the two-variable problem

$$\min \quad f(x) = x_1^2 + x_2^2$$
$$\text{subject to} \quad g_1(x) = x_2 - x_1^2 \geqslant 0$$
$$\text{and} \quad g_2(x) = x_2 + x_1 - 1 \geqslant 0.$$

The minimizing point and part of the feasible region for this problem are shown in Fig. 2.10. The Lagrangian function is

$$L(x, \lambda) = x_1^2 + x_2^2 - \lambda_1\{x_2 - x_1^2\} - \lambda_2(x_2 + x_1 - 1)$$

As before, consideration of the functions $\lambda(x)$ and $L\{x, \lambda(x)\}$ gives us no useful information. Turning to the dual problem, we find that

$$x_1(\lambda) = \tfrac{1}{2}\lambda_2/(1+\lambda_1)$$

and

$$x_2(\lambda) = \tfrac{1}{2}(\lambda_1 + \lambda_2)$$

The form of $L\{x(\lambda), \lambda\}$ is a little complicated (a foretaste of practical cases!), so let us try some specific values:

$\lambda = (0, 0)$ gives $x(\lambda) = (0, 0)$ and $L\{x(\lambda), \lambda\} = 0$

$\lambda = (0, 1)$ gives $x(\lambda) = (\tfrac{1}{2}, \tfrac{1}{2})$ and $L\{x(\lambda), \lambda\} = 0.5$

$\lambda = (0, 2)$ gives $x(\lambda) = (1, 1)$ and $L\{x(\lambda), \lambda\} = 0$

$\lambda = (1, 0)$ gives $x(\lambda) = (0, \tfrac{1}{2})$ and $L\{x(\lambda), \lambda\} = -0.25$

$\lambda = (1, 1)$ gives $x(\lambda) = (\tfrac{1}{4}, 1)$ and $L\{x(\lambda), \lambda\} = -0.125$

The largest of these values for $L\{x(\lambda), \lambda\}$ is indeed its maximum value (subject to $\lambda \geqslant 0$) and thus $\lambda^* = (0, 1)$ and $x^* = x(\lambda^*) = (\tfrac{1}{2}, \tfrac{1}{2})$ and $f(x^*) = L(x^*, \lambda^*) = \tfrac{1}{2}$. The first constraint is inactive, with $\lambda_1^* = 0$ and $g_1(x^*) > 0$, while constraint 2 is active, with $\lambda_2^* > 0$ and $g_2(x^*) = 0$.

Finally, in this section, we mention the important special case of *linear programming* problems, in which both $f(x)$ and $g(x)$ are linear functions of **x**. Suppose our problem is

$$\min f(x) \equiv \mathbf{c}^T\mathbf{x}$$

$$\text{subject to } g(x) \equiv \mathbf{Ax} - \mathbf{b} \geqslant \mathbf{0}$$

where **A** is a constant $m \times n$ matrix and **b** and **c** are constant vectors. The Lagrangian function is then

$$L(x,y) = \mathbf{c}^T\mathbf{x} - \mathbf{y}^T(\mathbf{Ax} - \mathbf{b})$$
$$= \mathbf{b}^T\mathbf{y} - \mathbf{x}^T(\mathbf{A}^T\mathbf{y} - \mathbf{c})$$

This is a case where the function $\mathbf{x}(y)$ which minimizes $L(x, y)$ is of little value, since each component $x_i(y)$ is $-\infty$, indeterminate, or $+\infty$ according as the coefficient of x_i in $L(x, y)$ is positive, zero, or negative (*cf.* $\lambda(x)$ in our first example). Thus we must turn to our other definition of the dual problem, which gives

$$\max L(x, y)$$

$$\text{subject to } \mathbf{y} \geqslant \mathbf{0}$$

$$\text{and } \nabla_x L(x, y) \equiv \mathbf{A}^T\mathbf{y} - \mathbf{c} = \mathbf{0}$$

But, when this last condition holds, we have

$$L(y) = \mathbf{b}^T \mathbf{y}$$

and this leads us to the usual form for the dual of a linear programming problem:

$$\max \mathbf{b}^T \mathbf{y}$$
$$\text{subject to } \mathbf{A}^T \mathbf{y} = \mathbf{c}, \mathbf{y} \geqslant \mathbf{0}$$

2.11 ALGORITHMS

In Sections 2.8 and 2.9 we discussed the properties which a function must possess at an optimizing point. The results obtained there provide us with tests which any such point must satisfy. Although this question of *identification* of solutions to optimization problems can cause difficulties when digital computers are used, most of the battle in structural optimization lies in devising efficient methods for the *location* of optima. In the simple examples, which we used to illustrate the theory, it was easy to differentiate the functions involved and solve the equations provided by the appropriate theorems. In most practical problems, with many variables and constraints, this procedure would require the solution of a large set of nonlinear equations, even supposing that we could determine which of the constraints were active at the solution. These difficulties usually require that some other method be used to locate optima and the most common procedures employ search techniques which recursively approach the optimizing point. We devote the next two sections to a general discussion of this type of method.

To clarify the ideas, consider the problem of finding a minimum value y^* for the function

$$y = f(x), y \in R^1, \mathbf{x} \in \mathbf{R}^n$$

We begin our search at a *starting point* $\mathbf{x}^{(1)} \in R^n$, $y^{(1)} = f\{x^{(1)}\} \neq y^*$ and, operating on this starting point according to some fixed set of rules, we arrive at a new point $\mathbf{x}^{(2)}$, $y = f\{x^{(2)}\} < y^{(1)}$. We can express the rules used to generate the new point by defining a *search function* A which maps any point of R^n onto another point of R^n and, in particular, maps $\mathbf{x}^{(1)}$ onto $\mathbf{x}^{(2)}$, i.e. $\mathbf{x}^{(2)} = \mathbf{A}\{x^{(1)}\}$. In the second stage, or *iteration*, of the search, we use $\mathbf{x}^{(2)}$, $y^{(2)}$ as starting point and arrive by means of the same set of rules at $\mathbf{x}^{(3)} = \mathbf{A}\{x^{(2)}\} \in R^n$, $y^{(3)} = f\{x^{(3)}\} < y^{(2)}$. The search continues in this manner, the kth iteration giving

$$\mathbf{x}^{(k+1)} = \mathbf{A}\{x^{(k)}\}, y^{(k+1)} = f(x^{(k+1)})$$

This procedure generates a sequence of points $x^{(k)}$; in most practical cases, this sequence is infinite and may therefore be written $\{x^{(k)}\}_1^\infty$. For the procedure to be of any value in solving our minimization problem, it is clearly necessary that

$$\mathbf{x}^{(\infty)} = \mathbf{x}^*$$

where

$$f(x^*) = y^*$$

In solving an actual problem, we cannot carry out the infinite number of operations needed to reach $\mathbf{x}^{(\infty)}$, so we must end our search after a finite number k of iterations, choosing k so as to achieve whatever accuracy we require. This accuracy requirement may be expressed in the form

$$|y^{(k+1)} - y^*| \leqslant \varepsilon'$$

where ε' is a suitably small positive number; since y^* is unknown, we must determine k by some other form of *termination criterion*, e.g.

$$|\nabla f\{x^{(k+1)}\}| \leqslant \varepsilon''$$

There are many possible types of criterion which may be used to decide when the iterative procedure has generated an acceptably close approximation to the optimizing point; most of these can be described by defining a function $B:R^n \to R^1$ and a positive number ε such that, if

$$B\{x^{(k+1)}\} \leqslant \varepsilon$$

then $\mathbf{x}^{(k+1)}$ is acceptably close to $\mathbf{x}^*$.

The entire process, outlined above, of starting from a given point and recursively generating a sequence of points until an acceptable approximation to the optimum has been located, is known as an *algorithm*. We may summarize the algorithmic process as follows:

Step 1   Choose a starting point $\mathbf{x}^{(1)} \in R^n, y^{(1)} = f\{x^{(1)}\}$ and choose a positive value for ε.
Step 2   Set $k = 0$.
Step 3   Set $k = k + 1$.
Step 4   Set $\mathbf{x}^{(k+1)} = \mathbf{A}\{x^{(k)}\}, y^{(k+1)} = f\{x^{(k)}\} < y^{(k)}$.
Step 5   If $B\{x^{(k+1)}\} \leqslant \varepsilon$, go to step 6; otherwise return to step 3.
Step 6   Set $\mathbf{x}^* = \mathbf{x}^{(k+1)}, y^* = y^{(k+1)}$ and stop.

In many algorithms, the search function of step 4 is defined in terms of a *search direction* $\mathbf{s}^{(k)}$ and a step length h_k. Thus we first decide that $\mathbf{A}\{x^{(k)}\}$ shall be of the form

$$\mathbf{A}\{x^{(k)}\} = \mathbf{x}^{(k)} + h\mathbf{s}^{(k)}$$

where $\mathbf{s}^{(k)} \in R^n$ is a fixed vector (dependent on $\mathbf{x}^{(k)}$) and h an unknown real number, and then we choose a specific value h_k for h. Often the definition of an algorithm requires that the step length be chosen so that $h = h_k$ minimizes $f\{x^{(k)} + hs^{(k)}\}$; however, we must recognize that, in general, this would require an infinite number of operations and that the best we can do in practice is to minimize $f\{x^{(k)} + hs^{(k)}\}$ to within some specified accuracy. Cauchy's method of steepest descent for minimizing a differentiable function is a simple example of this type of algorithm;

the search direction is the negative gradient direction

$$\mathbf{s}^{(k)} = -\nabla f\{x^{(k)}\}$$

and the step length is chosen to 'minimize' $f\{x^{(k)} + h\mathbf{s}^{(k)}\}$.

As an illustration of the algorithmic method, we use Cauchy's method to minimize the function

$$f(x) = (x_2 - x_1)^2 + (1 - x_1)^2.$$

In this case, it is easy to write down a general expression for the gradient:

$$\nabla f(x) = \left(\frac{\partial f}{\partial x_1}, \frac{\partial f}{\partial x_2}\right)$$

$$= (4x_1 - 2x_2 - 2, -2x_1 + 2x_2)$$

and of course we could easily determine $\mathbf{x}^* = (1, 1)$ by solving

$$f(x^*) = 0$$

However, our purpose is to illustrate a type of method which can be used in much more complicated situations:

Step 1   We arbitrarily choose $x^{(1)} = (5, 5)$, $y^{(1)} = 16$ and set $\varepsilon = \frac{1}{2}$; we shall use the termination criterion
$$B(x) = |\nabla f(x)| \leq \varepsilon.$$

Step 2   Set $k = 0$.

Step 3   Set $k = 1 + k$.

Step 4 $\mathbf{x}^{(2)} = \mathbf{A}\{x^{(1)}\}$
$$= x^{(1)} - h_1 \nabla f\{x^{(1)}\}$$
$$= (5, 5) + h_1(-8, 0)$$
$$= (5 - 8h_1, 5)$$
where $h = h_1$ is chosen to minimize
$$f(5 - 8h, 5) = (8h)^2 + (8h - 4)^2$$
A little algebra shows that $h_1 = \frac{1}{4}$ is the minimizing value and hence
$$x^{(2)} = (3, 5), \quad y^{(2)} = 8 < y^{(1)}$$

Step 5 $B\{x^{(2)}\} = |\nabla f\{x^{(2)}\}|$
$$= \left[\left(\frac{\partial f}{\partial x_1}\right)^2 + \left(\frac{\partial f}{\partial x_2}\right)^2\right]^{1/2} \text{ at } x = (3, 5)$$
$$= 4$$

Since the termination criterion is not satisfied, we must return to step 3 for a further iteration. Further steps are shown diagrammatically in Fig. 2.12.

Although the above example is only intended to illustrate the mechanics of carrying out an algorithm, it also serves to introduce some of the numerical difficulties which must be overcome when real problems are to be solved. It is clear, for example, that the use of another termination criterion, such as

$$\max\{|x_1^{(k+1)} - x_1^{(k)}|, |x_2^{(k+1)} - x_2^{(k)}|\} \leqslant 0.5$$

could cause the algorithm to locate a different 'solution'. Another noteworthy feature is the zigzag path towards the optimum, shown clearly in Fig. 2.12. We may note that this problem can be solved in a single step by the Newton method, which will be described later.

2.12 EFFICIENCY OF ALGORITHMS

An algorithm *converges* from a given starting point $x^{(1)}$ if the sequence $\{x^{(k)}\}$ of points which it generates has as its limit the solution x^* of the problem to be solved. The set of starting points from which a given algorithm converges to the solution of a given problem is its *domain of convergence* for that problem. In assessing the usefulness of an algorithm for solving any class of optimization problems, we should like to know how wide its domain of convergence is for members of the class, and also how much work we must do to obtain a satisfactory approximation to x^* by means of the algorithm.

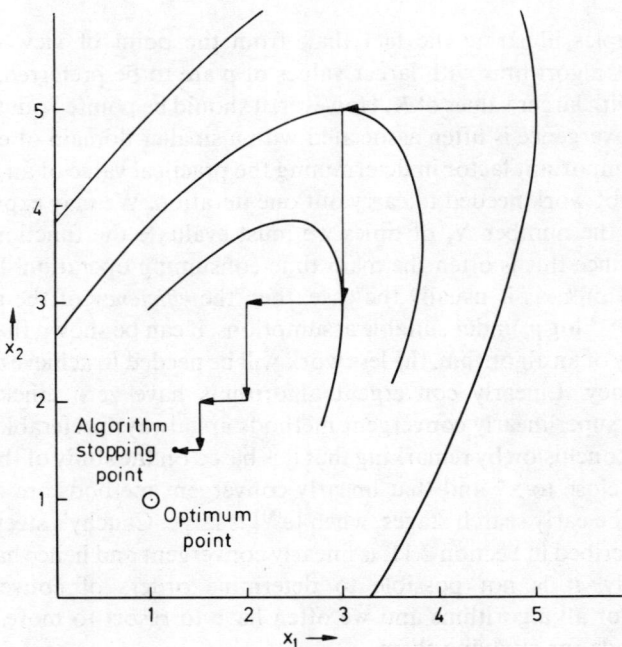

Fig. 2.12 Iteration history for the method of steepest descent in solving the problem   minimize $f(x) = (x_2 - x_1)^2 + (1 - x_1)^2$

Let $e^{(k)}$ denote the error $\mathbf{x}^* - \mathbf{x}^{(k)}$ in the kth iterative approximation to $\mathbf{x}^*$ generated by our algorithm. If there are real numbers $p \geqslant 1$ and $K \neq 0$ such that the sequence $\{|e^{(k+1)}|/|e^{(k)}|^p\}$ converges to K, then p is the *order of convergence* of the algorithm. If $p = 1$, the algorithm is said to possess *linear convergence*; if $p > 1$, the convergence is *superlinear*. Algorithms with an order of convergence of 2 are sometimes termed *quadratically convergent*; however, we shall reserve this term to describe algorithms which can locate exactly the minimum of a quadratic function in a finite number of iterations.

To illustrate the idea of order of convergence, consider the following three examples, in which we assume for simplicity that $|e^{(k+1)}| = K|e^{(k)}|^p$, which will be approximately true for large k.

(A) $K = 0.1, p = 1, |e^{(1)}| = 10^{-2}$
$|e^{(k)}| = \{10^{-2}, 10^{-3}, 10^{-4}, 10^{-5}, \ldots\}$

(B) $K = 1, p = 1.5, e^{(1)} = 10^{-2}$
$|e^{(k)}| = \{10^{-2}, 10^{-3}, 10^{-4.5}, 10^{-7.75}, \ldots\}$

(C) $K = 10, p = 2, e^{(1)} = 10^{-2}$
$|e^{(k)}| = \{10^{-2}, 10^{-3}, 10^{-5}, 10^{-9}, \ldots\}$

These examples illustrate the fact that, from the point of view of speed of convergence, algorithms with larger values of p are to be preferred, even when associated with larger values of K. However, it should be pointed out that a higher order of convergence is often associated with a smaller domain of convergence.

Another important factor in determining the practical value of an algorithm is the amount of work needed to carry out one iteration. We may express this, for example, as the number N_k of times we must evaluate the function $f(x)$ being optimized, since this is often the main time-consuming operation. If $N_k = N$ is independent of k, as is usually the case, then the *efficiency* of the algorithm is defined as $N^{-1} \log p$; under suitable assumptions, it can be shown that the higher the efficiency of an algorithm, the less work will be needed to achieve a solution of given accuracy. Linearly convergent algorithms have zero efficiency, which implies that superlinearly convergent methods are always preferable; we should temper this conclusion by remarking that it is based on the study of the behaviour of methods close to $\mathbf{x}^*$ and that linearly convergent methods are often highly effective in the early search stages, when $|e^{(k)}|$ is large. Cauchy's steepest descent method, described in Section 2.11, is linearly convergent and hence has effiency 0; unfortunately, it is not possible to determine orders of convergence and efficiencies for all algorithms and we often have to resort to more rough-and-ready methods for studying them.

A. J. M.

REFERENCES

1. T. M. Apostol, *Mathematical Analysis*, Addison-Wesley, 1957.
2. G. Birkhoff and S. Maclane, *A Survey of Modern Algebra*, The Macmillan Company, 1953.
3. T. M. Flett, *Mathematical Analysis*, McGraw-Hill, 1965.
4. W. Nef, *Linear Algebra*, McGraw-Hill, 1967.
5. W. I. Zangwill, *Nonlinear Programming: A Unified Approach*, Prentice Hall, 1969.
6. M. Avriel, *Nonlinear Programming: Analysis and Methods*, Prentice Hall, 1976.
7. D. M. Himmelblau, *Applied Nonlinear Programming*, McGraw-Hill, 1972.
8. O. Mangasarian, *Nonlinear Programming*, McGraw-Hill, 1969.

Foundations of Structural Optimization: A Unified Approach
Edited by A. J. Morris
© 1982 John Wiley & Sons Ltd.

Chapter 3

Elementary Concepts in Structural Optimization

3.1 INTRODUCTION

The concepts outlined in Chapter 2 have a general importance for the entire field of mathematical programming or general optimization. Using this framework attention can now be focused on structural optimization problems and, in particular, on the basic ideas on which solution algorithms can be established. Thus, in this chapter, some elementary concepts in the theory of optimal structural design are introduced.

In order to discuss the theory of structural optimization it is necessary to be able to analyse the structures being designed. For completeness the basic principles of structural analysis are introduced, including a brief description of the finite element method. However, this description is simply an outline designed more to refresh the readers' memories rather than attempting to substitute for the standard texts on analysis.

For the most part the theory is developed using pin-jointed frameworks progressing to finite element formulations in later sections. The aim is to introduce optimality criteria and duality theory as applied to structures in the simplest manner. Nevertheless, there is no loss of generality and the sound understanding of the ideas presented in this chapter will provide the reader with a sufficient grasp of the subject to allow an appreciation of the most advanced solution techniques available. It is a necessary building block to allow continued reading of the remaining chapters of the book.

3.2 STATICALLY DETERMINATE STRUCTURES

In starting an inquiry into structural optimization theory it is clearly convenient to begin with the simplest of all structures, the statically determinate pin-jointed framework. This offers two important advantages. Firstly, many of the equations and expressions essential for the development of a theoretical basis for structural optimization are exact for this case. Secondly, the formulations for structural response are particularly easy to manipulate and allow the development of the theory to be followed with ease.

Despite this simplicity the optimality conditions and the dual problems which are derived have wide applications and have formed the basis for some of the more important solution algorithms. Thus, a complete understanding of the principles illustrated in this section provides a sound base on which more complex formulations can be built. Indeed, it is possible to argue that the bulk of modern structural optimization theory can be developed using only these principles.

3.2.1 Analysis

As indicated in Chapter 1 analysis and structural optimization are intimately linked. Thus, before an attempt is made to define optimality or duality conditions for pin-jointed frameworks, the basic equations of structural analysis must be considered. In this section the elementary two-bar framework shown in Fig. 3.1 is used for illustrative purposes but the equations derived have a more general implication.

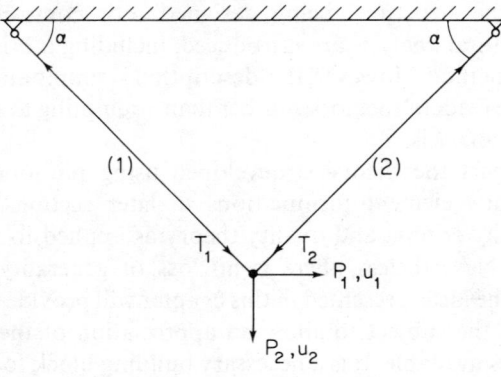

Fig. 3.1   A two-bar pin-jointed framework subject to two loads P_1, P_2 and with displacements u_1, u_2 and internal bar forces T_1, T_2

The two applied loads P_1, P_2 acting on the structure in Fig. 3.1 at the free node give rise to forces in the two bars denoted by T_1, T_2. Equilibrium between these internal and external forces generates the two equations:

$$\begin{bmatrix} P_1 \\ P_2 \end{bmatrix} = \begin{bmatrix} \cos\alpha & -\cos\alpha \\ \sin\alpha & \sin\alpha \end{bmatrix} \begin{bmatrix} T_1 \\ T_2 \end{bmatrix}$$

For a general structure with n bars and m nodes, this matrix equation becomes

$$\mathbf{P} = \mathbf{B}^T \mathbf{T}$$

where the vectors $\mathbf{P}$ and $\mathbf{T}$ have n elements and the matrix $\mathbf{B}$ is a square $n \times n$ matrix. The vector denoted by $\mathbf{T}$ represents the internal bar forces whilst the vector $\mathbf{P}$ is the applied loads. For known applied loads the corresponding internal forces can be found by solving this matrix equation to yield

$$\mathbf{T} = (\mathbf{B}^T)^{-1} \mathbf{P}$$

and in the special case of the two bar framework we have

$$\begin{bmatrix} T_1 \\ T_2 \end{bmatrix} = \frac{1}{2\cos\alpha \sin\alpha} \begin{bmatrix} \sin\alpha & \cos\alpha \\ -\sin\alpha & \cos\alpha \end{bmatrix} \begin{bmatrix} P_1 \\ P_2 \end{bmatrix}$$

Thus, for a statically determinate structure the internal bar forces depend only on the applied loads and the direction cosines of the individual bars. Similarly, for a general framework the bar stresses given by

$$\sigma_i = T_i / x_i$$

depend on the applied loads, the geometry of the structure and the bar cross-sectional areas $\mathbf{x} = (x_1, x_2, x_3, \ldots, x_n)$.

Having established a relationship between the internal stresses and the applied loads through equilibrium conditions we now want to examine compatibility and relate nodal displacements to the applied loads. Starting with the two-bar framework in Fig. 3.1 the application of Hooke's law and simple manipulation yields

$$\begin{bmatrix} T_1 \\ T_2 \end{bmatrix} = \begin{bmatrix} \dfrac{x_1 E}{l_1} & 0 \\ 0 & \dfrac{x_2 E}{l_2} \end{bmatrix} \begin{bmatrix} \cos\alpha & \sin\alpha \\ -\cos\alpha & \sin\alpha \end{bmatrix} \begin{bmatrix} u_1 \\ u_2 \end{bmatrix}$$

and thus the applied loads are related to the nodal displacements through the relationship

$$\begin{bmatrix} P_1 \\ P_2 \end{bmatrix} = \begin{bmatrix} \cos\alpha & -\cos\alpha \\ \sin\alpha & \sin\alpha \end{bmatrix} \begin{bmatrix} \dfrac{x_1 E}{l_1} & 0 \\ 0 & \dfrac{x_2 E}{l_2} \end{bmatrix} \begin{bmatrix} \cos\alpha & \sin\alpha \\ -\cos\alpha & \sin\alpha \end{bmatrix} \begin{bmatrix} u_1 \\ u_2 \end{bmatrix}$$

where l_1 and l_2 represent the lengths of the bars. For the general case this becomes

$$\mathbf{P} = \mathbf{B}^T \mathbf{D} \mathbf{B} \mathbf{u} = \mathbf{K} \mathbf{u}$$

where **D** is the generalized version of the corresponding Hooke's-law matrix, and the composite matrix **K** is called the stiffness matrix. The use of this stiffness matrix anticipates concepts which are required later when finite elements are used for structural analysis. We have now derived the matrices required to evaluate displacements in terms of internal loads and thus, for a general framework,

$$\mathbf{u} = (\mathbf{B}^T)^{-1}\mathbf{D}^{-1}\mathbf{T}$$

In this case the vector **u** gives all the displacements at all nodes and it is more usual in optimization to be interested in a specific displacement u_j (say). In order to extract the required component, the vector **u** may be multiplied by a vector **e** which contains zero elements except for the jth component which contains a 1. Thus

$$\mathbf{e}_{(j)}^T = \{0, 0, \ldots, 1, 0, 0, \ldots, 0\}$$

and

$$u_j = \mathbf{e}_{(j)}^T \mathbf{u} = \mathbf{e}^T B^{-1} \mathbf{D}^{-1} \mathbf{T}$$

or

$$u_j = \mathbf{t}^{(j)} \mathbf{D}^{-1} \mathbf{T}$$

where $\mathbf{t}^{(j)} = \mathbf{e}_{(j)}^T \mathbf{B}^{-1}$ and represents the internal forces in the bars due to the application of a unit, or dummy load, at u_j and acting in the direction of the displacement component. Expanding this matrix product we recover the familiar expression for calculating the magnitude of specific nodal displacements in pin-jointed frameworks,

$$u_j = \sum_{i=1}^n \frac{T_i l_i t_i^{(j)}}{E x_i}$$

where t_i^j represent the components of the vector $\mathbf{t}^{(j)}$ and l_i and x_i are again bar lengths and cross-sectional areas respectively.

3.2.2 Optimum Design of Strength-critical Frameworks

The most elementary optimum design problem for this class of structure consists in finding a set of bar cross-sectional areas which minimize the structural weight subject to limits on the allowable stresses in individual members. Although the problem is, in many aspects, trivial, it nevertheless forms a useful model for illustrating some of the concepts which play important rôles when more complex problems are considered.

The design problem requires that we find a vector $\mathbf{x}^* = (x_1^*, x_2^*, \ldots, x_n^*)^T$ for a structure of fixed layout which minimizes the weight

$$W = \sum_{i=1}^n \rho_i l_i x_i$$

Elementary Concepts in Structural Optimization

subject to the stress constraints

$$-\bar{\sigma}_i \leq \frac{T_i}{x_i} \leq \bar{\sigma}_i \qquad i = 1, 2, \ldots, n$$

where $\pm \bar{\sigma}_i$ represents the upper and lower bounds on the bar stresses and ρ_i is the specific weight for the ith bar. Because the structure is determinate each bar can be sized separately and the minimum value of the cross-sectional areas adequate to carry the applied loads are given by

$$x_i^* = \frac{T_i}{\bar{\sigma}_i} \qquad i = 1, 2, \ldots, n$$

If we take the vector **x** as a non-optimal set of cross-sectional areas then the above formulae can be rewritten in the form

$$x_i^* = \frac{\sigma_i}{\bar{\sigma}_i} x_i \qquad i = 1, 2, \ldots, n$$

which we can immediately recognize as the stress-ratioing formulae introduced in Chapter 1.

Returning to the simple two-bar problem shown in Fig. 3.1 and defining the angle α by

$$\alpha = \tan^{-1} 2$$

then the internal bar forces become

$$\begin{bmatrix} T_1 \\ T_2 \end{bmatrix} = P\sqrt{5} \begin{bmatrix} 2 \\ 1 \end{bmatrix}$$

If a minimum-weight design is now sought subject to limitations on the bar stresses of $\pm \bar{\sigma}$ then the constraints imposed on the design problem become

$$0 \geq g_1(x_1) = \frac{2P\sqrt{5}}{\bar{\sigma}} - x_1, \qquad 0 \geq g_2(x_2) = \frac{P\sqrt{5}}{\bar{\sigma}} - x_2$$

and this two-dimensional design problem is shown in Fig. 3.2. The problem is linear and the constraints are parallel to the axes defined by the design variables x_1 and x_2. It is clearly seen that each of these variables is associated with one and only one constraint and that the optimum design occurs at a vertex in design space. The optimum can, therefore, be found by seeking to simultaneously satisfy the design constraints rather than seeking to actually minimize the objective function. Although we are dealing with a determinate structure many of the features recur when indeterminate structures are designed for minimum weight. As a result many structural optimization algorithms concentrate on constraint satisfaction when seeking for optimum designs and many such designs do, indeed, turn out to be vertex solutions.

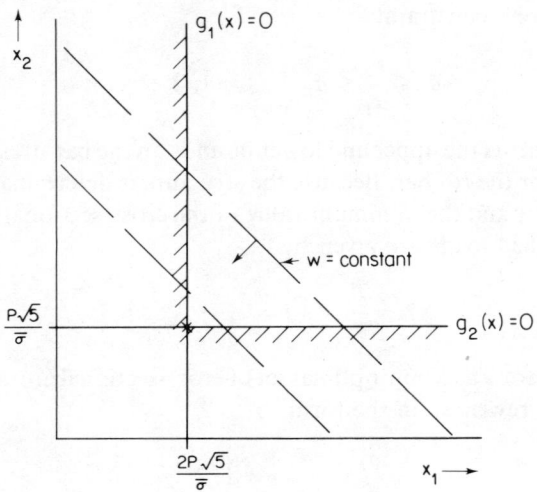

Fig. 3.2 Two-bar optimum design problém with stress constraints using cross-sectional areas as design variables

In later sections it is convenient to linearize the design constraints by using design variables defined as the reciprocals of the bar cross-sectional areas

$$z_i = \frac{1}{x_i} \qquad i = 1, 2, \ldots, n$$

The weight now becomes a nonlinear function given by

$$W = \sum_{i=1}^{n} \frac{\rho_i l_i}{z_i}$$

but the stress constraints remain linear,

$$-\bar{\sigma}_i \leqslant T_i z_i \leqslant \bar{\sigma}_i.$$

For the simple two-bar problem defined above the stress constraints are now given by

$$0 \geqslant g_1(z_1) = 2P\sqrt{5}\,z_1 - \bar{\sigma}, \qquad 0 \geqslant g_2(z_2) = P\sqrt{5}\,z_2 - \bar{\sigma}$$

and the resulting design problem is shown in Fig. 3.3.

The nonlinear nature of the transformed objective function is indicated by the curved constant weight lines which replace the straight lines of Fig. 3.2. For the current problem the introduction of such a nonlinearity is clearly a retrograde step but, as indicated later, the transformation has an important rôle in the solution of more complex structural optimization problems.

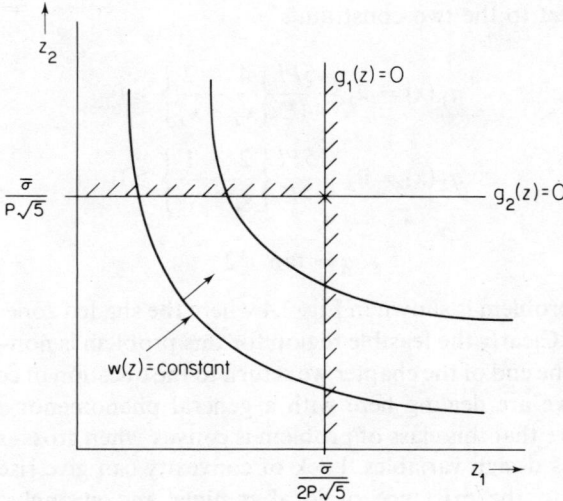

Fig. 3.3 Two-bar optimum design problem with stress constraints using the reciprocal of the cross-sectional areas as design variables

3.2.3 Optimum Design of Stiffness-critical Frameworks

Although the stress-constrained problem formed one of the starting points for the development of computer-based structural optimization methods, one of the major breakthroughs in developing practical techniques occurred when stiffness-constrained problems were studied. Techniques suitable for the solution of realistic design problems must be able to treat complex structures but once again many of the basic concepts and principles of the resulting algorithms are founded on the assumption that the structure is determinate. Thus we can continue our study of structural optimality theory by considering a statically determinate pin-jointed framework subject to constraints on specified nodal displacements.

3.2.3.1 Convexity

The first point to be addressed concerns the selection of the design variables which, for displacement-constrained statically determinate structures, has a direct bearing on the convexity of the problem. If cross-sectional areas are taken as the design variables then the problem may not be convex, as is easily illustrated by returning to the two-bar framework of Fig. 3.1. For such a layout we now consider the displacement-constrained problem where the minimum of the objective function for a structure with the same material in each bar

$$W(x) = \rho l_1 x_1 + \rho l_2 x_2$$

is sought subject to the two constraints

$$g_1(x) = \bar{u}_1 - \frac{5Pl}{4E}\left\{\frac{4}{x_1} - \frac{2}{x_1}\right\} \geq 0$$

$$g_2(x) = \bar{u}_2 - \frac{5Pl}{4E}\left\{\frac{2}{x_1} + \frac{1}{x_2}\right\} \geq 0$$

given that

$$\alpha = \tan^{-1} 2$$

This design problem is shown in Fig. 3.4 where the shaded zone represents the feasible region. Clearly the feasible region for this problem is non-convex. In the problem set at the end of the chapter we return to the question of convexity and it is found that we are dealing here with a general phenomenon since it is not possible to prove that this class of problem is convex when cross-sectional areas are employed as design variables. Lack of convexity can give rise to numerical problems due to the existence of local minima and strongly suggests that alternative design variables should be used. Before leaving this problem it should be emphasized that a non-convex feasible region need not give rise to a non-

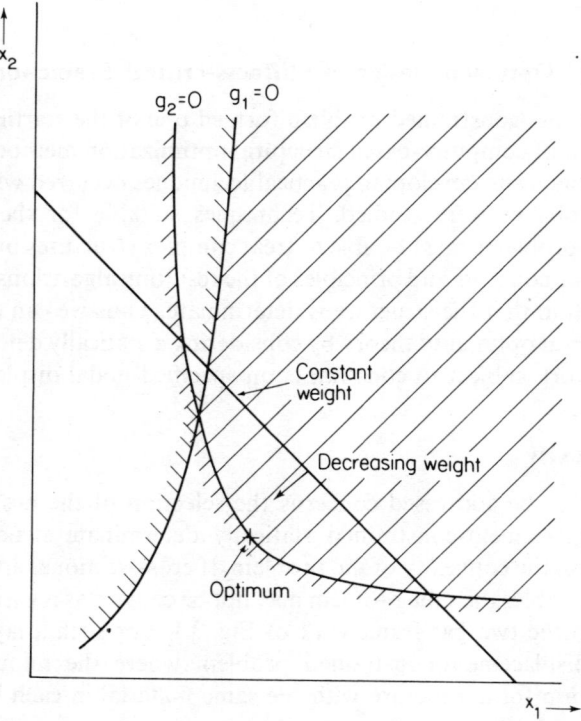

Fig. 3.4 Two-bar optimum design problem with displacement constraints using cross-sectional areas as design variables

Elementary Concepts in Structural Optimization

convex design problem since convexity of the overall problem is a function of both the constraints and the objective function. However, such a combination is fortuitous and non-convexity of the feasible region is best avoided.

In seeking a solution to this problem we can take the hint given in the last section and use the reciprocal of the cross-sectional areas as design variables. With this transformation the two-bar displacement-constrained problem now becomes

$$\text{minimize} \quad W(z) = \rho l \left\{ \frac{1}{z_1} + \frac{1}{z_2} \right\}$$

subject to

$$g_1(z) = \bar{u}_1 - \frac{5Pl}{4E} \{ 4z_1 - 2z_2 \} \geq 0$$

$$g_2(z) = \bar{u}_2 - \frac{5Pl}{4E} \{ 2z_1 + z_2 \} \geq 0$$

where $z_i = 1/x_i$ ($i = 1, 2$). The shape of the feasible region has now changed as can be seen from the shaded area on Fig. 3.5. This diagram also shows that the nonlinear constraints of Fig. 3.4 are now linear whilst the linear objective function

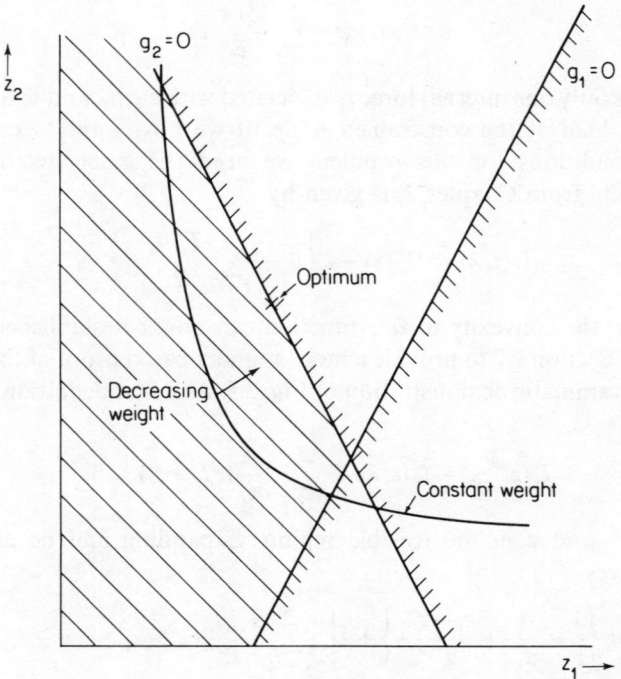

Fig. 3.5 Two-bar optimum design problem with displacement constraints using the reciprocal of the cross-sectional areas as design variables

is nonlinear. Although the essential nonlinear nature of the problem has not been removed the use of reciprocal variables has rendered the problem convex and produced linear constraints which later sections exploit to create efficient algorithms. The convexity property applies to minimum-weight design problems for statically determinate structures subject to displacement constraints and is referred to again in the problems at the end of this chapter.

3.2.3.2 The single displacement-constrained problem

Having considered convexity we now want to examine the optimality conditions associated with the displacement constrained problem. In order to do this we turn to a simple problem where the minimum weight is sought for a structure with the same material in each bar where there is only a single displacement constraint and a single loading condition. Employing reciprocal variables this optimum design problem requires finding a vector $\mathbf{z}^* = (z_1^*, z_2^*, \ldots, z_n^*)^T$ that

$$\text{minimizes} \quad W(z) = \sum_{i=1}^{n} \frac{\rho l_i}{z_i}$$

subject to (MWP1)

$$\bar{u} \geqslant u(z) \equiv \sum_{i=1}^{n} \frac{T_i l_i t_i}{E} z_i$$

where there is only one internal force t_i associated with each bar due to the action of a dummy load at the constrained node. If we now want to establish the optimality conditions for this problem we need the associated Lagrangian function which, from Chapter 2, is given by

$$L(z, \lambda) = W(x) - \lambda \left\{ \bar{u} - \sum_{i=1}^{n} \frac{T_i l_i t_i}{E} z_i \right\}$$

As a first step the convexity of this function can now be established using the definitions of Section 2.7 to provide a more soundly based proof of this property than the diagrammatic demonstration of Fig. 3.5. As a first definition we require that

$$L(z^+, \lambda) - L(z, \lambda) - \sum_{i=1}^{n} \frac{\partial L}{\partial z_i}(z_i^+ - z_i) \geqslant 0$$

for points $\mathbf{z}^+$ and $\mathbf{z}$ in the feasible region. Expanding and developing this expression gives

$$\sum_{i=1}^{n} \rho l_i \left\{ \frac{1}{z_i} - \frac{1}{z_i^+} + \frac{z_i - z_i^+}{z_i^{+2}} \right\} + \lambda \left[\sum_{i=1}^{n} \frac{T_i l_i t_i}{E} \{z_i - z_i^+ - (z_i - z_i^+)\} \right]$$

$$= \sum_{i=1}^{n} \frac{\rho l_i}{z_i z_i^{+2}} (z_i - z_i^+)^2 \geqslant 0$$

Elementary Concepts in Structural Optimization

because the design variables cannot be negative and, thus, the problem is convex. Alternatively, convexity can be established by noting that the Hessian matrix is given by

$$H = \text{diag}\left[\frac{2\rho l_i}{z_i^3}\right]$$

which is positive definite since each term in this diagonal matrix is positive. Thus, a vector z^* minimizing (MWP1) is unique and represents a global minimum for the problem. It should be noted that the convexity property and the consequential uniqueness of the minimizing vector z^* is not a function of the problem but of both the structure and the selection of the design variables.

Using the Lagrangian function as the basis for forming the optimality criterion we recall the Kuhn–Tucker optimality conditions defined in Chapter 2 then the necessary and sufficient optimality conditions for this convex problem are given by

$$\nabla_z L(z^*, \lambda^*) = 0,$$

$$\lambda^* \left\{ \sum_{i=1}^{n} \frac{T_i l_i t_i}{E} z_i^* - \bar{u} \right\} = 0$$

$$\lambda^* \geqslant 0, \qquad \bar{u} \geqslant u(z^*)$$

where the solution to these equations λ^*, z_i^* ($i = 1, 2, \ldots, n$) represent optimizing values. From the first of these equations,

$$\frac{\partial L}{\partial z_k} = \frac{\rho l_k}{z_k^{*2}} + \lambda^* T_k l_k t_k = 0 \qquad k = 1, 2, \ldots, n$$

or (3.1)

$$\left(\frac{1}{z_k^*}\right)^2 = \lambda^* \frac{T_k t_k}{\rho E} \qquad k = 1, 2, \ldots, n$$

which provide a first definition of the optimality criterion. These equations can be rearranged to provide the more illuminating form,

$$\sigma_k^{*+} \sigma_k^* = \frac{\rho E}{\lambda^*} \qquad k = 1, 2, \ldots, n$$

with

$$\sigma_k^{*+} = t_k^* z_k^* \qquad k = 1, 2, \ldots, n$$
$$\sigma_k^* = T z_k^*$$

If the Lagrangian multiplier λ^* is now interpreted as a pseudo-load, then

$$\tilde{\sigma}_k^* \sigma_k^* = \rho E = \text{constant} \qquad k = 1, 2, \ldots, n$$

The terms

$$\tilde{\sigma}_k^* = \lambda^* \sigma_k^{*+} \qquad k = 1, 2, \ldots, n$$

are the stresses in the bars of the framework due to the application of a load of magnitude λ^* in the direction of the constrained displacement. Since $\tilde{\sigma}_k^*$ is a stress due to a pseudo-load system and σ_k^* is a stress due to an actual load system then we can interpret $\tilde{\sigma}_k^* \sigma_k^*/E$ as the virtual strain energy of the bar k. Thus, for the problem (MWP 1), the optimum is achieved when the *virtual strain energy for each bar has the same value*.

This is a nice optimality theorem and is aesthetically pleasing but, unfortunately, does not permit us to identify the optimizing values for the design variables essential for a complete solution. To achieve such a solution we first observe that the single displacement constraint must be active at the optimum, hence

$$\bar{u} = u = \sum_{i=1}^{n} \frac{T_i l_i t_i}{E} z_i^*$$

By using expression (3.1) the optimizing design variables z^* can be replaced to give

$$\bar{u} = \sum_{i=1}^{n} \frac{T_i l_i t_i}{E} \bigg/ \sqrt{\frac{\lambda^* T_i t_i}{\rho E}} = \sum_{i=1}^{n} l_i \sqrt{\frac{T_i t_i \rho}{\lambda^* E}}$$

and the optimizing value for the Lagrangian multiplier can now be extracted;

$$\sqrt{\lambda^*} = \frac{1}{\bar{u}} \sum_{i=1}^{n} l_i \sqrt{\frac{T_i t_i \rho}{E}}$$

We are now poised to obtain the optimizing values for the design variables which are found by once more using (3.1) to provide the solution:

$$\frac{1}{z_i^*} = \left[\frac{1}{\bar{u}} \sum_{i=1}^{n} l_i \sqrt{\frac{\rho T_i t_i}{E}} \right] \sqrt{\frac{T_i t_i}{\rho E}} \qquad i = 1, 2, \ldots, n$$

and thus the optimum weight is given by

$$W_{(opt)} = \frac{\rho}{\bar{u} E} \left\{ \sum_{i=1}^{n} l_i \sqrt{T_i t_i} \right\}^2$$

The path followed has permitted a direct exploitation of the optimality criteria to achieve a complete optimal solution to the design problem (MWP 1). This is made possible because of the basic simplicity of the problem. Nevertheless, this simplicity provides a useful model for more complex indeterminate situations which is exploited later as the basis for a successful solution-seeking algorithm.

Although the optimality conditions are important, both in the theory of optimal structures and for algorithm development, duality theory has a significant rôle to play when practical computer programs are considered. As later sections show, many algorithms use the dual formulation as an essential part in the optimum-seeking process. In addition, the dual formulation offers the possibility of bounding the optimum weight and thereby provides the designer with some measure of the nearness of a specific design to the constrained minimum. Turning to the present problem, (MWP 1) represents the primal formulation and the dual problem is formulated using the procedure described in Section 2.10. Thus we seek to maximize the Lagrangian function

$$L(z, \lambda) = W(z) - \lambda \left\{ \bar{u} - \sum_{i=1}^{n} \frac{T_i l_i t_i}{E} z_i \right\}$$

subject to the constraints

$$\frac{\partial L(z, \lambda)}{\partial z_k} = -\frac{\rho l_k}{z_k^2} + \frac{\lambda T_k l_k t_k}{E} = 0 \qquad k = 1, 2, \ldots, n$$

over the variables z, λ and where $L(z, \lambda)$ is now the dual objective function. A modified form for the problem can be obtained by observing that each of the constraint equations can be multiplied by the appropriate design variable and summed to give

$$\sum_{k=1}^{n} z_k \frac{\partial L(z, \lambda)}{\partial z_k} = -W(z) + \lambda \sum_{k=1}^{n} \frac{T_k l_k t_k}{E} z_k = 0$$

This expression can now be used to substitute for the weight term $W(z)$ in the Lagrangian function to give

$$L(z, \lambda) = \lambda \left\{ 2 \sum_{i=1}^{n} \frac{T_i l_i t_i}{E} z_i - \bar{u} \right\}$$

which can now be interpreted as an alternative dual objective function. The dual problem which can now be defined as finding vectors z^*, λ^* which

$$\text{maximizes} \quad \lambda \left\{ 2 \sum_{i=1}^{n} \frac{T_i l_i t_i}{E} z_i - \bar{u} \right\}$$

subject to the constraints

$$\frac{\rho l_i}{z_i^2} = \frac{\lambda T_i l_i t_i}{E} \qquad i = 1, 2, \ldots, n$$

or, recalling the definition for the displacement u, the dual objective can be written in yet a further alternative form:

$$L(z, \lambda) = \lambda(2u - \bar{u})$$

In order to illustrate a dual problem we take the particularly trivial case shown in Fig. 3.6 of a bar of length l and cross-sectional area x subject to a single load P applied at the free end. The minimum design problem requires minimizing the bar weight

$$W = \frac{\rho l}{z}$$

subject to the constraint that displacement at the free end

$$u = \frac{lPz}{E} \leqslant \bar{u}.$$

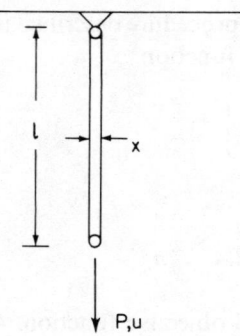

Fig. 3.6 A bar extended a distance u by an applied load P

This definition represents the primal problem shown in Fig. 3.7(a) with feasible region being to the left of the constraint $u = \bar{u}$. The dual problem, from the above discussion, is defined as

$$\text{maximize} \quad L(z, \lambda) = \frac{\rho l}{z} - \lambda \left(\bar{u} - \frac{P l z}{E} \right)$$

subject to the dual constraint

$$\frac{\partial L}{\partial z} = \frac{\rho l}{z^2} + \frac{\lambda P l}{E} = 0$$

and having selected a particularly simple problem we can use the dual constraint to solve for z, i.e. $z(\lambda) = \sqrt{\rho E / \lambda P}$ and obtain a dual problem which is now an unconstrained optimization problem. This new problem requires finding a value

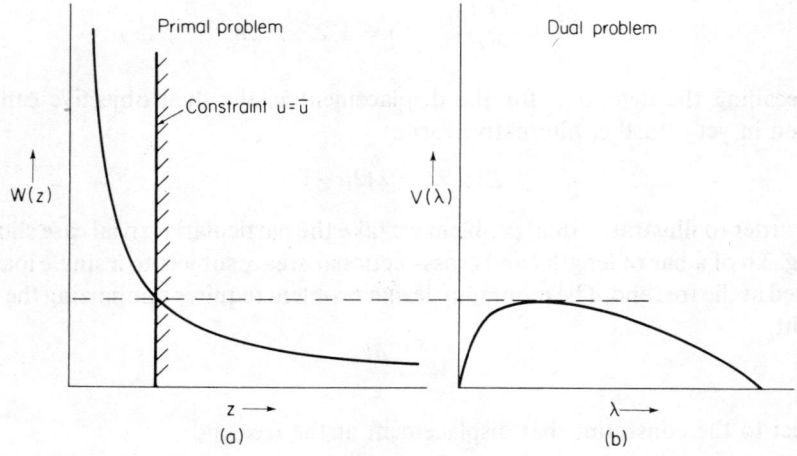

Fig. 3.7 Primal and dual problem for the single-bar minimum weight design problem

Elementary Concepts in Structural Optimization

λ^* for the Lagrangian multiplier which

$$\text{maximizes} \quad L(z(\lambda), \lambda) = V(\lambda) = 2l\sqrt{\frac{P\rho}{E}}\lambda^{1/2} - \lambda\bar{u}$$

and a plot of the function $V(\lambda)$ is shown in Fig. 3.7(b). Both primal and dual problems can be solved easily to give optimizing values

$$z^* = \frac{\bar{u}E}{Pl}, \quad \lambda^* = \frac{P\rho l^2}{\bar{u}^2 E}$$

giving

$$W(z^*) = V(\lambda^*) = \frac{P\rho l^2}{\bar{u}E}$$

and we may observe, in passing, that the sensitivity of the minimum weight to changes in the value of the constraint limit is given by

$$\frac{\partial W^*}{\partial \bar{u}} = -\frac{P\rho l^2}{\bar{u}^2 E} = -\lambda^*$$

Comparing this expression with that given in Section 2.9 we observe that a negative sign has appeared due to the use of reciprocal variables.

Although the bounding of the optimum by feasible values for the primal and dual problems is clearly seen by comparing Figs 3.7(a) and 3.7(b), it is helpful to calculate bounds in a more formal manner similar to the procedures adopted in certain of the major optimization programs. If we somewhat arbitrarily take a feasible value of the primal variable $z = 4\bar{u}E/5Pl$ then structural weight $W(z)$ is

$$\frac{5}{4}\frac{P\rho l^2}{\bar{u}E} = \frac{5}{4}W(z^*)$$

The dual variable associated with this value of z is found by solving the single dual constraint to give

$$\lambda(z) = \frac{25}{16}\frac{P\rho l^2}{\bar{u}^2 E}$$

thus

$$V(\lambda) = \frac{15}{16}\frac{P\rho l^2}{\bar{u}E} = \frac{15}{16}V(\lambda^*).$$

The bound is then provided by observing that

$$W(z) > W(z^*) = V(\lambda^*) > V(\lambda)$$

3.2.3.3 Multiple displacement-constrained problem

In turning to the minimum-weight design problem for a statically determinate pin-jointed framework subject to multiple displacement constraints we extend the theory discussed in the last section and are brought into contact with some of the mathematical complexities encountered when optimum designs for practical structures are sought. Because the single displacement constraint characterizes the solution of the structural problem the optimizing values for the design variables and the Lagrangian multiplier are directly available. Additional constraints destroy this attractive feature and render direct evaluation impossible so that a solution must now be sought by means of an optimum-seeking algorithm.

The problem being addressed is one of finding a vector of design variables $\mathbf{z}^* = (z_1^*, z_2^*, \ldots, z_n^*)^T$:

$$\text{minimize} \quad W(z) = \sum_{i=1}^{n} \frac{\rho l_i}{z_i} \quad \text{(MWP2)}$$

subject to

$$\bar{u}_j \geq u_j(z) = \sum_{i=1}^{n} \frac{T_i l_i t_i^{(j)}}{E} z_i$$

where the terms have their previous definitions and $t_i^{(j)}$ represents the force in bar i due to the application of a unit load of node j in the direction of u_j. The associated Lagrangian function is derived in the usual manner:

$$L(z, \lambda) = W(z) - \sum_{j=1}^{m} \lambda_j \left(\bar{u}_j - \sum_{i=1}^{n} \frac{T_i l_i t_i^{(j)} z_i}{E} \right)$$

and the Kuhn–Tucker sufficient optimality conditions become

$$\nabla_z L(z^*, \lambda^*) = 0$$

$$\lambda_j^* \left(\bar{u}_j - \sum_{i=1}^{n} \frac{T_i l_i t_i^{(j)} z_i^*}{E} \right) = 0 \quad j = 1, 2, \ldots, m$$

$$\lambda_j^* \geq 0, \quad \bar{u}_j \geq u_j(z^*) \quad j = 1, 2, \ldots, m$$

By following the arguments advanced in the case of a single constraint these conditions can be re-interpreted to provide an optimality condition which requires satisfaction of the expression

$$\sum_{j=1}^{m} \frac{C_{ij}}{\text{vol}_i} = \text{constant}$$

with

$$C_{ij} = \tilde{\sigma}_{ij}^* \frac{\sigma_i^*}{E} \quad (i \text{ not summed})$$

and $\tilde{\sigma}_{ij}^*$ now represents the stress in bar i due to the application of a load of magnitude λ_j^* and vol_i represents the volume of the ith bar. Again we are dealing with strain energy density terms but on this occasion the formulation is not so neat as in the single constraint case and we cannot now directly solve the optimality conditions to provide optimizing values for the inverse variables z_i^* ($i = 1, 2, \ldots, n$) and the Lagrangian multipliers λ_j^* ($j = 1, 2, \ldots, m$).

Although the avenue to a direct solution is closed we can, as has become our custom, exploit the optimality criterion to create the associated dual. Thus, the dual problem entails finding vectors $\mathbf{z}^*$, $\boldsymbol{\lambda}^*$ which

$$\text{maximize} \quad L(z, \lambda) = W(z) + \sum_{j=1}^{m} \lambda_j (u_j(z) - \bar{u}_j)$$

subject to

$$\nabla_z L(z, \lambda) = 0 \qquad \lambda_j \geq 0 \qquad j = 1, \ldots, m$$

or

$$\frac{\rho_k l_k}{z_k^2} - \sum_{j=1}^{m} \lambda_j \frac{T_k l_k t_k^{(j)}}{E} = 0 \qquad k = 1, 2, \ldots, n$$

Multiplying each equation by z_k and summing gives

$$\sum_{k=1}^{n} \frac{\rho_k l_k}{z_k} = W(z) = \sum_{k=1}^{n} \sum_{j=1}^{m} \lambda_j \frac{T_k l_k t_i^{(j)}}{E} z_k$$

Thus, the dual problem becomes

$$\text{maximize} \quad 2W(z) - \sum_{j=1}^{m} \lambda_j \bar{u}_j \qquad (3.2)$$

subject to

$$\nabla_z L(z, \lambda) = 0 \qquad \lambda_j \geq 0 \qquad j = 1, \ldots, m$$

This form of the dual mixing primal variables and Lagrangian multipliers (or dual variables) in the objective function is not particularly pleasing. However, there is nothing we can do about this unsatisfactory state but if, by chance, all the constraints of the problem are active some further progress can be made. In this situation $\bar{u}_j = u_j(z)$ ($j = 1, 2, \ldots, m$), $W(z) = \sum_{j=1}^{m} \lambda_j \bar{u}_j$ and the dual becomes

$$\text{minimize} \quad \sum_{j=1}^{m} \lambda_j \bar{u}_j$$

subject to

$$\frac{\rho_k l_k}{z_k^2} - \sum_{j=1}^{m} \lambda_j \frac{T_k l_k t_k^{(j)}}{E} = 0 \qquad (k = 1, \ldots n)$$

which provides a dual objective function in terms of the variables λ_j ($j = 1, 2, \ldots, m$)

Another version of the dual problem can be used if a feasible set of design variables $\tilde{z}_k$ ($k = 1, \ldots, n$) are available. These may be substituted into (3.2) to generate the alternative linear dual problem

$$\text{minimize} \quad \sum_{j=1}^{m} \lambda_j \bar{u}_j \qquad (3.3)$$

subject to
$$\frac{\rho_k l_k}{\tilde{z}_k^2} - \sum_{j=1}^{m} \lambda_j \frac{T_k l_k t_k^{(i)}}{E} = 0 \quad k = 1, \ldots n$$

$$\lambda_j \geq 0 \, (j = 1, 2, \ldots, m).$$

However, the solution to this problem does not generate the optimum value of the weight but only provides a lower bound estimate. An upper bound is given by computing the feasible design provided by the design variables $\tilde{z}_k$ ($k = 1, \ldots, n$) and in this way the optimum may be bracketed.

Example The framework shown in Fig. 3.8 is subject to the vertical load P at nodes 2 and 3. If all the bars of the structure are made from the same material find:
(i) the minimum weight configuration when the structure is subject to the single constraint $u > u_3$;
(ii) the Kuhn–Tucker conditions and the dual problem when a minimum-weight design is sought subject to the constraints $u > u_2, u > u_3$, and, thereby, bound the minimum weight.

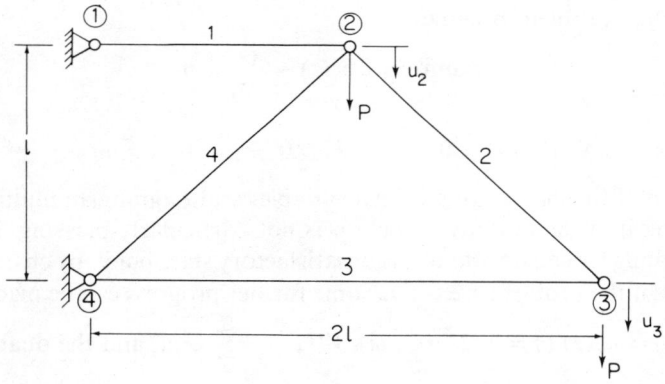

Fig. 3.8 Four-bar statically determinate framework

Case (i) Static equilibrium of the system requires that

$$T_1 = 3P, T_2 = \sqrt{2}P, T_3 = P, T_4 = -2\sqrt{2}P$$

and the tensions due to a unit load applied at node 3 in the direction of u_3:

$$t_1 = 2, t_2 = \sqrt{2}, t_3 = 1, t_4 = -\sqrt{2}$$

Using these expressions the problem becomes

$$\text{minimize the weight} \quad W(z) = \frac{\rho l}{z_1} + \frac{\sqrt{2}\rho l}{z_2} + \frac{2\rho l}{z_3} + \frac{\sqrt{2}\rho l}{z_4}$$

subject to the constraint

$$\bar{u} \geq u_3(z) = \frac{2\rho l}{E}(3z_1 + \sqrt{2}z_2 + z_3 + 2\sqrt{2}z_4)$$

The Lagrangian associated with this problem is given by

$$L(z, \lambda) = W(z) + \lambda \left\{ \frac{2Pl}{E}(3z_1 + \sqrt{2}z_2 + z_3 + 2\sqrt{2}z_4) - \bar{u} \right\}$$

and the Kuhn–Tucker sufficiency conditions are

$$\frac{\partial L}{\partial z_1} = \frac{\rho l}{z_1^{*2}} + \frac{6Pl}{E}\lambda^* = 0$$

$$\frac{\partial L}{\partial z_2} = -\frac{\sqrt{2}\rho l}{z_1^{*2}} + \frac{2\sqrt{2}Pl}{E}\lambda^* = 0$$

$$\frac{\partial L}{\partial z_3} = -\frac{2\rho l}{z_3^{*2}} + \frac{2Pl}{E}\lambda^* = 0$$

$$\frac{\partial L}{\partial z_4} = -\frac{\sqrt{2}\rho l}{z_4^{*2}} + \frac{4\sqrt{2}Pl}{E}\lambda^* = 0$$

$$\lambda^*(u_3(z^*) - \bar{u}) = 0, \quad \bar{u} \geq u_3(z^*), \quad \lambda^* \geq 0$$

Using the results of our earlier analysis the optimal value of the Lagrangian multiplier we have

$$\sqrt{\lambda^*} = \frac{1}{\bar{u}} \sum_{i=1}^{4} l_i \sqrt{\frac{T_i t_i \rho}{E}} = \frac{l}{\bar{u}} \sqrt{\frac{P\rho}{E}} (\sqrt{6} + 4 + 2\sqrt{2}) = 9.2779 \frac{l}{\bar{u}} \sqrt{\frac{P\rho}{E}}$$

From this the optimal values of the design values and, hence, the optimizing cross-sectional areas, are given by

$$\frac{1}{z_1^*} = x_1^* = 22.7262 \frac{Pl}{E\bar{u}}$$

$$\frac{1}{z_2^*} = x_2^* = \frac{x_1^*}{\sqrt{3}}; \quad \frac{1}{z_3^*} = x_3^* = \frac{x_1^*}{\sqrt{6}}; \quad \frac{1}{z_4^*} = x_4^* = \frac{2x_1^*}{\sqrt{6}}$$

with the optimum weight

$$W_{\text{opt}} = 63.681 \rho \frac{Pl^2}{E\bar{u}}$$

Case (ii) We now wish to solve the same problem but with the additional constraint $\bar{u} \geq u_2(z)$. Thus the problem is one of

$$\text{minimizing the weight} \quad W(z) = \frac{\rho l}{z_1} + \frac{\sqrt{2}\rho l}{z_2} + \frac{2\rho l}{z_3} + \frac{\sqrt{2}\rho l}{z_4}$$

subject to
$$\bar{u} \geq u_2(z) = \frac{Pl}{E}(3z_1 + 4\sqrt{2}z_4)$$
$$\bar{u} \geq u_3(z) = \frac{2Pl}{E}(3z_1 + \sqrt{2}z_2 + z_3 + 2\sqrt{2}z_4)$$

The associated Lagrangian is slightly enlarged from case (i); thus
$$L(z, \lambda) = W(z) + \lambda_1 \left\{ \frac{Pl}{E}(3z_1 + 4\sqrt{2}z_4) - \bar{u} \right\}$$
$$+ \lambda_2 \left\{ \frac{2Pl}{E}(3z_1 + \sqrt{2}z_2 + z_3 + 2\sqrt{2}z_4) - \bar{u} \right\}$$

and the Kuhn–Tucker sufficient conditions are
$$\frac{\partial L}{\partial z_1} = -\frac{\rho l}{z_1^{*2}} + 3\lambda_1^* \frac{Pl}{E} + 6\lambda_2^* \frac{Pl}{E} = 0$$
$$\frac{\partial L}{\partial z_2} = -\frac{\sqrt{2}\rho l}{z_2^{*2}} + 2\sqrt{2}\lambda_2^* \frac{Pl}{E} = 0$$
$$\frac{\partial L}{\partial z_3} = -\frac{2\rho l}{z_3^{*2}} + 2\lambda_2^* \frac{Pl}{E} = 0$$
$$\frac{\partial L}{\partial z_4} = -\frac{\sqrt{2}\rho l}{z_4^*} + 4\sqrt{2}\lambda_1^* \frac{Pl}{E} + 2\sqrt{2}\lambda_2^* \frac{Pl}{E} = 0$$
$$\lambda_1^*(u_2(z) - \bar{u}) = \lambda_2^*(u_3(z) - \bar{u}) = 0$$
$$\bar{u} \geq u_2(z), \quad \bar{u} \geq u_3(z), \quad \lambda_1^* \geq 0, \quad \lambda_2^* \geq 0$$

Clearly this problem cannot be solved in the direct manner which is employed in the solution of case (i). However, some progress can be made by going to the final part of the problem and looking at the associated dual problem. As we know, creating the dual requires minimizing the Lagrangian function over the primal design variables which is equivalent to the differential Kuhn–Tucker equations, hence
$$\frac{\partial L}{\partial z_i} = 0 \quad i = 1, 2, 3, 4$$

Multiplying each equation by z_i and summing gives a new expression for the weight function, i.e.
$$\sum_{i=1}^{4} z_i \frac{\partial L}{\partial z_i} = -W(z) + \lambda_1 u_2(z) + \lambda_2 u_3(z) = 0$$

This we can substitute into the Lagrangian function to give the dual objective

function
$$2W(z) - \bar{u}(\lambda_1 + \lambda_2)$$
and the dual problem requires the maximization of this function subject to the constraints
$$\frac{\partial L}{\partial z_i} = 0 \quad i = 1, 2, 3, 4$$

The dual formulation can now be used to obtain a lower bound on the minimum weight since the problem is convex. We now assume that the second constraint only is active, i.e.
$$\bar{u} > u_2(z), \quad \bar{u} = u_3(z)$$
and thus the Lagrangian multiplier λ_1 is equal to zero. The other multiplier λ_2 must be non-zero but, unfortunately, there is no straightforward procedure for deriving a suitable value. At this point we shall simply guess a number and so we assume
$$\lambda_2 = 100\rho \frac{Pl^2}{E\bar{u}^2}, \quad \lambda_1 = 0$$
and solving the constraints of the dual problem gives
$$\frac{1}{z_1} = x_1 = \frac{Pl}{E\bar{u}} \sqrt{600} = 24.4949 \frac{Pl}{E\bar{u}}$$
$$\frac{1}{z_2} = x_2 = \frac{x_1}{\sqrt{3}}, \quad \frac{1}{z_3} = x_3 = \frac{x_1}{\sqrt{6}}, \quad \frac{1}{z_4} = x_4 = \frac{2x_1}{\sqrt{6}}$$

For these values the primal objective function, that is the weight of the structure, is given by
$$W(z) = 82.7792\rho \frac{Pl^2}{E\bar{u}}$$
and the dual objective function is
$$2W(z) - \lambda_2 \bar{u} = \rho \frac{Pl^2}{E\bar{u}} \{165.5583 - 100\} = 65.5583\rho \frac{Pl^2}{E\bar{u}}$$

Therefore the minimum weight can now be bracketed by the primal and dual objective functions to give
$$82.7792\rho \frac{Pl^2}{E\bar{u}} \geq W(z^*) \geq 65.5583\rho \frac{Pl^2}{E\bar{u}}$$

An alternative lower bound to the solution of this problem is provided by the solution to part (i) but this gives a less satisfactory lower estimate of the optimum weight than the above dual.

3.3 STATICALLY INDETERMINATE STRUCTURES

After having examined the statically indeterminate pin-jointed framework the next obvious step is to consider the same type of structure where the layout gives rise to a statically indeterminate framework. We are now entering the realm of realistic design since this type of framework is often used to represent, at least in an idealistic form, the structure of certain types of telecommunications and power transmission towers. The additional complexity of the indeterminate framework both increases the difficulty associated with the analysis of the structure and destroys the convexity of the problem even when reciprocal variables are employed. In this respect statically indeterminate pin-jointed frameworks provide a preview of the type of problem which is encountered when practical structures, modelled by finite elements, are considered from a structural optimization. Nevertheless, the formulation of the optimum design problem is relatively straightforward when compared to the full finite element model involving a variety of different element types.

Thus, we consider the optimum design problem for statically indeterminate pin-jointed frameworks both as a precursor of more complicated design problems and, for certain special cases, as a design problem in its own right.

3.3.1 Analysis

The basic expressions used in Section 3.2.1 for the analysis of statically determinate structures apply in the case of statically indeterminate pin-jointed frameworks with the exception that the resulting matrices are no longer square and, thus, certain of the operations in Section 3.2.1 are now inadmissible. Once again the basic principles are demonstrated by means of a simple example and for this purpose we augment the structure of Fig. 3.1 to the three-bar framework of Fig. 3.9.

Equilibrium between the integral forces T_i ($i = 1, 2, 3$) and the externally applied forces P_j ($j = 1, 2$) gives the matrix equation

$$\begin{bmatrix} P_1 \\ P_2 \end{bmatrix} = \begin{bmatrix} \cos\alpha & 0 & -\cos\alpha \\ \sin\alpha & 1 & \sin\alpha \end{bmatrix} \begin{bmatrix} T_1 \\ T_2 \\ T_2 \end{bmatrix}$$

or, using the notation of Section 3.2.1, for a general pin-jointed framework,

$$\mathbf{P} = \mathbf{B}^\mathrm{T}\mathbf{T}$$

where the matrix $\mathbf{B}$ is non-square. Thus, for a set of applied external loads one cannot directly obtain the values for the internal bar forces. These internal forces

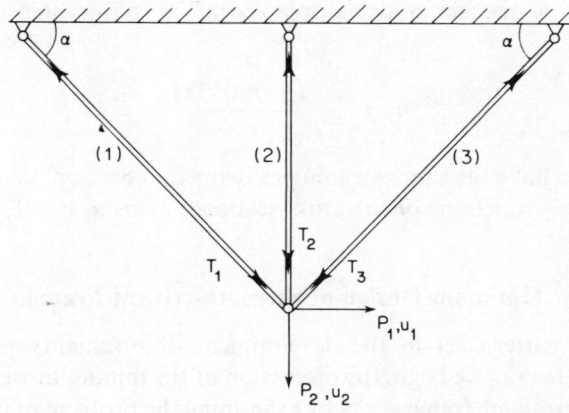

Fig. 3.9 A three-bar pin-jointed framework subject to two loads P_1, and P_2 and with displacements u_1, u_2 and internal bar forces T_1, T_2, T_3

can also be described in terms of the nodal displacements u_1, u_2 in the form

$$\begin{bmatrix} T_1 \\ T_2 \\ T_3 \end{bmatrix} = \begin{bmatrix} \dfrac{x_1 E_1}{l_1} & 0 & 0 \\ 0 & \dfrac{x_2 E_2}{l_2} & 0 \\ 0 & 0 & \dfrac{x_3 E_3}{l_3} \end{bmatrix} \begin{bmatrix} \cos\alpha & \sin\alpha \\ 0 & 1 \\ -\cos\alpha & \sin\alpha \end{bmatrix} \begin{bmatrix} u_1 \\ u_2 \end{bmatrix}$$

and again using the notation of Section 3.2.1 this may be written in the general form

$$\mathbf{T} = \mathbf{D}\mathbf{B}^T\mathbf{u}$$

and hence

$$\mathbf{P} = \mathbf{B}^T\mathbf{D}\mathbf{B}\mathbf{u} = \mathbf{K}\mathbf{u}$$

where $\mathbf{K}$ is again the stiffness matrix. To obtain the nodal displacements for a given set of applied loads requires the inversion of the stiffness matrix or the evolution of the simultaneous linear equations. If a specific nodal displacement u_j is required then

$$u_j = \mathbf{e}_{(j)}^T \mathbf{K}^{-1} \mathbf{P} = \mathbf{e}_{(j)}^T \mathbf{K}^{-1} \mathbf{B}\mathbf{T}$$

and if $\mathbf{e}_{(j)}^T$ is regarded as a dummy load applied at the nodal displacement

$$u_j = \hat{\mathbf{u}}_{(j)} \mathbf{B}^T \mathbf{T}$$

where $\hat{\mathbf{u}}_{(j)}$ are the nodal displacements due to the application of this dummy load. Continuing with the algebraic manipulation we arrive at the previously derived expression for a specific nodal displacement in pin-jointed indeterminate frameworks, i.e.

$$u_j = \sum_{i=1}^{n} \frac{T_i(x)l_i t_i^{(j)}(x)}{Ex_i}$$

where the terms have the same meaning as defined in Section 3.2.1 but with the bars forces being functions of the cross-sectional areas x_i ($i = 1, 2, \ldots, n$).

3.3.2 Optimum Design of Strength-critical Frameworks

Following the pattern set in the development of optimality theory for determinate frameworks we begin the discussion of the minimum-weight design of statically indeterminate frameworks by examining the problem of finding a set of cross-sectional areas which optimize the structure subject to stress constraints. This is identical in form to the statically determinate design problem of Section 3.2.1 and again requires

$$\text{minimizing} \quad W = \sum_{i=1}^{n} \rho l_i x_i$$

subject to

$$-\bar{\sigma}_i \leq \frac{T_i(x)}{x_i} \leq \bar{\sigma}_i \qquad i = 1, 2, \ldots, n$$

where the terms have their previously defined meaning.

In order to present the optimality criteria for this class of problem we select the alternative formulation using reciprocal variables so that the minimum-weight design problem requires finding variables $z_i^* = 1/x_i^*$ ($i = 1, 2, \ldots, n$) which

$$\text{minimize} \quad W(z) = \sum_{i=1}^{n} \frac{\rho l_i}{z_i} \qquad \text{(MWP3)}$$

subject to

$$-\bar{\sigma}_i \leq T_i(z)z_i \leq \bar{\sigma}_i \qquad i = 1, 2, \ldots, n$$

The associated Lagrangian function for this problem is, from Section 2.10, given by

$$L(z, \lambda, \mu) = \sum_{i=1}^{n} \frac{\rho l_i}{z_i} - \sum_{i=1}^{n} \lambda_i(\bar{\sigma}_i - T_i z_i) - \sum_{i=1}^{n} \mu_i(\bar{\sigma}_i + T_i z_i)$$

with $\lambda_i, \mu_i \geq 0$ ($i = 1, 2, \ldots, n$). Unlike the statically determinate framework this function is not necessarily convex and, thus, the optimality criteria establish only a local optimum. With this restriction the Kuhn–Tucker necessary and sufficient conditions for (MWP3) are

$$\nabla_z L(z^*, \lambda^*, \mu^*) = 0$$
$$\lambda_i^*(\bar{\sigma}_i - T_i z_i^*) = 0 \qquad i = 1, 2, \ldots, n \qquad (3.4)$$
$$\mu_i^*(\bar{\sigma}_i + T_i z_i^*) = 0 \qquad i = 1, 2, \ldots, n$$
$$\lambda_i^*, \mu_i^* \geq 0, \qquad \bar{\sigma}_i - T_i z_i^* \geq 0, \qquad \bar{\sigma}_i + T_i z_i^* \geq 0$$

and the vectors z^*, λ^*, μ^* therefore define the optimum. Clearly a specific bar cannot be constrained in both tension and compression at the same time, thus, for an integer i, either $\lambda_i^* > 0$, $\mu_i^* = 0$ or $\lambda_i^* = 0$, $\mu_i^* > 0$.

Although the Kuhn–Tucker conditions provide a mathematical description of the optimum it is possible to re-interpret these on a more illuminating structural basis providing certain assumptions are made. We begin by taking the statically determinate condition for the bar forces such that the derivatives of these terms vanish. Thus the first set of equations in (3.4) which are obtained by differentiating the Lagrangian are given in the form, either

$$-\frac{\rho l_i}{z_i^{*2}} + \lambda_i^* T_i = 0$$

or

$$-\frac{\rho l_i}{z_i^{*2}} - \mu_i^* T_i = 0$$

depending upon whether the tensile or compressive constraint applies. To reduce the number of equations we assume, for convenience, that the tensile constraint is active and then multiply the appropriate equations in (3.4) by z_i^* to give

$$-\frac{\rho l_i}{z_i^*} + \mu_i^* T_i z_i^* = 0$$

or

$$\rho = \frac{\lambda_i^* \bar{\sigma}_i}{\text{vol}_i^*} = \text{constant}$$

where we have appealed to the second set of equations above which require satisfaction of the constraints at the local optimum. The term vol_i^* represents the volume of bar i at the optimizing point. This formulation of the optimality conditions is similar to the constant strain energy density condition obtained for the displacement-constrained statically determinate framework.

The dual formulation for (MWP 3) using this simplification can be found from Section 2.10 which then requires that we

maximize $L(z, \lambda, \mu)$

subject to $\qquad\qquad\qquad\qquad\qquad\qquad\qquad\qquad\qquad\qquad\qquad(3.5)$

$$\nabla_z L(z, \lambda, \mu) = -\frac{\rho_i l_i}{z_i^2} + \lambda_i T_i - \mu_i T_i = 0 \qquad i = 1, 2, \ldots, n$$
$$\lambda_i, \mu_i \geq 0$$

A more compact form for the dual objective function is sought by multiplying each of the constraints in (3.5) by z_i and then summing the equations to give

$$-\sum_{i=1}^{n} \frac{\rho_i l_i}{z_i} + \sum_{i=1}^{n} \lambda_i T_i z_i - \sum_{i=1}^{n} \mu_i T_i z_i = 0$$

Substituting this into the objective function or (3.5) gives a new dual problem:

$$\text{maximize} \quad 2\sum_{i=1}^{n} \frac{\rho_i l_i}{z_i} + \sum_{i=1}^{n} \lambda_i \bar{\sigma}_i + \sum_{i=1}^{n} \mu_i \bar{\sigma}_i$$

subject to (3.6)

$$-\frac{\rho_i l_i}{z_i^2} + \lambda_i T_i - \mu_i T_i = 0$$

$$\lambda_i, \mu_i \geq 0$$

Whilst this is certainly a valid dual for the reduced problem it does not have a particularly pleasing form since both the primal variables z_i and the dual variables or Lagrangian multipliers λ_i, μ_i are present in the dual objective function. The preferred situation is to have only dual variables present in the objective function and to achieve this we must exploit all the possibilities inherent in the mathematical structure of the statically determinate structure. Thus, we follow Hemp [1] and introduce an alternative simplification which assumes that bar forces are fixed by equilibrium conditions only and that the compatibility conditions are ignored. In this form the structural optimization problem assumes a design philosophy equivalent to a lower bound plastic design. The minimum-weight design problem now becomes one of finding a vector $\mathbf{z}^* = (z_1^*, z_2^*, \ldots, z_n^*)^T$ which

$$\text{minimizes} \quad W(z) = \sum_{i=1}^{n} \frac{\rho l_i}{z_i}$$

subject to the equilibrium conditions

$$\sum_{i=1}^{n} B_{ji} T_i = P_j \quad j = 1, 2, \ldots, m$$

and the constraints

$$-\bar{\sigma}_i \leq z_i T_i(z) \leq \bar{\sigma}_i$$

where the framework has n bars and m nodes. Using the notation introduced earlier, P_j are the applied nodal loads and B_{ji} are the direction cosines of the bars making up the framework. The problem is now linear and following the usual practice with this class of problem we introduce the slack variables α_i, β_i ($i = 1, 2, \ldots, n$) through the expressions

$$T_i + 2\alpha_i = \bar{\sigma}_i/z_i$$
$$-T_i + 2\beta_i = \bar{\sigma}_i/z_i$$

The problem is now transformed and becomes one of

$$\text{minimizing} \quad W(\alpha, \beta) = \sum_{i=1}^{n} \frac{\rho l_i(\alpha_i + \beta_i)}{\bar{\sigma}_i}$$

subject to

$$\sum_{i=1}^{n} B_{ji}(\beta_i - \alpha_i) = P_j \quad j = 1, 2, \ldots, m$$

where

$$\alpha_i, \beta_i \geqslant 0 \quad i = 1, 2, \ldots, n$$

and the associated Lagrangian function is

$$L(\alpha, \beta, \lambda, \mu, v) = \sum_{i=1}^{n} \frac{\rho_i l_i(\alpha_i + \beta_i)}{\bar{\sigma}_i} + \sum_{j=1}^{m} \lambda_j \left\{ P_j - \sum_{j=1}^{n} B_{ji}(\beta_j - \alpha_j) \right\}$$

$$+ \sum_{i=1}^{n} \mu_i \alpha_i + \sum_{i=1}^{n} v_i \beta_i$$

with

$$\mu_i, v_i \geqslant 0 \quad i = 1, 2, \ldots, n$$

and where the multipliers λ_j ($j = 1, 2, \ldots, m$) are not necessarily positive since the corresponding equilibrium equations are equality constraints. Following what is now our normal practice we obtain the dual problem in the form:

$$\text{maximize} \quad L(\alpha, \beta, \lambda, \mu, v)$$

subject to

$$\frac{\partial L}{\partial \alpha_k}(\alpha, \beta, \lambda, \mu, v) = \frac{\rho_k l_k}{\bar{\sigma}_k} + \sum_{j=1}^{m} \lambda_j B_{jk} + \mu_k = 0 \quad k = 1, 2, \ldots, n$$

$$\frac{\partial L}{\partial \beta_k}(\alpha, \beta, \lambda, \mu, v) = \frac{\rho_k l_k}{\bar{\sigma}_k} - \sum_{j=1}^{m} \lambda_j B_{jk} + v_k = 0 \quad k = 1, 2, \ldots, n$$

Multiplying these constraints by α_i or β_i appropriately and summing gives

$$\sum_{i=1}^{n} \frac{\rho_i l_i(\alpha_i + \beta_i)}{\bar{\sigma}_i} = \sum_{j=1}^{m} \lambda_j \sum_{i=1}^{n} B_{ji}(\beta_i - \alpha_i) - \sum_{i=1}^{n} \mu_i \alpha_i - \sum_{i=1}^{n} v_i \beta_i$$

Substituting this into the Lagrangian function forming the new dual objective function and taking account of the positivity of the multipliers μ, v provides the dual problem

$$\text{maximize} \quad \sum_{j=1}^{m} \lambda_j P_j$$

subject to

$$\frac{\rho_k l_k}{\bar{\sigma}_k} \geqslant -\sum_{j=1}^{m} \lambda_j B_{jk} \quad k = 1, 2, \ldots, n$$

$$\frac{\rho_k l_k}{\bar{\sigma}_k} \geqslant \sum_{j=1}^{m} \lambda_j B_{jk} \quad k = 1, 2, \ldots, n$$

$$\lambda_j \geqslant 0 \quad j = 1, 2, \ldots, m$$

and we immediately observe that the objective function has only one set of variables the Lagrangian multipliers or dual variables λ_j ($j = 1, 2, \ldots, m$). We may once more seek after structural insights and to this end we now interpret the dual variables as virtual nodal displacements u_j ($j = 1, 2, \ldots, m$). This interpretation transforms the dual problem to a maximum work principle where we seek the largest value for the work term

$$\sum_{j=1}^{n} P_j u_j$$

subject to limitations on the virtual strains

$$\varepsilon_i = \frac{1}{l_i} \sum_{j=1}^{m} u_j B_{ji}$$

We conclude this section by looking at the three-bar framework shown in Fig. 3.9 and seeking a minimum-weight design subject to a limitation on the stress level in each of $\pm \bar{\sigma}$. As indicated in later sections this is a particularly useful problem but is solved here both to illustrate the Hemp technique and to obtain comparison results for later use.

Example Find the minimum-weight design for the three-bar framework in Fig. 3.9 with $P_1 = P_2 = P$ and $-\bar{\sigma} \leq T_i/z_i \leq \bar{\sigma}$.

Using the slack variables α_i, β_i ($i = 1, 2, 3$) the problem becomes

$$\text{minimize} \quad W(\alpha, \beta) = \frac{\rho l}{\bar{\sigma}} \{ (\alpha_1 + \beta_1) \sqrt{2} + (\alpha_2 + \beta_2) + (\alpha_3 + \beta_3) \sqrt{2} \}$$

subject to

$$(\beta_1 - \alpha_1) - (\beta_2 - \alpha_2) = P\sqrt{2}$$
$$(\beta_1 - \alpha_1) + (\beta_2 - \alpha_2)\sqrt{2} + (\beta_3 - \alpha_3) = P\sqrt{2}$$

and to impose symmetry,

$$(\beta_1 + \alpha_1) = (\beta_3 - \alpha_3)$$

The form of the problem suggests that T_1 and T_2 are positive whilst T_3 is negative, and noting

$$(\beta_i - \alpha_i) = T_i \quad i = 1, 2, 3$$

suggests the solution $\alpha_3, \beta_1, \beta_2 \neq 0$; $\alpha_1, \alpha_2, \beta_3 = 0$. Thus $\beta_1 = P/\sqrt{2}$, $\beta_2 = P$, $\alpha_3 = P/\sqrt{2}$, which gives

$$T_1 = P/\sqrt{2}, \quad T_2 = P, \quad T_3 = P/\sqrt{2}$$

and cross-sectional areas

$$x_1 = P/\bar{\sigma}\sqrt{2}, \quad x_2 = P/\bar{\sigma}, \quad x_3 = P/\bar{\sigma}\sqrt{2}$$

with an optimum weight $W = 3\rho l P/\bar{\sigma}$.

3.3.3 Displacement-constrained Statically Indeterminate Structures

Although the statically indeterminate frameworks are more complex than determinate frameworks from the viewpoint of analysis, where the optimum design of displacement-constrained problems is concerned, the basic mathematical formulation for the two types of structures is identical. Thus we examine the multiple displacement-constrained problem in outline and refer the reader to Section 3.2.3 for the full elaboration of the theory.

Using reciprocal variables the design problem requires finding a vector $\mathbf{z}^* = (z_1^*, z_2^*, \ldots, z_n^*)^T$ which

$$\text{minimizes} \quad W(z) = \sum_{i=i}^{n} \frac{\rho l_i}{z_i}$$

subject to

$$\bar{u}_j \geqslant u_j = \sum_{i=1}^{n} \frac{T_i(z) l_i t_i^{(j)}(z)}{E} z_i \quad j = 1, 2, \ldots, m$$

with the terms defined in sections 3.3.1. The associated Lagrangian function is identical to that for the multiple displacement-constrained statically determinate framework and is, therefore, given by

$$L(z, \lambda) = W(z) - \sum_{j=1}^{m} \lambda_j \left(\bar{u}_j - \sum_{i=1}^{n} \frac{T_i(z) l_i t_i^{(j)}(z)}{E} z_i \right)$$

and similarly the Kuhn–Tucker conditions are identical to the conditions for the determinate structure:

$$\nabla_z L(z^*, \lambda^*) = 0$$

$$\lambda_j^* \left(\bar{u}_j - \sum_{i=1}^{n} \frac{T_i(z) l_i t_i^{(j)}(z)}{E} z_i \right) = 0 \quad j = 1, 2, \ldots, m$$

$$\lambda_j^* \geqslant 0, \quad \bar{u}_j \geqslant u_j(z^*) \quad i = 1, 2, \ldots n$$

At first sight it may appear that obtaining the derivatives of the Lagrangian function with respect to the reciprocal design variables is now complicated by the dependence of the bar forces on these variables in the definition of the displacement terms. However, this turns out not to be the case as can be shown by looking at the form of these derivatives:

$$\frac{\partial u_j}{\partial z_i} = \frac{T_i(z) l_i t_i^{(j)}(z)}{E} + \sum_{k=1}^{n} \frac{\partial T_k(z)}{\partial z_i} \frac{l_k t_k^{(j)}(z)}{E} z_k + \sum_{k=1}^{n} \frac{T_k(t) l_k}{E} \frac{\partial t_i^{(j)}(z)}{\partial z_i} z_k$$

and recalling that the terms

$$\sum_{k=1}^{n} \frac{\partial T_k(z)}{\partial z_i} \frac{l_k t_i^{(j)}(z)}{E} z_k, \quad \sum_{k=1}^{n} \frac{T_k(z) l_k}{E} \frac{\partial t_k^{(j)}}{\partial z_i}(z) z_k$$

are self-equilibrating fields summing to zero. Thus, such terms merely represent a shift of the load path as the design is changed and the derivatives of the displacements reduce to

$$\frac{\partial u_j}{\partial z_i} = \frac{T_i(z) l_i t_i^{(j)}(z)}{E}$$

Because of this simplification much of the duality theory developed in Section 3.2.3 for the multiple constrained statically determinate pin-jointed frameworks also applies for indeterminate pin-jointed frameworks.

In particular, problem (3.2) represents the dual associated with (MWP3) on the assumption that the bar forces T and t are now functions of the design variables. This is also a nonlinear problem and as complex to solve as the original design problem (MWP3). If we want to exploit the duality principle to obtain bounds, then some economies are required in order to reduce the computational effort. Following the example of Section 3.2.3 a set of feasible design variables $\tilde{z}_i$ ($i = 1, 2, \ldots, n$) can be taken and a linearized dual problem created identical in form to (3.3). In the present case the bar forces due to the application of the actual and unit loads would need to be calculated. Thus the linearized dual becomes

$$\text{minimize} \sum_{j=1}^{m} \lambda_j \bar{u}_j$$

$$\text{subject to} \sum_{j=1}^{m} \frac{\lambda_j \tilde{T}_i l_i \tilde{t}_i^{(j)}}{E} = \frac{\rho_i l_i}{\tilde{z}_i^2} \quad i = 1, 2, \ldots, n$$

$$\lambda_j \geq 0 \quad (j = 1, 2, \ldots, m)$$

It should be emphasized that solutions to this problem do not give guaranteed lower bounds on the optimal weight except for values of $\tilde{z}$ sufficiently close to the optimum. Nevertheless, the linearized dual has a useful role to play in monitoring the performance of some of the solution algorithms introduced in later chapters which require major computation effort in calculating derivatives [2].

3.4 GENERAL STRUCTURES

As a final step in this gradual introduction to the theory of optimal structures we turn away from simple frameworks to more complex configurations employing the finite element analysis method. This section contains a brief introduction to this analysis method which is sufficient for our purposes, leaving a more detailed explanation to standard texts on the subject [3, 4, 5]. As in earlier sections derivatives with respect to design variables are required which make certain simplifying assumptions necessary. These are considered in some detail below and again in Chapter 4 where a more careful discussion is presented together with specific examples and complex constraints. On this basis the theory can be developed following the same outline as that used in the discussion of optimality and duality theory for frame structures.

3.4.1 Finite Element Analysis and Derivatives

In order to establish the basic relationships we begin at the level of the individual element and observe that for the simpler finite elements the displacement field within the element is defined in terms of nodal displacements **u**,

$$\mathbf{d} = \mathbf{Nu}$$

where **N** represents the appropriate shape function. This vector of nodal displacements would be replaced by a vector which may include rotations for more complex finite elements used to model beams, plates or shells.

The vector of strains ε is defined by

$$\varepsilon = \mathbf{Bu}$$

where the matrix **B** is obtained by operating on the shape function matrix. The relationship between stress and strain is given by

$$\sigma = \mathbf{E}\varepsilon = \mathbf{Du} \tag{3.7}$$

where the vector σ may correspond to stress or stress resultants depending on the structural type and **E** is a symmetric matrix of elastic constants which is also dependent on the structural model.

Following finite element theory these matrix equations can be employed to derive the element stiffness matrix

$$\mathbf{k} = \int_{\text{vol}} \mathbf{B}^T \mathbf{E} \mathbf{B} \, dv$$

where the integration is taken over the volume of the individual finite element. At this stage it is necessary to pause and consider the influence of structural optimization on the form of stiffness matrix. As we have seen the Kuhn–Tucker conditions and duality both require derivatives of the constraints. For a finite elemental model this requires differentiation of the stiffness matrix with respect to the design variables. It is important for practical solution algorithms that these derivatives are constants; an assumption implicit in the discussion of statically indeterminant structures and exact for the determinant case. If we consider thickness as the design variable then we are seeking stiffness matrices of the form

$$\mathbf{k} = x \int_{\text{area}} \mathbf{B}^T \mathbf{E} \mathbf{B} \, dA$$

where the integration is taken over a plane surface within the element and x represents the element thickness. This type of assumption implies that we are primarily dealing with membrane finite elements though other structural models can be incorporated if alternative design variables are taken as demonstrated in Chapters 4 and 8. For a membrane element we have a set of nodal forces **p** which are connected to the nodal displacements through the stiffness matrix to form

$$\mathbf{p} = \mathbf{k}\mathbf{u}$$

The entire structure may be modelled by assembling these individual element stiffness matrices into the global stiffness matrix **K** thus,

$$\mathbf{P} = \mathbf{K}\mathbf{U} \qquad (3.8)$$

where **P** is the vector of all the loads applied to the structure at the node points, and the vector **U** now includes all the nodal displacements for the entire structure. Since the loads are known and the rigid body or support conditions can be prescribed the displacements can be found by inverting the stiffness matrix to give the solution

$$\mathbf{U} = \mathbf{K}^{-1}\mathbf{P}$$

Element stresses can then be calculated at the element level by using the appropriate matrix operations outlined above.

If we take the thickness of each element as a discrete design variable we can now evaluate the derivatives of both stresses and displacements with respect to individual design variables. We begin by differentiating the load vector with respect to a specific design variable x_i and obtain from (3.8),

$$\frac{\partial \mathbf{P}}{\partial x_i} = 0 = \frac{\partial \mathbf{K}}{\partial x_i}\mathbf{U} + \mathbf{K}\frac{\partial \mathbf{U}}{\partial x_i}$$

or

$$\frac{\partial \mathbf{U}}{\partial x_i} = -\mathbf{K}^{-1}\frac{\partial \mathbf{K}}{\partial x_i}\mathbf{U}$$

This gives the rate of change of all displacements with respect to the design variables x_i whereas a constraint in a structural optimization problem normally applies to a specific displacement. In order to extract the derivative with respect to such a specified displacement u_j (say) we premultiply the derivative vector by the vector $\mathbf{e}_{(j)}$ introduced in Section 3.3.1. Thus, we now have

$$\frac{\partial \mathbf{U}_j}{\partial x_i} = -\mathbf{e}^t_{(j)}\mathbf{K}^{-1}\frac{\partial \mathbf{K}}{\partial x_i}\mathbf{U}$$

and recalling that the global stiffness matrix is assembled from individual element matrices, which are linear functions of the design variables, this expression can be rewritten in the form

$$\frac{\partial u_j}{\partial x_i} = -\hat{\mathbf{u}}_i^{(j)t}\frac{\mathbf{k}_i}{x_i}\mathbf{u}_i$$

The element stiffness matrix $\mathbf{k}_i$ corresponds to the ith element, $\mathbf{u}_i$ are the nodal displacements of this element under the action of the applied design loads, and $\hat{\mathbf{u}}_i^{(j)}$ are the displacements of the same nodes due to the action of a unit load applied at node j in the direction of u_j. In effect the terms in $\hat{\mathbf{u}}_i^{(j)}$ are extracted from a vector defined by the operation $\mathbf{e}^t_{(j)} \mathbf{K}^{-1}$ and specific examples are given in Chapter 4.

Stress derivatives are obtained by employing the definition of the element stress vector in terms of the element nodal displacements (3.7),

$$\frac{\partial \sigma_l}{\partial x_i} = \mathbf{D} \frac{\partial \mathbf{u}_l}{\partial x_i} \qquad (3.9)$$

where we have introduced the notation that σ_l and $\mathbf{u}_l$ are the stress and displacement vectors for the lth element. Thus (3.9) represents the derivatives of the stresses associated with the lth finite element with respect to the ith design variable which, in this case, represents the thickness of the ith finite element. Following the principles used in the derivation of displacement derivatives, (3.9) becomes

$$\frac{\partial \sigma_l}{\partial x_i} = -\mathbf{D} \tilde{\mathbf{u}}_i^{(l)\mathrm{T}} \frac{\mathbf{k}_i}{x_i} \mathbf{u}_i \qquad (3.10)$$

As before $\mathbf{k}_i$ is the stiffness matrix of the ith element, $\mathbf{u}_i$ the nodal displacements of the element corresponding to the applied loads and $\mathbf{u}_i^{(l)}$ is now a matrix of nodal displacements for the ith element due to action of unit loads applied at each of the nodes of the lth element.

Having derived the basic derivative formulations using a finite element structural model it can be seen that the computational procedure for obtaining the relevant terms is relatively straightforward. Sets of unit loads are required corresponding to the terms for which derivatives are sought and these must be processed in the same way as the design loads to generate the nodal displacement fields. In practice on the computer these would be processed at the same time as the design loads and would appear as extra right-hand sides during the solution process. Thus the matrix $\mathbf{P}$ in (3.8) would consist of columns representing both the design loads and the dummy loads.

3.4.2 Stress- and Displacement-constrained Finite Element Structural Models

We can now turn to the problem of designing a minimum-weight structure modelled by simple finite elements and subject to stress and displacement constraints. Initially it is assumed that the design variables correspond to individual element thickness x_i ($i = 1, 2, \ldots, n$) and the design problem can be defined as one of finding vector $\mathbf{x}^* = \{x_1^*, x_2^*, \ldots, x_n^*\}^\mathrm{T}$ which

$$\text{minimizes} \quad W(x) = \sum_{i=1}^{n} \omega_i x_i$$

$$\text{subject to} \quad \bar{u}_j > u_j(x) \quad j = 1, 2, \ldots, m$$

$$\phi_i^1 \geqslant \phi_i^2(x) \quad i = 1, 2, \ldots, n$$

where ω_i represents the weight/unit thickness of the finite element. The terms u_j are the m nodal displacement which are constrained to be less than $\bar{u}_j$ and $\phi_i^2(x)$ is a yield criterion, such as that Von Mise or Tresca, defined in terms of the stresses in the ith element. If we focus on the Von Mise criterion then an economic expression defining the function in terms of element stresses has been used by Fleury in the form

$$\phi_i^2(x) = \sigma^T V \sigma$$

where the matrix V depends on the finite element. For example, in the case of a membrane element with stresses described in terms of orthogonal cartesian coordinates the stress vector is given by

$$\sigma^T = \{\sigma_x, \sigma_y, \tau_{xy}\}$$

and

$$V = \begin{bmatrix} 1 & -\frac{1}{2} & 0 \\ -\frac{1}{2} & 1 & 0 \\ 0 & 0 & 3 \end{bmatrix}$$

where for the simplest elements the stresses are constant throughout the element and in more complex cases the stress vector is often evaluated at a specified point within the element.

Guided by the lessons learned in the design of both determinate and indeterminate frameworks, we redefine the design problem in terms of the inverse of the finite element thicknesses. Thus the optimum design problem now becomes one of finding a vector $z_i^* = 1/x_i^*$ $(i = 1, 2, \ldots, n)$ which

$$\text{minimizes} \quad W(z) = \sum_{i=1}^n \frac{\omega_i}{z_i}$$

(MWP4)

$$\text{subject to} \quad \begin{array}{ll} \bar{u}_j \geqslant u_j(z) & j = 1, 2, \ldots, m \\ \bar{\phi}_i^2 \geqslant \phi_i^2(z) & i = 1, 2, \ldots, n \end{array}$$

Although this formulation again destroys the linear form of the objective function, it gives rise to constraints which are more nearly linear in the sense that linearizing constraints in terms of the inverse variables, to allow the use of some of the algorithms introduced later, gives rise to smaller errors than linearization using element thickness as the design variable.

Following the usual route in obtaining optimality conditions the Lagrangian for (MWP4) is formed

$$L(x, \lambda, \mu) = W(z) + \sum_{j=1}^m \lambda_j(u_j(z) - \bar{u}_j) + \sum_{i=1}^n \mu_i(\phi_i^2(z) - \bar{\phi}_i^2)$$

and, thus, the Kuhn–Tucker optimality are

$$\nabla_z L(z^*, \lambda^*, \mu^*) = 0$$
$$\lambda_j^*(u_j(z^*) - \bar{u}_j) = 0 \quad j = 1, 2, \ldots, m$$
$$\mu_i^*(\phi_i^2(z^*) - \bar{\phi}_i^2) = 0 \quad i = 1, 2, \ldots, n$$
$$\lambda_j^* \geqslant 0 \quad \mu_i^* \geqslant 0, \quad \bar{u}_j \geqslant u_j(z^*), \phi_i^2 \geqslant \phi_i^2(z^*)$$

The derivatives are now with respect to the inverse variables z and are related to the derivatives in terms of the element thickness given in Section 3.4.1 through the transformation

$$\frac{\partial}{\partial z} = -\frac{1}{z^2}\frac{\partial}{\partial x}$$

Thus, the derivatives required in the first part of the Kuhn–Tucker conditions for (MWP4) become from Section 3.4.2,

$$\frac{\partial u_j}{\partial z_i} = \tilde{\mathbf{u}}_i^{(j)\mathrm{T}}\frac{\mathbf{k}_i}{z_i}\mathbf{u}_i$$

$$\frac{\partial(\phi_l^2(z))}{\partial z_i} = 2\frac{\partial \boldsymbol{\sigma}_i^\mathrm{T}}{\partial z_i}\mathbf{V}\boldsymbol{\sigma}_l$$

with

$$\frac{\partial \boldsymbol{\sigma}_i^\mathrm{T}}{\partial z_i} = \mathbf{D}\tilde{\mathbf{u}}_i^{(l)\mathrm{T}}\frac{\mathbf{k}_i}{z_i}\mathbf{u}_i$$

Specializing (MWP4) to the case where there are no stress constraints and only a single displacement constraint, the differential Kuhn–Tucker conditions become

$$-\frac{\omega_i}{z_i^{*2}} + \lambda^*\tilde{\mathbf{u}}_i^\mathrm{T}\frac{\mathbf{k}_i}{z_i^*}\mathbf{u}_i = 0 \qquad i = 1, 2, \ldots, n$$

where $\tilde{\mathbf{u}}_i$ are the displacements of the nodes of the ith element due to a unit load at the point where the single displacement constraint is applied. As before, $\mathbf{u}_i$ are the displacements at the same nodes due to the actual load and the matrix k_i is the element stiffness matrix of the ith element. Following the procedure adopted in the case of determinate structures and interpreting the Lagrangian multiplier λ^* as a pseudo-load gives

$$z_i^*\hat{\mathbf{u}}_i^\mathrm{T}\mathbf{k}_i\mathbf{u}_i = \omega_i$$

with $\hat{\mathbf{u}}_i$ now the nodal displacements associated with a load of a magnitude equal to the value of the multiplier λ^*. Interpreting the left-hand side as a virtual strain energy density we observe that the optimality condition requires that this attains a previously specified constant value. This optimality condition recovers the principle established in the case of determinate and indeterminate frameworks.

The Kuhn–Tucker optimality conditions for the more general problem are not exploited here but are considered in the chapters related to algorithm development. The more advanced methods of solution for problems of the type exemplified by (MWP4) rely on using these conditions either in full or in part to generate effective up-date formulae. These conditions also form the basis of the associated dual problem through the procedure developed in Chapter 2 and used in the earlier part of this chapter. Development of the relevant forms is left to the problems set at the end of the chapter.

3.4.3 Minimum-weight Design Subject to Buckling Constraints

We conclude the chapter by briefly looking at a different class of design where the solution of the structural problem has the additional complexity of requiring the calculation of eigenvalues and eigenvectors. The simplest problems of this type involve structures where linear stability or vibrational response are critical. In this section attention is focused on the linear stability problem and, because of the mathematical similarity of the governing equations, this also indirectly covers the case where free vibrational modes of the structure are critical. The forced linear vibration problem is similar to the standard displacement analysis covered earlier and a return to this problem is made in the problems set at the end of the chapter. Following the practice adopted in this chapter an elementary outline is presented leaving a more detailed examination of certain aspects of the problem to the next chapter. However, no attempt is made to cover the complex interaction and detailed buckling problems which occur in many real design situations.

The linear stability of the buckling problem is associated with structures consisting of 'slender' bars or 'thin-walled' components as illustrated by the cylinder problem in Chapter 1. Buckling requires an interaction between membrane and bending forces which is achieved in the finite element method by the introduction of the geometric stiffness matrix $\mathbf{K}_G$. This matrix is independent of the elastic properties of the structural material and is a function of both the element geometry and membrane forces. The linear stability of the structure is determined by solving the eigenvalue problem

$$(\mathbf{K} + \mu \mathbf{K}_G)\boldsymbol{\eta} = 0$$

where μ is the eigenvalue. The eigenvalues μ_j are given by

$$\mu_j = -\frac{\boldsymbol{\eta}_j^T \mathbf{K} \boldsymbol{\eta}_j}{\boldsymbol{\eta}_j^T \mathbf{K}_G \boldsymbol{\eta}_j}$$

where $\boldsymbol{\eta}_j$ is the buckling mode. The buckling load of the structure is given by $\mathbf{P}_{cr} = \mu \mathbf{P}$ where μ is the lowest eigenvalue and $\mathbf{P}$ a set of applied loads.

The design problem requires seeking a minimum-weight structure which can carry the applied loads $\mathbf{P}$ without buckling and, thus, generates the constraint $\mu \leqslant 1$. Because the stability problem is an interaction between bending and membrane action the linearity of the stiffness matrix with respect to element thickness does not apply. Although this causes some difficulty it can be surmounted in certain cases; for example, linearity may be preserved in the case of certain simple beam configurations if beam width is used as the design variable. Thus, linearity of the stiffness matrix with respect to the design variables is assumed to exist for the present case even though this represents a very restricted case.

Thus the design problem becomes one of finding a vector of design variables $\mathbf{x}^* = \{x_1^*, x_2^*, \ldots, x_n^*\}^T$ which minimize the weight

$$W(x) = \sum_{i=1}^{n} \omega_i x_i \qquad \text{(MWP5)}$$

Elementary Concepts in Structural Optimization

subject to the constraint

$$\mu \leqslant 1$$

Following what is now the normal course of action the associated Lagrangian function is defined in the form

$$L(x, \mu, \lambda) = W(x) + \lambda(\mu - 1)$$

which leads to the Kuhn–Tucker optimality conditions

$$\nabla_x L(x^*, \mu^*, \lambda^*) = 0$$
$$\lambda^*(\mu^* - 1) = 0$$
$$\lambda^* \geqslant 1, \mu^* \leqslant 1$$

Before developing these conditions the expression for the derivatives of the eigenvalues with respect to the design variables must be examined. To this purpose we follow the path normally used in deriving Rayleigh's quotient and introduce the term

$$\boldsymbol{\eta}^T (\mathbf{K} + \mu \mathbf{K}_G) \boldsymbol{\eta} = 0$$

Now taking derivatives,

$$2 \left(\frac{\partial \boldsymbol{\eta}}{\partial x_i} \right)^T (\mathbf{K} + \mu \mathbf{K}_G) \boldsymbol{\eta} + \boldsymbol{\eta}^T \mathbf{k}_i \boldsymbol{\eta} / x_i$$

$$+ \frac{\partial \mu}{\partial x_i} \boldsymbol{\eta}^T \mathbf{K}_G \boldsymbol{\eta} + \mu \boldsymbol{\eta}^T \frac{\partial \mathbf{K}_G}{\partial x_i} \boldsymbol{\eta} = 0$$

and if the assumption is made that the geometric stiffness matrix is independent of the design variables than $\partial \mathbf{K}_G / \partial x_i = 0$ and

$$\frac{\partial \mu}{\partial x_i} = \frac{-\boldsymbol{\eta}^T \mathbf{k}_i \boldsymbol{\eta} / x_i}{\boldsymbol{\eta}^T \mathbf{K}_G \boldsymbol{\eta}}$$

Armed with this expression the optimality conditions can now be written in the form

$$\frac{\partial L}{\partial x_i}(x, \mu, \lambda) = \omega_i - \lambda^* \frac{\boldsymbol{\eta}^T \mathbf{k}_i \boldsymbol{\eta} / x_i^*}{\boldsymbol{\eta}^T \mathbf{K}_G \boldsymbol{\eta}} = 0 \qquad i = 1, 2, \ldots, n$$

$$\lambda^*(\mu^* - 1) = 0$$
$$\lambda^* \geqslant 1, \mu^* \leqslant 1$$

Thus the optimality criterion for this problem is given by

$$\text{constant} = \frac{\boldsymbol{\eta}^T \mathbf{k}_i \boldsymbol{\eta} / x_i^*}{\omega_i \boldsymbol{\eta}^T \mathbf{K}_G \boldsymbol{\eta}} \qquad i = 1, 2, \ldots, n$$

and as pointed out by Khot et al. [6] this can be interpreted as implying that an optimum structure is one in which the ratio of the strain energy density to the mass density associated with the normalized buckling mode is the same for all the elements.

The dual problem associated with (MWP5) requires that we

$$\text{maximize} \quad L(x, \lambda, \mu) = W(x) + \lambda(\mu - 1)$$

$$\text{subject to} \quad \frac{\partial L}{\partial x_i} = \omega_i - \lambda \frac{\boldsymbol{\eta}^T \mathbf{k}_i \boldsymbol{\eta}/x_i}{\boldsymbol{\eta}^T \mathbf{K}_G \boldsymbol{\eta}} = 0 \quad i = 1, 2, \ldots, n$$

multiplying each of the constraint equations by the appropriate design variable and the resulting equations are summed to give

$$\sum_{i=1}^{n} x_i \frac{\partial L}{\partial x_i} = W(x) - \mu \frac{\boldsymbol{\eta}^T \mathbf{K} \boldsymbol{\eta}}{\boldsymbol{\eta}^T \mathbf{K}_G \boldsymbol{\eta}} = 0$$

$$\text{or} \quad W(x) = \mu \lambda$$

Using this last expression the dual problem becomes

$$\text{maximize} \quad -\lambda$$

$$\text{subject to} \quad \frac{\partial L}{\partial x_i} = \omega_i - \lambda \frac{\boldsymbol{\eta}^T \mathbf{k}_i \boldsymbol{\eta}/x_i}{\boldsymbol{\eta}^T \mathbf{K}_G \boldsymbol{\eta}} = 0$$

which represents a rather satisfactory form for the dual to the buckling constrained minimum-weight design problem.

PROBLEMS

1 Examine convexity for statically determinate structures–minimum weight design subject to displacement constraints for: (a) cross-sectional areas; (b) inverse variables.
2 Examine convexity for pin-jointed statically indeterminate structure–minimum weight design subject to displacement constraints.
3 For a single displacement-constrained statically determinate pin-jointed framework, show that for the special case of a single load applied at the node and in the direction of the constrained displacement a local optimum is achieved when all bars are *equally stressed*.
4 For statically indeterminate structures re-derive the optimality criterion in the presence of gauge constraints.
5 Repeat the analyses of Section 3.3.2 by deriving the optimality criterion for a structure modelled by the finite element method subject to displacement, stress and gauge constraints.
6 Derive appropriate optimality criteria and dual problems for the minimum weight design of a structure modelled by the finite element method subjected to a forced vibration with constraints on the amplitude of the displacements. This problem will require the introduction of the dynamic stiffness matrix.

7 Repeat question 6 for the case of or a freely vibrating structure subject only to the constraint that the natural frequency of the structure avoids a specified frequency.
8 Repeat question 7 and add gauge constraints.

A.J.M.

REFERENCES

1. W. S. Hemp, *Optimum Structures*, Oxford University Press, 1973.
2. A. J. Morris, P. Bartholomew and J. Dennis, A computer based system for structural design, analysis and optimisation, *Proceedings AGARD Symposium on 'The Use of Computers as a Design Tool'*, Proceedings No. 280, 1980.
3. R. H. Gallagher, *Finite Element Analysis Fundamentals*, Prentice-Hall, 1975.
4. O. C. Zienkiewicz, *The Finite Element Method* (3rd edn), McGraw-Hill, 1979.
5. B. Irons and S. Ahmad, *Techniques of Finite Elements*, Ellis Horwood, 1980.
6. N. S. Khot, V. B. Venkayya and L. Berke, Optimum structural design with stability constraints, *Int. J. Num. Met. Engg.*, **10**, pp. 1097–1114, 1976.

Foundations of Structural Optimization: A Unified Approach
Edited by A. J. Morris
© 1982 John Wiley & Sons Ltd.

Chapter 4
Finite Elements and Optimization

4.1 INTRODUCTION

In the previous chapter we looked at the theory of structural optimization using finite elements. We now re-examine the use of finite elements more carefully. The simplest way of using a finite element program for structural optimization is in a 'black box' mode. In this mode the finite element program is used to obtain displacements, stresses, vibration frequencies, or any other response quantities. The user of the finite element program then needs to provide an interface for calculating constraints and their derivatives. It is the calculation of the derivatives of constraints which is the major problem and is discussed here in detail. In order to introduce a fresh perspective from that adopted in Chapter 3 a slightly different notation is used here, but confusion is avoided by comparing the two notations at appropriate places in the text.

It is possible to obtain derivative information by a finite difference approach. This means running the finite element analysis for a nominal value of a design variable and then repeating the run for a slightly perturbed value of that design variable. Such a method, however, often proves to be very expensive computationally, especially when the number of design variables is large. It is also possible for specific structural configurations to develop algorithms and software for calculating such derivatives directly (e.g. Ref. [1]). So far, however, such algorithms seem to be problem dependent and may not be suitable for general finite element programs.

At the present it seems that for combining generality with computational efficiency the best approach is to modify and augment the finite element package for providing derivative (also called sensitivity) calculations. The present chapter is mainly devoted to a discussion of this approach.

For improved flexibility it is usually desirable to make the routines that calculate the derivatives as modular as possible. One way of achieving this goal is through the use of a data base for communication between the derivative modules and the main finite element program. An example of the use of such an approach is the PARS (Programs for Analysis and Resizing of Structures [2]) program.

PARS was written as a companion code to the SPAR [3] general finite element programs and like SPAR it consists of a series of programs or processors that communicate through a data base [4].

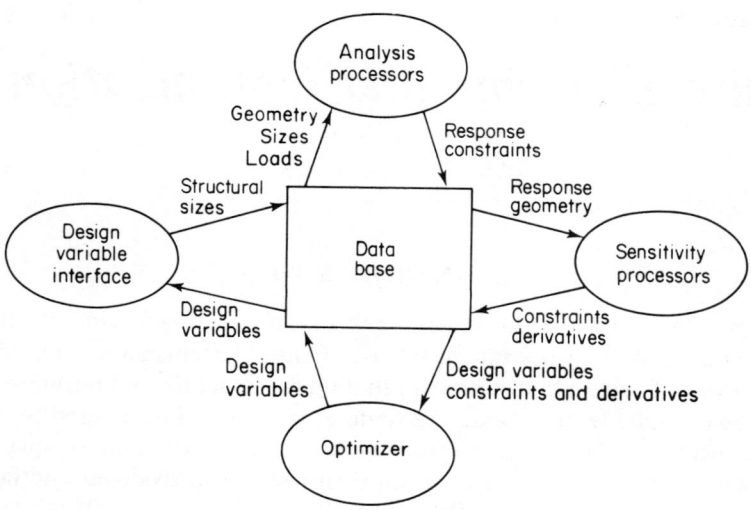

Fig. 4.1 PARS architecture

Figure 4.1 shows schematically the coupling of a finite element program with an optimization program achieved in PARS. The various programs of PARS are grouped in four categories for this purpose. The modules on the top include the SPAR analysis processors for calculating displacement, stresses and vibration frequencies plus additional processors for calculating flutter speeds. For a single analysis or for a trial-and-error type design these are the only modules that need be used.

For systematic optimization the first thing which is needed is a definition of design variables. The left group of modules in Fig. 4.1 represents an interface between design variables (which are typically structural sizes) and parameters of the finite element model. Such an interface is needed because, in general, one does not want to have a one-to-one correspondence between the structure and the finite element model. That is, often one structural element is discretized by several finite elements and sometimes one finite element represents contributions of several structural elements.

The right group of modules in Fig. 4.1 represents the derivatives or sensitivity processors and the bottom one represents the optimizer. In a typical run of a system the user defines the design variables and their initial values using the design variable interface, then the analysis processors calculate response and constraints, and the sensitivity processors calculate the derivatives of the response and the constraints. The optimizer in PARS uses the constraints and their derivatives to obtain approximate expressions for the constraints. These approximate expressions are used for conducting a search in the design variable space for improved values of the design variables. The design variable interface is now used to update the finite element model, and the system is ready for another go through the loop. All the information generated by the various processors resides in the data base so that the user can go back and display the response or the sensitivities during any stage of the process.

The above example shows an overall picture of a typical scheme of an interaction of a finite element program with an optimization package. The rest of this chapter is concerned with the algorithms used for obtaining derivatives of constraints. The material presented in the following sections is based on Refs [5] to [8].

4.2 CALCULATION OF SENSITIVITIES FOR STATIC LOADS

The optimization of structures under static loading typically requires derivatives of displacement or stress constraints. Because element stresses may be easily written as explicit functions of the nodal displacements, a general form of the constraint is

$$g(u, x) \geqslant 0 \qquad (4.1)$$

where $\mathbf{u}$ is the displacement vector and x is a design variable. The derivatives of g with respect to the design variable x may be written as

$$\frac{dg}{dx} = \frac{\partial g}{\partial x} + \mathbf{l}^T \frac{d\mathbf{u}}{dx} \qquad (4.2)$$

where $\mathbf{l}$ is a vector such that l_i is $\partial g/\partial u_i$.

The first term in Equation (4.2) is usually zero or very easy to obtain, so we discuss only the computation of the second term. This second term is obtained from the equations of equilibrium

$$\mathbf{K}\mathbf{u} = \mathbf{f} \qquad (4.3)$$

where $\mathbf{K}$ is the global stiffness matrix and $\mathbf{f}$ is the load vector. Differentiating Equation (4.3) with respect to the design variable, we get

$$\mathbf{K}\frac{d\mathbf{u}}{dx} = \frac{d\mathbf{f}}{dx} - \frac{d\mathbf{K}}{dx}\mathbf{u} \qquad (4.4)$$

Premultiplying Equation (4.4) by $\mathbf{l}^T \mathbf{K}^{-1}$, we obtain

$$\mathbf{l}^T \frac{d\mathbf{u}}{dx} = \mathbf{l}^T \mathbf{K}^{-1} \left[\frac{d\mathbf{f}}{dx} - \frac{d\mathbf{K}}{dx}\mathbf{u} \right]. \qquad (4.5)$$

Numerically, the calculation of $\mathbf{l}^T (d\mathbf{u}/dx)$ in Equation (4.5) may be performed in two different ways. The first, labelled herein method I, consists of solving Equation (4.4) for $d\mathbf{u}/dx$ and then taking the inner product with $\mathbf{l}$. The second approach, labelled herein method II, defines an adjoint variable λ which is the solution of the system

$$\mathbf{K}\lambda = \mathbf{l} \tag{4.6}$$

and then rewrites Equation (4.5) as

$$\mathbf{l}^T \frac{d\mathbf{u}}{dx} = \lambda^T \left[\frac{d\mathbf{f}}{dx} - \frac{d\mathbf{K}}{dx}\mathbf{u} \right] \tag{4.7}$$

where use has been made of the symmetry of $\mathbf{K}$.

Method II, introduced in Chapter 3, is also known as the dummy load method because the vector $\mathbf{l}$ is often described as a dummy load and we now observe that $\mathbf{l}$ is identical to the vector $\mathbf{e}_{(j)}$ used in Chapter 3. When the constraint g in Equation (4.1) is a simple upper limit on a single displacement component, the dummy force vector also has only a single non-zero component corresponding to the constrained displacement component.

Both method I and method II require the solution of a system of equations as the major component of computational effort. We must note, however, that the matrix $\mathbf{K}$ of the equations is already available in a factored form from the solution of Equation (4.3) for the displacements. The solution for $d\mathbf{u}/dx$ or λ is therefore much cheaper than the original solution of Equation (4.3). The difference between the computational effort associated with method I and that associated with method II is significant when we have multiple constraints and multiple design variables. Method I requires the solution of Equation (4.4) once for each design variable, while method II requires the solution of Equation (4.6) once for each constraint. Consequently, method I is more efficient when the number of design variables is smaller than the number of constraints. Method II is more efficient when the number of design variables is larger than the number of constraints.

4.3 EXAMPLE—CANTILEVER BEAM SUBJECT TO A DISPLACEMENT CONSTRAINT

Figure 4.2 shows a two-element cantilever beam under an end load. Assume that it is required to reduce the displacement at the tip with a minimum of weight penalty. Assuming that the weight penalty is proportional to $\Delta I_1 + \Delta I_2$, which beam element should be reinforced?

We adopt the following notation for this problem:

w = lateral displacement of the beam
θ = rotation of the beam = dw/dx
E = Young's modulus

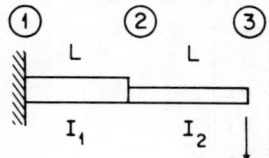

Fig. 4.2 Design of a beam for specified displacement

The problem is simple enough so that an analytical solution based on elementary beam theory is easily obtained:

$$w_{tip} = \left(\frac{7}{3}\right)\frac{PL^3}{EI_1} + \frac{PL^3}{3EI_2}$$

so that

$$\frac{\partial w_3}{\partial I_1} = -\left(\frac{7}{3}\right)\frac{PL^3}{EI_1^2}$$

$$\frac{\partial w_3}{\partial I_2} = -\frac{PL^3}{3EI_2^2}$$

It is therefore more effective to reinforce beam 1 if

$$I_1 \leqslant \sqrt{7}I_2 = 2.646 I_2$$

otherwise it is more efficient to reinforce beam 2.

The finite element solution is based on a standard cubic beam element. The element stiffness matrix is

$$\mathbf{k} = \left(\frac{EI}{L^3}\right)\begin{bmatrix} 12 & 6L & -12 & 6L \\ 6L & 4L^2 & -6L & 2L^2 \\ -12 & -6L & 12 & -6L \\ 6L & 2L^2 & -6L & 4L^2 \end{bmatrix}$$

so that the global stiffness matrix is

$$\mathbf{K} = \left(\frac{E}{L^3}\right)\begin{bmatrix} 12(I_1+I_2) & -6L(I_1-I_2) & -12I_2 & 6I_2L \\ & 4L^2(I_1+I_2) & -6I_2L & 2I_2L^2 \\ & \text{symm} & 12I_2 & -6I_2L \\ & & & 4I_2L^2 \end{bmatrix}$$

where the global degrees of freedom are $[w_2, \theta_2, w_3, \theta_3]$. The load vector $\mathbf{f}$ is $\mathbf{f}^T = [0, 0, P, 0]$ and the solution obtained by solving Equation (4.2.3) is

$$\mathbf{u} = \left(\frac{P}{E}\right)\begin{bmatrix} 5L^3/6I_1 \\ 3L^2/2I_1 \\ 7L^3/3I_1 + L^3/3I_2 \\ 3L^2/2I_1 + L^2/2I_2 \end{bmatrix}$$

For the derivative calculation we need to calculate $(\partial \mathbf{K}/\partial I_1)\mathbf{u}$ and $(\partial \mathbf{K}/\partial I_2)\mathbf{u}$.

$$\frac{\partial \mathbf{K}}{\partial I_1}\mathbf{u} = \left(\frac{E}{L^3}\right)\begin{bmatrix} 12 & -6L & 0 & 0 \\ & 4L^2 & 0 & 0 \\ & \text{symm} & 0 & 0 \\ & & & 0 \end{bmatrix}\begin{bmatrix} w_2 \\ \theta_2 \\ w_3 \\ \theta_3 \end{bmatrix}$$

$$= \left(\frac{E}{L^3}\right)\begin{bmatrix} 12w_2 - 6L\theta_2 \\ -6Lw_2 + 4L^2\theta_2 \\ 0 \\ 0 \end{bmatrix} = \frac{P}{I_1}\begin{bmatrix} 1 \\ L \\ 0 \\ 0 \end{bmatrix}$$

$$\frac{\partial \mathbf{K}}{\partial I_2}\mathbf{u} = \left(\frac{E}{L^3}\right)\begin{bmatrix} 12 & 6L & -12 & 6L \\ & 4L^2 & -6L & 2L^2 \\ & \text{symm} & 12 & -6L \\ & & & 4L^2 \end{bmatrix}\begin{bmatrix} w_2 \\ \theta_2 \\ w_3 \\ \theta_3 \end{bmatrix}$$

$$= \frac{E}{L^3}\begin{bmatrix} 12w_2 + 6L\theta_2 - 12w_3 + 6L\theta_3 \\ 6Lw_2 + 4L^2\theta_2 - 6Lw_3 + 2L^2\theta_3 \\ -12w_2 - 6L\theta_2 + 12w_3 - 6L\theta_3 \\ 6Lw_2 + 2L^2\theta_2 - 6Lw_3 + 4L^2\theta_3 \end{bmatrix} = \frac{P}{I_3}\begin{bmatrix} -1 \\ -L \\ 1 \\ 0 \end{bmatrix}$$

Method I

$$\frac{\partial \mathbf{u}}{\partial I_1} = \mathbf{K}^{-1}\left[\frac{\partial \mathbf{f}}{\partial I_1} - \frac{\partial \mathbf{K}}{\partial I_1}\mathbf{u}\right]$$

or

$$\frac{\partial}{\partial I_1}\begin{bmatrix} w_2 \\ \theta_2 \\ w_3 \\ \theta_3 \end{bmatrix} = -\mathbf{K}^{-1}\begin{bmatrix} P/I_1 \\ LP/I_1 \\ 0 \\ 0 \end{bmatrix} = -\frac{PL^2}{EI_1^2}\begin{bmatrix} 5L/6 \\ 3/2 \\ 7L/3 \\ 3/2 \end{bmatrix}$$

so

$$\frac{\partial w_3}{\partial I_1} = -\frac{7PL^3}{3EI_1^2}$$

Similarly,

$$\frac{\partial \mathbf{u}}{\partial I_2} = \mathbf{K}^{-1}\left[\frac{\partial \mathbf{f}}{\partial I_2} - \frac{\partial \mathbf{K}}{\partial I_2}\mathbf{u}\right]$$

or

$$\frac{\partial}{\partial I_2} \begin{bmatrix} w_2 \\ \theta_2 \\ w_3 \\ \theta_3 \end{bmatrix} = -\mathbf{K}^{-1} \begin{bmatrix} -P/I_2 \\ -PL/I_2 \\ P/I_2 \\ 0 \end{bmatrix} = -\frac{PL^2}{EI_2^2} \begin{bmatrix} 0 \\ 0 \\ L/3 \\ 1/2 \end{bmatrix}$$

so

$$\frac{\partial w_3}{\partial I_2} = -\frac{PL^3}{3EI_2^2}.$$

Method II

In our case,

$$\mathbf{e}_{(3)}^T = [0, 0, 1, 0] \qquad (w_3 = \mathbf{e}^T u)$$

where $\mathbf{e}_{(3)}^T$ is defined in Section 3.2.1,

$$\lambda = \mathbf{K}^{-1}\mathbf{e} = \mathbf{K}^{-1} \begin{bmatrix} 0 \\ 0 \\ 1 \\ 0 \end{bmatrix} = \left(\frac{L^2}{E}\right) \begin{bmatrix} 5L/6I_1 \\ 3/2I_1 \\ 7L/3I_1 + L/3I_2 \\ 3/2I_1 + 1/2I_2 \end{bmatrix}$$

$$\frac{\partial w_3}{\partial I_1} = -\lambda^T \frac{\partial \mathbf{k}}{\partial I_1} \mathbf{u} = \lambda^T \frac{P}{I_1} \begin{bmatrix} 1 \\ L \\ 0 \\ 0 \end{bmatrix} = -\frac{7PL^3}{3EI_1^2}$$

$$\frac{\partial w_3}{\partial I_2} = -\lambda^T \frac{\partial \mathbf{k}}{\partial I_2} \mathbf{u} = -\frac{PL^3}{3EI_2^2}$$

4.4 SENSITIVITY CALCULATION OF VIBRATION FREQUENCIES AND MODES

The finite element equation for the ith natural frequency ω_i is

$$\mathbf{K}\mathbf{a}_i - \omega_i^2 \mathbf{M}\mathbf{a}_i = 0 \tag{4.8}$$

where $\mathbf{M}$ is the mass matrix and $\mathbf{a}_i$ the vibration mode. The mode is assumed to be normalized so that

$$\mathbf{a}_i^T \mathbf{M} \mathbf{a}_i = 1 \tag{4.9}$$

Differentiating Equations (4.8) and (4.9) with respect to a design variable x, we

get

$$(\mathbf{K} - \omega_i^2 \mathbf{M})\frac{d\mathbf{a}_i}{dx} - \frac{d(\omega_i^2)}{dx}\mathbf{M}\mathbf{a}_i = -\left(\frac{d\mathbf{K}}{dx} - \omega_i^2 \frac{d\mathbf{M}}{dx}\right)\mathbf{a}_i \quad (4.10)$$

$$\mathbf{a}_i^T \frac{d\mathbf{M}}{dx}\mathbf{a}_i + 2\mathbf{a}_i^T \mathbf{M}\frac{d\mathbf{a}_i}{dx} = 0 \quad (4.11)$$

where use is made of the symmetry of the mass matrix. Combining Equations (4.10) and (4.11),

$$\begin{bmatrix} \mathbf{K} - \omega_i^2 \mathbf{M} & -\mathbf{M}\mathbf{a}_i \\ -\mathbf{a}_i^T \mathbf{M} & 0 \end{bmatrix} \begin{bmatrix} \dfrac{d\mathbf{a}_i}{dx} \\ \dfrac{d\omega_i^2}{dx} \end{bmatrix} = \begin{bmatrix} -\left(\dfrac{d\mathbf{K}}{dx} - \omega_i^2 \dfrac{d\mathbf{M}}{dx}\right)\mathbf{a}_i \\ -\tfrac{1}{2}\mathbf{a}_i^T \dfrac{d\mathbf{M}}{dx}\mathbf{a}_i \end{bmatrix} \quad (4.12)$$

The system (4.12) may be solved for the derivatives of the frequency and the mode. Care must be taken in the solution process because the principal minor $\mathbf{K} - \omega_i^2 \mathbf{M}$ is singular (see Ref. [7] for several solution strategies).

If only the derivatives of the frequency are required the calculation is much simpler. Premultiplying Equation (4.10) by $\mathbf{a}_i^T$, we get

$$\frac{d(\omega_i^2)}{dx} = \mathbf{a}_i^T \left(\frac{d\mathbf{K}}{dx} - \omega_i^2 \frac{d\mathbf{M}}{dx}\right)\mathbf{a}_i \quad (4.13)$$

4.5 EXAMPLE—A BEAM SUBJECT TO A FREQUENCY CONSTRAINT

Figure 4.3 shows a beam composed of two elements. Initially the two elements are identical but the fundamental frequency is below a given limit. Which element should be stiffened to increase the fundamental frequency? We assume that for each beam the total mass $m = \alpha \sqrt{I}$ where α is a given constant.

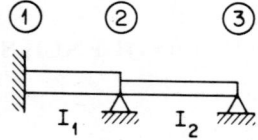

Fig. 4.3 Design of a beam for a specified vibration frequency

The stiffness matrix of a beam is given in the previous example. The global stiffness matrix corresponding to the unrestrained degrees of freedom $[\theta_2, \theta_3]$ is

$$\mathbf{K} = \frac{E}{L}\begin{bmatrix} 4(I_1 + I_2) & 2I_2 \\ 2I_2 & 4I_2 \end{bmatrix}$$

The consistent mass matrix of a beam element is

$$\mathbf{m} = \frac{m}{420} \begin{bmatrix} 156 & 22L & 54 & -13L \\ & 4L^2 & 13L & -3L^2 \\ & \text{symm} & 156 & -22L \\ & & & 4L^2 \end{bmatrix}$$

and the global mass matrix

$$\mathbf{M} = \frac{L^2}{420} \begin{bmatrix} 4(m_1+m_2) & -3m_2 \\ -3m_2 & 4m_2 \end{bmatrix} = \frac{L^2 \alpha}{420} \begin{bmatrix} 4(\sqrt{I_1}+\sqrt{I_2}) & -3\sqrt{I_2} \\ -3\sqrt{I_2} & 4\sqrt{I_2} \end{bmatrix}$$

4.5.1 Calculation of Vibration Mode and Frequency

Initially $I_1 = I_2$ so the eigenvalue problem is

$$\left| \frac{EI}{L} \begin{bmatrix} 8 & 2 \\ 2 & 4 \end{bmatrix} - \frac{\omega^2 \alpha \sqrt{I} L^2}{420} \begin{bmatrix} 8 & -3 \\ -3 & 4 \end{bmatrix} \right| = 0$$

Define

$$\lambda = \frac{\omega^2 \alpha L^3}{420 E \sqrt{I}}$$

then

$$\begin{vmatrix} 8-8\lambda & 2+3\lambda \\ 2+3\lambda & 4-4\lambda \end{vmatrix} = 0$$

$$23\lambda^2 - 76\lambda + 28 = 0$$

$$\lambda_1 = 0.422, \quad \lambda_2 = 2.88$$

The fundamental frequency is given by

$$\omega_1^2 = \frac{420 E \sqrt{I} \lambda_1}{\alpha L^3} = \frac{177 E \sqrt{I}}{\alpha L^3}$$

The corresponding mode satisfies

$$\begin{bmatrix} 4.62 & 3.27 \\ 3.27 & 2.31 \end{bmatrix} \begin{pmatrix} \theta_2 \\ \theta_3 \end{pmatrix} = \begin{pmatrix} 0 \\ 0 \end{pmatrix}$$

$$\theta_2 = -0.707 \theta_3$$

Also the normalization condition, by equation (4.10), is

$$\frac{\alpha \sqrt{I} L^2}{420} (8\theta_2^2 - 6\theta_2 \theta_3 + 4\theta_3^2) = 1$$

The vibration mode is

$$[\theta_2, \theta_3] = \frac{10.6}{(I\alpha^2)^{1/4}L}[-0.707 \quad 1]$$

4.5.2 Derivative Calculation

$$\frac{\partial \mathbf{K}}{\partial I_1} - \omega^2 \frac{\partial \mathbf{M}}{\partial I_1} = \frac{E}{L}\begin{bmatrix} 4 & 0 \\ 0 & 0 \end{bmatrix} - \left(\frac{177E\sqrt{I}}{\alpha L^3}\right)\left(\frac{L^2\alpha}{420\sqrt{I}}\right)\begin{bmatrix} 2 & 0 \\ 0 & 0 \end{bmatrix}$$

$$= \frac{E}{L}\begin{bmatrix} 3.16 & 0 \\ 0 & 0 \end{bmatrix}$$

$$\frac{\partial \mathbf{K}}{\partial I_2} - \omega^2 \frac{\partial \mathbf{M}}{\partial I_2} = \frac{E}{L}\begin{bmatrix} 4 & 2 \\ 2 & 4 \end{bmatrix} - \left(\frac{177E\sqrt{I}}{\alpha L^3}\right)\left(\frac{L^2\alpha}{420\sqrt{I}}\right)\begin{bmatrix} 2 & -1.5 \\ -1.5 & 2 \end{bmatrix}$$

$$= \frac{E}{L}\begin{bmatrix} 3.16 & 2.63 \\ 2.63 & 3.16 \end{bmatrix}$$

$$\frac{\partial \omega_1^2}{\partial I_1} = \frac{112}{\alpha L^2 \sqrt{I}}[-0.707 \quad 1]\frac{E}{L}\begin{bmatrix} 3.16 & 0 \\ 0 & 0 \end{bmatrix}\begin{bmatrix} -0.707 \\ 1 \end{bmatrix} = \frac{177E}{\alpha\sqrt{I}L^3}$$

$$= \frac{\omega_1^2}{I}$$

$$\frac{\partial \omega_1^2}{\partial I_2} = \frac{122}{\alpha L^2 \sqrt{I}}[-0.707 \quad 1]\frac{E}{L}\begin{bmatrix} 3.16 & 2.63 \\ 2.63 & 3.16 \end{bmatrix}\begin{bmatrix} -0.707 \\ 1 \end{bmatrix}$$

$$= \frac{122E}{\alpha\sqrt{I}L^2}$$

$$= \frac{0.64\omega_1^2}{I}$$

Result: It is more effective to stiffen beam 1.

4.6 DERIVATIVES OF TRANSIENT RESPONSE

The finite element equations resulting from a transient response problem may usually be written as

$$\mathbf{C}\dot{\mathbf{a}} + \mathbf{f}(x, a, t) = 0 \\ \mathbf{a}(0) = \mathbf{a}_0 \quad (4.14)$$

where $\mathbf{a}$ is the response vector, t is time, x is a design variable and a dot denotes derivative with respect to time. An example of such a system is the linear heat

transfer equation
$$\mathbf{C}\dot{\mathbf{T}} + \mathbf{K}\mathbf{T} - \mathbf{Q} = \mathbf{0} \qquad (4.15)$$
where $\mathbf{C}$ is the capacitance matrix, $\mathbf{K}$ is the conductivity matrix, and $\mathbf{Q}$ is a heat load vector.

The constraints that may be imposed on the response usually take the form of
$$g(x, a, t) \geq 0 \qquad 0 \leq t \leq t_f \qquad (4.16)$$
For actual computation the constraint has to be discretized at a series of time points
$$g_i = g(x, a, t_i) \geq 0 \qquad i = 1, \ldots, n_t \qquad (4.17)$$

It is often desirable to avoid having the large number of constraints implied by equation (4.17). There are several ways of removing the time dependence of the constraint without replacing it by several constraints.

(i) Equivalent exterior constraint:
$$\bar{g}(v, a) = \int_0^{t_f} [g(x, a, t) - |g(x, a, t)|] dt \geq 0 \qquad (4.18)$$
$\bar{g}$ is equal to zero in the entire feasible domain. Therefore, this constraint formulation is appropriate with methods which move outside the feasible domain or along its boundaries but not with interior methods such as interior penalty function methods.

(ii) Equivalent interior constraint:
$$\bar{g}(x, a) = \left[\int_0^{t_f} dt/g(x, a, t) \right]^{-1} \geq 0 \qquad (4.19)$$

This form of the constraint is defined only in the feasible domain. It is therefore not suitable for exterior methods.

(iii) Critical point constraint:
$$\bar{g}(x, a) = \min_{0 \leq t \leq t_f} g(x, a, t) \geq 0 \qquad (4.20)$$

This form of the constraint is based on following only the most critical point of the constraint. When g has several local minima it is advisable to follow all of them to prevent discontinuities in $\bar{g}$.

All three constraint forms may be written as
$$\bar{g}(x, a) = \int_0^{t_f} p(x, t, a) dt \geq 0 \qquad (4.21)$$

For the critical point constraint, for example,

$$p(x, t, a) = g(x, a, t)\delta(t - t_m) \tag{4.22}$$

where t_m is the point where $g(x, a, t)$ is miminal and δ is the Dirac delta function.
The derivative of $\bar{g}$ is obtained by differentiating equation (4.21):

$$\frac{d\bar{g}}{dx} = \int_0^{t_f} \left[\frac{\partial p}{\partial x} + \frac{\partial p}{\partial a}\frac{\partial a}{\partial x}\right] dt \tag{4.23}$$

As in the static case we can calculate the derivative in a straightforward manner, or by using an adjoint variable. The two techniques are again labelled method I and method II.

Method I

Method I consists of calculating the derivative of the response $\partial \mathbf{a}/\partial x$ and substituting it into equation (4.23). We obtain $\partial \mathbf{a}/\partial x$ by differentiating equation (4.14) to obtain

$$\mathbf{C}\frac{\partial \dot{\mathbf{a}}}{\partial x} + \mathbf{K}_t \frac{\partial \mathbf{a}}{\partial x} = \mathbf{F} \tag{4.24}$$

where

$$\mathbf{F} = -\frac{\partial \mathbf{f}}{\partial x} - \frac{\partial \mathbf{C}}{\partial x}\dot{\mathbf{a}} \tag{4.25}$$

and $\mathbf{K}_t$ is a matrix that

$$K_{tij} = \frac{\partial f_i}{\partial a_j} \tag{4.26}$$

Equation (4.24) is a system of differential equations which is linear but has time-dependent coefficients. If the solution technique is implicit it will be somewhat cheaper to solve than equation (4.14) for a. However, if equations (4.14) and (4.24) are solved with an explicit algorithm the amount of effort is comparable. Equation (4.24) has to be solved as many times as we have design variables.

Method II

Define an adjoint variable λ, multiply it by equation (4.24) and integrate;

$$\int_0^{t_f} \lambda^T \left[\mathbf{C}\frac{\partial \dot{\mathbf{a}}}{\partial x} + \mathbf{K}_t \frac{\partial \mathbf{a}}{\partial x} - \mathbf{F}\right] dt = 0 \tag{4.27}$$

Integrating by parts,

$$\lambda^T \mathbf{C} \frac{\partial \mathbf{a}}{\partial x}\bigg|_0^{t_f} + \int_0^{t_f} \left[\lambda^T(\mathbf{K}_t - \mathbf{C}) - \dot{\lambda}^T \mathbf{C}\right]\frac{\partial \mathbf{a}}{\partial x} dt = \int_0^{t_f} \lambda^T \mathbf{F} dt \tag{4.28}$$

We now require λ to satisfy

$$(\mathbf{K}_t - \dot{\mathbf{C}})^T \lambda - \mathbf{C}^T \dot\lambda = \left(\frac{\partial p}{\partial a}\right)^T \quad \lambda(t_f) = 0 \tag{4.29}$$

then

$$\int_0^{t_f} \left(\frac{\partial p}{\partial a}\right)\left(\frac{\partial a}{\partial x}\right) dt = \int_0^{t_f} \lambda^T \mathbf{F}\, dt \tag{4.30}$$

so that

$$\frac{d\bar{g}}{dx} = \int_0^{t_f} \left[\frac{\partial p}{\partial x} + \lambda^T \mathbf{F}\right] dt \tag{4.31}$$

Equation (4.29) is a system of linear ordinary differential equations which are integrated backwards (from $t = t_f$ to $t = 0$). This system has to be solved once for each constraint rather than once for each design variable. The choice of method I or method II is as in the static case based on the ratio of number of constraints to the number of design variables. Equation (4.29) takes a simpler form in the case of the critical point constraint:

$$(\mathbf{K}_t - \dot{\mathbf{C}})^T \lambda - \mathbf{C}^T \dot\lambda = \left(\frac{\partial g}{\partial a}\right)\delta(t - t_m) \quad \lambda(t_f) = 0 \tag{4.32}$$

or

$$(\mathbf{K}_t - \dot{\mathbf{C}})^T \lambda - \mathbf{C}^T \dot\lambda = 0 \quad \lambda^T(t_m) = \left(\frac{\partial g}{\partial a}\right)\mathbf{C}^{-1} \tag{4.33}$$

4.7 Example A multiwall titanium insulation layer protecting an aluminium structural layer which is proposed as a candidate for future space transportation systems [9] is used as an example. The panel shown in Fig. 4.4 consists of a series of flat layers interleaved with dimpled sheets. The layers must be closely spaced to prevent free convection in the air in the dimples. The dimpled construction reduce metal conduction and the use of a large number of layers reduces radiative heat transfer. The panel is subjected to a heat load $q(t)$ and radiates heat to space which is assumed to be at 0 K.

Lateral heat transfer is neglected, and the initial temperature of the insulation panel and of the aluminium panel it protects is 38 °C. The temperature in the aluminium panel is required to remain below 177 °C. A design study was conducted to determine whether a lighter insulation panel can be obtained by using a variable rather than a constant number of layers per unit thickness. The total thickness of the multiwall panel (h) was fixed by the requirement of interchangeability with the present ceramic insulation panels on the shuttle. This total thickness was divided into two regions. The design variables were the thickness of the first region and the number of titanium layers (both dimpled and flat) in each region. The only constraint on temperature was of 177 °C maximum aluminium temperature.

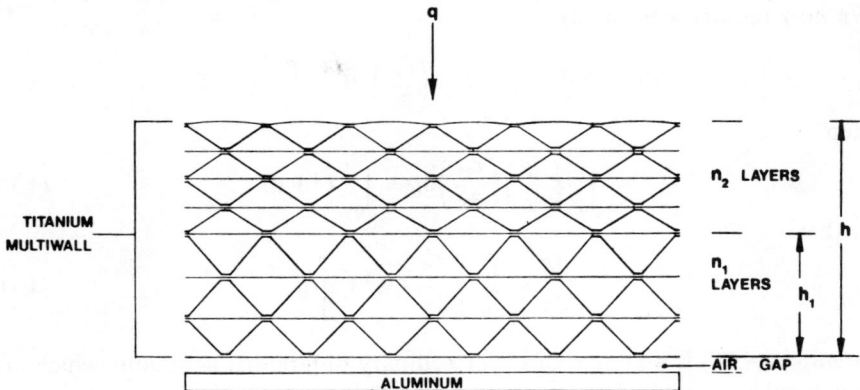

Fig. 4.4 Multiwall insulator

A one-dimensional finite element model was generated for the multiwall and aluminium panels. Two noded linear temperature elements were used; three in each region of the multiwall, one for the aluminium and one for the air gap between the insulation and aluminium for a total of eight elements and nine nodes. Because of the one-dimensional nature of the problem, implicit algorithms are at an advantage. Two algorithms were employed for the solution; a central difference (Crank–Nicholson) variable time step implicit algorithm and a forward difference (Euler) explicit algorithm.

The critical point constraint formulation was used to monitor the aluminium temperature constraint. The transient response and the derivatives of the constraint with respect to the three design variables were computed using both algorithms and are summarized in Table 4.1. It is seen that for this one-

Table 4.1 Comparison of integration times

	Explicit (Euler) CPU time* (average time step) (s)	Implicit (Crank–Nicholson) CPU time* (average time step) (s)
Analysis	27.5 (0.173)	1.7 (15)
Three derivatives finite differences	82.0 (0.173)	5.1 (15)
Three derivatives Method I	117.4 (0.173)	4.7 (7)
Three derivatives Method II	28.3 (0.173)	1.0 (12)

* Cyber-173

dimensional problem the implicit algorithm is much more efficient than the explicit one. For the analysis it required 1.7 CPU seconds on the CDU Cyber-173 compared to 27.5 seconds for the explicit algorithm. The reason for the difference is the time step that was restricted to 0.173 seconds for the explicit algorithm because of stability. The accuracy of the solution with the larger time step (15 seconds) of the implicit algorithm was about four significant digits. For comparison the transient analysis problem was run also with the SPAR program thermal analyser which employs an explicit integration scheme. SPAR required 51 seconds of CPU time with an average time step of 0.2 second.

R.H.

REFERENCES

1. E. J. Haug, Jr., K. C. Pan and T. D. Streeter, A computational method for optimal structural design, II: Continuous problems, *International Journal for Numerical Methods in Engineering*, **9**, 649–67, 1975.
2. R. T. Haftka and B. Prasad, Programs for analysis and resizing of complex structures, *Computers and Structures*, **10**, 323–30, 1979.
3. W. D. Whetstone, SPAR structural analysis system reference manual, *NASA CR-158970*, 1978.
4. G. L. Giles and R. T. Haftka, SPAR data handling utilities, *NASA TM-78701*, 1978.
5. J. S. Arora and E. J. Haug, Efficient methods of optimal structural design, *Journal of the Engineering Mechanics Division, ASCE*, **104** (EM3) 663–80, June 1978.
6. T. T. Feng, J. S. Arora and E. J. Haug, Optimal structural design under dynamic loads, *International Journal for Numerical Methods in Engineering*, **II**, 39–52, 1977.
7. C. Cardani and P. Mantegazza, Calculation of eigenvalue and eigenvector derivatives for algebraic flutter and divergence problems, *AIAA Journal*, **17** (4), 408–12, 1979.
8. R. T. Haftka, Techniques for thermal sensitivity analysis, *International Journal for Numerical Methods in Engineering* (in press).
9. L. R. Jackson, Multiwall TPS, in Recent advances in structures for hypersonic flight, *NASA CP-2065*, 1978.

Foundations of Structural Optimization: A Unified Approach
Edited by A. J. Morris
© 1982 John Wiley & Sons Ltd.

Chapter 5

Optimality Criterion Methods in Structural Optimization

5.1 INTRODUCTION

Having introduced the basic theory and the influence of the finite element, we now begin our discussion of solution algorithms starting with methods based on an optimality criterion to design a minimum-weight structure. These methods can be called indirect methods, since the objective is to obtain a design that satisfies a certain specified criterion and by so doing indirectly minimize the weight of the structure. The criterion may be intuitive or derived mathematically, based on the nature of the problem. We concentrate on a mathematical criterion derived by differentiating the Lagrangian with respect to the design variables. In the optimality criterion methods the criterion is derived for the dominant type of constraint imposed on the structure, and that criterion is used to develop the algorithm. The methods are iterative in nature due to the essential nonlinearity of the constraints and the statical indeterminacy of the structure. In deriving the optimality criterion and developing the algorithm, full use is made of the knowledge of the behaviour of the constraints imposed on the structure. The algorithms are efficient, because they are specifically developed for structural optimization and generally treat the constraints which weakly affect the behaviour of the structure as passive constraints.

The early discussion on the optimality criterion methods as applied to a discretized structure for different types of constraints may be found in Refs [1] through [4]. The application of the method to the stress-displacement constrained problem, with a variation in the recurrence relations and the methods used to estimate the Lagrange multipliers is discussed in Refs [5] through [20]. The general instability of the structure as a constraint on the structure is considered in Refs [21] and [22]. The optimization of a structure with dynamic loads is discussed in Refs [23] through [25]. These are a few of the references in the general field of structural optimization based on the optimality criterion approach. Additional references may be found in the recent survey paper (Ref.

[26]), and a detailed review of the state-of-the-art in Ref. [27]. In Ref. [28] the relationship between the different optimality criterion based algorithms is presented. The relationship between the optimality criterion and mathematical programming methods is presented in Ref. [29]. The relationship between the structural optimization algorithms based on an optimality criterion and the projection method is discussed in Refs [30] and [31]. Some of the computer programs based on the optimality criterion to design minimum-weight structures are documented in Refs [32] through [36].

The constraints imposed on the structure may include the maximum allowable stress in each element, the displacement limits at one or more locations, system stability, dynamic stiffness, local element buckling etc. In addition to these there would be limitations on the minimum and maximum sizes of the elements. An optimality criterion can be derived that includes all these constraints, and it may be desirable to find a design that satisfies this criterion. However, to develop an efficient algorithm based on such a criterion and effectively handle all types of constraints would be impractical and generally unnecessary. For most problems it is developing the algorithm that is more difficult than deriving the optimality criterion. In the case of most structures it is likely that one can predict the type of constraint which will be the most active at the optimum and use the algorithm based on that constraint. Then one can treat all other constraints as passive constraints. It is highly unlikely that all types of constraints will be active at the optimum. Sometimes this point of view may not be correct and will not give us an absolute minimum weight design. However, it gives a near minimum weight design, and the corresponding optimization algorithm will be efficient and easy to use for a structure with a large number of design variables. Even with this approximation to the overall problem, when the total number of constraints of the same type are large it is advantageous to make additional approximations in order to reduce the computational effort.

The optimality criterion derived for all the constraints imposed on the structure is equivalent to the Kuhn–Tucker conditions of nonlinear mathematical programming. However, in deriving the optimality criterion and the corresponding structural optimization algorithm, some of the constraints are treated as side constraints in order to simplify the algorithm. A good example of this is the minimum and maximum size limits on the design variables. These constraints generally are not included in the constraint equations and do not enter the optimality criterion. The optimality criterion derived for structural optimization for a particular type of constraint gives information on the distribution of energy in the structure necessary to have a minimum weight design. The nature of the energy depends upon the type of constraint.

The optimization procedure may be divided into two major steps. These are: (1) the analysis of the structure and (2) the redistribution of the material so that the weight of the structure is reduced and the active constraints are satisfied. Each of these has been considered seperately in previous chapters and now these two steps are repeated alternately until a satisfactory design is obtained. The analysis of the discretized structure is performed by the finite element method. The redistribution of the material is carried out by using a recurrence relation. The recurrence relation modifies the design variables so that in the design space, the initial design is moved towards a design that satisfies the optimality criterion. The recurrence relation contains two sets of unknown terms. The first set is related to the gradient of the constraints and the second set is the Lagrange multipliers. It is necessary to determine these unknowns before the recurrence relation can be used.

The efficiency and the convergence behaviour of the algorithm depend on: (1) the recurrence relation used to modify the design variables; (2) the nature of the approximations made to derive the mathematical expression for the unknowns in the recurrence relations; and (3) how these unknowns are determined.

In deriving the optimality criterion and the algorithm based on it, we will assume that the structure is discretized into a number of elements, and that it is suitable for finite element analysis. In common with our previous approach it is assumed that the only design variables are the cross-sectional areas of the bar elements and the thicknesses of the plate elements. The geometry of the structure and the loads applied to it are fixed.

In this chapter we discuss the different methods developed to optimize structures with constraints on displacements, stresses, minimum sizes and general instability. The detailed discussion is primarily devoted to constraints on stresses and displacements, since there are the basic constraints in any structure. At first the equations of the displacement method of finite element analysis are reviewed and then the optimization methods are presented. The effect of the reciprocal design variable on the optimality criterion and the optimization algorithms is also discussed. The general optimization procedure and the different approaches one can use to optimize a structure are then presented. The chapter is concluded with illustrative problems indicating the effect of the different algorithm, a summary and an appendix. The appendix contains details of a three-bar truss problem solved by using different algorithms.

5.2 BASIC EQUATIONS OF ANALYSIS

5.2.1 Displacement Method

The basic equations of finite element analysis are given in Chapters 3 and 4 but some extension of the terminology is required to develop the optimization algorithm, and this is summarized in this section.

Recall that the load vector **P** is related to the displacement **u** by the equation

$$\mathbf{Ku} = \mathbf{P} \qquad (5.1)$$

where **K** is the global stiffness matrix of the structure. The strain energy V_i stored in the ith element is given by

$$V_i = \mathbf{u}_i^T \mathbf{k}_i \mathbf{u}_i \tag{5.2}$$

where $\mathbf{u}_i$ is the displacement vector associated with the ith element and $\mathbf{k}_i$ is the ith element stiffness matrix. The ith element strain vector $\boldsymbol{\varepsilon}_i$ is related to $\mathbf{u}_i$ by

$$\boldsymbol{\varepsilon}_i = \mathbf{B}_i \mathbf{u}_i \tag{5.3}$$

where $\mathbf{B}_i$ is the strain-displacement matrix associated with the ith element.

The relation between the stress vector $\boldsymbol{\sigma}_i$ and the strain vector $\boldsymbol{\varepsilon}_i$ in the ith element is given by

$$\boldsymbol{\sigma}_i = \mathbf{E}_i \boldsymbol{\varepsilon}_i \tag{5.4}$$

where $\mathbf{E}_i$ is the stress–strain matrix of the elastic constants.

Using Equations (5.3) and (5.4), a relation between $\boldsymbol{\sigma}_i$ and $\mathbf{u}_i$ can be written as

$$\boldsymbol{\sigma}_i = \mathbf{D}_i \mathbf{u}_i \tag{5.5}$$

where $\mathbf{D}_i$ is the stress-displacement matrix and is given by

$$\mathbf{D}_i = \mathbf{E}_i \mathbf{B}_i \tag{5.6}$$

The displacement u_j at the jth node in the structure previously defined in Chapter 3 is now given by

$$u_j = \sum_{i=1}^{n} \frac{Q_{ij}}{x_i} \tag{5.7}$$

or

$$u_j = \sum_{i=1}^{n} Q_{ij} z_i \tag{5.8}$$

where n is the number of elements, and z_i is the reciprocal design variable related to the direct design variable x_i by

$$z_i = \frac{1}{x_i} \tag{5.9}$$

In Equations (5.7) and (5.8), Q_{ij} is the flexibility coefficient and is given by

$$Q_{ij} = x_i \mathbf{u}_i^T \mathbf{k}_i \mathbf{u}_i^{(j)} \tag{5.10}$$

or

$$Q_{ij} = \frac{1}{z_i} \mathbf{u}_i^T \mathbf{k}_i \mathbf{u}_i^{(j)} = \frac{1}{z_i} \mathbf{u}_i^{(j)T} \mathbf{k}_i \mathbf{u}_i \tag{5.11}$$

where $\mathbf{u}_i^{(j)}$ is the virtual displacement at node i associated with the virtual load vector $\mathbf{P}^{(j)}$, and the full vector displacement vector is defined by

$$\mathbf{u}^{(j)} = \mathbf{u}^T \mathbf{P}^{(j)}$$

If the virtual load vector consists of unit loads then the vector $\mathbf{P}^{(j)}$ reduces to one which is equivalent to the vector $\mathbf{e}_{(j)}$ defined in Section 3.2.1.

In the case of a bar structure it is shown in Chapter 3 that the deflection at a point is given by the well known relation

$$u_j = C_j = \sum_{i=1}^{n} \frac{T_i t_i^{(j)} l_i}{x_i E_i} \qquad (5.12)$$

where C_j represents a more generalized form for a structural response and the numerical value of the forces $t_i^{(j)}$ is different from the same terms defined in Chapter 3 if the value of the virtual loads is different from 1.

The flexibility coefficient Q_{ij} can be written as

$$Q_{ij} = \frac{T_i t_i^{(j)} l_i}{E_i} \qquad (5.13)$$

where T_i and $t_i^{(j)}$ are the forces in the ith bar due to the applied load P and the virtual load $P^{(j)}$ respectively.

5.2.2 Scaling of the Design

In an optimization algorithm, it is convenient to obtain a feasible design after each iteration. This helps keep track of the reduction in the weight of the structure after each iteration and also helps to pick the most active constraints. A feasible design can be obtained by scaling the design in order to satisfy the specified constraints. The design variable x_i can be written as

$$x_i = \Lambda \alpha_i \qquad (5.14)$$

where Λ is the scaling parameter. In Equation (5.14) α_i is the relative value of the design variable x_i. The scaling parameter Λ is the same for all the elements. Using the definition of x_i given in Equation (5.14), Equation (5.1) can be written as

$$\Lambda \mathbf{K}' \mathbf{u} = \mathbf{P} \qquad (5.15)$$

where $\mathbf{K}'$ is the stiffness matrix of the structure obtained by using the relative design vector α. If $\mathbf{u}'$ is the relative displacement vector, then the equilibrium equation can also be written as

$$\mathbf{K}' \mathbf{u}' = \mathbf{P} \qquad (5.16)$$

Comparing Equations (5.15) and (5.16), the relationship between the actual displacement vector $\mathbf{u}$ and the relative displacement vector $\mathbf{u}'$ can be written as

$$\mathbf{u}' = \Lambda \mathbf{u} \qquad (5.17)$$

Similarly, expressing the strain-displacement relation and the stress-strain relation in terms of the relative design variable α, the relationship between the relative strains and stresses can be obtained. This gives

$$\varepsilon'_i = \Lambda \varepsilon_i \qquad (5.18)$$
$$\sigma'_i = \Lambda \sigma_i \qquad (5.19)$$

where ε'_i and σ'_i are the relative strains and stresses associated with the ith element. It is seen from Equations (5.17) through (5.19), that one can select the parameter Λ to satisfy the displacement constraint at a node point or the stress constraint in an element. The Λ corresponding to the most critical constraint will satisfy all other constraints also. The design with the actual values of the design variables satisfying all constraints is given by Equation (5.14), where Λ is determined for the most critical constraint. The simple scaling procedure discussed here can be used when the stiffness matrix is a linear function of the design variable. In other cases it is required to iterate in order to obtain an acceptable scaling parameter. It is possible that for certain problems the scaling procedure cannot be used.

If the scaling parameter Λ is introduced in the stability equation from Section 3.4.3, we obtain

$$[\mathbf{K}' + \mu' \mathbf{K}'_G] \eta' = 0 \qquad (5.20)$$

where

$$\mu' = \frac{\mu}{\Lambda} \qquad (5.21)$$

and

$$\eta' = \Lambda \eta \qquad (5.22)$$

Using Equation (5.21), we can scale the design so that the buckling load of the structure is equal to the applied load.

5.3 DISPLACEMENT CONSTRAINTS

The displacement constraints are fundamental to the development of an optimization algorithm that uses the displacement method of finite element analysis. The algorithms for other types of constraints are similar to that of the displacement constrained problem. In the case of a stress-constrained problem, for example, an effective algorithm can be developed only by replacing the stress constraints by the equivalent displacement constraints. We will first consider the problem of designing a structure with a single displacement constraint and then discuss multiple constraints. In all the equations in this and in subsequent sections, the index i refers to an element and it goes from 1 to n, where n is the number of elements.

5.3.1 Single Displacement Constraint

5.3.1.1 Optimality criterion

The weight of the structure can be written as

$$W(x) = \sum_{i=1}^{n} \rho_i x_i l_i \tag{5.23}$$

where ρ_i is the specific weight, x_i is the design variable and l_i is a function of the geometry of an element. The product $x_i l_i$ represents the volume of the element. For a bar element x_i is the cross-sectional area and for a plate element it is the thickness. Equation (5.23) assumes that the weight of the structure is a linear function of the design variable x_i.

A single displacement constraint can be written as

$$g_1 = C_1 - \overline{C}_1 = 0 \tag{5.24}$$

or

$$g_1 = \sum_{i=1}^{n} \frac{Q_{i1}}{x_i} - \overline{C}_1 = 0 \tag{5.25}$$

or

$$g_1 = \mathbf{u}^T \mathbf{u}^{(1)} - \overline{C}_1 = 0 \tag{5.26}$$

where $\overline{C}_1$ is the limiting value of the displacement.

In this form the design problem is identical to that defined in Section 3.2.3 as (MWP1).

Using Equations (5.23) and (5.25), the Lagrangian for a single displacement constrained problem can be written as

$$L(x, \lambda) = \sum_{i=1}^{n} \rho_i x_i l_i + \lambda_1 \left(\sum_{i=1}^{n} \frac{Q_{i1}}{x_i} - \overline{C}_1 \right) \tag{5.27}$$

Differentiating this equation with respect to the design variable x_i by recalling (3.1) and setting it equal to zero gives

$$\rho_i l_i - \lambda_1 \frac{Q_{i1}}{x_i^2} = 0 \tag{5.28}$$

This equation can be rewritten as

$$x_i^2 = \lambda_1 \frac{Q_{i1}}{\rho_i l_i} \tag{5.29}$$

or

$$1 = \lambda_1 \frac{Q_{i1}}{x_i^2 \rho_i l_i} \tag{5.30}$$

or

$$x_i = \lambda_1 \frac{Q_{i1}}{x_i \rho_i l_i} \tag{5.31}$$

or

$$1 = \lambda_1 \frac{e_{i1}}{\rho_i} \tag{5.32}$$

In Equation (5.32), e_{i1} is the virtual strain energy density. Equations (5.28) through (5.32) represent the optimality criterion for a single constrained problem. Equation (5.32) states that in the optimum structure, the ratio of virtual strain energy density to specific weight is the same for all the elements.

The flexibility coefficients Q_{i1} are constant for a statically determinate structure. In the case of a statically indeterminate structure, they are functions of the design variables. In the derivation of the optimality criterion above, it seems that we have treated Q_{i1} as a constant with respect to the design variable. However, even without this assumption an identical criterion can be derived. Consider the definition of the displacement constraint as given in Equation (6.26). Using this equation and Equation (5.1) we can write

$$C_1 = \mathbf{u}^T \mathbf{K} \mathbf{u}^{(1)} \tag{5.33}$$

where $\mathbf{u}^{(1)}$ is the virtual displacement vector associated with the applied virtual load vector $P^{(1)}$. Differentiating Equation (5.33) with respect to x_i by using the results of Section 3.4.1 gives

$$\frac{\partial C_1}{\partial x_i} = -\frac{\mathbf{u}_i^T \mathbf{k}_i \mathbf{u}_i^{(1)}}{x_i} = -\frac{Q_{i1}}{x_i^2} \tag{5.34}$$

This derivation of the gradient of the constraint shows that in deriving the optimality criterion it is not necessary to assume Q_{i1} as a constant.

5.3.1.2 Recurrence relations

The optimality criterion derived in the last section is valid only at the optimum and has to be converted into a recurrence relation so that it can be used in an optimization algorithm. A recurrence relation can be written by multiplying both sides of Equation (5.30) by A_i^r and taking the rth root. This gives

$$x_i^{k+1} = x_i^k \left(\lambda_1 \frac{Q_{i1}}{x_i^2 \rho_i l_i} \right)_k^{1/r} \tag{5.35}$$

where $k+1$ and k are introduced to indicate the iteration numbers. The parameter r in Equation (5.35) determines the step size. We will call this an exponential recurrence relation. Equation (5.35) can also be written as

$$x_i^{k+1} = x_i^k \left(1 + \left(\lambda_1 \frac{Q_{i1}}{x_i^2 \rho_i l_i} - 1 \right) \right)_k^{1/r} \tag{5.36}$$

In this equation $\lambda_1 (Q_{i1}/x_i^2 \rho_i l_i)$ is equal to unity at the optimum, and thus near the optimum $(\lambda_1 (Q_{i1}/x_i^2 \rho_i l_i) - 1)$ is small compared to unity. Therefore using the binomial theorem to expand the right side of Equation (5.36) and retaining only the linear terms gives

$$x_i^{k+1} = x_i^k \left(1 + \frac{1}{r}\left(\lambda_1 \frac{Q_{i1}}{x_i^2 \rho_i l_i} - 1\right)\right)_k \qquad (5.37)$$

This equation we will call a linear recurrence relation. In this equation $(\lambda_1 (Q_{i1}/x_i^2 \rho_i l_i) - 1)$ is the error in satisfying the optimality criterion.

The exponential and linear recurrence relations can also be expressed in terms of the virtual strain energy density e_{i1} used to write the optimality criterion in Equation (5.32). In order to use the recurrence relation it is required to select the proper step size parameter. For most problems it has been found that $r = 2$ is generally adequate. Increasing the value of r reduces the step size.

The recurrence relations contain two unknowns, the flexibility coefficients Q_{i1} and the Lagrange multiplier λ_1. The coefficients Q_{i1} can be determined by using Equation (5.10). The Lagrange multiplier λ_1 can be obtained by using the relations derived in the next section.

5.3.1.3 Expressions for the Lagrange multiplier

The design variable x_i using Equation (5.29) can be written as

$$x_i = \sqrt{\lambda_1 \frac{Q_{i1}}{\rho_i l_i}} \qquad (5.38)$$

The Lagrange multiplier λ_1 can be evaluated by substituting Equation (5.38) in Equation (5.25) and solving for λ_1. This gives, as previously defined in Section 3.2.3,

$$\sqrt{\lambda_1} = \frac{\sum_{i=1}^{n} \sqrt{Q_{i1} \rho_i l_i}}{\overline{C}_1} \qquad (5.39)$$

We can derive two more expressions for the Lagrange multiplier by using the optimality criterion and the constraint equations. Substituting Equation (5.31) in Equation (5.23) and using Equation (5.23), the weight of the structure is given by

$$W = \lambda_1 \overline{C}_1 \qquad (5.40)$$

or

$$\lambda_1 = \frac{W}{\overline{C}_1} \qquad (5.41)$$

Equation (5.40) states that at the optimum the weight of the structure is equal to the product of the Lagrange multiplier λ_1 and the limiting value of the constraint $\overline{C}_1$.

Another expression for the Lagrange multiplier can be obtained by writing the optimality criterion as

$$\frac{1}{x_i} = \lambda_1 \frac{Q_{i1}}{\rho_i l_i x_i^3} \quad (5.42)$$

and substituting it in Equation (5.23). This gives

$$\lambda_1 = \frac{\overline{C}_1}{\sum_{i=1}^{n} Q_{i1}^2 / \rho_i l_i x_i^3} \quad (5.43)$$

Eliminating λ_1 from Equations (5.40) and (5.43), the weight of the structure at the optimum is given by

$$W = \frac{\overline{C}_1^2}{\sum_{i=1}^{n} Q_{i1}^2 / \rho_i l_i x_i^3} \quad (5.44)$$

The Lagrange multiplier λ_1 in Equations (5.35) and (5.37) can be eliminated by using one of the expressions for λ_1 derived above. However, in the case of the single constrained problem, if the structure can be scaled by using Equations (5.14) through (5.19), a relative value of the Lagrange multiplier can be used. Therefore, one can set λ_1 equal to unity. The expressions for the Lagrange multipliers derived in this section are strictly valid only at the optimum.

5.3.1.4 Effect of the passive elements on the Lagrange multipliers

In deriving the expressions for the Lagrange multipliers for the single displacement constraint we have assumed that all the design variables can be modified by using the recurrence relation. In a practical design problem there are always constraints on the minimum and maximum sizes of the elements. These elements or other elements, whose sizes are not modified during the iterations by using the recurrence relation, are called passive elements. And it is necessary to modify the algorithm to account for these elements. When the exponential recurrence relation is used, if the virtual strain energy associated with any element is negative, then that element is sized by some other criterion, and it becomes passive. The elements whose sizes are governed by the recurrence relation can be called active elements. At the optimum the optimality criterion is satisfied only by the active elements.

The contribution to the weight of the structure and the constraint can be divided into two parts, one due to the active and the other due to the passive elements. Thus the weight of the structure and the constraint equation can be written as

$$W = \sum_{i=1}^{n_1} \rho_i x_i l_i + W^* \quad (5.45)$$

and
$$g_1 = \sum_{i=1}^{n_1} \frac{Q_{i1}}{x_i} + C_1^* - \bar{C}_1 = 0 \tag{5.46}$$

where W^* is the weight of the passive elements and C_1^* is the contribution of the passive elements to the constraint. In Equations (5.45) and (5.46), n_1 is the number of active elements. This change in the definition of the weight of the structure and the constraint affects the expression for the Lagrange multiplier. With this change Equations (5.39), (5.41) and (5.43) can be written respectively as

$$\sqrt{\lambda_1} = \frac{\sum_{i=1}^{n_1} \sqrt{Q_{i1} \rho_i l_i}}{\bar{C}_1 - C_1^*} \tag{5.47}$$

$$\lambda_1 = \frac{W - W^*}{\bar{C}_1 - C_1^*} \tag{5.48}$$

$$\lambda_1 = \frac{\bar{C}_1 - C_1^*}{\sum_{i=1}^{n_1} Q_{i1}^2 / \rho_i l_i x_i^3} \tag{5.49}$$

Hence, when there are passive elements we should use Equations (5.47), (5.48) and (5.49) instead of (5.39), (5.41), and (5.43) respectively.

In the case of a single displacement constrained problem the explicit expression for the Lagrange multiplier does not have a significant effect on the algorithm. However, in the multiple displacement constrained problem, the method used to determine the Lagrange multipliers plays a major role in the optimization algorithm.

5.3.2 Multiple Displacement Constraints

The method discussed in the last section to design a minimum-weight structure would give a minimum weight design only if: (1) a structure is subjected to a single displacement constraint, or (2) only one constraint is active at the optimum. The active constraints may be defined as those which are satisfied as equality constraints at the optimum. The multiple constrained problem is much more difficult to solve than the single constrained problem. The main reason for this is that a simple explicit expression cannot be derived for the Lagrange multipliers. Also it is necessary to develop the logic to predict a probable set of active constraints during the iterations. If all the active constraints are not taken into consideration, the convergence behaviour is affected. If all the constraints imposed on the structure are considered potentially active, then the algorithm becomes inefficient from the point of view of computational effort. This is particularly true when there are a large number of inequality constraints imposed on the structure.

If the structure is subjected to more than one loading condition, the loading conditions can be treated as multiple constraints. So in order to simplify the notation we will not consider the multiple loading conditions in defining the problem.

The Lagrangian for the multiple displacement constrained problem can be written as

$$L(x, \lambda) = \sum_{i=1}^{n} \rho_i x_i l_i + \sum_{j=1}^{m} \lambda_j g_j \qquad (5.50)$$

where

$$g_j = C_j - \overline{C}_j \leqslant 0 \qquad j = 1, \ldots, m \qquad (5.51)$$

and

$$C_j = \sum_{i=1}^{n} \frac{Q_{ij}}{x_i} \qquad (5.52)$$

or

$$C_j = \mathbf{u}^T \mathbf{u}^{(j)} \qquad (5.53)$$

In Equations (5.50) and (5.51), g_j are the inequality constraints, λ_j are the Lagrange multipliers and m is the number of inequality constraints.

The optimality criterion can be derived by differentiating the Lagrangian with respect to the design variables x_i and setting the resulting equations to zero. This gives

$$\rho_i l_i - \sum_{j=1}^{m} \lambda_j \frac{Q_{ij}}{x_i^2} = 0 \qquad (5.54)$$

where $\lambda_j g_j = 0$. For the active constraints $\lambda_j > 0$ and $g_j = 0$, and for the passive constraints $\lambda_j = 0$ and $g_j \neq 0$. The optimality criterion in Equation (5.54) can be written in different forms as follows:

$$x_i^2 = \sum_{j=1}^{m} \lambda_j \frac{Q_{ij}}{\rho_i l_i} \qquad (5.55)$$

or

$$x_i = \sum_{j=1}^{m} \lambda_j \frac{Q_{ij}}{\rho_i l_i x_i} \qquad (5.56)$$

or

$$1 = \sum_{j=1}^{m} \lambda_j \frac{Q_{ij}}{\rho_i l_i x_i^2} \qquad (5.57)$$

or

$$1 = \sum_{j=1}^{m} \lambda_j \frac{e_{ij}}{\rho_i} \qquad (5.58)$$

or

$$1 = \frac{u_i^T k_i \sum_{j=1}^{m} \lambda_j u_i^{(j)}}{\rho_i l_i x_i} \quad (5.59)$$

or

$$1 = \frac{\bar{e}_{ij}}{\rho_i} \quad (5.60)$$

where

$$\bar{e}_{ij} = \frac{u_i^T k_i \sum_{j=1}^{m} \lambda_j u_i^{(j)}}{x_i l_i} \quad (5.61)$$

The optimality criterion as written in Equations (5.58) and (5.60) has a physical interpretation. According to Equation (5.58), at the optimum the weighted sum of the ratio of virtual strain energy density to mass density is equal to unity for all the elements, where the weighting parameters are the Lagrange multipliers. The virtual strain energy density in each element is due to a virtual load vector associated with each distinct constraint. Equation (5.60) which is also an optimality criterion states that the virtual strain energy to mass density is equal to unity for all elements where a single virtual load vector equal to $\sum_{j=1}^{m} \lambda_j P^{(j)}$ is applied to the structure. In using the recurrence relation based on Equation (5.57) or (5.58), it is required to determine the virtual strain energy in each element due to m virtual loads. But if the recurrence relation based on Equation (5.59) is used, it is only necessary to consider one virtual load vector in order to determine the virtual strain energy in an element. However, it is required to know the actual or relative values of the Lagrange multipliers before a single virtual load vector can be assembled. This is not always possible as we will see when we discuss the methods to determine the Lagrange multipliers.

The n optimality conditions and the m constraints must be satisfied by the optimum design. These are $(m + n)$ nonlinear equations that have to be solved in order to determine the n design variables and the m Lagrange multipliers. The optimization algorithm solves these nonlinear equations by an iterative scheme. In an iterative method the design variables are modified by using the recurrence relation so that the optimality criterion is satisfied. The Lagrange multipliers are evaluated on the basis that the constraint conditions are satisfied when the design variables are changed. Since the Lagrange multipliers and the design variables are functions of each other, any change in one nonlinearly affects the other, and this necessitates the use of an iterative method.

5.3.2.1 Recurrence relations for multiple constraints

The recurrence relations for the multiple displacement constrained problem can also be divided into two categories: (1) an exponential form, and (2) a linear form, as we have done for the single displacement constrained problem. The method of deriving them is similar to the one we used to write Equations (5.36) and (5.37) for the single constrained problem.

An exponential recurrence relation can be written by multiplying Equation (5.57) by A_i^r and taking the rth root. This gives

$$x_i^{k+1} = x_i^k \left(\sum_{j=1}^{m} \lambda_j \frac{Q_{ij}}{\rho_i l_i x_i^2} \right)_k^{1/r} \qquad (5.62)$$

where $k+1$ and k are the iteration numbers. In Equation (5.62) the parameter r determines the step size. Using the same argument given to derive Equation (5.37), we can write a linear recurrence relation for the multiple displacement constrained problem. This gives

$$x_i^{k+1} = x_i^k \left(1 + \frac{1}{r} \left(\sum_{j=1}^{m} \lambda_j \frac{Q_{ij}}{\rho_i l_i x_i^2} - 1 \right) \right)_k \qquad (5.63)$$

This equation can also be written as

$$x_i^{k+1} = x_i^k + \frac{1}{r} \left(\sum_{j=1}^{m} \lambda_j \frac{Q_{ij}}{\rho_i l_i x_i} - x_i \right)_k \qquad (5.64)$$

In this equation the term in parenthesis is the correction needed to the design variable A_i in order to satisfy the optimality criterion as defined in Equation (5.56). Thus we can say that the objective of modifying the design variable by using the recurrence relation is to move the current design towards a design that satisfies the optimality criterion. When the step size parameter $r = 1$, Equations (5.62) and (5.63) reduce to the same equation. For the multiple constrained problem, as in the case of a single constraint, $r = 2$ can be considered as the normal step size, but it is required for some problems to increase it to reduce the step size. In the exponential recurrence relation the design variable is modified by multiplying it by a quantity which is equal to unity at the optimum, and in the linear recurrence relation the design variable is modified by adding a quantity which is equal to zero at the optimum. Equations (5.62) and (5.63) can also be written by using the optimality criterion defined by Equations (5.58) and (5.60).

5.3.2.2 Methods to determine the Lagrange multipliers

In order to use any one of the recurrence relations the coefficients Q_{ij} and the Lagrange multipliers have to be known. The coefficients Q_{ij} can be determined by using Equation (5.10). The equations to determine the Lagrange multipliers can be derived by considering the effect of a change in the design variables or Lagrange multipliers on the constraint equations. We will derive in this section different methods, some rigorous and some not rigorous. The rigorous approach gives equations which are reliable but needs more computational effort. The nonrigorous methods are simple and need less computational effort, but give algorithms which may not lead to a minimum or are slow to converge. In discussing the different methods we will point out their advantages and disadvantages and their relationship to one another.

5.3.2.2.1 Recurrence relations for the Lagrange multipliers

A recurrence relation to estimate the Lagrange multipliers can be written by assuming that all the constraints in Equation (5.51) are equality constraints. This gives

$$C_j = \overline{C}_j \tag{5.65}$$

Multiplying both sides of this relation by λ_j^b and taking the bth root, a recurrence relation can be written as

$$\lambda_j^{k+1} = \lambda_j^k \left(\frac{C_j}{\overline{C}_j}\right)_k^{1/b} \tag{5.66}$$

where k refers to the iteration number and the parameter b controls the step size. The advantages of using this recurrence relation are:
1. An equivalent single virtual load vector can be used to determine the virtual strain energy in an element, and individual values of Q_{ij} for different constraints need not be determined.
2. It is not necessary to predict which constraints are potentially active. Repeated use of Equation (5.66) reduces the value of a Lagarange multiplier corresponding to a passive constraint, and its contribution to the virtual load vector is reduced after each iteration.
3. The computational effort required to determine the Lagrange multipliers and the virtual strain energy in an element is minimal as compared to other methods.

The disadvantages are:
1. Initial values of the Lagrange multipliers have to be assumed.
2. Convergence to the minimum weight design in slow and large oscillations sometimes occur in the scaled weight of the structure.

The recurrence relation in Equation (5.66) can be written as

$$\lambda_j^{k+1} = \lambda_j^k \left(1 + \left(\frac{\overline{C}_j}{C_j} - 1\right)\right)_k^{-1/b} \tag{5.67}$$

Since at the optimum $\overline{C}_j/C_j$ is nearly equal to unity, the difference $(\overline{C}_j/C_j - 1)$ is small as compared to unity. We can therefore expand Equation (5.67) by using the binomial theorem and retain only the linear terms. This gives

$$\lambda_j^{k+1} = \lambda_j^k \left(\frac{b+1}{b} - \frac{1}{b}\frac{\overline{C}_j}{C_j}\right)_k \tag{5.68}$$

This equation can also be written as

$$\bar{C}_j - C_j^k = bC_j^k\left(1 - \frac{\lambda_j^{k+1}}{\lambda_j^k}\right) \tag{5.69}$$

We will show in the next section that this linear recurrence relation for the Lagrange multipliers is an approximation to a set of linear equations that can be used to determine the Lagrange multipliers.

5.3.2.2.2. Linear equations to determine the Lagrange multipliers

A set of equations to determine the Lagrange multipliers can be obtained by considering the change in the constraint due to a change in the design variable x_i. The change in the jth constraint can be written as

$$\Delta g_j = g_j(x + \Delta x) - g_j(x) \tag{5.70}$$

$$= \sum_{i=1}^{n} \frac{\partial g_j}{\partial x_i} \Delta x_i \tag{5.71}$$

If the change in the design variable is assumed to be such that the constraints at point $(x + dx)$ are satisfied i.e. $g_j(x + dx) = 0$, then using Equation (5.51), Equation (5.71) can be written as

$$\bar{C}_j - C_j^k = \sum_{i=1}^{n} \frac{\partial g_j}{\partial x_i} \Delta x_i^k \tag{5.72}$$

where k refers to the iteration number. Differentiating Equation (5.52) with respect to the design variable x_i and assuming that Q_{ij} is a constant gives

$$\frac{\partial g_j}{\partial x_1} = -\frac{Q_{ij}}{x_i^2} \tag{5.73}$$

The change in the design variable from the kth to the $(k+1)$th iteration is given by

$$\Delta x_i^k = x_i^{k+1} - x_i^k \tag{5.74}$$

Substituting x_i^{k+1} from Equation (5.63) in this equation gives

$$\Delta x_i^k = \frac{1}{r}\left(\sum_{j=1}^{m} \lambda_j \frac{Q_{ij}}{\rho_i l_i x_i^2} - 1\right)_k x_i^k \tag{5.75}$$

Using Equations (5.73) and (5.75), Equation (5.72) can be written as

$$\sum_{p=1}^{m} \lambda_p^{k+1}\left(\sum_{i=1}^{n} \frac{Q_{ij}Q_{ip}}{\rho_i l_i x_i^3}\right)_k = r(C_j^k - \bar{C}_j) + \sum_{i=1}^{n}\left(\frac{Q_{ij}}{x_i}\right)_k \tag{5.76}$$

Since $\sum_{i=1}^{n}(Q_{ij}/x_i)_k = C_j^k$, Equation (5.76) can be written as

$$\sum_{p=1}^{m} \lambda_p^{k+1}\left(\sum_{i=1}^{n} \frac{Q_{ij}Q_{ip}}{\rho_i l_i x_i^3}\right)_k = (r+1)C_j^k - r\bar{C}_j \tag{5.77}$$

These are m equations corresponding to the m displacement constraints, which can be used to determine the Lagrange multipliers in the iterative algorithm. At each iteration it is required to use only those equations corresponding to the active constraints, giving positive Lagrange multipliers. This can generally be achieved by considering only those constraints which are closest to the constraint surface and by eliminating the equations yielding negative Lagrange multipliers.

When there are passive elements whose sizes are not governed by the optimality criterion, we have to separate the contribution of those elements in the summation of Equation (5.76). This gives

$$\sum_{p=1}^{m} \lambda_p^{k+1} \left(\sum_{i=1}^{n_1} \frac{Q_{ij}Q_{ip}}{\rho_i l_i x_i^3} \right)_k = r(C_j^k - \bar{C}_j) + \sum_{i=1}^{n_1} \left(\frac{Q_{ij}}{x_i} \right)_k$$

$$- r \sum_{i=n_1+1}^{n} \left(\frac{Q_{ij}}{x_i^2} \right) \Delta x_i^k(p) \quad (5.78)$$

where n_1 is the number of active elements and $\Delta x_i^k(p) = x_i^k(p) - x_i^k$. $x_i^k(p)$ is the size of a passive element dictated by considerations other than the optimality criterion. Generally the passive elements are those whose sizes are governed by a minimum or a maximum size. In the case of a problem with stress and displacement constraints, if we decide to treat the stress constraints as passive constraints, then the elements whose sizes are governed by stresses will be included in the passive category.

The advantages of using Equations (5.77) and (5.78) to determine the Lagrange multipliers are:
1. It is not required to assume initial values of the Lagrange multipliers, since the Lagrange multipliers are evaluated by solving the equations.
2. Since the equations contain coupling terms which take into account the interdependence of the different constraints, the Lagrange multipliers are more realistic. This is particularly true in the case of a structure where the constraints are sensitive to design changes.
3. The convergence behaviour is generally better than other approaches.

The disadvantages are:
1. A single virtual-load vector cannot be assembled since the Lagrange multipliers are not known before the flexibility coefficients Q_{ij} corresponding to the constraints are determined.
2. The computational effort to determine the flexibility coefficients and to assemble the coefficients of the Lagrange multipliers in Equations (5.76), (5.77), and (5.78) substantially increases with an increase in the number of potentially active constraints.
3. It is necessary to solve a set of simultaneous equations.
4. Some scheme has to be used in order to determine the potentially active constraints and to eliminate those equations corresponding to the passive constraints.

At the optimum the active constraints are satisfied as equality constraints, i.e. $C_j = \overline{C}_j$ and Equation (5.77) becomes

$$\sum_{p=1}^{m} \lambda_p^{k+1} \left(\sum_{i=1}^{n} \frac{Q_{ij}Q_{ip}}{\rho_i l_i x_i^3} \right) = \overline{C}_j \quad (5.79)$$

This relation, even though it is valid only at the optimum, can be used to estimate the Lagrange multipliers. Equation (5.79) can also be derived directly by using the optimality criterion and the constraint equations. The optimality criterion can be written as

$$\mathbf{I}_n = \mathbf{Q}_{n \times m} \lambda_m \quad (5.80)$$

The constraint equations can be written as

$$\mathbf{F}_{m \times n} \mathbf{I}_n = \overline{\mathbf{C}}_m \quad (5.81)$$

In Equations (5.80) and (5.81), I is a vector with all the elements equal to unity. Using these equations we can write

$$\mathbf{F}_{m \times n} \mathbf{Q}_{n \times m} \lambda_m = \overline{\mathbf{C}}_m \quad (5.82)$$

or

$$\mathbf{R}_{m \times m} \lambda_m = \overline{\mathbf{C}}_m \quad (5.83)$$

where $\mathbf{R}_{m \times m} = \mathbf{F}_{m \times n} \mathbf{Q}_{n \times m}$.
In Equation (5.83) the elements of R are given by

$$R_{jp} = \sum_{i=1}^{n} \frac{Q_{ij}Q_{ip}}{\rho_i l_i x_i^3} \quad (5.84)$$

Comparing Equation (5.79) and (5.83) it is seen that they are identical.
Equation (5.79) can also be derived by writing the optimality criterion as

$$\frac{1}{x_i} = \sum_{j=1}^{m} \lambda_j \frac{Q_{ij}}{\rho_i l_i x_i^3} \quad (5.85)$$

and substituting it into the constraint equation

$$\sum_{i=1}^{n} \frac{Q_{ij}}{x_i} = \overline{C}_j \quad (5.86)$$

The recurrence relation for the Lagrange multipliers given by Equation (5.68) can be shown to be an approximation to the linear equations in Equation (5.77). The matrix multiplying the vector λ in Equation (5.77) is square. If we neglect the off-diagonal terms, then these equations become

$$\lambda_j^{k+1} \sum_{i=1}^{n} \left(\frac{Q_{ij}Q_{ij}}{\rho_i l_i x_i^3} \right)_k = (r+1)C_j^k - r\overline{C}_j \quad (5.87)$$

These equations are uncoupled and assume that the constraints are independent of one another. With this assumption, and using the optimality criterion,

$(Q_{ij}/\rho_i l_i x_i^2)_k$ in Equation (5.87) can be replaced by $1/\lambda_j^k$. Also recalling that $\sum_{i=1}^n (Q_{ij}/x_i)_k = C^k$, Equation (5.87) can be written as

$$\frac{\lambda_j^{k+1}}{\lambda_j^k} C_j^k = (r+1)C_j^k - r\overline{C}_j \tag{5.88}$$

or

$$\lambda_j^{k+1} = \lambda_j^k \left((r+1) - r\frac{\overline{C}_j}{C_j} \right)_k \tag{5.89}$$

Comparing this equation with Equation (5.68) it is seen that they would be identical if

$$\frac{1}{b} = r \tag{5.90}$$

This shows that Equation (5.68), which is a linearized form of Equation (5.66), is an approximation to Equation (5.77).

5.3.2.2.3 Newton–Raphson method

An iterative algorithm to solve the optimality criterion and the constraint equations can be developed by considering the change in a constraint due to a change in the Lagrange multipliers. At this point we are anticipating the discussion of algorithmic methods given in Chapter 8 but in order to obtain a suitable solution technique we must turn to the application of Newton methods. However, we only introduce these aspects necessary for the solution of our displacement constrained structures problem and leave a more general discussion to later chapters. The change in a constraint can be written as

$$\Delta g_j = g_j(\lambda + \Delta\lambda) - g_j(\lambda) = \sum_{j=1}^m \frac{\partial g_j}{\partial \lambda} \Delta\lambda \tag{5.91}$$

Since in the Newton–Raphson method the change $\Delta\lambda$ is selected so as to satisfy the condition $g_j(\lambda + \Delta\lambda) = 0$, using Equation (5.91) an iterative relation can be written as

$$-\phi \mathbf{g}^k = \mathbf{H}\{\lambda^{k+1} - \lambda^k\} \tag{5.92}$$

or

$$\lambda^{k+1} = \lambda^k - \phi \mathbf{H}^{-1} \mathbf{g}^k \tag{5.93}$$

where $\mathbf{H}^{-1}$ is the inverse of the Hessian $\mathbf{H}$ whose elements are

$$H_{jp} = \frac{\partial g_j}{\partial \lambda_p} \tag{5.94}$$

and ϕ is a parameter introduced to control the step size.

Differentiating Equation (5.52) with respect to λ_i, recalling that λ_j is related to x_i by Equation (5.55) and assuming Q_{ij} as constants, we can write

$$\frac{\partial g_j}{\partial \lambda_p} = -\frac{1}{2} \sum_{i=1}^n \frac{Q_{ij} Q_{ip}}{\rho_i l_i x_i^3} \tag{5.95}$$

Equation (5.93) can be used to update the initially assumed Lagrange multipliers. The summation in Equation (5.95) is carried out only over the active elements. The iterative process in this method consists of using Equation (5.93) to estimate the Lagrange multipliers and Equation (5.53) to modify the design variables alternately until the constraint equations are satisfied within a specified tolerance. The disadvantages of this method are:
1. It is essential to assume initial values of the Lagrange multipliers.
2. The passive constraints cannot be easily separated from the active constraints.
3. If too many iterations are performed to satisfy the constraints, the design might move away from the region where the coefficient Q_{ij} are valid.

Even though the Newton–Raphson algorithm has a different form than the linear equations in Equation (5.77), it can be shown that the two algorithms are related. Equation (5.93) can be written as

$$\mathbf{H}\lambda^{k+1} = \mathbf{H}\lambda^k - \phi \mathbf{g}^k \tag{5.96}$$

Using Equation (5.95), this equation can be written as

$$\sum_{p=1}^{m} \lambda_p^{k+1} \sum_{i=1}^{n} \left(\frac{Q_{ij}Q_{ip}}{\rho_i l_i x_i^3}\right)_k = \sum_{p=1}^{m} \lambda_p^k \sum_{i=1}^{n} \left(\frac{Q_{ij}Q_{ip}}{\rho_i l_i x_i^3}\right)_k + 2\phi g_j^k \tag{5.97}$$

Substituting x_i^2 from Equation (5.53) into Equation (5.52) and rearranging, one obtains

$$g_j^k = \sum_{p=1}^{m} \lambda_p^k \sum_{i=1}^{n} \left(\frac{Q_{ij}Q_{ip}}{\rho_i l_i x_i^3}\right)_k - \overline{C}_j \tag{5.98}$$

Using this equation and recalling that $g_j^k = (C_j^k - \overline{C}_j)$, Equation (5.97) can be written as

$$\sum_{p=1}^{m} \lambda_p^{k+1} \sum_{i=1}^{n} \left(\frac{Q_{ij}Q_{ip}}{\rho_i l_i x_i^3}\right)_k = (2\phi + 1)C_j^k - 2\phi \overline{C}_j \tag{5.99}$$

This equation is identical to Equation (5.77), for $2\phi = r$.

The advantage of using Equation (5.99) in the Newton–Raphson method instead of Equation (5.93) is that it is not necessary to assume the initial values of the Lagrange multipliers. Equation (5.99) can be solved at each iteration. This approach we will refer to as the modified Newton–Raphson method.

5.4 STRESS CONSTRAINTS

An algorithm can be developed to design a structure with stress constraints by using two approaches. The first approach is to specify the constraint as a function of the stress in an element, and the second is to convert the stress constraint into an equivalent displacement constraint. These two approaches, depending on the approximations used to derive the criterion, lead to different algorithms. For

5.4.1 Fully Stressed Design (FSD) Method

The stress constraint in the ith element can be written as

$$g_i = \left(\frac{\sigma_i}{\bar{\sigma}_i} - 1\right) \leq 0 \tag{5.100}$$

where σ_i is the actual stress and $\bar{\sigma}_i$ is the maximum allowable stress in the ith element. The number of constraints is equal to the number of elements. The stress in the ith element is given by

$$\sigma_i = \frac{T_i}{x_i} \tag{5.101}$$

where T_i is the force in the ith element. Using Equations (5.23), (5.100) and (5.101), the Lagrangian for the stress constrained problem can be written as

$$L(x, \lambda) = \sum_{i=1}^{n} \rho_i x_i l_i + \sum_{i=1}^{n} \lambda_i \left(\frac{T_i}{x_i \bar{\sigma}_i} - 1\right) \tag{5.102}$$

Differentiating this equation with respect to the design variable x_i and setting it equal to zero gives

$$\rho_i l_i - \lambda_i \frac{T_i}{x_i^2 \bar{\sigma}_i} + \frac{1}{x_i \bar{\sigma}_i} \sum_{p=1}^{n} \lambda_p \frac{\partial T_p}{\partial x_i} = 0 \tag{5.103}$$

In this equation the term $\partial T_p/\partial x_i$, which is the gradient of the force in a bar, cannot be explicitly written. In the case of a determinate structure, this gradient is zero, since the force in a bar is independent of the areas of the elements. If we make the assumption that $\sum_{p=1}^{n} \lambda_p \partial T_p/\partial x_i = 0$, which is not true for an indeterminate structure, Equation (5.103) becomes

$$\rho_i l_i - \lambda_i \frac{T_i}{x_i^2 \bar{\sigma}_i} = 0 \tag{5.104}$$

or

$$x_i = \sqrt{\lambda_i} \sqrt{\frac{T_i}{\bar{\sigma}_i \rho_i l_i}} \tag{5.105}$$

Now if the stress constraints imposed on the structure are assumed to be satisfied as equality constraints, then using Equations (5.100) and (5.101) we can write

$$\frac{T_i}{x_i} = \bar{\sigma}_i \tag{5.106}$$

Substituting Equation (5.105) in this equation and solving for λ_i gives

$$\sqrt{\lambda_i} = \sqrt{\frac{T_i \rho_i l_i}{\bar{\sigma}_i}} \tag{5.107}$$

Substituting this relation in Equation (5.105) gives

$$x_i = \frac{T_i}{\bar{\sigma}_i} \qquad (5.108)$$

This equation states that at the optimum the area of the element is equal to the force in the bar divided by the maximum allowable stress in the bar, or the stress in an element is equal to the maximum allowable stress in that element. This is the well-known fully stressed design optimality criterion. A recurrence relation based on this criterion can be written as

$$x_i^{k+1} = \left(\frac{T_i}{\bar{\sigma}_i}\right)_k \qquad (5.109)$$

At the optimum $T_i = x_i \bar{\sigma}_i$. Substituting this in Equation (5.105) we can write

$$\lambda_i = x_i \rho_i l_i$$
$$= \text{weight of the element} \qquad (5.110)$$

This also shows that the weight of an optimum structure is equal to the sum of the Lagrange multipliers.

The use of the fully stressed design algorithm, if applied to an indeterminate structure, is an approximation, since for this structure

$$\sum_{p=1}^{n} \lambda_p \frac{\partial T_p}{\partial x_i} \neq 0$$

Therefore the FSD algorithm does not necessarily give a minimum-weight design for an indeterminate structure, and this is particularly true when the allowable stress in all the elements is not the same. The design obtained by the fully stressed design algorithm for an indeterminate structure with unequal stresses sometimes not only gives a non-optimum design but also gives a design with a bad distribution of material and an inefficient load path.

5.4.2 Stress Constraint Equivalent Displacement Constraint

The stress in the ith bar element can be written as

$$\sigma_i = \begin{bmatrix} D_{11} & D_{12} \end{bmatrix}_i \begin{bmatrix} U^1 \\ U^2 \end{bmatrix}_i \qquad (5.111)$$

where U^1 and U^2 are the longitudinal displacement of the two nodes defining the ith bar. In Equation (5.111), $D_{11} = -E_i/l_i$ and $D_{12} = E_i/l_i$ where E_i is the Young's modulus and l_i is the length of the bar. The stress constraint in the ith element can now be written as

$$g_i = \sigma_i - \bar{\sigma}_i \leqslant 0 \qquad (5.112)$$
$$= (D_{11} U^1 + D_{12} U^2)_i - \bar{\sigma}_i \leqslant 0 \qquad (5.113)$$

Thus we have replaced the stress constraint in an element by an equivalent constraint on the linear combination of the displacements in the longitudinal direction at the two nodes connecting the element. The constraint equation can now be written defining

$$g_j = \sigma_j - \bar{\sigma}_j < 0 \qquad j = 1, \ldots, n \tag{5.114}$$

or

$$g_j = \sum_{i=1}^{n} \frac{R_{ij}}{x_i} - \bar{\sigma}_j \leqslant 0 \tag{5.115}$$

or

$$g_j = \mathbf{u}^T \mathbf{P}^{(j)} - \bar{\sigma}_j \leqslant 0 \tag{5.116}$$

where

$$R_{ij} = x_i \mathbf{u}_i^T \mathbf{k}_i \mathbf{u}^{(j)}{}_i \tag{5.117}$$

There will be n constraints equal to the number of elements in a bar structure. In Equation (5.117) $\mathbf{u}^{(j)}{}_i$ is the virtual displacement vector associated with the ith element due to the virtual load vector $P^{(j)}$, which is equivalent to the forces $D_{11}(= -E_j/l_j)$ and $D_{12}(= E_j/l_j)$ acting in the longitudinal direction at the two nodes at the ends of jth element. Using Equations (5.23) and (5.115) the Lagrangian for the stress-constrained problem can be written as

$$L(x, \lambda) = \sum_{i=1}^{n} \rho_i x_i l_i + \sum_{j=1}^{n} \lambda_j \left(\sum_{i=1}^{n} \frac{R_{ij}}{x_i} - \bar{\sigma}_j \right) \tag{5.118}$$

Differentiating this relation with respect to the design variable x_i and setting the result equal to zero gives

$$\rho_i l_i - \sum_{j=1}^{n} \lambda_j \frac{R_{ij}}{x_i^2} = 0 \tag{5.119}$$

or

$$1 = \sum_{j=1}^{n} \lambda_j \frac{R_{ij}}{\rho_i l_i x_i^2} \tag{5.120}$$

where

$$\lambda_j \geqslant 0 \quad \text{and} \quad \lambda_j \bar{\sigma}_j = 0 \tag{5.121}$$

This optimality criterion derived for the stress constraints has the same form as the optimality criterion in Equation (5.57) for the multiple displacement constrained problem. Hence following the same procedure and logic we can derive the recurrence relations and the equations to determine the Lagrange multipliers. The recurrence relations for the stress constrained problem equivalent to Equations (5.62) and (5.63) are

$$x_i^{k+1} = x_i^k \left(\sum_{j=1}^{n} \lambda_j \frac{R_{ij}}{\rho_i l_i x_i^2} \right)_k^{1/r} \tag{5.122}$$

and

$$x_i^{k+1} = x_i^k \left(1 + \frac{1}{r} \left(\sum_{j=1}^{n} \lambda_j \frac{R_{ij}}{\rho_i l_i x_i^2} - 1 \right) \right)_k \tag{5.123}$$

Similarly, the equations to determine the Lagrange multipliers equivalent to Equations (5.66), (5.68), (5.69), (5.77), (5.78) and (5.79) respectively are

$$\lambda_j^{k+1} = \lambda_j^k \left(\frac{\sigma_j}{\bar{\sigma}_j}\right)_k^{1/b} \quad (5.124)$$

$$\lambda_j^{k+1} = \lambda_j^k \left(\frac{b+1}{b} - \frac{1}{b}\frac{\bar{\sigma}_j}{\sigma_j}\right)_k \quad (5.125)$$

$$\bar{\sigma}_j - \sigma_j^k = b\sigma_j^k \left(1 - \frac{\lambda_j^{k+1}}{\lambda_j^k}\right)_k \quad (5.126)$$

$$\sum_{p=1}^{n} \lambda_p^{k+1} \sum_{i=1}^{n} \left(\frac{R_{ij}R_{ip}}{\rho_i l_i x_i^3}\right)_k = (r+1)\sigma_j^k - r\bar{\sigma}_j \quad (5.127)$$

$$\sum_{p=1}^{n_1} \lambda_p^{k+1} \sum_{i=1}^{n_1} \left(\frac{R_{ij}R_{ip}}{\rho_i l_i x_i^3}\right)_k = r(\sigma_j^k - \bar{\sigma}_j)$$
$$+ \sum_{i=1}^{n_1} \left(\frac{R_{ij}}{x_i}\right)_k - r \sum_{i=n_1+1}^{n} \frac{R_{ij}}{x_i^2} \Delta x_i^k(p)$$
$$(5.128)$$

$$\sum_{p=1}^{n} \lambda_p^{k+1} \sum_{i=1}^{n} \left(\frac{R_{ij}R_{ip}}{\rho_i l_i x_i^3}\right)_k = \bar{\sigma}_j \quad (5.129)$$

In the case of a stress-constrained problem, the number of constraints is equal to the total number of stress constraints imposed on the structure. For a bar structure the number of constraints will be equal to the number of elements. If the structure is idealized with membrane elements and stress constraints are imposed on σ_x, σ_y, σ_{xy}, then there will be three constraints associated with each element.

In Equation (5.129), if the off-diagonal terms multiplying the vector of Lagrange multipliers are neglected, then this equation can be approximated as

$$\lambda_p^{k+1} \sum_{i=1}^{n} \left(\frac{R_{ip}R_{ip}}{\rho_i l_i x_i^3}\right)_k = \bar{\sigma}_p \quad (5.130)$$

In addition, if we assume that the virtual strain energy in an element is only due to the virtual load in that element, which is true for a statically determinate structure, then Equation (5.130) becomes

$$\lambda_i^{k+1} \left(\frac{R_{ii}R_{ii}}{\rho_i l_i x_i^3}\right)_k = \bar{\sigma}_i \quad (5.131)$$

Since in this equation we are assuming that the elements are not interdependent, $R_{ii} = T_i l_i / E_i$. In addition, if we assume that the elements are fully stressed i.e. $x_i = T_i / \bar{\sigma}_i$, a simple expression for the Lagrange multiplier can be written as

$$\lambda_i^{k+1} = \frac{T_i}{\bar{\sigma}_i^2}\left(\frac{\rho_i E_i^2}{l_i}\right) \tag{5.132}$$

It has been found that this simple expression leads to a correct near minimum weight design for the case of different allowable stresses in a structure, if the stresses in the elements at the optimum are equal to the allowable stresses in the elements.

The Newton–Raphson algorithm derived for the multiple displacement constrained problem can also be derived for the stress constrained problem by using the same procedure. The iterative relation would be

$$\lambda^{k+1} = \lambda^k - \phi \mathbf{H}^{-1} \mathbf{g}^k \tag{5.133}$$

where

$$H_{jp} = -\frac{1}{2}\sum_{i=1}^{n_1} \frac{R_{ij}R_{ip}}{\rho_i l_i x_i^3} \tag{5.134}$$

The optimality criterion for the stress constraints, Equation (5.120), can also be written as

$$1 = \sum_{j=1}^{n} \frac{\lambda_j \mathbf{u}_i^T \mathbf{k}_i \mathbf{u}_i^{(j)}}{x_i \rho_i l_i} \tag{5.135}$$

or

$$1 = \frac{\mathbf{u}_i^T \mathbf{k}_i \sum_{j=1}^{n} \lambda_j \mathbf{u}_i^{(j)}}{\rho_i x_i l_i} \tag{5.136}$$

In Equation (5.136), if the resultant displacements $\sum_{j=1}^{n} \lambda_j \mathbf{u}_i^{(j)}$ associated with the ith element due to the virtual-load system are equal to $\mathbf{u}_i$ due to the applied load, then the optimality criterion becomes

$$1 = \frac{\mathbf{u}_i^T \mathbf{k}_i \mathbf{u}_i}{\rho_i l_i x_i} \tag{5.137}$$

The assumption that $\sum_{j=1}^{n} \lambda_j \mathbf{u}^{(j)}_i = \mathbf{u}_i$ is equivalent to assuming that the product of the Lagrange multiplier and the virtual load applied to each element is equal to the force in the element due to the load P. Equation (5.137) can also be written as

$$1 = \frac{V_i}{\rho_i x_i l_i} = \frac{e_i}{\rho_i} \tag{5.138}$$

where V_i is the strain energy stored in the element and e_i is the strain energy density.

This criterion states that at the optimum the ratio of the strain energy density to the mass density is the same for all the elements. The criterion corresponds to a constraint on the generalized stiffness, and can be derived by replacing the virtual-load vector $P^{(i)}$ by the actual load vector P in Section 5.3.1.

A recurrence relation based on Equation (5.138) can be written as

$$x_i^{k+1} = x_i^k \left(\frac{e_i}{\rho_i}\right)_k^{1/r} \tag{5.139}$$

This recurrence relation can be used as an approximation to Equation (5.122) to design a structure with stress constraints only. In this case also if the maximum allowable stress in the different elements is not the same, it may not lead to a minimum-weight design. The advantage of Equation (5.139) is that it is not required to calculate the virtual strain energies in the elements and the Lagrange multipliers. A linear form of Equation (5.139) can be written as

$$x_i^{k+1} = x_i^k \left(1 + \frac{1}{r}\left(\frac{e_i}{\rho_i} - 1\right)\right)_k \tag{5.140}$$

5.5 ALGORITHM WITH THE RECIPROCAL DESIGN VARIABLE

In Sections 5.3 and 5.4 we have used the direct design variable x_i to define the objective function and the constraint equations. The recurrence relations and the equations to determine the Lagrange multipliers were expressed in terms of x_i. Now we will see how the optimality criterion and the algorithm are affected by defining the problem in terms of the reciprocal design variable z_i. The reciprocal design variable z_i and the direct design variable x_i are related by Equation (5.9). We will derive the equations only for the multiple displacement constrained problem, since the conclusions from this can be readily extended to the stress constrained problem.

5.5.1 Optimality Criterion

The weight of the structure can be written as

$$W(z) = \sum_{i=1}^{n} \frac{\rho_i l_i}{z_i} \tag{5.141}$$

The constraint equations in terms of z_i can be written as

$$g_j = C_j - \overline{C}_j \leq 0 \qquad j = 1, \ldots, m \tag{5.142}$$

where

$$C_j = \sum_{i=1}^{n} Q_{ij} z_i \tag{5.143}$$

Using Equations (5.23) and (5.142), the Lagrangian can be written as

$$L(z, \lambda) = \sum_{i=1}^{n} \frac{\rho_i l_i}{z_i} + \sum_{j=1}^{m} \lambda_j \left(\sum_{i=1}^{n} Q_{ij} z_i - \overline{C}_j\right) \tag{5.144}$$

Differentiating this equation with respect to the design variables z_i and setting the result equal to zero gives the optimality criterion,

$$-\frac{\rho_i l_i}{z_i^2} + \sum_{j=1}^{m} \lambda_j Q_{ij} = 0 \qquad (5.145)$$

or

$$1 = \frac{\rho_i l_i}{\sum_{j=1}^{m} \lambda_j Q_{ij} z_i^2} \qquad (5.146)$$

where $\lambda_j \geqslant 0$ and $\lambda_j g_j = 0$.
Equation (5.146) can also be written as

$$1 = \frac{\rho_i}{\sum_{j=1}^{m} \lambda_j e_{ij}} \qquad (5.147)$$

where e_{ij} is the virtual strain energy density. Comparing the optimality criterion for the direct design variable in Equation (5.58), with Equation (5.147), it is seen that they are equivalent, even though one is the reciprocal of the other.

5.5.2 Recurrence Relations

Applying the same logic used to write the exponential recurrence relation for x_i, (Equation (5.62)), the recurrence relation for z_i can be written by using the optimality criterion in Equation (5.146). This gives

$$z_i^{k+1} = z_i^k \left(\sum_{j=1}^{m} \frac{\lambda_j Q_{ij} z_i^2}{\rho_i l_i} \right)_k^{-1/r} \qquad (5.148)$$

This recurrence relation is equivalent to Equation (5.62) since one can be obtained from the other by using Equation (5.9).

The linear recurrence relation for the reciprocal design variable can be obtained by expanding Equation (5.148) using the binomial theorem and retaining only the linear terms. This gives

$$z_i^{k+1} = z_i^k \left(1 - \frac{1}{r} \left(\sum_{j=1}^{m} \frac{\lambda_j Q_{ij} z_i^2}{\rho_i l_i} - 1 \right) \right)_k \qquad (5.149)$$

This equation can be expressed in terms of x_i by using Equation (5.9). This gives

$$x_i^{k+1} = x_i^k \left(1 - \frac{1}{r} \left(\sum_{j=1}^{m} \lambda_j \frac{Q_{ij}}{\rho_i l_i x_i^2} - 1 \right) \right)_k^{-1} \qquad (5.150)$$

Comparing this equation with the linear recurrence relation for x_i in Equation (5.63), it is seen that they are not equivalent.

5.5.3 Evaluation of the Lagrange Multipliers

The set of equations to determine the Lagrange multipliers can be obtained by considering a change in the constraint due to a change in the design variable z_j. The change in a constraint can be written as

$$\Delta g_j = g_j(z + \Delta z) - g_j(z) \tag{5.151}$$

$$= \sum_{i=1}^{n} \frac{\partial g_j}{\partial z_i} \Delta z_i \tag{5.152}$$

Using Equations (5.142) and (5.149) and remembering that $g_j(z + \Delta z) = 0$, Equation (5.152) reduces to

$$\sum_{p=1}^{m} \lambda_p^{k+1} \left(\sum_{i=1}^{n} \frac{Q_{ij} Q_{ip} z_i^3}{\rho_i l_i} \right) = r(C_j^k - \bar{C}_j) + \sum_{i=1}^{n} (Q_{ij} z_i)_k \tag{5.153}$$

Comparing this relation with Equation (5.76) it is seen that they are equivalent and one can be obtained from the other by using the relationship between x_i and z_i. Similarly, it can be shown that the iterative relations for the Newton–Raphson method for z_i are also equivalent to those of the direct design variable x_i (Equation (5.93)).

The equations equivalent to Equation (5.148) and (5.149) for the stress-constrained problem can be written as

$$z_i^{k+1} = z_i^k \left(\sum_{j=1}^{n} \frac{\lambda_j R_{ij} z_i^2}{\rho_i l_i} \right)_k^{-1/r} \tag{5.154}$$

$$z_i^{k+1} = z_i^k \left(1 - \frac{1}{r} \left(\sum_{j=1}^{n} \lambda_j \frac{R_{ij} z_i^2}{\rho_i l_i} - 1 \right) \right)_k \tag{5.155}$$

It is seen from the equations derived in this section that the definition of the problem in terms of the reciprocal design variable affects the linear recurrence relation. However, the optimality criterion, the exponential recurrence relation and the equation to determine the Lagrange multipliers are not changed so as to affect the behaviour of the algorithm.

5.6 SYSTEM STABILITY CONSTRAINT

We now return to the problem of devising an algorithm for the solution of structural designs subject to constraints on the elastic stability of the system and begin by recapitulating the theory introduced in Section 3.4.3. The constraint equation for the linear static buckling of a structure can be written as

$$g_j = \mu_j - \bar{\mu} \tag{5.156}$$

where $\bar{\mu}$ is the lowest critical load factor and μ_j is defined in Section 3.4.3. The Lagrangian for the constraint on static stability can be written as

$$L(x, \lambda) = \sum_{i=1}^{n} \rho_i x_i l_i + \sum_{j=1}^{m} \lambda_j (\mu_j - \bar{\mu}) \tag{5.157}$$

The optimality criterion is obtained by differentiating equation (5.157) with respect to x_i and setting the result equal to zero. This gives

$$\rho_i l_i + \sum_{j=1}^{m} \lambda_j \frac{\partial \mu_j}{\partial x_i} = 0 \tag{5.158}$$

The gradient of μ_j given in Section 3.4.3 is obtained from

$$\eta_j^T \mathbf{K} \eta_j + \mu_j \eta_j^T \mathbf{K}_G \eta_j = 0 \tag{5.159}$$

by differentiating it with respect to x_i. This gives

$$\frac{\partial \mu_j}{\partial x_i} \eta_j^T \mathbf{K}_G \eta_j = \frac{1}{x_i} \eta_j^T \mathbf{k}_i \eta_j - \mu_j \eta_j^T \frac{\partial}{\partial x_i} \mathbf{K}_G \eta_j$$
$$- 2 \frac{\partial}{\partial x_i} \eta_j^T [\mathbf{K} \eta_j + \mu_j \mathbf{K}_G \eta_j] \tag{5.160}$$

The second and third term on the right side of this equation are equal to zero, and therefore we can write

$$\frac{\partial \mu_j}{\partial x_i} = -\frac{1}{x_i} \frac{\eta_j^T \mathbf{k}_i \eta_j}{\eta_j^T \mathbf{K}_G \eta_j} \tag{5.161}$$

Substituting this equation in Equation (5.158), the optimality criterion can be written as

$$1 = \sum_{j=1}^{m} \bar{\lambda}_j \frac{\bar{Q}_{ij}}{\rho_i x_i^2 l_i} \tag{5.162}$$

where

$$\bar{Q}_{ij} = x_i \eta_j^T \mathbf{k}_i \eta_j \tag{5.163}$$

and

$$\bar{\lambda}_j = \frac{\lambda_j}{\eta_j^T \mathbf{K}_G \eta_j} \tag{5.164}$$

Composing Equations (5.163) with Equation (5.57) it is seen that the optimality criterion for the system stability constraint problem has the same form as the optimality criterion for the displacement constraint problem. Equation (5.163) can also be written as

$$1 = \sum_{j=1}^{m} \bar{\lambda}_j \frac{q_{ij}}{\rho_i} \tag{5.165}$$

where

$$q_{ij} = \frac{\bar{Q}_{ij}}{x_i^2 l_i} \tag{5.166}$$

is the energy density in an element in the buckled mode. The exponential recurrence relation and the linear recurrence relation for the system stability constraint problem are given by

$$x_i^{k+1} = x_i^k \left(\sum_{j=1}^{m} \bar{\lambda}_j \frac{\bar{Q}_{ij}}{\rho_i l_i x_i^2} \right)_k^{1/r} \tag{5.167}$$

and

$$x_i^{k+1} = x_i^k \left(1 + \frac{1}{r} \left(\sum_{j=1}^{m} \bar{\lambda}_j \frac{\bar{Q}_{ij}}{\rho_i l_i x_i^2} - 1 \right) \right)_k \tag{5.168}$$

Equations to evaluate the Lagrange multipliers λ_i can be written by using the procedure discussed for the multiple displacement-constrained problem.

5.7 EFFECT OF STRUCTURAL SYMMETRY ON THE ALGORITHM

In the previous sections when deriving the optimality criterion and developing the algorithm, we had not taken into consideration the possible symmetry of the structure. The symmetry may be due to the nature of the constraints imposed on the structure and also due to the multiple loading conditions. We had mentioned that the multiple loading conditions can be treated as multiple constraints. A substantial amount of computational effort can be saved by making use of the summetry of the structure in the analysis phase and particularly in the optimization phase of the algorithm. A modification to the optimization algorithm becomes particularly essential when the structure is idealized with a large number of elements, the number of loading conditions is more than one, and the number of active constraints is large.

We will consider a very simple case and illustrate what modifications to the algorithm are required. Consider a symmetric structure subjected to multiple displacement constraints. Let us assume that there are five constraints which are active at the kth iteration. The exponential recurrence relation can be written as

$$x_i^{k+1} = x_i^k \left(\sum_{j=1}^{5} \lambda_j \frac{Q_{ij}}{\rho_i l_i x_i^2} \right)^{1/r} \tag{5.169}$$

Now because of the symmetry of the constraints let us assume that $\lambda_1 = \lambda_5 = \tilde{\lambda}_1$ and $\lambda_2 = \lambda_4 = \tilde{\lambda}_2$ and $\lambda_3 = \tilde{\lambda}_3$. The recurrence relation (Equation (5.169)) can now be written as

$$x_i^{k+1} = x_i^k \left(\tilde{\lambda}_1 \left(\frac{Q_{i1} + Q_{i5}}{\rho_i l_i x_i^2} \right) + \tilde{\lambda}_2 \left(\frac{Q_{i2} + Q_{i4}}{\rho_i l_i x_i^2} \right) + \tilde{\lambda}_3 \frac{Q_{i3}}{\rho_i l_i x_i^2} \right)^{1/r} \tag{5.170}$$

The linear Equation (5.77) to determine the Lagrange multipliers can be written as

$$Q^*_{5 \times 5} \lambda_5 = C^*_5 \tag{5.171}$$

where Q_{ij}^* are the coefficients of the square matrix multiplying the λ vector in Equation (5.77) and C^* is the right side of Equation (5.77). These quantities are evaluated for five constraints. In Equation (5.171) $Q_{ij}^* = Q_{ji}^*$. Replacing the λ's by the $\bar{\lambda}$'s and remembering that in Equation (5.171) because of symmetry the first equation is the same as the fifth equation and the second equation is the same as the fourth, we can write

$$\alpha_1(Q_{11}^* + Q_{15}^*) + \alpha_2(Q_{12}^* + Q_{14}^*) + \alpha_3(Q_{13}^*) = C_1^*$$
$$\alpha_1(Q_{21}^* + Q_{25}^*) + \alpha_2(Q_{22}^* + Q_{24}^*) + \alpha_3(Q_{23}^*) = C_2^* \quad (5.172)$$
$$\alpha_1(Q_{31}^* + Q_{35}^*) + \alpha_2(Q_{32}^* + Q_{34}^*) + \alpha_3(Q_{33}^*) = C_3^*$$

These three equations are not symmetric. Taking into consideration the symmetry of the constraints on the coefficients Q_{ij}^*, Equation (5.172) can be written as

$$(Q_{11}^* + Q_{15}^*)\alpha_1 + 2Q_{12}^*\alpha_2 + Q_{13}^*\alpha_3 = C_1^*$$
$$2Q_{12}^*\alpha_1 + (Q_{22}^* + Q_{24}^*)\alpha_2 + Q_{23}^*\alpha_3 = C_2^* \quad (5.173)$$
$$2Q_{13}^*\alpha_1 + 2Q_{23}^*\alpha_2 + Q_{33}^*\alpha_3 = C_3^*$$

Thus because of the symmetry we have to solve three equations instead of five. In addition, Equation (5.171) would require evaluation of fifteen Q_{ij}^* coefficients compared to eight coefficients for Equation (5.173). For this small problem this modification does not appear to be a substantial reduction in the computational effort. However, for a structure with a large number of design variables and potentially active constraints, it would save considerable CP time. The modification to algorithm discussed here can be extended to other types of constraints, to other symmetric conditions and also to other algorithms discussed previously.

It is interesting to note that if in a structure all the p Lagrange multipliers associated with all the active constraints are equal, because of the symmetry, all p equations will reduce to a single equation. This will require evaluation of only one Lagrange multiplier. Thus an apparently multiple constraint problem is reduced to an equivalent single constraint problem. For such a problem an algorithm based on the single dominant constraint would give the same design as obtained by using the multiple constraint algorithm. The seventy-two bar tower truss problem considered in Refs. 1, 2, 29 and other references fall under this category.

5.8 OPTIMIZATION PROCEDURE

In Sections 5.3 through 5.7 we have derived the optimality criterion, the different recurrence relations to modify the design variables, and the equations to estimate the Lagrange multipliers. We will discuss in this section how different algorithms can be developed based on these equations and what options we have in selecting an algorithm.

The main steps in the optimization of a structure are:
1. Assign initial sizes to all the elements. For all illustrative problems we have assumed equal sizes.
2. Analyse the structure using the displacement method of analysis.
3. Determine the stresses in the elements.
4. Scale the design to satisfy all the constraints. The scaled design is a feasible design. After scaling, at least one constraint will be on the constraint surface.
5. Separate the constraints into potentially active and passive categories.
6. Separate the elements into active and passive categories.
7. Determine the flexibility coefficients corresponding to the potentially active constraints.
8. Evaluate the Lagrange multipliers.
9. Modify the design variables using a recurrence relation.
10. Terminate the iterations if the specified criterion is satisfied or go to step 2. The terminating criterion may be the number of iterations or the percentage difference in the weight of the structure between the consecutive iterations.

The first four steps are common to all the algorithms. The subsequent steps differ depending upon: (a) the division of active and passive constraints; (b) the value of the parameters selected to control the step size; (c) the method used to determine the Lagrange multipliers; (d) the recurrence relation used to modify the design variables; and (e) the number of times the design variables are modified before reanalysing the structure. We will discuss in detail the different options available under these categories.

5.8.1 Active and Passive Constraints

The active constraints or potentially active constraints are those which are included in the constraint equations. The flexibility coefficients and the Lagrange multipliers associated with the active constraints enter the optimality criterion and the recurrence relation. The remaining constraints which are imposed on the structure are called passive constraints, and these are satisfied by the simple approach of scaling the individual design variables. In the case of a displacement and stress constraint problem, we can use one of the following approaches to design a structure:
1. Treat all the displacements as well as the stress constraints as potentially active constraints, determine the Lagrange multipliers associated with all the active constraints and use them in the recurrence relation. In this case only the elements whose sizes are governed by minimum or maximum size constraints would be passive.
2. Consider only the displacement constraints as potentially active and all the stress constraints as passive. Here, the stress constraints are not directly included in the constraint equations. Elements in which the stresses are greater than the maximum allowable stress are sized by using the FSD algorithm.

Optimality Criterion Methods in Structural Optimization 131

Thus the size of an element is a maximum of the sizes obtained by: (i) the recurrence relation for the displacement constraints; (ii) the FSD algorithm; or (iii) the minimum size constraint. Any element whose size is not equal to the one obtained by using a recurrence relation is characterized as a passive element. Since the number of elements and which elements are passive, affect the equations used to determine the Lagrange multipliers and consequently the recurrence relation, it is required to keep track of the passive elements and update the list whenever any change occurs.

3. Treat only one or some dominant displacement constraints as potentially active and consider the remaining displacement constraints and all the stress constraints as passive.
4. Treat all the displacement constraints as being independent of each other, and the stress constraints as passive constraints. In this case the size of an element is a maximum of the sizes obtained by: (i) the recurrence relation for the single displacement constraint applied to each displacement constraint separately; (ii) the FSD algorithm, or (iii) the minimum size constraint. This method is known as the envelop method.

For a structure subjected to stress constraints only, we can use one of the following approaches:

1. Treat all the stress constraints as potentially active, determine the Lagrangian multipliers associated with the active constraints, and use them in the recurrence relation. The passive elements in this case will be the elements whose sizes are governed by the minimum size limits.
2. Treat all the stress constraints as active, but approximate the virtual load system with the actual load system. In this case the Lagrange multipliers are not determined. The elements are sized on the basis of the strain energy stored in the element due to the applied load.
3. Treat all the stress constraints as active, but use the FSD algorithm.

5.8.2 Step Size

The convergence behaviour depends on the parameter r in the recurrence relation and also on the equations used to determine the Lagrange multipliers. We have introduced the parameter b in the recurrence relation to estimate the Lagrange multipliers and the parameter ϕ in the Newton–Raphson iterative method. We have derived a relationship between b, ϕ, and r. In the case of structural optimization $r = 2$ is generally adequate. However, for some problems, depending on the behaviour of the constraints, it is required to increase r in order to reduce the step size and get better convergence. The parameter r can be kept constant through all the iterations or can be changed after each iteration based on certain criteria. A good indication of the proper selection of r is a reduction in the weight of the structure after each iteration. Another approach in controlling the step size is to generate an intermediate design vector whenever the weight of structure is more than the previous iteration. This can be done by taking the average of the design variables in the previous and the present iteration. For example, if the weight of the structure with the design variables x_i^{k+1} is greater than the one with x_i^k, then the intermediate design variable would be $(x_i^{k+1} + x_i^k)/2$. This process can be continued until one obtains a design with a weight less than the weight of the previous iteration.

5.8.3 Lagrange Multipliers

When the structure is subjected to inequality constraints, we use only those Lagrange multipliers which are positive. A negative Lagrange multiplier is permissible only if the specific constraint is to be enforced as an equality constraint. The Lagrange multipliers can be determined by one of the following methods.
1. Linear simultaneous equations
   Equations (5.78), (5.79) for the displacement constraints.
   Equation (5.128) for the stress constraints.
2. Newton–Raphson iteration
   Equation (5.93) for the displacement constraints.
   Equation (5.133) for the stress constraints.
3. Exponential and linear recurrence relations
   Equations (5.66) and (5.68) for the displacement constraints.
   Equations (5.124) and (5.125) for the stress constraints.
4. Approximate relations
   Equation (5.87) for the displacement constraints.
   Equations (5.130), (5.131), (5.132) for the stress constraints.

Under the last category we can include a number of degenerate equations obtained from the first method. For the first two cases and some equations of the fourth case, it is required to calculate the coefficients Q_{ij} or R_{ij} before the Lagrange multipliers can be determined. In the third case the Lagrange multipliers are estimated based on the ratio of the actual value of the constraint to the limiting value of the constraint, and the coefficients Q_{ij} or F_{ij} need not be known before the λ's are estimated. The advantages and the disadvantages of using these methods were discussed in the section where they were derived.

5.8.4 Recurrence Relations

In the case of the displacement constraint problem there are basically three distinct relations. These are:
1. Exponential relation for x_i or z_i (Equation (5.62) or (5.148)).
2. Linear relation for x_i (Equation (5.63) or (5.64)).
3. Linear relation for z_i (Equation (5.149) or (5.150)).

We can use any one of these relations. In these relations we can keep the step size parameter r constant or change it after each iteration. All these relations need the evaluation of the Lagrange multipliers.

For the stress constraint problem we can use one of the following:
1. Exponential relation for x_i or z_i (Equation (5.122) or (5.154)).
2. Linear relation for x_i (Equation (5.123)).
3. Linear relation for z_i (Equation (5.155)).
4. FSD algorithm (Equation (5.109)).
5. Exponential relation for x_i based on the strain energy distribution (Equation (5.139)).
6. Linear relation for x_i based on the strain energy distribution (Equation (5.140)).

The first three of these relations require evaluation of the Lagrange multipliers before they can be used. The fourth requires knowing the force in the element and the last two use the strain energy stored in an element. For the last three the Lagrange multipliers are not evaluated, only the information from the applied load vector is used.

5.8.5 Design Variables

The flexibility coefficients Q_{ij} used to evaluate the Lagrange multipliers and modify the design variables are calculated with the assumed values of the design variables. When we modify the design variables, the previously calculated flexibility coefficients are no longer valid. We have two choices. One is to reanalyse the structure and calculate the new flexibility coefficients. And the other choice is to assume that the flexibility coefficients are not substantially affected with small changes in the design variables and reuse them. The second option would require a smaller number of analyses. However, one has to be careful that the design does not move away from the region where the flexibility coefficients are no longer valid.

5.9 ILLUSTRATIVE PROBLEMS

In this section results are presented for sample problems which are solved by using some of the algorithms and the optimization procedure discussed in the previous sections. Some of the problems selected for presentation are standard problems and have been used by several investigators to study the performance of their algorithms. The first problem is a ten-bar truss. It is designed to satisfy stress-displacement constraints and also different allowable stresses in the elements. The second structure is a 200-bar truss subjected to five loading conditions. This structure is designed with constraints, stresses and displacements. The third structure is a cantilever box-beam with the top and bottom skins made of a layered composite material. The box-beam is designed to satisfy stress constraints and local plate buckling. The fourth structure is a composite wing structure which is designed to satisfy stress constraints in all the elements and a twist constraint at the free end. The last structure is a three-dimensional dome designed to satisfy the constraint on system stability. For all the problems, unless otherwise mentioned, the sizes of all the elements are equal for the first iteration. The CP time given for some problems is for the CDC CYBER 74/175.

In problems where the linear equations (Equations (5.78) or (5.128)) are solved to determine the Lagrange multipliers, the equations were first arranged in the order of the degree of activity of the constraints. The first equation corresponds to the constraint which has been closest to the constraint surface, and the last equation corresponds to the constraint that was farthest from the constraint surface. The rearranged equations were solved by an iterative method. In the first iteration only one equation was solved, assuming that there was only one constraint, and during the subsequent iterations a new equation was added each time. This process was continued until all the equations corresponding to all the potentially active constraints were solved.

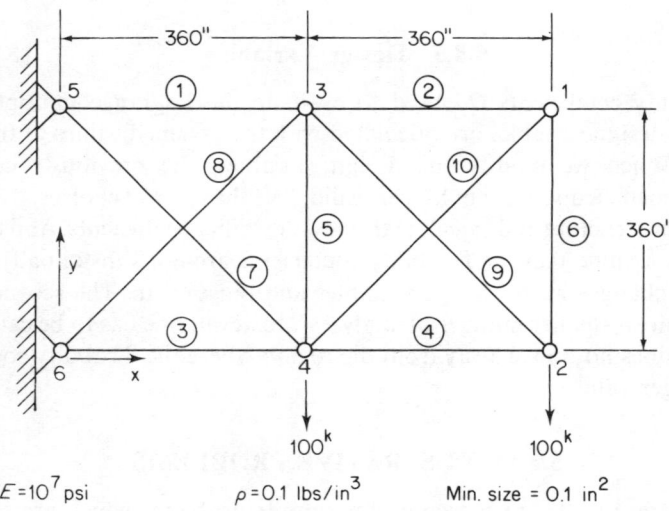

Fig. 5.1 Ten-bar truss

Example 1 Ten-bar truss (displacement and stress constraints

The truss shown in Fig. 5.1 was designed to satisfy a stress limit of 25 ksi in all the elements and a displacement limit of ± 2.0 in at all the node points in the x and y direction.

Case I The truss is designed by considering stress constraints as well as displacement constraints as potentially active constraints. The three recurrence relations used to modify the design variables were: (1) Case Ia, the exponential relation for x_i or z_i (Equations (5.62), (5.132), (5.148), (5.154)); (2) Case Ib, the linear relation for x_i (Equations (5.63) and (5.124); (3) Case Ic, the linear relation for z_i (Equations (5.149) and (5.155)). The Lagrange multipliers for the three cases are determined by solving the linear simultaneous equations given in Equations (5.78) and (5.128). The potentially active constraints were determined at each iteration. For the first iteration all the constraints satisfying the conditions $C_j/\overline{C}_j \geq \delta = 0.5$ for the scaled design were assumed to be potentially active. The maximum value of $C_j/\overline{C}_j$ is equal to unity for the scaled design. In subsequent iterations $C_j/\overline{C}_j$ was determined based upon the activity of the constraints in the previous iteration. For example, δ for the $(k+1)$th iteration is equal to $C_j/\overline{C}_j$ of the kth iteration that includes all the active constraints. In addition, a check was made so that there was at least one nonactive constraint included in the potentially active constraints. This was essential in order to make sure that all the active constraints were used in the recurrence relation. The step size parameter r for all three cases was set equal to 2. The structure was analysed after each iteration.

Table 5.1 Iteration history for ten-bar truss with stress and displacement constraints—Case I

Iteration no.	Case Ia Exponential relation		Case Ib Linear relation—x_i		Case Ic Linear relation—z_i		Total time in seconds
	Weight (lb)	Active constraints	Weight (lb)	Active constraints	Weight (lb)	Active constraints	
1	8266.1	5	8266.1	5	8266.1	5	0.09
2	6034.3	5	6666.8	5	6911.9	5	0.21
3	5824.3	5	6196.3	5	6438.4	5	0.33
4	5691.6	5	5946.9	5	6194.4	5	0.44
5	5574.6	5	5759.1	5	6018.7	5	0.55
6	5448.0	5	5627.3	5	5876.9	5	0.67
7	5305.7	2,5	5516.0	5	5749.5	5	0.78
8	5191.0	2,5	5400.5	5	5627.7	5	0.89
9	5083.5	2	5278.2	2,5	5510.6	5	1.01
10	6069.3	2,5,5*	5185.1	2,5	5401.5	5	1.13
11	5087.8	2	5118.3	2,5	5305.6	2,5	1.27
12	5070.1	—	5091.2	2,5	5228.0	2,5	1.41
13			5078.2	2,5	5169.2	2,5	1.54
14			5076.6	2,5	5134.1	2,5	1.68
15					5095.3	2,5	1.81
16					5085.0	2,5	1.95
17					5078.4	2,5	2.08
18					5076.6	2,5	2.23

Note: Active constraints   2   Vertical displacement at Node 1.
    5   Vertical displacement at Node 2.
    5* Element number 5.

The iteration history for Cases Ia through Ic is given in Table 5.1. The table also contains the active constraints at each iteration and the total time needed to complete the iterations for Case Ic. For Case Ia, after the 12th iteration, the weight of the structure was found to oscillate at weights greater than 5070.1 lbs. This behaviour was due to the large step size. If the step size parameter r had been increased after the 12th iteration, in order to reduce the step size, the additional iterations would lead to a design with a weight of 5060.85 lbs. The active constraints at the optimum for this design were the vertical displacement at node 1 and the stress in member 5. For Case Ia, at the 10th iteration, the scaled weight of the structure jumped to 6069.3 lbs, because the stress in element 5 suddenly became active. The iterations for Cases Ib and Ic lead to a design with a minimum weight of 5076.6 lbs with the vertical displacements at nodes 1 and 3 being active and equal to 2.0 in. The stress constraint for Cases Ib and Ic did not become active. Both of the designs satisfy the optimality criterion. The design with a weight of 5076.6 lbs is a relative minimum. The iteration history for the three cases is shown in Fig. 5.2. The computer time shown in this figure is for Case Ic. The total time required to complete 12 iterations for Case Ia and 14 iterations for Case Ib was 1.78 and 1.88 seconds, respectively.

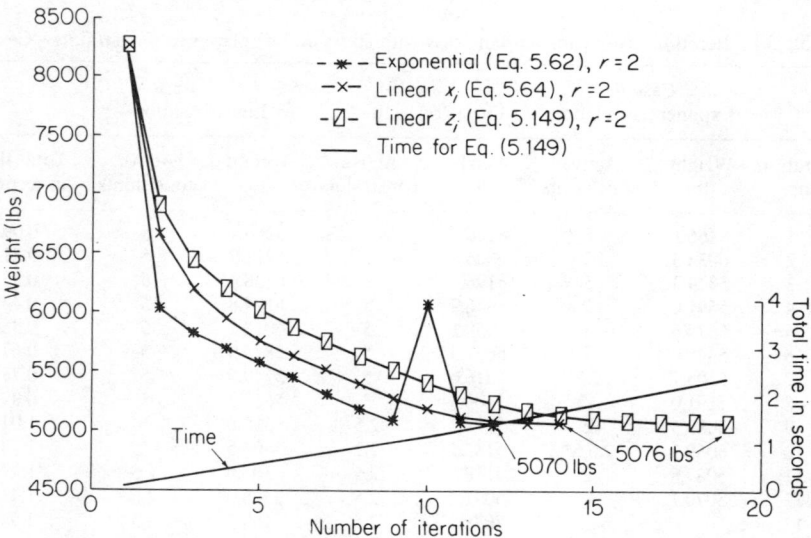

Fig. 5.2 Iteration history for ten-bar truss with stress–displacement constraints—Case I

Case II The truss is designed by using the exponential recurrence relation with the step size parameter r set equal to 4 after each analysis of the structure. The Lagrange multipliers were determined by using Equations (5.78) and (5.128). For

Optimality Criterion Methods in Structural Optimization   137

Case IIa, the structure was reanalysed after each iteration. For Case IIb, a maximum of 10 subiterations were allowed before reanalysing the structure with a criterion set to satisfy the constraints equal to 10^{-7}. During the subiterations the step size parameter was doubled, in order to reduce the step size, whenever the scaled weight of the structure was greater than the previous lower weight design within the subiterations. We have referred to this approach in the last section as the modified Newton–Raphson approach. The iteration history and the total time needed to complete the specified number of iterations are given in Table 5.2. Case IIa and Case IIb gave minimum-weight designs of 5076.6 lbs and 5060.8 lbs respectively. However, the approach of using repeated subiterations (Case IIb) needed substantially more computer time.

Table 5.2 Iteration history for ten-bar truss with stress–displacement constraints—Case II

Iteration no.	Case IIa		Case IIb	
	Weight (lb)	Total time in seconds	Weight (lb)	Total time in seconds
1	8266.1	0.09	8266.1	0.43
2	6423.2	0.21	6017.7	0.92
3	6009.5	0.34	5818.9	1.39
4	5856.1	0.45	5689.0	1.82
5	5780.1	0.58	5572.6	2.36
6	5719.4	0.70	5445.3	2.90
7	5660.4	0.81	5303.1	3.50
8	5599.3	0.93	5198.9	4.16
9	5536.6	1.04	5105.3	4.95
10	5469.9	1.16	5079.6	5.59
11	5398.4	1.23	5074.3	6.59
12	5320.4	1.41	5064.0	7.21
13	5232.3	1.56	5062.4	7.96
14	5155.0	1.69	5061.2	8.17
15	5076.9	1.84	5061.0	9.46
16	5076.6	2.00	5060.8	10.20

The iteration history for Case II is also shown in Fig. 5.3. The Lagrange multipliers and the areas of the members for the two minimum-weight designs are given in Table 5.3, and are designated as Design 1 and Design 2.

Example 2   Ten-bar truss (stress constraints)

The truss shown in Fig. 5.1 was designed to satisfy different stress constraints in the elements.

Case I   The maximum allowable stress for all the elements was 25 ksi except for element 9 where the maximum allowable stress was increased to 50 ksi. The FSD

Table 5.3 Minimum-weight designs of ten-bar truss

Design 1		Design 2		Design 3		
Member	Area	Member	Area	Member	Area	Lagrange multiplier
1	30.7297	1	30.5210	1	7.9000	248.0
2	0.1000	2	0.1000	2	0.1000	72.0
3	23.9407	3	23.1999	3	8.1000	392.0
4	14.7331	4	15.2229	4	3.9000	87.2
5	0.1000	5	0.1000	5	0.1000	0
6	0.1000	6	0.5514	6	0.1000	72.0
7	8.3406	7	7.4572	7	5.7983	319.2
8	20.9510	8	21.0364	8	5.5154	114.7
9	20.8358	9	21.5284	9	3.6770	0
10	0.1000	10	0.1000	10	0.1414	101.8
Weight	5076.66 lb	Weight	5060.85 lb	Weight	1497.6 lb	

Active constraints	Y-deflection at Node 1	Y-deflection at Node 2	Y-deflection at Node 1	Element 5
Lagrange multiplier	1638.0	885.3	2442.0	136.8

Note: Elements with zero Lagrange multipliers are inactive constraints

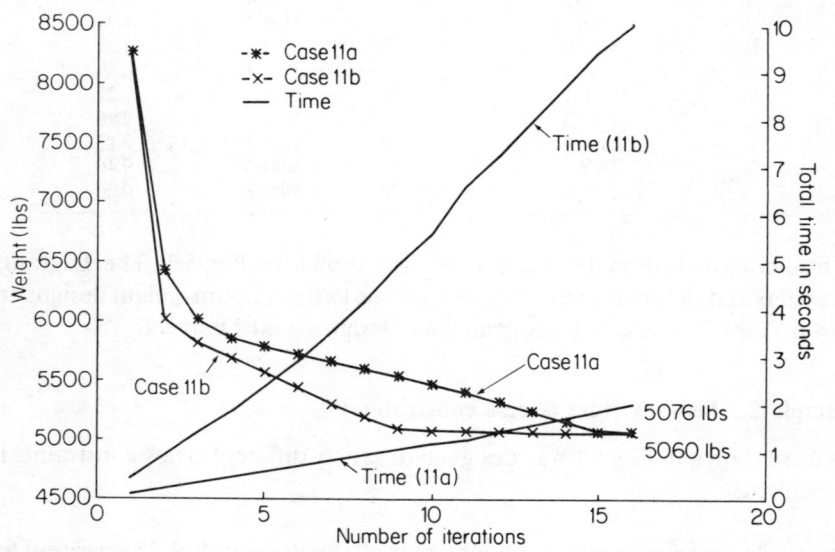

Fig. 5.3 Iteration history for ten-bar truss with stress–displacement constraints—Case II

Optimality Criterion Methods in Structural Optimization

design for this case is not the correct minimum-weight design. The minimum weight obtained by FSD algorithm was 1725 lbs.

The Lagrange multipliers for this case were obtained by solving Equation (5.128) and all ten stress constraints were assumed to be potentially active in all the iterations. The structure was reanalysed after each iteration. The parameter r was set equal to 2 for all three recurrence relations, the exponential relation (Equation (5.122)), the linear relation for x_i (Equation (5.123)) and the linear relation for z_i (Equation (5.155)). The parameter $r = 2$ was found to be too large to obtain proper convergence for the exponential relation. Therefore, these results are not given here. The iteration history for the other two cases, Case Ia, (Equation (5.123)) and Case Ib (Equation (5.175)) is given in Table 5.4. This table contains the active constraints at each iteration and also the total time required to complete the specified number of iterations. The iteration history is also shown in Fig. 5.4. The minimum-weight design for this problem is 1497.6 lbs with the stresses in elements 1, 2, 3, 4, 6, 7, 8, 10 equal to 25 ksi. The stress at the optimum in element 9 is 37.5 ksi, and this is not an active constraint. The cross-sectional areas of all the elements and the Lagrange multipliers associated with the minimum weight design are given in Table 5.3 as Design 3.

Table 5.4 Iteration history for ten-bar truss with stress constraints—Case I

	Case Ia Linear relation—x_i			Case Ib Linear relation—z_i		
Iteration no.	Weight (lb)	Active constraints	Total time in seconds	Weight (lb)	Active constraints	Total time in seconds
1	3434.9	1,3,7	0.18	3434.9	1,3,7	0.19
2	2444.3		0.36	2541.7		0.38
3	1835.7	1,2,3,4,	0.61	2103.3		0.57
4	2144.5	6,7,10	0.78	1804.6	1,2,3,4,	0.79
5	1796.1		0.97	1686.5	6,7,10	1.00
6	1758.7		1.15	1637.9		1.18
7	1591.1		1.35	1594.4		1.38
8	1556.2	1,2,3,4,	1.55	1572.9	1,2,3,4,	1.58
9	1532.5	6,7,8,10	1.75	1558.3	6,7,8,10	1.77
10	1524.7		1.96	1547.1		1.96
11	1518.9		2.12	1537.9		2.16
12	1514.2		2.39	1530.3		2.35
13	1510.2		2.62	1524.6		2.55
14	1506.3		2.35	1520.3		2.76
15	1504.0	1,2,3,4,	3.07	1516.2		2.97
16	1501.6	6,7,8,10	3.32	1512.7		3.20
17	1499.7		3.57	1509.6		3.42
18	1498.1		3.84	1506.9		3.65
19				1504.5		3.86
20				1502.5		4.09
21				1500.3		4.31
22				1499.3		4.55

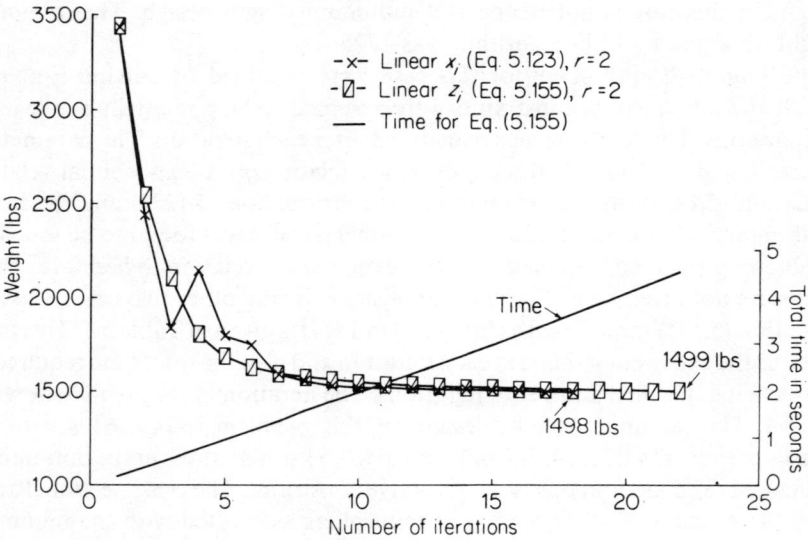

Fig. 5.4 Iteration history for ten-bar truss with stress constraints—Case I

Case II The problem solved was the same as Case I above except that the Lagrange multipliers were determined by using the recurrence relation in Equation (5.124). The step size parameter r was set equal to 2, and the parameter b in Equation (5.124) was set equal to 0.5 (See Equation (5.90)). The three recurrence relations used to modify the design variables were the exponential relation (Equation (5.122)), the linear relation for x_i (Equation (5.123)) and the linear relation for z_i (Equation (5.155)). The Lagrange multipliers for the first iteration were assumed to be proportional to the forces in the bars. All Lagrange multipliers were normalized so that the maximum value of the Lagrange multiplier was equal to unity. The structure was reanalysed each time the design variables were modified using the recurrence relations. The exponential relation gave a minimum-weight design of 1500.6 lbs after 13 iterations (see Table 5.5). Use of both of the linear relations caused oscillations in the weight of the structure and did not give a good minimum weight design. Therefore the problem was resolved by using the three recurrence relations. However, in the first iteration the design variables were modified by using the FSD algorithm (Equation (5.109)). The iteration history for the three cases, IIa (Equation (5.122)), IIb (Equation (5.123)) and IIc (Equation (5.155)) is given in Table 5.5. The table also contains the total computer time required to complete the specified number of iterations. This time is substantially less than that given in Table 5.4, where Equation (5.128) was used to determine the Lagrange multipliers. The iteration history for the three cases is also shown in Fig. 5.5. Previous experience

Optimality Criterion Methods in Structural Optimization 141

Table 5.5 Iteration history for ten-bar truss with stress constraints—Case II

Iteration no.	Case IIa Exp. relation	Case IIa Exp. relation	Case IIb Linear relation x_i	Case IIc Linear relation z_i	Total time in seconds
1	3434.9	3434.9	3434.9	3434.9	0.08
2	1969.1	1812.7*	1812.7*	1812.7*	0.16
3	1636.5	1942.1	1830.6	1801.6	0.24
4	1673.1	1668.1	1673.6	1734.0	0.35
5	1653.9	1678.3	1687.8	1633.5	0.40
6	1557.3	1600.5	1734.9	1742.9	0.48
7	1553.6	1567.3	1625.5	1774.1	0.56
8	1538.4	1554.0	1551.3	1631.7	0.64
9	1526.0	1537.7	1542.8	1597.1	0.72
10	1516.6	1525.6	1532.1	1559.7	0.79
11	1509.4	1516.1	1521.5	1553.5	0.88
12	1504.1	1509.3	1514.3	1530.9	0.96
13	1500.6	1505.7	1509.2	1523.0	1.05
14		1502.4	1505.8	1515.6	1.13
15		1500.7	1504.2	1514.1	1.21
16		1499.3	1503.2	1512.6	1.29
17			1502.0	1510.6	1.37
18				1508.7	1.46
19				1506.3	1.54
20				1505.0	1.61
21				1503.2	1.70

* FSD Design.

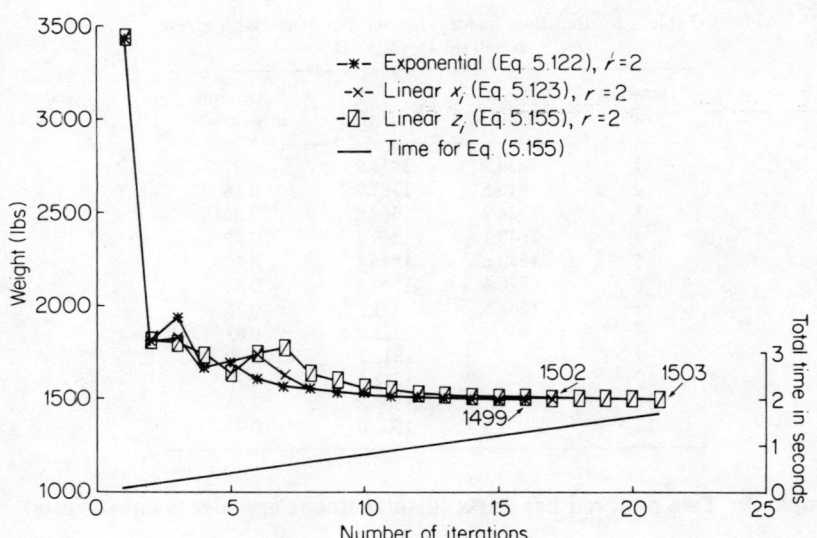

Fig. 5.5 Iteration history for ten-bar truss with stress constraints—Case II

with Equation (5.124) has shown that with the exponential recurrence relation with $r = 2$ and the parameter b increased to 4 to reduce the step size, a design with a weight 1497.6 lbs can be obtained. However, this needs a substantially larger number of iterations.

Case III The ten-bar truss with stress constraints was solved with a maximum allowable stress in element 9 equal to 37.5 ksi and a maximum allowable stress in all other elements equal to 25 ksi. The stress in element 9 was selected on the basis that its stress does not increase beyond 37.5 ksi for the minimum-weight design. The structure was designed by using two approaches. In Case IIIa, the FSD algorithm (Equation (5.109)) was used. And in Case IIIb, the design variables were modified by using the exponential relation (Equation (5.122)) and the Lagrange multipliers were determined by using Equation (5.132). This equation assumes that there is no coupling effect and that the elements are fully stressed. The iteration history for the two cases and the computer time for Case IIIb are given in Table 5.6. The FSD algorithm gives a fully stressed design which is nonoptimum. The use of Equation (5.122) to determine the Lagrange multipliers for Case IIIb gave a design with a weight of 1502.8 lbs which is fully stressed but closer to the real optimum weight of 1497.6 lbs. If the allowable stress in element 9 was increased to 50 ksi, the use of Equation (5.122) gives a design with a weight of 1621 lbs, which is in between the optimum design (1497.6 lbs) and the design (1725 lbs) obtained by using the FSD algorithm. The iteration history for Case IIIa and IIIb is also given in Fig. 5.6.

Table 5.6   Iteration history for ten-bar truss with stress constraints—Case III

Iteration no.	Case IIIa	Case IIIb	Total time in seconds
1	3434.9	3434.9	0.07
2	1815.5	1787.9	0.16
3	1726.2	1667.6	0.24
4	1648.1	1569.2	0.32
5	1590.6	1545.4	0.40
6	1570.4	1539.2	0.48
7	1568.5	1531.1	0.55
8		1523.9	0.63
9		1517.3	0.70
10		1512.6	0.78
11		1504.8	0.86
12		1502.0	0.94

Example 3   Two hundred bar truss (displacement and stress constraints)

The 200-bar truss shown in Fig. 5.7 is subjected to five loading conditions as follows:

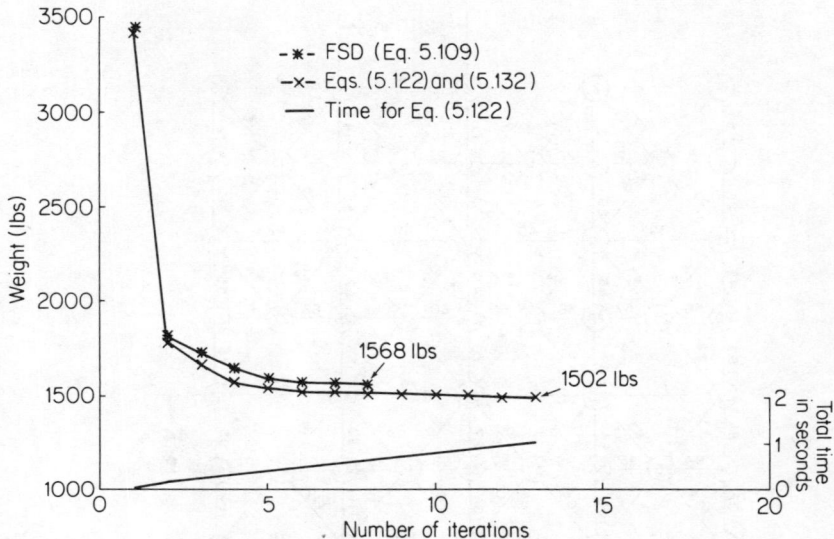

Fig. 5.6 Iteration history for ten-bar truss with stress constraints—Case III

1. A load of 1000 lbs in the positive X direction applied at nodes 1, 6, 15, 20, ..., 71.
2. A load of 1000 lbs in the negative X direction applied at nodes 5, 14, 19, ..., 75.
3. A load of 10 000 lbs in the negative Y direction at nodes 1, 2, 3, 4, 5, 6, 8, 10, 12, 14, 15, 16, ..., 73, 74, 75.
4. Loading conditions 1 and 3.
5. Loading conditions 2 and 3.

The maximum allowable stress in all the elements is equal to ± 10 ksi. The displacement limit in the X and Y directions at all the nodes is ± 0.5 in.

The structure was designed to satisfy stress and displacement constraints. Two approaches were used. In the first approach the constraints on the displacements and the stresses were both treated as potentially active constraints, i.e., the Lagrange multipliers associated with the stress as well as the displacement constraints were determined. In the second approach, the stress constraints were treated as passive constraints, i.e. no Lagrange multipliers or gradients of the constraints associated with stresses in the elements were evaluated. The Lagrange multipliers corresponding to only the displacement constraints were used in the recurrence relation. The linear recurrence relation for x_i (Equation (5.63) and (5.123)) were used. The effect of using other recurrence relations may be found in Ref. [31].

144 Foundations of Structural Optimization: A Unified Approach

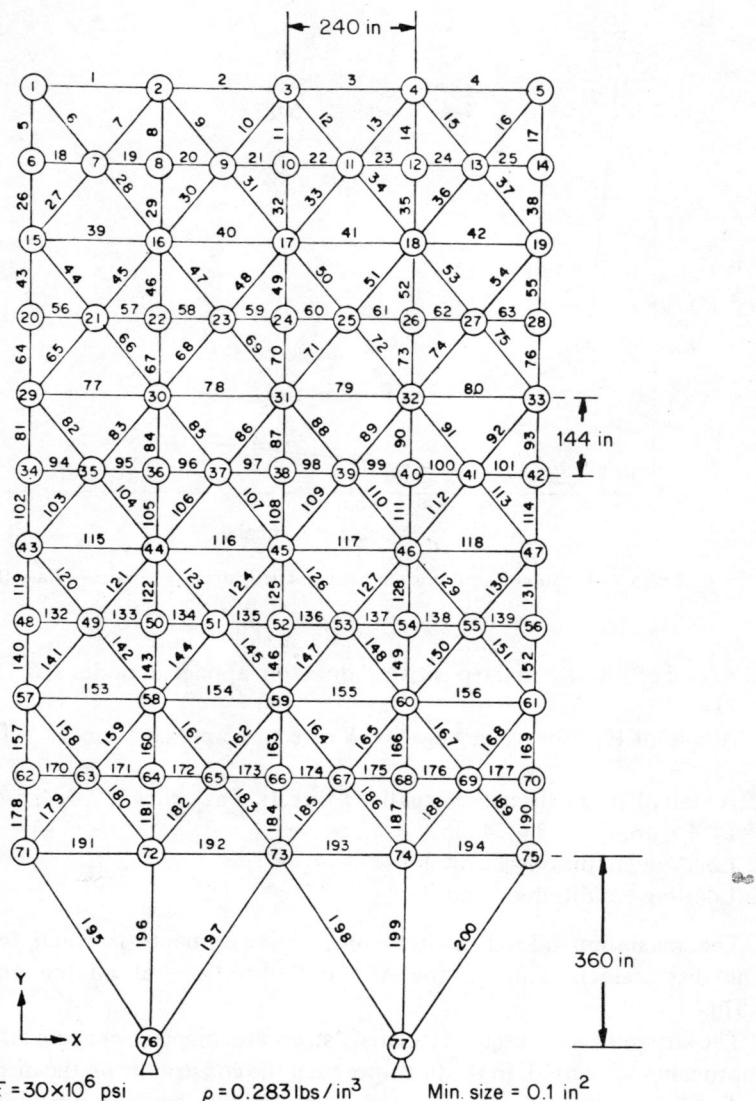

Fig. 5.7 200-Bar truss

The step size parameter r was set equal to 2, and the Lagrange multipliers were determined by using the linear equations (Equations (5.78), (5.128)). The structure was reanalysed for both cases after each iteration. The symmetry of the structure was not taken into consideration in order to reduce the loading conditions or the number of design variables. The number of potentially active constraints is determined by using the method discussed in Example 1. The iteration history for the two cases, Case I (stress + displacements) and Case II (displacements), is given in Table 5.7. The table also gives the CP time required for each iteration. The average CP time to analyse the structure was found to be 2.4 seconds. The difference between the time given in the table and 2.4 seconds is the time required to complete an optimization phase of the iteration. The optimization phase includes evaluation of the constraint gradients, assembling the elements of Equation (5.78) or (5.128), solving the equations to determine the Lagrange multipliers and modifying the design variables by using the recurrence relation. The iteration history for the two cases is also given in Fig. 5.8.

Table 5.7 Iteration history for 200-bar truss

	Case 1				Case II			
		Number of active constraints		Time per iteration in seconds		Number of active constraints		Time per iteration in seconds
Iteration no.	Weight	Disp.	Stress		Weight	Disp	Stress	
1	144 769	0	2	3.05	144 769			
2	86 154	0	4	3.49	42 258*	4	–	3.44
3	60 149	0	4	3.49	36 277	3	–	3.69
4	47 166	2	0	3.97	36 038	3	–	5.75
5	36 968	3	2	4.29	40 273	2	–	3.95
6	33 026	5	0	4.39	38 949	3	–	6.13
7	31 013	5	0	4.33	30 436	5	–	5.86
8	30 260	5	1	4.70	30 976	7	–	7.35
9	29 803	5	3	6.62	29 285	9	–	10.90
10	29 676	6	16	15.54	29 579	–	–	–
11	29 607	5	17	14.26				
12	29 563	5	17	14.80				
20	29 245	5	17	10.03	* FSD Design.			
30	29 121	5	17	–	Average time for analysis of the structure—2.4 seconds.			
40	29 015	5	17	–				
50	28 941	5	17	–				

The weight of the optimum design for this structure is 28 858.4 lbs. The areas of the elements associated with this design may be found in Ref. [31].

Example 4 Cantilever box-beam

The structure shown in Fig. 5.9 was idealized with membrane quadrilateral elements, shear panels and bar elements (posts). The quadrilateral membrane elements in the top and bottom skins consisted of four layers with fibres in the 0°, 90°, +45° and −45° directions. The 0° fibres are parallel to the length of the beam. The idealized structure consisted of 18 quadrilaterals, 18 shear panels and

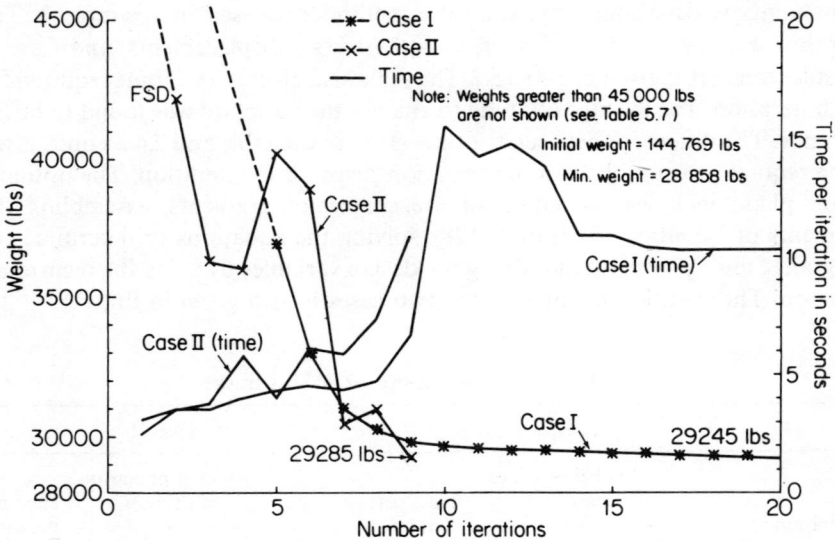

Fig. 5.8 Iteration history for 200-bar truss with stress and displacement constraints

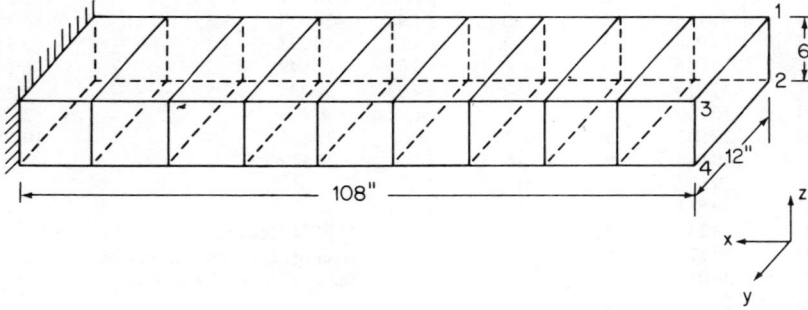

Fig. 5.9 Cantilever box-beam

18 posts. The six quadrilaterals near the tip of the box beam consisted of four layers of graphite epoxy. The six quadrilateral elements in the middle consisted of boron epoxy in the 0° direction and the remaining three layers of graphite epoxy. The six quadrilateral elements closer to the fixed end consisted of all layers of boron epoxy. The shear panels and posts have different material properties. The composite elements were designed by using the maximum stress criteria. The material properties used for this example are given in Table 5.8. The reason for selecting diverse properties for the different elements was to illustrate the versatility of the computer program. The program used to solve the problem is documented in Reference [33].

Table 5.8 Material properties for Example 4

	Graphite-epoxy	Boron-epoxy
E_{11}	18.5×10^6 psi	32.0×10^6 psi
E_{22}	1.6×10^6 psi	3.5×10^6 psi
v_{12}	0.208	0.25
v_{21}	0.0203	–
G	0.65×10^6 psi	0.93×10^6 psi
Density	0.055 lbs/in^3	0.0725 lbs/in^3
Thickness of laminae	0.0052	0.0052
Allowable stress fibre direction		
tension	139.0 kips/in^2	166.0 kips/in^2
compression	92.4 kips/in^2	86.0 kips/in^2
Transverse direction		
tension	4.95 kips/in^2	6.0 kips/in^2
compression	29.7 kips/in^2	11.86 kips/in^2
Shear stress	4.68 kips/in^2	3.95 kips/in^2

Shear panels
  $E_1 = 30 \times 10^6$ psi, $v = 0.3$, allowable stress $= 8.0$ kips/in^2.
  $\rho = 0.28$ lbs/in^3.

Posts
  $E_1 = 10.5 \times 10^6$ psi, $v = 0.3$, allowable stress $= 25$ kips/in^2.
  $\rho = 0.1$ lbs/in^3.

The box was subjected to two loading conditions. In the first loading condition 1000 lbs was applied at nodes 1 through 4 in the negative z direction. In the second loading condition 500 lbs was applied at nodes 2 and 4 in the negative y direction. The elements were designed by using the recurrence relation (Equation (5.139)) based on the strain energy stored in the element. In order to take into consideration the effect of the two loading conditions, the sum of the strain energies for both loading conditions was used. In the first iteration the number of layers in all four fibre directions for all the elements was assumed to be equal. The beam was first designed to satisfy the stress constraints. Then the design was modified to prevent local buckling of the elements. This was achieved by using the simple expression for the buckling of simply supported orthotropic plates (see Ref. [33]). The thickness of the plate elements was increased to satisfy the buckling equations, and the whole structure was reanalysed to determine the modified in-plane forces. This process was continued until there was no further increase in the thickness of the plate elements. Since the buckling load of the plate can be further increased by increasing the thickness of the $\pm 45°$ layers instead of $0°$ or $90°$ layers, only the thickness of these layers was primarily increased. The convergence behaviour of the box-beam is given in Table 5.9. The minimum weight of a design satisfying only the stress constraints was 26.45 lbs, and for the design with stress and local buckling constraints was 34.40 lbs. The distribution of the number of plies in the $0°$, $90°$, and $\pm 45°$ direction is given in Figs. 5.10 and 5.11.

Table 5.9 Iteration history for cantilever box-beam

Iteration no.	Weight (lb)	Iteration no.	Weight (lb)
1	153.20	10	29.27
2	74.83	11	26.97
3	32.71	12	26.50
4	30.48	13	26.46
5	29.27	14	26.45
6	29.34	15	32.08*
7	29.62	16	33.90*
8	29.56	17	34.26*
9	29.50	18	34.36*
		19	34.40*

* Modified for local buckling.
CP time for each iteration (1–14) 3.5 seconds.
CP time for each iteration (15–18) 4.9 seconds.

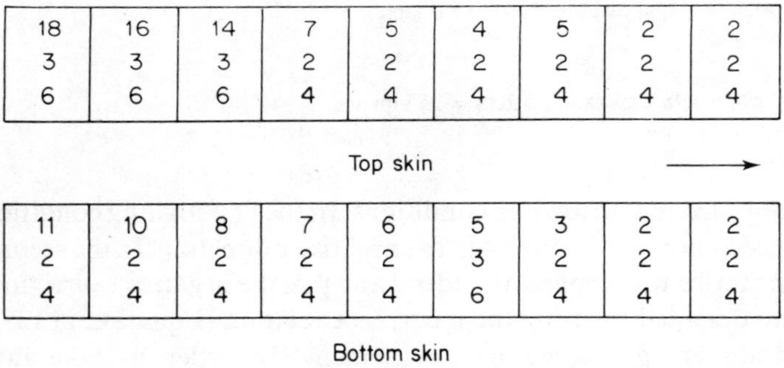

18	16	14	7	5	4	5	2	2	0°
3	3	3	2	2	2	2	2	2	90°
6	6	6	4	4	4	4	4	4	±45°

Top skin ⟶

11	10	8	7	6	5	3	2	2
2	2	2	2	2	3	2	2	2
4	4	4	4	4	6	4	4	4

Bottom skin

Fig. 5.10 Optimum design (stress constraints)

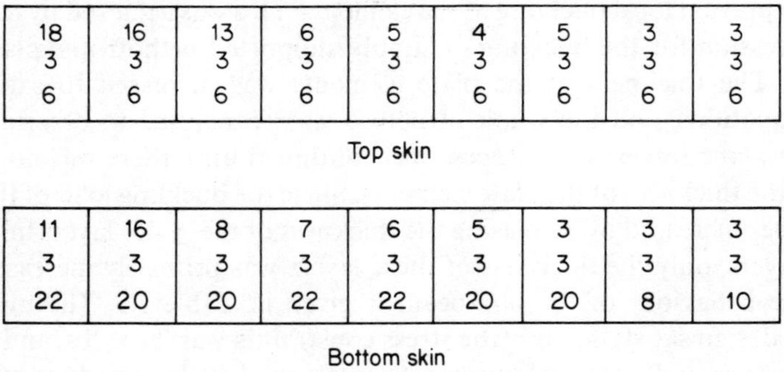

18	16	13	6	5	4	5	3	3
3	3	3	3	3	3	3	3	3
6	6	6	6	6	6	6	6	6

Top skin

11	16	8	7	6	5	3	3	3
3	3	3	3	3	3	3	3	3
22	20	20	22	22	20	20	8	10

Bottom skin

Fig. 5.11 Optimum design (stress and local buckling constraints)

The number of plies in the $+45°$ and $-45°$ direction is equal. The number of plies in each layer was obtained by dividing the thickness of each layer by the ply thickness, and then rounding the result to the next higher integer. The number of plies in the top skin for both cases is nearly the same, but the number of plies in the bottom skin, which is subjected to compressive loads, has substantially changed. The change in the number of plies in the $0°$ and $90°$ direction is due to scaling the design variables and rounding the thickness in order to make it equal to a multiple of a single ply thickness.

In the optimum design for the stress constraint problem it is interesting to see that the distribution of the number of plies in the top and bottom skin near the root is not the same. The distribution of layer thicknesses for the first ten iterations, until the weight reached 29.27 lbs, was symmetric, but with additional iterations the weight was reduced to 26.45 lbs and the symmetry was lost. This may be due to unsymmetric loading conditions.

Example 5 Wing structure

The finite element model of the wing structure (Ref. [37]) shown in Fig. 5.12 was designed to satisfy: (1) stress constraints, (2) negative twist constraints (a wash-

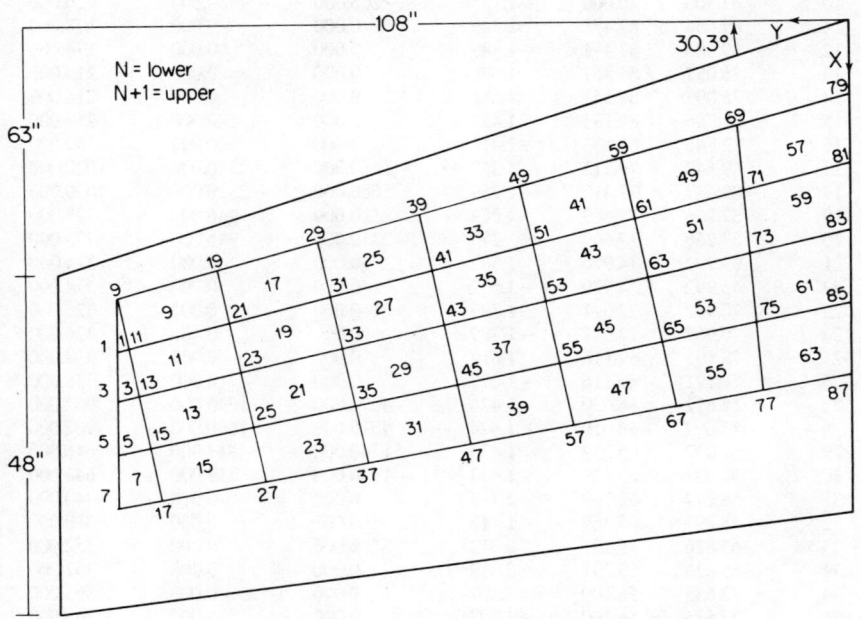

Fig. 5.12 Model of wing

out condition), and (3) positive twist constraints (a wash-in condition). The structure was subjected to a single loading condition. The coordinates of the node points and the loads applied at each node are given in Table 5.10. The structural model has 88 nodes and 158 members. The top and bottom skins were idealized by membrane quadrilaterals and triangles. The aluminium spars and ribs were idealized with shear panels and posts. The skin elements consist of four layers with fibres in $0°$, $90°$, $+45°$ and $-45°$ directions. The number of plies in the $+45°$ and $-45°$ are not necessarily equal. The $0°$ fibres are parallel to the direction of the middle spar as defined by connecting nodes 4 and 84 in the top skin or 3 and 83 in the bottom skin. The elastic constants and the allowable

Table 5.10 Coordinates of nodes and applied loads

JOINT	$-X$	$-Y$	$-Z$	FORCE$-X$	FORCE$-Y$	FORCE$-Z$
1	70.833	90.000	1.313	0.000	0.000	29.000
2	70.833	90.000	-1.313	0.000	0.000	29.000
3	78.167	90.000	1.500	-2800.000	-6960.000	1130.000
4	78.167	90.000	-1.500	2800.000	6960.000	1130.000
5	85.500	90.000	1.313	0.000	0.000	90.900
6	85.500	90.000	-1.313	0.000	0.000	90.900
7	92.833	90.000	1.125	-9870.000	-9780.000	1130.000
8	92.833	90.000	-1.125	9870.000	9780.000	1130.000
9	63.500	90.000	1.125	205.000	-7380.000	926.000
10	63.500	90.000	-1.125	-205.000	7380.000	926.000
11	69.686	87.471	1.349	0.000	0.000	178.000
12	69.686	87.471	-1.349	0.000	0.000	178.000
13	76.097	84.851	1.586	0.000	0.000	214.000
14	76.097	84.851	-1.586	0.000	0.000	214.000
15	82.746	82.133	1.427	0.000	0.000	253.000
16	82.746	82.133	-1.427	0.000	0.000	253.000
17	89.647	79.312	1.259	-5680.000	2320.000	1020.000
18	89.647	79.312	-1.259	5680.000	-2320.000	1020.000
19	57.266	77.669	1.279	2310.000	-946.000	723.000
20	57.266	77.669	-1.279	-2310.000	946.000	723.000
21	63.992	74.920	1.532	0.000	0.000	314.000
22	63.992	74.920	-1.532	0.000	0.000	314.000
23	70.962	72.071	1.799	0.000	0.000	326.000
24	70.962	72.071	-1.799	0.000	0.000	326.000
25	78.191	69.116	1.617	0.000	0.000	338.000
26	78.191	69.116	-1.617	0.000	0.000	338.000
27	85.692	66.050	1.424	-4070.000	1660.000	902.000
28	85.692	66.050	-1.424	4070.000	-1660.000	902.000
29	51.032	65.339	1.433	1740.000	-713.000	646.000
30	51.032	65.339	-1.433	-1740.000	713.000	646.000
31	58.297	62.369	1.715	0.000	0.000	340.000
32	58.297	62.369	-1.715	0.000	0.000	340.000
33	65.826	59.291	2.012	0.000	0.000	352.000
34	65.826	59.291	-2.012	0.000	0.000	352.000
35	73.635	56.100	1.807	0.000	0.000	365.000
36	73.635	56.100	-1.807	0.000	0.000	365.000
37	81.738	52.787	1.590	-4250.000	1740.000	974.000

JOINT	$-X$	$-Y$	$-Z$	FORCE-X	FORCE-Y	FORCE-Z
38	81.738	52.787	−1.590	4250.000	−1740.000	974.000
39	44.799	53.008	1.587	1820.000	−743.000	694.000
40	44.799	53.008	−1.587	−1820.000	743.000	694.000
41	52.603	49.818	1.898	0.000	0.000	365.000
42	52.603	49.818	−1.898	0.000	0.000	365.000
43	60.691	46.512	2.225	0.000	0.000	378.000
44	60.691	46.512	−2.225	0.000	0.000	378.000
45	69.079	43.083	1.997	0.000	0.000	392.000
46	69.079	43.083	−1.997	0.000	0.000	392.000
47	77.784	39.525	1.756	−4440.000	1820.000	1050.000
48	77.784	39.525	−1.756	4440.000	−1820.000	1050.000
49	38.565	40.678	1.742	1890.000	−773.000	742.000
50	38.565	40.678	−1.742	−1890.000	773.000	742.000
51	46.908	37.267	2.082	0.000	0.000	390.000
52	46.908	37.267	−2.082	0.000	0.000	390.000
53	55.555	33.732	2.438	0.000	0.000	404.000
54	55.555	33.732	−2.438	0.000	0.000	404.000
55	64.523	30.067	2.187	0.000	0.000	420.000
56	64.523	30.067	−2.187	0.000	0.000	420.000
57	73.830	26.262	1.922	−4640.000	1900.000	1120.000
58	73.830	26.262	−1.922	4640.000	−1900.000	1120.000
59	32.331	28.347	1.896	2290.000	−937.000	883.000
60	32.331	28.347	−1.896	−2290.000	937.000	883.000
61	41.214	24.716	2.265	0.000	0.000	413.000
62	41.214	24.716	−2.265	0.000	0.000	413.000
63	50.420	20.953	2.651	0.000	0.000	391.000
64	50.420	20.953	−2.651	0.000	0.000	391.000
65	59.967	17.050	2.376	0.000	0.000	368.000
66	59.967	17.050	−2.376	0.000	0.000	368.000
67	69.876	13.000	2.088	−3030.000	1240.000	804.000
68	69.876	13.000	−2.088	3030.000	−1240.000	804.000
69	25.166	14.173	2.073	3070.000	−520.000	1040.000
70	25.166	14.173	−2.073	−3070.000	520.000	1040.000
71	35.583	12.304	2.446	0.000	0.000	433.000
72	35.583	12.304	−2.446	0.000	0.000	433.000
73	46.181	10.403	2.827	0.000	0.000	370.000
74	46.181	10.403	−2.827	0.000	0.000	370.000
75	56.964	8.469	2.502	0.000	0.000	304.000
76	56.964	8.469	−2.502	0.000	0.000	304.000
77	67.938	6.500	2.169	−1370.000	262.000	446.000
78	67.938	6.500	−2.169	1370.000	−262.000	446.000
79	18.000	0.000	2.250	1730.000	0.000	544.000
80	18.000	0.000	−2.250	−1730.000	0.000	544.000
81	30.000	0.000	2.625	0.000	0.000	224.000
82	30.000	0.000	−2.625	0.000	0.000	224.000
83	42.000	0.000	3.000	0.000	0.000	190.000
84	42.000	0.000	−3.000	0.000	0.000	190.000
85	54.000	0.000	2.625	0.000	0.000	155.000
86	54.000	0.000	−2.625	0.000	0.000	155.000
87	66.000	0.000	2.250	−715.000	0.000	226.000
88	66.000	0.000	−2.250	715.000	0.000	226.000

strengths for graphite-epoxy and aluminium are given in Table 5.11. The maximum stress criterion was used to design the composite elements. For the first iteration for all three cases the relative sizes of the bar elements and the thicknesses of the plate element were 1.0 to 0.1, respectively. The percentage of plies in the four fibre directions for the initial design were assumed to be equal.

Table 5.11 Properties and allowable stresses for wing structure

Properties	Graphite-epoxy	Aluminium
E_{11} (psi)	18.5×10^6	10.5×10^6
E_{22} (psi)	1.6×10^6	10.5×10^6
v_{12}	0.25	0.3
Shear modulus	0.65×10^6	4.038×10^6
Density (lb/in^3)	0.055	0.1
Lamina thickness (in)	0.0052	
Allowable stresses (ksi)		
F_x (tension)	139.0	45
F_x (compression)	86.0	45
F_y (tension)	–	45
F_y (compression)	–	45
F_{xy}	46.8	25.9

(a) *Stress constraint design* The structure was designed by using the recurrence relation (Equation (5.139)) based on the strain energy in each layer and in each element. The initial weight of the structure, with an equal percentage of plies in the four fibre directions, was 312.84 lbs, and the minimum weight design obtained after 14 iterations was 45.45 lbs. The iteration history is given in Table 5.12. The weight of the structure after 7 iterations was found to be 45.63 lbs. The design was scaled after each iteration to obtain a feasible design. The distribution of the number of plies in the four fibre directions is shown in Fig. 5.13. The tip deflections at nodes 7 and 9 are given in Table 5.13. The tip twists through an

Table 5.12 Iteration history for wing structure stress constraints

Iteration no.	Weight (lb)	Iteration no.	Weight (lb)
1	312.84	8	45.75
2	112.18	9	45.71
3	68.37	10	45.59
4	59.89	11	45.49
5	55.78	12	45.49
6	55.78	13	45.51
7	45.63	14	45.45

Optimality Criterion Methods in Structural Optimization 153

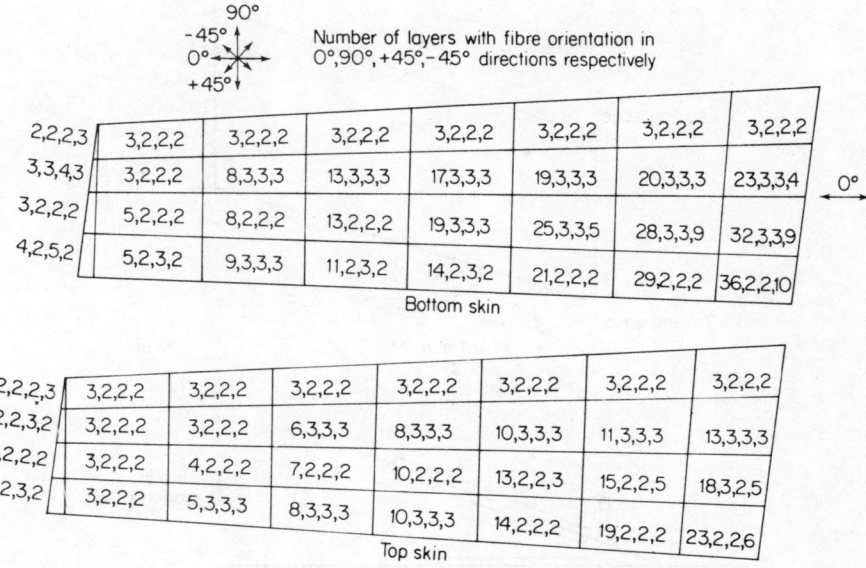

Fig. 5.13 Optimum design of wing (stress constraints)

Table 5.13 Deflections at tip for the stress constraint design

Node 7 (trailing edge) Direction (in)			Node 9 (leading edge) Direction (in)			Twist (°)
X	Y	Z	X	Y	Z	
−0.06	−0.19	13.43	−0.04	−0.16	10.62	−5.45

angle $-5.458°$. This is the angle between the lines joining nodes 9 and 7 before and after loading.

(b) *Twist constraint designs* The wing structure was designed to satisfy two twist constraints to illustrate the use of the algorithm. The two angles selected for this are $-7°$ (wash-out) and $+2°$ (wash-in) (see Fig. 5.14). The structure was designed by using Equation (5.37) for a single displacement constraint where the virtual load consisted of a unit couple. The stress constraints in the elements were treated as passive constraints. The iteration history for the wash-out and wash-in cases is shown in Figs. 5.15 and 5.16 respectively. After each iteration the structure was scaled to satisfy the stress and twist constraints. The two designs in Figs. 5.15 and 5.16 are referred to as the 'stress design' and the 'twist design'. In the 'stress design' the stresses in all the elements are less than or equal to the maximum

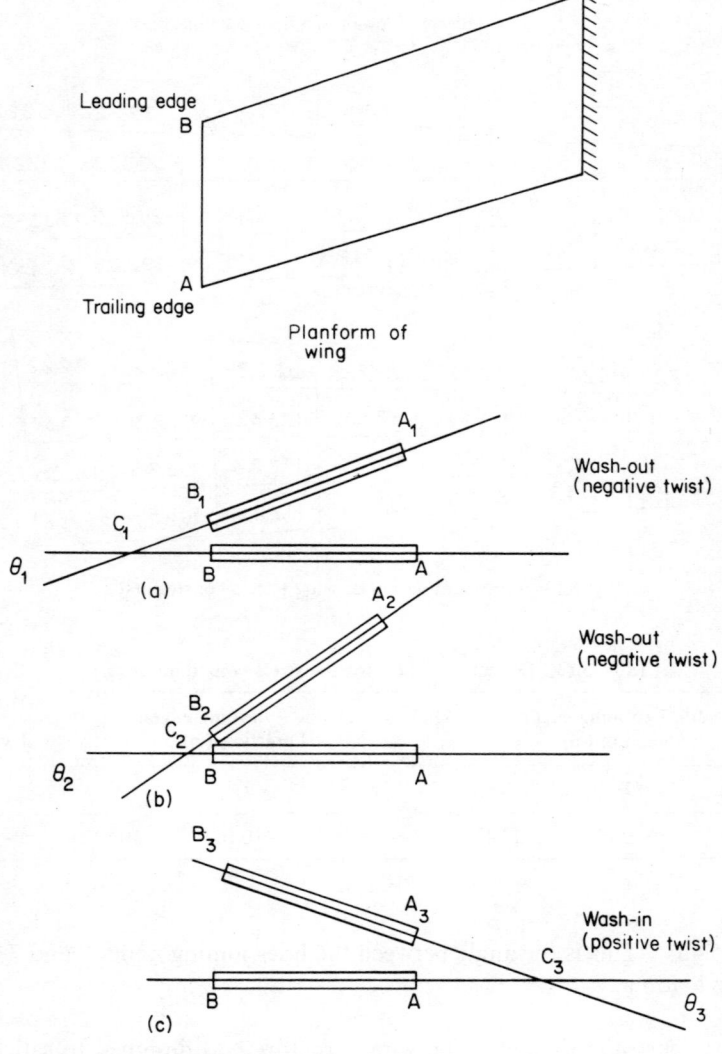

Fig. 5.14 Definition of twist

allowable stress, and in the 'twist design' the twist of the tip of the wing is equal to the prescribed value.

For the wash-out case the lowest design weight of 71.86 lbs was obtained in the 11th iteration. The weight after the 4th iteration was 72.22 lbs, and it did not improve significantly with additional iterations. The displacements at nodes 7 and 9 for the least weight design are given in Table 5.14, and the distribution of the plies in each element is given in Fig. 5.17.

Table 5.14 Deflections at tip for the twist constraint design (wash-out 7°)

Node 7 (trailing edge) Direction (in)			Node 9 (leading edge) Direction (in)			Twist (°)
X	Y	Z	X	Y	Z	
−0.09	−0.15	11.68	−0.08	−0.10	8.07	−7.00

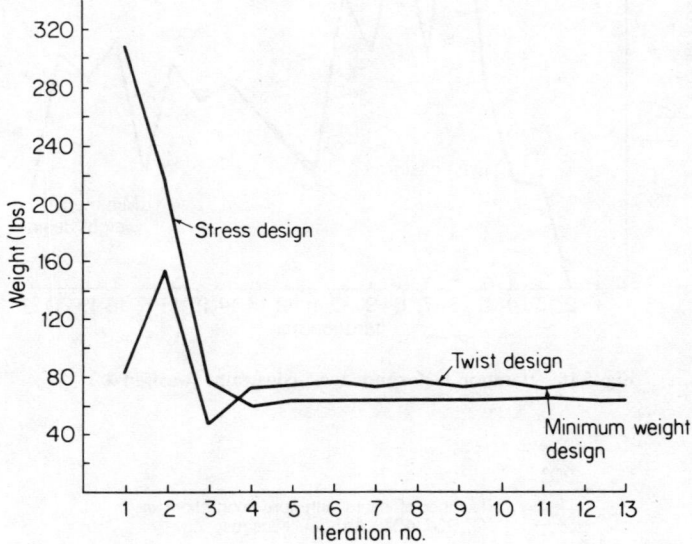

Fig. 5.15 Iteration history for twist constraints (wash-out −7°)

In the wash-in case the twist at the tip for the initial design was negative, and this design could not be scaled to achieve the +2° twist. From the 2nd to the 8th iteration the twist of the stress design was found to increase until it reached +2°. In the subsequent iterations for all cases, the twist designs were acceptable since they satisfied the stress constraints also. From Fig. 5.16 it is seen that the weight of the structure oscillates and does not reach a stable position in the design space. The least weight design with a weight of 105.54 lbs was obtained at the 21st iteration. The oscillations may be due to the large value of the step size parameter, $r = 2$, selected for this problem.

The displacements at nodes 7 and 9 for this design are given in Table 5.15. Figure 5.18 shows the distribution of the plies in each element. It is seen from this figure that there is a heavy concentration of 0° plies only on one side of the wing. A design with an even distribution of plies would have resulted for both cases if the 0° plies were placed at an angle instead of parallel to the line joining nodes 3 and 83.

156 *Foundations of Structural Optimization: A Unified Approach*

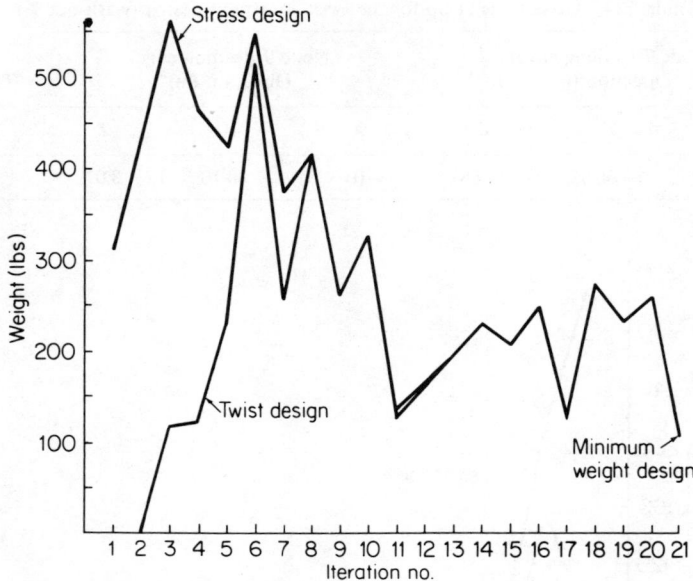

Fig. 5.16 Iteration history for twist constraint (wash-in 2°)

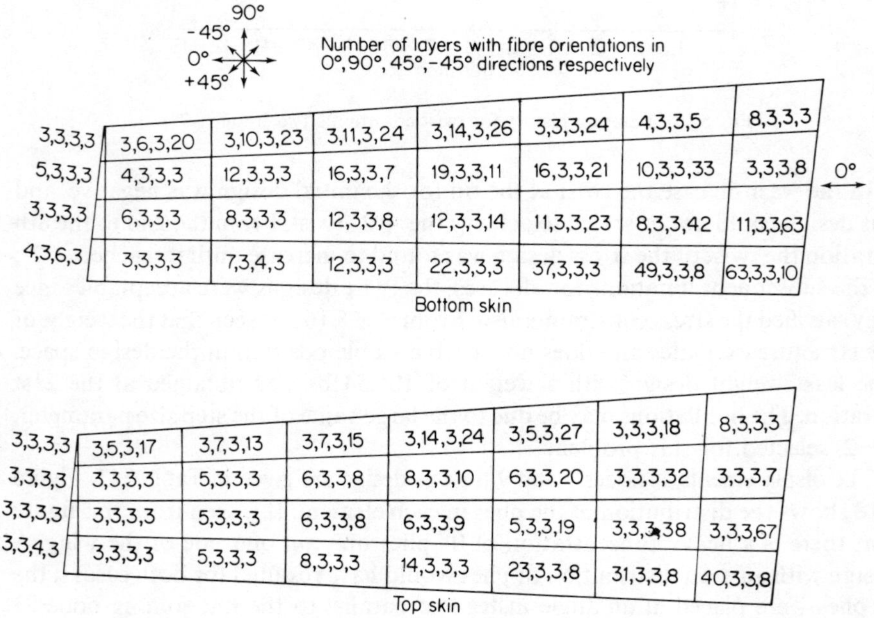

Fig. 5.17 Optimum design of wing (twist constraint, wash-out −7°)

Optimality Criterion Methods in Structural Optimization 157

Table 5.15 Deflections at tip for the twist constraint design (Wash-In 2°)

Node 7 (trailing edge) Direction (in)			Node 9 (leading edge) Direction (in)			Twist (°)
X	Y	Z	X	Y	Z	
0.06	−0.10	4.49	0.04	−0.11	5.51	2.00

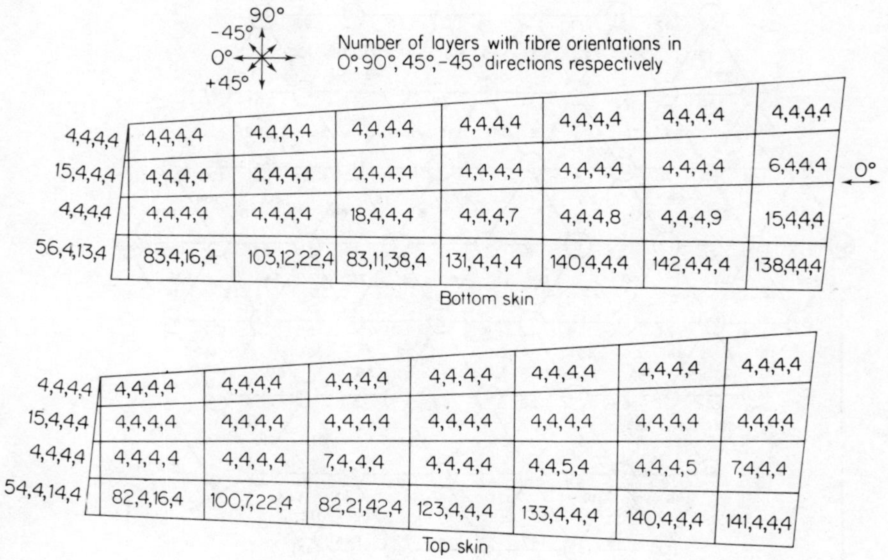

```
                90°
           -45°  ↑
             0° ←✳→  Number of layers with fibre orientations in
           +45°  ↓    0°, 90°, 45°, -45° directions respectively
```

Bottom skin:

4,4,4,4	4,4,4,4	4,4,4,4	4,4,4,4	4,4,4,4	4,4,4,4	4,4,4,4	4,4,4,4
15,4,4,4	4,4,4,4	4,4,4,4	4,4,4,4	4,4,4,4	4,4,4,4	4,4,4,4	6,4,4,4
4,4,4,4	4,4,4,4	4,4,4,4	18,4,4,4	4,4,4,7	4,4,4,8	4,4,4,9	15,4,4,4
56,4,13,4	83,4,16,4	103,12,22,4	83,11,38,4	131,4,4,4	140,4,4,4	142,4,4,4	138,4,4,4

Top skin:

4,4,4,4	4,4,4,4	4,4,4,4	4,4,4,4	4,4,4,4	4,4,4,4	4,4,4,4	4,4,4,4
15,4,4,4	4,4,4,4	4,4,4,4	4,4,4,4	4,4,4,4	4,4,4,4	4,4,4,4	4,4,4,4
4,4,4,4	4,4,4,4	4,4,4,4	7,4,4,4	4,4,4,4	4,4,5,4	4,4,4,5	7,4,4,4
54,4,14,4	82,4,16,4	100,7,22,4	82,21,42,4	123,4,4,4	133,4,4,4	140,4,4,4	141,4,4,4

Fig. 5.18 Optimum design of wing (twist constraint, wash-in 2°)

Example 6 Dome structure

The dome structure shown in Fig. 5.19 was optimized to satisfy system stability. The structure was subjected to a load of 1000 lbs applied in the vertical direction at each node point. The structure has 61 nodes and 132 members. The dome is supported at all the node points on the boundary and all the joints are hinged. The structure was idealized with bar elements carrying axial load only.

The structure was designed by using the recurrence relation (Equation (5.167)) which uses the strain energy in an element in the buckled mode associated with the lowest buckling load of the structure. The stress constraints in the elements were treated as passive constraints. The structure was designed by using two values of the step size parameter, $r = 2$ (design I) and $r = 10$ (design II). The iteration history for the two cases is given in Tables 5.16 and 5.17 respectively. The minimum-weight designs are given in Table 5.18.

158 Foundations of Structural Optimization: A Unified Approach

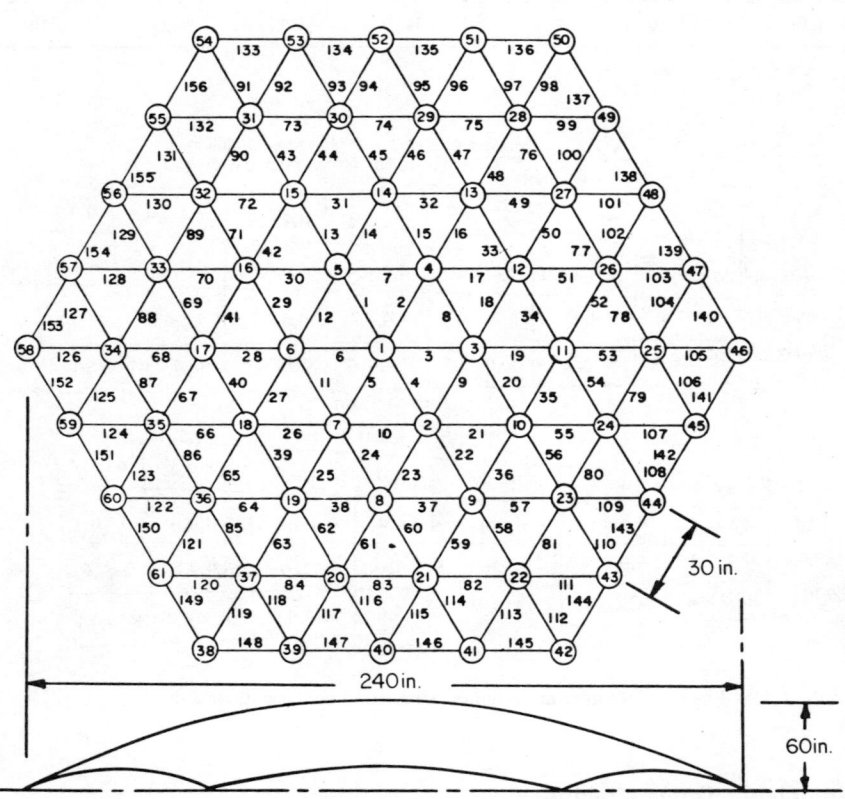

Fig. 5.19 Geodesic dome

In these tables 'Design B' is the scaled design satisfying the stress constraints in the elements. The buckling load of 'Design E' is equal to the applied load. It is seen that for all the designs the system buckling is the active constraint, and the stress constraint, because of the high allowable, does not become active.

The step size parameter $r = 2$ was found to be too large to obtain good convergence. Therefore the step size was reduced by generating the intermediate design vector whenever the scaled weight of the structure after each iteration was greater than the previous lower weight design (see discussion in Section 5.8). The intermediate designs obtained by this procedure are indicated by† in Table 5.16. In the case of the step size parameter $r = 10$, the minimum-weight design was

Table 5.16 Iteration history of dome structure for $m = 2$ and intermediate design vectors (Design I)

Cycle no.	Design B			Design E
	Wt (lb)	Buckling load*	B. load Wt	Wt (lb)
1	16.583	0.1877	0.0113	87.836
2	91.767	0.6073	0.0066	151.086
3	19.503	0.2679	0.0137	72.774†
4	45.307	0.4551	0.0100	99.537
5	25.144	0.3084	0.0122	81.511
6	21.708	0.3198	0.0147	67.859

* Load in kips over each node point
† Weight corresponding to intermediate design vectors
Initial design—Area of all members = 0.2160 in^2

Table 5.17 Iteration history of dome structure for $m = 10$ (Design II)

Cycle no.	Design B			Design E
	Wt (lb)	Buckling load*	B. load Wt	Wt (lb)
1	16.583	0.1887	0.0113	87.836
2	21.157	0.2949	0.0139	71.729
3	24.512	0.3655	0.0149	67.059

* Load in kips over each node point
Initial design—Area of all members = 0.2160 in^2

Table 5.18 Optimum design of Geodesic dome

Elements	Design I area (in^2)	Design II area (in^2)
1, 2	0.2016	0.2136
7	0.2805	0.2648
13, 16	0.2566	0.2469
14, 15	0.1775	0.1863
31, 32	0.1819	0.1936
43, 48	0.1391	0.1378
44, 47	0.2095	0.2134
45, 46	0.1760	0.1847
73, 75	0.1245	0.1115
74	0.1753	0.1837
91, 98	0.1261	0.1125
92, 97	0.1355	0.1336
93, 96	0.1253	0.1119
94, 95	0.1231	0.1079

Because of the symmetry only the areas of elements within one sector are given.

obtained after three iterations. The weights of the optimum designs obtained for the two step size parameters are nearly equal. However, from Table 5.18 it is seen that the two designs are not identical.

5.10 SUMMARY AND CONCLUSIONS

The minimum-weight design must satisfy the optimality criterion derived for the specific type of constraints imposed on the structure. The criterion states that a quantity

$$\psi = \sum_{j=1}^{m} \lambda_j \frac{F_{ij}}{\rho_i l_i x_i^2}$$

associated with all the elements in the structure is equal to unity. The quantity F_{ij} depends on the nature of the constraints. In the case of displacement, stress and stability constraints F_{ij} is equal to Q_{ij}, R_{ij} and $\overline{Q}_{ij}$ respectively and is defined in Equations (5.10), (5.117) and (5.161). F_{ij} is a function of the energy stored in the element, and the nature of the energy depends upon the type of constraints.

The algorithms based on an optimality criterion are iterative and use the recurrence relation derived from the optimality criterion to modify the design variables. There are basically three distinct relations which we have derived from the optimality criterion. These are: (1) the exponential relation; (2) the linear relation for x_i; and (3) the linear relation for z_i. Many other recurrence relations can be written from these three by selecting different values of the step size parameter r. Each of the recurrence relations contains the quantity ψ. Use of the recurrence relation moves the initial design to the optimum design satisfying the optimality criterion. When the condition $\psi = 1$ is satisfied, the recurrence relation does not change the design variables. However, in the optimization algorithm based on the optimality criterion some of the constraints are treated as passive and the step size parameter is generally kept constant, therefore the scaled weight of the structure reaches a certain minimum value, and then it increases. The real minimum weight lies closer to the lowest weight design obtained by using the algorithm. The real minimum-weight design satisfying the optimality criterion can be obtained from the lowest weight with additional iterations but with a smaller step size. The difference between the real minimum and the lowest weight design is generally small. This behaviour is particularly true when the stress constraints dominate.

The recurrence relations can only be used after determining the coefficients F_{ij} and the Lagrange multipliers. The coefficients F_{ij} are related to the gradients of the constraints. The number of Lagrange multipliers depends upon the number of active constraints. The active constraints are defined as those which are associated with the positive Lagrange multipliers, and these constraints are very close to the constraint surface. The number of active constraints changes with the iterations. Initially the number is small, but as the design approaches the optimum the number increases. The constraints sometimes switch from the active category to the passive category depending upon their activity as the optimization process proceeds from one iteration to the next. A prior knowledge of the active constraints at the optimum is not necessary. At each iteration one has to determine the set of active constraints.

We have presented different approaches to determine the Lagrange multipliers. These approaches depend on: (1) the solution to the linear equations; (2) the Newton–Raphson iterations; and (3) the use of the recurrence relations or approximations to the linear equations. In the first approach, as well as in the second, the coupling of the constraints on the overall problem is taken into consideration. The last approach ignores the interdependence of the various constraints imposed on the structure. The first approach requires the evaluation of the coefficients F_{ij} associated with all the elements corresponding to all the potentially active constraints. Since evaluation of these coefficients and assembling the matrix multiplying the Lagrange multipliers in the linear equations needs substantial CP time, it is necessary to adopt an efficient approach in order to pick the potentially active constraints. This can be achieved by selecting the constraints which are within a certain distance from the constraint surface and the knowledge of active constraints in the previous iteration. The most efficient method would be if the potentially active constraints are equal to the active constraints. In the case of a large structure such as the 200-bar truss in Example 3 of the last section, when the potentially active constraints become greater than about 6, the CP time required for the optimization phase became larger than the analysis phase. The use of linear equations to determine the Lagrange multipliers is the most suitable approach if there is a strong coupling between the constraints.

Since we have not presented any results using the Newton–Raphson method to determine the Lagrange multipliers, we will not discuss the advantages or disadvantages of this approach. However, in using this approach the coefficients Q_{ij} need to be evaluated for all the elements and constraints, and it is necessary to assume the initial values of the Lagrange multipliers. Also in the Newton–Raphson approach it is generally difficult to eliminate the equations associated with the passive constraints.

The last approach, based on using simple relations to determine the Lagrange multipliers, is computationally efficient. It requires less CP time to perform the optimization phase of the algorithm. With this approach it is possible to combine all the virtual loads associated with all the constraints into one equivalent virtual load, since the Lagrange multipliers can be estimated before evaluating the coefficients Q_{ij} or R_{ij}. Particularly for the stress-constrained problem, if the initial values of the Lagrange multipliers are assumed to be proportional to the forces in the bars and the exponential recurrence relation is used to modify the design variables, use of Equation (5.124) leads to the near optimum weight design with the least computational effort. However, if the constraints imposed on the structure are mixed, such as displacements and stresses, then this algorithm needs a large number of iterations and the scaled weight of the structure may violently oscillate. The approximate relation (Equation (5.132)) to determine the Lagrange multiplier for the FSD design, gives a near minimum weight design if the minimum weight design is a FSD. The FSD algorithm (Equation (5.108)) does not lead to the correct minimum weight design if there are large differences in the maximum allowable stresses in the elements.

In the optimization of a structure it may be desirable to obtain a design that satisfies the theoretical optimality criterion. However, in a practical design problem, true optimum may not always be feasible due to manufacturing considerations. In many practical cases it may be useful to obtain a design which is a near minimum weight design with a good distribution of material, then the design satisfying the theoretical optimality criterion. For many problems such a design can be obtained by including only a small number of constraints in the potentially active category and treating the other constraints as passive constraints. A good example of this is the displacement-stress constraint problem. For such a problem it is advantageous to treat all the stress constraints as passive constraints instead of including them in the constraint equations and evaluating their gradients and associated Lagrange multipliers. Since the number of active stress constraints is large compared to the displacement constraints, this simplification saves substantial CP time. This and other simple relations proposed in the previous sections to determine the Lagrange multiplier minimizes the total computational effort needed to design an optimum structure.

N.S.K.

REFERENCES

1. V. B. Venkayya, N. S. Khot and L. Berke, Application of optimality criteria approaches to automated design of large practical structures, *Second Symp. Struct. Opt., AGARD-CP-123*, Milan, Italy, 1973.
2. L. Berke and N. S. Khot, Use of optimality criteria methods for large scale systems, *AGARD Lecture Series No. 70, Struct. Opt.*, 1974.
3. N. S. Khot, V. B. Venkayya and L. Berke, Experience with minimum weight design of structures using optimality criteria methods, *2nd Int. Conf. Vehicle Struct. Mech.*, Southfield, Michigan, 1977.
4. J. Kiusalaas, Minimum weight design of structures via optimality criteria, *NASA TN D-7115*, 1972.
5. V. B. Venkayya, N. S. Khot and V. S. Reddy, Energy distribution in an optimum structural design, *AFFDL-TR-68-156*, 1969.
6. L. Berke, An efficient approach to the minimum weight design of deflection limited structures, *AFFDL-TM-70-4-FDTR*, 1970.
7. V. B. Venkayya, Design of optimum structures, *J. Computers and Struct.*, **1**, 265–309, 1971.
8. I. C. Taig and R. I. Kerr, Optimization of aircraft structures with multiple stiffness requirements, *Second Symp. Struct. Opt., AGARD-CP-123*, Milan, Italy, 1973.

9. R. A. Gellatly and L. Berke, Optimality-criterion algorithms, *Optimum Structural Design*, edited by R. M. Gallagher and O. C. Zienkiewicz, John Wiley & Sons, London, 1973.
10. J. C. Nagteggal, A new approach to optimal design of elastic structures, *Computational Methods in Applied Mechanics and Engineering*, **2**, 255–64, Feb. 1973.
11. N. S. Khot, V. B. Venkayya, C. D. Johnson and V. A. Tischler, Optimization of fiber reinforced composite structures, *Int. J. Solids Structures*, **9**, 1225–36, Pergamon Press, Sept. 1973.
12. N. S. Khot, V. B. Venkayya and L. Berke, Optimum design of composite structures with stress and deflection constraints, *AIAA Paper No. 75-141*, presented at AIAA 13th Aerospace Sciences Meeting, Pasadena, California, 1975.
13. R. A. Gellatly and L. Berke, Optimal structural design, *AFFDL-TR-70-165*, Air Force Flight Dynamics Laboratory, Wright-Patterson AFB, Ohio, 1975.
14. D. Rizzi, Optimization of multi-constrained structures based on optimality criteria, *Proc. AIAA/ASME/SAE 17th Struct., Structural Dynamics and Materials Conf.*, King of Prussia, Pennsylvania, 448–62, 1976.
15. M. W. Dobbs and R. B. Nelson, Application of optimality criteria to automated structural design, *AIAA J.*, **14**, 1436–43, 1976.
16. S. A. Segenreich, N. A. Zouain and J. Herskovits, An optimality criteria method based on slack variables concept for large scale structural optimization, *Proc. Symp. Applications of Computer Meth. In Engng.* (Ed. C. Wellford, Jr.), University of Southern California, 563–75, 1977.
17. F. Austin, A rapid optimization procedure for structures subjected to multiple constraints, *Paper No. 77-374*, AIAA/ASME/SAE 18th Struct., Structural Dynamics and Materials Conf., San Diego, California, 1977.
18. G. V. Rao, C. P. Shore and R. Narayanaswami, An optimality criterion for resizing heated structures with temperature constraints, *NASA TN D-8525*, 1977.
19. M. R. Khan, K. D. Willmert and W. A. Thornton, A new optimality criterion method for large scale structures, *AIAA/ASME 19th SDM Conf.*, Bethesda, Md., 47–58, 1978.
20. L. Berke and N. S. Khot, A simple virtual strain energy method to fully stress design structures with dissimilar stress allowables and material properties, *AFFDL-TM-77-28-FBR*,
21. J. Kiusalaas, Optimum design of structures with buckling constraints, *Int. J. Solids and Struct.*, **9**, 863–78, 1973.
22. N. S. Khot, V. B. Venkayya and L. Berke, Optimum structural design with stability constraints, *Int. J. Num. Meth. Engng.*, **10**, 1097–114, 1976.
23. V. B. Venkayya, N. S. Khot, V. A. Tischler and R. F. Taylor, Design of optimum structures for dynamic loads, *Third Conf. Matrix Meth. Struct. Mech.*, Wright-Patterson AFB, Ohio, 619–58, 1971.
24. V. B. Venkayya and N. S. Khot, Design of optimum structures to impulse type loading, *AIAA J.*, **13**, 989–94, 1975.
25. J. Kiusalaas and R. Shaw, An algorithm for optimal design with frequency constraints, *Int. J. for Num. Meth. in Engg.*, **13**(2), Dec. 1978.
26. V. B. Venkayya, Structural optimization: A review and some recommendations, *Int. J. for Num. Meth. in Engg.*, **13**(2), Dec. 1978.
27. N. S. Khot, Optimization of aerospace structures to satisfy strength, displacement, and stability requirements—A review and assessment of the state-of-the art, *AFFDL-TM-79-81-FBR*.
28. N. S. Khot, L. Berke and V. B. Venkayya, Comparisons of optimality criteria algorithms for minimum weight design of structures, *AIAA J.*, **17**, 182–90, 1979.

29. C. Fleury and G. Sander, Structural optimization by finite element, *L. T. A. S. Report SA-58*, University of Leige, 1978.
30. N. S. Khot, L. Berke and V. B. Venkayya, Minimum weight design of structures by the optimality criterion and projection method, *AIAA 79-0720*, Paper presented at AIAA/ASME/ASCE/AHS 20th Structures, Structural Dynamics and Material Conference, St. Louis, MO, April 1979.
31. N. S. Khot, Algorithms based on optimality criteria to design minimum weight structures, to be published in the Special Issue *Optimum Design of Large Structural Systems, Engineering Optimization Journal*, 1980.
32. R. A. Gellatly, D. M. Dupree and L. Berke, OPTIM II: A magic compatible large scale automated minimum weight design program, *AFFDL-TR-74-97*, I and II, 1974.
33. N. S. Khot, Computer program (OPTCOMP) for optimization of composite structures for minimum weight design, *AFFDL-TR-76-149*, 1977.
34. V. B. Venkayya and V. A. Tischler, OPTSTAT—A computer program for optimal design of structures subjected to static loads, *AFFDL-TR-79-* (in preparation).
35. J. Kiusalaas and G. B. Reddy, DESAP 2—A structural design program with stress and buckling constraints, *NASA CR-2797 to 2799* (3 vols), National Aeronautics and Space Administration, Washington, D.C., 1977.
36. G. Isakson and H. Pardo, ASOP-3: A program for the minimum-weight design of structures subjected to strength and deflection constraints, *AFFDL-TR-76-157*.
37. N. S. Khot, V. B. Venkayya, L. Berke and K. Schrader, Optimum design of composite wing structures with twist constraint for aeroelastic tailoring, *AFFDL-TR-76-117*, Dec. 1976.

APPENDIX

In order to illustrate some aspect of the algorithms described in this chapter we now look in detail at some simple problems. A three-bar truss shown in Fig. 1(A)

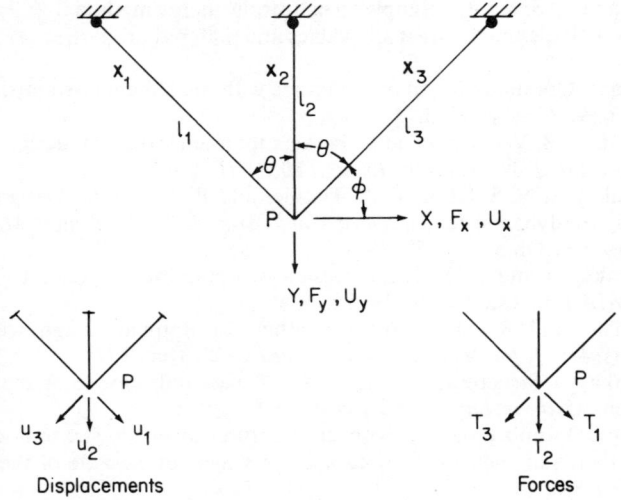

Fig. 1A   Three-bar truss

with the variation of loads and constraints will be used to illustrate the application of the optimality criterion approach to design a minimum-weight structure. The problem is small and most of the required calculations can be done by using a calculator.

The horizontal and vertical displacements of the node P are denoted by U_x and U_y. The forces applied at the node in the x and y directions are F_x and F_y respectively. The displacements U_1, U_2 and U_3 of the point P in the directions parallel to the length of the three bars are given by

$$U_1 = U_x \cos\phi + U_y \cos\theta$$
$$U_2 = U_y \quad \quad (1A)$$
$$U_3 = -U_x \cos\phi + U_y \cos\theta$$

The bar forces are given by

$$T_1 = \frac{x_1 E U_1}{l_1} \quad T_2 = \frac{x_2 E U_2}{l_2} \quad T_3 = \frac{x_3 E U_3}{l_3} \quad (2A)$$

where x_1, x_2 and x_3 are the cross-sectional areas of the three elements, and l_1, l_2 and l_3 are their respective lengths.

The stresses in the elements can be written as

$$\sigma_1 = \frac{T_1}{x_1} \quad \sigma_2 = \frac{T_2}{x_2} \quad \sigma_3 = \frac{T_3}{x_3} \quad (3A)$$

The equilibrium equations can be written by taking a summation of the forces at node P. This gives

$$T_1 \cos\phi - T_3 \cos\phi = F_x$$
$$\quad \quad (4A)$$
$$T_1 \cos\theta + T_2 + T_3 \cos\theta = F_y$$

Using Equations (1A) and (2A), these equations can be written as

$$\begin{bmatrix} K_i & K_1 \\ K_2 & K_2 \end{bmatrix} \begin{Bmatrix} U_x \\ U_y \end{Bmatrix} = \begin{Bmatrix} F_x \\ F_y \end{Bmatrix} \quad (5A)$$

where

$$K_1 = \left(\frac{x_1}{l_1} + \frac{x_2}{l_3}\right) E \cos^2\phi$$

$$K_2 = \left(\frac{x_1}{l_1} - \frac{x_3}{l_3}\right) E \cos\phi \cos\theta \quad (6A)$$

$$K_3 = \left(\frac{x_1}{l_1} + \frac{x_3}{l_3}\right) E \cos^2\theta + \frac{A_2}{l_2} E$$

Equations (5A) are the load-displacement relations and are equuivalent to Equation (5.1). Solving Equation (5A), the nodal displacements are given by

$$U_x = \frac{K_2 F_y - K_3 F_x}{K_2^2 - K_1 K_3}$$

$$U_y = \frac{K_2 F_x - K_1 F_y}{K_2^2 - K_1 K_3}$$ (7A)

The displacements U_x and U_y are also be given by

$$U_x = \sum_{i=1}^{3} \frac{T_i t_i^1 l_i}{x_i E}; \quad U_y = \sum_{i=1}^{3} \frac{T_i t_i^2 l_i}{x_i E}$$ (8A)

where t_i^1 and t_i^2 are the forces in the ith bar, due to a virtual load of unit magnitude, applied at node P in the x and y directions respectively.

The equilibrium equations for the virtual load system $\{f_x, f_y\}$ can be written by replacing the vector $\{F_x, F_y\}$ in Equation (5A) by $\{f_x, f_y\}$. If the displacements and forces due to the virtual load system are denoted by lowercase letters, then the expressions for these quantities, can be written by replacing the capital letters by the lowercase letters for the forces and displacements, in Equations (1A), (2A) and (7A).

Weight of the three-bar truss is given by

$$W = \rho(x_1 l_1 + x_2 l_2 + x_3 l_3)$$ (9A)

The solutions of the problems and the associated algorithms are discussed in subsequent pages. The details of the iterative steps are self-explanatory and follow the steps given in Section 5.8.

Problem 1   A Single Displacement Constraint

This problem is solved by using four different algorithms. All algorithms lead to the same minimum-weight design; however, the number of iterations required for each algorithm is not the same. This is a single constrained problem with only one constraint active for all iterations.

Problem 1(a)

The design variables are modified by using the recurrence relation of Equation (5.35):

$$x_i^{k+1} = x_i^k \left(\lambda_1 \frac{Q_{i1}}{\rho_i l_i x_i^2} \right)^{1/r}_k$$

with the step size parameter $r = 2$. The Lagrange multiplier is determined from

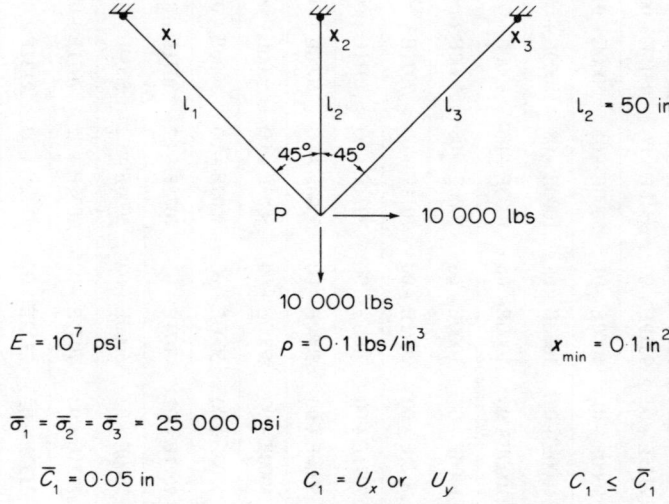

Fig. 2A Problem 1—A single displacement constraint

Equation (5.47):

$$\sqrt{\lambda_1} = \frac{\sum\limits_{i=1}^{n_1} \sqrt{Q_{i1} \rho_i l_i}}{\overline{C}_1 - C_1^*}$$

The areas x_1, x_2 and x_3 at the beginning of each iteration are given in rows 1 through 3, and the scaled areas are given in rows 44 through 46. The scaled weight after each iteration is given in row 48. The Lagrange multiplier λ_1 is calculated in row 71.

It is seen from row 48 that the minimum weight is 113.7 lbs and the associated cross-sectional areas are $x_1 = 15.07$ in² $x_2 = 0.1$ in² and $x_3 = 0.9318$ in². The deflection in the x direction at node P is the active constraint, and λ_1 at the optimum is 2141.0. This solution satisfies Equation (5.48) which gives the relation between λ_1, $W_{\min}$, W^*, $\overline{C}_1$ and C_1^*.

		CYCLE 1	CYCLE 2	CYCLE 3	CYCLE 4	CYCLE 5	CYCLE 6	CYCLE 7	CYCLE 8
1	x_1	.1000E+01	.1644E+02	.1500E+02	.1507E+02	.1507E+02	.1507E+02	.1507E+02	.1507E+02
2	x_2	.1000E+01	.1000E+00	.1000E+00	.1000E+00	.1000E+00	.1000E+00	.1000E+00	.1000E+00
3	x_3	.1000E+01	.1058E+02	.9783E+00	.9383E+00	.9325E+00	.9319E+00	.9318E+00	.9318E+00
4	$\bar{C}_1$	.5000E−01	.5000E−01	.5000E−01	.5000E−01	.5000E−01	.5000E−01	.5000E−01	.5000E−01
5	E	.1000E+08	.1000E+08	.1000E+08	.1000E+08	.1000E+08	.1000E+08	.1000E+08	.1000E+08
6	F_x	.1000E+06	.1000E+06	.1000E+06	.1000E+06	.1000E+06	.1000E+06	.1000E+06	.1000E+06
7	F_y	.1000E+06	.1000E+06	.1000E+06	.1000E+06	.1000E+06	.1000E+06	.1000E+06	.1000E+06
8	$\cos\theta$	.7071E+00	.7071E+00	.7071E+00	.7071E+00	.7071E+00	.7071E+00	.7071E+00	.7071E+00
9	$\cos\phi$	.7071E+00	.7071E+00	.7071E+00	.7071E+00	.7071E+00	.7071E+00	.7071E+00	.7071E+00
10	$\cos^2\theta$	.5000E+00	.5000E+00	.5000E+00	.5000E+00	.5000E+00	.5000E+00	.5000E+00	.5000E+00
11	$\cos^2\phi$	.5000E+00	.5000E+00	.5000E+00	.5000E+00	.5000E+00	.5000E+00	.5000E+00	.5000E+00
12	$\cos\theta\cdot\cos\phi$	.5000E+00	.5000E+00	.5000E+00	.5000E+00	.5000E+00	.5000E+00	.5000E+00	.5000E+00
13	l_1	.7071E+02	.7071E+02	.7071E+02	.7071E+02	.7071E+02	.7071E+02	.7071E+02	.7071E+02
14	l_2	.5000E+02	.5000E+02	.5000E+02	.5000E+02	.5000E+02	.5000E+02	.5000E+02	.5000E+02
15	l_3	.7071E+02	.7071E+02	.7071E+02	.7071E+02	.7071E+02	.7071E+02	.7071E+02	.7071E+02
16	$x_1 E/l_1$	.1414E+06	.2324E+07	.2121E+07	.2131E+07	.2132E+07	.2132E+07	.2132E+07	.2132E+07
17	$x_2 E/l_2$	.2000E+06	.2000E+05	.2000E+05	.2000E+05	.2000E+05	.2000E+05	.2000E+05	.2000E+05
18	$x_3 E/l_3$	.1414E+06	.1496E+07	.1384E+06	.1327E+06	.1319E+06	.1318E+06	.1318E+06	.1318E+06

19	$\dfrac{x_1 E}{l_1} + \dfrac{x_3 E}{l_3}(16+17)$	.2828E+06	.3820E+07	.2260E+07	.2264E+07	.2264E+07	.2264E+07	.2264E+07
20	$\dfrac{x_1 E}{l_1} - \dfrac{x_3 E}{l_3}(16-17)$	0.	.8284E+06	.1983E+07	.1998E+07	.2000E+07	.2000E+07	.2000E+07
21	$K_1 = (11)(19)$	.1414E+06	.1910E+07	.1130E+07	.1132E+07	.1132E+07	.1132E+07	.1132E+07
22	$K_2 = (12)(20)$	0.	.4142E+06	.9915E+06	.9992E+06	.9999E+06	.1000E+07	.1000E+07
23	$K_3 = (10)(19)+17$	.3414E+06	.1930E+07	.1150E+07	.1152E+07	.1152E+07	.1152E+07	.1152E+07
24	$\dfrac{K_2 \cdot F_y - K_3 \cdot F_x}{(22)(7) - (23)(6)}$	$-.3414E+11$	$-.1516E+12$	$-.1584E+11$	$-.1527E+11$	$-.1519E+11$	$-.1518E+11$	$-.1518E+11$
25	$\dfrac{K_2 \cdot K_2 - K_1 \cdot K_3}{(22)(22) - (21)(23)}$	$-.4828E+11$	$-.3515E+13$	$-.3161E+12$	$-.3054E+12$	$-.3038E+12$	$-.3036E+12$	$-.3036E+12$
26	$U_x = 24/25$	.7071E+00	.4312E−01	.5009E−01	.4999E−01	.5000E−01	.5000E−01	.5000E−01
27	$\dfrac{K_2 \cdot F_x - K_1 \cdot F_y}{(22)(6) - (21)(7)}$	$-.1414E+11$	$-.1496E+12$	$-.1384E+11$	$-.1327E+11$	$-.1319E+11$	$-.1318E+11$	$-.1318E+11$
28	$U_y = 27/25$	.2929E+00	.4255E−01	.4377E−01	.4345E−01	.4342E−01	.4341E−01	.4341E−01
29	$U_1 = (26)(9) + (28)(8)$	.7071E+00	.6058E−01	.6637E−01	.6607E−01	.6606E−01	.6605E−01	.6605E−01
30	$U_2 = U_y (28)$	.2929E+00	.4255E−01	.4377E−01	.4345E−01	.4342E−01	.4341E−01	.4341E−01
31	$U_3 = -(26)(9) + (28)(8)$	$-.2929E+00$	$-.4023E-03$	$-.4474E-02$	$-.4630E-02$	$-.4656E-02$	$-.4658E-02$	$-.4659E-02$
32	$T_1 = \dfrac{x_1 E}{l_1} U_1 (16)(29)$	.1000E+06	.1408E+06	.1408E+06	.1408E+06	.1408E+06	.1408E+06	.1408E+06
33	$T_2 = \dfrac{x_2 E}{l_2} U_2 (17)(30)$	.5858E+05	.8511E+03	.8754E+03	.8689E+03	.8683E+03	.8682E+03	.8682E+03

		CYCLE 1	CYCLE 2	CYCLE 3	CYCLE 4	CYCLE 5	CYCLE 6	CYCLE 7	CYCLE 8
34	$T_3 = \dfrac{x_3 \bar{E}}{l_3} U_3$ (18)(31)	$-.4142E+05$	$-.6018E+03$	$-.6190E+03$	$-.6144E+03$	$-.6140E+03$	$-.6139E+03$	$-.6139E+03$	$-.6139E+03$
35	$\sigma_1 = \dfrac{T_1}{x_1} = \dfrac{32}{1}$	$.1000E+06$	$.8568E+04$	$.9386E+04$	$.9344E+04$	$.9342E+04$	$.9341E+04$	$.9341E+04$	$.9341E+04$
36	$\sigma_2 = \dfrac{T_2}{x_2} = \dfrac{33}{2}$	$.5858E+05$	$.8511E+04$	$.8754E+04$	$.8689E+04$	$.8683E+04$	$.8682E+04$	$.8682E+04$	$.8682E+04$
37	$\sigma_3 = \dfrac{T_3}{x_3} = \dfrac{34}{3}$	$-.4142E+05$	$-.5689E+02$	$-.6327E+03$	$-.6548E+03$	$-.6584E+03$	$-.6588E+03$	$-.6589E+03$	$-.6589E+03$
38	$\Lambda(\sigma_1) = \dfrac{\sigma_1}{\bar{\sigma}}$	$.4000E+01$	$.3427E+00$	$.3755E+00$	$.3738E+00$	$.3737E+00$	$.3736E+00$	$.3736E+00$	$.3736E+00$
39	$\Lambda(\sigma_2) = \dfrac{\sigma_2}{\bar{\sigma}}$	$.2343E+01$	$.3404E+00$	$.3501E+00$	$.3476E+00$	$.3473E+00$	$.3473E+00$	$.3473E+00$	$.3473E+00$
40	$\Lambda(\sigma_3) = \dfrac{\sigma_3}{\bar{\sigma}}$	$-.1657E+01$	$-.2276E-02$	$-.2531E-01$	$-.2619E-01$	$-.2634E-01$	$-.2635E-01$	$-.2635E-01$	$-.2635E-01$
41	$\Lambda(U_x) = \dfrac{U_x}{\bar{c}_1} = \dfrac{26}{4}$	$.1414E+02$	$.8625E+00$	$.1002E+01$	$.9999E+00$	$.1000E+01$	$.1000E+01$	$.1000E+01$	$.1000E+01$
42	$\Lambda(U_y) = \dfrac{U_y}{\bar{c}_1} = \dfrac{28}{4}$	$.5858E+01$	$.8511E+00$	$.8754E+00$	$.8689E+00$	$.8683E+00$	$.8682E+00$	$.8682E+00$	$.8682E+00$
43	$\Lambda(\max)$ (38 – 42)	$.1414E+02$	$.8625E+00$	$.1002E+01$	$.9999E+00$	$.1000E+01$	$.1000E+01$	$.1000E+01$	$.1000E+01$
44	$x_1^* = \Lambda \cdot x_1$ (43)(1)	$.1414E+02$	$.1418E+02$	$.1503E+02$	$.1507E+02$	$.1507E+02$	$.1507E+02$	$.1507E+02$	$.1507E+02$
45	$x_2^* = \Lambda \cdot x_2$ (43)(2)	$.1414E+02$	$.8625E-01$	$.1002E+00$	$.9999E-01$	$.1000E+00$	$.1000E+00$	$.1000E+00$	$.1000E+00$
46	$x_3^* = \Lambda \cdot x_3$ (43)(3)	$.1414E+02$	$.9123E+01$	$.9802E+00$	$.9382E+00$	$.9325E+00$	$.9319E+00$	$.9318E+00$	$.9318E+00$

47	VOLUME = $\sum x_i l_i$	.2707E+04	.1652E+04	.1137E+04	.1137E+04	.1137E+04	.1137E+04	.1137E+04	.1137E+04
48	WT = (47)(ρ)	.2707E+03	.1652E+03	.1137E+03	.1137E+03	.1137E+03	.1137E+03	.1137E+03	.1137E+03
49	f_x	.1000E+01	.1000E+01	.1000E+01	.1000E+01	.1000E+01	.1000E+01	.1000E+01	.1000E+01
50	f_y	0.	0.	0.	0.	0.	0.	0.	0.
51	$\dfrac{K_2 \cdot f_y - K_3 \cdot f_x}{(22)(50) - (23)(49)}$	$-.3414E+06$	$-.1930E+07$	$-.1150E+07$	$-.1152E+07$	$-.1152E+07$	$-.1152E+07$	$-.1152E+07$	$-.1152E+07$
52	$u_x = 51/25$	.7071E−05	.5491E−06	.3638E−05	.3771E−05	.3792E−05	.3794E−05	.3794E−05	.3794E−05
53	$\dfrac{K_2 \cdot f_x - K_1 \cdot f_y}{(22)(49) - (21)(50)}$	0.	.4142E+06	.9915E+06	.9992E+06	.9999E+06	.1000E+07	.1000E+07	.1000E+07
54	$u_y = 53/25$	0.	$-.1178E−06$	$-.3137E−05$	$-.3271E−05$	$-.3292E−05$	$-.3294E−05$	$-.3294E−05$	$-.3294E−05$
55	$U_1 = (52)(9) + (54)(8)$	.5000E−05	.3049E−06	.3542E−06	.3535E−06	.3536E−06	.3536E−06	.3536E−06	.3536E−06
56	$u_2 = u_y (54)$	0.	$-.1178E−06$	$-.3137E−05$	$-.3271E−05$	$-.3292E−05$	$-.3294E−05$	$-.3294E−05$	$-.3294E−05$
57	$u_3 = (52)(9) + (54)(8)$	$-.5000E−05$	$-.4716E−06$	$-.4790E−05$	$-.4980E−05$	$-.5009E−05$	$-.5012E−05$	$-.5012E−05$	$-.5012E−05$
58	$t_1 = \dfrac{x_1 u_1 E}{l_1} (16)(55)$	.7071E+00	.7088E+00	.7515E+00	.7534E+00	.7537E+00	.7537E+00	.7537E+00	.7537E+00
59	$t_2 = \dfrac{x_2 u_2 E}{l_2} (17)(56)$	0.	$-.2357E−02$	$-.6273E−01$	$-.6543E−01$	$-.6583E−01$	$-.6588E−01$	$-.6589E−01$	$-.6589E−01$
60	$t_3 = \dfrac{x_3 u_3 E}{l_3} (18)(57)$	$-.7071E+00$	$-.7054E+00$	$-.6628E+00$	$-.6608E+00$	$-.6606E+00$	$-.6605E+00$	$-.6605E+00$	$-.6605E+00$
61	$Q_{11} = \dfrac{T_1 t_1 l_1}{E}$	.5000E+00	.7058E+00	.7482E+00	.7501E+00	.7504E+00	.7504E+00	.7504E+00	.7504E+00
62	$Q_{21} = \dfrac{T_2 t_2 l_2}{E}$	0.	$-.1003E−04$	$-.2746E−03$	$-.2843E−03$	$-.2858E−03$	$-.2860E−03$	$-.2860E−03$	$-.2860E−03$

		CYCLE 1	CYCLE 2	CYCLE 3	CYCLE 4	CYCLE 5	CYCLE 6	CYCLE 7	CYCLE 8
63	$Q_{31} = \dfrac{T_3 f_3 l_3}{E}$	.2071E+00	.3002E−02	.2901E−02	.2871E−02	.2868E−02	.2867E−02	.2867E−02	.2867E−02
64	Passive element	.2000E+01	.2000E+01	.2000E+01	.2000E+01	.2000E+01	.2000E+01	.2000E+01	.2000E+01
65	$C_1^* = \sum \dfrac{Q_{i1}}{x_i}(P)$	0.	−.1163E−03	−.2740E−02	−.2843E−02	−.2858E−02	−.2860E−02	−.2860E−02	−.2860E−02
66	$(Q_{11} \cdot \rho \cdot l_1)^{1/2}$	.1880E+01	.2234E+01	.2300E+01	.2303E+01	.2303E+01	.2304E+01	.2304E+01	.2304E+01
67	$(Q_{21} \cdot \rho \cdot l_2)^{1/2}$	0.	0.	0.	0.	0.	0.	0.	0.
68	$(Q_{31} \cdot \rho \cdot l_3)^{1/2}$	.1210E+01	.1457E+00	.1432E+00	.1425E+00	.1424E+00	.1424E+00	.1424E+00	.1424E+00
69	66 + 67 + 68	.3090E+01	.2380E+01	.2443E+01	.2446E+01	.2446E+01	.2446E+01	.2446E+01	.2446E+01
70	$\lambda_1^{1/2} = 69/4 - 65$	.6181E+02	.4748E+02	.4633E+02	.4628E+02	.4627E+02	.4627E+02	.4627E+02	.4627E+02
71	$\lambda_1 = (70)(70)$	.3820E+04	.2255E+04	.2146E+04	.2142E+04	.2141E+04	.2141E+04	.2141E+04	.2141E+04
72	$\lambda_1 \cdot Q_{11}/x_1^{*2} \cdot \rho \cdot l_1$	.1351E+01	.1120E+01	.1005E+01	.1001E+01	.10000E+01	.10000E+01	.10000E+01	.10000E+01
73	$\lambda_1 \cdot Q_{21}/x_2^{*2} \cdot \rho \cdot l_2$	0.	−.6079E+00	−.1174E+02	−.1218E+02	−.1224E+02	−.1225E+02	−.1225E+02	−.1225E+02
74	$\lambda_1 \cdot Q_{31}/x_3^{*2} \cdot \rho \cdot l_3$	.5595E+00	.1150E−01	.9164E+00	.9880E+00	.9986E+00	.9998E+00	.1000E+01	.1000E+01
75	$x_1^{k+1} = x_1^* \cdot (72)^{1/2}$	.1644E+02	.1500E+02	.1507E+02	.1507E+02	.1507E+02	.1507E+02	.1507E+02	.1507E+02
76	$x_2^{k+1} = x_2^* \cdot (73)^{1/2}$	0.	0.	0.	0.	0.	0.	0.	0.
77	$x_3^{k+1} = x_3^* \cdot (74)^{1/2}$	.1058E+02	.9783E+00	.9383E+00	.9325E+00	.9319E+00	.9318E+00	.9318E+00	.9318E+00
78	x_1	.1644E+02	.1500E+02	.1507E+02	.1507E+02	.1507E+02	.1507E+02	.1507E+02	.1507E+02
79	$x_2 (x_{min})$	.1000E+00	.1000E+00	.1000E+00	.1000E+00	.1000E+00	.1000E+00	.1000E+00	.1000E+00
80	x_3	.1058E+02	.9783E+00	.9383E+00	.9325E+00	.9319E+00	.9318E+00	.9318E+00	.9318E+00

Problem 1(b)

The design variables are modified by using the recurrence relation of Equation (5.35):

$$x_i^{k+1} = x_i^k \left(\lambda_i \frac{Q_{i1}}{\rho_i l_i x_i^2} \right)_k^{1/r}$$

with the step size parameter $r = 2$. The Lagrange multiplier is determined from Equation (5.49):

$$\lambda_i = \frac{\overline{C}_1 - C_1^*}{\sum_{i=1}^{n} Q_{i1}^2 / \rho_i l_i x_i^3}$$

The arrangement of the rows is the same as for Problem 1(a). The number of iterations required to obtain the minimum weight is nearly equal to that for Problem 1(a).

		CYCLE 1	CYCLE 2	CYCLE 3	CYCLE 4	CYCLE 5	CYCLE 6	CYCLE 7	CYCLE 8
1	x_1	.1000E+01	.1554E+02	.1423E+02	.1509E+02	.1507E+02	.1507E+02	.1507E+02	.1507E+02
2	x_2	.1000E+01	.1000E+00	.1000E+00	.1000E+00	.1000E+00	.1000E+00	.1000E+00	.1000E+00
3	x_3	.1000E+01	.1000E+02	.9539E+00	.9627E+00	.9340E+00	.9322E+00	.9318E+00	.9318E+00
4	$\bar{C}_1$	.5000E−01	.5000E−01	.5000E−01	.5000E−01	.5000E−01	.5000E−01	.5000E−01	.5000E−01
5	E	.1000E+08	.1000E+08	.1000E+08	.1000E+08	.1000E+08	.1000E+08	.1000E+08	.1000E+08
6	F_x	.1000E+06	.1000E+06	.1000E+06	.1000E+06	.1000E+06	.1000E+06	.1000E+06	.1000E+06
7	F_y	.1000E+06	.1000E+06	.1000E+06	.1000E+06	.1000E+06	.1000E+06	.1000E+06	.1000E+06
8	$\cos\theta$	.7071E+00	.7071E+00	.7071E+00	.7071E+00	.7071E+00	.7071E+00	.7071E+00	.7071E+00
9	$\cos\phi$	.7071E+00	.7071E+00	.7071E+00	.7071E+00	.7071E+00	.7071E+00	.7071E+00	.7071E+00
10	$\cos^2\theta$	.5000E+00	.5000E+00	.5000E+00	.5000E+00	.5000E+00	.5000E+00	.5000E+00	.5000E+00
11	$\cos^2\phi$	.5000E+00	.5000E+00	.5000E+00	.5000E+00	.5000E+00	.5000E+00	.5000E+00	.5000E+00
12	$\cos\theta\cdot\cos\phi$	.5000E+00	.5000E+00	.5000E+00	.5000E+00	.5000E+00	.5000E+00	.5000E+00	.5000E+00
13	l_1	.7071E+02	.7071E+02	.7071E+02	.7071E+02	.7071E+02	.7071E+02	.7071E+02	.7071E+02
14	l_2	.5000E+02	.5000E+02	.5000E+02	.5000E+02	.5000E+02	.5000E+02	.5000E+02	.5000E+02
15	l_3	.7071E+02	.7071E+02	.7071E+02	.7071E+02	.7071E+02	.7071E+02	.7071E+02	.7071E+02
16	$x_1 E/l_1$	.1414E+06	.2197E+07	.2012E+07	.2134E+07	.2131E+07	.2132E+07	.2132E+07	.2132E+07
17	$x_2 E/l_2$	.2000E+06	.2000E+05	.2000E+05	.2000E+05	.2000E+05	.2000E+05	.2000E+05	.2000E+05
18	$x_3 E/l_3$	.1414E+06	.1414E+07	.1349E+06	.1361E+06	.1321E+06	.1318E+06	.1318E+06	.1318E+06

19	$\dfrac{x_1 E}{l_1} + \dfrac{x_3 E}{l_3}(16+17)$	.2828E+06	.3612E+07	.2147E+07	.2270E+07	.2263E+07	.2264E+07	.2264E+07	.2264E+07
20	$\dfrac{x_1 E}{l_1} - \dfrac{x_3 E}{l_3}(16-17)$	0.	.7832E+06	.1877E+07	.1998E+07	.1999E+07	.2000E+07	.2000E+07	.2000E+07
21	$K_1 = (11)(19)$	.1414E+06	.1806E+07	.1073E+07	.1135E+07	.1132E+07	.1132E+07	.1132E+07	.1132E+07
22	$K_2 = (12)(20)$	0.	.3916E+06	.9385E+06	.9989E+06	.9994E+06	.1000E+07	.1000E+07	.1000E+07
23	$K_3 = (10)(19)+17$	.3414E+06	.1826E+07	.1093E+07	.1155E+07	.1152E+07	.1152E+07	.1152E+07	.1152E+07
24	$K_2 \cdot F_y - K_3 \cdot F_x$ $(22)(7)-(23)(6)$	$-.3414E+11$	$-.1434E+12$	$-.1549E+11$	$-.1561E+11$	$-.1521E+11$	$-.1518E+11$	$-.1518E+11$	$-.1518E+11$
25	$K_2 \cdot K_2 - K_1 \cdot K_3$ $(22)(22)-(21)(23)$	$-.4828E+11$	$-.3144E+13$	$-.2929E+12$	$-.3132E+12$	$-.3041E+12$	$-.3037E+12$	$-.3036E+12$	$-.3035E+12$
26	$U_x = 24/25$	.7071E+00	.4562E−01	.5289E−01	.4985E−01	.5001E−01	.5000E−01	.5000E−01	.5000E−01
27	$K_2^2 \cdot F_x - K_1 \cdot F_y$ $(22)(6)-(21)(7)$	$-.1414E+11$	$-.1414E+12$	$-.1349E+11$	$-.1361E+11$	$-.1321E+11$	$-.1318E+11$	$-.1318E+11$	$-.1318E+11$
28	$U_y = 27/25$	.2929E+00	.4499E+01	.4606E−01	.4346E−01	.4344E−01	.4341E−01	.4341E−01	.4341E−01
29	$U_1 = (26)(9)+(28)(8)$	.7071E+00	.6407E−01	.6997E−01	.6598E−01	.6608E−01	.6605E−01	.6605E−01	.6605E−01
30	$U_2 = U_y(28)$	.2929E+00	.4499E−01	.4606E−01	.4346E−01	.4344E−01	.4341E−01	.4341E−01	.4341E−01
31	$U_3 = -(26)(9)+(28)(8)$	$-.2929E+00$	$-.4499E−03$	$-.4828E−02$	$-.4515E−02$	$-.4651E−02$	$-.4657E−02$	$-.4659E−02$	$-.4659E−02$
32	$T_1 = \dfrac{x_1 E}{l_1} U_1(16)(29)$	.1000E+06	.1408E+06	.1408E+06	.1408E+06	.1408E+06	.1408E+06	.1408E+06	.1408E+06
33	$T_2 = \dfrac{x_2 E}{l_2} U_2(17)(30)$	.5858E+05	.8997E+03	.9212E+03	.8693E+03	.8687E+03	.8683E+03	.8682E+03	.8682E+03

		CYCLE 1	CYCLE 2	CYCLE 3	CYCLE 4	CYCLE 5	CYCLE 6	CYCLE 7	CYCLE 8
34	$T_3 = \dfrac{x_3 E}{l_3} U_3 (18)(31)$	$-.4142E+05$	$-.6362E+03$	$-.6514E+03$	$-.6147E+03$	$-.6143E+03$	$-.6140E+03$	$-.6139E+03$	$-.6139E+03$
35	$\sigma_1 = \dfrac{T_1}{x_1} = 32/1$	$.1000E+06$	$.9061E+04$	$.9895E+04$	$.9332E+04$	$.9345E+04$	$.9341E+04$	$.9341E+04$	$.9341E+04$
36	$\sigma_2 = \dfrac{T_2}{x_2} = \dfrac{33}{2}$	$.5858E+05$	$.8997E+04$	$.9212E+04$	$.8693E+04$	$.8687E+04$	$.8683E+04$	$.8682E+04$	$.8682E+04$
37	$\sigma_3 = \dfrac{T_3}{x_3} = \dfrac{34}{3}$	$-.4142E+05$	$-.6362E+02$	$-.6829E+03$	$-.6385E+03$	$-.6577E+03$	$-.6586E+03$	$-.6589E+03$	$-.6589E+03$
38	$\Lambda(\sigma_1) = \dfrac{\sigma_1}{\bar\sigma}$	$.4000E+01$	$.3624E+00$	$.3958E+00$	$.3733E+00$	$.3738E+00$	$.3737E+00$	$.3737E+00$	$.3737E+00$
39	$\Lambda(\sigma_2) = \dfrac{\sigma_2}{\bar\sigma}$	$.2343E+01$	$.3599E+00$	$.3685E+00$	$.3477E+00$	$.3475E+00$	$.3473E+00$	$.3473E+00$	$.3473E+00$
40	$\Lambda(\sigma_3) = \dfrac{\sigma_3}{\bar\sigma}$	$-.1657E+01$	$-.2545E-02$	$-.2731E-01$	$-.2554E-01$	$-.2631E-01$	$-.2635E-01$	$-.2635E-01$	$-.2636E-01$
41	$\Lambda(U_x) = \dfrac{U_x}{\bar c_1} \times \dfrac{26}{4}$	$.1414E+02$	$.9125E+00$	$.1058E+01$	$.9970E+00$	$.1000E+01$	$.1000E+01$	$.1000E+01$	$.1000E+01$
42	$\Lambda(U_y) = \dfrac{U_y}{\bar c_1} \dfrac{28}{4}$	$.5858E+01$	$.8997E+00$	$.9212E+00$	$.8693E+00$	$.8687E+00$	$.8683E+00$	$.8682E+00$	$.8682E+00$
43	$\Lambda(\max)(38-42)$	$.1414E+02$	$.9125E+00$	$.1058E+01$	$.9970E+00$	$.1000E+01$	$.1000E+01$	$.1000E+01$	$.1000E+01$
44	$x_1^* = \Lambda \cdot x_1(43)(1)$	$.1414E+02$	$.1418E+02$	$.1505E+02$	$.1504E+02$	$.1507E+02$	$.1507E+02$	$.1507E+02$	$.1507E+02$
45	$x_2^* = \Lambda \cdot x_2(43)(2)$	$.1414E+02$	$.9125E-01$	$.1058E+00$	$.9970E-01$	$.1000E+00$	$.1000E+00$	$.1000E+00$	$.1000E+00$
46	$x_3^* = \Lambda \cdot x_3(43)(3)$	$.1414E+02$	$.9124E+01$	$.1009E+01$	$.9598E+00$	$.9342E+00$	$.9322E+00$	$.9318E+00$	$.9318E+00$
47	$\text{VOLUME} = \sum x_i l_i$	$.2707E+04$	$.1652E+04$	$.1141E+04$	$.1137E+04$	$.1137E+04$	$.1137E+04$	$.1137E+04$	$.1137E+04$

48	$WT = (47)(\rho)$		.2707E+03	.1652E+03	.1141E+03	.1137E+03	.1137E+03	.1137E+03	.1137E+03	.1137E+03
49	f_x	.1000E+01	.1000E+01	.1000E+01	.1000E+01	.1000E+01	.1000E+01	.1000E+01	.1000E+01	.1000E+01
50	f_y	0.	0.	0.	0.	0.	0.	0.	0.
51	$K_2 f_y - K_3 f_x$ $(22)(50) - (23)(49)$	$-.3414E+06$	$-.1826E+07$	$-.1093E+07$	$-.1155E+07$	$-.1152E+07$	$-.1152E+07$	$-.1152E+07$	$-.1152E+07$
52	$U_x = 51/25$	.7071E−05	.5808E−06	.3733E−05	.3688E−05	.3787E−05	.3793E−05	.3794E−05	.3794E−05
53	$K_2 f_x - K_1 f_y$ $(22)(49) - (21)(50)$	0.	.3916E+06	.9385E+06	.9989E+06	.9994E+06	.1000E+07	.1000E+07	.1000E+07
54	$u_y = 53/25$	0.	$-.1246E−06$	$-.3204E−05$	$-.3189E−05$	$-.3287E−05$	$-.3293E−05$	$-.3294E−05$	$-.3294E−05$
55	$u_1 = (52)(9) + (54)(8)$	.5000E−05	.3226E−06	.3740E−06	.3525E−06	.3536E−06	.3536E−06	.3536E−06	.3536E−06
56	$u_2 = U_y (54)$	0.	$-.1246E−06$	$-.3204E−05$	$-.3189E−05$	$-.3287E−05$	$-.3293E−05$	$-.3294E−05$	$-.3294E−05$
57	$u_3 = -(52)(9) + (54)(8)$	$-.5000E−05$	$-.4988E−06$	$-.4905E−05$	$-.4863E−05$	$-.5002E−05$	$-.5011E−05$	$-.5012E−05$	$-.5012E−05$
58	$t_1 = \dfrac{x_1 u_1 E}{l_1}$ (16)(55)	.7071E+00	.7089E+00	.7524E+00	.7522E+00	.7536E+00	.7537E+00	.7537E+00	.7537E+00
59	$t_2 = \dfrac{x_2 U_2 E}{l_2}$ (17)(56)	0.	$-.2491E−02$	$-.6408E−01$	$-.6378E−01$	$-.6573E−01$	$-.6586E−01$	$-.6588E−01$	$-.6589E−01$
60	$t_3 = \dfrac{x_3 U_3 E}{l_3}$ (18)(57)	$-.7071E+00$	$-.7053E+00$	$-.6618E+00$	$-.6620E+00$	$-.6606E+00$	$-.6605E+00$	$-.6605E+00$	$-.6605E+00$
61	$Q_{11} = \dfrac{T_1 t_1 l_1}{E}$	.5000E+00	.7057E+00	.7490E+00	.7489E+00	.7503E+00	.7504E+00	.7504E+00	.7504E+00
62	$Q_{21} = \dfrac{T_2 t_2 l_2}{E}$	0.	$-.1121E−04$	$-.2952E−03$	$-.2772E−03$	$-.2855E−03$	$-.2859E−03$	$-.2860E−03$	$-.2860E−03$
63	$Q_{31} = \dfrac{T_3 t_3 l_3}{E}$	.2071E+00	.3173E−02	.3048E−02	.2877E−02	.2869E−02	.2868E−02	.2867E−02	.2867E−02

	Passive elements	CYCLE 1	CYCLE 2	CYCLE 3	CYCLE 4	CYCLE 5	CYCLE 6	CYCLE 7	CYCLE 8
64		.2000E+01	.2000E+01	.2000E+01	.2000E+01	.2000E+01	.2000E+01	.2000E+01	.2000E+01
65	$C_1^* = \sum \frac{Q_{t_1}}{x_i}(P)$	0.	$-.1228E-03$	$-.2791E-02$	$-.2781E-02$	$-.2854E-02$	$-.2859E-02$	$-.2860E-02$	$-.2860E-02$
66	$Q_{11}^2/\rho \cdot l_1 \cdot x_1^{*3}$	.1250E$-$04	.2471E$-$04	.2328E$-$04	.2330E$-$04	.2326E$-$04	.2325E$-$04	.2325E$-$04	.2325E$-$04
67	$Q_{21}^2/\rho \cdot l_2 \cdot x_2^{*3}$	0.	0.	0.	0.	0.	0.	0.	
68	$Q_{31}^2/\rho \cdot l_3 \cdot x_3^{*3}$	.2145E$-$05	.1874E$-$08	.1279E$-$05	.1324E$-$05	.1428E$-$05	.1436E$-$05	.1437E$-$05	.1437E$-$05
69	66+67+68	.1465E$-$04	.2472E$-$04	.2456E$-$04	.2462E$-$04	.2468E$-$04	.2469E$-$04	.2469E$-$04	.2469E$-$04
70	$\bar{C}_1 - C_1^*$	.5000E$-$01	.5012E$-$01	.5279E$-$01	.5278E$-$01	.5285E$-$01	.5286E$-$01	.5286E$-$01	.5286E$-$01
71	$\lambda_1 = 70/69$	.3414E+04	.2028E+04	.2150E+04	.2144E+04	.2141E+04	.2141E+04	.2141E+04	.2141E+04
72	$\lambda_1 \cdot Q_{11}/x_1^{*2} \cdot \rho \cdot l_1$	.1207E+01	.1007E+01	.1005E+01	.1003E+01	.1000E+01	.1000E+01	.1000E+01	.1000E+01
73	$\lambda_1 \cdot Q_{21}/x_2^{*2} \cdot \rho \cdot l_2$	0.	$-.5460E+00$	$-.1134E+02$	$-.1196E+02$	$-.1222E+02$	$-.1224E+02$	$-.1225E+02$	$-.1225E+02$
74	$\lambda_1 \cdot Q_{31}/x_3^{*2} \cdot \rho \cdot l_3$	.5000E+00	.1093E$-$01	.9101E+00	.9470E+00	.9957E+00	.9993E+00	.9999E+00	.1000E+01
75	$x_1^{k+1} = x_1^* \cdot (72)^{1/2}$	.1554E+02	.1423E+02	.1509E+02	.1507E+02	.1507E+02	.1507E+02	.1507E+02	.1507E+02
76	$x_2^{k+1} = x_2^* \cdot (73)^{1/2}$	0.	0.	0.	0.	0.	0.	0.	
77	$x_3^{k+1} = x_3^* \cdot (74)^{1/2}$	.1000E+02	.9539E+00	.9627E+00	.9340E+00	.9322E+00	.9318E+00	.9318E+00	.9318E+00
78	x_1	.1554E+02	.1423E+02	.1509E+02	.1507E+02	.1507E+02	.1507E+02	.1507E+02	.1507E+02
79	$x_2(x_{\min})$	.1000E+00	.1000E+00	.1000E+00	.1000E+00	.1000E+00	.1000E+00	.1000E+00	.1000E+00
80	x_3	.1000E+02	.9539E+00	.9627E+00	.9340E+00	.9322E+00	.9318E+00	.9318E+00	.9318E+00

Problem 1(c)

The design variables are modified by using the recurrence relation of Equation (5.37):

$$x_i^{k+1} = x_i^k \left(1 + \frac{1}{r}\left(\lambda_i \frac{Q_{i1}}{x_i^2 \rho_i l_i} - 1\right)\right)_k$$

with the step size parameter $r = 2$. The Lagrange multiplier is determined from Equation (5.49):

$$\lambda_1 = \frac{\overline{C}_1 - C_1^*}{\sum_{i=1}^{n_1} \frac{Q_{i1}^2}{\rho_i l_i x_i^3}}$$

The convergence is slower than for Problems 1(a) or 1(b). After eight iterations the minimum weight is 113.7 lbs, but comparing the areas of the members with problem 1(a), it is seen that the solution has not as yet converged. An additional 2 or 3 iterations are needed.

		CYCLE 1	CYCLE 2	CYCLE 3	CYCLE 4	CYCLE 5	CYCLE 6	CYCLE 7	CYCLE 8
1	x_1	.1000E+01	.1561E+02	.1603E+02	.1554E+02	.1441E+02	.1456E+02	.1485E+02	.1503E+02
2	x_2	.1000E+01	.7071E+01	.2849E+01	.5800E+00	.1000E+00	.1000E+00	.1000E+00	.1000E+00
3	x_3	.1000E+01	.1061E+02	.7473E+01	.4973E+01	.2858E+01	.1600E+01	.1101E+01	.9620E+00
4	$\bar{C}_1$	.5000E−01	.5000E−01	.5000E−01	.5000E−01	.5000E−01	.5000E−01	.5000E−01	.5000E−01
5	E	.1000E+08	.1000E+08	.1000E+08	.1000E+08	.1000E+08	.1000E+08	.1000E+08	.1000E+08
6	F_x	.1000E+06	.1000E+06	.1000E+06	.1000E+06	.1000E+06	.1000E+06	.1000E+06	.1000E+06
7	F_y	.1000E+06	.1000E+06	.1000E+06	.1000E+06	.1000E+06	.1000E+06	.1000E+06	.1000E+06
8	$\cos\theta$	.7071E+00	.7071E+00	.7071E+00	.7071E+00	.7071E+00	.7071E+00	.7071E+00	.7071E+00
9	$\cos\phi$	.7071E+00	.7071E+00	.7071E+00	.7071E+00	.7071E+00	.7071E+00	.7071E+00	.7071E+00
10	$\cos^2\theta$	.5000E+00	.5000E+00	.5000E+00	.5000E+00	.5000E+00	.5000E+00	.5000E+00	.5000E+00
11	$\cos^2\phi$	.5000E+00	.5000E+00	.5000E+00	.5000E+00	.5000E+00	.5000E+00	.5000E+00	.5000E+00
12	$\cos\theta\cdot\cos\phi$	.5000E+00	.5000E+00	.5000E+00	.5000E+00	.5000E+00	.5000E+00	.5000E+00	.5000E+00
13	l_1	.7071E+02	.7071E+02	.7071E+02	.7071E+02	.7071E+02	.7071E+02	.7071E+02	.7071E+02
14	l_2	.5000E+02	.5000E+02	.5000E+02	.5000E+02	.5000E+02	.5000E+02	.5000E+02	.5000E+02
15	l_3	.7071E+02	.7071E+02	.7071E+02	.7071E+02	.7071E+02	.7071E+02	.7071E+02	.7071E+02
16	$x_1 E/l_1$	.1414E+06	.2207E+07	.2267E+07	.2198E+07	.2038E+07	.2059E+07	.2101E+07	.2126E+07
17	$x_2 E/l_2$	.2000E+06	.1414E+07	.5698E+06	.1160E+06	.2000E+05	.2000E+05	.2000E+05	.2000E+05
18	$x_3 E/l_3$	.1414E+06	.1500E+07	.1057E+07	.7033E+06	.4041E+06	.2263E+06	.1557E+06	.1360E+06

19	$\dfrac{x_1 E}{l_1} + \dfrac{x_3 E}{l_3}(16+17)$	.2828E+06	.3707E+07	.3324E+07	.2902E+07	.2442E+07	.2286E+07	.2257E+07	.2262E+07
20	$\dfrac{x_1 E}{l_1} - \dfrac{x_3 E}{l_3}(16-17)$	0.	.7071E+06	.1210E+07	.1495E+07	.1634E+07	.1833E+07	.1945E+07	.1990E+07
21	$K_1 = (11)(19)$	.1414E+06	.1854E+07	.1662E+07	.1451E+07	.1221E+07	.1143E+07	.1128E+07	.1131E+07
22	$K_2 = (12)(20)$	0.	.3536E+06	.6050E+06	.7474E+06	.8170E+06	.9165E+06	.9725E+06	.9950E+06
23	$K_3 = (10)(19)+17$	.3414E+06	.3268E+07	.2232E+07	.1567E+07	.1241E+07	.1163E+07	.1148E+07	.1151E+07
24	$K_2 \cdot F_y - K_3 \cdot F_x$ $(22)(7) - (23)(6)$	−.3414E+11	−.2914E+12	−.1627E+12	−.8193E+11	−.4241E+11	−.2463E+11	−.1757E+11	−.1560E+11
25	$K_2 \cdot K_2 - K_1 \cdot K_3$ $(22)(22) - (21)(23)$	−.4828E+11	−.5932E+13	−.3343E+13	−.1714E+13	−.8481E+12	−.4888E+12	−.3498E+12	−.3119E+12
26	$U_x = 24/25$	.7071E+00	.4913E−01	.4866E−01	.4779E+01	.5001E−01	.5038E−01	.5025E−01	.5004E+01
27	$K_2 \cdot F_x - K_1 \cdot F_y$ $(22)(6) - (21)(7)$	−.1414E+11	−.1500E+12	−.1057E+12	−.7033E+11	−.4041E+11	−.2263E+11	−.1557E+11	−.1360E+11
28	$U_y = 27/25$	.2929E+00	.2529E−01	.3162E−01	.4103E−01	.4765E−01	.4629E−01	.4453E−01	.4362E−01
29	$U_1 = (26)(9) + (28)(8)$	.7071E+00	.5262E−01	.5676E−01	.6280E−01	.6906E−01	.6835E−01	.6702E−01	.6623E−01
30	$U_2 = U_y(28)$	.2929E+00	.2529E−01	.3162E−01	.4103E−01	.4765E−01	.4629E−01	.4453E−01	.4362E−01
31	$U_3 = -(26)(9) + (28)(8)$	−.2929E+00	−.1686E−01	−.1205E−01	−.4785E−02	−.1668E−02	−.2893E−02	−.4043E−02	−.4535E−02
32	$T_1 = \dfrac{x_1 E}{l_1} U_1 (16)(29)$	.1000E+06	.1161E+06	.1287E+06	.1381E+06	.1407E+06	.1408E+06	.1408E+06	.1408E+06
33	$T_2 = \dfrac{x_2 E}{l_2} U_2 (17)(30)$	.5858E+05	.3576E+05	.1801E+05	.4759E+04	.9530E+03	.9258E+03	.8906E+03	.8724E+03

		CYCLE 1	CYCLE 2	CYCLE 3	CYCLE 4	CYCLE 5	CYCLE 6	CYCLE 7	CYCLE 8
34	$T_3 = \dfrac{x_3 E}{l_3} U_3 (18)(31)$	$-.4142E+05$	$-.2529E+05$	$-.1274E+05$	$-.3365E+04$	$-.6739E+03$	$-.6546E+03$	$-.6297E+03$	$-.6169E+03$
35	$\sigma_1 = \dfrac{T_1}{x_1} = \dfrac{32}{1}$	$.1000E+06$	$.7441E+04$	$.8028E+04$	$.8882E+04$	$.9766E+04$	$.9667E+04$	$.9478E+04$	$.9366E+04$
36	$\sigma_2 = \dfrac{T_2}{x_2} = \dfrac{33}{2}$	$.5858E+05$	$.5057E+04$	$.6323E+04$	$.8205E+04$	$.9530E+04$	$.9258E+04$	$.8906E+04$	$.8724E+04$
37	$\sigma_3 = \dfrac{T_3}{x_3} = \dfrac{34}{3}$	$-.4142E+05$	$-.2384E+04$	$-.1705E+04$	$-.6766E+03$	$-.2358E+03$	$-.4091E+03$	$-.5718E+03$	$-.6413E+03$
38	$\Lambda(\sigma_1) = \dfrac{\sigma_1}{\sigma}$	$.4000E+01$	$.2977E+00$	$.3211E+00$	$.3553E+00$	$.3907E+00$	$.3867E+00$	$.3791E+00$	$.3746E+00$
39	$\Lambda(\sigma_2) = \dfrac{\sigma_2}{\sigma}$	$.2343E+01$	$.2023E+00$	$.2529E+00$	$.3282E+00$	$.3812E+00$	$.3703E+00$	$.3562E+00$	$.3490E+00$
40	$\Lambda(\sigma_3) = \dfrac{\sigma_3}{\sigma}$	$-.1657E+01$	$-.9536E-01$	$-.6818E-01$	$-.2707E-01$	$-.9433E-02$	$-.1637E-01$	$-.2287E-01$	$-.2565E-01$
41	$\Lambda(U_x) = \dfrac{U_x}{\bar{c}_1} = \dfrac{26}{4}$	$.1414E+02$	$.9825E+00$	$.9732E+00$	$.9558E+00$	$.1000E+01$	$.1008E+01$	$.1005E+01$	$.1001E+01$
42	$\Lambda(U_y) = \dfrac{U_y}{\bar{c}_1} = \dfrac{28}{4}$	$.5858E+01$	$.5057E+00$	$.6323E+00$	$.8205E+00$	$.9530E+00$	$.9258E+00$	$.8906E+00$	$.8724E+00$
43	$\Lambda(\max)(38 - 42)$	$.1414E+02$	$.9825E+00$	$.9732E+00$	$.9558E+00$	$.1000E+01$	$.1008E+01$	$.1005E+01$	$.1001E+01$
44	$x_1^* = \Lambda \cdot x_1 (43)(1)$	$.1414E+02$	$.1533E+02$	$.1560E+02$	$.1486E+02$	$.1441E+02$	$.1467E+02$	$.1493E+02$	$.1504E+02$
45	$x_2^* = \Lambda \cdot x_2 (43)(2)$	$.1414E+02$	$.6948E+01$	$.2773E+01$	$.5544E+00$	$.1000E+00$	$.1008E+00$	$.1005E+00$	$.1001E+00$
46	$x_3^* = \Lambda \cdot c_3 (43)(3)$	$.1414E+02$	$.1042E+02$	$.7273E+01$	$.4754E+01$	$.2858E+01$	$.1612E+01$	$.1107E+01$	$.9626E+00$
47	$\text{VOLUME} = \sum x_i l_i$	$.2707E+04$	$.2169E+04$	$.1756E+04$	$.1414E+04$	$.1226E+04$	$.1157E+04$	$.1139E+04$	$.1137E+04$

48	$WT = (47)(\rho)$		.2707E+03	.2169E+03	.1756E+03	.1414E+03	.1226E+03	.1157E+03	.1139E+03	.1137E+03
49	f_x		.1000E+01	.1000E+01	.1000E+01	.1000E+01	.1000E+01	.1000E+01	.1000E+01	.1000E+01
50	f_y		0.	0.	0.	0.	0.	0.	0.	0.
51	$K_2 \cdot f_y - K_3 \cdot f_x$ $(22)(50) - (23)(49)$		$-.3414E+06$	$-.3268E+07$	$-.2232E+07$	$-.1567E+07$	$-.1241E+07$	$-.1163E+07$	$-.1148E+07$	$-.1151E+07$
52	$U_x = 51/25$		.7071E$-$05	.5509E$-$06	.6676E$-$06	.9139E$-$06	.1463E$-$05	.2379E$-$05	.3283E$-$05	.3691E$-$05
53	$K_2 f_x - K_1 \cdot f_y$ $(22)(49) - (21)(50)$		0.	.3536E+06	.6050E+06	.7474E+06	.8170E+06	.9165E+06	.9725E+06	.9950E+06
54	$U_y = 53/25$		0.	$-.5960E-07$	$-.1810E-06$	$-.4360E-06$	$-.9633E-06$	$-.1875E-05$	$-.2781E-05$	$-.3191E-05$
55	$U_1 = (52)(9) + (54)(8)$		.5000E$-$05	.3474E$-$06	.3441E$-$06	.3379E$-$06	.3536E$-$06	.3562E$-$06	.3553E$-$06	.3538E$-$06
56	$u_2 = U_y(54)$		0.	$-.5960E-07$	$-.1810E-06$	$-.4360E-06$	$-.9633E-06$	$-.1875E-05$	$-.2781E-05$	$-.3191E-05$
57	$u_3 = -(52)(9) + (54)(8)$		$-.5000E-05$	$-.4317E-06$	$-.6000E-06$	$-.9545E-06$	$-.1716E-05$	$-.3008E-05$	$-.4288E-05$	$-.4866E-05$
58	$t_1 = \dfrac{x_1 u_1 E}{l_1}(16)(55)$		.7071E+00	.7667E+00	.7800E+00	.7429E+00	.7207E+00	.7336E+00	.7464E+00	.7522E+00
59	$t_2 = \dfrac{x_2 u_2 E}{l_2}(17)(56)$		0.	$-.8429E-01$	$-.1031E-00$	$-.5057E-01$	$-.1927E-01$	$-.3750E-01$	$-.5561E-01$	$-.6381E-01$
60	$t_3 = \dfrac{x_3 u_3 E}{l_3}(18)(57)$		$-.7071E+00$	$-.6475E+00$	$-.6342E+00$	$-.6713E+00$	$-.6935E+00$	$-.6806E+00$	$-.6678E+00$	$-.6620E+00$
61	$Q_{11} = \dfrac{T_1 t_1 l_1}{E}$		.5000E+00	.6296E+00	.7098E+00	.7252E+00	.7173E+00	.7302E+00	.7431E+00	.7490E+00
62	$Q_{21} = \dfrac{T_2 t_2 l_2}{E}$		0.	$-.1507E-01$	$-.9289E-02$	$-.1203E-02$	$-.9181E-04$	$-.1736E-03$	$-.2476E-03$	$-.2784E-03$
63	$Q_{31} = \dfrac{T_3 t_3 l_3}{E}$		.2071E+00	.1158E+00	.5712E$-$01	.1597E$-$01	.3305E$-$02	.3150E$-$02	.2974E$-$02	.2888E$-$02

	Passive elements	CYCLE 1	CYCLE 2	CYCLE 3	CYCLE 4	CYCLE 5	CYCLE 6	CYCLE 7	CYCLE 8
64									
65	$C_1^* = \sum \frac{Q_{i1}}{x_i}(P)$	$.2000E+01$	$.2000E+01$	$.2000E+01$	$.2000E+01$	$.2000E+01$	$.2000E+01$	$.2000E+01$	$.2000E+01$
		$0.$	$0.$	$0.$	$0.$	$-.9179E-03$	$-.1723E-02$	$-.2464E-02$	$-.2782E-02$
66	$Q_{11}^2/\rho \cdot l_1 \cdot x_1^{*3}$	$.1250E-04$	$.1555E-04$	$.1876E-04$	$.2268E-04$	$.2429E-04$	$.2387E-04$	$.2347E-04$	$.2330E-04$
67	$Q_{21}^2/\rho \cdot l_2 \cdot x_2^{*3}$	$0.$	$.1355E-06$	$.8096E-06$	$.1700E-05$	$0.$	$0.$	$0.$	$0.$
68	$Q_{31}^2/\rho \cdot l_3 \cdot x_3^{*3}$	$.2145E-05$	$.1675E-05$	$.1199E-05$	$.3360E-06$	$.6614E-07$	$.3350E-06$	$.9225E-06$	$.1322E-05$
69	$66+67+68$	$.1465E-04$	$.1736E-04$	$.2077E-04$	$.2471E-04$	$.2436E-04$	$.2421E-04$	$.2440E-04$	$.2462E-04$
70	$\bar{C}_1 - C_1^*$	$.5000E-01$	$.5000E-01$	$.5000E-01$	$.5000E-01$	$.5092E-01$	$.5172E-01$	$.5246E-01$	$.5278E-01$
71	$\lambda_1 = 70/69$	$.3414E+04$	$.2880E+04$	$.2407E+04$	$.2023E+04$	$.2090E+04$	$.2136E+04$	$.2151E+04$	$.2144E+04$
72	$\lambda_1 \cdot Q_{11}/x_1^{*2} \cdot \rho \cdot l_1$	$.1207E+01$	$.1091E+01$	$.9927E+00$	$.9400E+00$	$.1020E+01$	$.1025E+01$	$.1014E+01$	$.1003E+01$
73	$\lambda_1 \cdot Q_{21}/x_2^{*2} \cdot \rho \cdot l_2$	$0.$	$-.1799E+00$	$-.5816E+00$	$-.1584E+01$	$-.3836E+01$	$-.7305E+01$	$-.1055E+02$	$-.1192E+02$
74	$\lambda_1 \cdot Q_{31}/x_3^{*2} \cdot \rho \cdot l_3$	$.5000E+00$	$.4342E+00$	$.3676E+00$	$.2023E+00$	$.1196E+00$	$.3662E+00$	$.7383E+00$	$.9449E+00$
75	$x_1^{k+1} = x_1^*(1+0.5(72-1))$	$.1561E+02$	$.1603E+02$	$.1554E+02$	$.1441E+02$	$.1456E+02$	$.1485E+02$	$.1503E+02$	$.1507E+02$
76	$x_2^{k+1} = x_2^*(1+0.5(73-1))$	$.7071E+01$	$.2849E+01$	$.5800E+00$	$-.1620E+00$	$-.1418E+00$	$-.3177E+00$	$-.4797E+00$	$-.5464E+00$
77	$x_3^{k+1} = x_3^*(1+0.5(74-1))$	$.1061E+02$	$.7473E+01$	$.4973E+01$	$.2858E+01$	$.1600E+01$	$.1101E+01$	$.9620E+00$	$.9361E+00$
78	x_1	$.1561E+02$	$.1603E+02$	$.1554E+02$	$.1441E+02$	$.1456E+02$	$.1485E+02$	$.1503E+02$	$.1507E+02$
79	x_2	$.7071E+01$	$.2849E+01$	$.5800E+00$	$.1000E+00$	$.1000E+00$	$.1000E+00$	$.1000E+00$	$.1000E+00$
80	x_3	$.1061E+02$	$.7473E+01$	$.4973E+01$	$.2858E+01$	$.1600E+01$	$.1101E+01$	$.9620E+00$	$.9361E+00$

Problem 1(d)

The design variables are modified by using the recurrence relation of Equation (5.150):

$$x_i^{k+1} = \frac{x_i^k}{\left(1 - \frac{1}{r}\left(\lambda_1 \frac{Q_{i1}}{\rho_i l_i x_i^2} - 1\right)\right)_k}$$

with the step size parameter $r = 2$. The Lagrange multiplier is determined from Equation (5.49):

$$\lambda_1 = \frac{\bar{C}_1 - C_1^*}{\sum_{i=1}^{n_1} \frac{Q_{i1}^2}{\rho_i l_i x_i^3}}$$

This algorithm is equivalent to the algorithm based on the use of the reciprocal design variable. The convergence for this algorithm is slower than for all the three previous problems. After eight iterations the minimum weight is 123.0 lbs and needs additional iterations for the solution to converge to the minimum.

		CYCLE 1	CYCLE 2	CYCLE 3	CYCLE 4	CYCLE 5	CYCLE 6	CYCLE 7	CYCLE 8
1	x_1	.1000E+01	.1578E+02	.1618E+02	.1618E+02	.1585E+02	.1529E+02	.1467E+02	.1413E+02
2	x_2	.1000E+01	.9428E+01	.5855E+01	.3485E+01	.1944E+01	.9912E+00	.4486E+00	.1739E+00
3	x_3	.1000E+01	.1131E+02	.8733E+01	.6813E+01	.5343E+01	.4183E+01	.3245E+01	.2477E+01
4	$\bar{C}_1 = \bar{C}_2$	.5000E−01	.5000E−01	.5000E−01	.5000E−01	.5000E−01	.5000E−01	.5000E−01	.5000E−01
5	E	.1000E+08	.1000E+08	.1000E+08	.1000E+08	.1000E+08	.1000E+08	.1000E+08	.1000E+08
6	F_x	.1000E+06	.1000E+06	.1000E+06	.1000E+06	.1000E+06	.1000E+06	.1000E+06	.1000E+06
7	F_y	.1000E+06	.1000E+06	.1000E+06	.1000E+06	.1000E+06	.1000E+06	.1000E+06	.1000E+06
8	$\cos\theta$	.7071E+00	.7071E+00	.7071E+00	.7071E+00	.7071E+00	.7071E+00	.7071E+00	.7071E+00
9	$\cos\phi$	.7071E+00	.7071E+00	.7071E+00	.7071E+00	.7071E+00	.7071E+00	.7071E+00	.7071E+00
10	$\cos^2\theta$	.5000E+00	.5000E+00	.5000E+00	.5000E+00	.5000E+00	.5000E+00	.5000E+00	.5000E+00
11	$\cos^2\phi$	.5000E+00	.5000E+00	.5000E+00	.5000E+00	.5000E+00	.5000E+00	.5000E+00	.5000E+00
12	$\cos\theta\cdot\cos\phi$	.5000E+00	.5000E+00	.5000E+00	.5000E+00	.5000E+00	.5000E+00	.5000E+00	.5000E+00
13	l_1	.7071E+02	.7071E+02	.7071E+02	.7071E+02	.7071E+02	.7071E+02	.7071E+02	.7071E+02
14	l_2	.5000E+02	.5000E+02	.5000E+02	.5000E+02	.5000E+02	.5000E+02	.5000E+02	.5000E+02
15	l_3	.7071E+02	.7071E+02	.7071E+02	.7071E+02	.7071E+02	.7071E+02	.7071E+02	.7071E+02
16	$x_1 E/l_1$	.1414E+06	.2231E+07	.2289E+07	.2288E+07	.2241E+07	.2163E+07	.2074E+07	.1998E+07
17	$x_2 E/l_2$	.2000E+06	.1886E+07	.1171E+07	.6971E+06	.3889E+06	.1982E+06	.8971E+05	.3478E+05
18	$x_3 E/l_3$	.1414E+06	.1600E+07	.1235E+07	.9634E+06	.7556E+06	.5915E+06	.4589E+06	.3504E+06

19	$\dfrac{x_1 E}{l_1} + \dfrac{x_3 E}{l_3}(16+17)$	.2828E+06	.3831E+07	.3524E+07	.3252E+07	.2997E+07	.2754E+07	.2533E+07	.2349E+07
20	$\dfrac{x_1 E}{l_1} - \dfrac{x_3 E}{l_3}(16-17)$	0.	.6310E+06	.1054E+07	.1325E+07	.1486E+07	.1571E+07	.1615E+07	.1648E+07
21	$K_1 = (11)(19)$	.1414E+06	.1916E+07	.1762E+07	.1626E+07	.1498E+07	.1377E+07	.1267E+07	.1174E+07
22	$K_2 = (12)(20)$	0.	.3155E+06	.5268E+06	.6623E+06	.7428E+06	.7857E+06	.8077E+06	.8239E+06
23	$K_3 = (10)(19)+17$	.3414E+06	.3801E+07	.2933E+07	.2323E+07	.1887E+07	.1575E+07	.1356E+07	.1209E+07
24	$\dfrac{K_2 \cdot F_y - k_3 \cdot F_x}{(22)(7)-(23)(6)}$	$-$.3414E+11	$-$.3486E+12	$-$.2406E+12	$-$.1661E+12	$-$.1144E+12	$-$.7898E+11	$-$.5486E+11	$-$.3852E+11
25	$\dfrac{K_2 \cdot K_2 - K_1 \cdot K_3}{(22)(22)-(21)(23)}$	$-$.4828E+11	$-$.7182E+13	$-$.4890E+13	$-$.3338E+13	$-$.2276E+13	$-$.1552E+13	$-$.1065E+13	$-$.7409E+12
26	$U_x = 24/25$	.7071E+00	.4854E$-$01	.4920E$-$01	.4975E$-$01	.5028E$-$01	.5087E$-$01	.5149E$-$01	.5198E$-$01
27	$\dfrac{K_2 \cdot F_x - K_1 \cdot F_y}{(22)(6)-(21)(7)}$	$-$.1414E+11	$-$.1600E+12	$-$.1235E+12	$-$.9634E+11	$-$.7556E+11	$-$.5915E+11	$-$.4589E+11	$-$.3504E+11
28	$U_y = 27/25$	.2929E+00	.2228E$-$01	.2526E$-$01	.2887E$-$01	.3320E$-$01	.3810E$-$01	.4307E$-$01	.4729E$-$01
29	$U_1 = (26)(9)+(28)(8)$	.7071E+00	.5007E$-$01	.5265E$-$01	.5559E$-$01	.5903E$-$01	.6291E$-$01	.6686E$-$01	.7019E$-$01
30	$U_2 = U_y (28)$	.2929E+00	.2228E$-$01	.2526E$-$01	.2887E$-$01	.3320E$-$01	.3810E$-$01	.4307E$-$01	.4729E$-$01
31	$U_3 = -(26)(9)+(28)(8)$	$-$.2929E+00	$-$.1857E$-$01	$-$.1693E$-$01	$-$.1477E$-$01	$-$.1208E$-$01	$-$.9029E$-$02	$-$.5954E$-$02	$-$.3319E$-$02
32	$T_1 = \dfrac{x_1 E}{l_1} U_1 (16)(29)$	.1000E+06	.1117E+06	.1205E+06	.1272E+06	.1323E+06	.1361E+06	.1387E+06	.1403E+06
33	$T_2 = \dfrac{x_2 E}{l_2} U_2 (17)(30)$	.5858E+05	.4201E+05	.2958E+05	.2012E+05	.1291E+05	.7553E+04	.3864E+04	.1645E+04

		CYCLE 1	CYCLE 2	CYCLE 3	CYCLE 4	CYCLE 5	CYCLE 6	CYCLE 7	CYCLE 8
34	$T_3 = \dfrac{x_3 E}{l_3} U_3$ (18)(31)	$-.4142E+05$	$-.2971E+05$	$-.2091E+05$	$-.1423E+05$	$-.9129E+04$	$-.5341E+04$	$-.2732E+04$	$-.1163E+04$
35	$\sigma_1 = \dfrac{T_1}{x_1} = \dfrac{32}{1}$	$.1000E+06$	$.7082E+04$	$.7446E+04$	$.7862E+04$	$.8348E+04$	$.8897E+04$	$.9456E+04$	$.9927E+04$
36	$\sigma_2 = \dfrac{T_2}{x_2} = \dfrac{33}{2}$	$.5858E+05$	$.4456E+04$	$.5051E+04$	$.5773E+04$	$.6640E+04$	$.7620E+04$	$.8614E+04$	$.9457E+04$
37	$\sigma_3 = \dfrac{T_3}{x_3} = \dfrac{34}{3}$	$-.4142E+05$	$-.2626E+04$	$-.2395E+04$	$-.2089E+04$	$-.1709E+04$	$-.1277E+04$	$-.8420E+03$	$-.4694E+03$
38	$\Lambda(\sigma_1) = \dfrac{\sigma_1}{\sigma}$	$.4000E+01$	$.2833E+00$	$.2979E+00$	$.3145E+00$	$.3339E+00$	$.3559E+00$	$.3782E+00$	$.3971E+00$
39	$\Lambda(\sigma_2) = \dfrac{\sigma_2}{\sigma}$	$.2343E+01$	$.1782E+00$	$.2021E+00$	$.2309E+00$	$.2656E+00$	$.3048E+00$	$.3445E+00$	$.3783E+00$
40	$\Lambda(\sigma_3) = \dfrac{\sigma_3}{\sigma}$	$-.1657E+01$	$-.1050E+00$	$-.9579E-01$	$-.8354E-01$	$-.6835E-01$	$-.5108E-01$	$-.3368E-01$	$-.1878E-01$
41	$\Lambda(U_x) = \dfrac{U_x}{c_1} = \dfrac{26}{4}$	$.1414E+02$	$.9707E+00$	$.9841E+00$	$.9950E+00$	$.1006E+01$	$.1017E+01$	$.1030E+01$	$.1040E+01$
42	$\Lambda(U_y) = \dfrac{U_y}{c_1} = \dfrac{28}{4}$	$.5858E+01$	$.4456E+00$	$.5051E+00$	$.5773E+00$	$.6640E+00$	$.7620E+00$	$.8614E+00$	$.9457E+00$
43	$\Lambda(\max)(38-42)$	$.1414E+02$	$.9707E+00$	$.9841E+00$	$.9950E+00$	$.1006E+01$	$.1017E+01$	$.1030E+01$	$.1040E+01$
44	$x_1^* = \Lambda \cdot x_1 (43)(1)$	$.1414E+02$	$.1531E+02$	$.1593E+02$	$.1610E+02$	$.1594E+02$	$.1556E+02$	$.1510E+02$	$.1469E+02$
45	$x_2^* = \Lambda \cdot x_2 (43)(2)$	$.1414E+02$	$.9152E+01$	$.5762E+01$	$.3468E+01$	$.1955E+01$	$.1008E+01$	$.4619E+00$	$.1808E+00$
46	$x_3^* = \Lambda \cdot x_3 (43)(3)$	$.1414E+02$	$.1098E+02$	$.8594E+01$	$.6779E+01$	$.5373E+01$	$.4256E+01$	$.3341E+01$	$.2576E+01$

47	VOLUME $= \sum x_i l_i$	.2707E+04	.2317E+04	.2022E+04	.1791E+04	.1605E+04	.1452E+04	.1327E+04	.1230E+04
48	WT $= (47)(\rho)$	.2707E+03	.2317E+03	.2022E+03	.1791E+03	.1605E+03	.1452E+03	.1327E+03	.1230E+03
49	f_x	.1000E+01	.1000E+01	.1000E+01	.1000E+01	.1000E+01	.1000E+01	.1000E+01	.1000E+01
50	f_y	0.	0.	0.	0.	0.	0.	0.	0.
51	$K_2 \cdot f_y - K_3 \cdot f_x$ $(22)(50) - (23)(49)$	−.3414E+06	−.3801E+07	−.2933E+07	−.2323E+07	−.1887E+07	−.1575E+07	−.1356E+07	−.1209E+07
52	$U_x = 51/25$	.7071E−05	.5293E−06	.5998E−06	.6969E−06	.8292E−06	.1015E−05	.1273E−05	.1632E−05
53	$K_2 \cdot f_x - K_1 \cdot f_y$ $(22)(49) - (21)(50)$	0.	.3155E+06	.5268E+06	.6623E+06	.7428E+06	.7857E+06	.8077E+06	.8239E+06
54	$U_y = 53/25$	0.	−.4393E−07	−.1077E−06	−.1984E−06	−.3264E−06	−.5061E−06	−.7581E−06	−.1112E−05
55	$u_1 = (52)(9) + (54)(8)$	.5000E−05	.3432E−06	.3479E−06	.3518E−06	.3556E−06	.3597E−06	.3641E−06	.3676E−06
56	$u_2 = u_y(54)$	0.	−.4393E−07	−.1077E−06	−.1984E−06	−.3264E−06	−.5061E−06	−.7581E−06	−.1112E−05
57	$u_3 = -(52)(9) + (54)(8)$	−.5000E−05	−.4053E−06	−.5003E−06	−.6324E−06	−.8171E−06	−.1075E−05	−.1436E−05	−.1940E−05
58	$t_1 = \dfrac{x_1 u_1 E}{l_1} (16)(55)$	.7071E+00	.7657E+00	.7963E+00	.8049E+00	.7969E+00	.7781E+00	.7552E+00	.7345E+00
59	$t_2 = \dfrac{x_2 u_2 E}{l_2} (17)(56)$	0.	−.8284E−01	−.1262E+00	−.1383E+00	−.1269E+00	−.1003E+00	−.6801E−01	−.3867E−01
60	$t_3 = \dfrac{x_3 u_3 E}{l_3} (18)(57)$	−.7071E+00	−.6485E+00	−.6179E+00	−.6093E+00	−.6174E+00	−.6362E+00	−.6590E+00	−.6798E+00
61	$Q_{11} = \dfrac{T_1 t_1 l_1}{E}$	.5000E+00	.6049E+00	.6786E+00	.7239E+00	.7454E+00	.7487E+00	.7406E+00	.7284E+00

	CYCLE 1	CYCLE 2	CYCLE 3	CYCLE 4	CYCLE 5	CYCLE 6	CYCLE 7	CYCLE 8
62 $Q_{21} = \dfrac{T_2 t_2 l_2}{E}$	0.	−.1740E−01	−.1866E−01	−.1392E−01	−.8192E−02	−.3789E−02	−.1314E−02	−.3180E−03
63 $Q_{31} = \dfrac{T_3 t_3 l_3}{E}$	.2071E+00	.1362E+00	.9138E+01	.6130E−01	.3985E−01	.2403E−01	.1273E−01	.5590E−02
64 Passive elements	.2000E+01	.2000E+01	.2000E+01	.2000E+01	.2000E+01	.2000E+01	.2000E+01	.2000E+01
65 $C_1^* = \sum \dfrac{Q i_1}{x i}(P)$	0.	0.	0.	0.	0.	0.	0.	0.
66 $Q_{11}^2/\rho \cdot l_1 \cdot x_1^{*3}$	.1250E−04	.1441E−04	.1612E−04	.1777E−04	.1941E−04	.2104E−04	.2251E−04	.2368E−04
67 $Q_{21}^2/\rho \cdot l_2 \cdot x_2^{*3}$	0.	.7900E−07	.3639E−06	.9286E−06	.1795E−05	.2800E−05	.3503E−05	.3423E−05
68 $Q_{31}^2/\rho \cdot l_3 \cdot x_3^{*3}$	.2145E−05	.1981E−05	.1860E−05	.1706E−05	.1448E−05	.1059E−05	.6146E−06	.2586E−06
69 66+67+68	.1465E−04	.1647E−04	.1834E−04	.2040E−04	.2266E−04	.2490E−04	.2663E−04	.2736E−04
70 $\bar{C}_1 - C_1^*$	.5000E−01	.5000E−01	.5000E−01	.5000E−01	.5000E−01	.5000E−01	.5000E−01	.5000E−01
71 $\lambda_1 = 70/69$	.3414E+04	.3136E+04	.2726E+04	.2451E+04	.2207E+04	.2008E+04	.1878E+04	.1828E+04
72 $\lambda_1 \cdot Q_{11}/x_1^{*2} \cdot \rho \cdot l_1$	.1207E+01	.1108E+01	.1031E+01	.9683E+00	.9160E+00	.8781E+00	.8620E+00	.8726E+00
73 $\lambda_1 \cdot Q_{21}/x_2^{*2} \cdot \rho \cdot l_2$	0.	−.1262E+00	−.3063E+00	−.5672E+00	−.9456E+00	−.1497E+01	−.2312E+01	−.3557E+01
74 $\lambda_1 \cdot Q_{31}/x_3^{*2} \cdot \rho \cdot l_3$	.5000E+00	.4850E+00	.4769E+00	.4624E+00	.4308E+00	.3768E+00	.3028E+00	.2178E+00
75 $x_1^{*k+1} = x_1^*/(1 - 0.5(72 - 1))$	.1578E+02	.1618E+02	.1618E+02	.1585E+02	.1529E+02	.1467E+02	.1413E+02	.1381E+02
76 $x_2^{*k+1} = x_2^*/(1 - 0.5(73 - 1))$	.9428E+01	.5855E+01	.3485E+01	.1944E+01	.9912E+00	.4486E+00	.1739E+00	.5515E−01

77	$x_3^{k+1} = x_3^k/(1-0.5(74-1))$	.1131E+02	.8733E+01	.6813E+01	.5343E+01	.4183E+01	.3245E+01	.2477E+01	.1852E+01
78	x_1	.1578E+02	.1618E+02	.1618E+02	.1585E+02	.1529E+02	.1467E+02	.1413E+02	.1381E+00
79	x_2	.9428E+01	.5855E+01	.3485E+01	.1944E+01	.9912E+00	.4486E+00	.1739E+00	.1000E+00
80	x_3	.1131E+02	.8733E+01	.6813E+01	.5343E+01	.4183E+01	.3245E+01	.2477E+01	.1852E+01

Problem 2   Multiple Displacement Constraints

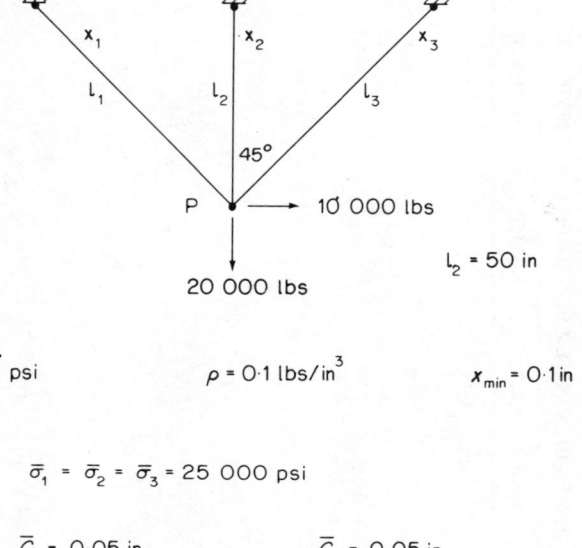

$E = 10^7$ psi $\rho = 0.1$ lbs/in^3 $x_{min} = 0.1$ in

$\bar{\sigma}_1 = \bar{\sigma}_2 = \bar{\sigma}_3 = 25\,000$ psi

$\bar{C}_1 = 0.05$ in $\bar{C}_2 = 0.05$ in

$C_1 = U_x$ $C_2 = U_y$

$C_1 \leq \bar{C}_1$ $C_2 \leq \bar{C}_2$

Fig. 3A   Problem 2—Multiple displacement constraints

This problem is solved by using three different algorithms. All algorithms lead nearly to the same minimum-weight design, but the convergence behaviour is not the same. The two contraints are active throughout the iterative history.

Problem 2(a)

The design variables are modified by using the recurrence relation of Equation (5.63):

$$x_i^{k+1} = x_i^k \left(1 + \frac{1}{r}\left(\sum_{j=1}^{2} \lambda_j \frac{Q_{ij}}{\rho_i l_i x_i^2} - 1\right)\right)_k$$

with the step size parameter $r = 2$. The Lagrange multipliers are determined from Equation (5.78):

$$\begin{bmatrix} A & B \\ B & C \end{bmatrix} \begin{Bmatrix} \lambda_1 \\ \lambda_2 \end{Bmatrix} = \begin{Bmatrix} R_1 \\ R_2 \end{Bmatrix}$$

where

$$A = \sum \frac{Q_{i1}Q_{i1}}{\rho_i l_i x_i^3} \quad \text{(active elements)}$$

$$B = \sum \frac{Q_{i1}Q_{i2}}{\rho_i l_i x_i^3} \quad \text{(active elements)}$$

$$C = \sum \frac{Q_{i2}Q_{i2}}{\rho_i l_i x_i^3} \quad \text{(active elements)}$$

$$R_1 = 2(C_1^* + B_1 - \overline{C}_1) + B_1$$

$$R_2 = 2(C_2^* + B_2 - \overline{C}_2) + B_2$$

$$C_1^* = \sum \frac{Q_{i1}}{x_i} \quad \text{(passive)}$$

$$C_2^* = \sum \frac{Q_{i2}}{x_i} \quad \text{(passive)}$$

$$B_1 = \sum \frac{Q_{i1}}{x_i} \quad \text{(active)}$$

$$B_2 = \sum \frac{Q_{i2}}{x_i} \quad \text{(active)}$$

The Lagrange multipliers are given by

$$\lambda_1 = \frac{BR_2 - CR_1}{B^2 - AB}$$

$$\lambda_2 = \frac{BR_1 - AR_2}{B^2 - AB}$$

where

$$Q_{i1} = \frac{T_i t_i^1 l_i}{E} \quad \text{and} \quad Q_{i2} = \frac{T_i t_i^2 l_i}{E}$$

t_i^1 and t_i^2 are the forces in the bars due to a unit load applied at node P in the x and y direction respectively.

The weight of the structure after eight iterations was 151.1 lbs. One more iteration was needed to reach the optimum design. The optimum weight of the structure was 150.7 lbs with $x_1 = 14.14$ in^2, $x_2 = 10.0$ in^2 and $x_3 = 0.1$ in^2. For the optimum design, $\lambda_1 = 1000$ and $\lambda_2 = 2000$. This design satisfies the relationshp between λ_i, $\overline{C}_i$ and W (min) written for the multiple constraint problem.

		CYCLE 1	CYCLE 2	CYCLE 3	CYCLE 4	CYCLE 5	CYCLE 6	CYCLE 7	CYCLE 8
1	x_1	.1000E+01	.1475E+02	.1359E+02	.1423E+02	.1411E+02	.1414E+02	.1414E+02	.1414E+02
2	x_2	.1000E+01	.8195E+01	.1047E+02	.9738E+01	.1004E+02	.9995E+01	.1000E+02	.1000E+02
3	x_3	.1000E+01	.7675E+01	.4423E+01	.2388E+01	.1218E+01	.6134E+00	.3068E+00	.1534E+00
4	$\bar{C}_1 = \bar{C}_2$	.5000E−01	.5000E−01	.5000E−01	.5000E−01	.5000E−01	.5000E−01	.5000E−01	.5000E−01
5	E	.1000E+08	.1000E+08	.1000E+08	.1000E+08	.1000E+08	.1000E+08	.1000E+08	.1000E+08
6	F_x	.1000E+06	.1000E+06	.1000E+06	.1000E+06	.1000E+06	.1000E+06	.1000E+06	.1000E+06
7	F_y	.2000E+06	.2000E+06	.2000E+06	.2000E+06	.2000E+06	.2000E+06	.2000E+06	.2000E+06
8	$\cos\theta$	.7071E+00	.7071E+00	.7071E+00	.7071E+00	.7071E+00	.7071E+00	.7071E+00	.7071E+00
9	$\cos\phi$	.7071E+00	.7071E+00	.7071E+00	.7071E+00	.7071E+00	.7071E+00	.7071E+00	.7071E+00
10	$\cos^2\theta$	.5000E+00	.5000E+00	.5000E+00	.5000E+00	.5000E+00	.5000E+00	.5000E+00	.5000E+00
11	$\cos^2\phi$	.5000E+00	.5000E+00	.5000E+00	.5000E+00	.5000E+00	.5000E+00	.5000E+00	.5000E+00
12	$\cos\theta.\cos\phi$	.5000E+00	.5000E+00	.5000E+00	.5000E+00	.5000E+00	.5000E+00	.5000E+00	.5000E+00
13	l_1	.7071E+02	.7071E+02	.7071E+02	.7071E+02	.7071E+02	.7071E+02	.7071E+02	.7071E+02
14	l_2	.5000E+02	.5000E+02	.5000E+02	.5000E+02	.5000E+02	.5000E+02	.5000E+02	.5000E+02
15	l_3	.7071E+02	.7071E+02	.7071E+02	.7071E+02	.7071E+02	.7071E+02	.7071E+02	.7071E+02
16	$x_1 E/l_1$	.1414E+06	.2086E+07	.1922E+07	.2012E+07	.1995E+07	.2000E+07	.2000E+07	.2000E+07
17	$x_2 E/l_2$	.2000E+06	.1639E+07	.2095E+07	.1948E+07	.2007E+07	.1999E+07	.2000E+07	.2000E+07
18	$x_3 E/l_3$	.1414E+06	.1085E+07	.6255E+06	.3376E+06	.1723E+06	.8675E+05	.4339E+05	.2170E+05

19 $\dfrac{x_1 E}{l_1} + \dfrac{x_3 E}{l_3}(16+17)$	.2828E+06	.3171E+07	.2547E+07	.2350E+07	.2167E+07	.2087E+07	.2043E+07	.2022E+07	
20 $\dfrac{x_1 E}{l_1} - \dfrac{x_3 E}{l_3}(16-17)$	0.	.1000E+07	.1296E+07	.1674E+07	.1823E+07	.1914E+07	.1957E+07	.1978E+07	
21 $K_1 = (11)(19)$	.1414E+06	.1586E+07	.1274E+07	.1175E+07	.1084E+07	.1044E+07	.1122E+07	.1011E+07	
22 $K_2 = (12)(20)$	0.	.5002E+06	.6481E+06	.8372E+06	.9114E+06	.9568E+06	.9783E+06	.9892E+06	
23 $K_3 = (10)(19)+17$	.3414E+06	.3225E+07	.3368E+07	.3122E+07	.3091E+07	.3043E+07	.3022E+07	.3011E+07	
24 $\dfrac{K_2 \cdot F_y - K_3 \cdot F_x}{(22)(7) - (23)(6)}$	−.3414E+11	−.2224E+12	−.2072E+12	−.1448E+12	−.1268E+12	−.1129E+12	−.1065E+12	−.1033E+12	
25 $\dfrac{K_2 K_2 - K_1 \cdot K_3}{(22)(22) - (21)(23)}$	−.4828E+11	−.4863E+13	−.3870E+13	−.2967E+13	−.2519E+13	−.2260E+13	−.2130E+13	−.2065E+13	
26 $U_x = 24/25$	.7071E+00	.4574E−01	.5355E−01	.4880E−01	.5034E−01	.4996E−01	.5000E−01	.5000E−01	
27 $\dfrac{K_2 \cdot F_x - K_1 \cdot F_y}{(22)(6) - (21)(7)}$	−.2828E+11	−.2671E+12	−.1899E+12	−.1512E+12	−.1256E+12	−.1130E+12	−.1065E+12	−.1033E+12	
28 $U_y = 27/25$	.5858E+00	.5493E−01	.4907E−01	.5097E−01	.4986E−01	.5002E−01	.5000E−01	.5000E−01	
29 $U_1 = (26)(9) + (28)(8)$	.9142E+00	.7118E−01	.7256E−01	.7055E−01	.7085E−01	.7070E−01	.7071E−01	.7071E−01	
30 $U_2 = U_y(28)$	.5858E+00	.5493E−01	.4907E−01	.5097E−01	.4986E−01	.5002E−01	.5000E−01	.5000E−01	
31 $U_3 = -(26)(9) + (28)(8)$	−.8579E−01	.6497E−02	−.3164E−02	.1536E−02	−.3414E−03	.4499E−04	−.2171E−05	.6921E−06	
32 $T_1 = \dfrac{x_1 E}{l_1} U_1 (16)(29)$	.1293E+06	.1485EE+06	.1394E+06	.1419E+06	.1414E+06	.1414E+06	.1414E+06	.1414E+06	
33 $T_2 = \dfrac{x_2 E}{l_2} U_2 (17)(30)$	.1172E+06	.9003E+05	.1028E+06	.9927E+05	.1001E+06	.9999E+05	.1000E+06	.1000E+06	

		CYCLE 1	CYCLE 2	CYCLE 3	CYCLE 4	CYCLE 5	CYCLE 6	CYCLE 7	CYCLE 8
34	$T_3 = \dfrac{x_3 E}{l_3} U_3 (18)(31)$	$-.1213E+05$	$.7052E+04$	$-.1979E+04$	$.5187E+03$	$-.5882E+02$	$.3902E+01$	$-.9422E-01$	$.1502E-01$
35	$\sigma_1 = \dfrac{T_1}{x_1} = \dfrac{32}{1}$	$.1293E+06$	$.1007E+05$	$.1026E+05$	$.9977E+04$	$.1002E+05$	$.9998E+04$	$.1000E+05$	$.1000E+05$
36	$\sigma_2 = \dfrac{T_2}{x_2} = \dfrac{33}{2}$	$.1172E+06$	$.1099E+05$	$.9814E+04$	$.1019E+05$	$.9972E+04$	$.1000E+05$	$.1000E+05$	$.1000E+05$
37	$\sigma_3 = \dfrac{T_3}{x_3} = \dfrac{34}{3}$	$-.1213E+05$	$.9188E+03$	$-.4474E+03$	$.2172E+03$	$-.4828E+02$	$.6362E+01$	$-.3071E+00$	$.9788E-01$
38	$\Lambda(\sigma_1) = \dfrac{\sigma_1}{\bar{\sigma}}$	$.5172E+01$	$.4027E+00$	$.4105E+00$	$.3991E+00$	$.4008E+00$	$.3999E+00$	$.4000E+00$	$.4000E+00$
39	$\Lambda(\sigma_2) = \dfrac{\sigma_2}{\bar{\sigma}}$	$.4686E+01$	$.4394E+00$	$.3926E+00$	$.4078E+00$	$.3989E+00$	$.4002E+00$	$.4000E+00$	$.4000E+00$
40	$\Lambda(\sigma_3) = \dfrac{\sigma_3}{\bar{\sigma}}$	$-.4853E+00$	$.3675E-01$	$-.1790E-01$	$.8690E-02$	$-.1931E-02$	$.2545E-03$	$-.1228E-04$	$.3915E-05$
41	$\Lambda(U_x) = \dfrac{U_x}{\bar{c}_1} = \dfrac{26}{4}$	$.1414E+02$	$.9148E+00$	$.1071E+01$	$.9760E+00$	$.1007E+01$	$.9992E+00$	$.1000E+01$	$.1000E+01$
42	$\Lambda(U_y) = \dfrac{U_y}{\bar{c}_1} = \dfrac{28}{4}$	$.1172E+02$	$.1099E+01$	$.9814E+00$	$.1019E+01$	$.9972E+00$	$.1000E+01$	$.1000E+01$	$.1000E+01$
43	$\Lambda(\max)(38-42)$	$.1414E+02$	$.1099E+01$	$.1071E+01$	$.1019E+01$	$.1007E+01$	$.1000E+01$	$.1000E+01$	$.1000E+01$
44	$x_1^* = \Lambda \cdot x_1 (43)(1)$	$.1414E+02$	$.1620E+02$	$.1455E+02$	$.1450E+02$	$.1420E+02$	$.1415E+02$	$.1414E+02$	$.1414E+02$
45	$x_2^* = \Lambda \cdot x_2 (43)(2)$	$.1414E+02$	$.9003E+01$	$.1122E+02$	$.9927E+01$	$.1011E+02$	$.9999E+02$	$.1000E+02$	$.1000E+02$
46	$x_3^* = \Lambda \cdot x_3 (43)(3)$	$.1414E+02$	$.8431E+01$	$.4737E+01$	$.2434E+01$	$.1227E+01$	$.6137E+00$	$.3068E+00$	$.1534E+00$

47	VOLUME $= \sum x_i l_i$		.2707E+04	.2192E+04	.1925E+04	.1694E+04	.1596E+04	.1544E+04	.1522E+04	.1511E+04
48	WT $= (47)(\rho)$		.2707E+03	.2192E+03	.1925E+03	.1694E+03	.1596E+03	.1544E+03	.1522E+03	.1511E+03
49	f_x^1		.1000E+01	.1000E+01	.1000E+01	.1000E+01	.1000E+01	.1000E+01	.1000E+01	.1000E+01
50	f_y^1		0.	0.	0.	0.	0.	0.	0.	0.
51	$K_2 f_x^1 - K_3 f_x^1$ $(22)(50) - (23)(49)$		$-.3414E+06$	$-.3225E+07$	$-.3368E+07$	$-.3122E+07$	$-.3091E+07$	$-.3043E+07$	$-.3022E+07$	$-.3011E+07$
52	$U_x^1 = 51/25$		.7071E$-$05	.6631E$-$06	.8704E$-$06	.1052E$-$05	.1227E$-$05	.1347E$-$05	.1419E$-$05	.1458E$-$05
53	$K_2 f_x^1 - K_1 f_y^1$ $(22)(49) - (21)(50)$		0.	.5002E+06	.6481E+06	.8372E+06	.9114E+06	.9568E+06	.9783E+06	.9892E+06
54	$u_y^1 = 53/25$		0.	$-.1029E-06$	$-.1675E-06$	$-.2821E-06$	$-.3618E-06$	$-.4235E-06$	$-.4592E-06$	$-.4790E-06$
55	$u_1^1 = (52)(9) + (54)(8)$		.5000E$-$05	.3962E$-$06	.4970E$-$06	.5446E$-$06	.6118E$-$06	.6527E$-$06	.6783E$-$06	.6922E$-$06
56	$u_2^1 = U_y (54)$		0.	$-.1029E-06$	$-.1675E-06$	$-.2821E-06$	$-.3618E-06$	$-.4235E-06$	$-.4592E-06$	$-.4790E-06$
57	$u_3^1 = -(52)(9) + (54)(8)$		$-.5000E-05$	$-.5416E-06$	$-.7339E-06$	$-.9435E-06$	$-.1123E-05$	$-.1252E-05$	$-.1328E-05$	$-.1370E-05$
58	$t_1^1 = \dfrac{x_1 u_1^1 E}{l_1} (16)(55)$		.7071E+00	.8263E+00	.9552E+00	.1096E+01	.1221E+01	.1306E+01	.1357E+01	.1385E+01
59	$t_2^1 = \dfrac{x_2 U_2^1 E}{l_2} (17)(56)$		0.	$-.1686E+00$	$-.3508E+00$	$-.5495E+00$	$-.7262E+00$	$-.8465E+00$	$-.9185E+00$	$-.9580E+00$
60	$t_3^1 = \dfrac{x_3 U_3^1 E}{l_3} (18)(57)$		$-.7071E+00$	$-.5879E+00$	$-.4590E+00$	$-.3186E+00$	$-.1936E+00$	$-.1086E+00$	$-.5762E-01$	$-.2972E-01$
61	$Q_{11} = \dfrac{T_1 t_1^1 l_1}{E}$		.6465E+00	.8675E+00	.9418E+00	.1100E+01	.1220E+01	.1306E+01	.1357E+01	.1385E+01

	CYCLE 1	CYCLE 2	CYCLE 3	CYCLE 4	CYCLE 5	CYCLE 6	CYCLE 7	CYCLE 8
62 $Q_{21} = \dfrac{T_2 l_2^1 l_2}{E}$	0.	.7589E−01	−.1803E+00	−.2727E+00	−.3634E+00	−.4232E+00	−.4593E+00	−.4790E+00
63 $Q_{31} = \dfrac{T_3 l_3^1 l_3}{E}$	.6066E−01	−.2932E−01	.6424E−02	−.1168E−02	.8051E−04	−.2996E−05	.3839E−07	−.3156E−06
64 Passive elements	0.	0.	0.	0.	0.	0.	0.	0.
65 $C_1^* = \sum \dfrac{Q\lambda_1}{x_i}(P)$	0.	0.	0.	0.	0.	0.	0.	0.
66 f_x^2	0.	0.	0.	0.	0.	0.	0.	0.
67 f_y^2	.1000E+01	.1000E+01	.1000E+01	.1000E+01	.1000E+01	.1000E+01	.1000E+01	.1000E+01
68 $\dfrac{K_2 f_y^2 - K_3 \cdot f_x^2}{(22)(67)-(23)(66)}$	0.	.5002E+06	.6481E+06	.8372E+06	.9114E+06	.9568E+06	.9783E+06	.9892E+06
69 $U_x^2 = 68/25$	0.	−.1029E−06	−.1675E−06	−.2821E−06	−.3618E−06	−.4235E−06	−.4592E−06	−.4790E−06
70 $\dfrac{K_2 f_x^2 - K_1 \cdot f_y^2}{(22)(66)-(21)(67)}$	−.1414E+06	−.1586E+07	−.1274E+07	−.1175E+07	−.1084E+07	−.1044E+07	−.1022E+07	−.1011E+07
71 $U_y^2 = 70/25$	.2929E−05	.3261E−06	.3291E−06	.3959E−06	.4302E−06	.4618E−06	.4796E−06	.4895E−06
72 $U_1^2 = (69)(9)+(71)(8)$	.2071E−05	.1578E−06	.1143E−06	.8046E−07	.4836E−07	.2715E−07	.1440E−07	.7429E−06
73 $U_2^2 = U_y^2$	.2929E−05	.3261E−06	.3291E−06	.3959E−06	.4302E−06	.4618E−06	.4796E−06	.4895E−06
74 $U_3^2 = -(69)(9)+(71)(8)$	.2071E−05	.3033E−06	.3511E−06	.4795E−06	.5600E−06	.6260E−06	.6639E−06	.6848E−06
75 $t_1^2 = \dfrac{x_1 U_2^2 E}{l_1}(16)(72)$	.2929E+00	.3292E+00	.2196E+00	.1619E+00	.9650E−01	.5430E−01	.2881E−01	.1486E−01
76 $t_2^2 = \dfrac{x_2 U_2^2 E}{l_2}(17)(73)$	.5858E+00	.5344E+00	.6894E+00	.7711E+00	.8635E+00	.9232E+00	.9593E+00	.9790E+00

77 $t_3^2 = \frac{x_3 U_3^2 E}{l_3}$ (18)(74)	.2929E+00	.3292E+00	.2196E+00	.1619E+00	.9650E-01	.5430E-01	.2881E-01	.1486E-01
78 $Q_{12} = \frac{T_1 t_1^2 l_1}{E}$	.2678E+00	.3456E+00	.2166E+00	.1625E+00	.9646E-01	.5430E-01	.2881E-01	.1486E-01
79 $Q_{22} = \frac{T_2 t_2^2 l_2}{E}$	.3431E+00	.2406E+00	.3543E+00	.3827E+00	.4321E+00	.4616E+00	.4796E+00	.4895E+00
80 $Q_{32} = T_3 t_3^2 l_3 / E$	-.2513E-01	.1642E-01	-.3073E-02	.5937E-03	-.4013E-04	.1498E-05	-.1919E-07	.1578E-08
81 $C_2^* = \sum \frac{Q \lambda_2}{x^i}$ (P)	0.	0.	0.	0.	0.	0.	0.	0.
82 $B_1 = \sum \frac{Q \lambda_1}{x_i}$ (Active)	.5000E-01	.4164E-01	.5000E-01	.4787E-01	.5000E-01	.4994E-01	.5000E-01	.5000E-01
83 $B_2 = \sum \frac{Q_{i2}}{x_i}$ (Active)	.4142E-01	.5000E-01	.4582E-01	.5000E-01	.4952E-01	.5000E-01	.5000E-01	.5000E-01
84 $R_1 = 2(C_1^* + B_1 - \bar{C}_1) + B_1$	.5000E-01	.2491E-01	.5000E-01	.4361E-01	.5000E-01	.4981E-01	.5000E-01	.5000E-01
85 $R_2 = 2(C_2^* + B_2 - \bar{C}_2) + B_2$	.2426E-01	.5000E-01	.3747E-01	.5000E-01	.4856E-01	.5000E-01	.4999E-01	.5000E-01
86 $A = \sum \frac{Q_{i1} Q_{i1}}{\rho_i l_i x_i^3}$ (Active)	.2108E-04	.2680E-04	.4537E-04	.7128E-04	.9906E-04	.1209E-03	.1342E-03	.1417E-03
87 $B = \sum \frac{Q_{i1} Q_{i2}}{\rho_i l_i x_i^3}$ (Active)	.8579E-05	.4852E-05	.2795E-06	-.1306E-04	-.2463E-04	-.3554E-04	-.4209E-04	-.4586E-04
88 $C = \sum \frac{Q_{i2} Q_{i2}}{\rho_i l_i x_i^3}$ (Active)	.1194E-04	.1990E-04	.1996E-04	.3117E-04	.3665E-04	.4276E-04	.4604E-04	.4793E-04
89 $\begin{array}{c} B \cdot B - A \cdot B \\ (87)(87) - (86)(87) \end{array}$	-.1781E-09	-.5098E-09	-.9054E-09	-.2051E-08	-.3024E-08	-.3907E-08	-.4406E-08	-.4690E-08

		CYCLE 1	CYCLE 2	CYCLE 3	CYCLE 4	CYCLE 5	CYCLE 6	CYCLE 7	CYCLE 8
90	$B \cdot R_2 - C \cdot R_1$ (87)(85) − (81)(84)	−.3890E−06	−.2531E−06	−.9874E−06	−.2013E−05	−.3029E−05	−.3907E−05	−.4406E−05	−.4690E−05
91	$\lambda_1 = 90/89$	.2184E+04	.4964E+03	.1091E+04	.9811E+03	.1002E+04	.9999E+03	.1000E+04	.1000E+04
92	$B \cdot R_1 - A \cdot R_2$ (87)(84) − (86)(85)	−.8253E−07	−.1219E−05	−.1686E−05	−.4134E−05	−.6042E−05	−.7815E−05	−.8813E−05	−.9380E−05
93	$\lambda_2 = 92/89$	.4633E+03	.2392E+04	.1862E+04	.2015E+04	.1998E+04	.2000E+04	.2000E+04	.2000E+04
94	IF (93) < 0 $\lambda_1 = R_1/A$	.2184E+04	.4964E+03	.1091E+04	.9811E+03	.1002E+04	.9999E+03	.1000E+04	.1000E+04
95	IF (93) < 0 $\lambda_2 = 0.0$	.4633E+03	.2392E+04	.1862E+04	.2015E+04	.1998E+04	.2000E+04	.2000E+04	.2000E+04
96	$\lambda_1 \cdot Q_{11} + \lambda_2 Q_{12}$ (94)(61) + (95)(78)	.1536E+04	.1257E+04	.1430E+04	.1406E+04	.1415E+04	.1414E+04	.1414E+04	.1414E+04
97	$\lambda_1 Q_{21} + \lambda_2 Q_{22}$ (94)(62) + (95)(79)	.1590E+03	.5377E+03	.4632E+03	.5036E+03	.4994E+03	.5000E+03	.5000E+03	.5000E+03
98	$\lambda_1 Q_{31} + \lambda_2 Q_{32}$ (94)(63) + (95)(80)	.1208E+03	.2471E+02	.1283E+01	.5006E−01	.4479E−03	.1499E−05	.7505E−09	.1750E−10
99	$96/(l_1 \cdot x_1^{*2} \rho)$	.1086E+01	.6773E+00	.9553E+00	.9455E+00	.9916E+00	.9987E+00	.1000E+01	.1000E+01
100	$97/(l_2 \cdot x_2^{*2} \cdot \rho)$	.1590E+00	.1327E+01	.7362E+00	.1022E+01	.9781E+00	.1000E+01	.9999E+00	.1000E+01
101	$98/(l_3 \cdot x_3^{2} \cdot \rho)$	.8543E−01	.4916E−01	.8088E−02	.1195E−02	.4209E−04	.5631E−06	.1127E−08	.1052E−09
102	$x_1^{k+1} = x_1^{*}(1 + 0.5(99-1))$	.1475E+02	.1359E+02	.1423E+02	.1411E+02	.1414E+02	.1414E+02	.1414E+02	.1414E+02
103	$x_2^{k+1} = x_2^{*}(1 + 0.5(100-1))$	.8195E+01	.1047E+02	.9738E+01	.1004E+02	.9995E+01	.1000E+02	.1000E+02	.1000E+02
104	$x_3^{k+1} = x_3^{*}(1 + 0.5(101-1))$	.7675E+01	.4423E+01	.2388E+01	.1218E+01	.6134E+00	.3068E+00	.1534E+00	.7671E−01

Problem 2(b)

The design variables are modified by using Equation (5.150):

$$x_i^{k+1} = \frac{x_i^k}{\left(1 - \frac{1}{r}\left(\sum_{j=1}^{2} \lambda_j \frac{Q_{ij}}{\rho_i l_i x_i^2} - 1\right)\right)_k}$$

with the step size parameter $r = 2$. The Lagrange multipliers are determined by using the same equations as for Problem 2(a) (see Equations (5.76) and (5.153)). From row 48 it is seen that the weight of the structure after eight iterations is 156.0 lb, and a few more iterations are required in order to reach the minimum weight of 150.7 lb. The algorithm used for this problem is equivalent to the one derived by using the reciprocal design variable. This algorithm is generally slower to converge than the one used for Problem 2(a).

	CYCLE 1	CYCLE 2	CYCLE 3	CYCLE 4	CYCLE 5	CYCLE 6	CYCLE 7	CYCLE 8
1 x_1	.1000E+01	.1478E+02	.1407E+02	.1416E+02	.1414E+02	.1414E+02	.1414E+02	.1414E+02
2 x_2	.1000E+01	.9956E+01	.1012E+02	.9970E+01	.1001E+02	.9999E+01	.1000E+02	.1000E+02
3 x_3	.1000E+01	.9704E+01	.6363E+01	.4279E+01	.2857E+01	.1907E+01	.1271E+01	.8474E+00
4 $\bar{C}_1 = \bar{C}_2$	.5000E−01	.5000E−01	.5000E−01	.5000E−01	.5000E−01	.5000E−01	.5000E−01	.5000E−01
5 E	.1000E+08	.1000E+08	.1000E+08	.1000E+08	.1000E+08	.1000E+08	.1000E+08	.1000E+08
6 F_x	.1000E+06	.1000E+06	.1000E+06	.1000E+06	.1000E+06	.1000E+06	.1000E+06	.1000E+06
7 F_y	.2000E+06	.2000E+06	.2000E+06	.2000E+06	.2000E+06	.2000E+06	.2000E+06	.2000E+06
8 $\cos\theta$	.7071E+00	.7071E+00	.7071E+00	.7071E+00	.7071E+00	.7071E+00	.7071E+00	.7071E+00
9 $\cos\phi$	.7071E+00	.7071E+00	.7071E+00	.7071E+00	.7071E+00	.7071E+00	.7071E+00	.7071E+00
10 $\cos^2\theta$	.5000E+00	.5000E+00	.5000E+00	.5000E+00	.5000E+00	.5000E+00	.5000E+00	.5000E+00
11 $\cos^2\phi$	.5000E+00	.5000E+00	.5000E+00	.5000E+00	.5000E+00	.5000E+00	.5000E+00	.5000E+00
12 $\cos\theta\cdot\cos\phi$	.5000E+00	.5000E+00	.5000E+00	.5000E+00	.5000E+00	.5000E+00	.5000E+00	.5000E+00
13 l_1	.7071E+02	.7071E+02	.7071E+02	.7071E+02	.7071E+02	.7071E+02	.7071E+02	.7071E+02
14 l_2	.5000E+02	.5000E+02	.5000E+02	.5000E+02	.5000E+02	.5000E+02	.5000E+02	.5000E+02
15 l_3	.7071E+02	.7071E+02	.7071E+02	.7071E+02	.7071E+02	.7071E+02	.7071E+02	.7071E+02
16 $x_1 E/l_1$	.1414E+06	.2090E+07	.1990E+07	.2003E+07	.1999E+07	.2000E+07	.2000E+07	.2000E+07
17 $x_2 E/l_2$	.2000E+06	.1991E+07	.2024E+07	.1994E+07	.2001E+07	.2000E+07	.2000E+07	.2000E+07
18 $x_3 E/l_3$	.1414E+06	.1372E+07	.8999E+06	.6052E+06	.4041E+06	.2696E+06	.1798E+06	.1198E+06

19 $\frac{x_1 E}{l_1} + \frac{x_3 E}{l_3}$ (16+17)		.2828E+06	.3462E+07	.2890E+07	.2608E+07	.2403E+07	.2270E+07	.2180E+07	.2120E+07
20 $\frac{x_1 E}{l_1} - \frac{x_3 E}{l_3}$ (16−17)		0.		.7173E+06	.1090E+07	.1398E+07	.1595E+07	.1731E+07	.1820E+07	.1880E+07
21 $K_1 = (11)(19)$		.1414E+06	.1731E+07	.1445E+07	.1304E+07	.1202E+07	.1135E+07	.1090E+07	.1060E+07
22 $K_2 = (12)(20)$		0.	.3586E+06	.5452E+06	.6990E+06	.7976E+06	.8653E+06	.9101E+06	.9401E+06
23 $K_3 = (10)(19)$		.3414E+06	.3722E+07	.3469E+07	.3298E+07	.3203E+07	.3135E+07	.3090E+07	.3060E+07
24 $\frac{K_2 \cdot F_y - K_3 \cdot F_x}{(22)(7) - (23)(6)}$		−.3414E+11	−.3005E+12	−.2378E+12	−.1900E+12	−.1608E+12	−.1404E+12	−.1270E+12	−.1180E+12
25 $\frac{K_2 \cdot K_2 - K_1 \cdot K_3}{(22)(22) - (21)(23)}$		−.4828E+11	−.6315E+13	−.4715E+13	−.3813E+13	−.3213E+13	−.2809E+13	−.2539E+13	−.2360E+13
26 $U_x = 24/25$		.7071E+00	.4759E−01	.5044E−01	.4984E−01	.5004E−01	.4999E−01	.5000E−01	.5000E−01
27 $\frac{K_2 \cdot F_x - K_1 \cdot F_y}{(22)(6) - (21)(7)}$		−.2828E+11	−.3103E+12	−.2345E+12	−.1909E+12	−.1606E+12	−.1405E+12	−.1270E+12	−.1180E+12
28 $U_y = 27/25$		.5858E+00	.4915E−01	.4973E−01	.5008E−01	.4998E−01	.5000E−01	.5000E−01	.5000E−01
29 $U_1 = (26)(9) + (28)(8)$		.9142E+00	.6840E−01	.7083E−01	.7065E−01	.7073E−01	.7071E−01	.7071E−01	.7071E−01
30 $U_2 = U_y (28)$		.5858E+00	.4915E−01	.4973E−01	.5008E−01	.4998E−01	.5000E−01	.5000E−01	.5000E−01
31 $U_3 = -(26)(9) + (28)(8)$		−.8579E−01	.1104E−02	−.4985E−03	.1682E−03	−.4717E−04	.1060E−04	−.1181E−05	.6806E−06
32 $T_1 = \frac{x_1 E}{l_1} U_1 (16)(29)$		.1293E+06	.1429E+06	.1410E+06	.1415E+06	.1414E+06	.1414E+06	.1414E+06	.1414E+06
33 $T_2 = \frac{x_2 E}{l_2} U_2 (17)(30)$		.1172E+06	.9786E+05	.1006E+06	.9986E+05	.1000E+06	.1000E+06	.1000E+06	.1000E+06

	CYCLE 1	CYCLE 2	CYCLE 3	CYCLE 4	CYCLE 5	CYCLE 6	CYCLE 7	CYCLE 8
34 $T_3 = \frac{x_3 E}{l_3} U_3 (18)(31)$	$-.1213E+05$	$.1515E+04$	$-.4486E+03$	$.1018E+03$	$-.1906E+02$	$.2858E+01$	$-.2123E+00$	$.8157E-01$
35 $\sigma_1 = \frac{T_1}{x_1} = \frac{31}{1}$	$.1293E+06$	$.9673E+04$	$.1002E+05$	$.9991E+04$	$.1000E+05$	$.1000E+05$	$.1000E+05$	$.1000E+05$
36 $\sigma_2 = \frac{T_2}{x_2} = \frac{33}{2}$	$.1172E+06$	$.9829E+04$	$.9946E+04$	$.1002E+05$	$.9995E+04$	$.1000E+05$	$.1000E+05$	$.1000E+05$
37 $\sigma_3 = \frac{T_3}{x_3} = \frac{34}{4}$	$-.1213E+05$	$.1561E+03$	$-.7050E+02$	$.2379E+02$	$-.6671E+01$	$.1499E+01$	$-.1670E+00$	$.9625E-01$
38 $\Lambda(\sigma_1) = \frac{\sigma_1}{\bar{\sigma}}$	$.5172E+01$	$.3869E+00$	$.4007E+00$	$.3997E+00$	$.4001E+00$	$.4000E+00$	$.4000E+00$	$.4000E+00$
39 $\Lambda(\sigma_2) = \frac{\sigma_2}{\bar{\sigma}}$	$.4686E+01$	$.3932E+01$	$.3979E+00$	$.4006E+00$	$.3998E+00$	$.4000E+00$	$.4000E+00$	$.4000E+00$
40 $\Lambda(\sigma_3) = \frac{\sigma_3}{\bar{\sigma}}$	$-.4853E+00$	$.6244E-02$	$-.2820E-02$	$.9516E-03$	$-.2669E-03$	$.5995E-04$	$-.6680E-05$	$.3850E-05$
41 $\Lambda(U_x) = \frac{U_x}{C_1} = \frac{26}{4}$	$.1414E+02$	$.9517E+00$	$.1009E+01$	$.9968E+00$	$.1001E+01$	$.9998E+00$	$.1000E+01$	$.1000E+01$
42 $\Lambda(U_y) = \frac{U_y}{C_1} = \frac{28}{4}$	$.1172E+02$	$.9829E+00$	$.9946E+00$	$.1002E+01$	$.9995E+00$	$.1000E+01$	$.1000E+01$	$.1000E+01$
43 $\Lambda(\max)(38-42)$	$.1414E+02$	$.9829E+00$	$.1009E+01$	$.1002E+01$	$.1001E+01$	$.1000E+01$	$.1000E+01$	$.1000E+01$
44 $x_1^* = \Lambda \cdot x_1 (43)(1)$	$.1414E+02$	$.1452E+02$	$.1420E+02$	$.1419E+02$	$.1415E+02$	$.1414E+02$	$.1414E+02$	$.1414E+02$
45 $x_2^* = \Lambda \cdot x_2 (43)(2)$	$.1414E+02$	$.9786E+01$	$.1021E+02$	$.9986E+01$	$.1002E+02$	$.1000E+02$	$.1000E+02$	$.1000E+02$
46 $x_3^* = \Lambda \cdot x_3 (43)(3)$	$.1414E+02$	$.9539E+01$	$.6419E+01$	$.4286E+01$	$.2860E+01$	$.1907E+01$	$.1271E+01$	$.8474E+00$

47 VOLUME $= \sum x_i l_i$		.2707E+04	.2191E+04	.1968E+04	.1805E+04	.1704E+04	.1635E+04	.1590E+04	.1560E+04
48 WT $= (47)(\rho)$		.2707E+03	.2191E+03	.1968E+03	.1805E+03	.1704E+03	.1635E+03	.1590E+03	.1560E+03
49 f_x^1		.1000E+01	.1000E+01	.1000E+01	.1000E+01	.1000E+01	.1000E+01	.1000E+01	.1000E+01
50 f_y^1		0.	0.	0.	0.	0.	0.	0.	0.
51 $\dfrac{K_2 \cdot f_y^1 - K_3 \cdot f_x^1}{(22)(50) - (23)(49)}$		−.3414E+06	−.3722E+07	−.3469E+07	−.3298E+07	−.3203E+07	−.3135E+07	−.3090E+07	−.3060E+07
52 $U_x^1 = 51/25$		.7071E−05	.5895E−06	.7356E−06	.8650E−06	.9969E−06	.1116E−05	.1217E−05	.1297E−05
53 $\dfrac{K_2 \cdot f_x^1 - K_1 \cdot f_y^1}{(22)(49) - (21)(50)}$		0.	.3586E+06	.5452E+06	.6990E+06	.7976E+06	.8653E+06	.9101E+06	.9401E+06
54 $U_y^1 = 53/25$		0.	−.5679E−07	−.1156E−06	−.1833E−06	−.2482E−06	−.3081E−06	−.3584E−06	−.3984E−06
55 $U_1^1 = (52)(9) + (54)(8)$		.5000E−05	.3766E−06	.4384E−06	.4820E−06	.5294E−06	.5713E−06	.6070E−06	.6353E−06
56 $U_2^1 = U_y(54)$		0.	−.5679E−07	−.1156E−06	−.1833E−06	−.2482E−06	−.3081E−06	−.3584E−06	−.3984E−06
57 $U_3^1 = -(52)(9) + (54)(8)$		−.5000E−05	−.4570E−06	−.6019E−06	−.7413E−06	−.8805E−06	−.1007E−05	−.1114E−05	−.1199E−05
58 $t_1^1 = \dfrac{x_1 U_1^1 E}{l_1}(16)(55)$		.7071E+00	.7871E+00	.8726E+00	.9656E+00	.1058E+01	.1143E+01	.1214E+01	.1271E+01
59 $t_2^1 = \dfrac{x_2 U_2^1 E}{l_2}(17)(56)$		0.	−.1131E+00	−.2340E+00	−.3655E+00	−.4968E+00	−.6160E+00	−.7168E+00	−.7968E+00
60 $t_3^1 = \dfrac{x_3 U_3^1 E}{l_3}(18)(57)$		−.7071E+00	−.6271E+00	−.5417E+00	−.4486E+00	−.3558E+00	−.2715E+00	−.2002E+00	−.1437E+00
61 $Q_{11} = \dfrac{T_1 t_1^1 l_1}{E}$		.6465E+00	.7955E+00	.8698E+00	.9663E+00	.1058E+01	.1143E+01	.1214E+01	.1271E+01

		CYCLE 1	CYCLE 2	CYCLE 3	CYCLE 4	CYCLE 5	CYCLE 6	CYCLE 7	CYCLE 8
62	$Q_{21} = \dfrac{T_2 t_2^1 l_2}{E}$	0.	$-.5533E-01$	$-.1177E+00$	$-.1825E+00$	$-.2485E+00$	$-.3080E+00$	$-.3584E+00$	$-.3984E+00$
63	$Q_{31} = \dfrac{T_3 t_3^1 l_3}{E}$	$.6066E-01$	$-.6718E-02$	$.1718E-02$	$-.3230E-03$	$.4796E-04$	$-.5486E-05$	$.3006E-06$	$-.8286E-07$
64	Passive elements								
65	$C_1^* = \sum \dfrac{Qi_1}{x_1}$ (P)	0.	0.	0.	0.	0.	0.	0.	0.
66	f_x^2	0.	0.	0.	0.	0.	0.	0.	0.
67	f_y^2	$.1000E+01$	$.1000E+01$	$.1000E+01$	$.1000E+01$	$.1000E+01$	$.1000E+01$	$.1000E+01$	$.1000E+01$
68	$\dfrac{K_2 f_y^2 - K_3 f_x^2}{(22)(67) - (23)(66)}$	0.	$.3586E+06$	$.5452E+06$	$.6990E+06$	$.7976E+06$	$.8653E+06$	$.9101E+06$	$.9401E+06$
69	$U_2^2 = 68/25$	0.	$-.5679E-07$	$-.1156E-06$	$-.1833E-06$	$-.2482E-06$	$-.3081E-06$	$-.3584E-06$	$-.3984E-06$
70	$\dfrac{K_2 f_x^2 - K_1 f_y^2}{(22)(66) - (21)(67)}$	$-.1414E+06$	$-.1731E+07$	$-.1445E+07$	$-.1304E+07$	$-.1202E+07$	$-.1135E+07$	$-.1090E+07$	$-.1060E+07$
71	$U_y^2 = 70/25$	$.2929E-05$	$.2741E-06$	$.3065E-06$	$.3420E-06$	$.3740E-06$	$.4041E-06$	$.4292E-06$	$.4492E-06$
72	$U_1^2 = (69)(9) + (71)(8)$	$.2071E-05$	$.1537E-06$	$.1349E-06$	$.1122E-06$	$.8893E-07$	$.6788E-07$	$.5006E-07$	$.3592E-07$
73	$U_2^2 = U_y^2$	$.2929E-05$	$.2741E-06$	$.3065E-06$	$.3420E-06$	$.3740E-06$	$.4041E-06$	$.4292E-06$	$.4492E-06$
74	$U_3^2 = (69)(9) + (71)(8)$	$.2071E-05$	$.2340E-06$	$.2985E-06$	$.3715E-06$	$.4400E-06$	$.5035E-06$	$.5569E-06$	$.5994E-06$
75	$t_1^2 = \dfrac{x_1 U_2^2 E}{l_1}$ (16)(72)	$.2929E+00$	$.3211E+00$	$.2686E+00$	$.2248E+00$	$.1778E+00$	$.1358E+00$	$.1001E+00$	$.7183E-01$
76	$t_2^2 = \dfrac{x_2 U_2^2 E}{l_2}$ (17)(73)	$.5858E+00$	$.5458E+00$	$.6202E+00$	$.6821E+00$	$.7486E+00$	$.8080E+00$	$.8584E+00$	$.8984E+00$

#	Formula									
77	$t_3^2 = \dfrac{x_3 U_3^2 E}{l_3}(18)(74)$	.2929E+00	.3211E+00	.2686E+00	.2248E+00	.1778E+00	.1358E+00	.1001E+00	.7183E−01	
78	$Q_{12} = \dfrac{T_1 t_1^2 l_1}{E}$	.2678E+00	.3246E+00	.2677E+00	.2250E+00	.1778E+00	.1358E+00	.1001E+00	.7183E−01	
79	$Q_{22} = \dfrac{T_2 t_2^2 l_2}{E}$	.3431E+00	.2671E+00	.3120E+00	.3405E+00	.3744E+00	.4040E+00	.4292E+00	.4492E+00	
80	$Q_{32} = T_3 t_3^2 l_3 / E$	−.2513E−01	.3440E−02	−.8520E−02	.1619E−03	−.2396E−04	.2743E−05	−.1503E−06	.4143E−07	
81	$C_2^* = \sum \dfrac{Q_{i2}}{x_i} (\text{P})$	0.	0.	0.	0.	0.	0.	0.	0.	
82	$B_1 = \sum \dfrac{Q_{i1}}{x_i} (\text{Active})$	.5000E−01	.4841E−01	.5000E−01	.4976E−01	.5000E−01	.4999E−01	.5000E−01	.5000E−01	
83	$B_2 = \sum \dfrac{Q_{i2}}{x_i} (\text{Active})$	.4142E−01	.5000E−01	.4930E−01	.5000E−01	.4993E−01	.5000E−01	.5000E−01	.5000E−01	
84	$R_1 = 2(C_1^* + B_1 - \overline{C}_1) + B_1$	.5000E−01	.4524E−01	.5000E−01	.4929E−01	.5000E−01	.4996E−01	.5000E−01	.5000E−01	
85	$R_2 = 2(C_2^* + B_2 - \overline{C}_2) + B_2$	.2426E−01	.5000E−01	.4790E−01	.5000E−01	.4980E−01	.5000E−01	.5000E−01	.5000E−01	
86	$A = \sum \dfrac{Q_{i1} Q_{i1}}{P_i l_i x_i^3} (\text{Action})$	.2108E−04	.2987E−04	.4000E−04	.5295E−04	.6820E−04	.8424E−04	.9937E−04	.1125E−03	
87	$B = \sum \dfrac{Q_{i1} Q_{i2}}{P_i l_i x_i^3} (\text{Action})$	.8579E−05	.8761E−05	.4598E−05	−.1715E−05	−.9125E−05	−.1713E−04	−.2469E−04	−.3123E−04	
88	$C = \sum \dfrac{Q_{i2} Q_{i2}}{P_i l_i x_i^3} (\text{Active})$	.1194E−04	.2009E−04	.2186E−04	.2580E−04	.2948E−04	.3357E−04	.3734E−04	.4062E−04	
89	$B \cdot B - A \cdot B$ $(87)(87) - (86)(87)$	−.1781E−09	−.5233E−09	−.8534E−09	−.1363E−08	−.1927E−08	−.2534E−08	−.3101E−08	−.3592E−08	

	CYCLE 1	CYCLE 2	CYCLE 3	CYCLE 4	CYCLE 5	CYCLE 6	CYCLE 7	CYCLE 8
$B \cdot R_2 - C \cdot R_1$ 90 (87)(85) − (88)(84)	−.3890E−06	−.4707E−06	−.8729E−06	−.1357E−05	−.1928E−05	−.2534E−05	−.3101E−05	−.3592E−05
91 $\lambda_1 = 90/89$	.2184E+04	.8995E+03	.1023E+04	.9958E+04	.1001E+04	.9999E+03	.1000E+04	.1000E+04
$B \cdot R_1 - A \cdot R_2$ 92 (87)(84) − (86)(85)	−.8253E−07	−.1097E−05	−.1686E−05	−.2732E−05	−.3853E−05	−.5068E−05	−.6203E−05	−.7185E−05
93 $\lambda_2 = 92/89$	.4633E+03	.2097E+04	.1976E+04	.2004E+04	.1999E+04	.2000E+04	.2000E+04	.2000E+04
IF (93) < 0 94 $\lambda_1 = R_1/A$	.2184E+04	.8995E+03	.1023E+04	.9958E+03	.1001E+04	.9999E+03	.1000E+04	.1000E+04
IF (93) < 0 95 $\lambda_2 = 0.0$	.4633E+03	.2097E+04	.1976E+04	.2004E+04	.1999E+04	.2000E+04	.2000E+04	.2000E+04
$\lambda_1 \cdot Q_{11} + \lambda_2 Q_{12}$ 96 (94)(61) + (95)(78)	.1536E+04	.1396E+04	.1419E+04	.1413E+04	.1414E+04	.1414E+04	.1414E+04	.1414E+04
$\lambda_1 Q_{21} + \lambda_2 Q_{22}$ 97 (94)(62) + (95)(79)	.1590E+03	.5102E+03	.4962E+03	.5007E+03	.4999E+03	.5000E+03	.5000E+03	.5000E+03
$\lambda_1 Q_{31} + \lambda_2 Q_{32}$ 98 (94)(63) + (95)(80)	.1208E+03	.1171E+01	.7377E−01	.2740E−02	.7266E−04	.1278E−05	.5766E−08	.7242E−09
99 $96/(l_1 \cdot x_1^{*2} \cdot \rho)$	.1086E+01	.9360E+00	.9955E+00	.9931E+00	.9991E+00	.9997E+00	.1000E+01	.1000E+01
100 $97/(l_2 \cdot x_2^{*2} \cdot \rho)$	.1590E+00	.1066E+01	.9527E+01	.1004E+01	.9965E+01	.1000E+01	.9999E+01	.1000E+01
101 $98/(l_3 \cdot x_3^{*2} \cdot \rho)$	.8543E−01	.1820E−02	.2532E−03	.2110E−04	.1256E−05	.4973E−07	.5046E−09	.1426E−09
102 $x_1^{k+1} = x_1^*/(1 - 0.5(99-1))$	.1478E+02	.1407E+02	.1416E+02	.1414E+02	.1414E+02	.1414E+02	.1414E+02	.1414E+02
103 $x_2^{k+1} = x_2^*/(1 - 0.5(100-1))$	.9956E+01	.1012E+02	.9970E+01	.1001E+02	.9999E+01	.1000E+02	.1000E+02	.1000E+02
104 $x_3^{k+1} = x_3^*/(1 - 0.5(101-1))$	.9704E+01	.6363E+01	.4279E+01	.2857E+01	.1907E+01	.1271E+01	.8474E+00	.5650E+00

Problem 2(c)

In this problem the relative design vector α_i (see Equation (5.14)) is used. This is modified by using the relation

$$\alpha_i^{k+1} = \alpha_i^k \left(\frac{Q_{i1}}{\alpha_1^2 \rho_i l_i} \right)_k^{1/2}$$

and the Lagrange multipliers are estimated from Equation (5.66):

$$\lambda_j^{k+1} = \lambda_j^k \left(\frac{C_j}{\bar{\bar{C}}_j} \right)_k^{1/2} \qquad j = 1, 2$$

For the first iteration $\lambda_1 = \lambda_2 = 1.0$. The Lagrange multipliers and the relative design variables α_i are normalized after each modification so that the maximum α_i or λ_j is equal to unity. A single virtual load vector with $f_x = \lambda_1.1$ and $f_y = \lambda_2.1$ is used.

After eight iterations the weight of the structure is 150.8 lbs (see row 48), but x_3 for this design is $0.0999\,\text{in}^2$, which should be equal to $0.1\,\text{in}^2$. The minimum weight design is nearly the same as the one for Problem 2(a).

		CYCLE 1	CYCLE 2	CYCLE 3	CYCLE 4	CYCLE 5	CYCLE 6	CYCLE 7	CYCLE 8
1	α_1	.1000E+01	.1000E+01	.1000E+01	.1000E+01	.1000E+01	.1000E+01	.1000E+01	.1000E+01
2	α_2	.1000E+01	.7286E+00	.3693E+00	.9169E+00	.8486E+00	.7267E+00	.7065E+00	.7067E+00
3	α_3	.1000E+01	.1971E+00	.2317E−01	.3232E−01	.5870E−02	.6079E−02	.6890E−02	.7065E−02
4	$\bar{C}_1 = \bar{C}_2$	.5000E−01	.5000E−01	.5000E−01	.5000E−01	.5000E−01	.5000E−01	.5000E−01	.5000E−01
5	E	.1000E+08	.1000E+08	.1000E+08	.1000E+08	.1000E+08	.1000E+08	.1000E+08	.1000E+08
6	F_x	.1000E+06	.1000E+06	.1000E+06	.1000E+06	.1000E+06	.1000E+06	.1000E+06	.1000E+06
7	F_y	.2000E+06	.2000E+06	.2000E+06	.2000E+06	.2000E+06	.2000E+06	.2000E+06	.2000E+06
8	$\cos\theta$	.7071E+00	.7071E+00	.7071E+00	.7071E+00	.7071E+00	.7071E+00	.7071E+00	.7071E+00
9	$\cos\phi$	.7071E+00	.7071E+00	.7071E+00	.7071E+00	.7071E+00	.7071E+00	.7071E+00	.7071E+00
10	$\cos^2\theta$	.5000E+00	.5000E+00	.5000E+00	.5000E+00	.5000E+00	.5000E+00	.5000E+00	.5000E+00
11	$\cos^2\phi$	.5000E+00	.5000E+00	.5000E+00	.5000E+00	.5000E+00	.5000E+00	.5000E+00	.5000E+00
12	$\cos\theta\cdot\cos\phi$	.5000E+00	.5000E+00	.5000E+00	.5000E+00	.5000E+00	.5000E+00	.5000E+00	.5000E+00
13	l_1	.7071E+02	.7071E+02	.7071E+02	.7071E+02	.7071E+02	.7071E+02	.7071E+02	.7071E+02
14	l_2	.5000E+02	.5000E+02	.5000E+02	.5000E+02	.5000E+02	.5000E+02	.5000E+02	.5000E+02
15	l_3	.7071E+02	.7071E+02	.7071E+02	.7071E+02	.7071E+02	.7071E+02	.7071E+02	.7071E+02
16	$x_1 E/l_1$	.1414E+06	.1414E+06	.1414E+06	.1414E+06	.1414E+06	.1414E+06	.1414E+06	.1414E+06
17	$x_2 E/l_2$	.2000E+06	.1457E+06	.7386E+05	.1834E+06	.1697E+06	.1453E+06	.1413E+06	.1413E+06
18	$x_3 E/l_3$	.1414E+06	.2788E+05	.3277E+04	.4571E+04	.8302E+03	.8597E+03	.9744E+03	.9992E+03

#	Formula										
19	$\dfrac{x_1 E}{l_1} + \dfrac{x_3 E}{l_3}(16+17)$		.2828E+06	.1693E+06	.1447E+06	.1460E+06	.1423E+06	.1423E+06	.1424E+06	.1424E+06	.1424E+06
20	$\dfrac{x_1 E}{l_1} - \dfrac{x_3 E}{l_3}(16-17)$	0.		.1135E+06	.1381E+06	.1369E+06	.1406E+06	.1406E+06	.1404E+06	.1404E+06	.1404E+06
21	$K_1 = (11)(19)$	.1414E+06		.8465E+05	.7235E+05	.7300E+05	.7113E+05	.7114E+05	.7120E+05	.7121E+05	
22	$K_2 = (12)(20)$	0.		.5677E+05	.6907E+05	.6843E+05	.7030E+05	.7028E+05	.7022E+05	.7021E+05	
23	$K_3 = (10)(19) + 17$	.3414E+06		.2304E+06	.1462E+06	.2564E+06	.2408E+06	.2165E+06	.2125E+06	.2126E+06	
24	$\dfrac{K_2 \cdot F_y - K_3 \cdot F_x}{(22)(7) - (23)(6)}$	$-.3414E+11$	$-.1168E+11$	$-.8068E+09$	$-.1195E+11$	$-.1002E+11$	$-.7592E+10$	$-.7205E+10$	$-.7213E+10$	
25	$\dfrac{K_2 \cdot K_2 - K_1 \cdot K_3}{(22)(22) - (21)(23)}$	$-.4828E+11$	$-.1628E+11$	$-.5807E+10$	$-.1403E+11$	$-.1219E+11$	$-.1046E+11$	$-.1020E+11$	$-.1021E+11$	
26	$U_x = 24/25$	.7071E+00	.7177E+00	.1389E+00	.8518E+00	.8225E+00	.7257E+00	.7065E+00	.7067E+00	
27	$\dfrac{K_2 \cdot F_x - K_1 \cdot F_y}{(22)(6) - (21)(7)}$	$-.2828E+11$	$-.1125E+11$	$-.7563E+10$	$-.7757E+10$	$-.7196E+10$	$-.7200E+10$	$-.7217E+10$	$-.7221E+10$	
28	$U_y = 27/25$	.5858E+00	.6913E+00	.1302E+01	.5528E+00	.5904E+00	.6883E+00	.7077E+00	.7075E+00	
29	$U_1 = (26)(9) + (28)(8)$	.9142E+00	.9963E+00	.1019E+01	.9932E+00	.9990E+00	.9998E+00	.1000E+01	.1000E+01	
30	$U_2 = U_y (28)$	.5858E+00	.6913E+00	.1302E+01	.5528E+00	.5904E+00	.6883E+00	.7077E+00	.7075E+00	
31	$U_3 = -(26)(9) + (28)(8)$	$-.8579E-01$	$-.1864E-01$	.8226E+00	$-.2114E+00$	$-.1641E+00$	$-.2647E+01$	.8225E$-$03	.5182E$-$03	
32	$T_1 = \dfrac{x_1 E}{l_1} U_1 (16)(29)$	.1293E+06	.1409E+06	.1441E+06	.1405E+06	.1413E+06	.1414E+06	.1414E+06	.1414E+06	
33	$T_2 = \dfrac{x_2 E}{l_2} U_2 (17)(30)$	.1172E+06	.1007E+06	.9619E+05	.1014E+06	.1002E+06	.1000E+06	.1000E+06	.1000E+06	

		CYCLE 1	CYCLE 2	CYCLE 3	CYCLE 4	CYCLE 5	CYCLE 6	CYCLE 7	CYCLE 8
34	$T_3 = \dfrac{x_3 E}{l_3} U_3 (18)(31)$	$-.1213E+05$	$-.5198E+03$	$.2695E+04$	$-.9663E+03$	$-.1363E+03$	$-.2276E+02$	$.8014E+00$	$.5177E+00$
35	$\sigma_1 = \dfrac{T_1}{x_1} = \dfrac{31}{1}$	$.1293E+06$	$.1409E+06$	$.1441E+06$	$.1405E+06$	$.1413E+06$	$.1414E+06$	$.1414E+06$	$.1414E+06$
36	$\sigma_2 = \dfrac{T_2}{x_2} = \dfrac{33}{2}$	$.1172E+06$	$.1383E+06$	$.2604E+06$	$.1106E+06$	$.1181E+06$	$.1377E+06$	$.1415E+06$	$.1415E+06$
37	$\sigma_3 = \dfrac{T_3}{x_3} = \dfrac{34}{3}$	$-.1213E+05$	$-.2636E+04$	$.1163E+06$	$-.2990E+05$	$-.2321E+05$	$-.3744E+04$	$.1163E+03$	$.7328E+02$
38	$\Lambda(\sigma_1) = \dfrac{\sigma_1}{\bar{\sigma}_1}$ (Abs)	$.5172E+01$	$.5636E+01$	$.5765E+01$	$.5618E+01$	$.5651E+01$	$.5656E+01$	$.5657E+01$	$.5657E+01$
39	$\Lambda(\sigma_2) = \dfrac{\sigma_2}{\bar{\sigma}_2}$ (Abs)	$.4686E+01$	$.5531E+01$	$.1042E+02$	$.4422E+01$	$.4723E+01$	$.5506E+01$	$.5662E+01$	$.5660E+01$
40	$\Lambda(\sigma_3) = \dfrac{\sigma_3}{\bar{\sigma}_3}$ (Abs)	$.4853E+00$	$.1055E+00$	$.4653E+01$	$.1196E+01$	$.9285E+00$	$.1498E+00$	$.4653E-02$	$.2931E-02$
41	$\Lambda(U_x) = \dfrac{U_x}{\bar{c}_1} = \dfrac{26}{4}$	$.1414E+02$	$.1435E+02$	$.2778E+01$	$.1704E+02$	$.1645E+02$	$.1451E+02$	$.1413E+02$	$.1413E+02$
42	$\Lambda(U_y) = \dfrac{U_y}{\bar{c}_1} = \dfrac{28}{4}$	$.1172E+02$	$.1383E+02$	$.2604E+02$	$.1106E+02$	$.1181E+02$	$.1377E+02$	$.1415E+02$	$.1415E+02$
43	$\Lambda(\max)(38-42)$	$.1414E+02$	$.1435E+02$	$.2604E+02$	$.1704E+02$	$.1645E+02$	$.1451E+02$	$.1415E+02$	$.1415E+02$
44	$x_1^* = \Lambda \cdot x_1 (43)(1)$	$.1414E+02$	$.1435E+02$	$.2604E+02$	$.1704E+02$	$.1645E+02$	$.1451E+02$	$.1415E+02$	$.1415E+02$
45	$x_2^* = \Lambda \cdot x_2 (43)(2)$	$.1414E+02$	$.1046E+02$	$.9619E+01$	$.1562E+02$	$.1396E+02$	$.1055E+02$	$.1000E+02$	$.1000E+02$
46	$x_3^* = \Lambda \cdot x_3 (43)(3)$	$.1414E+02$	$.2830E+01$	$.6034E+00$	$.5506E+00$	$.9656E-01$	$.8823E-01$	$.9752E-01$	$.9997E-01$

47	VOLUME $=\sum x_i l_i$	.2707E+04	.1738E+04	.2365E+04	.2024E+04	.1868E+04	.1560E+04	.1508E+04	.1508E+04
48	WT $= (47)(\rho)$	.2707E+03	.1738E+03	.2365E+03	.2024E+03	.1868E+03	.1560E+03	.1508E+03	.1508E+03
49	$\lambda_1^{k+1} = \lambda_1^k (41)^{1/2}$	.1000E+01	.3789E+01	.1667E+01	.1374E+01	.1675E+01	.1858E+01	.1882E+01	.1881E+01
50	$\lambda_2^{k+1} = \lambda_2^k (42)^{1/2}$	.1000E+01	.3718E+01	.5009E+01	.3325E+01	.3436E+01	.3710E+01	.3762E+01	.3762E+01
51	$\lambda_1^{k+1} = 49/\text{Max}(49,50)$	.1000E+01	.1000E+01	.3328E+00	.4131E+00	.4876E+00	.5007E+00	.5003E+00	.5000E+00
52	$\lambda_2^{k+1} = 50/\text{Max}(49, 50)$	.1000E+01	.9815E+00	.1000E+01	.1000E+01	.1000E+01	.1000E+01	.1000E+01	.1000E+01
53	$f_x^1 = \lambda_1^{k+1} \cdot 1 (51 \cdot 1)$	.1000E+01	.1000E+01	.3328E+00	.4131E+00	.4876E+00	.5007E+00	.5003E+00	.5000E+00
54	$f_y^1 = \lambda_2^{k+1} \cdot 1 (52 \cdot 1)$	.1000E+01	.9815E+00	.1000E+01	.1000E+01	.1000E+01	.1000E+01	.1000E+01	.1000E+01
55	$\dfrac{K_2 \cdot f_y^1 - K_3 \cdot f_x}{(22)(54) - (23)(53)}$	−.3414E+06	−.1746E+06	.2042E+05	−.3748E+05	−.4713E+05	−.3810E+05	−.3608E+05	−.3607E+05
56	$U_x = 55/25$	.7071E−05	.1073E−04	−.3515E−05	.2671E−05	.3867E−05	.3642E−05	.3538E−05	.3534E−05
57	$\dfrac{K_2 \cdot f_x^1 = k_1 \cdot f_y^1}{(22)(53) - (21)(54)}$	−.1414E+06	−.2631E+05	−.4936E+05	−.4473E+05	−.3685E+05	−.3595E+05	−.3607E+05	−.3611E+05
58	$U_y = 57/25$	.2929E−05	.1616E−05	.8500E−05	.3188E−05	.3023E−05	.3437E−05	.3537E−05	.3537E−05
59	$U_1^1 = (56)(9) + (58)(8)$	.7071E−05	.8730E−05	.3525E−05	.4143E−05	.4872E−05	.5006E−05	.5003E−05	.5000E−05
60	$U_2^1 = U_y$	.2929E−05	.1616E−05	.8500E−05	.3188E−05	.3023E−05	.3437E−05	.3537E−05	.3537E−05
61	$U_3^1 = -(56)(9) + (58)(8)$	−.2929E−05	−.6444E−05	.8496E−05	.3654E−06	−.5965E−06	−.1453E−06	−.9188E−09	.2640E−08
62	$t_1^1 = \dfrac{x_1 U_1^1 E}{l_1} (16)(59)$	.1000E+01	.1235E+01	.4985E+00	.5859E+00	.6891E+00	.7079E+00	.7075E+00	.7071E+00
63	$t_2^1 = \dfrac{x_2 U_2^1 E}{l_2} (17)(60)$	.5858E+00	.2355E+00	.6278E+00	.5845E+00	.5131E+00	.4995E+00	.4997E+00	.5000E+00

		CYCLE 1	CYCLE 2	CYCLE 3	CYCLE 4	CYCLE 5	CYCLE 6	CYCLE 7	CYCLE 8
64	$t_3^1 = \dfrac{x_3 U_3^1 E}{l_3}$ (18)(61)	$-.4142E+00$	$-.1797E+00$	$.2784E-01$	$.1670E-02$	$-.4952E-03$	$-.1249E-03$	$-.8952E-06$	$.2638E-05$
65	$Q_{11} = \dfrac{T_1 t_1^1 l_1}{E}$	$.9142E+00$	$.1230E+01$	$.5080E+00$	$.5819E+00$	$.6884E+00$	$.7078E+00$	$.7075E+00$	$.7071E+00$
66	$Q_{21} = \dfrac{T_2 t_2^1 l_2}{E}$	$.3431E+00$	$.1186E+00$	$.3020E+00$	$.2963E+00$	$.2571E+00$	$.2498E+00$	$.2499E+00$	$.2500E+00$
67	$Q_{31} = \dfrac{T_3 t_3^1 l_3}{E}$	$.3553E-01$	$.6603E-03$	$.5306E-03$	$-.1141E-04$	$.4771E-06$	$.2011E-07$	$-.5073E-11$	$.9656E-11$
68	$Q_{11}/\alpha_1^2 \cdot \rho \cdot l_1$	$.1293E+00$	$.1740E+00$	$.7184E-01$	$.8229E-01$	$.9735E-01$	$.1001E+00$	$.1001E+00$	$.1000E+00$
69	$Q_{21}/\alpha_2^2 \cdot \rho \cdot l_2$	$.6863E-01$	$.4470E-01$	$.4428E+00$	$.7048E-01$	$.7140E-01$	$.9462E-01$	$.1001E+00$	$.1001E+00$
70	$Q_{31}/\alpha_3^2 \cdot \rho \cdot l_3$	$.5025E-02$	$.2403E-02$	$.1398E+00$	$-.1545E-02$	$.1958E-02$	$.7694E-04$	$-.1511E-07$	$.2736E-07$
71	$\alpha_1^{k+1} = \alpha_1 (68)^{1/2}$	$.3596E+00$	$.4171E+00$	$.2680E+00$	$.2869E+00$	$.3120E+00$	$.3164E+00$	$.3163E+00$	$.3162E+00$
72	$\alpha_2^{k+1} = \alpha_2 (69)^{1/2}$	$.2620E+00$	$.1540E+00$	$.2457E+00$	$.2434E+00$	$.2267E+00$	$.2235E+00$	$.2235E+00$	$.2236E+00$
73	$\alpha_3^{k+1} = \alpha_3 (70)^{1/2}$	$.7089E-01$	$.9663E-02$	$.8662E-02$	$0.$	$.2598E-03$	$.5332E-04$	$0.$	$.1169E-05$
74	$\alpha_1 = 71/\text{Max}(71-73)$	$.1000E+01$	$.1000E+01$	$.1000E+01$	$.1000E+01$	$.1000E+01$	$.1000E+01$	$.1000E+01$	$.1000E+01$
75	$\alpha_2 = 72/\text{Max}(71-73)$	$.7286E+00$	$.3693E+00$	$.9169E+00$	$.8486E+00$	$.7267E+00$	$.7065E+00$	$.7067E+00$	$.7071E+00$
76	$\alpha_3 = 73/\text{Max}(71-73)$	$.1971E+00$	$.2317E-01$	$.3232E-01$	$0.$	$.8325E-03$	$.1685E-03$	$0.$	$.3695E-05$
77	α_1 ⎫ checked for x_min	$.1000E+01$	$.1000E+01$	$.1000E+01$	$.1000E+01$	$.1000E+01$	$.1000E+01$	$.1000E+01$	$.1000E+01$
78	α_2 ⎬ $\alpha_\text{min} = x_\text{min}$	$.7286E+00$	$.3693E+00$	$.9169E+00$	$.8486E+00$	$.7267E+00$	$.7065E+00$	$.7067E+00$	$.7071E+00$
79	α_3 ⎭ $A(\text{max})(43)$	$.1971E+00$	$.2317E-01$	$.3232E-01$	$.5870E+02$	$.6079E+02$	$.6890E+02$	$.7065E+02$	$.7067E-02$

Problem 3 Stress-constrained problem

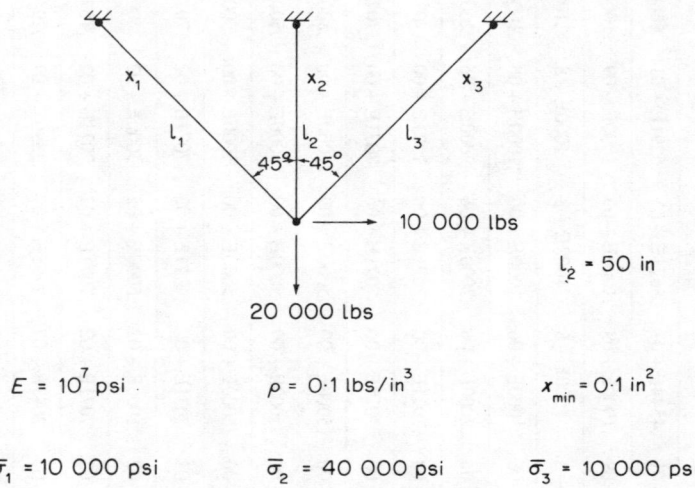

Fig. 4A Problem 3—Stress-constrained problem

This problem is solved by using three different algorithms. First the structure is designed by using the FSD method, then the other two approaches are used. It is seen that for this problem the FSD algorithm does not lead to the real minimum-weight design.

Problem 3(a)

The design variables are modified by Equation (5.109):

$$x_i^{k+1} = \left(\frac{T_i}{\bar{\sigma}_i}\right)_k$$

This is the fully stressed design algorithm. The weight of the structure with the same size for all the elements is 247.5 lbs. After the first iteration the weight reduced to 158.4 lbs. In the second iteration the weight increased to 246.7 lbs and then decreased with subsequent iterations until it reached 198.5 lbs. The cross-sectional areas for the design with a weight equal to 198.5 lbs are $x_1 = 21.07\ \text{in}^2$, $x_2 = 0.1\ \text{in}^2$ and $x_3 = 6.92\ \text{in}^2$. The stress in the second element equals 20 000 psi, and 10 000 psi in the other two elements. The size of x_2 is dictated by the minimum size requirement. This is not an optimum design (see Problem 3(b) or 3(c)).

		CYCLE 1	CYCLE 2	CYCLE 3	CYCLE 4	CYCLE 5	CYCLE 6	CYCLE 7	CYCLE 8
1	x_1	.1000E+01	.1293E+02	.1582E+02	.1725E+02	.1869E+02	.1976E+02	.2043E+02	.2081E+02
2	x_2	.1000E+01	.2929E+01	.1907E+01	.1403E+01	.8935E+00	.5126E+00	.2759E+00	.1433E+00
3	x_3	.1000E+01	.1213E+01	.1676E+01	.3103E+01	.4544E+01	.5621E+01	.6291E+01	.6666E+01
4	$\bar{C}_1 = \bar{C}_2$	.1000E+09	.1000E+09	.1000E+09	.1000E+09	.1000E+09	.1000E+09	.1000E+09	.1000E+09
5	E	.1000E+8	.1000E+8	.1000E+8	.1000E+8	.1000E+8	.1000E+8	.1000E+8	.1000E+8
6	F_x	.1000E+06	.1000E+06	.1000E+06	.1000E+06	.1000E+06	.1000E+06	.1000E+06	.1000E+06
7	F_y	.2000E+06	.2000E+06	.2000E+06	.2000E+06	.2000E+06	.2000E+06	.2000E+06	.2000E+06
8	$\cos\theta$	.7071E+00	.7071E+00	.7071E+00	.7071E+00	.7071E+00	.7071E+00	.7071E+00	.7071E+00
8	$\cos\phi$	.7071E+00	.7071E+00	.7071E+00	.7071E+00	.7071E+00	.7071E+00	.7071E+00	.7071E+00
10	$\cos^2\theta$	.5000E+00	.5000E+00	.5000E+00	.5000E+00	.5000E+00	.5000E+00	.5000E+00	.5000E+00
11	$\cos^2\phi$	.5000E+00	.5000E+00	.5000E+00	.5000E+00	.5000E+00	.5000E+00	.5000E+00	.5000E+00
12	$\cos\theta\cdot\cos\phi$	.5000E+00	.5000E+00	.5000E+00	.5000E+00	.5000E+00	.5000E+00	.5000E+00	.5000E+00
13	l_1	.7071E+02	.7071E+02	.7071E+02	.7071E+02	.7071E+02	.7071E+02	.7071E+02	.7071E+02
14	l_2	.5000E+02	.5000E+02	.5000E+02	.5000E+02	.5000E+02	.5000E+02	.5000E+02	.5000E+02
15	l_3	.7071E+02	.7071E+02	.7071E+02	.7071E+02	.7071E+02	.7071E+02	.7071E+02	.7071E+02
16	$x_1 E/l_1$	.1414E+06	.1828E+07	.2237E+07	.2439E+07	.2643E+07	.2795E+07	.2890E+07	.2943E+07
17	$x_2 E/l_2$	.2000E+06	.5858E+06	.3815E+06	.2806E+06	.1787E+06	.1025E+06	.5518E+05	.2866E+05
18	$x_3 E/l_3$	.1414E+06	.1716E+06	.2370E+06	.4389E+06	.6426E+06	.7950E+06	.8897E+06	.9427E+06

19	$\frac{x_1 E}{l_1} + \frac{x_3 E}{l_3}(16+7)$	.2828E+06	.2000E+07	.2474E+07	.2878E+07	.3285E+07	.3590E+07	.3779E+07	.3885E+07
20	$\frac{x_1 E}{l_1} - \frac{x_3 E}{l_3}(16-17)$	0.	.1657E+07	.2000E+07	.2000E+07	.2000E+07	.2000E+07	.2000E+07	.2000E+07
21	$K_1 = (11)(19)$	.1414E+06	.1000E+07	.1237E+07	.1439E+07	.1643E+07	.1795E+07	.1890E+07	.1943E+07
22	$K_2 = (12)(20)$	0.	.8284E+06	.1000E+07	.1000E+07	.1000E+07	.1000E+07	.1000E+07	.1000E+07
23	$K_3 = (10)(19)+17$	.3414E+06	.1586E+07	.1619E+07	.1719E+07	.1821E+07	.1897E+07	.1945E+07	.1971E+07
24	$\frac{K_2 \cdot F_y - K_3 \cdot F_x}{(22)(7) - (23)(6)}$	−.3414E+11	.7108E+10	.3815E+11	.2806E+11	.1787E+11	.1025E+11	.5518E+10	.2866E+10
25	$\frac{K_2 \cdot K_2 - K_1 \cdot K_3}{(22)(22) - (21)(23)}$	−.4828E+11	−.8995E+12	−.1002E+13	−.1474E+13	−.1992E+13	−.2406E+13	−.2675E+13	−.2830E+13
26	$U_x = 24/25$	.7071E+00	−.7902E−02	−.3807E−01	−.1903E−01	−.8972E−02	−.4261E−02	−.2063E−02	−.1013E−02
27	$\frac{K_2 \cdot F_x - K_1 \cdot F_y}{(22)(6) - (21)(7)}$	−.2828E+11	−.1172E+12	−.1474E+12	−.1878E+12	−.2285E+12	−.2590E+12	−.2779E+12	−.2885E+12
28	$U_y = 27/25$	.5858E+00	.1302E+00	.1471E+00	.1274E+00	.1147E+00	.1076E+00	.1039E+00	.1020E+00
29	$U_1 = (26)(9) + (28)(8)$	.9142E+00	.8651E−01	.7709E−01	.7662E−01	.7479E−01	.7311E−01	.7201E−01	.7139E−01
30	$U_2 = U_y (28)$	.5858E+00	.1302E+00	.1471E+00	.1274E+00	.1147E+00	.1076E+00	.1039E+00	.1020E+00
31	$U_3 = -(26)(9) + (28)(8)$	−.8579E−01	.9769E−01	.1309E+00	.1035E+00	.8748E−01	.7913E−01	.7493E−01	.7282E−01
32	$T_1 = \frac{x_1 E}{l_1} U_1 (16)(29)$	.1293E+06	.1582E+06	.1725E+06	.1869E+06	.1976E+06	.2043E+06	.2081E+06	.2101E+06
33	$T_2 = \frac{x_2 E}{l_2} U_2 (17)(30)$	.1172E+06	.7630E+05	.5611E+05	.3574E+05	.2050E+05	.1104E+05	.5733E−04	.2923E−04

		CYCLE 1	CYCLE 2	CYCLE 3	CYCLE 4	CYCLE 5	CYCLE 6	CYCLE 7	CYCLE 8
34	$T_3 = \dfrac{x_3 E}{l_3} U_3 \,(18)(31)$	$-.1213E+05$	$.1676E+05$	$.3103E+05$	$.4544E+05$	$.5621E+05$	$.6291E+05$	$.6666E+05$	$.6864E+05$
35	$\sigma_1 = \dfrac{T_1}{x_1} = \dfrac{32}{1}$	$.1293E+06$	$.1223E+05$	$.1090E+05$	$.1084E+05$	$.1058E+05$	$.1034E+05$	$.1018E+05$	$.1010E+05$
36	$\sigma_2 = \dfrac{T_2}{x_2} = \dfrac{33}{2}$	$.1172E+06$	$.2605E+05$	$.2942E+05$	$.2548E+05$	$.2295E+05$	$.2153E+05$	$.2078E+05$	$.2039E+05$
37	$\sigma_3 = \dfrac{T_3}{x_3} = \dfrac{34}{3}$	$-.1213E+05$	$.1382E+05$	$.1852E+05$	$.1464E+05$	$.1237E+05$	$.1119E+05$	$.1060E+05$	$.1030E+05$
38	$\Lambda(\sigma_1) = \dfrac{\sigma_1}{\bar{\sigma}_1}$ (Abs)	$.1293E+02$	$.1223E+01$	$.1090E+01$	$.1084E+01$	$.1058E+01$	$.1034E+01$	$.1018E+01$	$.1010E+01$
39	$\Lambda(\sigma_2) = \dfrac{\sigma_2}{\bar{\sigma}_2}$ (Abs)	$.2929E+01$	$.6512E+00$	$.7354E+00$	$.6369E+00$	$.5737E+00$	$.5382E+00$	$.5195E+00$	$.5098E+00$
40	$\Lambda(\sigma_3) = \dfrac{\sigma_3}{\bar{\sigma}_3}$ (Abs)	$.1213E+01$	$.1382E+01$	$.1852E+01$	$.1464E+01$	$.1237E+01$	$.1119E+01$	$.1060E+01$	$.1030E+01$
41	$\Lambda(U_x) = \dfrac{U_x}{\bar{c}_1} = \dfrac{26}{4}$	$.7071E-08$	$.7902E-10$	$.3807E-09$	$.1903E-09$	$.8972E-10$	$.4261E-10$	$.2863E-10$	$.1013E-10$
42	$\Lambda(U_y) = \dfrac{U_y}{\bar{c}_1} = \dfrac{28}{4}$	$.5858E-08$	$.1302E-08$	$.1471E-08$	$.1274E-08$	$.1147E-08$	$.1076E-08$	$.1039E-08$	$.1020E-08$
43	$\Lambda(\max)(38-42)$	$.1293E+02$	$.1382E+01$	$.1852E+01$	$.1464E+01$	$.1237E+01$	$.1119E+01$	$.1060E+01$	$.1030E+01$
44	$x_1^* = \Lambda \cdot x_1 \,(43)(1)$	$.1293E+02$	$.1786E+02$	$.2929E+02$	$.2525E+02$	$.2312E+02$	$.2212E+02$	$.2165E+02$	$.2143E+02$
45	$x_2^* = \Lambda \cdot x_2 \,(43)(2)$	$.1293E+02$	$.4046E+01$	$.3532E+01$	$.2054E+01$	$.1105E+01$	$.5736E+00$	$.2923E+00$	$.1476E+00$
46	$x_3^* = \Lambda \cdot x_3 \,(43)(3)$	$.1293E+02$	$.1676E+02$	$.3103E+02$	$.4544E+02$	$.5621E+02$	$.6291E+02$	$.6666E+02$	$.6864E+02$
47	VOLUME $= \sum x_i l_i$	$.2475E+04$	$.1584E+04$	$.2467E+04$	$.2210E+04$	$.2087E+04$	$.2037E+04$	$.2017E+04$	$.2000E+04$

48	$WT = (47)(\rho)$	.2475E+03	.1584E+03	.2467E+03	.2210E+03	.2087E+03	.2037E+03	.2017E+03	.2008E+03
49	$x_1^{k+1} = \dfrac{T_1}{\bar{\sigma}_1}\left(\dfrac{2}{\bar{\sigma}_1}\right)$	.1293E+02	.1582E+02	.1725E+02	.1869E+02	.1976E+02	.2043E+02	.2081E+02	.2101E+02
50	$x_2^{k+1} = \dfrac{T_2}{\bar{\sigma}_2}\left(\dfrac{33}{\bar{\sigma}_2}\right)$	.2929E+01	.1907E+01	.1403E+01	.8935E+00	.5126E+00	.2759E+00	.1433E+00	.1000E+00
51	$x_3^{k+1} = \dfrac{T_3}{\bar{\sigma}_3}\left(\dfrac{34}{\bar{\sigma}_3}\right)$	.1213E+01	.1676E+01	.3103E+01	.4544E+01	.5621E+01	.6291E+01	.6666E+01	.6864E+01

Note: If in rows 49–51 $x_i < x_{\min}$ then $x_i = x_{\min}$.

Problem 3(b)

In this problem the relative design variable α_i is used. It is modified by using the relation

$$\alpha_i^{k+1} = \alpha_i^k \left(\frac{Q_{i1}}{\rho_i l_i \alpha_i^2}\right)_k^{1/2}$$

The Lagrange multipliers are estimated by using Equation (5.124):

$$\lambda_j^{k+1} = \lambda_j^k \left(\frac{\sigma_j}{\bar{\sigma}_j}\right)_k^{1/2} \qquad j = 1, 2, 3$$

For the first iteration the λ_j's are assumed to be proportional to the forces in the bars. The relative design variables and the Lagrange multipliers are normalized after each iteration so that the maximum α_i or λ_j is equal to unity. A single virtual load vector is used with

$$f_x = \frac{E}{l_1} \cos\phi \, \lambda_1 \left(\frac{T_1}{\text{ABS}(T_1)}\right)$$
$$+ 0$$
$$- \frac{E}{l_3} \cos\phi \, \lambda_3 \left(\frac{T_3}{\text{ABS}(T_3)}\right)$$

and

$$f_y = \frac{E}{l_1} \cos\theta \, \lambda_1 \left(\frac{T_1}{\text{ABS}(T_1)}\right)$$
$$+ \frac{E}{l_2} \lambda_2 \left(\frac{T_2}{\text{ABS}(T_2)}\right)$$
$$+ \frac{E}{l_3} \cos\theta \, \lambda_3 \left(\frac{T_3}{\text{ABS}(T_3)}\right)$$

The ratio $T_i/\text{ABS}(T_i)$ is equal to $+1$ or -1 depending on whether the force in the bar is tension or compression.

From row 48 it is seen that the minimum weight of the structure is 128.6 lbs. This is achieved with five iterations. With additional iterations the weight increases. The cross-sectional areas for the minimum weight design are $x_1 = 14.22 \text{ in}^2$; $x_2 = 5.468 \text{ in}^2$; $x_3 = 0.1003 \text{ in}^2$. The Lagrange multipliers associated with this design are $\lambda_1 = 1.0, \lambda_2 = 0.086$ and $\lambda_3 = 0.01$. For convenience the flexibility coefficient is denoted by Q_{ij} instead of R_{ij}.

		CYCLE 1	CYCLE 2	CYCLE 3	CYCLE 4	CYCLE 5	CYCLE 6	CYCLE 7	CYCLE 8
1	α_1	.1000E+01	.1000E+01	.1000E+01	.1000E+01	.1000E+01	.1000E+01	.1000E+01	.1000E+01
2	α_2	.1000E+01	.9588E+00	.5781E+00	.4953E+00	.3844E+00	.3213E+00	.2842E+00	.2627E+00
3	α_3	.1000E+01	.7735E−02	.7085E−02	.7060E−02	.7050E−02	.7031E−02	.6112E−02	.4925E−02
4	$\bar{C}_1 = \bar{C}_2$	.1000E+09	.1000E+09	.1000E+09	.1000E+09	.1000E+09	.1000E+09	.1000E+09	.1000E+09
5	E	.1000E+08	.1000E+08	.1000E+08	.1000E+08	.1000E+08	.1000E+08	.1000E+08	.1000E+08
6	F_x	.1000E+06	.1000E+06	.1000E+06	.1000E+06	.1000E+06	.1000E+06	.1000E+06	.1000E+06
7	F_y	.2000E+06	.2000E+06	.2000E+06	.2000E+06	.2000E+06	.2000E+06	.2000E+06	.2000E+06
8	$\cos\theta$	.7071E+00	.7071E+00	.7071E+00	.7071E+00	.7071E+00	.7071E+00	.7071E+00	.7071E+00
9	$\cos\phi$	.7071E+00	.7071E+00	.7071E+00	.7071E+00	.7071E+00	.7071E+00	.7071E+00	.7071E+00
10	$\cos^2\theta$	.5000E+00	.5000E+00	.5000E+00	.5000E+00	.5000E+00	.5000E+00	.5000E+00	.5000E+00
11	$\cos^2\phi$	.5000E+00	.5000E+00	.5000E+00	.5000E+00	.5000E+00	.5000E+00	.5000E+0	.5000E+00
12	$\cos\theta \cdot \cos\phi$	.5000E+00	.5000E+00	.5000E+00	.5000E+00	.5000E+00	.5000E+00	.5000E+00	.5000E+00
13	l_1	.7071E+02	.7071E+02	.7071E+02	.7071E+02	.7071E+02	.7071E+02	.7071E+02	.7071E+02
14	l_2	.5000E+02	.5000E+02	.5000E+02	.5000E+02	.5000E+02	.5000E+02	.5000E+02	.5000E+02
15	l_3	.7071E+02	.7071E+02	.7071E+02	.7071E+02	.7071E+02	.7071E+02	.7071E+02	.7071E+02
16	$x_1 E/l_1$	.1414E+06	.1414E+06	.1414E+06	.1414E+06	.1414E+06	.1414E+06	.1414E+06	.1414E+06
17	$x_2 E/l_2$	.2000E+06	.1918E+06	.1156E+06	.9905E+05	.7689E+05	.6426E+05	.5684E+05	.5254E+05
18	$x_3 E/l_3$	.1414E+06	.1094E+04	.1002E+04	.9985E+03	.9971E+03	.9943E+03	.8644E+03	.6965E+03

		CYCLE 1	CYCLE 2	CYCLE 3	CYCLE 4	CYCLE 5	CYCLE 6	CYCLE 7	CYCLE 8
19	$\dfrac{x_1 E}{l_1} + \dfrac{x_3 E}{l_3}(16+17)$	.2828E+06	.1425E+06	.1424E+06	.1424E+06	.1424E+06	.1424E+06	.1423E+06	.1421E+06
20	$\dfrac{x_1 E}{l_1} - \dfrac{x_3 E}{l_3}(16-17)$	0.	.1403E+06	.1404E+06	.1404E+06	.1404E+06	.1404E+06	.1406E+06	.1407E+06
21	$K_1 = (11)(19)$	.1414E+06	.7126E+05	.7121E+05	.7121E+05	.7121E+05	.7121E+05	.7114E+05	.7106E+05
22	$K_2 = (12)(20)$	0.	.7016E+05	.7021E+05	.7021E+05	.7021E+05	.7021E+05	.7028E+05	.7036E+05
23	$K_3 = (10)\,(19)+17$	.3414E+06	.2630E+06	.1868E+06	.1703E+06	.1481E+06	.1355E+06	.1280E+06	.1236E+06
24	$\dfrac{K_2 \cdot F_y - K_3 \cdot F_x}{(22)(7) - (23)(6)}$	−.3414E+11	−.1227E+11	−.4641E+10	−.2984E+10	−.7670E+09	.4962E+09	.1258E+10	.1712E+10
25	$\dfrac{K_2 \cdot K_2 - K_1 \cdot K_3}{(22)(22) - (21)(23)}$	−.4828E+11	−.1382E+11	−.8375E+10	−.7195E+10	−.5616E+10	−.4716E+10	−.4166E+10	−.3832E+10
26	$U_x = 24/25$	.7071E+00	.8878E+00	.5542E+00	.4147E+00	.1366E+00	−.1052E+00	−.3019E+00	−.4468E+00
27	$\dfrac{K_2 \cdot F_x - K_1 \cdot F_y}{(22)(6) - (21)(7)}$	−.2828E+11	−.7235E+10	−.7221E+10	−.7221E+10	−.7221E+10	−.7220E+10	−.7201E+10	−.7176E+10
28	$U_y = 27/25$	.5858E+00	.5236E+00	.8622E+00	.1004E+01	.1286E+01	.1531E+01	.1728E+01	.1872E+01
29	$U_1 = (26)(9) + (28)(8)$	.9142E+00	.9980E+00	.1002E+01	.1003E+01	.1006E+01	.1008E+01	.1009E+01	.1008E+01
30	$U_2 = U_y(28)$	.5858E+00	.5236E+00	.8622E+00	.1004E+01	.1286E+01	.1531E+01	.1728E+01	.1872E+01
31	$U_3 = -(26)(9) + (28)(8)$	−.8579E+01	−.2576E+00	.2178E+00	.4164E+00	.8126E+00	.1157E+01	.1436E+01	.1640E+01
32	$T_1 = \dfrac{x_1 E}{l_1} U_1 (16)(29)$	.1293E+06	.1411E+06	.1416E+06	.1418E+06	.1422E+06	.1426E+06	.1427E+06	.1426E+06
33	$T_2 = \dfrac{x_2 E}{l_2} U_2 (17)(30)$	.1172E+06	.1004E+06	.9969E+05	.9941E+05	.9885E+05	.9837E+05	.9825E+05	.9838E+05

34 $T_3 = \frac{x_3 E}{l_3} U_3 (18)(31)$	−.1213E+05	−.2817E+03	.2183E+03	.4158E+03	.8102E+03	.1150E+04	.1241E+04	.1142E+04
35 $\sigma_1 = \frac{T_1}{x_1} = 32/1$	.1293E+06	.1411E+06	.1416E+06	.1418E+06	.1422E+06	.1426E+06	.1427E+06	.1426E+06
36 $\sigma_2 = \frac{T_2}{x_2} = \frac{33}{2}$	.1172E+06	.1047E+06	.1724E+06	.2007E+06	.2571E+06	.3062E+06	.3457E+06	.3745E+06
37 $\sigma_3 = \frac{T_3}{x_3} = \frac{34}{3}$	−.1213E+05	−.3643E+05	.3060E+05	.5889E+05	.1149E+06	.1636E+06	.2030E+06	.2319E+06
38 $\Lambda(\sigma_1) = \frac{\sigma_1}{\sigma}$	.1293E+02	.1411E+02	.1416E+02	.1418E+02	.1422E+02	.1426E+02	.1427E+02	.1426E+002
39 $\Lambda(\sigma_2) = \frac{\sigma_2}{\sigma}$	.2929E+01	.2618E+01	.4311E+01	.5018E+01	.6429E+01	.7655E+01	.8642E+01	.9362E+01
40 $\Lambda(\sigma_3) = \frac{\sigma_3}{\sigma}$	.1213E+01	.3643E+01	.3080E+01	.5889E+01	.1149E+02	.1636E+02	.2030E+02	.2319E+02
41 $\Lambda(U_x) = \frac{U_x}{\bar{c}_1} = \frac{26}{4}$	.7071E−08	.8878E−08	.5542E−08	.4147E−08	.1366E−08	.1052E−08	.3019E−08	.4468E−08
42 $\Lambda(U_y) = \frac{U_y}{\bar{c}_1} = \frac{28}{4}$	.5858E−08	.5236E−08	.8622E−08	.1004E−07	.1286E−07	.1531E−07	.1728E−07	.1872E−07
43 $\Lambda(\max)(38-42)$	.1293E+02	.1411E+02	.1416E+02	.1418E+02	.1422E+02	.1636E+02	.2030E+02	.2319E+02
44 $x_1^* = \Lambda \cdot x_1 (43)|(1)$	.1293E+02	.1411E+02	.1416E+02	.1418E+02	.1422E+02	.1636E+02	.2030E+02	.2319E+02
45 $x_2^* = \Lambda \cdot x_2 (43)|(2)$	.1293E+02	.1353E+02	.8188E+01	.7025E+01	.5468E+01	.5257E+01	.5770E+01	.6093E+01
46 $x_3^* = \Lambda \cdot x_3 (43)|(3)$	.1293E+02	.1092E+00	.1004E+00	.1001E+00	.1003E+00	.1150E+00	.1241E+00	.1142E+00
47 VOLUME $= \sum x_i l_i$	.2475E+04	.1682E+04	.1418E+04	.1361E+04	.1286E+04	.1428E+04	.1733E+04	.1953E+04

		CYCLE 1	CYCLE 2	CYCLE 3	CYCLE 4	CYCLE 5	CYCLE 6	CYCLE 7	CYCLE 8
48	$WT = (47)/(\rho)$	.2475E+03	.1682E+03	.1418E+03	.1361E+03	.1286E+03	.1428E+03	.1733E+03	.1953E+03
49	$\lambda_1^{k+1} = \lambda_1^k (38)^{1/2}$	.1000E+01	.3757E+01	.3764E+01	.3766E+01	.3771E+01	.3776E+01	.3777E+01	.3776E+01
50	$\lambda_2^{k+1} = \lambda_2^k (39)^{1/2}$	.9062E+00	.1466E+01	.8103E+00	.4823E+00	.3247E+00	.2382E+00	.1855E+00	.1502E+00
51	$\lambda_3^{k+1} = \lambda_3^k (40)^{1/2}$	.9383E+01	.1791E+00	.8367E−01	.5395E−01	.4856E−01	.5208E+01	.6215E−01	.7924E−01
52	$\lambda_1^{k+1} = 49/\text{Max}(49-51)$	.1000E+01	.1000E+01	.1000E+01	.1000E+01	.1000E+01	.1000E+01	.1000E+01	.1000E+01
53	$\lambda_2^{k+1} = 50/\text{Max}(49-51)$	.9062E+00	.3903E+00	.2153E+00	.1281E+00	.8610E−01	.6309E−01	.4910E−01	.3979E−01
54	$\lambda_3^{k+1} = 51/\text{Max}(49-51)$	.9383E−01	.4767E−01	.2223E−01	.1432E−01	.1288E−01	.1379E−01	.1645E−01	.2099E−01
55	$\dfrac{E}{l_1} \cdot \cos\phi \cdot \lambda_1 \cdot (\text{SIGN} - T_1)$	.1000E+06	.1000E+06	.1000E+06	.1000E+06	.1000E+06	.1000E+06	.1000E+06	.1000E+06
56	0.0	0.	0.	0.	0.	0.	0.	0.	0.
57	$-\dfrac{E}{l_3} \cdot \cos\phi \cdot \lambda_3 \cdot (\text{SIGN} = T_3)$	.9384E+04	.4767E+04	−.2223E+04	−.1432E+04	−.1288E+04	−.1379E+04	−.1646E+04	−.2099E+04
58	$\dfrac{E}{l_1} \cdot \cos\theta \cdot \lambda_1 \cdot (\text{SIGN} - T_1)$	.1000E+06	.1000E+06	.1000E+06	.1000E+06	.1000E+06	.1000E+06	.1000E+06	.1000E+06
59	$\dfrac{E}{l_2} \cdot \lambda_2 \cdot (\text{SIGN} - T_3)$	.1812E+06	.7805E+05	.4306E+05	.2561E+05	.1722E+05	.1262E+05	.9820E+04	.7958E+04
60	$\dfrac{E}{l_3} \cdot \cos\theta \cdot \lambda_3 \cdot (\text{SIGN} - T_3)$	−.9384E+04	−.4767E+04	.2223E+04	.1432E+04	.1288E+04	.1379E+04	.1646E+04	.2099E+04
61	$f_x^1 = 55 + 56 + 57$	.1094E+06	.1048E+06	.9778E+05	.9857E+05	.9871E+05	.9862E+05	.9836E+05	.9790E+05
62	$f_y^1 = 58 + 59 + 60$	.2718E+06	.1733E+06	.1453E+06	.1270E+06	.1185E+06	.1140E+06	.1115E+06	.1101E+06

#									
63	$K_2 \cdot f_y^1 - K_3 \cdot f_x$ $(22)(62) - (23)(61)$	$-.3735E+11$	$-.1540E+11$	$-.8068E+10$	$-.7862E+10$	$-.6298E+10$	$-.5356E+10$	$-.4754E+10$	$-.4357E+10$
64	$U_x = 63/65$	$.7735E+00$	$.1114E+01$	$.9633E+00$	$.1093E+01$	$.1121E+01$	$.1136E+01$	$.1141E+01$	$.1137E+01$
65	$K_2 \cdot f_x^1 - K_1 \cdot f_y$ $(22)(61) - (21)(62)$	$-.3845E+11$	$-.4997E+10$	$-.3481E+10$	$-.2126E+10$	$-.1508E+10$	$-.1193E+10$	$-.1018E+10$	$-.9320E+09$
66	$U_y = 65/25$	$.7962E+00$	$.3616E+00$	$.4156E+00$	$.2955E+00$	$.2685E+00$	$.2529E+00$	$.2443E+00$	$.2432E+00$
67	$U_1 = (64)(9) + (66)(8)$	$.1110E+01$	$.1044E+01$	$.9750E+00$	$.9817E+00$	$.9829E+00$	$.9818E+00$	$.9797E+00$	$.9759E+00$
68	$U_2 = U_y(66)$	$.7962E+00$	$.3616E+00$	$.4156E+00$	$.2955E+00$	$.2685E+00$	$.2529E+00$	$.2443E+00$	$.2432E+00$
69	$U_3 = (-64)(9) + (66)(8)$	$.1610E+01$	$-.5322E+00$	$-.3873E+00$	$-.5638E+00$	$-.6031E+00$	$-.6241E+00$	$-.6342E+00$	$-.6320E+00$
70	$t_1^1 = \dfrac{x_1 U_1 E}{l_1} \ (16)(67)$	$.1570E+06$	$.1476E+06$	$.1379E+06$	$.1388E+06$	$.1390E+06$	$.1389E+06$	$.1385E+06$	$.1380E+06$
71	$t_2^1 = \dfrac{x_2 U_2 E}{l_2} \ (17)(68)$	$.1592E+06$	$.6934E+05$	$.4806E+05$	$.2927E+05$	$.2064E+05$	$.1625E+05$	$.1389E+05$	$.1278E+05$
72	$t_3^1 = \dfrac{x_3 U_3 E}{l_3} \ (18)(69)$	$.2276E+04$	$-.5821E+03$	$-.3880E+03$	$-.5629E+03$	$-.6014E+03$	$-.6206E+03$	$-.5482E+03$	$-.4402E+03$
73	$Q_{11} = T_1 t_1^1 l_1 / E$	$.1435E+06$	$.1473E+06$	$.1381E+06$	$.1392E+06$	$.1398E+06$	$.1400E+06$	$.1398E+06$	$.1391E+06$
74	$Q_{21} = T_2 t_2^1 l_2 / E$	$.9328E+05$	$.3481E+05$	$.2395E+05$	$.1455E+05$	$.1020E+05$	$.7994E+04$	$.6821E+04$	$.6286E+04$
75	$Q_{31} = T_3 t_3^1 l_3 / E$	$-.1953E+03$	$.1160E+01$	$-.5988E+00$	$-.1655E+01$	$-.3445E+01$	$-.5048E+01$	$-.4810E+01$	$-.3556E+01$
76	$Q_{11}/\alpha_1^2 \cdot p \cdot l_1$	$.2029E+05$	$.2083E+05$	$.1953E+05$	$.1969E+05$	$.1977E+05$	$.1980E+05$	$.1977E+05$	$.1968E+05$
77	$Q_{21}/\alpha_2^2 \cdot p \cdot l_2$	$.1866E+05$	$.7573E+04$	$.1433E+05$	$.1186E+05$	$.1381E+05$	$.1549E+05$	$.1689E+05$	$.1821E+05$
78	$Q_{31}/\alpha_3^2 \cdot p \cdot l_3$	$-.2762E+02$	$.2742E+04$	$-.1687E+04$	$-.4698E+04$	$-.9802E+04$	$-.1444E+05$	$-.1821E+05$	$.2073E+05$

		CYCLE 1	CYCLE 2	CYCLE 3	CYCLE 4	CYCLE 5	CYCLE 6	CYCLE 7	CYCLE 8
79	$\alpha_1^{k+1} = \alpha_1(76)^{1/2}$	.1425E+03	.1443E+03	.1398E+03	.1403E+03	.1406E+03	.1407E+03	.1406E+03	.1403E+03
80	$\alpha_2^{k+1} = \alpha_2(77)^{1/2}$	.1366E+03	.8344E+02	.6921E+02	.5395E+02	.4518E+02	.3999E+02	.3694E+02	.3546E+02
81	$\alpha_3^{k+1} = \alpha_3(78)^{1/2}$	0.	.4050E+00	0.	0.	0.	0.	0.	0.
82	$\alpha_1 = 79/\text{Max}(79-81)$	.1000E+01	.1000E+01	.1000E+01	.1000E+01	.1000E+01	.1000E+01	.1000E+01	.1000E+01
83	$\alpha_2 = 80/\text{Max}(79-81)$	.9588E+00	.5781E+00	.4953E+00	.3844E+00	.3213E+00	.2842E+00	.2627E+00	.2528E+00
84	$\alpha_3 = 81/\text{Max}(78-81)$	0.	.2806E-02	0.	0.	0.	0.	0.	0.
85	α_1 ⎫ checked	.1000E+01	.1000E+01	.1000E+01	.1000E+01	.1000E+01	.1000E+01	.1000E+01	.1000E+01
86	α_2 ⎬ for x min	.9588E+01	.5781E+01	.4953E+00	.3844E+00	.3213E+00	.2842E+00	.2627E+00	.2528E+00
87	α_3 ⎭ $\alpha_{\min} = \dfrac{x\min}{\Delta(\max)}$ (43)	.7735E-02	.7085E+02	.7060E-02	.7050E-02	.7031E-02	.6112E-02	.4925E-02	.4312E-02

Problem 3(c)

In this problem the design variables are modified by using Equation (5.123);

$$x_i^{k+1} = x_i^k \left(1 + \frac{1}{r}\left(\sum_{j=1}^{2} \lambda_j \frac{Q_{ij}}{\rho_i l_i x_i^2} - 1\right)\right)_k$$

with step size parameter $r = 2$. The equations used for evaluating the Lagrange multipliers are the same as those of Problem 2(a). The stress constraints in elements 1 and 3 are assumed to be potentially active in order to reduce the size of the problem solution. The stress in element 2 is inactive, and assuming it to be potentially active would only increase the size of the problem. For element 1 the virtual load is

$$f_x^1 = \frac{E}{l_1} \cos\phi \left(\frac{T_1}{\text{ABS}(T_1)}\right)$$

$$f_y^1 = \frac{E}{l_1} \cos\theta \left(\frac{T_1}{\text{ABS}(T_1)}\right)$$

and for element 2 the virtual load is

$$f_x^2 = -\frac{E}{l_3} \cos\phi \left(\frac{T_3}{\text{ABS}(T_3)}\right)$$

$$f_y^2 = +\frac{E}{l_3} \cos\theta \left(\frac{T_3}{\text{ABS}(T_3)}\right)$$

The flexibility coefficients are denoted by Q_{ij} instead of R_{ij} for convenience. The weight of the optimum design is 126.06 lbs, and the cross-sectional areas of the three members are $x_1 = 14.24$ in^2, $x^2 = 4.929$ in^2 and $x_3 = 0.1$ in^2. The stresses in the three elements are 10 000 psi, 20 000 psi and 10 000 psi respectively. The stress constraint in element 1 is active for all iterations, but the stress constraint in element 3 becomes active only after the fourth iteration. The stress in element 2 is not active.

Comparing this design with the one obtained by using the FSD algorithm, it is seen that even though the stress distribution for the two designs is the same, the two designs are not the same. The design obtained by the FSD algorithm is non-optimum.

	CYCLE 1	CYCLE 2	CYCLE 3	CYCLE 4	CYCLE 5	CYCLE 6	CYCLE 7	CYCLE 8
1 x_1	.1000E+01	.1408E+02	.1485E+02	.1517E+02	.1502E+02	.1283E+02	.1428E+02	.1423E+02
2 x_2	.1000E+01	.1051E+02	.7952E+01	.5780E+01	.4047E+01	.4731E+01	.4858E+01	.4926E+01
3 x_3	.1000E+01	.6760E+01	.3404E+01	.1498E+01	.4377E+00	.2936E+00	.1000E+00	.1000E+00
4 $\bar{C}_1 = \bar{C}_2$	.1000E+14	.1000E+14	.1000E+14	.1000E+14	.1000E+14	.1000E+14	.1000E+14	.1000E+14
5 E	.1000E+08	.1000E+08	.1000E+08	.1000E+08	.1000E+08	.1000E+08	.1000E+08	.1000E+08
6 F_x	.1000E+06	.1000E+06	.1000E+06	.1000E+06	.1000E+06	.1000E+06	.1000E+06	.1000E+06
7 F_y	.2000E+06	.2000E+06	.2000E+06	.2000E+06	.2000E+06	.2000E+06	.2000E+06	.2000E+06
8 $\cos\theta$	.7071E+00	.7071E+00	.7071E+00	.7071E+00	.7071E+00	.7071E+00	.7071E+00	.7071E+00
9 $\cos\phi$	.7071E+00	.7071E+00	.7071E+00	.7071E+00	.7071E+00	.7071E+00	.7071E+00	.7071E+00
10 $\cos^2\theta$	.5000E+00	.5000E+00	.5000E+00	.5000E+00	.5000E+00	.5000E+00	.5000E+00	.5000E+00
11 $\cos^2\phi$	.5000E+00	.5000E+00	.5000E+00	.5000E+00	.5000E+00	.5000E+00	.5000E+00	.5000E+00
12 $\cos\theta \cdot \cos\phi$	.5000E+00	.5000E+00	.5000E+00	.5000E+00	.5000E+00	.5000E+00	.5000E+00	.5000E+00
13 l_1	.7071E+02	.7071E+02	.7071E+02	.7071E+02	.7071E+02	.7071E+02	.7071E+02	.7071E+02
14 l_2	.5000E+02	.5000E+02	.5000E+02	.5000E+02	.5000E+02	.5000E+02	.5000E+02	.5000E+02
15 l_3	.7071E+02	.7071E+02	.7071E+02	.7071E+02	.7071E+02	.7071E+02	.7071E+02	.7071E+02
16 $x_1 E/l_1$	.1414E+06	.1991E+07	.2100E+07	.2145E+07	.2125E+07	.1814E+07	.2020E+07	.2012E+07
17 $x_2 E/l_2$	.2000E+06	.2101E+07	.1590E+07	.1156E+07	.8095E+06	.9462E+06	.9716E+06	.9852E+06
18 $x_3 E/l_3$	.1414E+06	.9561E+06	.4813E+06	.2118E+06	.6190E+05	.4153E+05	.1414E+05	.1414E+05

19 $\frac{x_1 E}{l_1} + \frac{x_3 E}{l_3}(16+17)$			.2828E+06	.2947E+07	.2582E+07	.2357E+07	.2187E+07	.1855E+07	.2034E+07	.2026E+07
20 $\frac{x_1 E}{l_1} - \frac{x_3 E}{l_3}(16-17)$	0.	.1035E+07	.1619E+07	.1933E+07	.2063E+07	.1772E+07	.2006E+07	.1998E+07
21 $K_1 = (11)(19)$	.1414E+06	.1474E+07	.1291E+07	.1178E+07	.1093E+07	.9277E+06	.1017E+07	.1013E+07
22 $K_2 = (12)(20)$	0.	.5175E+06	.8095E+06	.9664E+06	.1031E+07	.8862E+06	.1003E+07	.9991E+06
23 $K_3 = (10)(19)+17$	.3414E+06	.3575E+07	.2881E+07	.2334E+07	.1903E+07	.1874E+07	.1989E+07	.1998E+07
24 $\frac{K_2 \cdot F_y - K_3 \cdot F_x}{(22)(7)-(23)(6)}$	−.3414E+11	−.2540E+12	−.1262E+12	−.4013E+11	.1601E+11	−.1015E+11	.1703E+10	−.2012E+08
25 $\frac{K_2 \cdot K_2 - K_1 \cdot K_3}{(22)(22)-(21)(23)}$	−.4828E+11	−.5000E+13	−.3064E+13	−.1816E+13	−.1017E+13	−.9531E+12	−.1017E+13	−.1027E+13
26 $U_x = 24/25$	.7071E+00	.5080E−01	.4120E−01	.2210E−01	−.1575E−01	.1065E−01	−.1675E−02	.1960E−04
27 $\frac{K_2 \cdot F_x - K_1 \cdot F_y}{(22)(6)-(21)(7)}$	−.2828E+11	−.2430E+12	−.1772E+12	−.1390E+12	−.1155E+12	−.9692E+11	−.1031E+12	−.1027E+12
28 $U_y = 27/25$	.5858E+00	.4859E−01	.5784E−01	.7653E−01	.1136E+00	.1017E+00	.1014E+00	.1001E+00
29 $U_1 = (26)(9)+(28)(8)$	.9142E+00	.7028E−01	.7003E−01	.6974E−01	.6922E−01	.7944E−01	.7053E−01	.7077E−01
30 $U_2 = U_y (28)$	.5858E+00	.4859E−01	.5784E−01	.7653E−01	.1136E+00	.1017E+00	.1014E+00	.1001E+00
31 $U_3 = -(26)(9)+(28)(8)$	−.8579E−01	−.1559E−02	−.1177E−01	.3849E−01	.9150E−01	.6438E−01	.7290E−01	.7075E−01
32 $T_1 = \frac{x_1 E}{l_1} U_1 (16)(29)$	.1293E+06	.1399E+06	.1471E+06	.1496E+06	.1471E+06	.1441E+06	.1425E+06	.1424E+06
33 $T_2 = \frac{x_2 E}{l_2} U_2 (17)(30)$	.1172E+06	.1021E+06	.9199E+05	.8847E+05	.9199E+05	.9622E+05	.9854E+05	.9859E+05

	CYCLE 1	CYCLE 2	CYCLE 3	CYCLE 4	CYCLE 5	CYCLE 6	CYCLE 7	CYCLE 8
34 $T_3 = \dfrac{x_3 E}{l_3} U_3$ (18)(31)	$-.1213E+05$	$-.1490E+04$	$.5665E+04$	$.8155E+04$	$.5663E+04$	$.2673E+04$	$.1031E+04$	$.1001E+04$
35 $\sigma_1 = \dfrac{T_1}{x_1} = \dfrac{32}{1}$	$.1293E+06$	$.9940E+04$	$.9903E+04$	$.9863E+04$	$.9790E+04$	$.1123E+05$	$.9974E+04$	$.1001E+05$
36 $\sigma_2 = \dfrac{T_2}{x_2} = \dfrac{33}{2}$	$.1172E+06$	$.9719E+04$	$.1157E+05$	$.1531E+05$	$.2273E+05$	$.2034E+05$	$.2028E+05$	$.2001E+05$
37 $\sigma_3 = \dfrac{T_3}{x_3} = \dfrac{34}{3}$	$-.1213E+05$	$-.2204E+03$	$.1664E+04$	$.5444E+04$	$.1294E+05$	$.9104E+04$	$.1031E+05$	$.1001E+05$
38 $\Lambda(\sigma_1) = \dfrac{\sigma_1}{\bar{\sigma}_1}$ (abs)	$.1293E+02$	$.9940E+00$	$.9903E+00$	$.9863E+00$	$.9790E+00$	$.1123E+01$	$.9974E+00$	$.1001E+01$
39 $\Lambda(\sigma_2) = \dfrac{\sigma_2}{\bar{\sigma}_2}$ (abs)	$.2929E+01$	$.2430E+00$	$.2892E+00$	$.3827E+00$	$.5682E+00$	$.5085E+00$	$.5071E+00$	$.5004E+00$
40 $\Lambda(\sigma_3) = \dfrac{\sigma_3}{\bar{\sigma}_3}$ (abs)	$.1213E+01$	$.2204E-01$	$.1664E+00$	$.5444E+00$	$.1294E+01$	$.9104E+00$	$.1031E+01$	$.1001E+01$
41 $\Lambda(U_x) = \dfrac{U_x}{\bar{c}_1} = \dfrac{26}{4}$	$.7071E-13$	$.5080E-14$	$.4120E-14$	$.2210E-14$	$-.1575E-14$	$.1065E-14$	$-.1675E-15$	$.1960E-17$
42 $\Lambda(U_y) = \dfrac{U_y}{\bar{c}_1} = \dfrac{28}{4}$	$.5858E-13$	$.4859E-14$	$.5784E-14$	$.7653E-14$	$.1136E-13$	$.1017E-13$	$.1014E-13$	$.1001E-13$
43 $\Lambda(\max)(38-42)$	$.1293E+02$	$.9940E+00$	$.9903E+00$	$.9863E+00$	$.1294E+01$	$.1123E+01$	$.1031E+01$	$.1001E+01$
44 $x_1^* = \Lambda \cdot x_1$ (43)(1)	$.1293E+02$	$.1399E+02$	$.1471E+02$	$.1496E+02$	$.1944E+02$	$.1441E+02$	$.1472E+02$	$.1424E+02$
45 $x_2^* = \Lambda \cdot x_2$ (43)(2)	$.1293E+02$	$.1044E+02$	$.7875E+01$	$.5701E+01$	$.5237E+01$	$.5315E+01$	$.5009E+01$	$.4930E+01$
46 $x_3^* = \Lambda \cdot x_3$ (43)(3)	$.1293E+02$	$.6720E+01$	$.3371E+01$	$.1477E+01$	$.5663E+00$	$.3299E+00$	$.1031E+00$	$.1001E+00$
47 VOLUME $= \sum x_i^* l_i$	$.2475E+04$	$.1987E+04$	$.1672E+04$	$.1447E+04$	$.1677E+04$	$.1308E+04$	$.1299E+04$	$.1261E+04$

48	$\mathbf{WT} = (47)(\rho)$	.2475E+03	.1987E+03	.1672E+03	.1447E+03	.1677E+03	.1308E+03	.1299E+03	.1261E+03
49	$f_x^1 = \dfrac{E}{l_1}\cos\phi \cdot (\text{sign}\cdot T_1)$	.1000E+06	.1000E+06	.1000E+06	.1000E+06	.1000E+06	.1000E+06	.1000E+06	.1000E+06
50	$f_y^1 = \dfrac{E}{l_1}\cos\theta \cdot (\text{sign}\cdot T_1)$	.1000E+06	.1000E+06	.1000E+06	.1000E+06	.1000E+06	.1000E+06	.1000E+06	.1000E+06
51	$\dfrac{K_2 \cdot f_y^1 - K_3 \cdot f_x^1}{(22)(50)-(23)(49)}$	−.3414E+11	−.3057E+12	−.2072E+12	−.1368E+12	−.8714E+11	−.9877E+11	−.9858E+11	−.9993E+11
52	$U_x^1 = 51/25$	.7071E+00	.6115E−01	.6762E−01	.7530E−01	.8572E−01	.1036E+00	.9696E−01	.9734E−01
53	$\dfrac{K_2 \cdot f_x^1 - K_1 \cdot f_y^1}{(22)(49)-(21)(50)}$	−.1414E+11	−.9561E+11	−.4813E+11	−.2118E+11	−.6190E+10	−.4153E+10	−.1414E+10	−.1414E+10
54	$U_y^1 = 53/25$	.2929E+00	.1912E−01	.1571E−01	.1166E−01	.6089E−02	.4357E−02	.1391E−02	.1377E−02
55	$U_1^1 = (52)(9) + (54)(8)$	.7071E+00	.5676E+00	.5892E−01	.6150E−01	.6492E−01	.7636E−01	.6955E−01	.6980E−01
56	$U_2^1 = U_y(54)$	.2929E+00	.1912E−01	.1571E−01	.1166E−01	.6089E−02	.4357E−02	.1391E−02	.1377E−02
57	$U_3^1 = (52)(9) + (54)(8)$	−.2929E+00	−.2972E−01	−.3670E−01	−.4500E−01	−.5631E−01	−.7020E−01	−.6758E−01	−.6785E−01
58	$t_1^1 = \dfrac{x_1 U_1^1 E}{l}$ (16)(55)	.1000E+06	.1130E+06	.1238E+06	.1319E+06	.1379E+06	.1385E+06	.1405E+06	.1405E+06
59	$t_2^1 = \dfrac{x_2 U_2^1 E}{l_2}$ (17)(56)	.5858E+05	.4018E+05	.2498E+05	.1348E+05	.4929E+04	.4123E+04	.1352E+04	.1357E+04
60	$t_3^1 = \dfrac{x_3 U_3^1 E}{l_3}$ (18)(57)	−.4142E+05	−.2841E+05	−.1767E+05	−.9533E+04	−.3485E+04	−.2915E+04	−.9557E+03	−.9596E+03
61	$Q_{11} = \dfrac{T_1 t_1^1 l_1}{E}$	.9142E+05	.1118E+06	.1287E+06	.1395E+06	.1435E+06	.1411E+06	.1415E+06	.1415E+06

	CYCLE 1	CYCLE 2	CYCLE 3	CYCLE 4	CYCLE 5	CYCLE 6	CYCLE 7	CYCLE 8
62 $Q_{21} = \dfrac{T_2 f_x^2 l_2}{E}$	.3431E+05	.2051E+05	.1149E+05	.5964E+04	.2267E+04	.1983E+04	.6659E+03	.6689E+03
63 $Q_{31} = \dfrac{T_3 f_x^2 l_3}{E}$	.3553E+04	.2994E+03	−.7077E+03	−.5497E+03	−.1396E+03	−.5511E+02	−.6967E+01	−.6789E+01
64 Passive elements	0.	0.	0.	0.	0.	0.	3	3
65 $C_1^* = \sum \dfrac{Q_{i1}}{x_i}$ (P)	0.	0.	0.	0.	0.	0.	−.6758E+02	−.6783E+02
66 $f_x^2 = -\dfrac{E}{l_3}\cdot\cos\phi\,(\mathrm{sign}\,T_3)$	.1000E+06	.1000E+06	.1000E+06	.1000E+06	.1000E+06	.1000E+06	.1000E+06	.1000E+06
67 $f_y^2 = \dfrac{E}{l_3}\cdot\cos\theta\,(\mathrm{sign}\,T_3)$	−.1000E+06	−.1000E+06	.1000E+06	.1000E+06	.1000E+06	.1000E+06	.1000E+06	.1000E+06
68 $\dfrac{K_2 f_x^2 - K_3 \cdot f_x^2}{(22)(67)-(23)(66)}$	−.3414E+11	−.4092E+12	.3691E+12	.3301E+12	.2934E+12	.2760E+12	.2991E+12	.2998E+12
69 $u_x^2 = 68/25$	.7071E+00	.8185E−01	−.1205E+00	−.1817E+00	−.2887E+00	−.2896E+00	−.2942E+00	−.2920E+00
70 $\dfrac{K_2 f_x^2 - K_1 f_y^2}{(22)(66)-(21)(67)}$	.1414E+11	.1991E+12	−.2100E+12	−.2145E+12	−.2125E+12	−.1814E+12	−.2020E+12	−.2012E+12
71 $U_y^2 = 70/25$	−.2929E+00	−.3982E−01	.6855E−01	.1181E−00	.2090E+00	.1903E+00	.1987E+00	.1960E+00
72 $U_1^2 = (69)(9) + (71)(8)$	.2929E+00	.2972E−01	−.3670E−01	−.4500E−01	−.5631E−01	−.7020E−01	−.6758E−01	−.6785E−01
73 $U_2^2 = U_y^2$	−.2929E+00	−.3982E−01	.6855E−01	.1181E+00	.2090E+00	.1903E+00	.1987E+00	.1960E+00
74 $U_3^2 = -(69)(9) + (71)(8)$	−.7071E+00	−.8604E−01	.1336E+00	.2120E+00	.3519E+00	.3394E+00	.3485E+00	.3451E+00
75 $t_1^2 = \dfrac{x_1 U_2^2 E}{l_1} - (16)(72)$	.4142E+05	.5917E+05	−.7709E+05	−.9651E+05	−.1196E+06	−.1273E+06	−.1365E+06	−.1365E+06

76 $t_2^2 = \dfrac{x_2 U_2^2 E}{l_2}$ (17)(73)		−.5858E+05	−.8367E+05	.1090E+06	.1365E+06	.1692E+06	.1801E+06	.1930E+06	.1931E+06
77 $t_3^2 = \dfrac{x_3 U_3^2 E}{l_3}$ (18)(74)	−.1000E+06	−.8226E+05	.6433E+05	.4491E+05	.2178E+05	.1409E+05	.4929E+04	.4880E+04
78 $Q_{12} = \dfrac{T_1 t_1^2 l_1}{E}$	.3787E+05	.5854E+05	−.8018E+05	−.1021E+06	−.1244E+06	−.1297E+06	−.1375E+06	−.1375E+06
79 $Q_{22} = \dfrac{T_2 t_2^2 l_2}{E}$	−.3431E+05	−.4272E+05	.5015E+05	.6038E+05	.7782E+05	.8663E+05	.9511E+05	.9518E+05
80 $Q_{32} = T_3 t_3^2 l_3 / E$	.8579E+04	.8668E+03	.2577E+04	.2589E+04	.8723E+03	.2664E+03	.3593E+02	.3452E+02
81 $C_2^* = \sum \dfrac{Q_{i2}}{x_i}$ (P)	0.	0.	0.	0.	0.	0.	.3485E+03	.3449E+03
82 $B_1 = \sum \dfrac{Q_{i1}}{x_i}$ (Active)	.1000E+05	.1000E+05	.1000E+05	.1000E+05	.7566E+04	.1000E+05	.9743E+04	.1007E+05
83 $B_2 = \sum \dfrac{Q_{i2}}{x_i}$ (Active)	.9384E+03	.2218E+03	.1681E+04	.5519E+04	.1000E+05	.8103E+04	.9651E+04	.9651E+04
84 $R_1 = 2(C_1^* + B_1 - \bar{\sigma}_1) + B_1$	.1000E+05	.1000E+05	.1000E+05	.1000E+05	.1000E+05	.2697E+04	.1000E+05	.9093E+04	.1007E+05
85 $R_2 = 2(C_2^* + B_2 - \bar{\sigma}_2) + B_2$	−.1718E+05	−.1933E+05	−.1496E+05	−.3442E+04	.1000E+05	.4310E+04	.9651E+04	.9643E+04
86 $A = \sum \dfrac{Q_{i1} Q_{i1}}{\rho_i l_i x_i^3}$ (Active)	.6567E+06	.7193E+06	.7922E+06	.8740E+06	.4185E+06	.9586E+06	.8877E+06	.9803E+06
87 $B = \sum \dfrac{Q_{i1} Q_{i2}}{\rho_i l_i x_i^3}$ (Active)	.1196E+06	.1841E+06	.2294E+06	−.2754E+06	−.1927E+06	−.6944E+06	−.7611E+06	−.8460E+06
88 $C = \sum \dfrac{Q_{i2} Q_{i2}}{\rho_i l_i x_i^3}$ (Active)	.2076E+06	.4978E+06	.1340E+07	.4670E+07	.9324E+07	.1107E+08	.1524E+08	.1605E+08

		CYCLE 1	CYCLE 2	CYCLE 3	CYCLE 4	CYCLE 5	CYCLE 6	CYCLE 7	CYCLE 8
89	$B \cdot B - A \cdot B$ $(87)(87) - (81)(87)$	$-.1221E+12$	$-.3242E+12$	$-.1009E+13$	$-.4006E+13$	$-.3864E+13$	$-.1013E+14$	$-.1295E+14$	$-.1501E+14$
90	$B \cdot R_2 - C \cdot R_1$ $(87)(85) - (81)(84)$	$-.4131E+10$	$-.8537E+10$	$-.9967E+10$	$-.4575E+11$	$-.2708E+11$	$-.1137E+12$	$-.1459E+12$	$-.1697E+12$
91	$\lambda_1 = 90/89$	$.3385E-01$	$.2633E-01$	$.9880E-02$	$.1142E-01$	$.7007E-02$	$.1122E-01$	$.1127E-01$	$.1130E-01$
92	$B \cdot R_1 - A \cdot R_2$ $(87)(84) - (86)(85)$	$.1248E+11$	$.1575E+11$	$.9556E+10$	$.2535E+09$	$-.4704E+10$	$-.1108E+11$	$-.1549E+11$	$-.1797E+11$
93	$\lambda_2 = 92/89$	$-.1023E+00$	$-.4858E-01$	$-.9472E-02$	$-.6330E-04$	$.1217E-02$	$.1093E-02$	$.1196E-02$	$.1197E-02$
94	IF (93) < 0 $\lambda_1 = R_1/A$	$.1523E-01$	$.1390E-01$	$.1262E-01$	$.1144E-01$	$.7007E-02$	$.1122E-01$	$.1127E-01$	$.1130E-01$
95	IF (93) < 0 $\lambda_2 = 0 \cdot 0$	$0.$	$0.$	$0.$	$0.$	$.1217E-02$	$.1093E-02$	$.1196E-02$	$.1197E-02$
96	$\lambda_1 \cdot Q_{11} + \lambda_2 Q_{12}$ $(94)(61) + (95)(78)$	$.1392E+04$	$.1555E+04$	$.1625E+04$	$.1596E+04$	$.8537E+04$	$.1442E+04$	$.1430E+04$	$.1434E+04$
97	$\lambda_1 Q_{21} + \lambda_2 Q_{22}$ $(94)(62) + (95)(79)$	$.5225E+03$	$.2852E+03$	$.1415E+03$	$.6824E+02$	$.1106E+03$	$.1170E+03$	$.1213E+03$	$.1215E+03$
98	$\lambda_1 Q_{31} + \lambda_2 Q_{32}$ $(94)(63) + (95)(80)$	$.5411E+02$	$.4162E+01$	$-.8933E+01$	$-.6290E+01$	$.8393E-01$	$-.3273E+00$	$-.3553E-01$	$-.3541E-01$
99	$96/(l_1 \cdot x_1^{*2} \cdot \rho)$	$.1178E+01$	$.1123E+01$	$.1062E+01$	$.1009E+01$	$.3195E+00$	$.9823E+00$	$.9329E+00$	$.1000E+01$
100	$97/(l_2 \cdot x_2^{*2} \cdot \rho)$	$.6252E+00$	$.5231E+00$	$.4678E+00$	$.4200E+00$	$.8067E+00$	$.8281E+00$	$.9670E+00$	$.9996E+00$
101	$98/(l_3 \cdot x_3^{*2} \cdot \rho)$	$.4578E-01$	$.1304E-01$	$-.1112E+00$	$-.4075E+00$	$.3701E-01$	$-.4253E+00$	$-.4727E+00$	$-.4999E+00$
102	$x_1^{k+1} = x_1^* (1 + 0.5(99 - 1))$	$.1408E+02$	$.1485E+02$	$.1517E+02$	$.1502E+02$	$.1283E+02$	$.1428E+02$	$.1423E+02$	$.1424E+02$
103	$x_2^{k+1} = x_2^* (1 + 0.5(100 - 1))$	$.1051E+02$	$.7952E+01$	$.5780E+01$	$.4047E+01$	$.4731E+01$	$.4858E+01$	$.4926E+01$	$.4929E+01$

104	$x_3^{k+1} = x_3^*(1 + 0.5(101-1))$	.6760E+01	.3404E+01	.1498E+01	.4377E+00	.2936E+00	.9480E−01	.2718E−01	.2503E−01
105	x_1 checked		.1408E+02	.1485E+02	.1517E+02	.1502E+02	.1283E+02	.1428E+02	.1423E+02	.1424E+02
106	x_2 for		.1051E+02	.7952E+01	.5780E+01	.4047E+01	.4731E+01	.4858E+01	.4926E+01	.4929E+01
107	x_3 x min		.6760E+01	.3404E+01	.1498E+01	.4377E+00	.2936E+00	.1000E+00	.1000E+00	.1000E+00

Foundations of Structural Optimization: A Unified Approach
Edited by A. J. Morris
© 1982 John Wiley & Sons Ltd.

Chapter 6

Optimality Criteria using a Force Method Analysis Approach

6.1 INTRODUCTION

In essence the techniques discussed in Chapter 5 are based on solving a displacement-constrained problem with stress constraints accommodated within this framework. This bias towards displacement-constrained problems occurs because the displacement finite element method is used as the underlying analysis technique. An alternative approach which offers a promise of being more suited to design problems subject to stress constraints is to abandon the displacement method for analysing the structure and turn instead to the force method. Although this approach is in an early stage of development it has both interest and promise and is presented here as a research concept.

The main drawback is that the finite element displacement method has dominated the field of analysis for many years and has been used almost exclusively in optimization work. The force method has largely been ignored in analysis and consequently has not been considered of general or particular relevance to optimization. Some recent use has been made of the force method [1, 2], but this was primarily to reduce the computational effort associated with iterative analysis. Earlier, the role of the force method was examined [3] to determine the possibility of a fundamental integration of the analysis and redesign philosophy. The preliminary results did indicate a potential for the rigorous incorporation of stress constraints. While extremely simple, the approach of using the force method concept to overcome the difficulties associated with strength optimization of redundant systems appears to be entirely novel. If the internal redundant forces are considered to be self-equilibrating external forces, the structure considered becomes effectively statically determinate. The optimality criteria for such a determinate structure can then be derived rigorously for either displacement or stress constraints. The compatibility conditions, associated with the redundancies, now become zero-valued quasi-external displacement constraints. Additional optimality criteria are derived to permit the consideration of the redundant forces as subsidiary variables.

Of course the problem of stress constraints can be transformed into one involving displacement constraints following the procedure of Chapter 5. While this approach is undoubtedly entirely valid, its extension to other types of finite elements may present a major problem.

If the force method is adopted and the optimality criteria generated for both stress and displacement constraints, our next problem is the solution of the resulting set of nonlinear equations. The most appropriate approach is to follow the example of Chapter 5 and use some form of Newton–Raphson technique. However, attempting to solve the linearized form of the total set of governing equations is ineffective due to the inability of the search technique to identify and distinguish between active and inactive constraints.

This difficulty may be largely overcome by the introduction of the dual problem of Chapter 3 which provides an estimate of the optimal population of active constraints. As we saw earlier, this approach is based upon a local linearization of the domain and leads to the selection of a full vertex corresponding to the satisfaction of the requirements that the number of constraints is equal to the number of variables. This problem can then be solved using the linear programming technique.

For stress-constrained problems in which this situation applies at the optimum, the use of the linear programming approach generates the optimal design in only one or two iterative steps. The validity of the optimality is demonstrated by checking that all Lagrangian multipliers are non-negative.

Historically, in test problems involving only stress and minimum-size constraints, the optimum designs determined have appeared to be full vertices, i.e. the number of constraints is equal to the number of variables. Even when it has not been possible to determine the optimum design rigorously, there have been indications that the optimum design tends to have a large number of active constraints. The number of active constraints is important since it affects significantly the strategy required. Therefore, for a number of variations on a 22-bar truss theme, analytic studies were undertaken to determine the optimal design and demonstrate rigorously the active constraint populations. It was shown that, depending on the values of certain parameters, some optimal designs had 22 constraints and others only 21.

For such problems, in which the optimum does not occur at a full vertex, the linear programming approach can still play a dominant role. The linear programming stage will lead to a full vertex design but one for which some Lagrangian multipliers are negative. These correspond to constraints which would not be active at the optimum design and hence should be discarded. For the reduced set of constraints, the Newton–Raphson method can be used satisfactorily. The key to the solution of the optimal system is a good reliable indication of which constraints are active at the optimum design. We return to this aspect of dual solution techniques in later chapters where the full implications of this technique are explored.

In the case when additional displacement constraints are considered the basic approach is essentially unaltered. Displacement constraints appear even in the stress-constraint case—as compatibility conditions. Thus, the mathematical formulation is readily extendible to represent real external displacement constraints. What is different is the uncertainty regarding which of these displacement constraints is active. By definition, all compatibility conditions are always active constraints; additional displacement constraints need not be active. Also, displacement constraints tend to dominate, sometimes to the exclusion of other types of constraints. Hence, a displacement-constrained optimum design may have a much sparser constraint population than a full vertex design. Nonetheless, the same strategy can be applied as for the purely stress-limited cases. But the constraint population predicted by the LP appears to be less valid and greater reliance has to be placed on constraint acquisition and discard algorithms which are incorporated in the full Newton–Raphson search.

6.2 TECHNICAL DISCUSSION

6.2.1 Optimality Criteria

The general problem is the minimization of the weight W of a structure subject to a number of inequality constraints of the form $C_j - \overline{C}_j \geqslant 0$ corresponding to various behavioural response characteristics of the structure.

While many types of responses can be constrained, for the purposes of this chapter attention is confined solely to displacements, stresses and minimum member sizes.

In developing optimality conditions our standard approach is to form a Lagrangian

$$L = W + \sum_j \lambda_j (C_j - \overline{C}_j) \tag{6.1}$$

where, as before, W is the merit function (weight) of the system to be minimized,

C_j is the value of some response characteristic which is constrained to be less than or equal to an allowable value $\overline{C}_j$

and λ_j is an undetermined Lagrangian multiplier.

Differentiation of Equation (6.1) with respect to the primary design variables once more will yield parts of the Kuhn–Tucker conditions which must be satisfied by the optimum design. Differentiation of Equation (6.1) with respect to the secondary variables (Lagrangian multipliers) yields explicitly the constraint conditions. One major problem in this, or any other, approach to structural optimization is determining which constraints are active (e.g. satisfied as equalities) and which are inactive (inequalities). This problem will be addressed in a later section.

We now wish to extend the Lagrangians and optimality conditions derived earlier and include stress, displacement and gauge constraints. In this chapter we make the same assumptions exploited earlier, namely that:

(1) the total weight is the sum of the component member weights;
(2) the weight of each member is a linear function of a single design variable.

As we have seen, these are fairly traditional assumptions for weight minimization problems. They assume that the effect of attachments and joints is ignored and attention is restricted to an essentially membrane-type of behaviour and that only membrane plates and pin-ended frames are used to model the structure.

The weight of the structure can then be written in the usual form

$$W = \sum_{i=1}^{m} w_i x_i. \tag{6.2}$$

For the constraints basically similar conditions must be satisfied:

(3) a constraint is applied to a behavioural characteristic of each individual member of the structure or on the linear sum of contributions arising from each member;
(4) each constraint or constraint contribution must be expressible as an explicit differentiable function of the design variables.

Condition (3) clearly permits the consideration of member stresses and minimum sizes directly and is consistent with the classical virtual force method for calculating displacements used in earlier chapters.

Condition (4) is of much more complex significance and has a dominant influence on the range of applicability of the Lagrangian approach to structural optimization.

For minimum size condition (4) imposes no special restrictions. To be consistent with other constraints, we express minimum size constraints in the form

$$\frac{\bar{x}_i}{x_i} \leq 1 \qquad i = 1, \ldots, n \tag{6.3}$$

where x_i is the current value of a member size
 $\bar{x}_i$ is its minimum allowable value
and n is the number of elements in the structural model.

The corresponding term in the Lagrangian is written

$$\sum_{i=1}^{n} \mu_i \left(\frac{\bar{x}_i}{x_i} - 1 \right)$$

where μ_i are the associated Lagrangian multipliers.

From (5.51, 5.52) the displacement constraint condition can be written in the form

$$\sum_{i=1}^{n} \frac{Q_{ij}}{x_i} \leq \overline{C}_j \qquad j = 1, \ldots, m$$

The corresponding terms in the Lagrangian are

$$\sum_{j=1}^{m} \lambda_j \left(\frac{1}{\overline{C}_j} \sum_{i=1}^{n} \frac{Q_{ij}}{x_i} - 1 \right)$$

where λ_j are the associated Lagrangian multipliers.

For stress constraints, the form is deceptively simple:

$$\sigma_i = \frac{T_i}{x_i} \tag{6.4}$$

As discussed above, this will not meet requirements of condition (4) directly for redundant systems, since T is a complex function of all the x_i. There is no corresponding simple term for derivatives of stress constraints equivalent to that for a displacement constraint. If it is assumed that T_i is invariant which, as we saw in Chapter 3, is exact for statically determinate structures only, the use of the optimality criteria approach will lead directly to a fully stressed design. One solution to the problem of finding a suitable derivative for stress constraints is to compute finite difference approximations but this is generally too computationally expensive. Another approach lies in the use of the force method which forms the central theme of this chapter and which is discussed in the next section.

The Lagrangian can now be written as

$$L = \sum_{i=1}^{n} w_i x_i + \sum_{i=1}^{n} \mu_i \left(\frac{\overline{x}_i}{x_i} - 1 \right) + \sum_{i=1}^{n} v_i \left(\frac{T_i}{\sigma_i x_i} - 1 \right) + \sum_{j=1}^{m} \lambda_j \left(\frac{1}{\overline{C}_j} \sum_{i=1}^{n} \frac{Q_{ij}}{x_i} - 1 \right)$$
(6.5)

Setting the differentials of Equation (6.5) with respect to x_i equal to zero to obtain the first part of the Kuhn–Tucker conditions leads to n nonlinear equations with the design variables x_i and the Lagrangian multipliers μ_i and λ_j as unknowns. Selection of a set of active constraint conditions will define: (a) which Lagrangian multipliers are zero—corresponding to inactive (inequality) constraints, and (b) provide a set of k nonlinear equations (where $k \leq n$) with design variables as unknowns.

If the set of active constraints is known accurately and $k = n$, the constraint equations can be solved using a Newton–Raphson approach to yield the optimum design. As will be discussed later, this happy set of circumstances does not occur too frequently. Either $k = n$ or, even more likely, the active set of constraints is unknown.

6.2.2 Fundamentals of the Force Method

The main effort in much of the early work in the field of structural optimization was directed towards the development of methods of redesign *per se*. The use of a finite element method of analysis was considered merely as an adjunct for determining stresses and displacements rapidly and with a minimum of computational effort. It was assumed that any other method of analysis (e.g. finite difference) would be equally suitable. There was no attempt (or indeed any real reason) to integrate the analysis and redesign philosophies in any real way. With the dominance of the displacement method of finite element analysis in the 1960s and 1970s, this method was fully developed and readily available for incorporation into any optimization program with a minimum of effort. The force method which had achieved initial prominence in the 1950s was effectively discarded because of some inherent disadvantages it was felt to display in comparison with the apparently simpler displacement method.

Use was made of the force method in earlier work on optimization [1] but this was primarily directed at the reduction of computational effort associated with iterative analysis. The analysis was not made an integral part of the optimization procedure. Advantage was taken of the form of the equations to be solved in the iterative analysis to reduce computational times considerably compared with displacement method techniques.

Turning now to the force method of analysis, this is based upon the overall enforcement of structural equilibrium and the subsequent satisfaction of compatibility. We fix attention once again on bar structures but as before the principles apply more generally. For a general finite element model of a structure, the set of overall equilibrium equations is given in Section 3.2:

$$\mathbf{P} = \mathbf{B}^T \mathbf{T} \tag{6.6}$$

which relate the externally applied loads $\mathbf{P}$ and the internal member forces $\mathbf{T}$.

The internal force distribution $\mathbf{T}$ is assumed to be the sum of two component distributions. The first component is one which is in static equilibrium only with the applied loading system while the second component arises from internally self-equilibrating force systems of undetermined magnitudes.

Expressed mathematically,

$$\mathbf{T} = \mathbf{b}_0 + \mathbf{b}_1 \mathbf{X} \tag{6.7}$$

where $\mathbf{b}_0$ is in static equilibrium with the applied loading
and $\mathbf{b}_1$ are unit values of self-equilibrating force systems whose magnitudes $\mathbf{X}$ must be determined.

Although both $\mathbf{b}_0$ and $\mathbf{b}_1$ systems will individually violate compatibility conditions, their weighted linear combination will nevertheless ensure satisfaction of compatibility. The force distributions $\mathbf{b}_0$ and $\mathbf{b}_1$ are not generally unique, except in very simple structures. Their determination may require use of a concept referred to as a structure cutter.

From the definitions of the $\mathbf{b}_0$ matrix, it is clear that there must be zero forces in as many members as there are redundancies. These members can be considered as being cut in the basic (b_0) system[†]. The $\mathbf{b}_0$ matrix can be written as

$$\mathbf{b}_0 = \begin{bmatrix} \mathbf{b}'_0 \mathbf{P} \\ \mathbf{0} \end{bmatrix} \quad (6.8)$$

where $\mathbf{b}'_0$ are the values of the element forces due to unit values of $\mathbf{P}$. In the 'cut' members the only forces will be those due to the redundant systems. If a redundant system is assigned a unit magnitude in a 'cut' member, the $\mathbf{b}_1$ matrix may be written as

$$\mathbf{b}_1 = \begin{bmatrix} \mathbf{b}'_1 \\ \mathbf{I} \end{bmatrix} \quad (6.9)$$

where the partitioning is as in Equation (6.8). Combining Equations (6.8) and (6.9) into (6.7) yields

$$[\mathbf{T}] = \begin{bmatrix} \mathbf{b}'_0 & \mathbf{b}'_1 \\ \mathbf{0} & \mathbf{I} \end{bmatrix} \begin{bmatrix} \mathbf{P} \\ \mathbf{X} \end{bmatrix} \quad (6.10)$$

This matrix is square and invertible and its inverse may be written

$$\begin{bmatrix} \mathbf{P} \\ \mathbf{X} \end{bmatrix} = \begin{bmatrix} \mathbf{A}_1 & \mathbf{A}_2 \\ \mathbf{0} & \mathbf{I} \end{bmatrix} [\mathbf{T}] \quad (6.11)$$

Equation (6.11) corresponds to two equations, the second of which is trivial. The first is exactly Equation (6.7) which defines overall system equilibrium.

Thus, to generate the $\mathbf{b}_0$ and $\mathbf{b}_1$ matrices, one need only start with the equilibrium condition, Equation (6.7). From the rectangular matrix $\mathbf{A}$, a non-singular square matrix $\mathbf{A}_1$ is extracted using a suitable rank technique. The $\mathbf{b}_0$ and $\mathbf{b}_1$ matrices are then given by

$$\mathbf{b}_0 = \begin{bmatrix} \mathbf{A}_1^{-1} \\ \mathbf{0} \end{bmatrix}, \quad \mathbf{b}_1 = \begin{bmatrix} -\mathbf{A}_1^{-1} \mathbf{A}_2 \\ \mathbf{I} \end{bmatrix} \quad (6.12)$$

The extraction of $\mathbf{A}_1$ is not a unique process and can be influenced by a number of factors. The choice of the $\mathbf{b}_0$ matrix is of minor significance, but the $\mathbf{b}_1$ matrix will greatly affect the conditioning of equations to be solved in satisfaction of compatibility. A considerable effort [4] has been devoted to the development of structure cutters which improve conditioning but their study is outside the scope of the present discussion.

To enforce compatibility, the magnitudes of the redundant forces must be selected to ensure that the relative displacements of the cuts in the structure are set to zero. (An analogy to structural optimization wherein displacements are

---

[†] These cuts are not real but are expressed as such to give physical verisimilitude to the concept of compatibility violation.

constrained to a prescribed value can be observed.) To accomplish this the virtual-force method is used with the virtual-force distribution being given by the b_1 matrix.

In matrix notation, the compatibility condition is given by

$$b_1^T f S = 0 \qquad (6.13)$$

or

$$b_1^T f b_0 + b_1^T f b_1 X = 0$$

Hence

$$X = -(b_1^T f b_1)^{-1} (b_1^T f b_0) \qquad (6.14)$$

where f is the flexibility matrix and is the inverse of the stiffness matrix.

Equation (6.14) is the classic expression for the value of the redundancies. While the apparent complexity of Equation (6.14) may indicate computational cost, it must be pointed out that the order of the matrix $(b_1^T f b_1)$ is only equal to the number of redundancies—a number usually considerably smaller than the corresponding number of degrees of freedom encountered in the solution of a displacement method approach.

6.2.3 Role of the Force Method in Structural Optimization

In the force method analysis, the internal force distributions b_0, b_1 are determined purely from static considerations. Hence, apart from the final satisfaction of compatibility conditions, the entire analysis effectively treats only a statically determinate system. Reference to Equation (6.14) indicates very clearly that if the redundant forces X are regarded as part of the applied force set, the entire system is indeed statically determinate. Thus, instead of considering a redundant structure with an external load system P applied, the structure can equally be viewed as a determinate system with an external loading system $\begin{bmatrix} P \\ \overline{X} \end{bmatrix}$. The only additional requirement is that this structure must be designed to ensure that the displacements associated with the (unknown) forces X are constrained to have zero values. In an optimization sense, stress constraints may now be considered in a very simple manner because of the static determinancy generated by the augmented loading system.

Clearly this transformation from redundancy to determinancy cannot be accomplished without paying some penalty. Additional terms must be introduced into the mathematical formulation and the set of design variables expanded to include the values of the redundancies as unknowns.

The full set of optimality criteria and constraint conditions applicable to a force method formulation now contain the additional design variables and are presented in Fig. 6.1. In the derivation of the equations of Fig. 6.1, two types of displacement constraints have been included, those associated with the compatibility conditions and those with externally defined displacement limitations. The

Lagrangian

$$L = \sum_{i=1}^{n} w_i x_i + \sum_{i=1}^{n} \mu_i \left(\frac{x_i^*}{x_i} - 1\right) + \sum_{i=1}^{n} v_i \left[\left(\frac{B_0^i + \sum_{j=1}^{p} B_1^{ij} X_j}{x_i \sigma_i}\right) - 1\right]$$

$$+ \sum_{j=1}^{p} \lambda_j \left[\sum_{i=1}^{n} \left(B_0^i + \sum_{k=1}^{p} B_1^{ik} X_k\right) \overline{f_i} B_1^{ij} \frac{1}{x_i}\right].$$

Optimality criteria

$$\frac{\partial L}{\partial x_i} = 0 = w_i - \mu_i \frac{x_i^*}{x_i^2} - v_i \left(\frac{B_0^i + \sum_{j=1}^{p} B_i^{ij} X_j}{x_i^2 \sigma_i}\right)$$

$$- \sum_{j=1}^{p} \lambda_j \frac{1}{x_i^2} \left[\left(B_0^i + \sum_{k=1}^{p} B_1^{ik} X_k\right) \overline{f_i} B_1^{ij}\right] = f x_i$$

$$\frac{\partial L}{\partial X_j} = 0 = \sum_{l=1}^{n} v_i \frac{B_1^{ij}}{x_i \sigma_i} + \sum_{k=1}^{p} \lambda_k \sum_{i=1}^{n} \frac{B_1^{ij} B_1^{ik} \overline{f_i}}{x_i} = f X_j$$

Note: $\overline{f_i}$ = flexibility matrix

Constraint conditions

Minimum area

$$\frac{\partial L_i}{\partial \mu_i} = 0 = \left(\frac{x_i^*}{x_j} - 1\right) = f a_i$$

Maximum stress

$$\frac{\partial L}{\partial v_i} = 0 = \left[\left(\frac{B_0^i + \sum_{j=1}^{p} B_1^{ij} X_j}{x_i \sigma_i}\right) - 1\right] = f s_i$$

Compatibility

$$\frac{\partial L}{\partial \lambda_j} = 0 = \sum_{i=1}^{m} \left[\left(B_0^i + \sum_{k=1}^{p} B_1^{ik} X_b\right) \overline{f_1} B_1^{ij} \frac{1}{x_i}\right] = f \varepsilon_i$$

Fig. 6.1 Optimality criteria and constraint conditions

major difference between the two terms lies principally in the absence of the unity for the compatibility constraint value. Thus, the stress-constraint problem has been solved effectively by transformation into a type of displacement constraint. It is recognized that an alternate approach would have been to express each individual (bar) element stress constraint as a relative displacement constraint at the two ends. This idea has been used [5], but it is felt to have the disadvantage of requiring more computational effort. The definition of a simple relative displacement constraint criterion will not usually suffice for a multi-node plate element.

6.2.4 Equation Solution and Constraint Detection

In general, there are potentially many more constraints than variables and the actual number of active constraints is a small subset of the total number present. An inactive constraint is represented by a zero value of the corresponding Lagrangian multiplier and the term associated with it vanishes from Equation (6.5). An active constraint corresponds to a non-zero multiplier.

Consider the first two equations of Fig. 6.1, e.g. L and $\partial L/\partial x_i$. Examining the form of L and $\partial L/\partial x_i$, they can be written in the abbreviated forms

$$L = \sum_i w_i x_i + \sum_k \lambda_k \left(\sum_i \frac{K_i}{x_i} - 1 \right) \tag{6.15}$$

$$\frac{\partial L}{\partial x_i} = w_i - \sum_k \lambda_k \frac{K_i}{x_i^2} \tag{6.16}$$

where λ_k represents all multipliers
and K_i represents the remaining terms

which, as we have seen in earlier chapters, constitutes the dual formulation of the original design problem. Forming the usual linear combination, the following relationship can be formed:

$$L + \sum_i x_i \frac{\partial L}{\partial x_i} = 2 \sum_i w_i x_i - \sum_k \lambda_k \tag{6.17}$$

For the optimum structure all terms in L, except the first, vanish. Hence at the optimum $L^* = W^*$, when an asterisk indicates an optimal value. Also, for the optimum structure $\partial L/\partial x = 0$. Therefore, Equation (6.17) reduces to

$$W^* = 2W^* - \sum \lambda_k^*$$

or

$$W^* = \sum \lambda_k^* \tag{6.18}$$

Thus, at the optimum the weight of the structure is given by the sum of the active constraint Lagrangian multipliers†.

Since the form of the nonlinear equations to be solved is well defined, the Newton–Raphson (N–R) method provides a most practical solution technique.

† It is to be noted that Lagrangian multipliers associated with compatibility constraints do not contribute to summation of multipliers which equals the weight. This is due to the absence of the unity which appears in the behavioural constraints and which survives in the combination of equations (6.16) and (6.17).

In an N–R formulation derivatives of all the functions are required, but as the entire approach via the force method results in explicit terms of relatively simple algebraic form, this is a straightforward matter. The calculation of these derivative terms is exact and may assist in convergence, but more importantly, their evaluation does not require finite difference or similar calculations which would involve repeated structural analyses.

For the present set of nonlinear equations consisting of optimality criteria and constraints, the linearized N–R equations are given in Fig. 6.2. All the derivatives are explicitly expressible in terms of the constituent matrices and are listed in Fig. 6.3.

Optimality criteria

$$\frac{\partial fx_i}{\partial x_i}\delta x_i + \sum_{j=1}^{p}\frac{\partial fx_i}{\partial X_j}\delta X_j + \frac{\partial fx_i}{\partial \mu_j}\delta \mu_i + \frac{\partial fx_i}{\partial v_i}\delta v_i + \sum_{j=1}^{p}\frac{\partial fx_i}{\partial \lambda_j}\delta \lambda_j = -fx_i$$

$$\sum_{i=1}^{n}\frac{\partial fX_j}{\partial x_i}\delta x_i + \sum_{i=1}^{n}\frac{\partial fX_j}{\partial v_i}\delta v_i + \sum_{k=1}^{p}\frac{\partial fX_j}{\partial \lambda_k}\delta \lambda_k = -fX_j$$

Constraint conditions

$$\frac{\partial fa_i}{\partial x_i}\delta x_i = -fa_i$$

$$\frac{\partial fs_i}{\partial x_i}\delta x_i + \sum_{j=1}^{p}\frac{\partial fs_i}{\partial X_j}\delta X_j = -fs_i$$

$$\sum_{i=1}^{n}\frac{\partial f\varepsilon_j}{\partial x_i}\delta x_i + \sum_{k=1}^{p}\frac{\partial f\varepsilon_j}{\partial X_k}\delta X_k = -f\varepsilon_j$$

Fig. 6.2 Linearized Newton–Raphson equations

In simple test problems, where the active constraints population was known from the start, the N–R procedure converges rapidly on the optimal solution. In a small problem with three structural elements and two displacement type of constraints, the envelope method required 54 iterations to achieve a converged solution. The N–R approach converged in five iterations on a design slightly lighter (by 0.3%) than that obtained by the envelope method.

6.2.5 Linear Programming

While the N–R solution technique combined with the force method provides the potential for the rigorous incorporation of stress constraints, it is clear that the major problem of detecting active constraints still remains. In order to overcome this problem we turn to the dual formation [6] to furnish the information necessary to make a detection of the active constraints. While the usual direct

$$\frac{\partial fx_i}{\partial x_i} = \frac{2\mu_i x_i^*}{x_i^3} + 2v_i \left(\frac{B_0^i + \sum\limits_{j=1}^{p} B_1^{ij} X_j}{x_i^3 \sigma_i} \right)$$

$$+ \sum_{j=1}^{p} 2\lambda_j \frac{1}{x_i^3} \left[\left(B_0^i + \sum_{k=1}^{p} B_1^{ik} X_k \right) \bar{f} B_1^{ij} \right]$$

$$= \frac{2(\bar{w}_i - fx_i)}{x_i}$$

$$\frac{\partial fx_i}{\partial X_j} = -\frac{v_i B_1^{ij}}{x_i^2 \sigma_i} - \sum_{k=1}^{p} \lambda_k \frac{B_1^{ik} B_1^{ij} \bar{f}_i}{x_i^2} = \frac{\partial fX_j}{\partial x_i}$$

$$\frac{\partial fx_i}{\partial \mu_i} = -\frac{x_i^*}{x_i^2} = \frac{\partial fa_i}{\partial x_i}$$

$$\frac{\partial fx_i}{\partial v_i} = -\frac{B_0^i + \sum\limits_{j=1}^{p} B_1^{ij} X_j}{x_i^2 \sigma_i} = \frac{\partial fs_i}{\partial x_i}$$

$$\frac{\partial fx_i}{\partial \lambda_j} = -\frac{\left(B_0^i + \sum\limits_{k=1}^{p} B_1^{ik} X_k \right) \bar{f}_i B_1^{ij}}{x_i^2} = \frac{\partial f\varepsilon_j}{\partial x_i}$$

$$\frac{\partial fX_j}{\partial v_i} = \frac{B_1^{ij}}{x_i \sigma_i} = \frac{\partial fs_i}{\partial X_j}$$

$$\frac{\partial fX_i}{\partial \lambda_k} = \sum_{i=1}^{n} B_1^{ik} B_1^{ij} \bar{f}_i \frac{1}{x_i} = \frac{\partial f\varepsilon_j}{\partial X_k}$$

Fig. 6.3 Newton–Raphson derivatives

duality approach is entirely valid, the same linearization can be achieved in a more direct manner.

We demonstrated in Equation (6.18) that the optimum weight is the sum of the Lagrangian multipliers associated with all non-zero value constraints. That means the optimum weight is explicitly independent of the (primal) design variables. If we take a problem which has n elements with p redundancies and constraints on stresses and minimum sizes, then, selecting design variables which satisfy the constraint conditions either as equalities or inequalities, in an arbitrary manner, the optimality criteria become a non-square set of $(n+p)$ linear equations with a set of unknowns $> (n+p)$.

Solution of this problem can now be accomplished using a standard linear programming technique, with the merit function being simply the sum of the individual multipliers. In the LP all Lagrangian multipliers, except those associated with the compatibility conditions, will be restrained to have values greater than or equal to zero. The merit function is minimized by a process which selects those variables having non-zero values (active constraints) and those which remain zero (inactive constraints). The result is a constraint population with as many constraints as there are variables.

The resulting new design may not, in general, satisfy all the actual constraints in the problem and hence may have to be adjusted, e.g. areas defined by stress constraints may be smaller than minimum allowable sizes. Dependent upon some simple test criteria, this new design may again be used as the starting point for another iteration of the LP. This cycle may be repeated until convergence occurs, in which case the optimum design has been generated.

In passing it is of interest to note what occurs in purely stress-constrained problems. If the force method is not used then all displacement- and compatibility-related terms vanish from the optimality criteria equations of Fig. 6.1. Examination of the remaining terms in the equations shows that the array uncouples into n individual problems, each involving μ_i and v_i only. The solution to this then indicates that either a μ_i (minimum area) or v_i (maximum stress) is present in each member, i.e. the classic FSD solution. However, in the present case, the presence of the compatibility-related terms associated with the force method formulation ensures avoidance of this pitfall.

The LP approach generates a potential constraint population at each iteration and appropriate non-zero values for the Lagrangian multipliers. Experience indicates that these values are of minor worth, but the population so defined is extremely relevant. The force method approach is ideally suited to the determination of a structure which satisfies a prescribed set of constraints. The proviso must be made that the number of constraints equals the number of variables, i.e. that the design being sought does lie at a full vertex in the design space. This is consistent with the design population prescribed by the LP solution technique which selects such a vertex as optimal in the linearized design space. The consequences of the optimal structure not occurring at a full vertex are discussed in subsequent sections.

In determining a structure which corresponds to a given (full) set of active constraints, the key lies in the derivation of the internal force distribution. An illustration of this is the usual method of generating an FSD via stress-ratio.

An initial guess is chosen, the structure analysed, and areas adjusted to bring stresses to their critical values and/or to their minimum sizes. The process is simple, but many analysis iterations may be required for convergence. This indeed is the method conventionally used in conjunction with a displacement analysis technique.

Assume now that the distribution of critical constraints is known with regard to members which are fully stressed and those with minimum areas. Using the force method approach, the compatibility equations are

$$(\mathbf{b}_1^T \mathbf{f} \mathbf{b}_0) + (\mathbf{b}_1^T \mathbf{f} \mathbf{b}_1)\mathbf{X} = \mathbf{0} \qquad (6.19)$$

Using a slightly different notation, Equation (6.19) can be written in the form

$$\sum_{i=1}^{n} b_1^{ij} \frac{\overline{f_i}}{x_i} \left(b_0^i + \sum_{k} b_1^{ik} X_k \right) = 0 \qquad j = 1, \ldots, n \qquad (6.20)$$

For members in which the stress is critical $(b_0 + \Sigma b_1^{ik} X_k)/x_i$ is replaced by σ_i, and for the other members x is set to $x^{\min}$. All x-values are now eliminated, and the only unknowns are the redundant forces X_k. Solution of the linear equations leads to the required internal force distribution which satisfies the prescribed FSD conditions—without iteration. The areas are obtained from the known stresses and the computed internal forces.

If the constraint distribution given by the LP routine is other than FSD-type, the equations to be solved become nonlinear and a member may be both fully stressed and have a minimum area. Consequently, other member sizes may be completely unrestrained. The presence of active displacement constraints will similarly leave some members undefined.

In such cases, the set of compatibility equations must be expanded by force–stress–minimum area relationships and/or displacement relationships. In either case the number of equations and variables increases to (redundancies + undefined areas) and the equations become nonlinear. At the time of writing a prototype Newton–Raphson technique has been developed incorporating the LP dual solution and has always proved to be entirely adequate for problems in which the defined constraint population corresponds to a feasible population.

6.2.5.1 Ten-bar truss problem

We now turn to illustrating this technique by means of some simple design problems and begin by discussing a case similar to that introduced in Section 5.9. The ten-bar truss shown in Fig. 6.4 is constructed of aluminium alloy throughout ($E = 10^7$ psi, $\rho = 0.1$ lb/in^3). The allowable stress in all bars except member 9 is $\pm 25\,000$ psi and $\bar{x} = 1.0$ in^2 for all bars. The allowable stress for member 9 has been varied in past problems from $\pm 25\,000$ up to $\pm 75\,000$ psi. It has been shown [7] that for $\sigma_9 \leqslant 37\,500$ psi, an FSD is optimum. For $\sigma_9 > 37\,500$ psi the optimum design (weighing 1497.6 lb) is independent of σ_9 and member 9 is neither critically stressed or at minimum size. Use of a stress-ratio approach leads to a totally different design weighing 1725.24 lb, an error of approximately 15% with a completely different distribution of element areas.

The output from the pilot computer program applied to the 10-bar problem with $\sigma_9 = \pm 75\,000$ psi, is shown in Table 6.1. Computer program development is discussed in a later section, but it is appropriate to mention here some of the salient features of these output results.

Optimality Criteria using a Force Method Analysis Approach

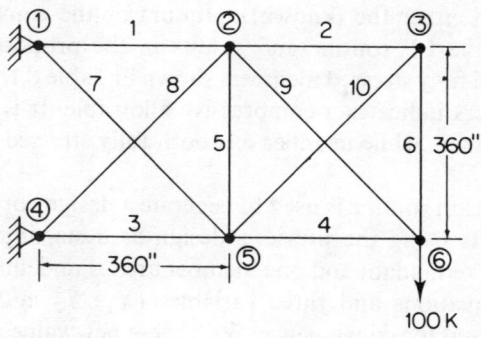

	x_{min} (in²)	σ_{max} (psi)
MEM 1–8, 10	0.1	25 000
MEM 9	0.1	75 000

Fig. 6.4   10-bar truss, basic structure

Table 6.1   10-bar truss—computer results

STRESS RATIO AREAS

| 1 | 0.78146E | 01 | 2 | 0.16050E | 01 | 3 | 0.81854E | 01 | 4 | 0.23950E | 01 | 5 | 0.14196E | 01 |
| 6 | 0.16050E | 01 | 7 | 0.59190E | 01 | 8 | 0.53946E | 01 | 9 | 0.11290E | 01 | 10 | 0.22698E | 01 |

X VALUES   1   −0.21033E 05   2   0.50183E 05

MEM	1	2	3	4	5	6	7	8	9	10
AREA	0	1	0	0	1	0	0	0	0	0
STRA	1	1	−1	1	0	1	1	−1	0	1

N–R AREAS

| 1 | 0.79000E | 01 | 2 | 0.10000E | 01 | 3 | 0.21000E | 00 | 4 | 0.39000E | 00 | 5 | 0.10000E | 01 |
| 6 | 0.10000E | 01 | 7 | 0.57983E | 01 | 8 | 0.55154E | 01 | 9 | 0.36769E | 01 | 10 | 0.14142 | 00 |

X VALUES   1   −0.25000E 04   2   0.25000E 04

LAGRANGIAN MULTIPLIERS

1	0.0		2	0.60001E	01	3	0.0		4	0.0		5	0:36000E	01
6	0.0		7	0.0		8	0.0		9	0.0		10	0.0	
11	0.22200E	03	12	0.66000E	02	13	0.35400E	03	14	0.78000E	02	15	0.69333E	02
16	0.66000E	02	17	0.42000E	03	18	0.15600E	03	19	0.0		20	0.13200E	03
21	0.69333E	02	22	−0.69333E	02									

ITER No. 1
SUM MEM WGT = 0.14976E 04   SUM LAM = 0.14976E 04

No starting point design was specified so the program automatically selects minimum areas as the default option. Experience indicates that such designs may be very poor approximations of optimal systems, hence one stress-ratio cycle is optionally permitted to generate a more-or-less feasible design. This design is still very different from either the (known) optimum or the converged FSD. This design is used in the LP routine and results in the proposed population of minimum areas and fully stressed members shown in Table 6.1. The negative sign on stressed members indicates a compressive allowable. It is to be noted that member 9 is undefined, while member 6 is both fully stressed and of minimum area.

A Newton–Raphson routine is used to generate a design corresponding to the specified constraints using the previous design as a starting point. Since the structure is doubly redundant and one member area is undefined, there are only three nonlinear equations and three variables (X_1, X_2 and x_9). Only three iterations are required for convergence. With these new values of the redundant forces, the known stresses, minimum-sized members and an area for member 9, the new design is generated and printed out. It can be seen that member 2 is at minimum size but the Lagrangian multiplier 14 indicates it is also stress-constrained. This is a rare situation which arises from the geometry and loading at Gridpoint 3, whereby members 2 and 6 are effectively linked together. The presence of surplus constraints has no influence on the problem or solution technique. A shortage of constraints, on the other hand, is of major consequence as will be discussed later. For this design, the corresponding non-zero Lagrangian multipliers can be obtained from the linear solution of the optimality criteria equations. In the print-out the first ten multipliers are those associated with minimum sizes, the next ten with stresses and the final two are the compatibility constraints.

Since all multipliers associated with non-zero type inequality constraints have positive values, the design is optimal. As an additional check, the sum of the multipliers may be computed and are found to equal the weight. This solution, with its extremely rapid convergence, is an excellent demonstration of the power of the LP/force method.

The next stage in the checking process involves the application of the program to a larger problem—the 22-bar double truss.

6.2.5.2 Twenty-two bar truss problem

The 22-bar truss problem is an extension of the ten-bar problem and consists of two ten-bar centilever frames connected at their extremities by a pair of hangers (Fig. 6.5). A single load of 100 000 lbs is applied vertically at the extremity of the hanger. While the modulus is held constant at 10^7 psi throughout, various allowable stresses and material densities are considered. The stress-ratio method is clearly inappropriate for this type of problem since it includes no reference whatever to material density which is a dominant factor in the optimal design. This problem has been considered previously in the literature with a certain amount of confusion [5].

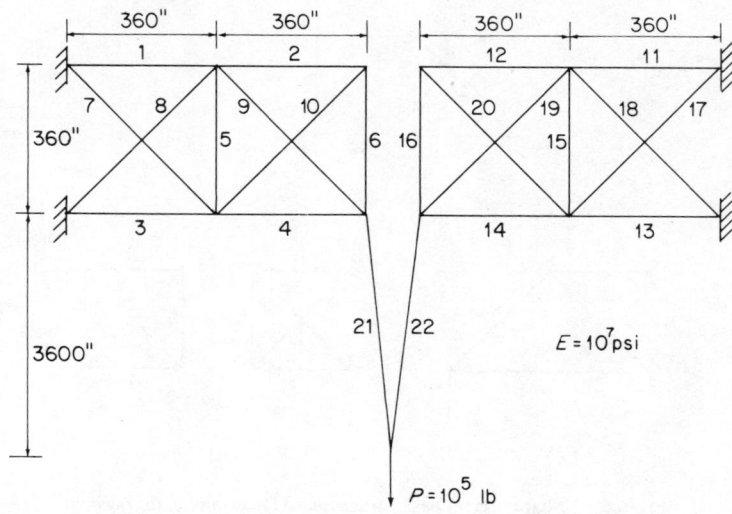

Fig. 6.5 22-bar truss

In this particular problem, selecting the correct set of active constraints is of prime importance and thus a more detailed study of the 22-bar problem is appropriate. Although the structure (in its various forms) is difficult to optimize using general-purpose optimization methods, the generation of optimal solutions analytically is relatively straightforward.

In all the cases of interest, the allowable stresses and densities are uniform throughout each of the 10-bar trusses which form the two principal halves of the structure. The approach to the analytical optimization is through substructuring. The individual 10-bar trusses are optimized parametrically in terms of the applied load and the two halves are coupled through a compatibility condition. This approach is only possible here due to the special geometry of the structure, although the substructuring concept has been proposed for general use in the optimization of complex structures [8].

In order to analyse this structure we break up the framework into two basic substructures, one of which is shown in Fig. 6.6. The constituent determinate (b_0) and redundant (b_1) systems for the basic structure are shown in Fig. 6.6(b) and (c). The choice of these systems is non-unique, but the members retained in the b_0-system are those which form the natural primary load-bearing structure. Provided the ratio between the applied load and the stress allowable is sufficiently high, the members 1, 3, 4, 8 and 9 present in the b_0-system are fully stressed. The remaining members 2, 5, 6, 7 and 10 have minimum areas. It is possible that the applied load could be so small that some or all of the primary b_0-system members would have to be set to their minimum areas instead of being fully stressed. In the present analysis, this situation does not arise. For the systems shown, the relevant

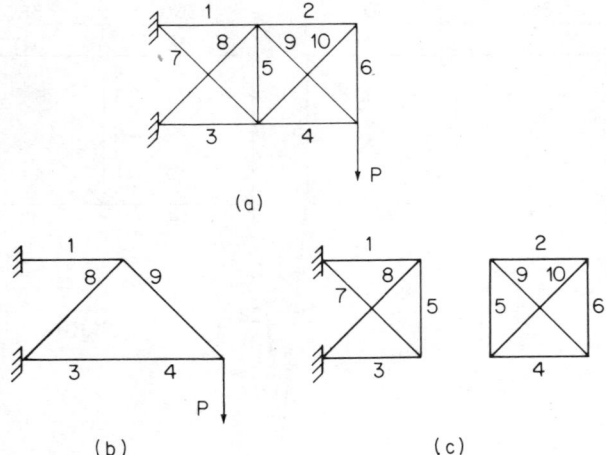

Fig. 6.6 10-bar truss (a) Basic structure (b) Determinate (b_0) system (c) Redundant (b_1) system

matrices are

$$\mathbf{b}_0 = P \begin{bmatrix} 2 \\ 0 \\ -1 \\ -1 \\ 0 \\ 0 \\ 0 \\ -\sqrt{2} \\ \sqrt{2} \\ 0 \end{bmatrix} \quad \mathbf{b}_1 = \begin{bmatrix} 1 & 0 \\ 0 & 1 \\ 1 & 0 \\ 0 & 1 \\ 1 & 1 \\ 0 & 1 \\ -\sqrt{2} & 0 \\ -\sqrt{2} & 0 \\ 0 & -\sqrt{2} \\ 0 & -\sqrt{2} \end{bmatrix} \quad \bar{\mathbf{f}} = \frac{L}{E} \begin{bmatrix} 1 \\ 1 \\ 1 \\ 1 \\ 1 \\ 1 \\ \sqrt{2} \\ \sqrt{2} \\ \sqrt{2} \\ \sqrt{2} \end{bmatrix} \quad (6.21)$$

where $\bar{\mathbf{f}}$ is the diagonal matrix of unit sized element flexibilities
 L is the length of a vertical or horizontal element ($= 360$ in.)
and E is the uniform elastic modulus.

With the information that members 1, 3, 4, 8 and 9 are fully stressed to the allowable value σ and the remaining members have a minimum area x^*, the usual compatibility conditions can be written as

$$\begin{aligned} (2\sqrt{2}+1)X_1 + X_2 &= -2\sigma x^* \\ X_1 + (3+2\sqrt{2})X_2 &= 3\sigma x^* \end{aligned} \quad (6.22)$$

Optimality Criteria using a Force Method Analysis Approach

Hence
$$X_1 = -0.68767264\sigma x^* = \alpha_1 \sigma x^*$$
and
$$X_2 = 0.63270460\sigma x^* = \alpha_2 \sigma x^*. \qquad (6.23)$$

The resultant internal force distribution **T** is

$$\mathbf{T} = \begin{bmatrix} 2P + \alpha_1 \sigma x^* \\ \alpha_2 \sigma x^* \\ -P + \alpha_1 \sigma x^* \\ -P + \alpha_2 \sigma x^* \\ (\alpha_1 + \alpha_2)\sigma x^* \\ \alpha_2 \sigma x^* \\ -\sqrt{2}\alpha_1 \sigma x^* \\ -\sqrt{2}P - \sqrt{2}\alpha_1 \sigma x^* \\ \sqrt{2}P - \sqrt{2}\alpha_2 \sigma x^* \\ -\sqrt{2}\alpha_2 \sigma x^* \end{bmatrix} \qquad (6.24)$$

The corresponding areas are

$$\begin{bmatrix} 2P/\sigma + \alpha_1 x^* \\ x^* \\ P/\sigma - \alpha_1 x^* \\ P/\sigma - \alpha_2 x^* \\ x^* \\ x^* \\ x^* \\ \sqrt{2}(P/\sigma + \alpha_1 x^*) \\ \sqrt{2}(P/\sigma - \alpha_2 x^*) \\ x^* \end{bmatrix} \qquad (6.25)$$

The weight of the optimized 10-bar truss is
$$W = [8P/\sigma + (3 + 2\sqrt{2} + 2\alpha_1 - 3\alpha_2)x^*]360\rho \qquad (6.26)$$
where ρ is material density.

Thus, the weight of each half-structure is expressible purely as a function of the load P at the free end. Since the total load on the structure is known, the problem is now reduced to a single-variable optimization which defines the sharing of the externally applied load between the left- and right-hand halves.

The problem is further simplified when the deflection of each half is computed. Using the unit virtual load method, only the members in the determinate b_0 system need be considered. Since these members are, by definition, fully stressed, their strains are independent of their areas, hence the deflection at the centre point of the two structures is independent of the applied load, i.e.

$$u = \sum_{1,3,4,8,9} \sigma \bar{f} b'_1 \qquad (6.27)$$

Note here that σ is intended to signify both positive and negative stresses which must be accounted for in performing the summation.

Using the values of Equation (6.21), the deflection of the 10-bar frame is

$$u = \frac{8L\sigma}{E} \qquad (6.28)$$

Since L and E are not variable in this problem, the deflection of each half-frame is directly proportional to the allowable stress only. Thus, when the allowable stresses are different on each side, the higher-stressed side will always deflect more, irrespective of which side carries the predominant share of the applied load.

In any particular case, the stresses are specified and the individual deflections are known (e.g. u_L and u_R). The ends of the hangers (members 21 and 22) are joined together. Hence, the difference in stress between the two hangers is

$$\Delta \sigma_H = \frac{8L}{L'}(\sigma_L - \sigma_R) \qquad (6.29)$$

where L' is the hanger length (= 3600 in.) and the subscripts H, L and R indicate hanger, LHS and RHS, respectively.

Finally, from an inspection of the problem parameters (allowable stress and density), it can be readily decided which half-structure should carry the major portion of the applied load. A logical corollary from this is the fact that the hanger on the lightly loaded side has a minimum area. It follows, therefore that the optimum structure has at least 21 active constraints and the existence, or otherwise, of the twenty-second depends on the values of the parameters.

To complete the optimization, it is assumed that the LHS always has the higher allowable stress. Initially it is taken that the stress/density ratio ensures that the major load is carried on the LHS. Substituting numerical values into Equation (6.26), the weights of the two halves can be written as

$$W_L = \left[\frac{8P_L}{\sigma_L} + 2.554\,968 x^*\right] 360 \rho_L$$

$$W_R = \left[\frac{8P_R}{\sigma_R} + 2.554\,968 x^*\right] 360 \rho_R \qquad (6.30)$$

The applied load of the 10^5 lb is the sum of the individual hanger loads, i.e.

$$P_L + P_R = 10^5 \qquad (6.31)$$

Hence

$$W_L = \left[-\frac{8P_R}{\sigma_L} + \frac{8 \times 10^5}{\sigma_L} + 2.554\,968 x^*\right] 360 \rho_L \qquad (6.32)$$

Since the major load is on the LHS, $x_R = x_H^*$ (x_H^* = minimum area of hanger) and

$$\sigma_{RH} = \frac{P_R}{x_H^*} \tag{6.33}$$

Because the LHS deflects more than the RHS, the stress in the RH hanger is less than that of the LH hanger.
Then

$$\sigma_{LH} = \sigma_{RH} - \Delta\sigma_H$$

$$= \frac{P_R}{x^*} - \frac{8L}{L'}(\sigma_L - \sigma_R) \tag{6.34}$$

The corresponding area x_L is given by

$$x_L = \frac{P_L}{\sigma_{LH}} = \frac{10^5 - P_R}{\frac{P_R}{x^*} - \frac{8L}{L'}(\sigma_L - \sigma_R)} \tag{6.35}$$

The total weight is

$$W_{tot} = W_L + W_R + W_{LH} + W_{RH} \tag{6.36}$$

While this expression is somewhat complicated algebraically, substitution of numerical values for known parameters simplifies it to the form

$$W_{tot} = C_1 + C_2 P_R + \frac{C_3}{(P_R - C_4)} \tag{6.37}$$

where C_1, C_2, C_3 and C_4 are numerical coefficients.

The weight of the structure is, therefore, a nonlinear function of the single variable P_R. The minimum weight may then be either a free minimum from Equation (6.37), or may be constrained. Differentiating Equation (6.37) with respect to P_R yields

$$\frac{\partial W_{tot}}{\partial P_R} = C_2 - \frac{C_3}{(P_R - C_4)^2} = 0$$

ie

$$P_R = C_4 + \left(\frac{C_3}{C_2}\right)^{1/2} \tag{6.38}$$

Since the RH hanger has minimum area, if its stress (Equation (6.33), computed using P_R given by Equation (6.38)) is greater than its allowable, then the RH hanger is at both minimum area and maximum stress and Equation (6.38) does not apply. In this case,

$$P_R = \sigma_{RH}^* x_H^* \tag{6.39}$$

and the optimum design has 22 active constraints. With the determination of P_R from either Equation (6.38) or Equation (6.39), the optimum design is completely defined. Both fully and partially constrained designs can therefore exist for stress-constrained problems. If the stress/density ratio transfers the loading to the RHS, a similar procedure is followed to obtain the optimum designs, and, depending on the values of the parameters, both 21 and 22 active constraint designs may occur.

During the study of this basic problem by various investigators, a number of different combinations of parameters have been used, labelled either case A or case B.

To attempt to bring some clarity into the situation, the eight cases given in Table 6.2 are defined. It is believed that these eight cases encompass the principal problems studied by various investigators. It is left to the reader to correlate the new and old designations. Using the analytical approach described in this section, the optimal solutions have been determined as in Table 6.3. With the determination that stress-constrained problems could exhibit less than fully constrained optimal designs, it is necessary to revise the previously developed approach to accommodate this situation.

Table 6.2 22-bar truss—computer results

Case	LHS			RHS			Hanger		
	σ_{ksi}	ρ	x^*	σ_{ksi}	ρ	x^*	σ_{ksi}	ρ	x^*
1	50	0.1	0.001	25	0.1	0.001	5000	0.1	0.001
2	30	0.3	0.001	25	0.1	0.001	5000	0.1	0.001
3	50	0.1	0.001	25	0.1	0.001	500	0.1	0.001
4	30	0.3	0.001	25	0.1	0.001	500	0.1	0.001
5	50	0.1	0.01	25	0.1	0.01	5000	0.1	0.001
6	30	0.3	0.01	25	0.1	0.01	5000	0.1	0.01
7	50	0.1	0.01	25	0.1	0.01	500	0.1	0.001
8	30	0.3	0.01	25	0.1	0.01	500	0.1	0.001

Table 6.3 22-bar truss—optimal solution of principal cases

Case	1	2	3	4	5	6	7	8
Weight	605.096	1202.174	654.04	1232.94	606.752	1312.764	655.70	1236.2
Number of constraints	21	21	22	22	21	21	22	22

6.2.6 CONSTRAINT DISCARD AND THE GENERALIZED NEWTON–RAPHSON TECHNIQUE

In Section 6.2.4 the Newton–Raphson technique is used for the solution of the full set of nonlinear equations consisting of the optimality criteria and all the potential constraints. This approach fails because of the difficulty of identifying meaningful sets of active constraints. If an incomplete set is selected, the resulting equations are linearly dependent and problems with singularity occur.

In the LP routine, the variables (Lagrangian multipliers) are restrained to have values greater than or equal to zero; negative multipliers are not permitted (except those associated with compatibility constraints). Subsequently, a new design is created which has active constraints corresponding to the population of positive Lagrangian multipliers generated by the LP routine. For this new design, a new set of Lagrangian multipliers can be generated by linear solution of the optimality criteria equations. If all the multipliers are positive, the design is optimal. The presence of one or more negative multipliers indicates that non-optimal constraints, associated with these negative multipliers, are present and we must seek a reduced set of potential active constraints for the optimum design. When the constraint population is known the full set of Newton equations must be reduced to provide the non-singular set to be solved.

If the structure has n elements, each of which has both stress and minimum-size constraints and there are m redundancies, the total number of equations available is $(3n+2m)$. The Newton equations can be arranged in the symmetric array shown in Fig. 6.7. If a member is known to be a minimum-size constraint, it is not strictly a variable. Hence the corresponding optimality criterion does not apply and can be deleted along with the associated variable, δ_i, which results in the elimination of the constraint equation associated with that minimum area. For the case where the minimum-area constraint equations for unconstrained members do not apply, the set of minimum-area constraint equations are deleted. Finally, constraint equations and Lagrangian multipliers for non-stress constrained members are inapplicable. This reduces the number of applicable equations to the more manageable order $(2m+n+s-a)$, where s is number of stress constraints and a is number of minimum-sized members.

When a feasible design has been generated by the LP section, the corresponding Lagrangian multipliers are determined. Constraints corresponding to any negative multipliers are discarded and the surviving nonlinear equations are solved iteratively using the full Newton–Raphson method.

As starting points, the current values of member sizes and redundancies are used. Since the equations are linear in the Lagrangian multipliers, the Newton solution will always generate the same final values for the multipliers, at each step, entirely independently of the initially selected guess values. The rate of convergence is therefore unaffected by selection of the starting point for the multipliers. It is controlled by the definition of the constraint population and is also influenced by the initial guesses for the areas.

In many linearized solutions of nonlinear problems, the early iterations have to be controlled to prevent instabilities occurring. The usual method, used here, is the imposition of move-limits on variables.

To test the extended computer program, yet another three-bar truss problem was devised. This one is relatively unique in that it has only two active constraints.

For a structure with four elements and two redundacies the Newton–Raphson array is populated as shown

	δx_1	δx_2	δx_3	δx_4	δX_1	δX_2	$\delta \mu_1$	$\delta \mu_2$	$\delta \mu_3$	$\delta \mu_4$	δv_1	δv_2	δv_3	δv_4	$\delta \lambda_1$	$\delta \lambda_2$
fx_1	×				×	×	×				×				×	×
fx_2		×			×	×		×				×			×	×
fx_3			×		×	×			×				×		×	×
fx_4				×	×	×				×				×	×	×
fx_1	×	×	×	×							×	×	×	×	×	×
fx_2	×	×	×	×							×	×	×	×	×	×
fa_1	×															
fa_2		×														
fa_3			×													
fa_4				×												
fs_1	×				×	×										
fs_2		×			×	×										
fs_3			×		×	×										
fs_4				×	×	×										
$f\varepsilon_1$	×	×	×	×	×	×										
$f\varepsilon_2$	×	×	×	×	×	×										

Condensation:
(1) All fa_i equations and $\delta\mu_i$ variables are deleted
(2) For each element without stress constraint, fs_i equation and δv_i variable are deleted

Fig. 6.7 Symmetric array—Newton equations

Details of the problem are given in Fig. 6.8. The program presents no difficulty in determining the optimal design.

With the development of the extended program, the 22-bar structures which had contributed so much to the detection of the non-fully constrained problems can now be solved. Of the eight problems tested, all except cases 1 and 3 executed successfully. Table 6.4 summarizes the numbers of steps required for convergence. In all a single stress-ratio step was used initially to produce a more-or-less feasible design. As can be seen from Table 6.4, the results generated correspond exactly to those given in Table 6.5.

Optimality Criteria using a Force Method Analysis Approach

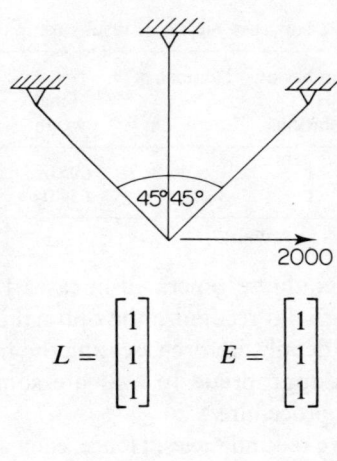

$$L = \begin{bmatrix} 1 \\ 1 \\ 1 \end{bmatrix} \quad E = \begin{bmatrix} 1 \\ 1 \\ 1 \end{bmatrix}$$

$$\sigma_{max} = \begin{bmatrix} =100 \\ \pm 100 \\ \pm 100 \end{bmatrix} \quad x_{min} = \begin{bmatrix} 0.001 \\ 1.0 \\ 0.001 \end{bmatrix}$$

$$\rho = \begin{bmatrix} 5.7403 \\ 1000 \\ 1 \end{bmatrix}$$

Fig. 6.8   3-bar truss with stress constraints

Table 6.4   22-bar truss—iteration history

Case number	Stress ratio iteration	Number of LP iterations	Number of N–R iterations	Final weight	Number of active constraints	Remarks
1	1	2				LP stages diverged
2	1	2	7	1202.17	21	
3	1	5				LP stages diverged
4	1	2		1232.9	22	Final solution is linear
5	1	2	4	606.75	21	Fastest convergence
6	1	2	11	1312.76	21	Slow N–R convergence
7	1	2		655.70	22	
8	1	2		1236.2	22	Final solution is linear

Table 6.5 22-bar truss—test case results using optimal input

Case number	Stress ratio	Number of LP iterations	Number of N–R iterations	Final weight	Number of active constraints	Remarks
1	0	1	4	605.094	21	Optimal
3	0	1		654.04	22	solution input

Although no solution could be generated in cases 1 and 3 using the general starting point, the program did recognize the optimal designs when supplied as input. Table 6.5 provides details. Before assessing the reasons for the difficulties with cases 1 and 3, it is appropriate to evaluate some of the computational requirements of the new procedure.

The 22-bar truss has five redundancies. Hence, each analysis requires solution of five equations. The stress-ratio requires one such analysis. The LP stage extracts the optimal solution from a set of $(22 + 5) = 27$ equations involving $(2 \times 22 + 5) = 49$ unknowns. Typically, the solution requires 40–50 row/column interchanges. The reduced N–R used between LP stages requires the iterative solution of $(5 + \text{NIQ})$ equations, where NIQ is the number of doubly defined elements. When the solution is linear, five equations are solved once.

In the major N–R, the number of symmetric equations solved was only 31, since ten stress constraints and eleven minimum areas were defined for the optimal population. The generation of all the terms in the array solved is simple, direct and requires no auxiliary analyses to generate approximate derivatives.

Turning to the two cases on which the program failed, there appears at first glance to be no obvious reason for the inability of the LP routines to generate feasible constraint distributions. It is true that these cases represent the most radical variations in allowable stresses coupled with extremely small minimum areas. Examination of the values of the Lagrangian multipliers does indicate a potential conditioning problem. The sum of the Lagrangian multipliers is the merit condition for the LP. In case 1, the multipliers range in value from 140.31 down to 0.002 76, which immediately suggests a potential problem. In the successful case 2, the corresponding range is 283.77 to 0.002 76, and no difficulty was encountered. It is, nevertheless, entirely possible that a conditioning problem does exist in the LP state. It may be necessary, for problems as extreme as those considered herein, to formulate the terms in the LP in double precision.

The 22-bar is discussed extensively in Ref. [9] where two cases are presented. From the results, these are believed to correspond to cases 1 and 2 presented here. Case 1 yields a final weight of 606 lb and case 2 a weight of 1203 lb. Both require a large number of iterations to converge. For case 2, an even lower weight of 1188 lb is quoted, but it is not clear as to whether this represents a truly feasible design.

The team of Bartholomew and Morris at the Royal Aircraft Establishment (RAE) has developed the STARS structural optimization system [10] based on a projected gradient method described in Chapter 8 and applied it to the 22-bar problem [11]. For case 2 a weight of 1204.38 lb is generated in 15 iterations. This is not a converged design, as illustrated by the duality gap between the above primal weight and a dual weight of 1200.66 lb. The optimization may have been terminated either due to a limit on the number of iterations or because there is practically no weight change between the last two iterations. The final design presented is close to the optimum but has a significant number of constraints (particularly minimum areas) which are not considered active.

That this situation should occur is puzzling since a comparison of the two methods shows very few differences. In each case, an LP step (or steps) is used to determine a constraint population followed by a N–R search. The STARS program allows both accumulation and discard of constraints. This strategy was not found necessary for the present problems but is used in later versions of the program. The statement is made by the RAE team that the number of active constraints is typically much smaller than the number of variables—a fact not really borne out in the current examples. The RAE solution requires the solution of an $m \times m$ matrix, where m is the number of active constraints. The $m \times m$ matrix is itself a triple product $(GH^{-1}G^T)$ where the H is a diagonal $n \times n$ matrix with n as the number of variables. Finally, it should be noted that G is a constraint derivative matrix. As is well known, if the constraints are stresses, treating the internal forces as invariants can lead to gross errors. To include the internal force redistribution effects in computing the derivatives either requires a finite difference method or the formulation given in Chapter 3. Both of these methods are computationally expensive. Owing to the known extreme nonlinearity of the problem, it is entirely possible that the convergence characteristics are strongly influenced by the accuracy of the constraint derivative matrix G.

With the force method, the constraint derivative matrix is always exact and is generated with a minimum of auxiliary calculation.

6.2.7 Displacement Constraints

The next stage is the formal incorporation of displacement constraints. In principle, this does not raise any new problems, since displacement constraints have already been treated in the form of the compatibility conditions.

The only obvious feature of external displacement constraints are that they are associated with non-zero constraint values and are inequality constraints which need not be satisfied as equalities of the optimum. Compatibility conditions are zero-valued and must be satisfied as equalities throughout.

The governing Lagrangian is modified as shown in Fig. 6.9. The figure also gives the modified forms of the derivatives (optimality criteria and constraints). In the modified equations, one new matrix parameter is introduced (B_D). This is the virtual force in each member of the structure arising from the (virtual) unit loads imposed on the structure to calculate the displacements. Strictly from the

Additional terms required for displacement constraints:
Lagrangian

$$L = \ldots + \sum_{l=1}^{q} \tau_l \left[\frac{1}{C_l} \sum_{i=1}^{n} \frac{\left(B_0^i + \sum_{k=1}^{b} B_1^{ik} X_k\right) \bar{f}_i B_D^{il}}{x_i} - 1 \right]$$

Optimality criteria

$$\frac{\partial L}{\partial x_i} = 0 = \ldots - \sum_{l=1}^{q} \tau_l \frac{1}{C_l x_i^2} \left[\left(B_0^i + \sum_{k=1}^{p} B_1^{ik} X_k\right) \bar{f}_i B_D^{il} \right]$$

$$\frac{\partial L}{\partial x_j} = 0 = \ldots + \sum_{l=1}^{q} \tau_l \frac{1}{C_l} \sum_{i=1}^{m} \frac{B_1^{ij} B_D^{il} \bar{f}_i}{x_i}$$

Displacement constraint

$$\frac{\partial L}{\partial \tau_l} = \frac{1}{C_l} \sum_{i=1}^{m} \frac{\left(B_0^i + \sum_{k=1}^{p} B_1^{ik} X_k\right) \bar{f}_i B_D^{il}}{x_i} - 1 = f_{Dl}$$

Fig. 6.9 Modified Lagrangian to include displacement constraints

definition of the virtual load method, B_D need only be a system in static equilibrium with the virtual load. In the present context it has been shown [2] in Chapter 3 that the B_D system must be the actual internal force distribution arising from the virtual force.

Because of the additional displacement-related terms, the linearized Newton–Raphson equations must also be expanded as indicated in Fig. 6.10. Naturally only active displacement constraints will be retained in the N–R equations to be solved. The total number of equations is given by $(2m + n + s - a + d)$, where d is the number of active displacement constraints. The additional term, d, does not necessarily mean an increase in the order of the equations since the total number of active constraints can still never exceed the number of variables. The upper limit on the number of equations is $(2n + 2m - 1)$ which would only occur when no minimum-size constraints existed.

Because of the similarity between the compatibility and displacement constraints, modification of the computer program was relatively straightforward. The B_D matrices necissitated the introduction of additional loading cases, but this concept had been used in the earlier OPTIM programs.

The new program was tested on the small-scale 3-bar truss problem, which is shown in Fig. 6.11. A few minor developmental problems were encountered and more limits were introduced on the early iterations to prevent initial transients

Optimality criteria

$$\frac{\partial fx_i}{\partial x_i}\delta x_i + \sum_{j=1}^{p}\frac{\partial fx_i}{\partial X_j}\delta X_j + \frac{\partial fx_i}{\partial \mu_i}\partial \mu_i + \frac{\partial fx_i}{\partial v_i}\delta v_i$$

$$+ \sum_{j=1}^{p}\frac{\partial fx_i}{\partial \lambda_j}\delta \lambda_j + \sum_{l=1}^{q}\frac{\partial fx_i}{\partial \tau_l}\delta \tau_l = -fx_i$$

$$\sum_{i=1}^{n}\frac{\partial fX_j}{\partial x_i}\partial x_i + \sum_{i=1}^{n}\frac{\partial fX_j}{\partial v_i}\delta v_i + \sum_{k=1}^{p}\frac{\partial fX_j}{\partial \lambda_k}\delta \lambda_k$$

$$+ \sum_{l=1}^{q}\frac{\partial fX_j}{\partial \tau_l}\delta \tau_l = -fX_j$$

Constraint conditions

$$\frac{\partial fa_i}{\partial x_i}\delta x_i = -fa_i; \quad \frac{\partial fs_i}{\partial x_i}\delta x_i + \sum_{j=1}^{p}\frac{\partial fs_i}{\partial X_j}\delta X_j = -fs_i$$

$$\sum_{i=1}^{n}\frac{\partial f\varepsilon_j}{\partial x_i}\delta x_i + \sum_{k=1}^{p}\frac{\partial f\varepsilon_j}{\partial X_k}\delta X_k = -f\varepsilon_j; \quad \sum_{l=1}^{n}\frac{\partial f_{Dl}}{\partial x_i}\delta x_i + \sum_{k=1}^{p}\frac{\partial f_{Dl}}{\partial X_k}\delta X_k = -f_{Dl}$$

Additional derivatives required are $\partial fx_i/\partial \tau_l = \partial f_{Dl}/\partial x_i$ and $\partial fX_j/\partial \tau_l$ ($= \partial f_{Dl}/\partial X_k$). Other derivatives must be modified to include terms for the displacement constraints. The form of all additional equations and terms corresponds to that used for compatibility.

Fig. 6.10 Modified Newton equations to include displacement constraints

producing unstable situations. Also, the ability to acquire and discard constraints was incorporated.

After each iteration, following the initial transients—estimated as dying out in two or three iterations—all the constraints associated with the design were examined. If any additional constraints had become activated, they were duly recognized and considered in subsequent steps. Should the number of recognized constraints equal or exceed the number of variables, the program then returns to the LP mode to select a new design. Discarding of constraints was based on the same criterion as used initially, i.e. non-positive Lagrangian multipliers.

The next problem tested is another varient of the 10-bar truss structure. This problem, used in Ref. [12], proved to be a rather difficult one to solve and no exact optimum solution has been published. The problem is shown in Fig. 6.12. The major difficulty arises from the specification of constraints on the vertical displacements at both the lower and upper nodes at the free end. If it is assumed,

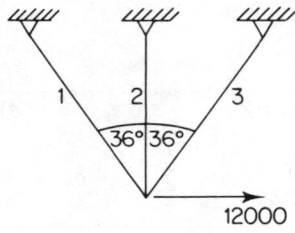

$$L = \begin{bmatrix} 125 \\ 100 \\ 125 \end{bmatrix} \quad \rho = \begin{bmatrix} 0.1 \\ 0.1 \\ 0.1 \end{bmatrix} \quad x_{\min} = \begin{bmatrix} 0.01 \\ 0.01 \\ 0.01 \end{bmatrix}$$

$$\sigma_{\max} = \begin{bmatrix} \pm 50 \\ \pm 50 \\ \pm 50 \end{bmatrix} \quad E = \begin{bmatrix} 10000 \\ 10000 \\ 10000 \end{bmatrix} \quad \delta x, y = \pm 0.05$$

Fig. 6.11 3-bar truss, with displacement constraints

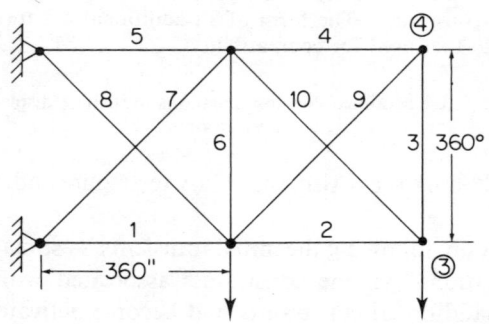

	$x_{\min}$ (in^2)	$\sigma_{\max}$ (psi)	δ_y max (in)
Mem 1–10	0.1	25 000	± 2.0 at 3 and 4

Fig. 6.12 10-bar truss with displacement constraints

say, that the lower node deflects more than the upper and the structure is optimized, using both FSD and the envelope method with the single displacement constraint, the upper node in the optimized system experiences an excessive displacement. Using the upper displacement as the constrained value reverses the situation and the lower displacement violates at the new optimum. Using both upper and lower displacements in the envelope method produces an unsatisfactory situation in which no converged solution is obtained. In the course of the iterations, which tend to become unstable at one stage, a very low weight design is generated. From an examination of this design it does appear, however, to be near optimum. Unfortunately, it has not been possible to determine exactly what the true optimum is and which constraints are active.

Using the current version of the program the above problem was tried and a number of difficulties encountered. It proved impossible to generate a feasible full vertex design based upon any population predicted by an LP solution. Although a number of iterations were permitted, resulting in a number of different LP populations, it was never possible to converge on any full vertex designs using the Newton–Raphson technique. This situation had been encountered previously for stress-limited problems. The solution adopted previously was to generate a fully stressed design (or a close approximation thereto) using a limited number of stress-ratio iterations. For this design, the Lagrangian multipliers were found and after discarding inactive constraints the full Newton–Raphson search was used. In the presence of potentially active displacement constraints, the FSD alone may result in a violated design. From expediency, it was decided to use this FSD, suitably scaled, to ensure satisfaction of the single dominant displacement constraint as the starting point for the Newton search. Consideration was also given to the use of an envelope method to generate a starting point, either using a single or multiple displacement constraint. This approach was not implemented in the present program for two main reasons. Firstly, the envelope method is computationally more complex than the stress-ratio FSD. Secondly, from the previous history with this particular problem, use of either of the envelope methods could be unreliable—a single constraint always failed and double constraints would not converge.

Using this scaled FSD as a starting point, the Newton–Raphson stage was entered. For a limited number of steps, convergence appeared to be satisfactory, moving in the direction of the assumed optimal design. But after additional constraints were captured, according to the built-in logic, the direction of convergence altered and rapidly became unstable. Table 6.6 provides details.

After considerable numerical experimentation, the problem appeared to reside in the accumulation of undesirable constraints during the search, with no satisfactory indications of which should be rejected. Thus, the problem again appeared to devolve on the question of constraint identification.

Table 6.6 10-bar truss with displacement constraints—computer results

STRESS RATIO AREAS
1 0.30373E 02 2 0.88870E 01 3 0.59555E 01 4 0.59555E 01 5 0.28997E 02
6 0.52675E 01 7 0.20017E 02 8 0.21963E 02 9 0.84222E 01 10 0.12568E 02
X VALUES 1 −0.13894E 02 2 0.30352E 02

MEM	1	2	3	4	5	6	7	8	9	10
AREA	0	0	1	0	0	1	0	0	0	0
STR	−1	1	0	0	1	0	−1	1	1	0

DISPL 1 2
ACTIVE 1 1
N–R AREAS
1 0.80000E 01 2 0.40000E 01 3 0.10000E 00 4 0.1000E 00 5 0.80000E 01
6 0.10000E 00 7 0.56568E 01 8 0.56568E 01 9 0.1000E 00 5 0.10000E 01
X VALUES 1 −0.48369E−03 2 0.43948E−06
N–R DID NOT CONVERGE, REENTER L–P

MEM	1	2	3	4	5	6	7	8	9	10
AREA	0	0	0	0	0	1	0	0	0	1
STR	−1	−1	0	0	0	0	−1	1	1	−1

DISPL 1 2
ACTIVE 1 1
N–R AREAS
1 0.11906E 02 2 0.10000E 00 3 0.35122E 06 4 0.10000E 00 5 0.14967E 01
6 0.16486E 00 7 0.13315E 00 8 0.11180E 02 9 0.57568E 01 10 0.10000E 00
X VALUES 1 −0.97646E 02 2 0.10177E 03
STRESS RATIO AREAS
1 0.29026E 02 2 0.14176E 02 3 0.36001E 00 4 0.36001E 00 5 0.28575E 02
6 0.36001E 00 7 0.20046E 02 8 0.20684E 02 9 0.36001E 00 10 0.20048E 02
X VALUES 1 −0.15550E 01 2 0.15536E 01
SINGLE DISPLACEMENT CONSTRAINT DOMINATES
SUM MEM WGT = 0.57355E 04   SUM LAM = 0.57355E 04
N–R AREAS
1 0.21374E 02 2 0.14281E 02 3 0.44131E 00 4 0.44131E 00 5 0.30355E 02
6 0.21601E 00 7 0.21989E 02 8 0.12410E 02 9 0.50402E 00 10 0.20197E 02
X VALUES 1 0.17070E 01 2 0.87457E 00
SUM MEM WGT = 0.5221102E 04   SUM LAM = 0.4374711E 04
N–R AREAS
1 0.20732E 02 2 0.15925E 02 3 0.26478E 00 4 0.26478E 00 5 0.30621E 02
6 0.15470E 00 7 0.22099E 02 8 0.74462E 01 9 0.30241E 00 10 0.22521E 02
X VALUES 1 0.43817E 01 2 −0.40262E 00
SUM MEM WGT = 0.5112699E 04   SUM LAM = 0.4502434E 04
N–R AREAS
1 0.22290E 02 2 0.15085E 02 3 0.15887E 00 4 0.15887E 00 5 0.30860E 02
6 0.10607E 00 7 0.21748E 02 8 0.54855E 01 9 0.185535E 00 10 0.21334E 02

Table 6.6 (*continued*)

X VALUES	1	0.46337E 01	2	−0.63535E 00					
SUM MEM WGT = 0.4953781E 04			SUM LAM = 0.4780637E 04						

N–R AREAS

1	0.20604E	02	2	0.15652E	02	3	0.10000E	00	4	0.10000E	00	5	0.31637E	02
6	0.14849	00	7	0.23173E	02	8	0.32913E	01	9	0.11132E	00	10	0.22135E	02

X VALUES	1	0.10773E 02	2	−0.10200E 01
SUM MEM WGT = 0.4936586E 04			SUM LAM = 0.4581691E 04	

N–R AREAS

1	0.22037E	02	2	0.15565E	02	3	0.10000E	00	4	0.10000E	00	5	0.30110E	02
6	0.18551E	00	7	0.21093E	02	8	0.46078E	01	9	0.15585E	00	10	0.22012E	02

X VALUES	1	0.86500E 01	2	−0.62550E 00
SUM MEM WGT = 0.4888543E 04			SUM LAM = 0.5920930E 04	

N–R AREAS

1	0.22552E	02	2	0.15364E	02	3	0.10000E	00	4	0.10000E	00	5	0.30453E	02
6	0.20382E	00	7	0.21233E	02	8	0.64510E	01	9	0.13928E	00	10	0.21728E	02

X VALUES	1	0.61123E 01	2	−0.23041E 00
SUM MEM WGT = 0.4998473E 04			SUM LAM = 0.5225891E 04	

N–R AREAS

1	0.22810E	02	2	0.14992E	02	3	0.10000E	00	4	0.10000E	00	5	0.30930E	02
6	0.17419E	00	7	0.21601E	02	8	0.65412E	01	9	0.10000E	00	10	0.21202E	02

X VALUES	1	0.52540E 01	2	−0.18723E 00
SUM MEM WGT = 0.5005059E 04			SUM LAM = 0.3111493E 04	

N–R AREAS

1	0.28306E	02	2	0.89953E	01	3	0.10000E	00	4	0.10000E	00	5	0.39001E	02
6	0.14835E	00	7	0.27401E	02	8	0.71342E	01	9	0.10000E	00	10	0.12721E	02

X VALUES	1	0.41969E 01	2	0.91289E−01
SUM MEM WGT = 0.5170379E 04			SUM LAM = 0.1882996E 04	

N–R AREAS

1	0.31648E	02	2	0.96492E	01	3	0.10000E	00	4	0.10000E	00	5	0.43846E	02
6	0.11804E	00	7	0.30818E	02	8	0.70263E	01	9	0.10000E	00	10	0.13646E	02

X VALUES	1	0.36351E 01	2	0.51553E−02
SUM MEM WGT = 0.5703102E 04			SUM LAM = 0.1138443E 04	

N–R AREAS

1	0.28042E	02	2	0.93600E	01	3	0.24000E	00	4	0.10000E	00	5	0.41884E	02
6	0.10000E	00	7	0.29597E	02	8	0.59015E	01	9	0.10000E	00	10	0.13236E	02

X VALUES	1	0.39562E 01	2	−0.14401E 00
SUM MEM WGT = 0.5356340E 04			SUM LAM = −0.5979176E 04	

6.3 CONCLUSIONS

The new approach to structural optimization via the use of the force method of analysis and a linear programming stage has achieved a major degree of success. Some of the classic optimization test problems have been reduced to almost trivial solutions. The 10-bar truss problem with non-uniform stresses is solved without iteration. On the larger 22-bar trusses, solutions have been achieved in the majority of cases, which are confirmed as being optimal by comparison with analytically generated results. The results provide a clear indication of the validity of the approach for the rigorous incorporation of stress constraints. Because of the relatively rapid convergences achieved, the iterative computational effort appears to be significantly lower than that required by alternate approaches. The key to the generation of the optimal design lies in the ability to predict or identify the active constraints. The linear programming stage incorporated in the developed approach does an excellent job of selecting dominant constraints through local linearization of the nonlinear domain, but, as in any nonlinear situation, does require either a good starting point and or a well-conditioned problem to solve.

In the study of the 22-bar problems, no solutions could be generated using the standard program, although the correctness of the optima were demonstrated by the program. In these cases, one of which was a full vertex design, the problem was felt to be attributable to conditioning. This could not be entirely proven, since some of the other 22-bar cases which were successfully solved might have appeared to be more poorly conditioned. The conditioning problem here was taken to be associated with the very large spreads in magnitude between the largest and smallest Lagrangian multipliers. The spread is 10^5 and since the merit function is the sum of the multipliers, there is potential for round-off error to obscure the situation and lead to erroneous definition of active constraints.

Naturally, the 22-bar problem is unrealistic being selected to exhibit certain characteristics on a relatively small scale and to be a test for optimization techniques. It is not clear whether the problem is really pathological—especially in its ranges of values of multipliers—or if it is truly characteristic of larger-scale problems. Some consideration has been given to transformation of the problem to try and eliminate the size disparity in numbers which are linearly combined. Simple scaling alone will not suffice, but no suitable transformation could be evolved.

The presence of very small, but non-zero, values for Lagrangian multipliers associated with real (active) constraints presents a hazard in the Newton–Raphson search procedure which allows for constraint acquisition and discard. The question of when a multiplier becomes zero and the corresponding constraint should be discarded is made extremely difficult, since such small values are readily masked by round-off error in large-scale problems.

With regard to displacement constraints, the fundamental concepts and approach seem to have been validated. Some small-scale problems were successfully solved. Difficulties with constraint detection on larger problems appear to remain. In purely stress and minimum-size problems, the number of active constraints equals or approaches the number of variables. The population defined by the LP stage is therefore a reliable indication of the final active constraints. In the presence of displacement constraints, this no longer holds true. The number of active constraints is frequently very much reduced and the LP is much less reliable in predicting population.

To overcome this situation, the expedient solution of selecting a single dominant displacement constraints was adopted to generate a starting design for the Newton search. This did not perform satisfactorily and placed too much reliance on the constraint acquisition/discard algorithms. It is clear that the method of selecting a starting design for the Newton search must be revised.

In all optimization research, there has always been a great deal of concern for computational efforts involved in the various approaches. Both the numbers of iterations and the cost per iteration are major contributing factors. In this force method work, the costs appear to compare most favourably with other methods. The force method, generally, requires the solution of a smaller array of equations than the displacement method for a given structure (redundancies versus displacements). The need for a structure cutter in the force method has a major cost impact in a single-shot analysis but this effect is reduced considerably in iterative situations. The LP stage is equivalent to the inversion of an array of order of the number of variables. The full Newton–Raphson search does require iterating with a larger set of equations. For the 22-bar trusses, 31 equations are required. Although this array is large, the computation of the individual terms is very straightforward. No auxiliary analyses are required to generate approximations to gradients of constraints. The absence of approximations throughout is considered to be a major factor in the good convergence characteristics observed. The size of the Newton array may be a problem in larger systems. The matrix is symmetric and sparse, so that the possibility of using efficient solution techniques exists.

R. A. G.
R. D. T.

REFERENCES

1. G. Thierauf and A. Topcu, Structural optimization using the force method, World Congress on Finite Element Methods in structural Mechanics, England, October 1975.
2. W. Lipp and G. Thierauf, Die Bedeutung des Kraft–und Weggrossenverfahrens fur die Optimierung von Tragwerken nach der Lagrange 'schen Multiplikatorenmethode', Tenth Congress, IABSE, Tokyo, September 1976.
3. R. A. Gellatly and L. Berke, A preliminary study of a new approach to the optimization of strength limited structures, *AFFDL-TM*-75-162-*FBR*, September 1975.
4. J. Pichard, FORMAT II–Second version of Fortran matrix abstraction technique: Engineering User Report, Vol. I, *AFFDL-TR*-66-207, Air Force Flight Dynamics Laboratory, WPAFB, Ohio, September 1965.
5. L. Berke and N. S. Khot, A simple virtual strain energy method to fully stress design structures with dissimilar stress allowables and material properties, *AFFDL-TM-77-28-FBR*, December 1977.

6. P. Bartholomew and A. J. Morris, A unified approach to fully-stressed design *Engineering Optimization*, **2**, 3–15, 1976.
7. L. Berke and N. S. Khot, Use of optimality criteria methods for large scale system *AGARD Lecture Services No 170*, on Structural Optimization, AGARD-LS-70, 1974.
8. J. S. Arora and A. K. Govil, An efficient method for optimal structural design by substructuring, *Comput. Structures*, **7(4)**, 507–15, 1977.
9. N. S. Khot, L. Berke and V. B. Venkayya, Minimum weight design of structures by the optimality criterion and projection method *20th Structures and Structural Dynamics Conference*, St Louis, Missouri, April 1979.
10. STARS, *RAE Structural Analysis and Redesign System User Guide*, Structures Department, RAE Farnborough, Hants, UK, May 1981.
11. P. Bartholomew, Private communication, 1978.
12. R. A. Gellatly and L. Berke, Optimal structural design *AFFDL-TR*-70-165, February 1971.

Foundations of Structural Optimization: A Unified Approach
Edited by A. J. Morris
© 1982 John Wiley & Sons Ltd.

Chapter 7
Introduction to Mathematical Programming Methods

7.1 INTRODUCTION

The aim of the four next chapters is twofold. On one hand we wish to introduce the reader to the numerous available methods of mathematical programming, in such a way that a wide variety of optimum design problems can be covered in this book. On the other hand, in the special but important case of optimal sizing problems, such as those considered in the preceding chapters, we want to present a unified approach where optimality criteria are seen to reside within the general framework of the mathematical programming approach to structural optimization.

In this chapter, a classification of mathematical programming problems is proposed, the primal and dual statements are again discussed, basic descent algorithms are examined, and several line-search techniques are studied in detail. Unconstrained and linearly constrained minimization algorithms are investigated in Chapter 8, while Chapter 9 is concerned with general nonlinear programming methods. Finally, Chapter 10 deals with the more specific problem of structural weight minimization for a finite element model with fixed geometry and material properties; several numerical examples are provided to support a comparison of various primal and dual solution schemes.

Most of the material on mathematical programming methods presented hereafter is based on the books of Fiacco and McCormick [1], Lasdon [2], Fox [3], Lootsma [4], Himmelblau [5], Luenberger [6], Gill and Murray [7] and Wolfe [8].

7.11 CLASSIFICATION OF MATHEMATICAL PROGRAMMING PROBLEMS

The mathematical programming problem studied in the next chapters can be written as follows: determine a vector $\mathbf{x}^* = (x_1^*, \ldots, x_n^*)^T$ that solves the problem

$$\text{minimize} \quad f(x) \tag{7.1}$$
$$\text{subject to} \quad h_j(x) \geq 0 \quad j = 1, \ldots, m \tag{7.2}$$
$$\bar{x}_i \geq x_i \geq \underline{x}_i \quad i = 1, \ldots, n \tag{7.3}$$

In this statement of the problem, lower and upper bound constraints (1.3) are considered apart from the general constraints (1.2) because these *side constraints* are very simple explicit functions and they can be treated separately in most mathematical programming algorithms. In particular, there is no need to introduce Lagrangian multipliers associated with the side constraints in order to establish the necessary optimality conditions for a minimum (see Section 2.9). The well-known Kuhn–Tucker conditions take the form

$$\frac{\partial f}{\partial x_1} - \sum_{j=1}^{m} \lambda_j \frac{\partial h_j}{\partial x_i} = 0 \quad \text{if} \quad \underline{x}_i < x_i < \bar{x}_i \tag{7.4}$$

$$\frac{\partial f}{\partial x_i} - \sum_{j=1}^{m} \lambda_j \frac{\partial h_j}{\partial x_i} > 0 \quad \text{if} \quad x_i = \underline{x}_i \tag{7.5}$$

$$\frac{\partial f}{\partial x_i} - \sum_{j=1}^{m} \lambda_j \frac{\partial h_j}{\partial x_i} < 0 \quad \text{if} \quad x_i = \bar{x}_i \tag{7.6}$$

where the λ_j's must be non-negative ($\lambda_j = 0$ if $h_j > 0$). Note that equality constraints $h_k(x) = 0$ (see Section 2.8) are not included here because each of them can be taken into account in an optimization algorithm by introducing two constraints $h_k(x) \geq 0$ and $h_k(x) \leq 0$.

In the context of structural optimization, the design variables x_i are the cross-sectional sizes of the structural members and the objective function is the structural weight, which is linear in the design variables (see Chapter 3). The main constraints are the behaviour constraints $h_j(x) \geq 0$, which impose limitations on quantities describing the structural response (stresses, displacements, frequencies, buckling loads, etc.), while the side constraints, which prescribe lower and upper bounds $\underline{x}_i$ and $\bar{x}_i$ to the member sizes, usually reflect fabrication and analysis validity considerations.

When the objective function $f(x)$ and all the constraint functions $\{h_j(x), j = 1, \ldots, m\}$ are linear, the problem stated in Equations (7.1)–(7.3) is called a *linear programming problem*. This special case has been treated extensively in the literature and well-established solution methods are now available on nearly all computer installations. Therefore we shall assume that we are provided with a standard linear programming problem-solver and shall not study it further in this book. If any of the problem functions $\{f, h_j\}$ is nonlinear, the problem is called a *nonlinear programming problem*. Solution methods are then much less standard and in fact are problem dependent.

The *unconstrained minimization problem*

$$\text{minimize } f(x) \text{ for } \mathbf{x} \in E^n \tag{7.7}$$

is a very important special case, because it is at the origin of the basic theory underlying nonlinear programming methods. Also, many efficient techniques for dealing with constrained problems consist in solving a sequence of unconstrained problems. Another interesting special case is that of the *linearly constrained minimization problem*. Indeed, methods for unconstrained problems can be readily adapted when linear constraints are considered. Moreover, some effective general-purpose optimization techniques proceed by transforming the original problem into a sequence of linearly constrained subproblems. This is especially true in modern structural weight-minimization methods (see Chapter 10).

When the general constraints (7.2) are absent in the problem statement, we are faced with a special linearly constrained problem:

$$\begin{aligned}&\text{minimize } f(x)\\&\text{subject to } \bar{x}_i \geqslant x_i \geqslant \underline{x}_i\end{aligned} \tag{7.8}$$

Because taking care solely of the side constraints is a very simple matter, we call it a *quasi-constrained problem*. In particular, the Kuhn–Tucker conditions (7.4)–(7.6) take the form:

$$\frac{\partial f}{\partial x_i} = 0 \quad \text{if} \quad \underline{x}_i < x_i < \bar{x}_i \tag{7.9}$$

$$\frac{\partial f}{\partial x_i} > 0 \quad \text{if} \quad x_i = \underline{x}_i \tag{7.10}$$

$$\frac{\partial f}{\partial x_i} < 0 \quad \text{if} \quad x_i = \bar{x}_i \tag{7.11}$$

The problem (7.1)–(7.3) is called a *convex programming problem* when the functions $\{f(x); -h_j(x), j = 1, \ldots, m\}$ are convex. The smoothness of the problem functions makes the problem well behaved, the feasible region is a convex set, and most importantly, any local solution is also global. Also, many of the duality results, which are so important in structural optimization, are strictly valid for convex problems only. Two interesting special cases of convex problems have been widely studied in the literature. *The quadratic programming problem* consists in minimizing a positive semidefinite quadratic form, subject to linear constraints:

$$\text{minimize } q(x) = \tfrac{1}{2}x^T A x - b^T x \tag{7.12}$$

$$\text{subject to } \sum_{i=1}^{n} c_{ij} x_i \geqslant \underline{c}_j \tag{7.13}$$

$$\bar{x}_i \geqslant x_i \geqslant \underline{x}_i \tag{7.14}$$

This problem mainly serves as a reference for establishing the convergence properties of the various algorithms available. In the unconstrained case, the quadratic problem is specially important and it leads to many fundamental concepts, such as those of conjugate directions, quadratic termination, etc. (see Section 8.3). The other special case in convex programming is that of a convex separable problem, for which duality results can be readily implemented into efficient algorithms. A *separable programming problem* is one that can be written

$$\text{minimize} \quad f(x) = \sum_{i=1}^{n} f_i(x_i) \tag{7.15}$$

$$\text{subject to} \quad h_j(x) = \sum_{i=1}^{n} h_{ji}(x_i) \geq 0 \quad j = 1, \ldots, m \tag{7.16}$$

$$\bar{x}_i \geq x_i \geq \underline{x}_i \tag{7.17}$$

where each function $\{f_i(x_i), h_{ji}(x_i)\}$ depends only on the single variable x_i. Separable functions have several computationally important properties. In particular, the Hessian matrix of such a function is diagonal.

7.2 PRIMAL AND DUAL PROBLEM STATEMENT

The important concept of duality is explored in Section 2.10 and it is used in Chapter 3 and other chapters in conjunction with structural optimization problems. In this section we again consider the dual problem, by taking now an algorithmic point of view. As indicated in the beginning of Section 7.1, only the main constraints (7.2) have to be associated with Lagrangian multipliers, or *dual variables*, $(\lambda_j, j = 1, \ldots, m)$, while the side constraints are treated separately. Let therefore X define the set of all primal points satisfying the side constraints (7.3), that is,

$$X = \{\mathbf{x} : \underline{x}_i \leq x_i \leq \bar{x}_i; i = 1, \ldots, n\} \tag{7.18}$$

and let Λ denote the set of all dual points satisfying the non-negatively conditions, that is,

$$\Lambda = \{\lambda : \lambda_j \geq 0; j = 1, \ldots, m\} \tag{7.19}$$

Corresponding to the *primal problem*

$$\begin{aligned}\text{minimize} \quad & f(x) \quad \text{for} \quad \mathbf{x} \in X \\ \text{subject to} \quad & h_j(x) \geq 0 \quad j = 1, \ldots, m\end{aligned} \tag{7.20}$$

there exists a unique dual problem if the function $f(x)$ is strictly convex and if the function $h_j(x)$ are concave.

Formally, following the procedure of Chapter 2 the solution of the dual problem can be obtained through a two-phase procedure as follows:

$$\max_{\lambda \in \Lambda} \min_{x \in X} L(x, \lambda) \tag{7.21}$$

where
$$L(x, \lambda) = f(x) - \sum_{j=1}^{m} \lambda_j h_j(x) \qquad (7.22)$$
is the Lagrangian function. Therefore the *dual problem* can be written

$$\begin{aligned} &\text{maximize} \quad l(\lambda) \\ &\text{subject to} \quad \lambda_j \geqslant 0 \qquad j = 1, \ldots, m \end{aligned} \qquad (7.23)$$

where
$$l(\lambda) = \min_{x \in X} L(x, \lambda) \qquad (7.24)$$
is defined as the dual function.

While the primal problem (7.20) involves n variables, m general constraints and $2n$ side constraints, the dual problem (7.23) involves m variables and m nonnegativity constraints. The dual problem is therefore quasi-unconstrained, and it can readily be solved by using, for example, the algorithm given at the end of Section 7.4. This requires the gradient of the dual function to be known. Fortunately, and this is perhaps the most interesting feature of the dual method approach, $\nabla l(\lambda)$ is extremely simple to compute, because it is given by the primal constraints

$$\frac{\partial l}{\partial \lambda_j} = -h_j[x(\lambda)] \qquad (7.25)$$

where $x(\lambda)$ denotes the primal point which minimizes $L(x, \lambda)$ over X for given λ (see Equation (7.24)). In term of this $\mathbf{x}(\lambda)$, the dual function can also be written

$$l(\lambda) = f[x(\lambda)] - \sum_{j=1}^{m} \lambda_j h_j[x(\lambda)] \qquad (7.26)$$

So, when a numerical maximization scheme is employed to solve the dual problem, the evaluation of the dual function requires the determination of the primal constraint values $h_j[x(\lambda)]$, so that the first derivatives (7.25) are available without additional computation.

Anticipating on the next chapters, we could also decide to employ a Newton-type method for solving the dual problem. This needs the second derivatives of the dual function to be evaluated. Let $\mathbf{N}$ represent the matrix of the primal constraint gradients, that is,

$$N = [\nabla h_1, \ldots, \nabla h_m] \qquad (7.27)$$

Then the Hessian of the dual function is given by

$$\nabla^2 l = -\mathbf{N}^T \tilde{\mathbf{G}}^{-1} \mathbf{N} \qquad (7.28)$$

where $\tilde{\mathbf{G}}$ is the Hessian matrix of the Lagrangian function, restricted to the free primal variables. A primal variable is said to be free if it has not taken on its lower or upper bound value. Introducing the set of indices

$$\tilde{I} = \{i : \underline{x}_i < x_i < \bar{x}_i\} \qquad (7.29)$$

the matrix $\tilde{\mathbf{G}}$ is thus given by

$$\tilde{G}_{ik} = \frac{\partial^2 L}{\partial x_i \partial x_k} \quad \text{for} \quad i, k \in \tilde{I}$$

and (7.30)

$$\tilde{G}_{ik} = 0 \quad \text{otherwise}$$

Of course, in (7.28), $\mathbf{N}$ and $\tilde{\mathbf{G}}$ depend on the dual variables through $\mathbf{x} = \mathbf{x}(\lambda)$.

To compute $l(\lambda)$, it is necessary to find the x that minimizes the Lagrangian, as formally stated in (7.24). For certain problems, this is not very difficult. For example, for the separable programming problem (7.15)–(7.17), the dual function reads as follow:

$$l(\lambda) = \sum_{i=1}^{n} \left\{ \min_{\underline{x}_i \leq x_i \leq \bar{x}_i} \left[f_i(x_i) - \sum_{j=1}^{m} \lambda_j h_{ji}(x_i) \right] \right\} \quad (7.31)$$

So a one-dimensional search in each of n components is all that is required to obtain the dual function value for given λ. This can be accomplished by restoring to line-search techniques (see Section 7.5). In some problems, the simplicity of each single-variable minimization problem appearing in (7.31) is such that it can be solved in closed form, yielding thus an explicit dual function. Furthermore, for a separable problem, the Hessian matrix (7.28) takes the form

$$\frac{\partial^2 l}{\partial \lambda_j \partial \lambda_k} = - \sum_{i \in \tilde{I}} \frac{n_{ij} n_{ik}}{G_{ii}} \quad (7.32)$$

where n_{ij} are the elements of N, namely $n_{ij} = \partial h_{ji}/\partial x_i$ and G_{ii} are the diagonal elements of G, namely $G_{ii} = \partial^2 L/\partial x_i^2$ (off-diagonal elements are zero, because of the separability). The summation in (7.32) is made only on the free variables (see Equation (7.29)). Hence it is apparent that while $l(\lambda)$ is continuously differentiable in all the feasible dual space, it is not, in general, twice continuously differentiable everywhere. Indeed, whenever the set $\tilde{I}$ is modified, $\nabla^2 l$ undergoes a discontinuity. This happens each time a free primal becomes fixed, or inversely.

To help fix ideas, we consider the two-variable problem examined at the end of Section 2.10, and we add side constraints on x_1 and x_2. Because we know from Section 2.10 that the optimal solution is $x_1^* = x_2^* = \frac{1}{2}$ with $\lambda_1 = 0$, $\lambda_2 = 1$, we simplify the problem and we ignore the first constraint from the very beginning. The problem then reads as follow:

minimize $f(x) = x_1^2 + x_2^2$ for $x \in X = \{x: \frac{1}{4} \leq x_i \leq 1; i = 1, 2\}$
subject to $h(x) = x_1 + x_2 - 1 \geq 0$

The problem is separable and thus we use the definition (7.31) of $l(\lambda)$. So we have to solve two single-variable minimization problems:

$$\min_{\frac{1}{4} \leq x_i \leq 1} \{x_i^2 - \lambda x_i\} \quad i = 1, 2$$

Solution of these problems show that the dual function takes on different explicit forms depending on the value of λ:

if $\lambda < 1/2$: $x_1 = x_2 = 1/4$ and $l(\lambda) = \lambda/2 + 1/8$
if $1/2 \leqslant \lambda \leqslant 2$: $x_1 = x_2 = \lambda/2$ and $l(\lambda) = \lambda - \lambda^2/2$
if $\lambda > 2$: $x_1 = x_2 = 1$ and $l(\lambda) = 2 - \lambda$

The dual space is partitioned into three subregions as indicated in Fig. 7.1. It can be seen that $l(\lambda)$ and $l'(\lambda)$ are continuous for any $\lambda > 0$. However, $l''(\lambda)$ is discontinuous at $\lambda = 1/2$ and $\lambda = 2$.

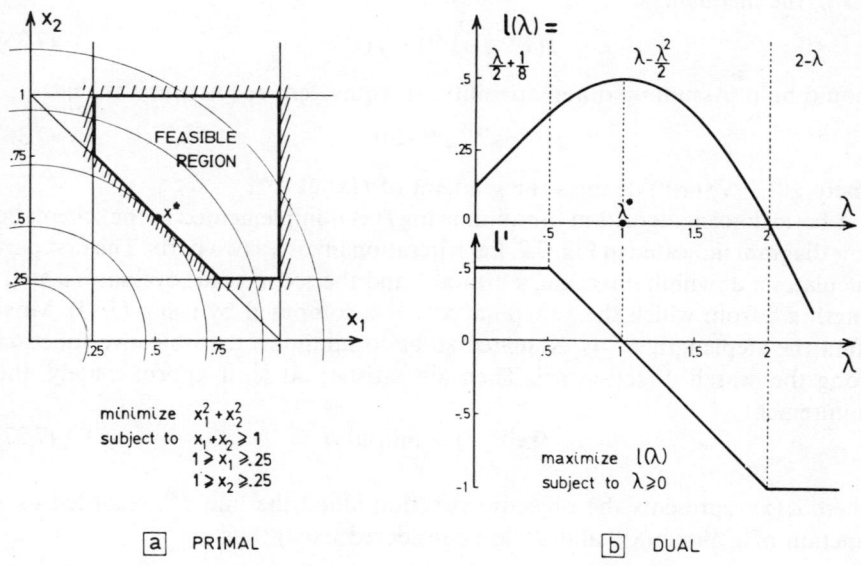

Fig. 7.1 Example of dual problem

In summary, then, the dual method formulation is interesting because it replaces the initial constrained problem with a quasi-unconstrained problem. In addition, the effective dimensionality of the dual problem, that is, the number of non-zero dual variables, is usually much lower than that of the primal problem. Indeed, in many problems of practical interest, the number of active constraints $h_j(x) = 0$ is small when compared with the number of primal variables.

7.3 DESCENT METHODS FOR MINIMIZATION

As previously mentioned in Section 2.11, most algorithms for solving a minimization problem are iterative. They require an initial estimate of the solution, $\mathbf{x}^{(0)}$, and then, for $k = 0, 1, 2, \ldots$, the kth iteration replaces $\mathbf{x}^{(k)}$ by

$\mathbf{x}^{(k+1)}$, which should be a better estimate of the solution. Furthermore, nearly all of the unconstrained minimization methods which are described in the sequel are descent methods, that is:

$$f(x^{(k+1)}) < f(x^{(k)}) \tag{7.33}$$

They usually involve sequential minimization of $f(x)$ along successive search directions $\mathbf{s}^{(k)}$, so that

$$\mathbf{x}^{(k+1)} = \mathbf{x}^{(k)} + \alpha^{(k)} \mathbf{s}^{(k)} \tag{7.34}$$

Clearly $\mathbf{s}^{(k)}$ must be a *downhill direction*, which means that for sufficiently small $\alpha > 0$, the inequality

$$f(\mathbf{x}^{(k)} + \alpha \mathbf{s}^{(k)}) < f(\mathbf{x}^{(k)}) \tag{7.35}$$

should hold. Assuming differentiability, an equivalent requirement is that

$$\mathbf{s}^{(k)\mathrm{T}} \mathbf{g}^{(k)} < 0 \tag{7.36}$$

where $\mathbf{g}^{(k)} = \nabla f(\mathbf{x}^{(k)})$ denotes the gradient of $f(x)$ at $\mathbf{x}^{(k)}$.

A basic descent algorithm for minimizing $f(x)$ is implemented in the schematic flow diagram indicated in Fig. 7.2. Each iteration involves two parts. The first part calculates a downhill direction, $\mathbf{s}^{(k)}$, at $\mathbf{x}^{(k)}$, and the second part evaluates a steplength $\alpha^{(k)}$ from which the new point $\mathbf{x}^{(k+1)}$ is computed by using (7.34). Most often the steplength $\alpha^{(k)}$ is estimated so as to minimize the objective function along the search direction $\mathbf{s}^{(k)}$. Then $\alpha^{(k)}$ satisfies, at least approximately, the requirement

$$f(x^{(k+1)}) = \min_{\alpha} \phi(\alpha) \tag{7.37}$$

where $\phi(\alpha)$ represents the objective function along the line $\mathbf{s}^{(k)}$, regarded as a function of α alone ($\mathbf{x}^{(k)}$ and $\mathbf{s}^{(k)}$ are considered fixed) thus:

$$\phi(\alpha) = f(\mathbf{x}^{(k)} + \alpha \mathbf{s}^{(k)}) \tag{7.38}$$

Following the arguments outlined in Section 2.11 each iteration in the above basic descent algorithm implies solving the equation

$$\phi'(\alpha) = 0 \tag{7.39}$$

where ϕ' is the first derivative of ϕ. From (7.38) it is apparent that·

$$\begin{aligned} \phi'(\alpha) &= \sum_{i=1}^{n} \frac{\partial f(\mathbf{x}^{(k)} + \alpha \mathbf{s}^{(k)})}{\partial x_i} \frac{\partial}{\partial \alpha}(x_i^{(k)} + \alpha s_i^{(k)}) \\ &= \mathbf{g}(\mathbf{x}^{(k)} + \alpha \mathbf{s}^{(k)})^{\mathrm{T}} \mathbf{s}^{(k)} \end{aligned} \tag{7.40}$$

Condition (7.39) and the definition (7.34) of $x^{(k+1)}$ show that the gradient $\mathbf{g}^{(k+1)}$ of f at $\mathbf{x}^{(k+1)}$ is orthogonal to the search direction $\mathbf{s}^{(k)}$:

$$\mathbf{g}^{(k+1)\mathrm{T}} \mathbf{s}^{(k)} = 0 \tag{7.41}$$

Introduction to Mathematical Programming Methods

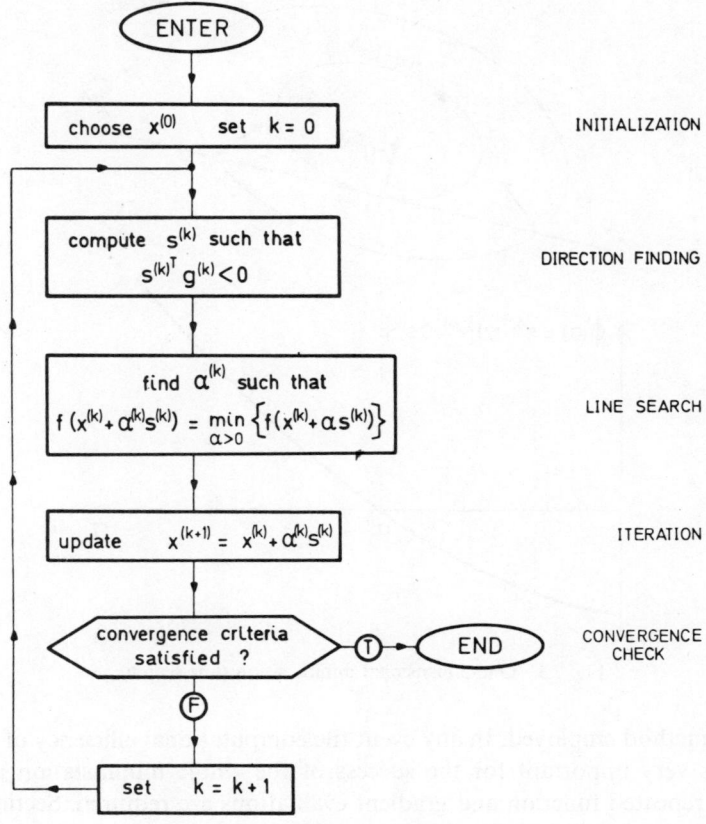

Fig. 7.2 Basic descent algorithm

This property is illustrated in Fig. 7.3. Note that because of (7.41), $s^{(k)}$ is tangent to the contour of f on which $f(x) = f(x^{(k+1)})$.

Note also that

$$\begin{aligned}\phi'(\alpha) < 0 &\quad \text{if} \quad \alpha < \alpha^{(k)} \\ \phi'(\alpha) > 0 &\quad \text{if} \quad \alpha > \alpha^{(k)}\end{aligned} \quad (7.42)$$

which are useful inequalities for trapping the optimal step length $\alpha^{(k)}$.

Solving the one-dimensional minimization problem (7.37), or, equivalently, the single-variable nonlinear Equation (7.39), is an important part of many descent methods. This process of finding $\alpha^{(k)}$ by estimating a minimum of $\phi(\alpha)$ is called *line search*. Most line searches are performed by using iterative numerical procedures which are terminated when some convergence criteria are satisfied. Line searches are therefore usually not exact, and their accuracy must be adapted to the type of

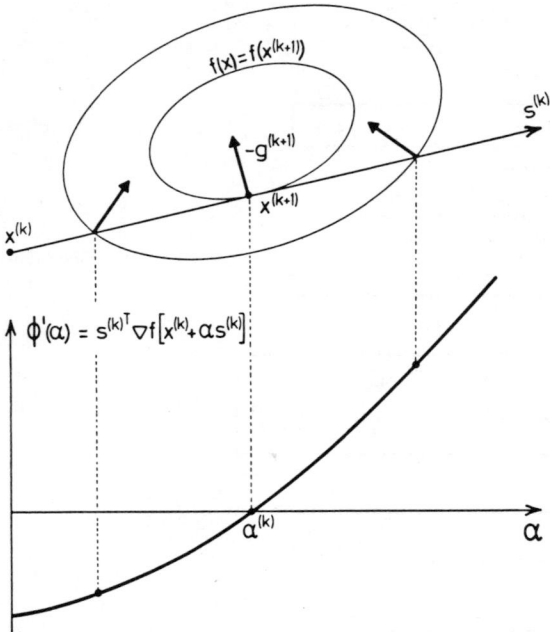

Fig. 7.3 One-dimensional minimization (line search)

descent method employed. In any event the computational efficiency of the line search is very important for the success of the whole minimization method, because repeated function and gradient evaluations are required. Section 7.5 is focused on some efficient line-search techniques.

Finally, it is worth noticing that the second derivative of $\phi(\alpha)$, if required in a line-search process, is given by

$$\phi''(\alpha) = \mathbf{s}^{(k)T} \mathbf{H}[x^{(k)} + \alpha \mathbf{s}^{(k)}] \mathbf{s}^{(k)} \tag{7.43}$$

where **H** denotes the Hessian of the objective function.

7.4 THE METHOD OF STEEPEST DESCENT

A fundamental method in nonlinear programming is the method of steepest descent, in which the downhill directions are taken as the negative gradient vectors. This method is not in itself very efficient, but provides the basis for all gradient methods. Furthermore, convergence properties can be established, which serve as a reference situation for other minimization algorithms. Because of these characteristics and also because its simple geometrical interpretation helps in developing an understanding of the basic phenomena, the method of steepest descent is discussed in this introductory chapter.

In view of the descent condition (7.36), a natural choice for a downhill direction $s^{(k)}$ in the basic descent algorithm schematized in Fig. 7.2, is $s^{(k)} = -g^{(k)}$, yielding $s^{(k)T}g^{(k)} < 0$ if $g^{(k)} \neq 0$. So, from the point $x^{(k)}$, we search along the direction of negative gradient to a minimum point $x^{(k+1)}$ on this line. This is the method of steepest descent, and Fig. 7.4 shows how it works. From the line search condition (7.41) it is apparent that the successive search directions are orthogonal. Also, it can be seen from Fig. 7.4 that the method of steepest descent should perform well on an objective function with nearly circular contours. On the other hand, convergence will be extremely slow if the contours are very elongated. These intuitive results are confirmed by studying the rate of convergence of the steepest-descent method when applied to quadratic problems.

Consider the quadratic function

$$q(x) = \tfrac{1}{2}x^T A x - b^T x \tag{7.44}$$

where A is a symmetric positive definite matrix with eigenvalues

$$0 < e = e_1 \leqslant e_2 \leqslant \ldots \leqslant e_n = E$$

Clearly, q is strictly convex and has a unique minimum point $x^* = A^{-1}b$. It can be shown that the method of steepest descent converges to x^* for any starting point $x^{(0)}$. The rate of convergence is linear and the convergence ratio is

$$\rho = \left(\frac{E-e}{E+e}\right)^2 = \left(\frac{r-1}{r+1}\right)^2 \tag{7.45}$$

where r is the condition number of the matrix A, that is, the ratio E/e of its largest to its smallest eigenvalue. The meaning of (7.45) is that each iteration will reduce the error in the objective function by more than a factor ρ. Note that $\rho \in (0, 1)$; and the smaller is ρ, the more rapid is the convergence. Clearly convergence is slowed as r increases, and is very rapid when r is close to unity, which can only happen if *all* the eigenvalues of A are close to each other.

The contours of $q(x)$ are n-dimensional ellipsoïds with axes in the direction of the n mutually orthogonal eigenvectors of the Hessian matrix A. The axis corresponding to the ith eigenvector has length proportional to $1/\sqrt{\lambda_i}$, where λ_i is the ith eigenvalue of A. Therefore, in view of (7.45), the speed of convergence in the steepest-descent method depends upon the ratio of the longest to the shortest principal axis of the elliptical contours of q, that is, it is primarily governed by the eccentricity of the ellipsoïds. For example, in the limiting case of unit condition number ($E = e$), corresponding to circular contours, convergence occurs in a single step. Conversely, the rate of convergence is slowed as the contours of q become more eccentric, which is intuitively obvious (see Fig. 7.4).

In the example of Section 2.11, the steepest-descent method was applied to the function $q(x) = (x_2 - x_1)^2 + (1 - x_1)^2$, whose Hessian matrix is

$$\nabla^2 q = \begin{bmatrix} 4 & -2 \\ -2 & 2 \end{bmatrix}$$

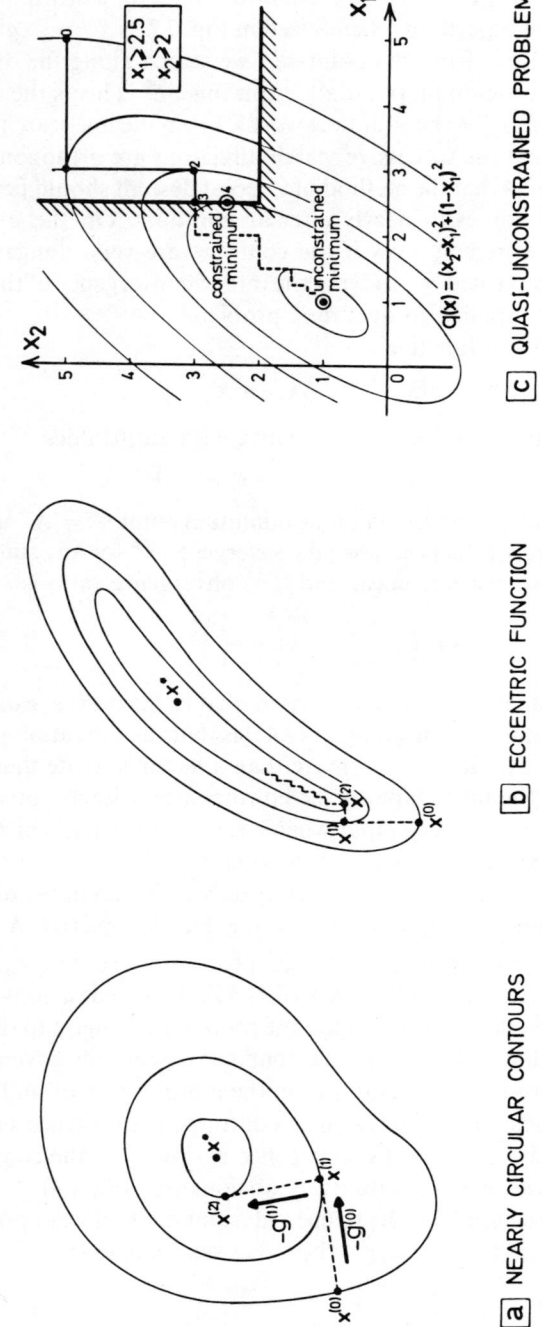

Fig. 7.4 The method of steepest descent (a) Nearly circular contours (b) Eccentric function (c) Quasi-unconstrained problem

It is easily seen that the two eigenvalues of this matrix are $e_1 = 0.76$ and $e_2 = 5.24$, yielding a condition number of 0.15 and a convergence ratio of 0.55. This means that the error in $q(x)$ should be about halved at each iteration. This theoretical prediction is in good agreement with the numerical results given in Fig. 2.12.

The quasi-unconstrained case

The steepest descent algorithm can be easily modifed so as to deal with the quasi-unconstrained problem (7.8). In view of the optimality conditions (7.9)–(7.11), an obvious choice for the search direction at iteration k is

$$\begin{aligned} s_i^{(k)} &= 0 && \text{if } x_i = \underline{x}_i \text{ and } g_i^{(k)} \geq 0 \\ s_i^{(k)} &= 0 && \text{if } x_i = \bar{x}_i \text{ and } g_i^{(k)} \leq 0 \\ s_i^{(k)} &= -g_i^{(k)} && \text{otherwise} \end{aligned} \qquad (7.46)$$

This direction-finding process allows us to decide whether or not an active side constraint must be kept active in the next iteration. We must also be capable of determining if an inactive side constraint will become active in the current iteration. This is done in the line-search process by replacing condition (7.37) with

$$f(x^{(k+1)}) = \min_{\alpha \leq \tilde{\alpha}} \phi(\alpha) \qquad (7.47)$$

where $\tilde{\alpha}$ is the maximal allowable step length that can be taken without violating the constraints. It is apparent that

$$\tilde{\alpha} = \min\{\underline{\alpha}, \bar{\alpha}\} \qquad (7.48)$$

with

$$\underline{\alpha} = \min_{s_i^{(k)} < 0} \left\{ \frac{\underline{x}_i - x_i^{(k)}}{s_i^{(k)}} \right\}$$

and

$$\bar{\alpha} = \min_{s_i^{(k)} > 0} \left\{ \frac{\bar{x}_i - x_i^{(k)}}{s_i^{(k)}} \right\} \qquad (7.49)$$

Now, if $\phi'(\tilde{\alpha}) > 0$, then the minimum of ϕ holds for $\alpha^* < \tilde{\alpha}$; thus set $\alpha^{(k)} = \alpha^*$, where α^* is computed by using a line-search technique. On the other hand, if $\phi'(\tilde{\alpha}) < 0$, then set $\alpha^{(k)} = \tilde{\alpha}$, because the minimum of ϕ holds for $\alpha^* > \tilde{\alpha}$, for which the new point $\mathbf{x}^{(k+1)}$ would violate at least one side constraint. With $\alpha^{(k)}$ a solution of (7.47), compute $\mathbf{x}^{(k+1)}$ from (7.34), and the kth iteration is complete (see Fig. 7.4 for a graphical interpretation of this algorithm).

This scheme for solving a quasi-unconstrained problem is interesting, because it gives a foretaste of the active-set strategy employed for handling more complex constraints. Also, it is a useful algorithm in itself. For example, it can be applied to the dual problem (7.23).

7.5 LINE-SEARCH TECHNIQUES

Most nonlinear programming methods considered in this book involve a sequence of one-dimensional problems, each of which corresponds to minimizing the objective function along a search direction (or to finding the point of intersection of a search direction with a constraint surface). It is therefore very important in practice to be capable of solving efficiently such single-variable problems. This is the purpose of the line-search techniques examined in this section.

It is assumed for simplicity that the function $\phi(\alpha)$ to be minimized (see Equation (7.38)) is unimodal, that is, it has a single relative minimum α^* (see Fig. 7.5). Furthermore, in most problems, the function being searched can be assumed to possess a certain degree of smoothness, and so many line-search procedures are based on curve-fitting techniques. Depending upon whether or not derivatives of the function can be measured, one or several points must be used to determine the fit, and a variety of line-search techniques can be devised.

7.5.1 One-point Pattern: Newton–Raphson Iteration

In a line-search technique that uses only one measurement point at each iteration, it is apparent that the first and second derivatives of the objective function must be available in order to compute $\phi'(\alpha)$ and $\phi''(\alpha)$ (see Equations (7.40) and (7.43)). Suppose therefore that at a point α_1, we evaluate the three numbers $\phi(\alpha_1)$, $\phi'(\alpha_1)$ and $\phi''(\alpha_1)$. It is then possible to construct a quadratic function $q(\alpha)$ which at α_1 agrees with $\phi(\alpha)$ up to second derivative:

$$q(\alpha) = \phi(\alpha_1) + \phi'(\alpha_1)(\alpha - \alpha_1) + \frac{1}{2}\phi''(\alpha_1)(\alpha - \alpha_1)^2 \qquad (7.50)$$

An estimate α_2 of the minimum point of ϕ can now be calculated by finding the vanishing point of the derivative of q. Thus setting

$$0 = q'(\alpha_2) = \phi'(\alpha_1) + \phi''(\alpha_1)(\alpha_2 - \alpha_1) \qquad (7.51)$$

we obtain

$$\alpha_2 = \alpha_1 - \frac{\phi'(\alpha_1)}{\phi''(\alpha_1)} \qquad (7.52)$$

This process can then be repeated at α_2. It is apparent from (7.52) that the new point α_2 does not depend on the value $\phi(\alpha_1)$. Therefore the method can be viewed as iteratively solving the equation $\phi'(\alpha) = 0$. It is in fact the well-known method of Newton–Raphson. It is illustrated in Fig. 7.5 both as a technique for minimizing $\phi(\alpha)$ and as a technique for solving $\phi'(\alpha) = 0$.

It can be shown that Newton's method has order-two convergence. However, because it demands second derivative evaluations this line-search technique is limited to special problems for which a simple explicit function $\phi''(\alpha)$ is available.

Introduction to Mathematical Programming Methods

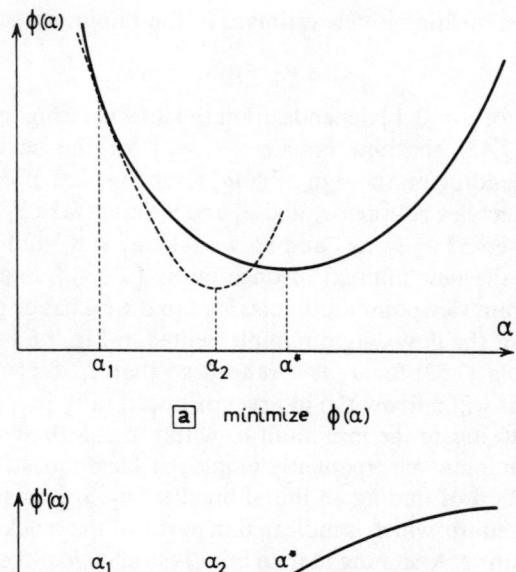

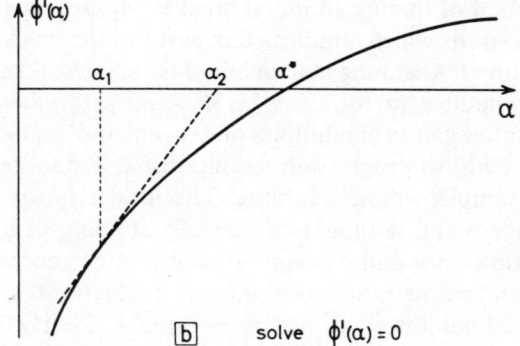

Fig. 7.5 Line search by Newton method

7.5.2 Two-point Pattern

In many practical optimization problems the second derivatives of the objective function are not available, or they can only be evaluated by spending a prohibitive computational effort. Therefore several line-search techniques have been devised to efficiently find the minimum of the function $\phi(\alpha)$ by using first-derivative information only. Two measurement points are then necessary to produce a polynomial fit.

Let us thus define an initial bracket $[\alpha_1, \alpha_2]$ in which we know that the minimum α^* lies, i.e. $\phi'(\alpha_1) \leq 0$ and $\phi'(\alpha_2) \geq 0$. Such a bracket is sometimes called an *interval of uncertainty*. An important class of line-search techniques consists of gradually reducing the interval of uncertainty until the minimum is trapped with sufficient accuracy according to a given convergence criterion. The general equation yielding a new estimate of the minimum is

$$\alpha_3 = \alpha_2 - \rho(\alpha_2 - \alpha_1) \tag{7.53}$$

where the value of $\rho \in [0, 1]$ depends upon the interpolating formula being used. According to (7.42), the new bracket $[\alpha'_1, \alpha'_2]$ for the next iteration can be determined depending on the sign of $\phi'(\alpha_3)$ (see Fig. 7.3). If $\phi'(\alpha_3) > 0$ then the minimum point α^* lies between α_1 and α_3 and we must take $\alpha'_1 = \alpha_1$ and $\alpha'_2 = \alpha_3$. If $\phi'(\alpha_3) < 0$ then $\alpha^* \in [\alpha_3, \alpha_2]$ and so we take $\alpha'_1 = \alpha_3$ and $\alpha'_2 = \alpha_2$. We then apply (7.53) to the new interval of uncertainty $[\alpha'_1, \alpha'_2]$, and so on. From the computer program viewpoint these rules lead to the exchange procedure which is given in part 3 of the flow diagram implemented in Fig. 7.6 (see test C). Clearly, since the formula (7.53) for α_3 is arranged so that $\alpha_1 \leqslant \alpha_3 \leqslant \alpha_2$ (because $0 \leqslant \rho \leqslant 1$), each refit will narrow the interval of uncertainty $[\alpha_1, \alpha_2]$ and successive refits allow us to locate the minimum to within the desired accuracy.

Before examining some frequently employed techniques for evaluating ρ in (7.53), the question of finding an initial bracket $[\alpha_1, \alpha_2]$ must be considered. A convenient procedure, which is indicated in part 1 of the block diagram shown in Fig. 7.6, is as follows. Assuming that an initial estimate h of the minimum α^* of ϕ is known, we compute $\phi'(\alpha)$ for $\alpha = h, 2h, 3h, \ldots, \alpha_1, \alpha_2$, where α_1 and α_2 are the first two consecutive values of multiples of h such that $\phi'(\alpha_1) < 0$ and $\phi'(\alpha_2) \geqslant 0$. Instead of this additive progression for increasing h, another rationale can be adopted, for example, simple doubling. The ideal strategy is of course very problem-dependent and it should be carefully studied for each new problem, because much time-consuming computation might be generated in this part of the minimization routine (numerous gradient evaluations).

Another crucial point is the choice of the initial h. Clearly if h is very large, so that $\phi'(h)$ is positive in the first pass through test A (Fig. 7.6), then the interpolation might have to be repeated many times before test B is satisfied. On the other hand, if h is too small, numerous traverses of loop 1 will be necessary before test A is satisfied. It is therefore apparent that the initial h should be already a good estimate of the minimum point α^*. Several methods for achieving this are available, and they depend largely upon the particular descent method with which the line-search technique is employed. For example, when employing Newton and quasi-Newton methods, $h = 1$ is usually a good estimate of α^* (see Chapter 8). In gradient methods, however, such an estimate is generally not acceptable, and the following technique is recommended. Let us first consider the quadratic function

$$\phi(\alpha) = \frac{1}{2}a\alpha^2 + b\alpha + c \tag{7.54}$$

Equating to zero the derivative

$$\phi'(\alpha) = a\alpha + b \tag{7.55}$$

Introduction to Mathematical Programming Methods

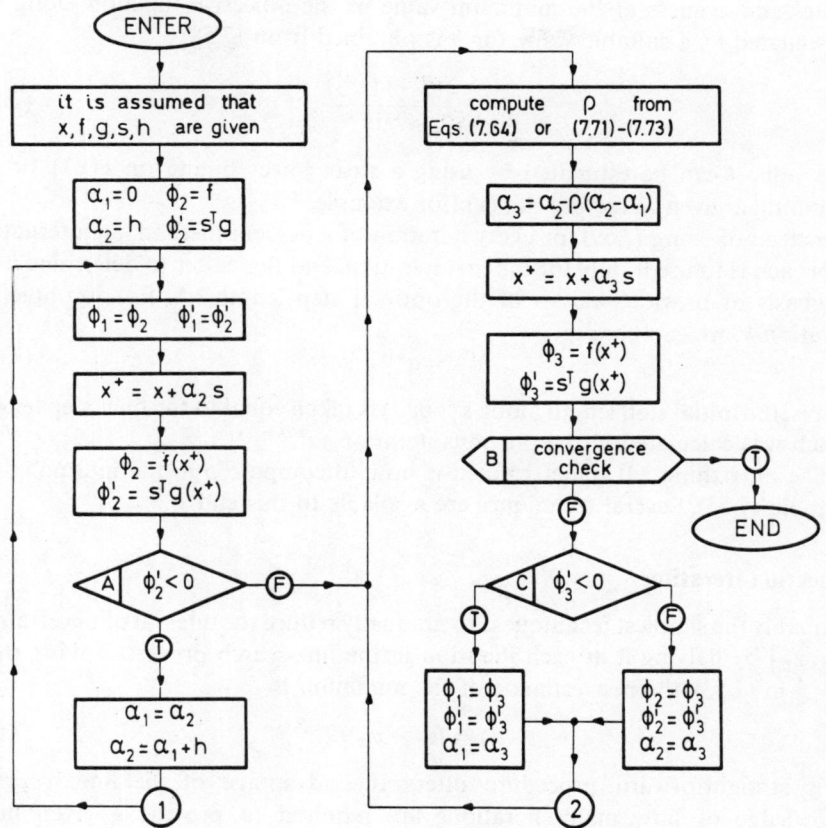

Fig. 7.6 Line search using a two-point pattern

we find the minimum of $\phi(\alpha)$ as

$$\alpha^* = -\frac{b}{a} \tag{7.56}$$

Therefore,

$$\phi(\alpha^*) = -\frac{b^2}{2a} + c \tag{7.57}$$

and so, because it follows from (7.55) and (7.54) that $b = \phi'(0)$ and $c = \phi(0)$, we have

$$\alpha^* = \frac{2(\phi(\alpha^*) - \phi(0))}{\phi'(0)} \tag{7.58}$$

In the kth iteration of a descent method, we know $\phi(0) = f(x^{(k)})$ and $\phi'(0) = \mathbf{g}(x^{(k)})^T \mathbf{s}^{(k)}$, where $s^{(k)}$ is the search direction (see Section 7.3). Using these values and a guess at the minimum value of the objective function along $\mathbf{s}^{(k)}$ (designated $\tilde{f}$), a suitable value for h is obtained from (7.58):

$$h = \frac{2(\tilde{f} - f(\mathbf{x}^{(k)}))}{\mathbf{g}(\mathbf{x}^{(k)})^{\mathrm{T}} \mathbf{s}^{(k)}} \tag{7.59}$$

The value $\tilde{f}$ can be estimated by using a strict lower bound on $f(\mathbf{x}^*)$, or by assuming a given reduction in $f(\mathbf{x})$ (for example, 10%).

Instead of using (7.59) for every iteration of a descent method, an alternative approach is to use it only for the first iteration, and thereafter to determine h on the basis of previous values of the optimal step length α^*. For example, at iteration k, we can choose

$$h^{(k)} = \alpha^{*(k-1)} \tag{7.60}$$

that is, the initial step length along $\mathbf{s}^{(k)}$ ($h^{(k)}$) is taken equal to the final step length which was calculated in the previous iteration ($\alpha^{*(k-1)}$).

The only thing left undetermined is how to compute ρ in the interpolating formula (7.53). Several techniques are available to this end.

Bisection iteration

Probably the simplest technique is to gradually reduce the interval of uncertainty $[\alpha_1, \alpha_2]$ by halving it at each iteration in the line-search process. Taking thus $\rho = \frac{1}{2}$ in (7.53), the new estimate of the minimum is

$$\alpha_3 = (\alpha_1 + \alpha_2)/2 \tag{7.61}$$

This straightforward procedure offers the advantage of yielding a prior knowledge of how many iterations are required to provide a given final uncertainty interval (for example, 20 bisection iterations lead to a reduction of the initial bracket by a factor of 10^6).

It is apparent from (7.61) that the bisection iteration is very crude and does not use fully the derivative information on $\phi(\alpha)$. It seems therefore desirable to exploit this information by resorting to curve fitting, that is, by approximating ϕ by a polynomial in α of degree two or three, and determining the minimum of the polynomial approximation analytically. That is considered in the next two paragraphs.

Method of false position

The method of Newton discussed in Section 7.5.1 is based on fitting a quadratic on the basis of information at a single point. By using two points, less information is required at each of them. Knowing $\phi(\alpha_1)$, $\phi'(\alpha_1)$, $\phi'(\alpha_2)$ it is possible to fit the quadratic

$$q(\alpha) = \phi(\alpha_1) + \phi'(\alpha_1)(\alpha - \alpha_1) + \frac{\phi'(\alpha_2) - \phi'(\alpha_1)}{\alpha_2 - \alpha_1} \frac{(\alpha - \alpha_1)^2}{2} \tag{7.62}$$

which has the same corresponding values. The new estimate α_3 can then be determined by finding the point where the derivative of q vanishes, namely

$$\alpha_3 = \alpha_2 - \phi'(\alpha_2)\frac{\alpha_2 - \alpha_1}{\phi'(\alpha_2) - \phi'(\alpha_1)} \tag{7.63}$$

This shows that the method of false position consists of choosing

$$\rho = \frac{\phi'(\alpha_2)}{\phi'(\alpha_2) - \phi'(\alpha_1)} \tag{7.64}$$

in the general interpolating formula (7.53).

Note that this method does not depend on values of ϕ directly. Therefore it can be regarded as a method for solving $\phi'(\alpha) = 0$. Viewed in this way, the method consists in linearly interpolating $\phi'(\alpha)$ between α_1 and α_2:

$$\phi'(\alpha) \simeq \frac{\alpha_1 \phi'(\alpha_2) - \alpha_2 \phi'(\alpha_1)}{\alpha_1 - \alpha_2} + \frac{\phi'(\alpha_1) - \phi'(\alpha_2)}{\alpha_1 - \alpha_2}\alpha \tag{7.65}$$

whose vanishing point is α_3, given in (7.63). It is also worth mentioning that the formula (7.63) can be interpreted as an approximation to Newton's method (see Equation (7.52)) where the second derivative is replaced by the difference of two first derivatives. The geometric interpretation of the method of false position is indicated in Fig. 7.7 (hence the name 'secant' method which is sometimes used).

It is significant that for a true quadratic, the result of the first iteration is theoretically the exact α^*. The rate of convergence of the method is more than linear. It can be shown that it is of order 1.618 which is the golden mean. Therefore convergence is usually quite rapid. Occasionally, however, the situation shown in Fig. 7.7(b) will happen: the minimum α^* is approached very slowly by point α_2, while point α_1 remain unchanged. A possible expedient consists of alternating a bisection iteration (7.61) with the false-position iteration (7.63), so that α_2 is moved closer to α^*.

Cubic interpolation

The next method hinges on approximating $\phi(\alpha)$ by a cubic polynomial

$$H(\alpha) = a\alpha^3 + b\alpha^2 + c\alpha + d \tag{7.66}$$

We assume that values of $\phi(\alpha_1), \phi'(\alpha_1), \phi(\alpha_2), \phi'(\alpha_2)$ are available with the customary condition $\phi'(\alpha_1) < 0$ and $\phi'(\alpha_2) > 0$. Those values permit identification of the four coefficients a, b, c, d. ($H(\alpha)$ is called the Hermite interpolating polynomial of degree three). From (7.66) it follows that

$$H'(\alpha) = 3a\alpha^2 + 2b\alpha + c \tag{7.67}$$
$$H''(\alpha) = 6a\alpha + 2b \tag{7.68}$$

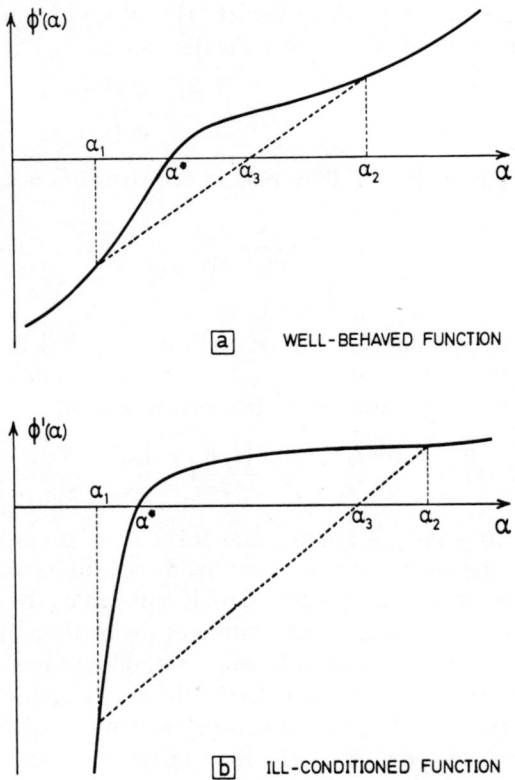

Fig. 7.7 Line search by the method of false position (a) Well-behaved function (b) Ill-conditioned function

The unique minimum point $\hat{\alpha}$ of H satisfies $H'(\alpha) = 0$ and $H''(\alpha) > 0$. This leads to the solution

$$\hat{\alpha} = \frac{-b + (b^2 - 3ac)^{1/2}}{3a} \qquad (7.69)$$

if a is non-zero and, if a is zero,

$$\hat{\alpha} = \frac{-c}{2b} \qquad (7.70)$$

When a is very small but non-zero, $\hat{\alpha}$ computed from (7.69) suffers from loss of significance. Therefore the following approach, which is applicable whether or not a is zero, is recommended:

Compute

$$S = \frac{3(\phi(\alpha_1) - \phi(\alpha_2))}{\alpha_2 - \alpha_1} + \phi'(\alpha_1) + \phi'(\alpha_2) \qquad (7.71)$$

and
$$R = (S^2 - \phi'(\alpha_1)\phi'(\alpha_2))^{1/2} \tag{7.72}$$
The new estimate α_3 is then given by the general interpolation formula (7.53) with
$$\rho = \frac{\phi'(\alpha_2) + R - S}{\phi'(\alpha_2) - \phi'(\alpha_1) + 2R} \tag{7.73}$$
Because $\phi'(\alpha_1) < 0$ and $\phi'(\alpha_2) > 0$, the square root in (7.72) is always defined. Moreover, it can be shown that the condition $H''(\alpha_3) > 0$ is satisfied and that the minimum α_3 lies between α_1 and α_2.

It can be proved that the order of convergence of the cubic interpolation method is two (and not three, although the method is exact for cubic functions).

7.5.3 Three-point Pattern: Quadratic Interpolation

After considering successively a one-point pattern, which requires second derivative evaluation, and several two-point patterns, where only the first derivatives must be computed, it can be argued that a three-point pattern using a quadratic fit can be envisioned that is based on the objective function values only. Thus, let α_1, α_2 and α_3 be distinct measurement points yielding $\phi_1 = \phi(\alpha_1)$, $\phi_2 = \phi(\alpha_2)$ and $\phi_3 = \phi(\alpha_3)$. The quadratic polynomial $q(\alpha)$ for which
$$q(\alpha_i) = \phi_i \qquad i = 1, 2, 3 \tag{7.74}$$
is easily seen to be given by
$$q(\alpha) = \sum_{i=1}^{3} \phi_i \frac{\prod_{j \neq i}(\alpha - \alpha_j)}{\prod_{j \neq i}(\alpha_i - \alpha_j)} \tag{7.75}$$
(Lagrange interpolating polynomial of degree two). The critical point of q is the point where the derivative q' vanishes:
$$\hat{\alpha} = \frac{1}{2} \frac{b_{23}\phi_1 + b_{31}\phi_2 + b_{12}\phi_3}{a_{23}\phi_1 + a_{31}\phi_2 + a_{12}\phi_3} \tag{7.76}$$
where $a_{ij} = \alpha_i - \alpha_j$ and $b_{ij} = \alpha_i^2 - \alpha_j^2$. In particular, if we choose $\alpha_1 = 0, \alpha_2 = h$ and $\alpha_3 = 2h$, where $h > 0$ is a preselected trial step, then
$$\hat{\alpha} = h \frac{4\phi_2 - 3\phi_1 - \phi_3}{4\phi_2 - 2\phi_1 - 2\phi_3} \tag{7.77}$$
Differentiating $q(\alpha)$ twice, it can be proved that $q''(\hat{\alpha}) > 0$ if and only if
$$\phi_3 + \phi_1 > 2\phi_2 \tag{7.78}$$

which means that the value of ϕ_2 must be below the line connecting ϕ_1 and ϕ_3 (see Fig. 7.8).

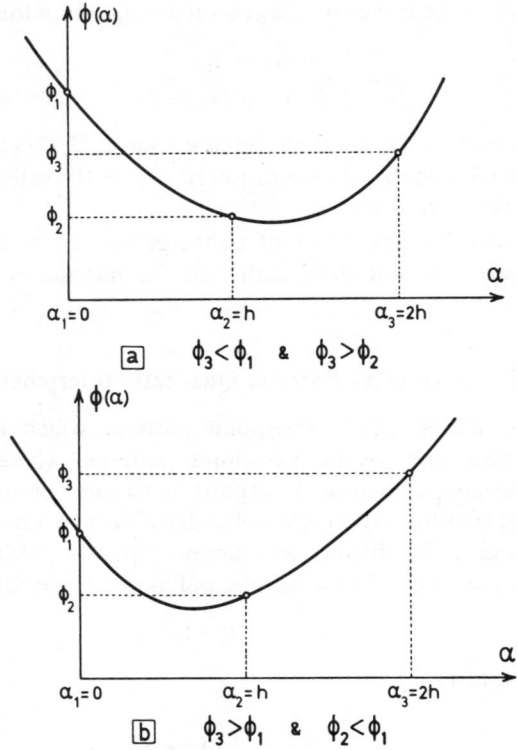

Fig. 7.8   Line search by quadratic interpolation

These formulas can be employed in several ways in order to devise efficient line-search techniques requiring only the evaluation of ϕ. Most existing methods comprise a bracketing section, where three acceptable points α_1, α_2 and α_3 are found, an interpolating section yielding a new estimate α_4 and a refining section, where a new set of interpolating points is selected from $\alpha_1, \alpha_2, \alpha_3$ and α_4. The interpolating section is then re-entered, followed by the refining section, and so on, until α_4 is considered as a sufficiently good estimate of the minimum point α^*.

A convenient implementation of such a three-point pattern is given in the schematic block diagram indicated in Fig. 7.9. In the bracketing section, it is assumed that an initial value for the step length h has been found, which should be ideally of the order of α^* (see Equations (7.59) and (7.60)). If $\phi(h) > \phi(0) \equiv \phi_1$, then set $\phi_3 = \phi(h)$, cut h in half and repeat this step until $\phi(h) < \phi_1$. Otherwise, set $\phi_2 = \phi(h)$, double h and repeat this step until $\phi(h) > \phi_2$. When entering the interpolating section, the value of h is such that $\phi_2 < \phi_1$ and $\phi_3 > \phi_2$ (with

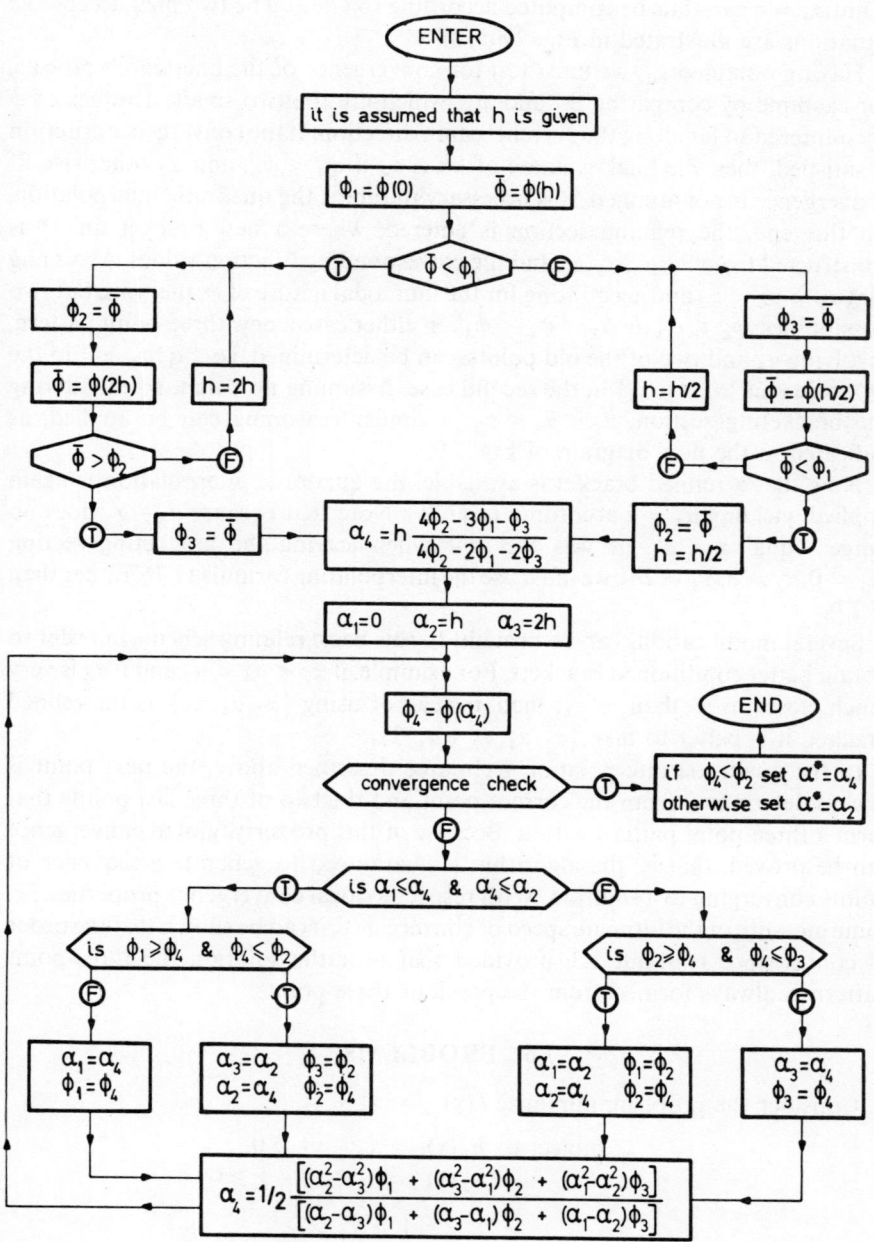

Fig. 7.9 Quadratic interpolation—three-point pattern

$\alpha_1 = 0, \alpha_2 = h, \alpha_3 = 2h$) and therefore (7.78) is automatically satisfied. The new point $\alpha_4 = \hat{\alpha}$ can thus be computed according to (7.77). The two final acceptable situations are illustrated in Fig. 7.8(a, b).

Having obtained α_4, we must test for convergence of the line-search process, for example by comparing ϕ_2 and ϕ_4, which are the two smallest values of ϕ encountered so far along the current search direction. If the convergence criterion is satisfied, then the final estimate of α^* is α_4 if $\phi_4 < \phi_2$, and α_2 otherwise. If convergence is not attained, it is necessary to repeat the quadratic interpolation. To this end, the refining section is entered, where a new bracket on α^* is constructed from $\alpha_1, \alpha_2, \alpha_3, \alpha_4$ and the corresponding function values. Assuming that $\alpha_1 < \alpha_4 < \alpha_2$ and accounting for the unimodal nature of ϕ, there are but two possibilities: $\phi_4 \leqslant \phi_2$ or $\phi_2 < \phi_4 < \phi_1$. In either case a new three-point pattern, involving α_4 and two of the old points, can be determined, i.e. $\{\alpha_1, \alpha_4, \alpha_2\}$ in the first case and $\{\alpha_4, \alpha_2, \alpha_3\}$ in the second case. Assuming now that, when entering the bracketing section, $\alpha_2 < \alpha_4 < \alpha_3$, a similar reasoning can be applied, as indicated in the flow diagram of Fig. 7.9..

Now that a refined bracket is available, the quadratic interpolation is again applied, yielding $\alpha_4 = \hat{\alpha}$ according to (7.76). Note that because $\alpha_3 - \alpha_2$ does no longer equal $\alpha_2 - \alpha_1$, as was the case when leaving the bracketing section ($\alpha_1 = 0, \alpha_2 = h, \alpha_3 = 2h$), we must use the interpolating formula (7.76) rather than (7.77).

Several modifications can be brought to this basic refining scheme in order to obtain better conditioned brackets. For example, if $\alpha_2 < \alpha_4 < \alpha_3$ and if α_4 is very much closer to α_2 than to α_3, then, instead of using $\{\alpha_2, \alpha_4, \alpha_3\}$ as the refined bracket, it is better to take $\{\alpha_2, \alpha_4, \alpha_3 + \alpha_4/2\}$.

In the quadratic interpolation technique described above, the next point is always determined from the current point and the two of three last points that form a three-point pattern with it. Because of this property, global convergence can be proved, that is, the algorithm is guaranteed to generate a sequence of points converging to a solution. With respect to local convergence properties, i.e. some measure of the ultimate speed of convergence, it can be shown that the order of convergence is about 1.3, provided that near the solution the three-point pattern is always formed from the previous three points.

7.6 PROBLEMS

1. Consider the problem: minimize $f(x) = x_1^2 + x_2^2$

$$\text{subject to } h_1(x) = x_2 - x_1^2 \geqslant 0$$
$$h_2(x) = x_2 + x_1 - 1 \geqslant 0$$
$$1 \geqslant x_1 \geqslant \frac{1}{4}$$
$$1 \geqslant x_2 \geqslant \frac{1}{4}$$

Write the dual of this problem (with only two dual variables). Show that the dual space is partioned into six subregions by four hyperplanes. What is the meaning of these subregions? What happens when passing from one region to another? Using (7.25) and (7.32), find an expression for the gradient and the Hessian of the dual function.

2. Find an explicit form for the steepest descent algorithm applied to a positive definite quadratic form. (*Hint*: find $\alpha^{(k)}$ such that $g^{(k)T}g^{(k+1)} = 0$.) Use this algorithm for minimizing the function

$$q(x) = x_1^2 + x_2^2 + x_1 x_2 - 3x_1$$

With the sequence of values obtained for the objective function, determine the convergence ratio. Compare it with the theorical estimate (7.45). Solve again the problem when the side constraints $x_1 \geqslant 0, x_2 \geqslant 0$ are added in.

3. Assume that the function

$$q(x) = \frac{c}{2}(x_1^2 + x_2^2) + x_1 x_2 - 3x + cy$$

is minimized by the method of steepest descent.
What are the acceptable values of c such that q has a unique minimum point?
What is the value of c such that one digit will be gained at each iteration?

4. Consider the iterative process

$$x^{(k+1)} = \frac{1}{2}\left(x^{(k)} + \frac{a}{x^{(k)}}\right) \qquad a > 0$$

To what does it converge? What is the rate of convergence?

5. Apply the method of false position and the cubic interpolation to the function

$$\phi(\alpha) = (\alpha + 1)(\alpha - 1)^2$$

with initial h computed from (7.58) with $\phi(\alpha^*) = 0$.
Repeat the exercise with the quadratic interpolation procedure.

C.F.

REFERENCES

1. A. V. Fiacco and G. P. McCormick, *Non-linear Programming: Sequential Unconstrained Minimization Techniques*, Wiley, New York, 1968.
2. L. S. Lasdon, *Optimization Theory for Large Systems*, Macmillan, New York, 1970.
3. R. L. Fox, *Optimization Methods for Engineering Design*, Addison-Wesley, Reading, 1971.
4. F. A. Lootsma (ed.), *Numerical Methods for Non-linear Optimization*, Academic Press, London, 1972.
5. D. M. Himmelblau, *Applied Nonlinear Programming*, McGraw-Hill, New York, 1972.

6. D. G. Luenberger, *Introduction to Linear and Nonlinear Programming*, Addison-Wesley, Reading, 1973.
7. P. E. Gill and W. Murray (eds), *Numerical Methods for Constrained Optimization*, Academic Press, London, 1974.
8. M. A. Wolfe, *Numerical Methods for Unconstrained Optimization—An Introduction*, Van Nostrand Reinhold, Wokingham, 1978.

Foundations of Structural Optimization: A Unified Approach
Edited by A. J. Morris
© 1982 John Wiley & Sons Ltd.

Chapter 8

Unconstrained and Linearly Constrained Minimization

8.1 INTRODUCTION

Chapter 7 provides the background to general optimization method and we now want to move onto solution algorithms. Unconstrained and linearly constrained problems form an important subject of nonlinear programming, because many general methods finally lead to a requirement for solving such problems (sequence of sub-problems). A wide range of very effective techniques for the solution of this type of problem have evolved as a result of a massive research effort over the last few decades. This chapter describes the techniques in this general area appropriate for the solution of structural optimization problems. We begin in Section 8.2 by considering Newton-type methods for unconstrained optimization. In order to provide for effective algorithms safeguarding procedures ensuring positive-definiteness and remedies against failure of Newton's method are discussed. Section 8.3 is focused on conjugate direction methods which exhibit good convergence properties at low storage cost when applied to quadratic as well as general functions. Section 8.4 is concerned with quasi-Newton methods, which construct approximations to the inverse Hessian. Rank 1 and rank 2 update formula are described, including the well-known DFP and BFGS formulas. Quasi-Newton methods are very efficient, at the price of relatively high core requirement.

In Section 8.5, we introduce the idea of generalized steepest descent direction by defining the concepts of metric and norm. This leads to unifying the various unconstrained minimization techniques previously discussed. Then, the classical gradient projection method for linear constraints is studied in some detail, with emphasis on the three main steps: leaving a binding constraint, adding a constraint to the active set, and finding an unconstrained minimum along a line (Section 8.6).

Section 8.7 is concerned with more sophisticated projection methods, working in non-Euclidian metrics. These can be considered as extensions of Newton-type and quasi-Newton methods to linearly constrained problems. In Section 8.8, the important question of determining a suitable active-set strategy is considered. The essential difficulty lies in deleting inactive constraints early enough to reduce the computational cost, but without inducing zigzagging. Finally, Section 8.9 shows how the side constraints can be treated separately from ordinary linear constraints, so that storage requirements and computing times are reduced.

8.2 NEWTON-TYPE METHODS FOR UNCONSTRAINED OPTIMIZATION

8.2.1 The Quadratic Approximation

Consider the Taylor expansion of $f(x)$ around the point $\mathbf{x}^{(k)}$:

$$f(x^{(k)} + s) = f(x^{(k)}) + \mathbf{g}^{(k)T}\mathbf{s} + \frac{1}{2}\mathbf{s}^T\mathbf{H}^{(k)}\mathbf{s} + O(|s|^3) \quad (8.1)$$

$$= q(s) + O(|s|^3)$$

where $q(s)$ is the quadratic local approximation to $f(x)$ at $\mathbf{x} = \mathbf{x}^{(k)}$:

$$q(s) = f^{(k)} + \mathbf{g}^{(k)T}\mathbf{s} + \frac{1}{2}\mathbf{s}^T\mathbf{H}^{(k)}\mathbf{s} \quad (8.2)$$

We can choose $\mathbf{s}$ as the solution of the local quadratic minimum problem

$$\min_{s} q(s) \quad (8.3)$$

The gradient of this function is $\mathbf{g}^{(k)} + \mathbf{H}^{(k)}\mathbf{s}$ and thus the minimum occurs at the point $\mathbf{s}^{(k)}$ satisfying the equation

$$\mathbf{H}^{(k)}\mathbf{s}^{(k)} = -\mathbf{g}^{(k)} \quad (8.4)$$

Provided that the local Hessian matrix is positive definite, we have

$$\mathbf{g}^{(k)T}\mathbf{s}^{(k)} = -\mathbf{s}^{(k)T}\mathbf{H}^{(k)}\mathbf{s}^{(k)} < 0 \quad (8.5)$$

from which results,

$$f(x^{(k)} + s^{(k)}) - f(x^{(k)}) = -\frac{1}{2}\mathbf{s}^{(k)T}\mathbf{H}^{(k)}\mathbf{s}^{(k)} + O(|s|^3). \quad (8.6)$$

Equation (8.6) shows that the function $f(x)$ is decreased if it is sufficiently well approximated in the vicinity of $\mathbf{x}^{(k)}$ by the quadratic approximation (8.2).

8.2.2 The Standard Newton Algorithm

Equation (8.4) is the basis of Newton's method for estimating a critical point of $f(x)$ which is embodied in the algorithm shown in Fig. 8.1.

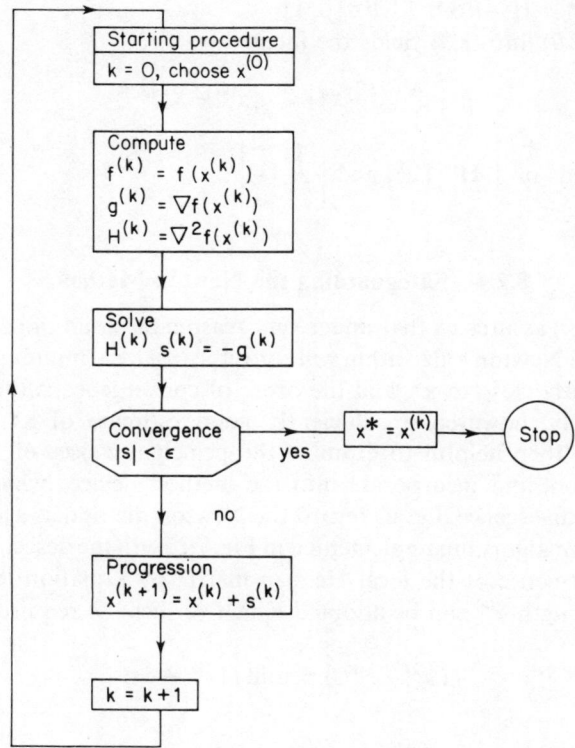

Fig. 8.1 Standard Newton algorithm

8.2.3 Order of Convergence of the Newton Method

It is a simple matter to show that if the sequence $\mathbf{x}^{(k)}$ generated from the algorithm converges to a local minimum $\mathbf{x}^*$, then its order of convergence is 2 and is, therefore quadratically convergent. From

$$\mathbf{x}^{(k+1)} = \mathbf{x}^{(k)} - [\mathbf{H}^{(k)}]^{-1} \mathbf{g}^{(k)} \tag{8.7}$$

and using the fact that $\mathbf{g}(x^*) = 0$, we have

$$|\mathbf{x}^{(k+1)} - \mathbf{x}^*| = |[\mathbf{H}^{(k)}]^{-1} \{\mathbf{H}^{(k)}(\mathbf{x}^{(k)} - \mathbf{x}^*) + \mathbf{g}(x^*) - \mathbf{g}^{(k)}\}| \tag{8.8}$$

A Taylor expansion of the right-hand side gives

$$\mathbf{H}^{(k)}(\mathbf{x}^{(k)} - \mathbf{x}^*) + \mathbf{g}(\mathbf{x}^*) - \mathbf{g}^{(k)} = -\frac{1}{2}[\mathbf{x}^{(k)} - \mathbf{x}^*]^T \frac{\partial \mathbf{H}}{\partial x}(\bar{x})[\mathbf{x}^{(k)} - \mathbf{x}^*]] \quad (8.9)$$

where $\bar{\mathbf{x}} = \theta \mathbf{x}^{(k)} + (1-\theta)\mathbf{x}^* \qquad \theta \in [0, 1]$
Substituting (8.9) into (8.8) yields the inequality

$$|\mathbf{x}^{(k+1)} - \mathbf{x}^*| \leqslant c |\mathbf{x}^{(k)} - \mathbf{x}^*|^2 \quad (8.10)$$

where c depends on $|[\mathbf{H}^{(k)}]^{-1}|$ and $\left|\frac{\partial \mathbf{H}}{\partial x}(\bar{x})\right|$.

8.2.4 Safeguarding the Newton Method

The result (8.10) assures us that under very reasonable conditions the sequence generated from Newton's algorithm will converge to a local minimum $\mathbf{x}^*$ of $f(x)$ if $\mathbf{x}^{(0)}$ is sufficiently close to $\mathbf{x}^*$, and the order of convergence is then at least 2.

Unfortunately, however, a sufficiently good estimate of $\mathbf{x}^*$ is often not available. It is then helpful to examine the principal causes of failure of the Newton method, and incorporate into the method devices which reduce the probability of divergence. Let us regard the Newton method as a special case of the basic descent algorithm implemented in Fig. 7.2, with the descent direction $\mathbf{s}^{(k)}$ defined in the metric of the local Hessian matrix by Equation (8.7). Then, an optimal step length $\alpha^{(k)}$ can be adopted which verifies the requirement

$$f(x^{(k)} + \alpha^{(k)}s) = \min_{\alpha} f(x^{(k)} + \alpha s) \quad (8.11)$$

or

$$\min_{\alpha} \{f(x^{(k)}) + \alpha \mathbf{s}^T \mathbf{g}^{(k)} + \frac{\alpha^2}{2} \mathbf{s}^T \mathbf{H}^{(k)} \mathbf{s} + \mathrm{O}(\alpha^3 |s|^3)\} \quad (8.12)$$

which yields the line-search equation

$$\mathbf{s}^T \mathbf{g}^{(k)} + \alpha \mathbf{s}^T \mathbf{H}^{(k)} \mathbf{s} + \mathrm{O}(\alpha^2 |s|^3) = 0 \quad (8.13)$$

with the particular choice (8.4) of the search direction

$$(\alpha - 1)\mathbf{g}^{(k)T}[\mathbf{H}^{(k)}]^{-1}\mathbf{g}^{(k)} + \mathrm{O}(\alpha^2 |s|^3) = 0 \quad (8.14)$$

it shows that the step length $\alpha = 1$ may not give a good estimate of the minimum if the quadratic approximation is not accurate enough in the vicinity of $\mathbf{x}^{(k)}$.

Therefore, it may happen that $f^{(k+1)} > f^{(k)}$ with Newton's standard algorithm for one of the following reasons:
1. $[\mathbf{H}^{(k)}]^{-1}$ exists and is positive definite, but $\mathbf{s}^{(k)}$ is so large that $f^{(k+1)} > f^{(k)}$.
2. If $[\mathbf{H}^{(k)}]^{-1}$ exists but is not positive definite, we may have $\mathbf{s}^{(k)T}\mathbf{g}^{(k)} < 0$. This means that $\mathbf{s}^{(k)}$ is no longer downhill: again, it may happen that $f^{(k+1)} > f^{(k)}$.

3. The case $s^{(k)T}g^{(k)} = 0$ is a limiting case of the preceding one. It corresponds to orthogonality between the search direction and the gradient, and cannot occur if $H^{(k)}$ is positive definite.
4. If $H^{(k)}$ is singular, then $s^{(k)}$ is not even defined, so that if further progress is to be made one needs an alternative means to construct $s^{(k)}$.

8.2.5 Remedies against Failure

8.2.5.1 Line search: generalized Newton method

If $[H^{(k)}]^{-1}$ remains positive definite, the series expansion (8.12) shows that the condition $f^{(k+1)} < f^{(k)}$ may always be fulfilled if a line search is adopted. The original Newton algorithm is thus modified by adding a line search before the progression step.

8.2.5.2 Remedies against positive indefiniteness

(a) Both failures due to orthogonality of the search direction with the gradient and to the singularity of the Hessian matrix can be overcome by replacing the Newton step by a steepest-descent iteration;
(b) failure due to fact that the search direction is uphill can be avoided by reversing the search direction.

These modifications lead to a first safeguarded Newton algorithm defined by the flowchart shown in Fig. 8.2.

An alternative against all the modes of failure due to positive indefiniteness of the Hessian matrix is to modify the Hessian matrix in order to preserve its positive definite character (see Fig. 8.3).

(a) In a method due originally to Greenstadt [3], a modified $\bar{H}^{(k)}$ is found by computing the eigenvalues
$(\mu_1, \ldots, \mu_n)$ $(\mu_i \leq \mu_j$ if $i \leq j)$ and eigenvectors
$(v_1, \ldots, v_n)$ of $H^{(k)}$, and setting

$$\bar{H}^{(k)} = \sum_{j=1}^{n} \bar{\mu}_j v_j v_j^T \tag{8.15}$$

where

$$\bar{\mu}_j = \max\{\delta, |\mu_j|\} \tag{8.16}$$

The parameter δ is introduced to avoid numerical difficulties in evaluating $s^{(k)}$ when $H^{(k)}$ is ill conditioned.

(b) Levenberg [4] and later on, Murray [5] have proposed modifications of Newton's method which are more efficient from a computational point of view. Both consist in modifying the Hessian matrix while computing its Choleski decomposition $H^{(k)} = L^{(k)}D^{(k)}L^{(k)T}$. In Murray's method, the diagonal $D^{(k)}$ is increased by a sufficient amount during the computation to

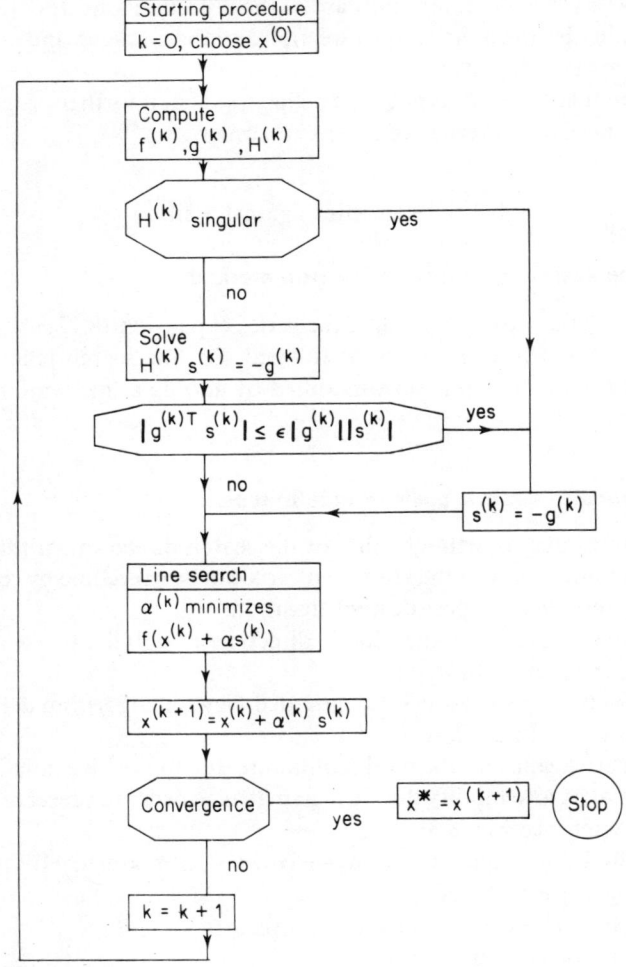

Fig. 8.2 Safeguarded Newton algorithm I

restore the positive definiteness of $\mathbf{H}^{(k)}$. Eventually a factorization

$$\mathbf{L}^{(k)} \mathbf{D}^{(k)} \mathbf{L}^{(k)T} = \mathbf{H}^{(k)} + \mathbf{E}^{(k)} = \bar{\mathbf{H}}^{(k)}$$

is obtained, where $\mathbf{E}^{(k)}$ is a diagonal matrix which vanishes if $\mathbf{H}^{(k)}$ is sufficiently positive definite: that is, $D_{ii}^{(k)} > \delta$.

In both methods, the selection of an appropriate δ is somewhat of an art. A small δ means that nearly singular matrices must be inverted, while a large δ means that the order two convergence may be lost.

Unconstrained and Linearly Constrained Minimization

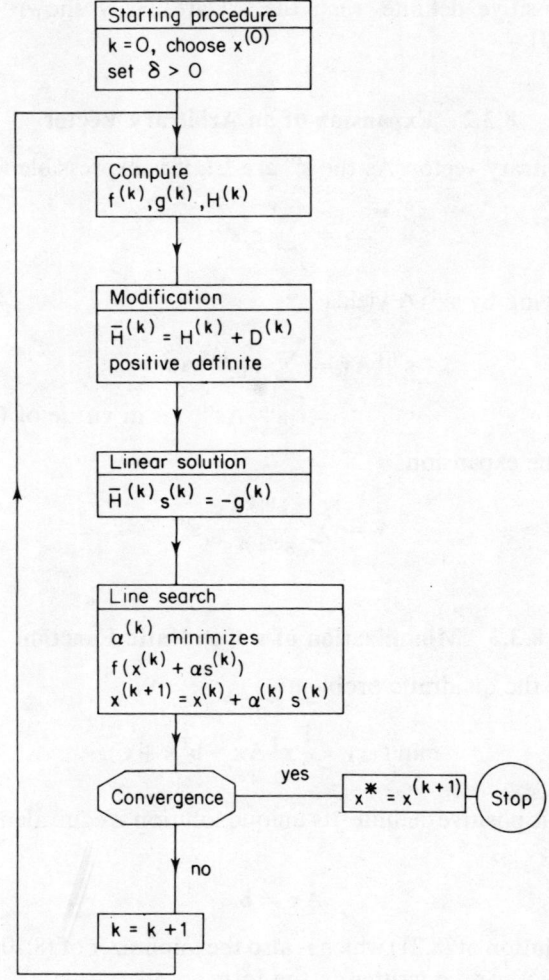

Fig. 8.3 Safeguard Newton algorithm II

8.3 CONJUGATE DIRECTION METHODS

8.3.1 Definition of Conjugate directions

Let **A** be a symmetric $n \times n$ matrix. Then, the vectors $\mathbf{s}^{(i)}$ ($i = 1, \ldots, n$) are A-conjugate if and only if

$$\mathbf{s}^{(i)\mathrm{T}} \mathbf{A} \mathbf{s}^{(j)} = 0 \qquad i \neq j \tag{8.17}$$

If $\mathbf{A}$ is also positive definite, then the $\mathbf{s}^{(i)}$ are easily shown to be linearly independent (LI).

8.3.2 Expansion of an Arbitrary Vector

Let $\mathbf{v}$ be an arbitrary vector. As the $\mathbf{s}^{(i)}$ are LI, it is expressible in the form

$$\mathbf{v} = \sum_{j=1}^{n-1} c_j \mathbf{s}^{(j)} \qquad (8.18)$$

and premultiplying by $\mathbf{s}^{(i)\mathrm{T}}\mathbf{A}$ yields

$$\mathbf{s}^{(i)\mathrm{T}}\mathbf{A}\mathbf{v} = \sum_{j=1}^{n-1} c_j \mathbf{s}^{(i)\mathrm{T}}\mathbf{A}\mathbf{s}^{(j)}$$
$$= c_i \mathbf{s}^{(i)\mathrm{T}}\mathbf{A}\mathbf{s}^{(i)} \qquad \text{in virtue of (8.17)}$$

Hence we get the expansion

$$\mathbf{v} = \sum_{i=0}^{n-1} \frac{\mathbf{s}^{(i)\mathrm{T}}\mathbf{A}\mathbf{v}}{\mathbf{s}^{(i)\mathrm{T}}\mathbf{A}\mathbf{s}^{(i)}} \mathbf{s}^{(i)} \qquad (8.19)$$

8.3.3 Minimization of a Quadratic Function

Let us consider the quadratic problem

$$\min_{x} f(x) = \frac{1}{2}\mathbf{x}^{\mathrm{T}}\mathbf{A}\mathbf{x} - \mathbf{b}^{\mathrm{T}}\mathbf{x} + \mathbf{c} \qquad (8.20)$$

with $\mathbf{A}$ symmetric positive definite. Its unique solution is equivalent to solving the linear problem

$$\mathbf{A}\mathbf{x} = \mathbf{b} \qquad (8.21)$$

Let $\mathbf{x}^*$ be the solution of (8.21) which is also the minimizer of (8.20). According to (8.18) and (8.19) it can be written in the form

$$\mathbf{x}^* = \sum_{i=0}^{n-1} \alpha_i \mathbf{s}^{(i)} \qquad (8.22)$$

in which the coefficients

$$\alpha_i = \frac{\mathbf{s}^{(i)\mathrm{T}}\mathbf{A}\mathbf{x}^*}{\mathbf{s}^{(i)\mathrm{T}}\mathbf{A}\mathbf{s}^{(i)}} = \frac{\mathbf{s}^{(i)\mathrm{T}}\mathbf{b}}{\mathbf{s}^{(i)\mathrm{T}}\mathbf{A}\mathbf{s}^{(i)}} \qquad (8.23)$$

The result (8.23) suggests that using the property of A orthogonality, the solution of (8.21) can be obtained from an iterative process of n steps in which the successive coefficients are evaluated without knowing $\mathbf{x}^*$. It yields the following algorithm.

8.3.4 The General Conjugate Direction Algorithm

The general conjugate algorithm is shown in Fig. 8.4.

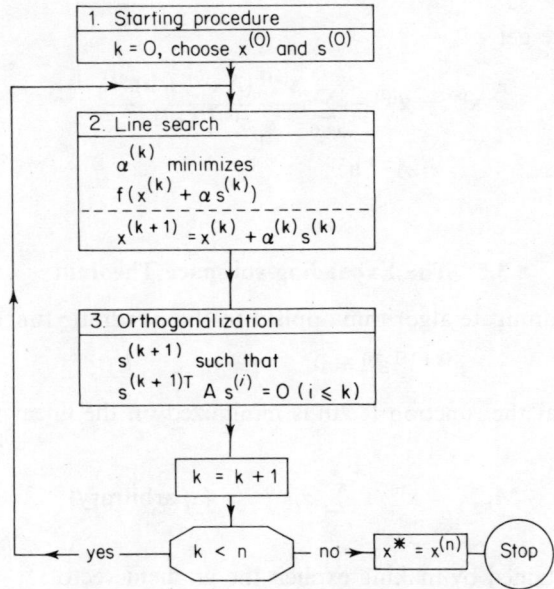

Fig. 8.4 The general conjugate direction algorithm for minimizing quadratic functions

It is easily shown that when applied to the quadratic function (8.20), the algorithm converges in n steps and is, thus, quadratically convergent. In the case of algorithms quadratic convergence is often called *quadratic termination* because this property can only be demonstrated in the terminal phase of the iteration history for non-quadratic functions.

Proof By step 2 of the algorithm, we have

$$\mathbf{x}^{(k+1)} = \mathbf{x}^{(0)} + \sum_{j=0}^{k} \alpha^{(j)} \mathbf{s}^{(j)} \qquad 0 \leqslant k < n$$

and, since $\alpha^{(k)}$ minimizes $f(x)$ along $\mathbf{s}^{(k)}$, we have also

$$\mathbf{s}^{(k)\mathrm{T}} \mathbf{g}^{(k+1)} = 0 \qquad (8.24)$$

The gradient of (8.20) evaluated at $\mathbf{x}^{(k+1)}$ is

$$\mathbf{g}^{(k+1)} = \mathbf{A}\mathbf{x}^{(k+1)} - \mathbf{b}$$

$$= \mathbf{A}\mathbf{x}^{(0)} - \mathbf{b} + \sum_{j=0}^{k} \alpha^{(j)} \mathbf{A}\mathbf{s}^{(j)}$$

and making use of (8.17) and (8.24) we obtain

$$\alpha^{(k)} = -\frac{\mathbf{s}^{(k)\mathrm{T}}(\mathbf{A}\mathbf{x}^{(0)} + \mathbf{b})}{\mathbf{s}^{(k)\mathrm{T}}\mathbf{A}\mathbf{s}^{(k)}}$$

For $k = n - 1$ we get

$$\mathbf{x}^{(n)} = \mathbf{x}^{(0)} - \sum_{j=0}^{n-1} \frac{\mathbf{s}^{(k)\mathrm{T}}(\mathbf{A}\mathbf{x}^{(0)} + \mathbf{b})\mathbf{s}^{(k)}}{\mathbf{s}^{(k)\mathrm{T}}\mathbf{A}\mathbf{s}^{(k)}}$$

$$= \mathbf{A}^{-1}\mathbf{b}$$

8.3.5 The Expanding-subspace Theorem

In the general conjugate algorithm applied to the quadratic function (8.20),

$$\mathbf{g}^{(k+1)\mathrm{T}}\mathbf{s}^{(i)} = 0 \qquad i = 0, \ldots, k \tag{8.25}$$

which means that the function (8.20) is minimized on the linear variety $\mathbf{M}_{k+1}$ defined by

$$\mathbf{M}_{k+1} = \mathbf{x}^{(0)} + \sum_{i=0}^{k} \alpha_i \mathbf{s}^{(i)} \qquad (\alpha_i \text{ arbitrary})$$

The proof is obtained by making explicit the gradient vector:

$$\mathbf{g}^{(k+1)\mathrm{T}}\mathbf{s}^{(i)} = (\mathbf{A}\mathbf{x}^{(k+1)} + \mathbf{b})^{\mathrm{T}}\mathbf{s}^{(i)}$$

$$= [\mathbf{A}(\mathbf{x}^{(i+1)} + \sum_{j=i+1}^{k} \alpha^{(j)}\mathbf{s}^{(j)}) + \mathbf{b}]^{\mathrm{T}}\mathbf{s}^{(i)}$$

$$= \mathbf{g}^{(i+1)\mathrm{T}}\mathbf{s}^{(i)}$$

The right-hand side vanishes by making use of the property (8.17) of orthogonality.

8.3.6 The Conjugate Gradient Algorithm

The conjugate gradient algorithm is a special case of the general conjugate direction algorithm, in which the set of conjugate directions is obtained from an orthogonalization of the successive gradients.

Suppose that the $\mathbf{s}^{(i)}$ have been obtained by reorthogonalization of the $\mathbf{g}^{(i)}$. As the vectors $[\mathbf{s}^{(1)}, \ldots, \mathbf{s}^{(i)}]$ and $[\mathbf{g}^{(1)}, \ldots, \mathbf{g}^{(i)}]$ span the same subspace of R_n, equation (8.25) is equivalent to

$$\mathbf{g}^{(k+1)\mathrm{T}}\mathbf{g}^{(i)} = 0 \qquad i = 0, \ldots, k \tag{8.26}$$

Equation (8.26) is the fundamental relation on which the conjugate gradient algorithm is based.

Indeed, at the $(k+1)$th iteration, $\mathbf{s}^{(k+1)}$ is computed from

$$\mathbf{s}^{(k+1)} = -\mathbf{g}^{(k+1)} + \sum_{i=0}^{k} \beta^{(i)} \mathbf{s}^{(i)}$$

where the $\beta^{(i)}$ are constructed to ensure orthogonality:

$$\beta^{(i)} = \frac{\mathbf{g}^{(k+1)\mathrm{T}} \mathbf{A} \mathbf{s}^{(i)}}{\mathbf{s}^{(i)\mathrm{T}} \mathbf{A} \mathbf{s}^{(i)}}$$

If use is made of the fact that

$$\mathbf{A}\mathbf{s}^{(i)} = \mathbf{g}^{(i+1)} - \mathbf{g}^{(i)}$$

we obtain $\beta_i = 0$ $(i < k)$, and thus

$$\mathbf{s}^{(k+1)} = -\mathbf{g}^{(k+1)} + \beta^{(k)} \mathbf{s}^{(k)} \tag{8.27}$$

with

$$\beta^{(k)} = \frac{\mathbf{g}^{(k+1)\mathrm{T}}[\mathbf{g}^{(k+1)} - \mathbf{g}^{(k)}]}{\mathbf{s}^{(k)\mathrm{T}}[\mathbf{g}^{(k+1)} - \mathbf{g}^{(k)}]} \tag{8.28}$$

The conjugate gradient algorithm is then as implemented in Fig. 8.5.

The formulas (8.26)–(8.27) for calculating the A-conjugate directions are due to Hestenes and Stiefel [6]. They may be further simplified by noting that:
1. The line search along $\mathbf{s}^{(k)}$ is exact so that

$$\mathbf{s}^{(k)\mathrm{T}} \mathbf{g}^{(k+1)} = 0$$

2. The function f is quadratic, so that by (8.26)

$$\mathbf{g}^{(k+1)\mathrm{T}} \mathbf{g}^{(k)} = 0$$

3. The line search along $\mathbf{s}^{(k-1)}$ is also exact, so that

$$\begin{aligned} \mathbf{s}^{(k)\mathrm{T}} \mathbf{g}^{(k)} &= [-\mathbf{g}^{(k)} + \beta^{(k-1)} \mathbf{s}^{(k-1)}]^{\mathrm{T}} \mathbf{g}^{(k)} \\ &= -\mathbf{g}^{(k)\mathrm{T}} \mathbf{g}^{(k)} \end{aligned}$$

Therefore:
(a) If hypotheses 1, 2 and 3 are simultaneously valid, one then obtains the widely used Fletcher–Reeves formula [7]:

$$\beta^{(k)} = \frac{\mathbf{g}^{(k+1)\mathrm{T}} \mathbf{g}^{(k+1)}}{\mathbf{g}^{(k)\mathrm{T}} \mathbf{g}^{(k)}} \tag{8.29}$$

(b) If only hypotheses 1 and 3 are valid, then

$$\beta^{(k)} = \frac{\mathbf{g}^{(k+1)\mathrm{T}} (\mathbf{g}^{(k+1)} - \mathbf{g}^{(k)})}{\mathbf{g}^{(k)\mathrm{T}} \mathbf{g}^{(k)}} \tag{8.30}$$

a formula due to Polak and Ribiere [8]
(c) If only hypotheses 1 and 2 are valid, the corresponding formula is attributed by Dixon [9] to Myers:

$$\beta^{(k)} = \frac{\mathbf{g}^{(k+1)\mathrm{T}} \mathbf{g}^{(k+1)}}{\mathbf{s}^{(k)\mathrm{T}} [\mathbf{g}^{(k+1)} - \mathbf{g}^{(k)}]} \tag{8.31}$$

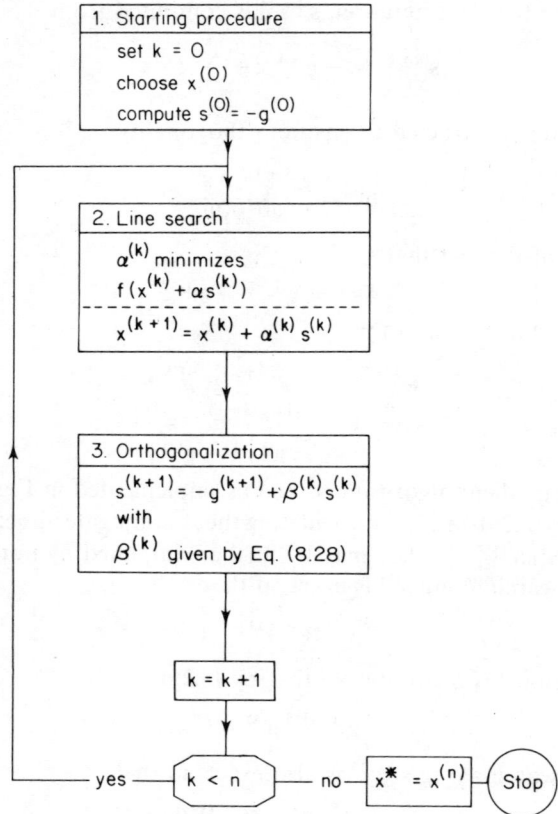

Fig. 8.5 The conjugate gradient algorithm for minimizing a quadaratic functions

8.3.7 Application to Non-quadratic Objective Functions

The family of conjugate gradient methods described above is applicable, with minor modifications, to the problem of finding the local minimizers of a non-quadratic objective function. Clearly, if $\mathbf{x}$ is sufficiently close to $\mathbf{x}^*$, then f is sufficiently well approximated by a quadratic function and the directions $\mathbf{s}^{(k)}$ generated by (8.26) and (8.27) become nearly A-conjugate.

We expect that $\mathbf{x}^{(n)}$ generated by n steps of the algorithm will be much closer to $\mathbf{x}^*$ than is $\mathbf{x}^{(0)}$, so that the quadratic approximation becomes much better in the vicinity of $\mathbf{x}^{(n)}$. This suggests a strategy called resetting, which consists of restarting the algorithm after n steps with $\mathbf{s}^{(0)} = -\mathbf{g}^{(0)}$.

An alternate strategy consists of continuing the iteration without resetting, but it has been found to be far less computationally efficient, in general, than the resetting strategy.

Hestenes and Stiefel [6] have obtained satisfactory results by resetting every $n+1$ iterations rather than every n, but the basis of this appears to be empirical.

Two other modifications have to be made to the algorithm if the objective function is non-quadratic:
  (a) If the line search is not accurate enough, it is no longer guaranteed that $\mathbf{g}^{(k+1)\mathrm{T}}\mathbf{s}^{(k)} < 0$. If it is the case, the algorithm has to be restarted.
  (b) A test of convergence has to be included to stop the algorithm.

The final algorithm is described by the flowchart in Fig. 8.6. It is difficult to advise the use of one formula (8.28)–(8.31) rather than another. The numerical experiments reported in [1] indicate that they all perform almost equally well if used in conjunction with a cubic line search.

Conjugate gradient methods are useful when the number of variables n is large, because of the relatively small amount of machine storage space required.

8.4 QUASI-NEWTON METHODS

8.4.1 Principle of Quasi-Newton Methods

We have seen that Newton's method for minimizing a function $f(x)$ consists in generating a sequence $\mathbf{x}^{(k)}$ from

$$\mathbf{x}^{(k+1)} = \mathbf{x}^{(k)} - [\mathbf{H}^{(k)}]^{-1}\mathbf{g}^{(k)} \qquad (8.32)$$

where $\mathbf{H}^{(k)}$ is the Hessian matrix evaluated at $\mathbf{x} = \mathbf{x}^{(k)}$. There are, however, a certain number of objections to Newton's method, the principal one being that each iteration requires the direct evaluation of $\mathbf{H}^{(k)}$ and the solution of the associated linear system (8.32). This main drawback of Newton's method provides motivation for constructing minimization algorithms in which the Hessian matrix is approximated from available quantities rather than calculated directly. The basis for such approximations is the quasi-Newton equation.

8.4.2 The Quasi-Newton Equation

In order to obtain an approximation to the Hessian matrix, let us consider a first-order expansion of the gradient of $f(x)$ in the vicinity of $\mathbf{x}^{(k+1)}$:

$$\mathbf{g}(x^{(k)}) = \mathbf{g}(x^{(k+1)}) + \mathbf{H}(x^{(k+1)})[\mathbf{x}^{(k)} - \mathbf{x}^{(k+1)}] + \Delta$$

where $\Delta \to 0$ as $\mathbf{x}^{(k)} \to \mathbf{x}^{(k+1)}$.

If we define thus

$$\mathbf{y}^{(k)} = \mathbf{g}^{(k+1)} - \mathbf{g}^{(k)}$$
$$\mathbf{s}^{(k)} = \mathbf{x}^{(k+1)} - \mathbf{x}^{(k)}$$

and neglect the second-order term Δ, we obtain

$$\mathbf{y}^{(k)} \simeq \mathbf{H}^{(k+1)}\mathbf{s}^{(k)}$$

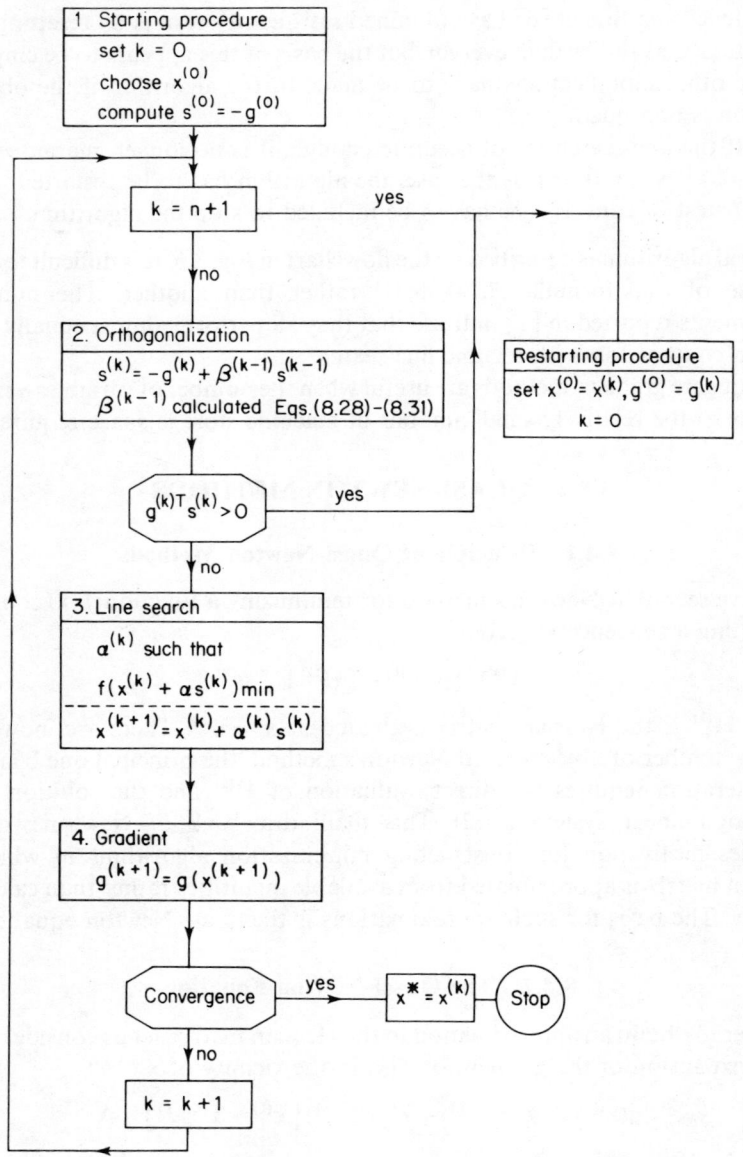

Fig. 8.6 Conjugate gradient algorithm for non-quadratic functions

Therefore, if we call $\mathbf{S}^{(k+1)}$ an approximation to the inverse Hessian matrix $[\mathbf{H}^{(k+1)}]^{-1}$, it should be constructed in order to verify the so-called quasi-Newton equation

$$\mathbf{S}^{(k+1)}\mathbf{y}^{(k)} = \mathbf{s}^{(k)} \quad k \geqslant 0 \tag{8.33}$$

The matrix $\mathbf{S}^{(k+1)}$ is easily computable from $\mathbf{S}^{(k)}$ if it is obtained by adding to $\mathbf{S}^{(k)}$ a correction term $\mathbf{C}^{(k)}$ which depends upon $\mathbf{S}^{(k)}$, $\mathbf{y}^{(k)}$ and $\mathbf{s}^{(k)}$ only:

$$\mathbf{S}^{(k+1)} = \mathbf{S}^{(k)} + \mathbf{C}^{(k)} \tag{8.34}$$

At the same time, the correction $\mathbf{C}^{(k)}$ should be constructed to preserve the symmetry and the positive definiteness of $\mathbf{S}^{(k)}$. Then, the vector $-\mathbf{S}^{(k)}\mathbf{g}^{(k)}$ is downhill at $\mathbf{x} = \mathbf{x}^{(k)}$ and there exists an $\alpha^{(k)} > 0$ such that

$$f(x^{(k)} - \alpha^{(k)} S^{(k)} g^{(k)}) < f(x^{(k)})$$

8.4.3 Quasi-Newton Algorithms

Collecting the ideas described above, the general quasi-Newton method for minimizing a function $f(x)$ is contained in the flow diagram given in Fig. 8.7. The various quasi-Newton algorithms will differ only by the choice of the matrix updating formula.

8.4.4 Rank-one Updates

To obtain a symmetric correction matrix $\mathbf{C}^{(k)}$, we investigate a correction in the form of a rank-one matrix

$$\mathbf{S}^{(k+1)} = \mathbf{S}^{(k)} + \beta^{(k)} \mathbf{z}^{(k)} \mathbf{z}^{(k)\mathrm{T}} \tag{8.35}$$

The vector $\mathbf{z}^{(k)}$ and the coefficient $\beta^{(k)}$ are selected so that (8.33) is satisfied:

$$\mathbf{s}^{(k)} = \mathbf{S}^{(k+1)} \mathbf{y}^{(k)} = \mathbf{S}^{(k)} \mathbf{y}^{(k)} + \beta^{(k)} \mathbf{z}^{(k)} \mathbf{z}^{(k)\mathrm{T}} \mathbf{y}^{(k)} \tag{8.36}$$

This yields

$$\mathbf{z}^{(k)} = \frac{\mathbf{s}^{(k)} - \mathbf{S}^{(k)} \mathbf{y}^{(k)}}{\beta^{(k)} \mathbf{z}^{(k)\mathrm{T}} \mathbf{y}^{(k)}} \tag{8.37}$$

Taking also the inner product of (8.37) with $\mathbf{y}^{(k)\mathrm{T}}$ we have

$$\beta^{(k)} [\mathbf{y}^{(k)\mathrm{T}} \mathbf{z}^{(k)}]^2 = \mathbf{y}^{(k)\mathrm{T}} \mathbf{s}^{(k)} - \mathbf{y}^{(k)\mathrm{T}} \mathbf{S}^{(k)} \mathbf{y}^{(k)} \tag{8.38}$$

Reintroducing (8.37) and (8.38) into (8.35) yields the symmetric rank-one formula [10] which verifies the quasi-Newton equation

$$\mathbf{S}^{(k+1)} = \mathbf{S}^{(k)} + \frac{[\mathbf{s}^{(k)} - \mathbf{S}^{(k)} \mathbf{y}^{(k)}][\mathbf{s}^{(k)} - \mathbf{S}^{(k)} \mathbf{y}^{(k)}]^{\mathrm{T}}}{\mathbf{y}^{(k)\mathrm{T}}[\mathbf{s}^{(k)} - \mathbf{S}^{(k)} \mathbf{y}^{(k)}]} \tag{8.39}$$

8.4.5 Application to the Quadratic Case

To demonstrate the quadratic convergence of the quasi-Newton method with the rank-one updates, let us consider its application to the quadratic function (8.20). By the very definition of the gradient $g = \mathbf{A}\mathbf{x} - \mathbf{b}$ we have

$$\mathbf{A}\mathbf{s}^{(k)} = \mathbf{y}^{(k)} \tag{8.40}$$

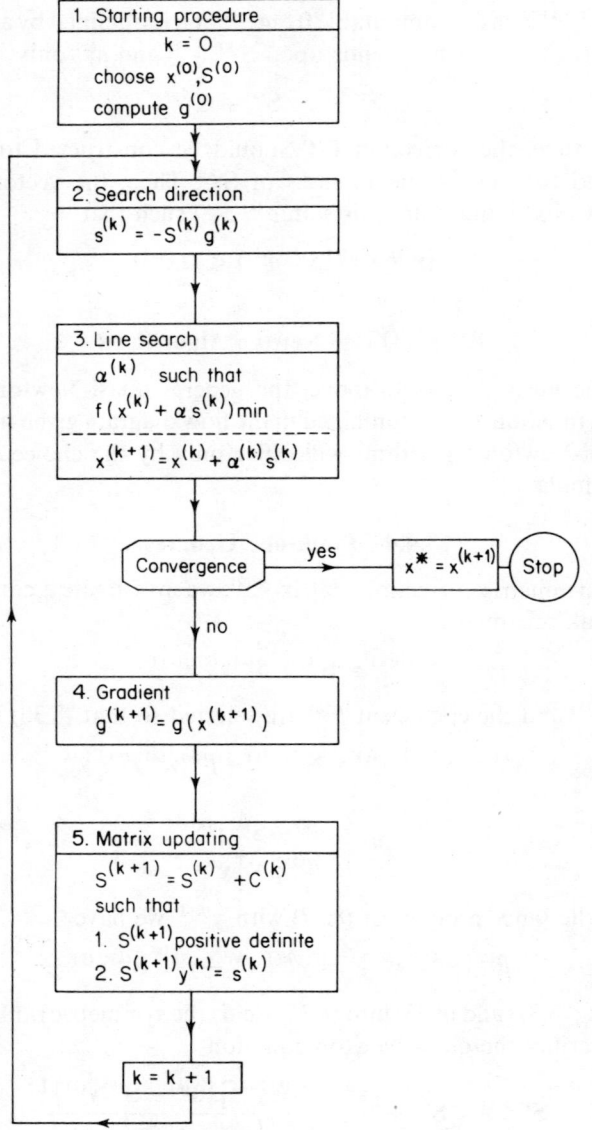

Fig. 8.7 The general quasi-Newton algorithm

Now, let us consider the sequence of estimates to $\mathbf{A}^{-1}$ generated by (8.39) with $\mathbf{S}^{(o)}$ positive definite: the proof of quadratic convergence lies on the fact that

$$\mathbf{S}^{(k)}\mathbf{y}^{(i)} = \mathbf{s}^{(i)} \qquad (i = 0, \ldots, k-1) \tag{8.41}$$

Since for $k = n$ we obtain the equality

$$\mathbf{S}^{(n)}\mathbf{y}^{(i)} = \mathbf{S}^{(n)}\mathbf{A}^{-1}\mathbf{s}^{(i)} = \mathbf{s}^{(i)} \qquad (i = 0, \ldots, n-1)$$

the $\mathbf{s}^{(i)}$ are thus eigenvectors of the matrix $\mathbf{S}^{(n)}\mathbf{A}^{-1}$ with a unit eigenvalue. Provided that they are linearly independent, we have

$$\mathbf{S}^{(n)} = \mathbf{A}^{-1} \tag{8.42}$$

The proof of (8.41) is obtained by induction:
For $k = 0$, the quasi-Newton equation yields

$$\mathbf{S}^{(1)}\mathbf{y}^{(0)} = \mathbf{s}^{(0)}$$

Suppose now that (8.41) holds for $k > 0$. Then, for $i = 0, \ldots, k-1$, equation (8.39) yields

$$\mathbf{S}^{(k+1)}\mathbf{y}^{(i)} = \mathbf{S}^{(k)}\mathbf{y}^{(i)} + \frac{[\mathbf{s}^{(k)} - \mathbf{S}^{(k)}\mathbf{y}^{(k)}][\mathbf{s}^{(k)} - \mathbf{S}^{(k)}\mathbf{y}^{(k)}]^T\mathbf{y}^{(i)}}{\mathbf{y}^{(k)T}[\mathbf{s}^{(k)} - \mathbf{S}^{(k)}\mathbf{y}^{(k)}]}$$

where, by making use of (8.40) and (8.41),

$$[\mathbf{s}^{(k)} - \mathbf{S}^{(k)}\mathbf{y}^{(k)}]^T\mathbf{y}^{(i)} = \mathbf{s}^{(k)T}\mathbf{y}^{(i)} - \mathbf{y}^{(k)T}\mathbf{S}^{(k)}\mathbf{y}^{(i)}$$
$$= \mathbf{s}^{(k)T}\mathbf{A}\mathbf{s}^{(i)} - \mathbf{s}^{(k)T}\mathbf{A}\mathbf{s}^{(i)} = 0$$

If use is also made of the quasi-Newton equation for $k+1$, we obtain

$$\mathbf{S}^{(k+1)}\mathbf{y}^{(i)} = \mathbf{s}^{(i)} \qquad (i = 0, \ldots, k)$$

It is worthwhile noticing that the quasi-Newton algorithm with the rank-one update (8.39), when applied to a quadratic function, does not involve any line search. It is thus expected that its application to non-quadratic cases dispenses also with an accurate line search, which makes it very attractive.

Unfortunately, the positive definite character of $\mathbf{S}^{(k)}$ may be lost in the iteration process. This may lead to a breakdown of the algorithm. Therefore, the rank-two updates which may be constructed in order to preserve the positive definiteness of $\mathbf{S}^{(k)}$ are generally preferred.

8.4.6 Rank-two Update: The Davidon–Fletcher–Powell (DFP) Method

In the DFP method Ref. [11] the updating formula is taken of the form

$$\mathbf{S}^{(k+1)} = \mathbf{S}^{(k)} + \beta \mathbf{s}^{(k)}\mathbf{s}^{(k)T} + \gamma [\mathbf{S}^{(k)}\mathbf{y}^{(k)}][\mathbf{S}^{(k)}\mathbf{y}^{(k)}]^T$$

and has to verify the quasi-Newton equation

$$\mathbf{S}^{(k+1)}\mathbf{y}^{(k)} = \mathbf{S}^{(k)}\mathbf{y}^{(k)} + \beta \mathbf{s}^{(k)}\mathbf{s}^{(k)T}\mathbf{y}^{(k)} + \gamma [\mathbf{S}^{(k)}\mathbf{y}^{(k)}]\mathbf{y}^{(k)T}\mathbf{S}^{(k)}\mathbf{y}^{(k)}$$
$$= \mathbf{s}^{(k)}$$

The equality above holds for the particular choice of coefficients β and γ which gives the DFP updating formula:

$$\mathbf{S}^{(k+1)}_{\text{DFP}} = \mathbf{S}^{(k)}_{\text{DFP}} + \frac{\mathbf{s}^{(k)}\mathbf{s}^{(k)T}}{\mathbf{s}^{(k)T}\mathbf{y}^{(k)}} - \frac{[\mathbf{S}^{(k)}_{\text{DFP}}\mathbf{y}^{(k)}][\mathbf{S}^{(k)}_{\text{DFP}}\mathbf{y}^{(k)}]^T}{\mathbf{y}^{(k)T}\mathbf{S}^{(k)}_{\text{DFP}}\mathbf{y}^{(k)}} \tag{8.43}$$

8.4.6.1 Positive definiteness

We prove first that if $\mathbf{S}^{(k)}$ is positive definite, then so is $\mathbf{S}^{(k+1)}$ if a line search is made to calculate $\mathbf{x}^{(k+1)}$.

To prove it, we note that for an arbitrary $\mathbf{x}$ we have

$$\mathbf{x}^T \mathbf{S}^{(k+1)} \mathbf{x} = \mathbf{x}^T \mathbf{S}^{(k)} \mathbf{x} + \frac{[\mathbf{x}^T \mathbf{s}^{(k)}]^2}{\mathbf{s}^{(k)T} \mathbf{y}^{(k)}} - \frac{[\mathbf{x}^T \mathbf{S}^{(k)} \mathbf{y}^{(k)}]^2}{\mathbf{y}^{(k)T} \mathbf{S}^{(k)} \mathbf{y}^{(k)}}$$

If a line search is assumed, then, with the optimal step length $\alpha^{(k)}$,

$$\mathbf{s}^{(k)T} \mathbf{y}^{(k)} = \mathbf{s}^{(k)T} (\mathbf{g}^{(k+1)} - \mathbf{g}^{(k)}) = -\mathbf{s}^{(k)T} \mathbf{g}^{(k)}$$
$$= \alpha^{(k)} \mathbf{g}^{(k)T} \mathbf{S}^{(k)} \mathbf{g}^{(k)} > 0$$

Let us set also

$$\mathbf{p} = [\mathbf{S}^{(k)}]^{1/2} \mathbf{x} \quad \text{and} \quad \mathbf{q} = [\mathbf{S}^{(k)}]^{1/2} \mathbf{y}$$

Making use of the Cauchy–Schwarz inequality we get

$$\mathbf{x}^T \mathbf{S}^{(k+1)} \mathbf{x} = \frac{(\mathbf{p}^T \mathbf{p})(\mathbf{q}^T \mathbf{q}) - (\mathbf{p}^T \mathbf{q})^2}{(\mathbf{q}^T \mathbf{p})^2} + \frac{[\mathbf{x}^T \mathbf{s}^{(k)}]^2}{\alpha^{(k)} \mathbf{g}^{(k)T} \mathbf{S}^{(k)} \mathbf{g}^{(k)}} > 0$$

Actually, any $\alpha^{(k)}$ which gives $\mathbf{s}^{(k)T} \mathbf{y}^{(k)} > 0$ can be used in the algorithm since the positive definite character of $\mathbf{S}^{(k)}$ will be preserved.

8.4.6.2 Conjugate directions and finite step convergence

Applied to a quadratic function, the DFP method has two fundamental properties:

(a) it generates conjugate directions

$$\mathbf{s}^{(i)T} \mathbf{A} \mathbf{s}^{(j)} = 0 \qquad i < j \leqslant k \qquad (8.44)$$

(b)
$$\mathbf{S}^{(k+1)} \mathbf{A} \mathbf{s}^{(i)} = \mathbf{s}^{(i)} \qquad 0 \leqslant i \leqslant k \qquad (8.45)$$

The property (8.45) implies that $\mathbf{S}^{(n)} = \mathbf{A}^{-1}$ so that convergence is obtained after n steps.

To prove it, we start from (8.40) to note that

$$\mathbf{S}^{(k+1)} \mathbf{A} \mathbf{s}^{(k)} = \mathbf{S}^{(k+1)} \mathbf{y}^{(k)} = \mathbf{s}^{(k)} \qquad (8.46)$$

The proof of (8.44) and (8.45) is obtained next by induction.

For $k = 0$, we have

$$\mathbf{S}^{(1)} \mathbf{A} \mathbf{s}^{(0)} = \mathbf{s}^{(0)}$$

and

$$\mathbf{s}^{(0)T} \mathbf{g}^{(1)} = 0$$

For $k > 0$, we can write

$$\mathbf{g}^{(k)} = \mathbf{g}^{(i+1)} + \mathbf{A}(\mathbf{s}^{(i+1)} + \ldots + \mathbf{s}^{(k-1)})$$

and from (8.44) for $i < k$,
$$s^{(i)T}g^{(k)} = s^{(i)T}g^{(i+1)} = 0 \qquad 0 \leqslant i < k$$
Hence, from (8.45),
$$s^{(i)T}AS^{(k)}g^{(k)} = 0$$
Thus, since $s^{(k)} = -\alpha^{(k)}s^{(k)}g^{(k)}$ with $\alpha^{(k)} > 0$, we obtain
$$s^{(i)T}As^{(k)} = 0 \qquad i \neq k$$
which proves that the DFP method generates conjugate directions. To obtain the result (8.45) which proves quadratic convergence, let us calculate $S^{(k+1)}As^{(i)}$. Making use of (8.44) and (8.45) for k after substitution into (8.43), we get
$$S^{(k+1)}As^{(i)} = s^{(i)} - \frac{[S^{(k)}y^{(k)}]y^{(k)T}S^{(k)}As^{(i)}}{y^{(k)T}S^{(k)}y^{(k)}}$$
The second term of the right-hand side vanishes since in virtue of (8.44), (8.45) and (8.46) we have
$$y^{(k)T}S^{(k)}As^{(i)} = y^{(k)T}s^{(i)} = s^{(k)T}As^{(i)} = 0$$

Since the $s^{(k)}$ are A-orthogonal and since f is minimized successively in these directions we see that the DFP method is a particular conjugate method. Furthermore, if the initial approximation $S^{(0)}$ is the identity matrix, the method becomes the conjugate gradient method.

Just as with the rank-one method, the vectors $s^{(i)}$ are the eigenvectors of $S^{(n)}A$ with a unit eigenvalue. They are necessarily linearly independent, since they are A-orthogonal. Therefore, $S^{(n)} = A^{-1}$, and breakdown of the method is no longer possible.

8.4.7 Inverse DFP Update

It is a simple matter to show that the updating formula (8.43) may also be inverted to approximate the Hessian matrix itself rather than its inverse. The resulting matrix $R^{(k)} = [S^{(k)}]^{-1}$ is calculated by the DFP update

$$R_{DFP}^{(k+1)} = \left(I - \frac{y^{(k)}s^{(k)T}}{y^{(k)T}s^{(k)}}\right)R_{DFP}^{(k)}\left(I - \frac{s^{(k)}y^{(k)T}}{y^{(k)T}s^{(k)}}\right)$$
$$+ \frac{y^{(k)}y^{(k)T}}{y^{(k)T}s^{(k)}} \tag{8.47}$$

8.4.8 Complementary updates: the Broyden–Fletcher–Goldfarb–Shanno (BFGS) Formulas

Complementary formulas to (8.43) and (8.47) are easily calculated by adopting for $R^{(k)}$ an updating formula analogous to (8.43) which verifies the quasi-Newton equation written in the form

$$R^{(k+1)}s^{(k)} = y^{(k)}$$

It is directly obtained from the DFP formula by inverting the roles of $s^{(k)}$ and $y^{(k)}$:

$$R_{BFGS}^{(k+1)} = R_{BFGS}^{(k)} + \frac{y^{(k)}y^{(k)T}}{s^{(k)T}y^{(k)}} - \frac{[R_{BFGS}^{(k)}s^{(k)}][R_{BFGS}^{(k)}s^{(k)}]^T}{s^{(k)T}R_{BFGS}^{(k)}s^{(k)}} \quad (8.48)$$

In inverse form, it is similarly obtained from the DFP formula (8.47):

$$S_{BFGS}^{(k+1)} = \left(I - \frac{s^{(k)}y^{(k)T}}{s^{(k)T}y^{(k)}}\right)S_{BFGS}^{(k)}\left(I - \frac{y^{(k)}s^{(k)T}}{s^{(k)T}y^{(k)}}\right)$$
$$+ \frac{s^{(k)}s^{(k)T}}{s^{(k)T}y^{(k)}} \quad (8.49)$$

According to several authors [12, 13], there is growing evidence that the BFGS is the best current update formula for use in unconstrained optimization. This is due to the fact that the eigenvalues of (8.49) are systematically larger than those of (8.43)

8.5 A GLOBAL VIEW OF UNCONSTRAINED MINIMIZATION

To conclude the description of the unconstrained minimization techniques, we give some unifying concepts by generalizing the idea of 'steepest descent'. When looking at a descent method, a question that naturally arises is: at a given point, what is the 'best' downhill direction? In other words, we seek the move direction which ensures the largest decrease, locally, of the objective function. The mathematical description of this problem implies minimizing the directional derivative of the objective function at the current point $x^{(k)}$:

$$\phi'(0) = s^T g^{(k)} \quad (8.50)$$

which is a first-order estimate of the decrease $f(x^{(k+1)}) - f(x^{(k)})$ (see Equation (7.40)). Note that $\phi'(0)$ is negative, because s must be a downhill direction (see Equation (7.36)).

If we attempt to minimize $s^T g^{(k)}$ as a function of s, we find that no bounded solution exists. However, we can pose the problem: which direction s is such that the maximum decrease in the linear function $s^T g^{(k)}$ is obtained for a unit step? In mathematical terms this gives the constrained problem:

$$\underset{s}{\text{minimize}} \quad s^T g^{(k)} \quad \text{with the condition} \quad |s| = 1 \quad (8.51)$$

Clearly the solution to this problem depends upon the way we measure the length of a vector, that is its norm $|s|$ defined in Section 2.3. The usual definition is the Euclidian norm $|s| = (s^T s)^{1/2}$. It can be generalized as follows:

$$|s|_M = (s^T M s)^{1/2} \quad (8.52)$$

where **M** is a symmetric positive-definite matrix which can be interpreted as our 'metric'. The symbol $|s|_M$ is often read 'length with respect to M'. Such a definition of length emphasizes some directions over other. In terms of the generalized norm (8.52), the Lagrangian function of the minimization problem (8.51) is

$$L(s, \mu) = \mathbf{s}^T \mathbf{g}^{(k)} + \mu(\mathbf{s}^T \mathbf{M} \mathbf{s} - 1) \tag{8.53}$$

where μ is a Lagrangian multiplier. Equating to zero the derivative with respect to s of this function, we obtain

$$\mathbf{M}\mathbf{s} = -\frac{1}{2\mu} \mathbf{g}^{(k)} \tag{8.54}$$

the value of μ following from the condition that $\mathbf{s}^T \mathbf{M} \mathbf{s} = 1$. However, because we are interested only in the direction of search, we can set

$$\mathbf{s}^{(k)} = -\mathbf{M}^{-1} \mathbf{g}^{(k)} \tag{8.55}$$

which is our 'best' downhill direction for the kth iteration of a descent method (note that the value of μ will be implicitly supplied in the line search).

Returning now to a Euclidian metric, i.e. $\mathbf{M} = \mathbf{I}$ in (8.55), it is seen that the locally best downhill direction is $\mathbf{s}^{(k)} = -\mathbf{g}^{(k)}$, which defines the classical method of steepest descent (see Section 7.4). Therefore it can be concluded that (8.55) yields a generalized direction of steepest descent in the metric defined by **M**. It should be noted that **M** is positive definite, $\mathbf{s}^{(k)}$ as given by (8.55) is a downhill direction, because condition (7.36) is satisfied ($\mathbf{s}^{(k)T} \mathbf{g}^{(k)} = -\mathbf{s}^{(k)T} \mathbf{M} \mathbf{s}^{(k)} < 0$ if $\mathbf{s}^{(k)} \neq 0$). Comparing (8.4) with (8.55) it is apparent that the method of Newton is in fact a method of steepest descent under the norm $(\mathbf{s}^T \mathbf{H}^{(k)} \mathbf{s})^{1/2}$, that is, in the metric of the Hessian matrix $\mathbf{H}^{(k)} = \nabla^2 f(x^{(k)})$. Also we can conclude that any modified Newton method, with some positive definite matrix $\bar{\mathbf{H}}^{(k)}$ replacing $\mathbf{H}^{(k)}$, is a steepest-descent method in the metric $\bar{\mathbf{H}}^{(k)}$ (see section 8.2.5). In this connection, comparing (8.33) with (8.55) shows that the quasi-Newton methods are obtained by setting $\mathbf{M}^{-1} = \mathbf{S}^{(k)}$. Because $\mathbf{S}^{(k)}$ is an approximation to the inverse Hessian, gradually updated in the iterative process, quasi-Newton methods can be viewed as methods of steepest descent in a metric that varies from point to point. Whence the appellation 'variable metric' method originally suggested by Davidon [18] for his method (DFP update). Finally, it is worth mentioning that the conjugate direction methods reviewed in Section 8.3 are also based on the idea of employing a non-Euclidian metric. Indeed, they use search directions that are orthogonal in the metric of the Hessian matrix

$$(\mathbf{s}_j^T \mathbf{H} \mathbf{s}_k = 0 \text{ for } j \neq k).$$

8.6 THE GRADIENT PROJECTION METHOD

In Section 7.4, we discussed an algorithm for solving a quasi-unconstrained problem, which is a special case of linearly constrained minimization. It was concluded that: (1) an active side constraint was kept satisfied by moving within the corresponding constraint hyperplane; (2) in some situations, we can leave a hyperplane to enter inside the feasible domain; (3) a maximal allowance step length must be computed in each line search. These three points will still be present in a general method for solving the problem

$$\text{minimize} \quad f(x) \tag{8.56}$$

$$\text{subject to} \quad \sum_{i=1}^{n} c_{ij} x_i \geqslant b_j \quad j = 1, \ldots, m \tag{8.57}$$

$$\bar{x}_i \geqslant x_i \geqslant \underline{x}_i \quad i = 1, \ldots, n \tag{8.58}$$

The projection methods considered hereafter are characterized by a move along downhill directions that are constrained to reside on the polyhedral boundary of the feasible domain. Therefore each point in the process is feasible and the value of the objective function constantly decreases. Furthermore, an estimate of the Lagrangian multipliers is generated at each iteration.

Let $x^{(k)}$ be a feasible point at which q linear constraints are active (i.e. $c_j^T x^{(k)} - b_j = 0$; $j = 1, \ldots, q$), the others being inactive ($c_j^T x^{(k)} - b_j > 0$; $j > q$). Let N_q denote the matrix composed of the gradients of active constraints:

$$N_q = [c_1 \ldots c_q] \tag{8.59}$$

Initially the side constraints (8.58) will be treated like regular linear constraints. Therefore the columns of N_q may contain vectors c_j that are simple base vectors (e.g. $(0, \ldots, 0, 1, 0, \ldots, 0)^T$).

However, it will be shown in Section 8.8 how to handle the side constraints in a more efficient way. Assuming regularity of the active constraints, N_q is a $n \times q$ matrix of rank $q < n$. The kth iteration must lead from $x^{(k)}$ to another feasible point $x^{(k+1)}$ according to the usual descent iteration (7.34). For notational convenience we shall henceforth omit the iteration index k and simply use the superscript $+$ to indicate a new point (at iteration $k+1$). Therefore the current iteration (7.34) becomes

$$x^+ = x + \alpha s \tag{8.60}$$

where α is the step length made along the search direction s. We want s to be a downhill direction, so we require $s^T g < 0$, where $g = \nabla f(x)$ (see Section 7.3). In addition, we wish s to lie in the intersection of the active constraint hyperplanes, so we require

$$N_q^T s = 0 \tag{8.61}$$

so that all currently active constraints remain active at x^+. The particular search direction that we shall use is the projection of the negative gradient onto the constraint intersection:

$$s = -P_q g \tag{8.62}$$

where $\mathbf{P}_q$ is an *orthogonal projection operator*. To find the form of the matrix $\mathbf{P}_q$ we notice that any vector $\mathbf{v}$ can be written as the difference between the projected vector $\mathbf{P}_q\mathbf{v}$ and a vector $\mathbf{N}_q\lambda$ orthogonal to the constraint intersection:

$$\mathbf{v} = \mathbf{P}_q\mathbf{v} - \mathbf{N}_q\lambda \qquad (8.63)$$

where $\lambda \in E^q$. Taking $\mathbf{v}$ as the negative gradient and using condition (8.61), it follows that

$$\mathbf{N}_q^T[\mathbf{P}_q\mathbf{g}] \equiv \mathbf{N}_q^T(\mathbf{g} - \mathbf{N}_q\lambda) = 0 \qquad (8.64)$$

Because $\mathbf{N}_q$ has rank q, we can solve this equation for λ and obtain

$$\lambda = (\mathbf{N}_q^T\mathbf{N}_q)^{-1}\mathbf{N}_q^T\mathbf{g} \qquad (8.65)$$

Now (8.62) can be written

$$\mathbf{s} = -\mathbf{P}_q\mathbf{g} = -\mathbf{g} + \mathbf{N}_q\lambda \qquad (8.66)$$

which shows that

$$\mathbf{P}_q = \mathbf{I} - \mathbf{N}_q(\mathbf{N}_q^T\mathbf{N}_q)^{-1}\mathbf{N}_q^T \qquad (8.67)$$

The direction $\mathbf{s}$ given by (8.66) is the 'projected gradient'. By construction the feasibility requirement (8.61) is satisfied. Furthermore, the descent condition is also fulfilled, because $\mathbf{s}^T\mathbf{g} = -\mathbf{s}^T\mathbf{s} < 0$ if $\mathbf{s} \neq 0$. On the other hand, if $\mathbf{s} = 0$, then from (8.66):

$$\mathbf{g} - \mathbf{N}_q\lambda = 0 \qquad (8.68)$$

Since $\mathbf{N}_q$ is made up of the active constraint gradient, (8.68) implies that the Kuhn–Tucker conditions are satisfied provided that the components of λ are all non-negative. The process is then terminated. Suppose now that $\mathbf{s} = 0$ and at least one component of λ is negative, say $\lambda_r < 0$. It is thus possible to find a new feasible search direction $\tilde{\mathbf{s}}$ by relaxing the corresponding inequality $\mathbf{c}_r^T\mathbf{x} - b_r \geq 0$ and projecting the negative gradient onto the intersection of the remaining $q - 1$ active constraints:

$$\tilde{\mathbf{s}} = -\mathbf{P}_{q-1}\mathbf{g} \qquad (8.69)$$

where $\mathbf{P}_{q-1}$ is the new projection matrix, which is computed from (8.67) with $\mathbf{N}_q$ replaced by $\mathbf{N}_{q-1}$ ($\mathbf{N}_{q-1}$ is simply the matrix $\mathbf{N}_q$ with column $\mathbf{c}_r$ deleted). The new vector $\tilde{\mathbf{s}}$ is a feasible down hill direction. Indeed, the constraint just left cannot be violated, because it can be shown that

$$\mathbf{c}_r^T\tilde{\mathbf{s}} = -\frac{1}{\lambda_r}\tilde{\mathbf{s}}^T\tilde{\mathbf{s}} > 0$$

This process of dropping a binding constraint at a point satisfying (8.68) is illustrated in Fig. 8.8(a).

From the foregoing developments, it appears that λ satisfying (8.68) can be identified as the vector of Lagrangian multipliers associated with the active constraints. When condition (8.68) is not fulfilled, then the λ_j computed from (8.65) are no longer the true Lagrangian multipliers. They are called *first-order estimates of the multipliers*.

We next consider selection of the step size α in (8.60), which leads to eventually adding a constraint to the active set. Iteration (8.60) must be performed so as to minimize the objective function along the direction $\mathbf{s}$, with the additional requirement that $\mathbf{x}^+$ must still be a feasible point. Therefore there exists a maximum allowable step length along s, which can be computed as follows:

$$\bar{\alpha} = \min_{j=q+1,\ldots,m} \{\alpha_j > 0 : \mathbf{x} + \alpha_j \mathbf{s} \text{ is feasible}\} \qquad (8.70)$$

It is easily seen that

$$\alpha_j = -\frac{\mathbf{c}_j^T \mathbf{x} - b_j}{\mathbf{c}_j^T \mathbf{s}} \quad j = q+1, \ldots, m \qquad (8.71)$$

which implies $\alpha_j > 0$ only if $\mathbf{c}_j^T \mathbf{s} < 0$. As indicated in Fig. 8.8(b), the α_j are the intercept distances to the constraint hyperplanes corresponding to previously inactive constraints. With $\bar{\alpha}$ known from (8.70), compute $\bar{\mathbf{x}} = \mathbf{x} + \bar{\alpha}\mathbf{s}$. If $\mathbf{s}^T \mathbf{g}(\bar{\mathbf{x}}) \leq 0$, $\bar{\mathbf{x}}$ is the minimum of $f(x)$ along $\mathbf{s}$, because for $\alpha > \bar{\alpha}$, at least one constraint would be violated. Thus set $\mathbf{x}^+ = \bar{\mathbf{x}}$ and add the newly encountered constraint to the active set. This means that the corresponding constraint gradient is added to the matrix $\mathbf{N}_q$ to form the $n \times (q+1)$ matrix $\mathbf{N}_{q+1}$. The associated projection matrix $\mathbf{P}_{q+1}$ is given by (8.67) with $\mathbf{N}_{q+1}$ replacing $\mathbf{N}_q$. The iterative process can now be repeated at $\mathbf{x}^+$. On the other hand, if $\mathbf{s}^T \mathbf{g}(\bar{\mathbf{x}}) > 0$, then there exists $\alpha^* \in [0, \bar{\alpha}]$ such that $\mathbf{x}^* = \mathbf{x} + \alpha^* \mathbf{s}$ is the minimum of $f(x)$ along $\mathbf{s}$ (see Fig. 8.8(c)). This value α^* can be determined by using a line-search technique (see Section 7.5). In this case, no new constraint has been added to the active set. Thus simply set $\mathbf{x}^+ = \mathbf{x}^*$ and repeat the iteration at $\mathbf{x}^+$.

The gradient projection algorithm previously described is summarized in the flow diagram of Fig. 8.9. The three basic schemes, namely leaving a constraint, adding a constraint and finding a minimum, are illustrated in Fig. 8.8. To further understand the method, let us apply it to the example problem of Section 7.2, namely, minimize $f(x) = x_1^2 + x_2^2$ subject to $x_1 + x_2 \geq 1$ (ignoring the side constraints for simplicity). We have $g = \nabla f = (2x_1, 2x_2)^T$, $\mathbf{c} = (1, 1)^T$ and $b = 1$. Starting from $\mathbf{x}^{(0)} = (2, 0)^T$, gives $\mathbf{g}(x^\circ) = (4, 0)^T$ and then $\mathbf{s} = (-4, 0)^T$, as no constraint is active at $\mathbf{x}^{(0)}$. The maximal allowable step length computed from (8.71) is $\bar{\alpha} = 1/4$ and therefore $\bar{\mathbf{x}} = \mathbf{x}^{(0)} + \bar{\alpha}\mathbf{s} = (1, 0)^T$. From $\mathbf{g}(\bar{\mathbf{x}}) = (2, 0)^T$, it follows that $\mathbf{s}^T \mathbf{g}(\bar{\mathbf{x}}) = -8$, so that the constraint has become active at the new iteration point $\mathbf{x}^{(1)} = (1, 0)^T$. Using (8.65) with $\mathbf{N} = (1, 1)^T$, we obtain $\lambda = 1$ and from (8.66), $\mathbf{s} = (-1, 1)^T$. The subsequent line search amounts to solving $\min_{\alpha} [(1 - \alpha)^2 + \alpha^2]$, which gives $\alpha = 1/2$ and hence $\mathbf{x}^{(2)} = (1/2, 1/2)^T$. It is easily verified

Unconstrained and Linearly Constrained Minimization

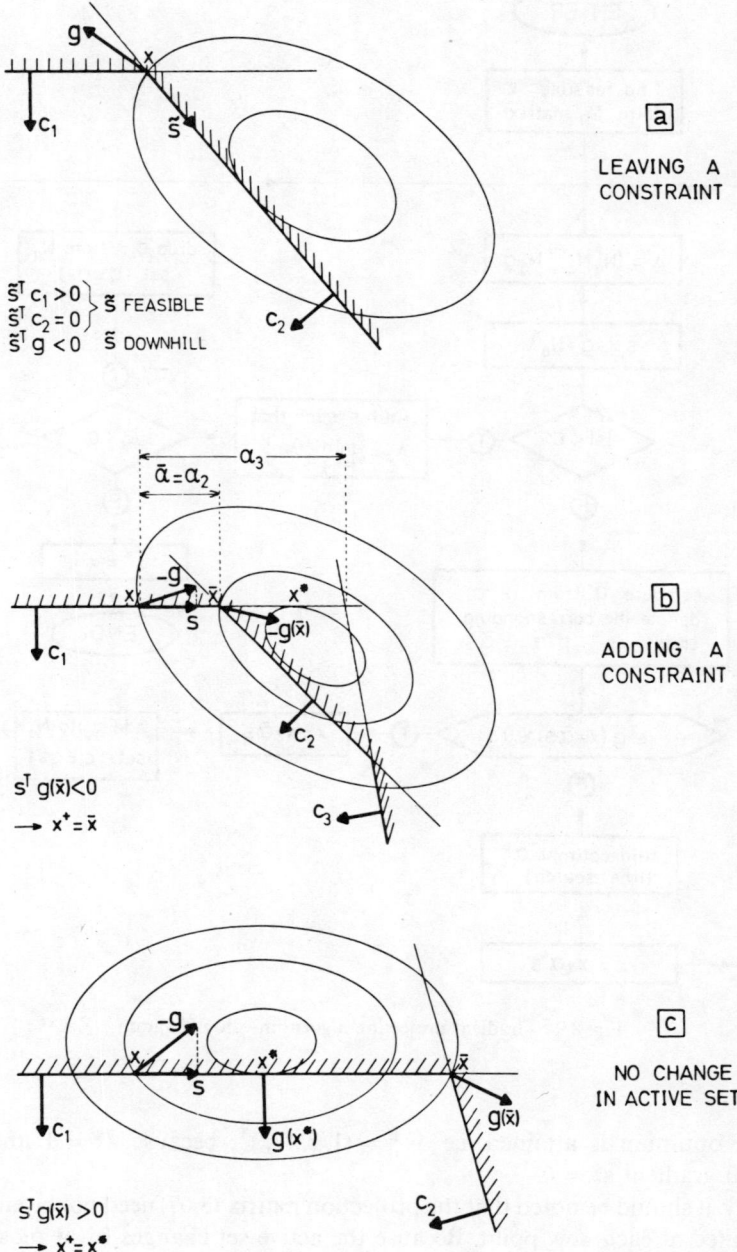

Fig. 8.8 Active-set strategy (a) Leaving a constraint (b) Adding a constraint (c) No change in active set

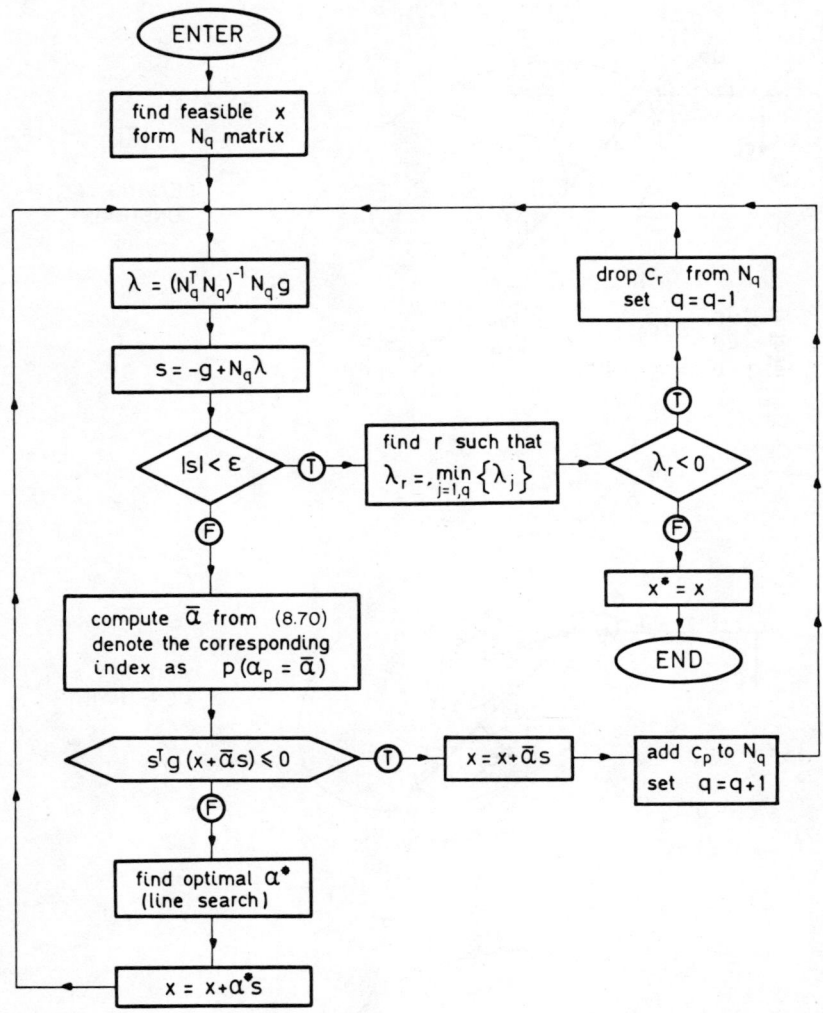

Fig. 8.9 Gradient projection algorithm—flow diagram

that the optimum is attained, i.e., $\mathbf{x}^* = (1/2, 1/2)^T$, because $\lambda^* = 1$ and the projected gradient $\mathbf{s}^* = 0$.

Finally, it should be noted that the projection matrix (8.67) need not be entirely re-evaluated at each new point. Because the active set changes by at most one constraint at a time, rather it is possible to compute each new projection matrix from the previous one by using the updating procedure on the basic matrix $(\mathbf{N}_q^T \mathbf{N}_q)^{-1}$ (see, for example, Rosen [14]).

8.7 FIRST- AND SECOND-ORDER PROJECTION METHODS

Just as the method of steepest descent, the gradient projection method, as implemented above, is largely inefficient, because of its slow convergence speed. It is therefore natural to consider Newton-type, quasi-Newton, or conjugate gradient methods, which implies defining adequate projection operators. To do this, we use the unifying concepts developed in Section 8.5. Assuming that q constraints are active at the current point $\mathbf{x}^{(k)}$, we pose the problem: which direction $\mathbf{s}$, lying in the intersection of the active constraint hyperplanes, is such that the maximum decrease in the linear approximation (8.50) to the objective function is obtained for a unit step? The problem analogous to (8.51) is (omitting again the index k):

$$\text{minimize} \quad \{\mathbf{s}^T\mathbf{g} + \mu(\mathbf{s}^T\mathbf{M}\mathbf{s} - 1)\} \tag{8.72}$$

$$\text{subject to} \quad \mathbf{N}_q^T\mathbf{s} = 0 \tag{8.73}$$

The Lagrangian of this problem is written

$$L(\mathbf{s}, \mu, \hat{\lambda}) = \mathbf{s}^T\mathbf{g} + \mu(\mathbf{s}^T\mathbf{M}\mathbf{s} - 1) - \mathbf{s}^T\mathbf{N}_q\hat{\lambda} \tag{8.74}$$

The stationarity conditions yield the search direction

$$\mathbf{s} = -\frac{1}{2\mu}\mathbf{M}^{-1}(\mathbf{g} - \mathbf{N}_q\hat{\lambda}) \tag{8.75}$$

The indeterminate Lagrangian multipliers $\hat{\lambda}$ can be obtained from the linear constraints (8.73):

$$\mathbf{N}_q^T\mathbf{M}^{-1}(\mathbf{g} - \mathbf{N}_q\hat{\lambda}) = 0 \tag{8.76}$$

Solving for $\hat{\lambda}$, introducing $\hat{\lambda}$ in (8.75) and taking $\mu = \frac{1}{2}$ for simplicity, the feasible search direction becomes

$$\mathbf{s} = -\hat{\mathbf{P}}_q\mathbf{M}^{-1}\mathbf{g} \tag{8.77}$$

where

$$\hat{\mathbf{P}}_q = \mathbf{I} - \mathbf{M}^{-1}\mathbf{N}_q(\mathbf{N}_q^T\mathbf{M}^{-1}\mathbf{N}_q)^{-1}\mathbf{N}_q^T \tag{8.78}$$

is an *oblique projection operator*, weighted by $\mathbf{M}^{-1}$, which projects vectors of E^n so that they are orthogonal to the space spanned by the columns of $\mathbf{N}_q$. $\hat{\mathbf{P}}_q$ can be thought of as a projection operator in a metric $\mathbf{M}$ rather than in a Euclidian metric.

Taking $\mathbf{M}$ as the Hessian $\mathbf{H}$ of the objective function, $\mathbf{s} = -\hat{\mathbf{P}}_q\mathbf{H}^{-1}\mathbf{g}$ can be viewed as the direction of steepest descent in the subspace defined by the active constraints, when lengths are measured under the norm $|s|_H = (\mathbf{s}^T\mathbf{H}\mathbf{s})^{1/2}$, that is, in the metric of the Hessian matrix. Also the matrix $\hat{\mathbf{P}}_q\mathbf{H}^{-1}$ can be considered as a projected inverse Hessian and consequently the direction $\mathbf{s} = -\hat{\mathbf{P}}_q\mathbf{H}^{-1}\mathbf{g}$ is the direction of the Newton method within the intersection of the active constraint hyperplanes. Note that if we adopt $\mathbf{M} = \mathbf{I}$, we recover the conventional gradient projection method of Section 8.6, which is a method of steepest descent in the Euclidian metric.

It is easily verified that for a quadratic problem with linear equality constraints, the descent direction (8.77) with $\mathbf{M} = \mathbf{H}$ (i.e. Newton direction) leads to the optimum in only one iteration, with a step length equatl to unity, exactly like Newton's method for unconstrained minimization (see Fig. 8.10). When applied to a general objective function subject to linear inequality constraints, the second-order projection algorithm proceeds iteratively just as the first-order projection algorithm of Section 8.6. The search for the maximal allowable step length $\bar{\alpha}$ is performed by using (8.70) (8.71), while the optimal step length α^* can be either achieved by resorting to a line-search technique, or it can be done approximately by adopting $\alpha^* = 1$. Instead of the first-order estimates (8.65) for the Lagrangian multipliers, we are now provided with second-order estimates,

$$\hat{\lambda} = (\mathbf{N}_q^T \mathbf{H}^{-1} \mathbf{N}_q)^{-1} \mathbf{N}_q^T \mathbf{H}^{-1} \mathbf{g} \tag{8.79}$$

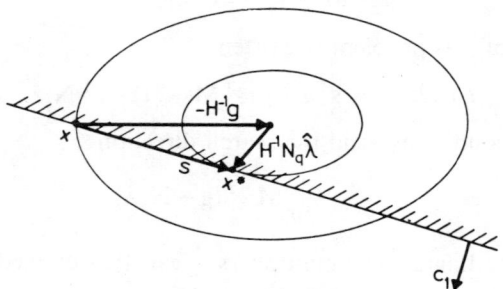

Fig. 8.10 Second-order projection method

in terms of which the search direction can be written as

$$\mathbf{s} = -\mathbf{H}^{-1}(\mathbf{g} - \mathbf{N}_q \hat{\lambda}) \tag{8.80}$$

This 'projected' Newton method is probably the most appropriate for solving linearly constrained problems, provided that the objective function Hessian matrix be readily evaluated and inverted at each new point in the iterative process. Note, however, that, in opposition with the unconstrained case, the Hessian is not necessarily positive definite at a constrained minimum†. Therefore techniques for enforcing positive definiteness are more likely to be required, and several modified Newton methods have been proposed, which are similar to the well known Levenberg–Marquardt safeguarding procedure discussed in Section 8.1.5 (see Gill and Murray [15], chs 2 and 5).

Another approach is to resort to quasi-Newton methods for approximating the Hessian matrix or its inverse on the basis of first-order information accumulated from the preceding steps. Putting $\mathbf{M}^{-1} = \mathbf{S}^{(k)}$ in (8.77), (8.78), the direction of

---

† The second-order optimality conditions imply positive definiteness of the Hessian of the Lagrangian restricted to the tangent plane, but not of the objective function Hessian itself.

steepest descent in the metric of $[\mathbf{S}^{(k)}]^{-1}$ is given by

$$\mathbf{s}^{(k)} = \mathbf{S}^{(k)}(\mathbf{N}_q \hat{\lambda}^{(k)} - \mathbf{g}^{(k)}) \qquad (8.81)$$

where

$$\hat{\lambda}^{(k)} = (\mathbf{N}_q^T \mathbf{S}^{(k)} \mathbf{N}_q)^{-1} \mathbf{N}_q^T \mathbf{S}^{(k)} \mathbf{g}^{(k)} \qquad (8.82)$$

In such a quasi-Newton method, the matrix $\mathbf{S}^{(k)}$ is an approximation to the inverse Hessian, which has to be gradually improved according to some update formula (see Section 8.4). An interesting possibility, exploited by Murtagh and Sargent [16], is to employ the rank-one formula (8.39), because this avoids the necessity of an exact one-dimensional minimization along $\mathbf{s}^{(k)}$.

A similar strategy, proposed by Goldfarb [17], is to mix the DFP update formula and the gradient projection algorithm as follows. The search direction at iteration k is given by

$$\mathbf{s}^{(k)} = -\mathbf{S}_q^{(k)} \mathbf{g}^{(k)} \qquad (8.83)$$

where the matrix $\mathbf{S}_q^{(k)}$ is computed according to the DFP update formula (8.43) as long as the active set remains unchanged (q active constraints). Indeed, as observed by Davidon himself in his original paper [18], if the initial inverse Hessian approximation $\mathbf{S}_q^{(0)}$ is orthogonal to the constraint gradients (e.g. $\mathbf{S}_q^{(0)} \equiv \mathbf{P}_q$ as given by Equation (8.67), then so are all subsequent approximations and consequently all search directions (8.83) will reside within the initial subspace of linear constraints. If a constraint, say $\mathbf{c}_j^T \mathbf{x} - b_j \geqslant 0$, is added to the active set, then the update formula, such that $\mathbf{S}_{q+1}^{(k+1)} \mathbf{c}_j = 0$, is

$$\mathbf{S}_{q+1}^{(k+1)} = \mathbf{S}_q^{(k)} - \frac{\mathbf{S}_q^{(k)} \mathbf{c}_j \mathbf{c}_j^T \mathbf{S}_q^{(k)}}{\mathbf{c}_j^T \mathbf{S}_q^{(k)} \mathbf{c}_j} \qquad (8.84)$$

Finally, if the jth constraint is dropped from the active set, then

$$\mathbf{S}_{q-1}^{(k+1)} = \mathbf{S}_q^{(k)} + \frac{\mathbf{P}_{q-1} \mathbf{c}_j \mathbf{c}_j^T \mathbf{P}_{q-1}}{\mathbf{c}_j^T \mathbf{P}_{q-1} \mathbf{c}_j} \qquad (8.85)$$

with

$$\mathbf{P}_{q-1} = \mathbf{I} - \mathbf{N}_{q-1} (\mathbf{N}_{q-1}^T \mathbf{N}_{q-1})^{-1} \mathbf{N}_{q-1}^T \qquad (8.86)$$

Note that the decision as whether or not to drop a constraint from the active set can be taken on the basis of the first-order estimates (8.65) of the Lagrangian multipliers. In this first-order projection method, the matrix $\mathbf{S}_q$ is an approximation to the matrix $\hat{\mathbf{P}}_q \mathbf{H}^{-1}$ where $\hat{\mathbf{P}}_q$ is the weighted projection operator defined in (8.78). If the objective function is a positive definite quadratic form and if $n - q$ successive iterations are performed with a constant set of q active constraints then, starting from $\mathbf{S}_q^{(0)} = \mathbf{P}_q$, $\mathbf{S}_q^{(n-q)}$ will be equal to $\hat{\mathbf{P}}_q \mathbf{H}^{-1}$. In addition, the search directions generated in these $n - q$ iterations are H-conjugate.

To close this section, it is worth giving a simple but effective way of improving the convergence speed of the gradient projection method. It merely consists of conjugating the projected gradient vectors, according to formula (8.27) with $\mathbf{g}^{(k)}$ replaced by $\mathbf{P}_q \mathbf{g}^{(k)}$. For a quadratic problem with q linear equality constraints, convergence is then achieved within $n - q$ iterations. Of course, in the general case, the conjugacy procedure must be reinitiated whenever the projection matrix is modified because of a change in the set of active constraints. This corresponds to throwing away useful information and thus constitutes a disadvantage with respect to the methods which approximate inverse Hessians.

8.8 THE ACTIVE-SET STRATEGY

An essential ingredient in any projection algorithm is the determination at each iteration of the set of active constraints whose intersection forms the basis for projection. The strategy for adding constraints to the basis is rather straightforward and is fixed by computing the step to the nearest constraint (see Equation (8.70)). It should be mentioned that some difficulties occur when the constraints are linearly dependent. There is then a risk of introducing a redundant constraint in the active set, which renders the projection matrix singular. It is, however, relatively simple to devise *ad hoc* strategies for avoiding degeneracy and subsequent cycling phenomena that can possibly arise.

Much less apparent are the strategies for determining which constraints to delete from the basis. The simplest strategy is that adopted in Section 8.6 and implemented in the flow diagram of Fig. 8.9. It consists of retaining all the currently active constraints until a minimum is found with respect to the corresponding subspace. Only at this point is the sign of the Lagrangian multipliers examined, which eventually leads to deleting constraints from the basis. It can be shown that this scheme will terminate at a strong local minimum after a finite number of basis changes. Unfortunately, this strategy suffers from a major drawback, in that it requires computing the minimum on each successive subspace. If the initial basis differs substantially from the optimal one, it is apparent that a prohibitive computational effort will be needed before reaching the final solution.

An alternative strategy is to compute estimates of the Lagrangian multipliers every iteration and move off the constraint with the most negative multiplier, if any. This is essentially the rule used in the simplex method of linear programming, in which case the estimates of the Lagrangian multipliers are exact. This is no longer true when a general function is minimized subject to linear constraints, and in fact the first-order estimates (8.65), and even the second-order estimates (8.79), of the multipliers tend to be very inaccurate when computed far from the minimum with respect to the current subspace†. This inaccuracy can be responsible for the phenomenon of 'zigzagging', in which a given constraint may be dropped from the basis and then reintroduced later, repeatedly until the sign of the corresponding multiplier is stabilized. Consequently progress to the solution

---

† The error is $O(h)$ for the first-order estimates and $O(h^2)$ for the second-order ones, where h is the distance to the minimum in the current basis.

is considerably slowed. In fact, global convergence cannot be proved, that is, this type of strategy is not guaranteed to terminate after a finite number of basis changes. However, several anti-zigzagging rules have proved to be quite effective in practice. For example, Zoutendijk [19] suggests the rule: 'if a constraint previously dropped from an active set is added again, it is retained in the active set until a constrained stationary point is attained'.

Yet another approach, which is somewhat more sophisticated, consists in deleting a constraint only if its removal will result in a sufficient decrease in the objective function. As a reference case, we consider the minimization of a positive definite quadratic form using Newton's method (see Equations (8.79), (8.80). Let Δf denote the expected reduction of $f(x)$ obtained by keeping the same active set and $\Delta \tilde{f}$ the reduction that might be achieved by deleting the jth constraint, which is associated to $\tilde{\lambda}_j < 0$. It can be shown that

$$\Delta f = \frac{1}{2} \mathbf{s}^T \mathbf{H} \mathbf{s} \qquad \text{(active set unchanged)}$$

and

$$\Delta \tilde{f} = \frac{1}{2} \mathbf{s}^T \mathbf{H} \mathbf{s} + \frac{1}{2} \frac{\hat{\lambda}_j^2}{u_j} \qquad (j\text{th constraint removed})$$

where u_j is the jth diagonal element of the matrix $(\mathbf{N}_q^T \mathbf{H}^{-1} \mathbf{N}_q)^{-1}$. Therefore the additional benefit that is gained when deleting the jth constraint is $\frac{1}{2} \hat{\lambda}_j^2 / u_j$. In the general case the foregoing results hold approximately and a suitable active-set strategy is thus as follows. Find the constraint for which $\frac{1}{2} \hat{\lambda}_j^2 / u_j$ is a maximum over the constraints which are candidates for deletion (i.e. $\hat{\lambda}_j < 0$) and drop the corresponding constraint only if the additional reduction in $f(x)$ exceeds γ times the reduction that would be obtained by keeping it in the active basis. Mathematically this constraint (say the qth) is deleted if

$$-\mathbf{s}^T \mathbf{g} = \mathbf{s}^T \mathbf{H} \mathbf{s} \leqslant \frac{1}{\gamma} \frac{\hat{\lambda}_q^2}{u_q} \qquad (8.87)$$

where γ is a positive constant. In the gradient projection method, the Hessian matrix is not computed and the first-order estimates (8.65) of the Lagrangian multipliers only are available. Substitute tests are however possible, which approximate the ideal test (8.87). For example, Rosen [14], to whom the above procedure can be traced back, employs the form

$$-\mathbf{s}^T \mathbf{g} = \mathbf{s}^T \mathbf{s} \leqslant \frac{1}{4} \frac{\lambda_q^2}{v_q} \qquad (8.88)$$

where v_q is the qth diagonal element of the matrix $(\mathbf{N}_q^T \mathbf{N}_q)^{-1}$ (i.e. $\mathbf{H} = \mathbf{I}$ and $\gamma = 4$).

The validity of the foregoing strategy is strongly dependent on the quality of the Lagrangian multiplier estimates. Therefore it is a wise decision to assure the consistency of the tests (8.87) or (8.88) by applying them only if

$$|\mathbf{s}^T \mathbf{g}| \leqslant \varepsilon \qquad (8.89)$$

where ε is a weak tolerance. This additional condition guarantees that the estimates of the multipliers are sufficiently accurate, because (8.89) can only be satisfied in the neighbourhood of a stationary point, where the first- and second-order multiplier estimates are known to be exact. Note that the active-set strategy outlined above becomes very close to the simple strategy presented at the outset, provided that γ is chosen large enough in the tests (8.87), (8.88) and ε small enough in (8.89). That is why this strategy is not very sensitive to the zigzagging phenomenon and is thus strongly recommended.

Finally, it should be mentioned that similar tests can be devised for the quasi-Newton iterations (8.81) of Murtagh–Sargent and (8.83) of Goldfarb. More detailed informations about this important subject can be found in Gill and Murray [15] and Fletcher [20], where several other practical rules are provided.

8.9 SPECIAL TREATMENT OF SIDE CONSTRAINTS

In problems with a large number of variables, many of which will take on a lower or upper bound value at the solution, it is desirable to use special procedures for handling the side constraints. The modifications brought in this section to the basic gradient projection algorithm, yield a significant reduction in dimension and consequently a great saving in computation time and storage space. Also effects of round-off errors, which might be important in large-scale problems, are reduced.

In the projection methods previously described, the side constraints (8.58) are treated exactly like ordinary linear constraints, namely, if an inequality of the form $x_i \geq \underline{x}_i$ or $x_i \leq \bar{x}_i$ becomes active, its gradient is added to the matrix $\mathbf{N}_q$ and the projection matrix $\mathbf{P}_q$ or $\hat{\mathbf{P}}_q$ is modified accordingly. However, because of the special nature of this gradient, either $(0, \ldots, 0, 1, 0, \ldots, 0)^T$ or $(0, \ldots, 0, -1, 0, \ldots, 0)^T$, several simplifications are possible. Suppose thus that the jth column of $\mathbf{N}_q$ is all zeros except for a 1 or -1 in row i. Let $\mathbf{A}$ be the same as $\mathbf{N}_q$ with the ith row and jth column deleted. Then it can be shown that $\hat{\mathbf{P}}_q = [\mathbf{I}_{n-1} - \mathbf{A}(\mathbf{A}^T\mathbf{A})^{-1}\mathbf{A}^T]$ is exactly the same as $\mathbf{P}_q$ if its ith row and column, which are zeros, are taken out and its dimension reduced to $(n-1)$. Based on the foregoing observation, the gradient projection method for solving problem (8.56)–(8.58) can be modified as follows. Let us write the matrix $\mathbf{N}_q$ of the active constraint gradients in the form:

$$\mathbf{N}_q = \begin{bmatrix} A & 0 & 0 \\ B & I & 0 \\ C & 0 & -I \end{bmatrix} \begin{matrix} \} \\ \} \\ \} \end{matrix} \begin{matrix} \tilde{n} \\ \underline{n} \\ \bar{n} \end{matrix} \Bigg\} \; n \qquad (8.90)$$

$$\underbrace{\tilde{m} \quad \underline{n} \quad \bar{n}}_{q}$$

where it is assumed that the variables are ordered so that the $\tilde{n}$ first variables are not bounded, the $\underline{n}$ following variables are fixed to their lower bound and the $\bar{n}$

remaining variables are fixed to their upper bound. Also the q active constraints are arranged so that the $\tilde{m}$ first constraints are full linear constraints (extracted from Equation (8.57)) and the $q - \tilde{m}$ remaining constraints are simple side constraints (extracted from Equation (8.58)). The objective function gradient is accordingly partitioned as $\mathbf{g}^T = (\tilde{\mathbf{g}}^T, \underline{\mathbf{g}}^T, \bar{\mathbf{g}}^T)$, the projected gradient as $\mathbf{s}^T = (\tilde{\mathbf{s}}^T, \underline{\mathbf{s}}^T, \bar{\mathbf{s}}^T)$ and the vector of Lagrangian multiplier estimates as $\lambda^T = (\tilde{\lambda}^T, \underline{\lambda}^T, \bar{\lambda}^T)$. Then it can be shown that

$$\tilde{\lambda} = (\mathbf{A}^T\mathbf{A})^{-1}\mathbf{A}^T\tilde{\mathbf{g}} \qquad \underline{\lambda} = \underline{\mathbf{g}} - \mathbf{B}\tilde{\lambda} \qquad \bar{\lambda} = -\bar{\mathbf{g}} + \mathbf{C}\tilde{\lambda} \qquad (8.91)$$

and

$$\tilde{\mathbf{s}} = -\tilde{\mathbf{g}} + \mathbf{A}\tilde{\lambda} \qquad \underline{\mathbf{s}} = \mathbf{0} \qquad \bar{\mathbf{s}} = \mathbf{0} \qquad (8.92)$$

Therefore computing $\tilde{\lambda}$ from the first equation (8.91) and

$$\mathbf{v} = -\mathbf{g} + \tilde{\mathbf{N}}\tilde{\lambda} \qquad (8.93)$$

where $\tilde{\mathbf{N}} = (\mathbf{A}^T, \mathbf{B}^T, \mathbf{C}^T)^T$, is all that is required at each iteration. The $\tilde{n}$ first components of $\mathbf{v}$ are the non-zero components of the projected gradient; the $\underline{n}$ following components are the negatives of the multipliers associated with the lower bound constraints $x_i = \underline{x}_i$; and the $\bar{n}$ remaining components of $\mathbf{v}$ are the multipliers associated with the upper bound constraints $x_i = \bar{x}_i$, that is, $\mathbf{v}^T = (\tilde{\lambda}^T, -\underline{\lambda}^T, \bar{\lambda}^T)$. In the computer program implementation, taking account of the side constraints can thus be achieved in a very simple way through row and column permutations in the $\mathbf{N}_q$ matrix when constraints are activated, so that the form (8.90) is kept at each stage of the process. When using second-order or variable-metric methods, such as those described in Section 8.7, similar modifications can be brought to bear on the basic projection schemes.

To fix ideas consider the problem

$$\begin{aligned}\text{minimize} \quad & f(x) = x_1^2 + x_2^2 + x_3^2 + x_4^2 \\ \text{subject to} \quad & x_1 + x_2 + x_3 + x_4 \leq 1 \\ & x_1 - x_2 - x_3 + x_4 \geq 2 \\ & x_3 \geq 2 \qquad x_4 \leq 0 \end{aligned}$$

We ask whether or not the point $\mathbf{x} = (1, -2, 2, 0)^T$ is a solution to this problem. The gradient at $\mathbf{x}$ is $\mathbf{g} = (2, -4, 4, 0)^T$ and it is easily seen that

$$\tilde{\mathbf{N}}^T = \begin{bmatrix} -1 & -1 & -1 & -1 \\ 1 & -1 & -1 & 1 \end{bmatrix} \qquad (8.94)$$

from which it follows that

$$\mathbf{A} = \begin{bmatrix} -1 & 1 \\ -1 & -1 \end{bmatrix}, \qquad (\mathbf{A}^T\mathbf{A})^{-1} = \begin{bmatrix} 1/2 & 0 \\ 0 & 1/2 \end{bmatrix}$$

Hence, using (8.91) with $\tilde{\mathbf{g}} = (2, -4)^T$, we obtain $\tilde{\lambda} = (1, 3)^T$. From (8.93) and (8.94), it then follows that $\mathbf{v} = (0, 0, -8, 2)^T$ which decomposes into $\tilde{\mathbf{s}} = (0, 0)^T$, $\underline{\lambda} = 8$ and $\bar{\lambda} = 2$. The projected gradient is zero and all Lagrangian multipliers $\tilde{\lambda}$,

λ and $\bar{\lambda}$ are positive. Therefore the given point $(1, -2, 2, 0)^T$ is indeed a solution to the problem. Because it is a convex problem, it is in fact the global solution.

Problems

1. Try to apply the basic Newton's method (Fig. 8.1) to the function
$$f(x) = 4x_1^2 + x_2^2 - x_1^2 x_2$$
with the following starting points $\mathbf{x}_A^{(0)} = (1, 1)^T$, $\mathbf{x}_B^{(0)} = (3, 4)^T$, $\mathbf{x}_C^{(0)} = (2, 0)^T$.

2. Apply Greenstadt's procedure (Eq. (8.15)) to the function defined in Problem 1, to determine $\mathbf{x}^{(1)}$ if $\mathbf{x}^{(0)} = (2, 0)^T$.

3. Use the conjugate gradient algorithm (Fig. 8.5) to minimize
$$q(x) = 2(x_1^2 + x_2^2 + x_1 x_2) - 3(x_1 + x_2)$$
Note that the line search can be done analytically.
Verify that $\beta^{(k)}$ given by (8.28), (8.29), (8.30) or (8.31) are equivalent for this problem.

4. Let $f(x) = \frac{1}{2}\mathbf{x}^T \mathbf{Q}\mathbf{x} - \mathbf{b}^T \mathbf{x}$ be defined on E^n with $\mathbf{Q}$ positive definite. Let $\mathbf{x}^{(1)}$ be a minimum point of f over a subspace of E^n containing the vector $\mathbf{s}$ and let $\mathbf{x}^{(2)}$ be the minimum of f over another subspace containing $\mathbf{s}$. Suppose $f(\mathbf{x}^{(1)}) < f(\mathbf{x}^{(2)})$. Show that $\mathbf{x}^{(1)} - \mathbf{x}^{(2)}$ is Q-conjugate to $\mathbf{S}$.

5. Repeat Problem 3 with the DFP method (Equation (8.43)). Verify that by taking $\mathbf{S}^{(0)} = \mathbf{I}$, the DFP method becomes identical to the conjugate gradient method.

6. Consider the problem
$$\text{minimize} \quad (x_1 - 2)^2 + (x_2 - 2)^2$$
$$\text{subject to} \quad x_1 + x_2 \leqslant 2$$
$$x_1 - x_2 \leqslant 1$$
$$0 \leqslant x_1 \leqslant 1.5$$
$$0 \leqslant x_2 \leqslant 1.5$$

Starting from $\mathbf{x}^{(0)} = (0, 1.5)^T$ (vertex), solve the problem by using:
(a) the basic gradient projection algorithm of Section 8.6, with six linear constraints.
(b) the modified projection relations of Section 8.9, with only two linear constraints (the side constraints being treated separately).

7. Consider the quadratic programming problem
$$\text{minimize} \quad f(x) = \frac{1}{2}\mathbf{x}^T \mathbf{Q}\mathbf{x} + \mathbf{q}^T \mathbf{x} + P$$
$$\text{subject to} \quad \mathbf{A}^T \mathbf{x} = \mathbf{b}$$

where $\mathbf{Q}$ is a positive definite matrix. Starting from a feasible point $\mathbf{x}^{(0)}$, which direction **s** and step length α are such that

$$\mathbf{x}^* = \mathbf{x}^{(0)} + \alpha \mathbf{s}$$

is the solution (one iteration only)?
Assuming now that $\mathbf{x}^{(0)}$ is not feasible, find $\mathbf{x}^*$ in at most two iterations.

8. Solve the problem

$$\text{minimize} \quad f(x) = x_1^2 + 4x_2^2$$
$$\text{subject to} \quad \frac{3}{5}x_1 + \frac{4}{5}x_2 \geq \frac{13}{5}$$

by using the Goldfarb algorithm, starting from $\mathbf{x}^{(0)} = (0, 5)^T$ and $\mathbf{H}^{(0)} = \mathbf{I}$ (see Equations (8.83)–(8.86)).

M.G.
C.F.

REFERENCES

Reference Books on Unconstrained Optimization

1. M. A. Wolfe, *Numerical Methods for Unconstrained Optimization*, Van Nostrand, 1978.
2. D. G. Luenberger, *Introduction to Linear and Nonlinear Programming*, Addison Wesley, 1973.

References to Papers

3. J. Greenstadt, On the relative efficiencies of gradient methods, *Mathematics of Computation*, **21**, 360–7, 1967.
4. K. Levenberg, A method for the solution of certain non-linear problems in least squares, *Quarterly Journal of Applied Mathematics*, **2**, 164–8, 1944.
5. W. Murray, Second derivative methods, in *Numerical Methods for Unconstrained Optimization*, (W. Murray, ed.) Academic Press, 1972.
6. M. R. Hestenes and E. Stiefel, Methods of conjugate gradients for solving linear systems, *Journal of Research of the National Bureau of Standards*, **49**, 409–36, 1952.
7. R. Fletcher and C. M. Reeves, Function minimization by conjugate gradients, *The Computer Journal*, **7**, 149–53, 1964.
8. E. Polak and G. Ribiere, Note sur la convergence de méthodes de directions conjuguées, *Rev. Française d'Informatique et de Recherche Opérationnelle*, **16**-R1, 35–43, 1969.
9. L. C. W. Dixon, Conjugate gradient algorithms: Quadratic termination without linear searches, The Hatfield Polytechnic, Numerical Optimization Centre, T.R. 38, 1972.
10. C. G. Broyden, Quasi-Newton methods and their application to function minimization, *Mathematics of Computation*, **21**, 368–81, 1967.
11. R. Fletcher and M. J. D. Powell, A rapidly convergent descent method for minimization, *The Computer Journal*, **6**, 163–8, 1963.
12. J. E. Dennis and J. J. More, Quasi-Newton methods, motivation and theory, *SIAM Review*, **19**, 46–89, 1977.

13. L. C. W. Dixon, The choice of step length, a crucial factor in the performance of variable metric algorithms, in *Numerical Methods for Nonlinear Optimization* (F. A. Lootsma, ed.), Academic Press, 1972.
14. J. B. Rosen, The gradient projection method for nonlinear programming—Part I: Linear constraints, *SIAM J.*, **8**(1), 181–217, 1960.
15. P. E. Gill and W. Murray, *Numerical Methods for Constrained Optimization*, Academic Press, London, 1974.
16. B. A. Murtagh and R. W. H. Sargent, A constrained minimization method with quadratic convergence, in *Optimization* (R. Fletcher, ed.), Academic Press, London, 1969, pp. 215–46.
17. D. Goldfarb, Extension of Davidon's variable metric method to maximization under linear inequality and equality constraints, *SIAM J. Appl. Math.*, **17**, 739–64, 1969.
18. W. Davidon, Variable metric methods for minimization, *AEC Res. and Devel. Rept. ANL*-5990, Argonne National Lab., Illinois, 1959.
19. G. Zoutendijk, *Methods of Feasible Directions*, Elsevier, Amsterdam, 1960.
20. R. Fletcher, Minimizing general functions subject to linear constraints, in *Numerical Methods for Non-Linear Optimization* (F. A. Lootsma, ed.), Academic Press, 1971, pp. 279–96.

Foundations of Structural Optimization: A Unified Approach
Edited by A. J. Morris
© 1982 John Wiley & Sons Ltd.

Chapter 9
General Nonlinear Programming Methods

9.1 INTRODUCTION

In this chapter general methods for solving the constrained nonlinear programming problem (7.1)–(7.3) are examined. The first three sections are concerned with transformation methods, in which the constrained problem is approximated by a sequence of unconstrained problems. In the barrier and penalty function methods, the approximation is accomplished by adding to the objective function a term that reflects the degree of satisfaction or violation of the constraints. In both methods the resulting unconstrained problems tend to become ill-conditioned as a solution point is approached. In the augmented Lagrangian function method, a penalty term is added to the Lagrangian function, rather than merely the objective function. The effect is that the ill-conditioning of the unconstrained problems can be avoided.

Another type of approach to a general nonlinear programming problem, discussed in Section 9.5, is to attack it directly by using sequential one-dimensional minimizations along usable–feasible directions. Because these methods work directly on the original primal problem by searching through the feasible region for an optimal solution, they are often referred to as primal methods. Projection methods and feasible-direction methods belong to that category.

Perhaps the most natural approach to a nonlinear problem is to replace it with a sequence of linear programming problems, and these are discussed in Section 9.5. Because they correspond to a rather natural process, these methods are appealing to the engineer. Also the availability of standard linear programming packages facilitates their practical implementation. Unfortunately, linearization techniques often require the introduction of artificial constraints in order to prevent convergence to a non-optimal vertex. Finally some recent and promising methods involving constraint linearization will be discussed (recursive quadratic programming.)

9.2 BARRIER FUNCTION METHODS

The barrier function approach proceeds by forming an auxiliary function whose minima are unconstrained inside the feasible region. The auxiliary function is defined so as to construct a barrier at the boundary of the feasible region, thus preventing violation of the constraints. By gradually removing the effect of the constraints in the auxiliary function through controlled changes in the value of a parameter, a sequence of unconstrained problems is generated, whose solutions are interior points converging to a minimum of the original constrained problem. Therefore barrier methods are often referred to as *interior-point unconstrained minimization techniques*. These methods are especially attractive for structural design applications, since they exhibit the reassuring feature that, should the algorithm be terminated prematurely, a feasible solution is nevertheless returned, which usually corresponds to a better design than the initial one.

We again disregard the side constraints of problem (7.1)–(7.3), assuming that they can be treated separately or that they are included into the general constraints $h_j(x) \geq 0$. Therefore the transformation methods described in the sequel will eventually construct quasi-unconstrained subproblems instead of purely unconstrained ones. The feasible region is defined as $R = \{x : h_j(x) \geq 0; j = 1, \ldots, m\}$ and the interior of the feasible region is written as $R_0 = \{x : h_j(x) > 0; j = 1, \ldots, m\}$. With these definitions in mind, the barrier function transformation is stated as

$$\phi(x, r) = f(x) + rB(x) \qquad r > 0 \tag{9.1}$$

where $B(x)$, the barrier function, is positive on R_0 and $B(x) \to \infty$ as x approaches the boundary of R. This implies R_0 being non-empty and so the barrier function approach is not a suitable transformation for problems involving equality constraints (even if each of them is written as two inequalities!). Frequently used barrier functions are the logarithmic function

$$B(x) = -\sum_{j=1}^{m} \ln[h_j(x)] \tag{9.2}$$

and the inverse function

$$B(x) = \sum_{j=1}^{m} \frac{1}{h_j(x)} \tag{9.3}$$

It can be proved under mild conditions that if $\{r^{(k)}\}$ is a monotonic decreasing sequence with $r^{(k)} \to 0$ as $k \to \infty$, then the solutions of the unconstrained problems

$$\underset{x}{\text{minimize}} \quad \phi(x, r^{(k)}) \tag{9.4}$$

initiated at an interior point, are interior points $\{\mathbf{x}^{(k)} = \mathbf{x}[r^{(k)}]\}$ converging to $\mathbf{x}^*$, a solution to the constrained problem. Furthermore:

$$\lim_{k \to \infty} \phi(x^{(k)}, r^{(k)}) = \lim_{k \to \infty} f(x^{(k)}) = f(x^*) \tag{9.5}$$

and $f(x^{(k)})$ is monotonic decreasing. An intuitive understanding of these theoretical results is as follows. If, for example, the jth constraint is active at x^*, then as $x \to x^*$, $h_j(x) \to 0$ and, from (9.2) or (9.3), it is apparent that $B(x) \to \infty$. However, if $r \to 0$ the growth of $B(x)$ is cancelled, so making possible the constraint value to be reduced, and x to approach x^*.

An interesting aspect of transformation methods is that estimates of the Lagrangian multipliers are available:

$$\lambda_j(r) = \frac{r_j}{h_j} \quad \text{and} \quad \lambda_j(r) = \frac{r_j}{h_j^2} \tag{9.6}$$

for the logarithmic and inverse barrier functions, respectively. It can be shown that

$$\lim_{k \to \infty} \lambda_j(r^{(k)}) = \lambda_j^* \tag{9.7}$$

where the λ_j^* are the Lagrangian multipliers satisfying the Kuhn–Tucker conditions at x^*. These estimates of Lagrangian multipliers, together with the corresponding constraint values, are helpful in identifying the set of active constraints.

The basic algorithm in the barrier function approach is as follows:
(i) select a monotonic decreasing sequence $\{r^{(k)}\} \to 0$ as $k \to \infty$; find $x^{(0)} \in R_0$ and set $k = 0$;
(ii) with $x^{(k)}$ as a starting point, minimize $\phi(x, r^{(k)})$ to find $x^{(k+1)} = x(r^{(k)})$;
(iii) if convergence criteria are not satisfied, set $k = k+1$ and return to (ii).

To help fix ideas, we consider the problem

$$\begin{aligned} \text{minimize} & \quad x_1^2 + x_2^2 \\ \text{subject to} & \quad x_1 + x_2 - 1 \geq 0 \end{aligned} \tag{9.8}$$

which has already been solved using the dual method approach (see Sections 2.10 and 7.2). Be employing the logarithmic barrier function (9.2), this problem is replaced with the following unconstrained problem:

$$\text{minimize} \quad \phi(x, r) = x_1^2 + x_2^2 - r \ln(x_1 + x_2 - 1) \tag{9.9}$$

From the optimality conditions $\partial \phi / \partial x_1 = 0$ and $\partial \phi / \partial x_2 = 0$, it follows that

$$2x_1 - \frac{r}{x_1 + x_2 - 1} = 0$$

$$2x_2 - \frac{r}{x_1 + x_2 - 1} = 0$$

which gives

$$\begin{cases} x_1 = x_2 \\ 4x_1^2 - 2x_1 - r = 0 \end{cases}$$

The solution of the problem is therefore

$$x_1(r) = x_2(r) = \frac{1 + \sqrt{1 + 4r}}{4} \tag{9.10}$$

The other root $(1 - \sqrt{1 + 4r})/4$ is ruled out since the inequality $x_1(r) + x_2(r) - 1 > 0$ must be satisfied. Finally, taking the limit for $r \to 0$, it is concluded that the solution is $x_1^* = x_2^* = \frac{1}{2}$. An estimate of the Lagrangian multiplier associated with the constraint is provided by (9.6), which yields

$$\lambda(r) = \frac{r}{x_1 + x_2 - 1} = \frac{1 + \sqrt{1 + 4r}}{2}$$

showing that

$$\lambda^* = \lim_{r \to 0} \lambda(r) = 1$$

The barrier function transformation provides a powerful way of solving general constrained minimization problems. Unfortunately this approach suffers from a number of undesirable features. These weaknesses are mainly of computational nature and they become worse when the controlling parameter $r^{(k)}$ decreases, that is when $\mathbf{x}(r^{(k)})$ approaches $\mathbf{x}^*$. The essential difficulty lies in the fact that the auxiliary function $\phi(x, r^{(k)})$ is easy to minimize as long as $r^{(k)}$ remains reasonably large; however when $r^{(k)}$ becomes small, the function is more and more difficult to minimize. This behaviour can be related to the ill-conditioned nature of the Hessian matrix. As explained in Section 7.4, it is important, in order to evaluate the difficulty of an unconstrained minimization problem, to determine the eigenvalue structure of the Hessian. For the auxiliary function (9.1), this structure becomes increasingly unfavourable as r decreases. Assuming that there are q active constraints at the solution $\mathbf{x}^*$ to the original constrained problem, then the Hessian of the auxiliary function, $\nabla^2 \phi(x, r)$, has q eigenvalues that vary with r^{-1} and thus tend to infinity as $r \to 0$. The other $n - q$ eigenvalues tend to finite positive limits. This implies that as r decreases the condition number of $\nabla^2 \phi(x, r)$ varies with r^{-1} and therefore problem (9.4) becomes less and less manageable. In other words, the function $\phi(x, r)$ is more and more eccentric as $r \to 0$, thereby slowing considerably the speed of convergence of any first-order minimization algorithm, for example, conjugate gradient or quasi-Newton methods.

In the previous example, the Hessian of the auxiliary function (9.9) is given by

$$\nabla^2 \phi(x, r) = \begin{bmatrix} 2 + \dfrac{r}{(x_1 + x_2 - 1)^2} & \dfrac{r}{(x_1 + x_2 - 1)^2} \\ \dfrac{r}{(x_1 + x_2 - 1)^2} & 2 + \dfrac{r}{(x_1 + x_2 - 1)^2} \end{bmatrix}$$

General Nonlinear Programming Methods

The two eigenvalues of this matrix, with $x_1(r)$ and $x_2(r)$ known from (9.10), are easily seen to be

$$e_1 = 2 \text{ and } e_2 = 4 + \frac{1}{r} + \frac{\sqrt{1+4r}}{r}$$

showing that, as $r \to 0$, the condition number (7.45) of the Hessian matrix varies with r^{-1}. Therefore the unconstrained problem (9.9) becomes increasingly ill-conditioned as r decreases (see Fig. 9.1).

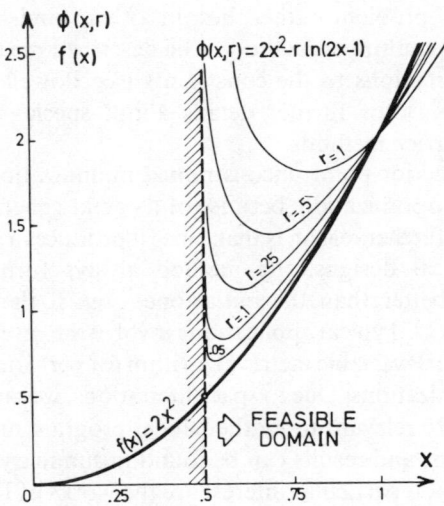

Fig. 9.1 Barrier function transformation

The question of knowing how the sequence $\{r^{(k)}\}$ should be chosen is thus very important, because it can seriously affect the computational effort required to find a solution. The conflict is clear: ultimately $r^{(k)}$ must be small enough to force the minimum $\mathbf{x}(r^{(k)})$ to approach the boundary of the feasible region, but large enough to enable the auxiliary function $\phi(x, r^{(k)})$ to be minimized without excessive difficulty. In practice a convenient choice is

$$r^{(k+1)} = r^{(k)} f \tag{9.11}$$

with f ranging from 0.1 to 0.5 depending upon the nature of the problem and the unconstrained minimization algorithm employed. We still have to supply an initial value for r. One possible approach is to choose $r^{(0)}$ so that neither $f(x^{(0)})$ nor $r^{(0)} B(x^{(0)})$ dominates in the auxiliary function $\phi(x^{(0)}, r^{(0)})$. So if $x^{(0)}$ is a reasonable guess to the solution, we could pick $r^{(0)}$ so that $r^{(0)} B(x^{(0)})$ approximately equals $f(x^{(0)})$. Note also that an essential requirement is that $\mathbf{x}^{(0)}$ should be a feasible starting point.

One idea that may be used for avoiding slow convergence is to resort to Newton's method, since its order-two convergence is unaffected by the poor eigenvalue structure. In applying the method, however, attention should be paid to the manner by which the ill-conditioned Hessian is inverted. Also Newton's method requires the second derivatives of the problem functions to be readily available, which is not often the case in practical problems. Other techniques have been devised in order to alleviate the difficulties inherent to the barrier function approach, for example, acceleration procedures, extrapolation schemes, and scaling processes. However it seems now that the best strategy to solve a general constrained problem rather lies in other kinds of transformation methods, such as the multiplier method to be described next, or methods based upon linear approximations to the constraints (see Powell [1]). We therefore refer to Refs [2, 3, 4] for further details about special techniques used in conjunction with barrier methods.

The concept of interior-point unconstrained minimization met considerable success in structural optimization, because of its great generality, reliability and relative simplicity. A further reason is that, since it produces a sequence of steadily improving non-critical designs, the method always furnishes a practically meaningful design, better than the initial one, even if the numerical process terminates prematurely. Typical applications involve the inverse barrier function (9.3) and use of the DFP variable metric algorithm for performing the sequence of unconstrained minimizations. Due to space limitations, we can only discuss those particularities that are relevant to mathematical programming algorithms. The structural implications and results can be found in summary form in the review papers of Refs [5, 6]. Of particular interest are the works of Kavlie and Moe [7], Cassis and Schmit [8] and Haftka and Starnes [9], later used extensively by Schmit and Miura [6]. In these works, effective techniques are devised to cope with part of the unconstrained minimization difficulties previously discussed. The auxiliary function $\phi(x, r)$ is not defined at infeasible points, and the simple expedient of setting its value to infinity is unlikely to be sufficient if this value is then required to determine an interpolating polynomial in the line-search process. Hence the idea of using an extended barrier function that is defined outside of the feasible region. In the usual inverse barrier function (9.3), the constraint repulsion terms $1/h_j(x)$ are defined only for $h_j(x) > 0$. The following linear *extended barrier function*:

$$B(x) = \sum_{j=1}^{m} b_j(x) \tag{9.12}$$

with

$$b_j(x) = \begin{cases} \dfrac{1}{h_j(x)} & \text{for } h_j(x) \geqslant \epsilon \\ \dfrac{2\epsilon - h_j(x)}{\epsilon^2} & \text{for } h_j(x) < \epsilon \end{cases} \tag{9.13}$$

is defined over an enlarged domain [7]. At the transition point with the usual barrier function (9.3), the modified function (9.12) satisfies continuity of both function value and first derivative. A rational procedure for determining ϵ, and hence the transition point, is set forth in Ref. [8]. It provides a relatively well-behaved extended barrier function (9.12), where the ordinary barrier function (9.3) becomes extremely awkward (see Fig. 9.2). In Ref. [9], a quadratic extended barrier function is introduced, which preserves the continuity of the second derivatives throughout the design space and is therefore well suited for second-order unconstrained minimization methods (i.e. Newton-type algorithms of Section 8.2). The quadratic extended barrier function is given by (9.12) with

$$b_j(x) = \begin{cases} \dfrac{1}{h_j(x)} & \text{for } h_j(x) \geqslant \epsilon \\ \dfrac{1}{\epsilon[(h_j(x)/\epsilon)^2 - 3(h_j(x)/\epsilon) + 3]} & \text{for } h^j(x) < \epsilon \end{cases} \qquad (9.14)$$

This quadratic extension of the barrier function is illustrated in Fig. 9.2, as well as the linear extension and the corresponding exact barrier function.

It should be mentioned that cubic [10] and variable [11] extended barrier functions have also been devised.

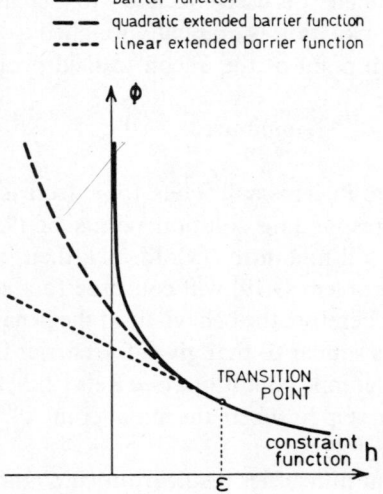

Fig. 9.2 Extended barrier functions

9.3 PENALTY FUNCTION METHODS

In contrast to the barrier function transformation, the penalty function transformation is defined so as to prescribe a high cost for violation of the constraints:

$$\psi(x, r) = f(x) + \frac{1}{r} P(x) \qquad r > 0 \tag{9.15}$$

where $P(x)$ is such that $P(x) \geq 0$ for all $x \in E^n$ and $P(x) = 0$ if and only if $x \in R$. $\psi(x, r)$ is defined on E^n and $\psi(x, r) \to \infty$ as constraint violation increases. Frequently used penalty functions are the quadratic loss function

$$P(x) = \sum_{j=1}^{m} [\min(0, h_j(x))]^2 \tag{9.16}$$

and the Zangwill loss function

$$P(x) = -\sum_{j=1}^{m} \min(0, h_j(x)) \tag{9.17}$$

It should be emphasized that this type of method can handle equality-constrained problems. An important penalty function transformation for such a problem is (assuming each h_j represents an equality constraint:

$$\psi(x, r) = f(x) + \frac{1}{r} \sum_{j=1}^{m} [h_j(x)]^2 \tag{9.18}$$

The controlling parameter r is used effectively to increase the magnitude of the penalty, i.e., constraint violation is gradually weighed as $r \to 0$. For small r, it is clear that the minimum point of the unconstrained problem

$$\underset{x}{\text{minimize}} \quad \psi(x, r) \tag{9.19}$$

will be in a region where $P(x)$ is small. Thus, for a decreasing sequence $\{r^{(k)}\}$, it is expected that the corresponding solution points of (9.19) will approach the feasible region R and will minimize $f(x)$. Ideally, then, as $r^{(k)} \to 0$, the solution point of the penalty problem (9.19) will converge to a solution of the original constrained problem. Therefore the behaviour of the penalty function transformation as $r^{(k)}$ decreases is similar to that given for barrier functions. Convergence can still be ensured under mild conditions (see Refs [2, 3]). The only difference in the basic algorithm is in step (i), where the initial point $x^{(0)}$ does no longer need to be feasible (see Section 9.1).

Again, computational difficulties result from $\psi(x, r)$ forming an increasingly steep-sided valley as the controlling parameter r decreases. It can be shown that the Hessian of the penalty function becomes ill-conditioned as $r \to 0$, just as the Hessian matrix of the barrier function. Also, there exist Lagrangian multiplier estimates. For the quadratic loss function (9.16), they are given by

$$\lambda_j^{(k)} = -\frac{2}{r^{(k)}} \min[0, h_j(x^{(k)})] \tag{9.20}$$

As a final observation, we note that in general the sequence $\mathbf{x}^{(k)}$ approaches $\mathbf{x}^*$ from outside the feasible region. Indeed, as $\mathbf{x}^{(k)} \to \mathbf{x}^*$, then since $\lambda^{(k)} \to \lambda^*$, all constraints that are active at $\mathbf{x}^*$ and have positive Lagrangian multipliers will be violated at $\mathbf{x}^{(k)}$ (see (9.20)). Therefore the penalty function transformation methods are also called *exterior-point unconstrained minimization techniques* (see Fig. 9.3 for a geometrical interpretation).

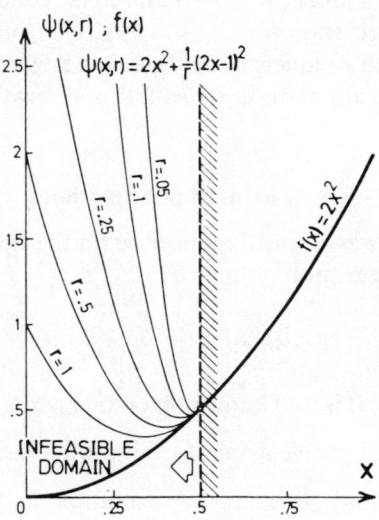

Fig. 9.3   Penalty function transformation

As an example, let us solve again problem (9.8) by assuming that the constraint imposed will be active at the optimum and that it can thus be viewed as an equality constraint. The penalty function transformation (9.18) is written

$$\psi(x, r) = x_1^2 + x_2^2 + \frac{1}{r}(x_1 + x_2 - 1)^2 \tag{9.21}$$

The solution of the unconstrained problem (9.19) is given by

$$x_1(r) = x_2(r) = \frac{1}{2+r} \tag{9.22}$$

and therefore

$$x_1^* = x_2^* = \lim_{r \to 0} \left(\frac{1}{2+r} \right) = \frac{1}{2}$$

It can also be seen from (9.20) that

$$\lambda^* = \lim_{r \to 0} \left(\frac{2}{2+r} \right) = 1$$

Finally, the Hessian matrix has the form

$$\nabla^2 \psi(x, r) = \begin{bmatrix} 2 + \dfrac{2}{r} & \dfrac{2}{r} \\ \dfrac{2}{r} & 2 + \dfrac{2}{r} \end{bmatrix}$$

Its eigenvalues are $e_1 = 2$ and $e_2 = 2 + 4/r$ and so the condition number $1 + 2/r$ is deteriorating when r decreases.

Because they generate a sequence of infeasible designs, penalty methods have received relatively little attention in structural optimization applications [12].

9.4 The multiplier methods

The multiplier method was originally conceived for the equality constrained case. Again the Lagrangian function

$$L(x, \lambda) = f(x) - \sum_{j=1}^{m} \lambda_j h_j(x) \qquad (9.23)$$

plays an important role. It is well known that if there exists some λ^* for which x^* solves the unconstrained problem $\min_x L(x, \lambda^*)$, while satisfying the constraints $\{h_j(x) = 0; \ j = 1, \ldots, m\}$, then x^* is a solution to the original problem. Therefore the problem

$$\begin{aligned} &\underset{x}{\text{minimize}} && L(x, \lambda) \text{ given suitable } \lambda && (9.24) \\ &\text{subject to} && h_j(x) = 0 \quad j = 1, \ldots, m \end{aligned}$$

is equivalent to the original problem in the sense that both problems have x^* as local minimum and λ^* as associated Lagrangian multiplier vector. Now consider solving the equivalent problem (9.24) by resorting to the penalty function transformation (9.18). This leads us to add the penalty term not to the objective function $f(x)$, but rather to the Lagrangian function $L(x, \lambda)$, thus forming the *augmented Lagrangian function* for the equality constrained problem [13, 14]:

$$\chi(x, \lambda, r) = f(x) - \sum_{j=1}^{m} \lambda_j h_j(x) + \frac{1}{r} \sum_{j=1}^{m} [h_j(x)]^2 \qquad (9.25)$$

Given values of λ and r, the multiplier method consists in applying an algorithm for unconstrained minimization to the function (9.25), yielding a point $x(\lambda, r)$. Then λ and r are adjusted for the next iteration so that $x(\lambda, r)$ converges to x^* as the iterations proceed.

The two following extreme cases are instructive:
1. $\lambda = 0$: then (9.25) reduces to the standard penalty function transformation (9.18). As r is gradually decreased, the solution is approached from the infeasible region and the unconstrained minimizations become ill-conditioned. Note that the points $\mathbf{x}(r)$ and $\mathbf{x}^*$ have to be different for $r \neq 0$, because the gradient of the function $\psi(x, r)$ is zero at $\mathbf{x}(r)$ but has the value $\nabla f(x^*)$ at $\mathbf{x}^*$.
2. $\lambda = \lambda^*$: that is, the optimal Lagrangian multipliers are known at the outset. Then it can be shown that minimizing $\chi(x, \lambda^*, r)$ gives the solution to the original problem for any value of $r > 0$. Note that by virtue of the first-order optimality conditions, the gradient of $\chi(x, \lambda^*, r)$ is zero at $\mathbf{x}^*$ for all values of r.

The foregoing observations suggest that, by updating the Lagrangian multipliers λ so that they approach λ^*, convergence may occur without the need for r to be very small. Thus the ill-conditioning associated with penalty methods can be avoided. To obtain the update formula for the λ_j, we compare the optimality conditions of $\chi(x, \lambda, r)$

$$\frac{\partial \chi}{\partial x_i}(x, \lambda, r) \equiv \frac{\partial f}{\partial x_i}(x) - \sum_{j=1}^{m} \left[\lambda_j - \frac{2}{r} h_j(x)\right] \frac{\partial h_j}{\partial x_i}(x) = 0 \qquad (9.26)$$

with the stationarity conditions at the optimal solution $(\mathbf{x}^*, \lambda^*)$:

$$\frac{\partial L}{\partial x_i}(x^*, \lambda^*) \equiv \frac{\partial f}{\partial x_i}(x^*) - \sum_{j=1}^{m} \lambda_j^* \frac{\partial h_j}{\partial x_i}(x^*) = 0 \qquad (9.27)$$

In the limit as $\mathbf{x}(\lambda, r)$ converges to $\mathbf{x}^*$, (9.26) should collapse to (9.27), which implies that

$$\lambda_j - \frac{2}{r} h_j(x) \to \lambda_j^* \quad \text{as} \quad \mathbf{x}(\lambda, r) \to \mathbf{x}^*$$

This important observation naturally suggests the following update formula:

$$\lambda_j^{(k+1)} = \lambda_j^{(k)} - \frac{2}{r^{(k)}} h_j(x^{(k)}) \qquad (9.28)$$

where k is the iteration index [13]. Using this adjustment of the Lagrangian multipliers accelerates the convergence of a sequence of unconstrained minimizations executed on $\chi(x, \lambda^{(k)}, r^{(k)})$. It can be shown that the convergence rate is superlinear if we let r decrease to zero. However, in order to avoid ill-conditioned problems, it is better to keep the response factor $r^{(k)}$ within an appropriate problem dependent lower bound $\underline{r}$ (see Fig. 9.4). The convergence rate is then linear, which is much better than the convergence rate of usual penalty methods [15].

The adjustment of λ can be presented very elegantly as a dual problem. From the saddle point condition

$$\chi(x^*, \lambda, r) \leqslant \chi(x^*, \lambda^*, r) \leqslant \chi(x, \lambda^*, r) \qquad (9.29)$$

it follows that the optimal pair $(\mathbf{x}^*, \lambda^*)$ could be obtained from first minimizing $\chi(x, \lambda, r)$ over $\mathbf{x}$ for given λ and r, and next maximizing $\chi(\bar{x}, \lambda, r)$ over λ where $\bar{x}$ is the solution of the minimization problem. Repeating this process by gradually reducing r yields the multiplier method, sometimes called 'primal–dual' method. The problem of adjusting λ is thus the problem of maximizing the dual function (see Sections 2.10 and 7.2):

$$l_r(\lambda) = \min_x \chi(x, \lambda, r) = \chi(\bar{x}, \lambda, r) \qquad (9.30)$$

Because the first derivatives of $l_r(\lambda)$ are given by minus the constraint values (see Equation (7.25)), it is apparent that the update formula (9.28) corresponds to a steepest ascent move with step size $2/r$ in the dual space:

$$\lambda^+ = \lambda + \frac{2}{r}\nabla l_r \qquad (9.31)$$

In place of the steepest ascent iteration, it is possible to consider the Newton iteration

$$\lambda^+ = \lambda - [\nabla^2 l_r]^{-1}\nabla l_r \qquad (9.32)$$

where the Hessian matrix of the dual function, $\nabla^2 l_r$, is given by (7.28). This correction formula provides a quadratic rate of convergence. However, in practice the simple steepest ascent iteration (9.31) is usually preferred, because it has already a good convergence rate, and it does not require the calculation of any derivatives. Note that quasi-Newton methods can also be considered. By employing second derivative approximations from the unconstrained calculations instead of $\nabla^2 l_r$ in (9.32), the sequence of values of λ converges to λ^* at a superlinear rate.

Therefore the multiplier method can be viewed as a kind of *primal–dual optimization method* with very limited search in the dual space for optimum Lagrangian multipliers.

As a computational example, we take again problem (9.9), for which the augmented Lagrangian function (9.25) has the form:

$$\chi(x, \lambda, r) = x_1^2 + x_2^2 - \lambda(x_1 + x_2 - 1) + \frac{1}{r}(x_1 + x_2 - 1)^2$$

This function is minimized for

$$x_1(\lambda, r) = x_2(\lambda, r) = \frac{1}{2}\frac{2 + r\lambda}{2 + r}$$

For $\lambda = 0$, we of course obtain the solution (9.22) corresponding to the penalty method, while, for $\lambda = \lambda^* = 1$, we have $x_i(\lambda^*, r) = x_i^* = \frac{1}{2}$ for any value of r. In Table 9.1, we show the results of the computation for the penalty method where

$\lambda^{(k)} = 0$ for all k and for the multiplier method with the update formula (9.28), namely, $\lambda^{(0)} = 0$ and, for subsequent k:

$$\lambda^{(k+1)} = \lambda^{(k)} - \frac{2}{r^{(k)}}(x_1^{(k)} + x_2^{(k)} - 1)$$

Table 9.1 Comparison of penalty and multiplier methods

$x_i^{(k)}$	$r^{(k+1)} = r^{(k)} \times 0.50$		$r^{(k+1)} = r^{(k)} \times 0.25$		$r^{(k+1)} = r^{(k)} \times 0.10$	
k	Penalty	multiplier	penalty	multiplier	penalty	multiplier
1	0.1000	0.1000	0.1000	0.1000	0.1000	0.1000
2	0.1667	0.2333	0.2500	0.3000	0.3571	0.3857
3	0.2500	0.3667	0.4000	0.4600	0.4808	0.4956
4	0.3333	0.4556	0.4706	0.4976	0.4980	0.5000
5	0.4000	0.4911	0.4923	0.5000	0.4998	
6	0.4444	0.4990	0.4981		0.5000	
7	0.4706	0.4999	0.4995			
8	0.4848	0.5000	0.4999			
9	0.4923		0.5000			
10	0.4961					
11	0.4981					
12	0.4990					
13	0.4995					
14	0.4998					
15	0.4999					
16	0.4999					
17	0.5000					

It can be observed from Table 9.1 that the number of unconstrained minimizations required for both methods decreases when the response factor $r^{(k)}$ is decreased at a faster rate (note that the effects of ill-conditioning cannot be felt in this closed form solution). As expected, the multiplier method performs much better than the penalty method. If now we decided to resort to the second-order update formula (9.32), we would obtain the solution in only two minimizations, for any values of r. This is not surprising, because the dual function (9.30) is quadratic for the problem under consideration, so that λ^* is attained in only one application of the update formula (9.32):

$$\lambda^{(1)} = \lambda^{(0)} - \frac{2 + r^{(0)}}{r^{(0)}}\left(\frac{2 + r^{(0)}\lambda^{(0)}}{2 + r^{(0)}} - 1\right) = 1$$

for any $r^{(0)}$ and $\lambda^{(0)}$.

The multiplier method can easily be extended to deal with inequality constraints by converting them to equality constraints through the use of slack variables [16]. It turns out that these additional variables finally disappear from the augmented Lagrangian function, which exhibits the following form:

$$\chi(x, \lambda, r) = f(x) + \frac{1}{r} \sum_{j=1}^{m} m_j(x, \lambda, r) \qquad (9.33)$$

with

$$m_j(x, \lambda, r) = \begin{cases} h_j^2(x) - r\lambda_j h_j(x) & \text{if } h_j(x) \leq \frac{r}{2}\lambda_j \\ -\left(\frac{r}{2}\lambda_j\right)^2 & \text{if } h_j(x) > \frac{r}{2}\lambda_j \end{cases} \qquad (9.34)$$

The update formula for the Lagrangian multipliers becomes:

$$\lambda_j^+ = \max\left[0, \lambda_j - \frac{2}{r} h_j(x)\right] \qquad (9.35)$$

It is interesting to note that the primal–dual nature of the multiplier method can be directly employed to find the form of the augmented Lagrangian function (9.33). Indeed, when inequality constraints $h_j(x) > 0$ are introduced, the corresponding Lagrangian multipliers λ_j must remain non-negative throughout the optimization process. The steepest ascent step in dual space is thus constrained, which leads to (9.35), or, equivalently (see Equation (7.46)):

$$\lambda_j^+ = \lambda_j - \frac{2}{r} \min\left[h_j(x), \frac{r}{2}\lambda_j\right] \qquad (9.36)$$

This observation suggests that the expression (9.25) for $\chi(x, \lambda, r)$ can still be used, provided that $h_j(x)$ is replaced with $\min[h_j(x), r/2\lambda_j]$:

$$\chi(x, \lambda, r) = f(x) - \sum_{j=1}^{m} \lambda_j \min\left[h_j(x), \frac{r}{2}\lambda_j\right] + \frac{1}{r} \sum_{j=1}^{m} \left\{\min\left[h_j(x), \frac{r}{2}\lambda_j\right]\right\}^2 \qquad (9.37)$$

It is easily shown that (9.37) reduces to (9.33) after some algebraic manipulations. It is important to realize that no contribution to $\chi(x, \lambda, r)$ arises from the inactive constraints associated with zero Lagrangian multipliers. This fact permits dynamic selection of the active constraint set. In practice, it has been observed that the correct set of active constraints usually emerges after only a few stages of unconstrained minimizations. It should also be mentioned that although the multiplier method tends in general to generate a sequence of infeasible primal points, it can also be used to produce a sequence of feasible points, provided that the initial Lagrangian multipliers are large enough. Finally, note that the augmented Lagrangian function (9.33) is continuous and has continuous first derivatives; however, its second derivatives are discontinuous along the boundaries $h_j(x) = r/2\lambda_j$. An implementation of the multiplier method is given in the flow diagram of Fig. 9.4.

General Nonlinear Programming Methods

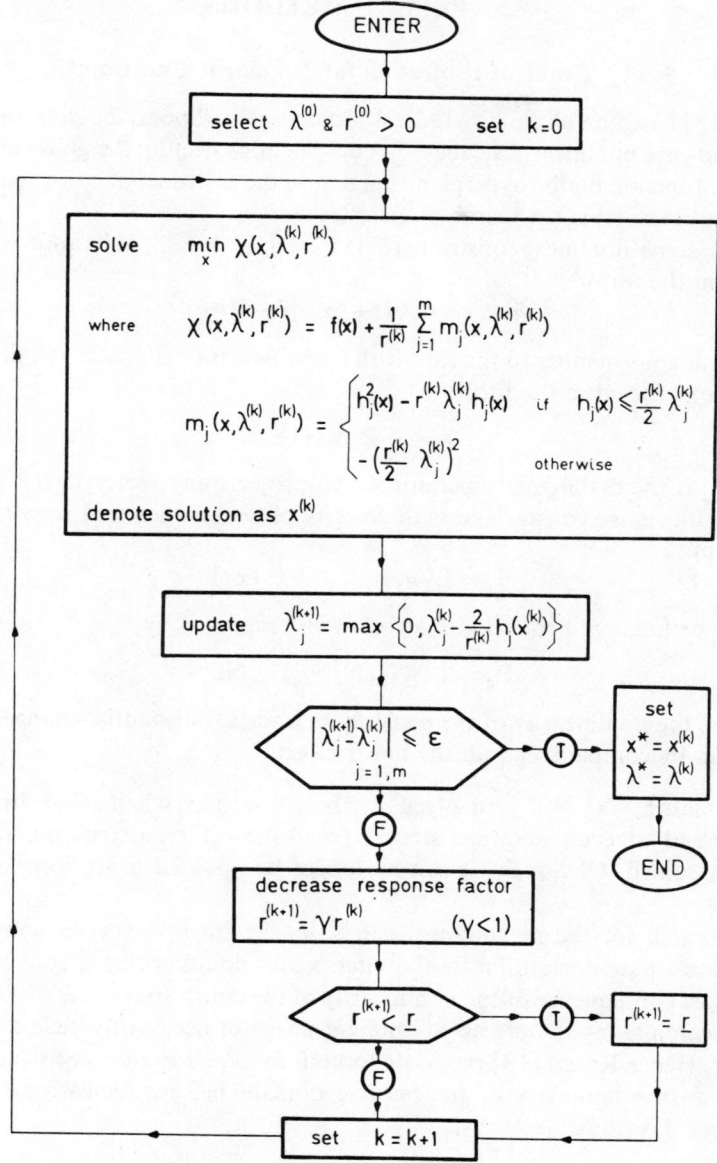

Fig. 9.4 Multiplier method—flow diagram

The multiplier method has not yet been extensively applied to structural optimization; however, Imai [17] employed this efficient method in the case of truss structures considering both configuration and sizing type design variables.

9.5 PRIMAL METHODS

9.5.1 Gradient Projection for Nonlinear Constraints

Rosen [18] has generalized his gradient projection method to the case where the constraints are nonlinear. The method consists in projecting the gradient of the objective function on the hyperplane tangent to the active set of constraints. The vector so obtained is the search direction.

Let the active nonlinear constraints $h_j(x) \geq 0$ $(j = 1, \ldots, q)$ be approximated linearly in the form

$$h_j(x) = h_j(\hat{x}) + (\mathbf{x} - \hat{\mathbf{x}})^T \nabla h_j \tag{9.38}$$

In an analogous manner to the case with linear constraints, the search direction can be constructed in the form

$$\mathbf{s} = -\mathbf{P}_q \nabla f(x) \tag{9.39}$$

where $\mathbf{P}_q$ is the orthogonal operator which projects any vector in the tangent plane to the active constraints. If the matrix of active linearized constraints is defined by

$$\mathbf{N}_q = [\nabla h_1(x) \ldots \nabla h_q(x)] \tag{9.40}$$

then the orthogonal projection matrix is still calculated by

$$\mathbf{P}_q = \mathbf{I} - \mathbf{N}_q (\mathbf{N}_q^T \mathbf{N}_q)^{-1} \mathbf{N}_q^T \tag{9.41}$$

However, the nonlinearity of the constraints leads to substantial changes in the algorithm by comparison with the linear case:

1. The matrix $(\mathbf{N}_q^T \mathbf{N}_q)^{-1}$ involved in the projection scheme has to be re-evaluated at each iteration step, even if the set of active constraints is not modified. Obviously, it can no longer be updated using recursion formulas.
2. The search for the maximum feasible step length involves the solution of nonlinear equations to locate the intersection points with the constraints.
3. The last consequence of the nonlinearity of the constraints is the fact that the line search along the projected gradient does not necessarily yield a feasible point. Hence Rosen [18] proposes to perform a *restoration step* which leads back to the boundary of the feasible domain before calculating the next descent direction.

To perform the restoration step, the linear approximation to the constraints (9.38) may be used: a progression is made along a direction normal to the constraints at the initial point as illustrated by Fig. 9.5.

According to Equations (9.38) and (9.40), a point $\mathbf{x}$ such that $h(x) = 0$ is generated by the iterative sequence

$$\mathbf{x}^{k+1} = \mathbf{x}^k - \mathbf{N}_q (\mathbf{N}_q^T \mathbf{N}_q)^{-1} h_q(x^k) \tag{9.42}$$

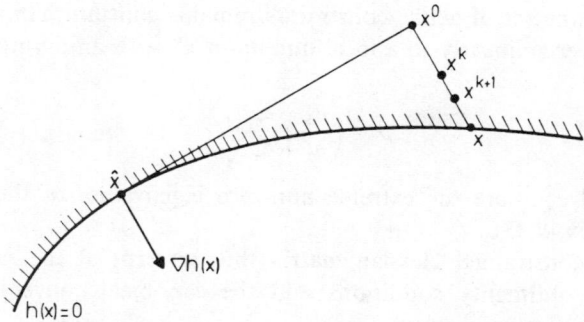

Fig. 9.5 Restoration phase in the gradient projection method for nonlinear constraints

The convergence of the restoration algorithm (9.42) depends on whether the point $\mathbf{x}^0$ is sufficiently close to the constraint surface, or not. And also, at each iteration of the restoration phase the condition $f(x^k) < f(\hat{x})$ has to remain verified. Hence, the effectiveness of the restoration phase depends on the length of the step performed in the minimization phase.

Finally, it is worth while noticing that in the restoration phase advantage is taken of the availability of the matrix $(\mathbf{N}_q^T \mathbf{N}_q)^{-1}$ which was needed in the minimization step.

Just as in the case of projection methods with linear constraints, more powerful minimization algorithms can be devised which are based on [19]:

(a) conjugate gradients;
(b) Newton iteration;
(c) quasi-Newton (variable metric) iteration.

In the gradient projection methods, the starting point of the iterative process must be a feasible point. Since the constraints are nonlinear, localizing such a point may be far from trivial. Generally, an iterative scheme such as (9.42) is used to locate a feasible point from a minimum problem in which the objective function is a measure of constraint violation.

9.5.1.1 Convergence of gradient projection methods

The global convergence of most primal methods cannot be attained unless adequate remedies are adopted to prevent the zigzag phenomenon. Most of the gradient projection schemes, in their original form, may be regarded as a steepest descent method applied on the hypersurface defined by the active constraints. Thus, they are characterized by a convergence rate similar to that of the steepest descent method which is dependent on the Hessian matrix of the Lagrangian function $L(x)$ restrained to the tangent plane to the active constraints [20]

$$\mathbf{H} = \mathbf{P}_q^T \nabla^2 \mathbf{L} \mathbf{P}_q \tag{9.43}$$

As long as the set of active constraints remains constant, almost all primal methods converge linearly to a local minimum x^* with an asymptotic convergence ratio

$$\rho = \left(\frac{e_1 - e_{n-q}}{e_1 + e_{n-q}}\right)^2 \qquad (9.44)$$

where e_1 and e_{n-q} are the extreme non-zero eigenvalues of the constrained Hessian matrix (9.43).

The same constrained Hessian matrix thus governs at the same time the second-order optimality conditions and the canonical convergence rate of gradient projection methods.

9.5.2 Methods of Feasible Directions

Numerous primal methods of minimization are based on the principle of feasible directions introduced initially by Zoutendijk [21]: according to this principle, the succesive iterations consist in a line search in a direction **s** starting from a feasible point $\hat{\mathbf{x}}$ and leading to a feasible point **x**:

$$\mathbf{x} = \hat{\mathbf{x}} + \alpha \mathbf{s} \qquad \alpha > 0 \qquad (9.45)$$

Such a direction which does not immediately leave the feasible domain is a *feasible direction*. The feasibility condition at a limit point where $q \leqslant m$ constraints are simultaneously active is written

$$\mathbf{s}^T \nabla h_j \geqslant 0 \qquad j = 1, \ldots, q \qquad (9.46)$$

with strict inequality for nonlinear constraints. Any vector satisfying (9.46) lies at least partly in the feasible domain, as indicated by Fig. 9.6.

Moreover, the feasible direction **s** is *usable* if it is a *direction of descent*, which means that it leads to a decrease of the objective function in the vicinity of **x**; the

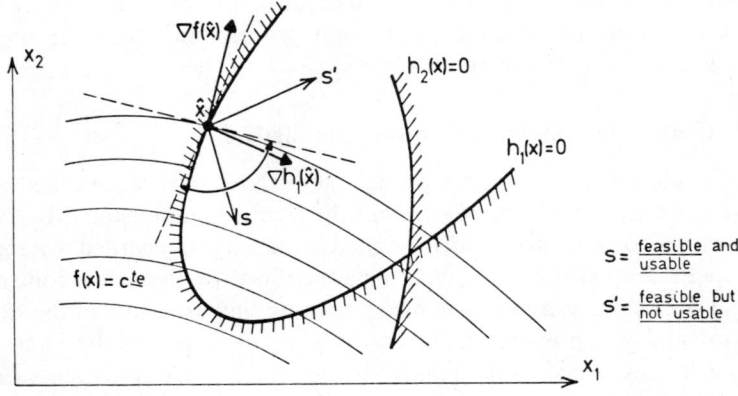

Fig. 9.6 Method of feasible directions: cone of feasible directions

condition of descent involves the gradient of the objective function

$$\mathbf{s}^T \nabla f(\hat{\mathbf{x}}) \leq 0 \qquad (9.47)$$

A method of feasible directions proceeds in two steps:
1. finding a feasible direction $\mathbf{s}$;
2. taking the step length α in order to find a new feasible point $\mathbf{x}$ which minimizes the objective function along the selected direction.

If the current point $\hat{\mathbf{x}}$ is not a local minimum, both inequalities (9.46) and (9.47) define a cone of feasible directions as shown in Fig. 9.6. Several feasible direction methods have been proposed which differ mainly in the direction-finding strategy.

9.5.2.1 Finding a feasible direction

Zoutendijk [21] has shown that the direction finding problem may be formulated as a linear programming problem in which the feasible direction leading to the maximum decrease of the objective function is a solution of

$$\max \beta$$

subject to
$$\mathbf{s}^T \nabla f + \beta \leq 0 \qquad (9.48)$$

$$\mathbf{s}^T \nabla h_j - \theta_j \beta \geq 0 \qquad j = 1, \ldots, q$$

some measure of the length $\mathbf{s}$ is bounded

where the θ_j are arbitrary positive constants.

Clearly, if $\beta_{\max} > 0$, the strict inequalities in (9.46) and (9.47) hold and the selected feasible direction is also a direction of descent. If $\beta_{\max} = 0$, the initial point $\hat{\mathbf{x}}$ is then a local minimum.

The positive constants θ_j can be fixed *a priori*, and measure the extent to which the feasible direction $\mathbf{s}$ is projected from the boundary of the feasible domain. Therefore, Zoutendijk calls them *push-off factors*.

The influence of the push-off factors is illustrated in Fig. 9.7.

1. If θ_j is taken close to zero, the feasible direction is essentially chosen so that $\mathbf{s}^T \nabla f + \beta \leq 0$. Therefore, the objective function is decreased rapidly but with a feasible direction which generally almost follows the boundary of the feasible domain. It is thus likely that the surface of active constraints will be hit very rapidly.
2. On the other hand, if θ_j is taken very large, the direction $\mathbf{s}$ tends to stay on the isolines of the objective function. The risk of running out the feasible domain is reduced, but is paid by lower decreases in the objective function.

Intermediate values of θ_j may lead to search directions for which, in the vicinity of $\hat{\mathbf{x}}$, the constraints and the objective function are decreased at similar rates.

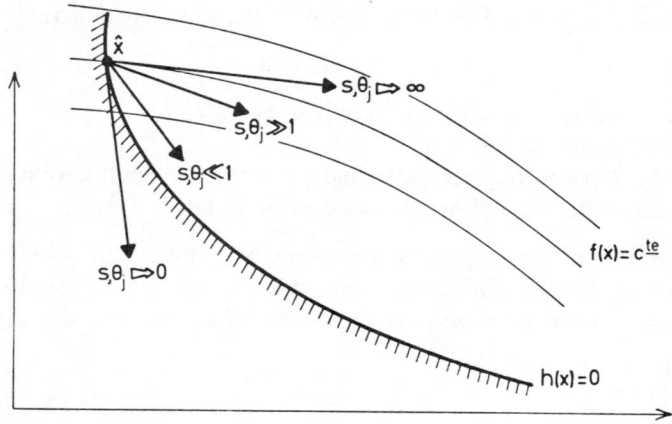

Fig. 9.7 Method of feasible directions: influence of push-off factors

9.5.2.2 Taking the step

Assuming that a strictly usable feasible direction has been obtained from Zoutendijk's or some other method, the problem of selecting the step size can still be challenging. There are two possibilities for the outcome of the line search:

(a) at the final point **x**, one or more of the constraints are active. The problem is then to find the maximum allowable step length. In particular, due to the nonlinearity of the constraints, one of the constraints where **s** has been evaluated may become active again.
(b) the final point **x** is an unconstrained minimum with respect to α, in which case taking the step reduces to a one-dimensional minimization.

Moreover, the zigzag phenomenon due to instability in the selection of active constraints may be very acute in the presence of nonlinear constraints. Therefore, care must be taken in the selection process of active constraints: zigzagging is prevented by the same techniques as in the presence of linear constraints.

9.6 LINEARIZATION METHODS

9.6.1 Recursive Linear Programming

The simplest method to solve a nonlinear programming problem is probably to transform it into a *sequence of linear programming problems* as follows:

$$\min_{x} \{f(\hat{x}) + (\mathbf{x} - \hat{\mathbf{x}})^T \nabla f(\hat{x})\}$$

subject to

$$h_j(\hat{x}) + (\mathbf{x} - \hat{\mathbf{x}})^T \nabla h_j(\hat{x}) \geq 0 \qquad j = 1, \ldots, m$$

(9.49)

where $\hat{x}$ is the point where the objective function and the constraints are linearized. The solution of this problem is a starting point to the next linearization.

However, this very attractive recursive method encounters strong limitations:

1. It does not converge to a local minimum unless the latter occurs at a vertex of the feasible domain. If the minimum is located somewhere else, which is the most frequent case, the process either converges to a non-optimal vertex, or it oscillates indefinitely between two vertices.
2 The problem admits a solution only if the number of constraints exceeds the number of variables.

Figure 9.8 illustrates this very rough linearization process.

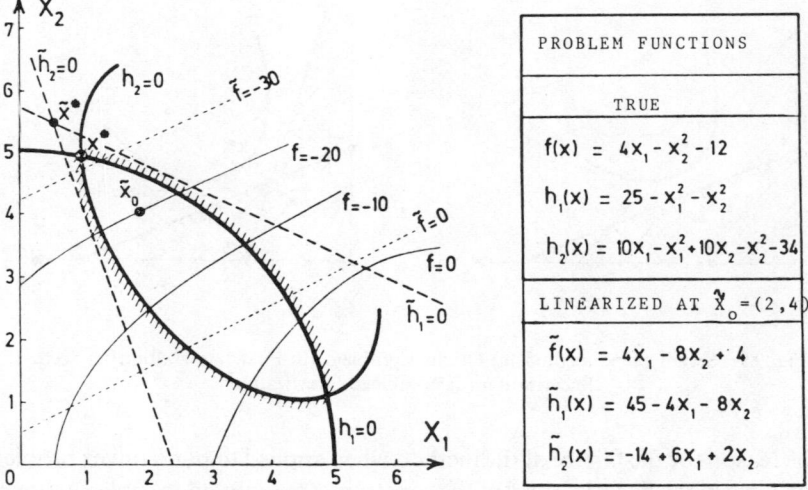

Fig. 9.8 Linearization of the objective function and constraints

9.6.2 Kelley's Cutting Plane Method [22]

The cutting plane method is a very successful variant of the recursive linear programming method when applied to convex problems. The principle of the method consists, as illustrated by Fig. 9.9, in generating a piecewise-linear envelope to the constraints. In Kelley's method, this envelope is obtained by adding a new linearized constraint after each solution of the linearized problem. The new constraint included into the problem corresponds to the most violated constraint after each linearization.

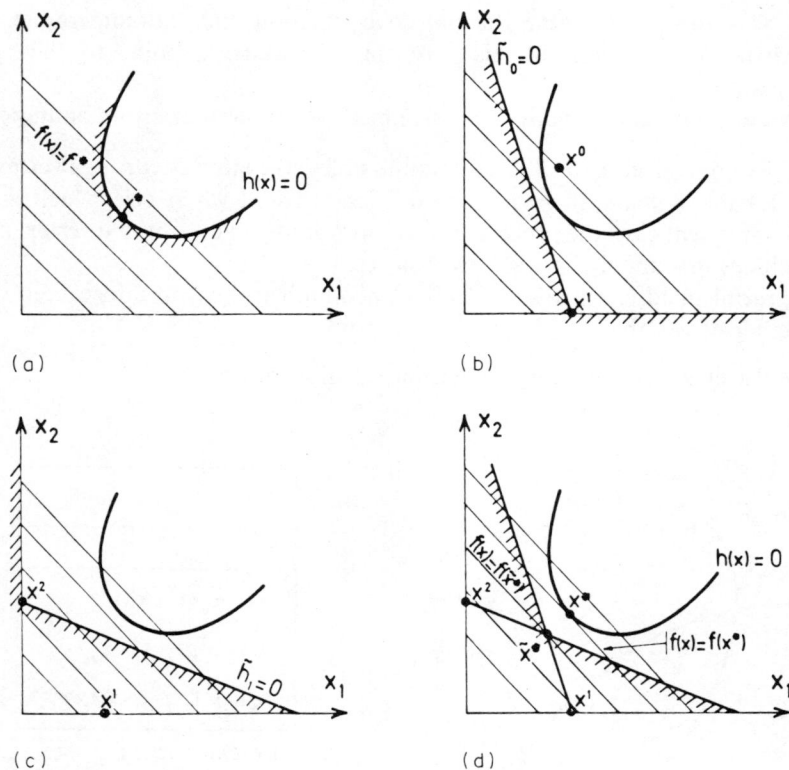

Fig. 9.9 Secant plane method (a) Original problem (b) First linearization (c) Second linearization (d) Combined linearization

The reason for the failure of the method when applied to non-convex problems is that some of the linearizations of the constraints may cut off feasible portions of the space that include the optimal solution of the original problem.

A proof of the global convergence of the method applied to convex problems can be given; however, the associated convergence rate is very poor. The method suffers from two other important drawbacks:
(a) as the number of linear constraints increases at each iteration step, the resulting linear programming problem can get very large;
(b) even in convex problems, the designs leading to the optimum are usually infeasible. This is a serious limitation from an engineering point of view.

9.6.3 Method of Approximation Programming [23]

The method of approximation programming is another interesting variant of the recursive linear programming method. It solves the linearized problem in the form (9.49), but with a limitation on the variation domain of the variables:

$$\hat{x}_i - \alpha_i \leqslant x_i \leqslant \hat{x}_i + \beta_i \tag{9.50}$$

where α and β are vectors of properly chosen positive constants, called *move limits*. After solution of the problem (9.49) with the additional constraints (9.50), the objective function and the constraints are again linearized, and the move limits are possibly modified.

The technique is illustrated in Fig. 9.10.

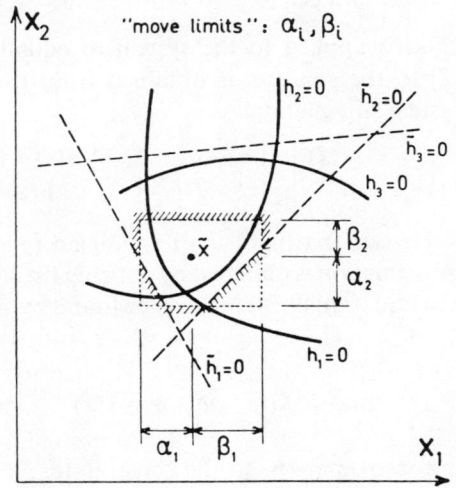

Fig. 9.10 Programming by approximation

9.6.4 Recursive Quadratic Programming Methods [24]

The use of a Lagrangian function is probably the basis of the most successful minimization methods with nonlinear constraints. Their success relies on the fact that account of the curvature of the constraints is taken through their quadratic approximation contained in the Lagrangian function.

Let us suppose for sake of simplicity that our minimum problem is subjected only to equality constraints

$$\begin{aligned}&\min_x f(x)\\ \text{subject to}\quad & \\ & h_j(x) = 0 \qquad j = 1, \ldots, q\end{aligned} \tag{9.51}$$

The extension to inequality constraints is then made using an active constraint strategy based on the magnitude of the Lagrangian multipliers.

The constrained problem (9.51) is treated by writing the stationarity conditions of the Lagrangian function

$$L(x, \lambda) = f(x) - \sum_j \lambda_j h_j(x) \tag{9.52}$$

which gives rise to the system of nonlinear equations

$$\begin{cases} \nabla f(x) - \sum_{j=1}^{q} \lambda_j \nabla h_j(x) = 0 \\ h_j(x) = 0 \quad j = 1, \ldots, q \end{cases} \tag{9.53}$$

If Newton's method is applied to the system of equations (9.53), a better approximation (x, λ) to the solution is obtained from the estimate $(\hat{x}, \hat{\lambda})$ by solving the linear system of equations

$$\begin{bmatrix} G(\hat{x}, \hat{\lambda}) & -N(\hat{x}) \\ -N^T(\hat{x}) & 0 \end{bmatrix} \begin{bmatrix} x - \hat{x} \\ \lambda - \hat{\lambda} \end{bmatrix} = \begin{bmatrix} -\nabla f(\hat{x}) + N(\hat{x})\hat{\lambda} \\ h(\hat{x}) \end{bmatrix} \tag{9.54}$$

where $G(x, \lambda)$ is the Hessian matrix of the Lagrangian function.

The quadratic approximation is obtained by noticing that the correction vector $\delta = (\hat{x} - x)$ that solves Equation (9.54) can also be found by solving the quadratic problem

$$\min_\delta \frac{1}{2} \delta^T G(\hat{x}, \hat{\lambda}) \delta + \delta^T \nabla f(\hat{x}) \tag{9.55}$$

subject to the linear approximations to the constraints

$$N^T(\hat{x})\delta + h(\hat{x}) = 0 \tag{9.56}$$

Equations (9.55)–(9.56) thus suggest transforming the initial problem (9.51) into a succession of quadratic programs in which the objective function is a second-order approximation of the Lagrangian calculated with the active constraints and where the constraints are replaced by their linear approximations (9.56).

A further simplification in the method is obtained if the second derivative matrix $G(\hat{x}, \hat{\lambda})$ is no longer evaluated at each step of the method, but simply obtained from successive quasi-Newton updates. Such a scheme using the BFGS update is outlined in Fig. 9.11.

PROBLEMS

1 Solve the problem

$$\begin{aligned} \text{minimize} \quad & x_1 + x_2 \\ \text{subject to} \quad & -x_1^2 + x_2 \geq 0 \\ & x_1 \geq 0 \end{aligned}$$

General Nonlinear Programming Methods

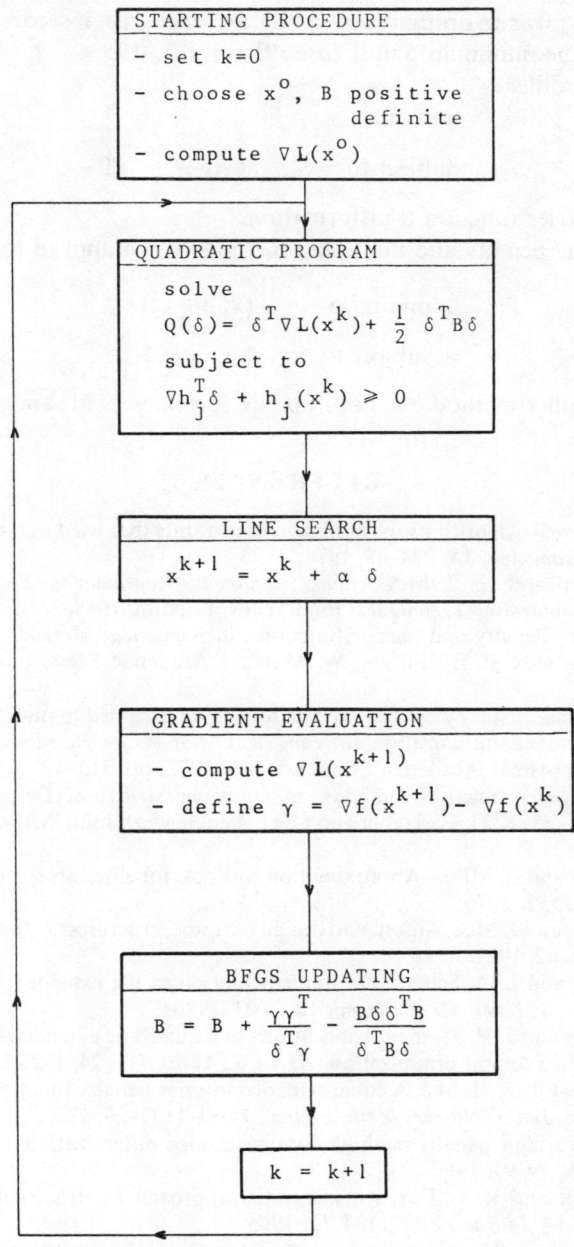

Fig. 9.11 Principle of recursive quadratic programming based on quasi-Newton updating

by using the logarithmic barrier function ransformation (9.2). Treat first the side constraint $x_1 \geq 0$ as an ordinary constraint, and then treat it separately. Draw the trajectory of the minimum points (use $r^{(k)} = 1, 0.5, 0.25, \ldots$).

2 Solve the problem

$$\text{minimize} \quad x_2$$
$$\text{Subject to} \quad x_2 - \sin x_1 - \frac{x_1}{2} \geq 0$$

by using a barrier function transformation.

3 Compare the penalty and multiplier methods when applied to the problem

$$\text{minimize} \quad \frac{1}{2}(x_1^2 + \frac{1}{3}x_2^2)$$
$$\text{subject to} \quad x_1 + x_2 = 1$$

For the multiplier method, try both update formulas (9.31) and (9.32).

REFERENCES

1. M. J. D. Powell, Algorithms for nonlinear constraints that use Lagrangian functions, *Math. Programming*, **14**, 224–48, 1978.
2. A. V. Fiacco and G. P. McCormick, *Nonlinear Programming: Sequential Unconstrained Minimization Techniques*, John Wiley, London, 1968.
3. D. M. Ryan, Penalty and barrier functions, in *Numerical Methods for Constrained Optimization* (eds. P. E. Gill and W. Murray), Academic Press, London, 1974, pp. 175–90.
4. F. A. Lootsma, A survey of methods for solving constrained minimization problems via unconstrained minimization, in *Numerical Methods for Non-linear Optimization*, (ed. F. A. Lootsma), Academic Press, London, 1972, pp. 313–47.
5. J. Moe, Penalty function methods, in *Optimum Structural Design—Theory and Applications* (eds R. H. Gallagher and O. C. Zienkiewicz), John Wiley, London, 1973, pp. 143–77.
6. L. A. Schmit and H. Miura, Approximation concepts for efficients structural synthesis, *NASA-CR-2552*, 1976.
7. W. C. Kavlie and J. Moe, Automated design of frame structures, *J. Struct. Div.*, ASCE, **97**(ST 1), 33–62, 1971.
8. J. H. Cassis and L. A. Schmit, On implementation of the extended interior penalty function, *Int. J. Num. Meth. Engng*, **10**, 3–23, 1976.
9. R. T. Haftka and J. H. Starnes, Applications of a quadratic extended interior penalty function for structural optimization, *AIAA J.*, **14**(6), 718–24, 1976.
10. B. Prasad and R. T. Haftka, A cubic extended interior penalty function for structural optimization, *Int. J. Numer. Meth. Engng.*, **14**(9), 1107–26, 1979.
11. B. Prasad, Variable penalty methods for constrained minimization, *Comp. & Maths. with Appl.*, **6**, 79–97, 1980.
12. L. A. Schmit and R. L. Fox, An integrated approach to structural synthesis and analysis, *AIAA Journal*, **3**(6), 1104–12, 1965.
13. M. R. Hestenes, Multiplier and gradient method, *J. Optimization Theory and Applications*, **4**(5), 303–20, 1969.
14. M. J. D. Powell, A method for nonlinear constraints in minimization problems, in *Optimization* (ed. R. Fletcher), Academic Press, New York, 1969, pp. 283–98.

15. D. P. Bertsekas, Multiplier methods: a survey, *Automatica*, **12**, 133–45, 1976.
16. R. T. Rockafellar, A dual approach to solving nonlinear programming problems by unconstrained optimization, *Math. Programming*, **5**, 354–73, 1973.
17. K. Imai, Configuration optimization of trusses by the multiplier method, PhD Thesis, *Report UCLA-ENG-7842*, 1978.
18. J. B. Rosen, The gradient projection method for nonlinear programming—Part II: nonlinear constraints, *SIAM Jnl.*, **9**, 514–32, 1961.
19. R. W. H. Sargent, Reduced-gradient and projection methods for nonlinear programming, in *Numerical Methods for Constrained Optimization*, (eds. Gill and Murray), Academic Press, pp. 149–74, 1974.
20. D. G. Luenberger, *Introduction to Linear and Nonlinear Programming*, Addison Wesley, 1965.
21. G. Zoutendijk, *Methods of Feasible Directions*, Elsevier, Amsterdam, 1960.
22. J. E. Kelley, The cutting-plane method for solving complex programs, *SIAM J.*, **8**(4), 703–12, 1960.
23. R. E. Griffith and R. A. Stewart, A nonlinear programming technique for the optimization of continuous processing systems, *Mangmt Sci.*, **7**, 379–92, 1961.
24. M. J. D. Powell, Algorithms for nonlinear constraints that use Lagrangian functions, *Math. Progr.*, **14**, 224–48, 1978.

Foundations of Structural Optimization: A Unified Approach
Edited by A. J. Morris
© 1982 John Wiley & Sons Ltd.

Chapter 10
Reconciliation of Mathematical Programming and Optimality Criteria Methods

10.1 INTRODUCTION

Essentially two main approaches have been explored in this book which claim to solve the structural optimization problem, namely, the mathematical programming (MP) methods of Chapters 8 and 9 and the optimality criteria (OC) approaches discussed in Chapter 5 and 6. In this chapter we show that these two classical schools of thought have now reached a stage where they use the same basic principles and that their apparent opposition can be interpreted in many respects as a conflict of primal versus dual methods.

The optimality criteria approach can be viewed as transforming the initial problem into a sequence of simple explicit problems in which the constraints are approximated from virtual work considerations. Solving the explicit subproblems via a dual solution scheme leads to a generalized optimality criterion, which inherently contains a mechanism for selecting the set of active constraints.

Then, based on a primal solution scheme, a mixed method is described, which permits a continuous transition between a strict mathematical programming method and a pure optimality criteria technique. The generalized optimality criterion is shown to be equivalent to a mathematical programming linearization method using the reciprocals of the design variables. Having established this connection some other relationships between mathematical programming and optimality criteria approaches are also discussed (approximation concepts approach, constraint gradients, stress ratioing).

After describing primal and dual optimization algorithms, several examples of application to various structures are offered. Each example specifically serves as a basis for providing further developments of structural optimization methods:

hybrid optimality criterion, treatment of frequency constraints, consideration of bending elements, introduction of discrete design variables, large-scale problems.

10.1.1 The Structural Optimization Problem

The structural optimization problem considered in this chapter consists of the weight minimization of a finite element model with fixed geometry and material properties. The design variables are the transverse sizes of the structural members, namely the cross-sectional areas of bar or beam elements and the thicknesses of membrane or plate elements. The constrained minimization problem has the mathematical form:

$$\text{minimize} \quad W(x) = \sum_{i=1}^{n} w_i x_i \quad (10.1)$$

$$\text{subject to} \quad h_j(x) \geqslant 0 \quad j = 1, \ldots, m \quad (10.2)$$

$$\bar{x}_i \geqslant x_i \geqslant \underline{x}_i \quad i = 1, \ldots, n \quad (10.3)$$

The objective function $W(x)$ is the structural weight. It is a linear function of the design variables x_i, because the weight coefficients w_i are constant quantities when geometry and material properties are fixed. The inequalities (10.2) represent behaviour constraints, which impose limitations on quantities describing the structural response, for example, the stresses and the displacements under multiple static loading cases, the natural frequencies, the buckling loads, etc. The design variables are also subjected to the side constraints (10.3), where $\underline{x}_i$ and $\bar{x}_i$ are lower and upper limits that reflect fabrication and analysis validity considerations.

The structural optimization problem (10.1)–(10.3) is a nonlinear mathematical programming problem to which standard minimization techniques can be applied (see Chapter 9). However, this problem exhibits some characteristics that make it complicated when practical structural design applications are considered. The main difficulty arises from the fact that the behaviour constraints (10.2) are in general implicit functions of the design variables and their precise numerical evaluation for a particular design requires a complete finite element analysis. Since the solution scheme is essentially iterative, it involves a large number of structural reanalyses. Therefore the computational cost often becomes prohibitive when large structural systems are dealt with.

In the last decade, two main approaches have been used to solve this problem. One is based on the many rigourous numerical methods of nonlinear mathematical programming (see Chapters 8, 9) The other uses the more intuitive concept of optimality criteria (see Chapters 5, 6). These approaches have often been opposed in the past and two corresponding schools were developed with apparently contradictory viewpoints. The advantages claimed for the mathematical programming methods are sound foundations, guaranteed convergence properties, and generality which permit consideration of any type of constraints. The essential disadvantage lies in the computing time, which increases rapidly with the size of the problem leading to unacceptable cost even for relatively simple problems. On the other hand, in optimality criteria approaches, the number of reanalysis cycles does not increase with the number of design variables and is generally very small. The main drawbacks are a lack of generality and of sound mathematical foundations, which explains the often unpredictable convergence properties and even convergence to non-optimal solutions.

Based on the results obtained in previous work [1–3], this chapter will attempt to reconcile these two schools of thought. It will be shown that the mathematical programming (MP) and optimality criteria (OC) approaches, far from being ineluctably opposed, have in fact converged toward the same method. This unifield approach proceeds by transforming the original problem (10.1)–10.3) into a sequence of explicit subproblems. The dual form of the explicit problem expressed in terms of the Lagrangian multipliers generalizes the conventional OC redesign relations. Its primal form in terms of the reciprocal design variables leads to a mixed method, which permits controlling the convergence of the optimization process through a continuous transition between pure MP and pure OC approaches.

10.2 THE GENERALIZED OPTIMALITY CRITERION (DUAL SOLUTION SCHEME)

Most of the optimality criteria techniques are based on the consideration of a statically determinate truss subject to stress and displacement constraints, in which case the behaviour constraints (10.2) can be written

$$h_j(x) \equiv \bar{u}_j - u_j(x) \geqslant 0 \tag{10.4}$$

where $\bar{u}_j$ denotes an upper bound to a response quantity $u_j(x)$ (stress, nodal displacement, relative displacement, etc.). As explained previously in this book, by using virtual-load considerations, the following explicit forms of the behaviour constraints can be generated:

$$\tilde{h}_j(x) \equiv \bar{u}_j - \sum_{i=1}^{n} \frac{Q_{ij}}{x_i} \geqslant 0 \tag{10.5}$$

where the 'flexibility coefficients' Q_{ij} defined by (5.7) are related to virtual energy densities in the structural members. They are constant for a statically determinate truss, so that problem (10.1)–(10.3) takes on an exact explicit form:

minimize $\quad W(x) = \sum_{i=1}^{n} w_i x_i \tag{10.6}$

subject to $\quad \bar{u}_j - \sum_{i=1}^{n} \frac{Q_{ij}}{x_i} \geqslant 0 \quad j = 1, \ldots, m \tag{10.7}$

$\quad\quad\quad\quad\quad \bar{x}_i \geqslant x_i \geqslant \underline{x}_i \quad i = 1, \ldots, n \tag{10.8}$

In the case of a statically determinate truss, the minimum-weight design can be obtained after one structural analysis as the solution of the explicit problem (10.6)–(10.8). However, for a general statically indeterminate structure, the coefficients Q_{ij} depend implicitly on the design variables and the redesign process must be employed iteratively. Therefore the OC approach can be interpreted as transforming the original problem (10.1)–(10.3) into a sequence of explicit approximate problems (10.6)–(10.8). The basic assumption is that internal force redistribution is moderate enough to ensure convergence.

Instead of solving each approximate problem by employing an MP algorithm, an alternative approach, which is typical of the OC philosophy, is to use its explicit character in order to express analytically the optimal design variables. This can be achieved through the use of the Kuhn–Tucker conditions (7.4)–(7.6), where the Lagrangian multipliers λ_j will be associated to the behaviour constraints (10.7) (and not to the side constraints). These conditions lead to a generalized OC yielding explicitly the design variables in terms of the Lagrangian multipliers:

free (active) design variables:

$$x_i = \left(\frac{1}{w_i}\sum_{j=1}^{m} Q_{ij}\lambda_j\right)^{1/2} \quad \text{if} \quad w_i \underline{x}_i^2 < \sum_{j=1}^{m} Q_{ij}\lambda_j < w_i \bar{x}_i^2 \quad (10.9)$$

fixed (passive) design variables:

$$x_i = \underline{x}_i \quad \text{if} \quad \sum_{j=1}^{m} Q_{ij}\lambda_j \leqslant w_i \underline{x}_i^2 \quad (10.10)$$

$$x_i = \bar{x}_i \quad \text{if} \quad \sum_{j=1}^{m} Q_{ij}\lambda_j \geqslant w_i \bar{x}_i^2 \quad (10.11)$$

where the λ_j must remain non-negative.

active constraints:

$$\lambda_j \geqslant 0 \quad \text{if} \quad \sum_{i=1}^{n} \frac{Q_{ij}}{x_i} = \bar{u}_j \quad (10.12)$$

inactive constraints:

$$\lambda_j = 0 \quad \text{if} \quad \sum_{i=1}^{n} \frac{Q_{ij}}{x_i} < \bar{u}_j \quad (10.13)$$

The subdivision of the design variables into active and passive groups is classical in the OC approaches [4].

When the optimal Lagrangian multipliers satisfying (10.12)–(10.13) are known—and, in fact, this is the central problem—the corresponding optimal design variables can be readily computed using the explicit OC relations

(10.9)–(10.11). Therefore the problem has been replaced with a new one, which is defined in terms of the Lagrangian multipliers only. To solve this new problem, the conventional OC techniques usually make the assumption that the set of active behaviour constraints is known *a priori*, thus avoiding the non-negativity constraints on the Lagrangian multipliers appearing in (10.12)–(10.13). An update procedure for the retained Lagrangian multipliers is then employed, so that the optimal design variables can be sought iteratively by coupling the update procedure and the explicit OC defined in (10.9)–(10.11). A frequently used technique [5, 6] is to solve the system of nonlinear equations $u_j(\lambda) = \bar{u}_j$ by resorting to the Newton–Raphson method, which leads to the following update procedure:

$$\lambda_j^{(k+1)} = \lambda_j^{(k)} + [H^{(k)}]^{-1} \tilde{h}^{(k)} \qquad (10.14)$$

where

$$H_{jk} = \frac{\partial u_j}{\partial \lambda_k} = -\frac{1}{2} \sum_i \frac{Q_{ij} Q_{ik}}{w_i x_i^3} \qquad (10.15)$$

The process consists thus in applying (10.14) recursively to obtain new estimates of the Lagrangian multipliers from which the design variables and the corresponding constraint values are computed by (10.9)–(10.11) and (10.5), respectively.

As first noted by Kiusalaas [7], the essential difficulties involved in applying these OC methods are those associated with identifying the correct active constraint set and the proper corresponding set of passive members. These difficulties were also recognized by Berke and Khot [8]. They were addressed with varying degrees of success in studies such as those reported in Refs [9, 10]. However, it was only with the advent of the dual formulation set forth in Refs [2, 11, 12] that these obstacles were conclusively overcome.

10.2.1 The Dual Formulation

Corresponding to the primal minimization problem (10.6)–(10.8), the dual maximization problem exhibits the following form (see Section 7.2):

$$\text{maximize} \quad l(\lambda) = \sum_{i=1}^{n} w_i x_i(\lambda) - \sum_{j=1}^{m} \lambda_j \tilde{h}_j [x(\lambda)] \qquad (10.16)$$

$$\text{subject to } \lambda_j \geq 0 \quad j = 1, \ldots, m \qquad (10.17)$$

Because the primal problem is separable, the expression (7.31) of the dual function can be employed to find $x_i(\lambda)$. It turns out that each one-dimensional minimization problem appearing in (7.31), namely:

$$\min_{\underline{x}_i \leq x_i \leq \bar{x}_i} \left[w_i x_i + \frac{1}{x_i} \sum_{j=1}^{m} \lambda_j Q_{ij} \right] \qquad (10.18)$$

can be solved in closed form, yielding explicitly the primal variables x_i in terms of the dual variables λ_j. As would be expected, $x_i(\lambda)$ is given by the OC equations (10.9)–(10.11). Therefore the dual function (10.16) is known explicitly and it can be readily maximized.

To relate the dual formulation with the OC techniques, we recall that the first derivatives of the dual function are given by minus the primal constraints (see Equation (7.25)), that is:

$$\frac{\partial l}{\partial \lambda_j} = -\tilde{h}_j(x) = \sum_{i=1}^{n} \frac{Q_{ij}}{x_i} - \bar{u}_j \qquad (10.19)$$

Hence it appears that (10.12)–(10.13) can be interpreted as the optimality conditions for the dual problem (see Equations (7.9)–(7.11)), namely:

$$\frac{\partial l}{\partial \lambda_j} = 0 \quad \text{if} \quad \lambda_j \geqslant 0 \qquad (10.20)$$

$$\frac{\partial l}{\partial \lambda_j} < 0 \quad \text{if} \quad \lambda_j = 0 \qquad (10.21)$$

Therefore, solving the system of equations (10.12)–(10.13) to find the optimal Lagrangian multipliers is equivalent to maximizing the dual function. In other words, the dual methods can be interpreted as using an update procedure for the Lagrangian multipliers, exactly like the OC techniques. It is worth mentioning that the matrix given in (10.15) can be identified as the Hessian of the dual function, i.e. $\mathbf{H} = \nabla^2 l$. This shows that the update procedure (10.14) corresponds to maximizing the dual function by the Newton method.

As explained in Section 7.2, the second derivatives of the dual function are discontinuous whenever an active design variable becomes passive, or inversely (the summation in (10.15) must actually be taken on the active variables only). From the relations (10.10)–(10.11) it is apparent that these discontinuities occur on hyperplanes in the dual space given by

$$\sum_{j=1}^{m} Q_{ij} \lambda_j = w_i \underline{x}_i^2 \qquad (10.22)$$

and

$$\sum_{j=1}^{m} Q_{ij} \lambda_j = w_i \bar{x}_i^2 \qquad (10.23)$$

Consequently the dual space is partitioned into several domains separated by these second-order discontinuity planes. In each domain the set of active design variables remains constant. However, when passing from one domain to another, across a discontinuity plane, this set is modified and the second derivatives of the dual function change abruptly.

From the foregoing interpretation of the dual problem, it is clear that the update procedure on the Lagrangian multipliers involved in OC approaches must be carefully devised so as to correctly select the sets of passive/active design variables and to identify the set of critical behaviour constraints. A simple numerical scheme such as (10.14)–(10.15), based on the discontinuous matrix **H**, is obviously not sufficient to avoid numerical difficulties (singularity of the **H** matrix, detection of zero Lagrangian multipliers, erratic convergence, etc.). Efficient dual algorithms that resolve these crucial difficulties will be discussed in Section 10.5. Since the dual maximization problem (10.16)–(10.17) is quasi-unconstrained and explicit, its exact solution can be generated at a low computational cost, which is comparable to that required by the recursive techniques of conventional OC approaches. The dual algorithms can handle a large number of inequality constraints and they intrinsically contain a rational scheme for identifying the active behaviour constraints through the non-negativity constraints on the Lagrangian multipliers (or dual variables). They also automatically sort out the active and passive design variable groups by using the explicit relationships between primal and dual variables.

Example A two-dimensional problem is used to illustrate graphically the concept of dual formulation. The two-bar truss schematized in Fig. 10.1 is subjected to two displacement constraints (plus the usual side constraints). With appropriate numerical values, the structural optimization problem takes on the following explicit form (a_1 and a_2 denoting here the bar cross-sectional areas):

$$\text{minimize} \quad W = a_1 + a_2$$

$$\text{subject to} \quad \frac{1}{a_1} + \frac{1}{a_2} \leq \frac{3}{2} \quad (10.24)$$

$$\frac{1}{a_1} - \frac{1}{a_2} \leq \frac{1}{2} \quad (10.25)$$

$$1 \leq a_1 \leq 2$$
$$1 \leq a_2 \leq 2$$

The design space is represented in Fig. 10.1. The optimum design corresponds to the point $(4/3, 4/3)^T$, where only one constraint is active (horizontal displacement $u = 3/2$). The dual problem statement involves two Lagrangian multipliers λ_1 and λ_2 associated with the behaviour constraints (10.24) and (10.25) respectively. The discontinuity planes defined in (10.22), (10.23) are illustrated in Fig. 10.2 and they correspond to

$$\lambda_1 + \lambda_2 = 1 \quad \lambda_1 + \lambda_2 = 4$$
$$\lambda_1 - \lambda_2 = 1 \quad \lambda_1 - \lambda_2 = 4 \quad (10.26)$$

They subdivide the dual space into six domains. In each of them, the expression of the dual function in terms of the dual variables λ_1, λ_2 is different and is given in

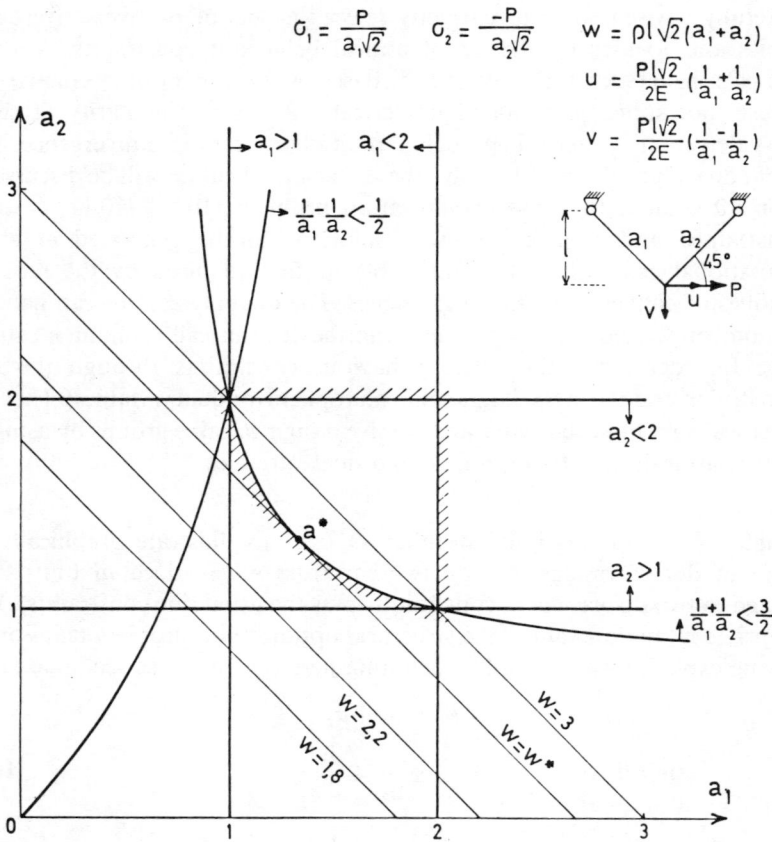

Fig. 10.1 Two-bar truss—primal problem

explicit form in Table 10.1. The maximum in the feasible dual space occurs in the subdomain II, for $\lambda_1 = 16/9$ and $\lambda_2 = 0$, corresponding to $l(\lambda_1, \lambda_2) = 8/3$. A one-dimensional maximization is illustrated in Fig. 10.3 by assuming that the search direction $(1, 0)^T$ has been selected at the dual point $(0, \frac{1}{2})^T$. The continuity properties of the dual function and its derivatives along a line are also indicated.

10.3 THE MIXED METHOD (PRIMAL SOLUTION SCHEME)

In contrast to the intuitive OC techniques, the MP approach attacks directly the nonlinear problem (10.1)–(10.3) by making use of classical minimization techniques. As indicated in Chapter 9, such an approach requires the frequent reanalysis of the structure in order to evaluate the behaviour constraints and their gradients at each new step in the iterative optimization process. In this section, we

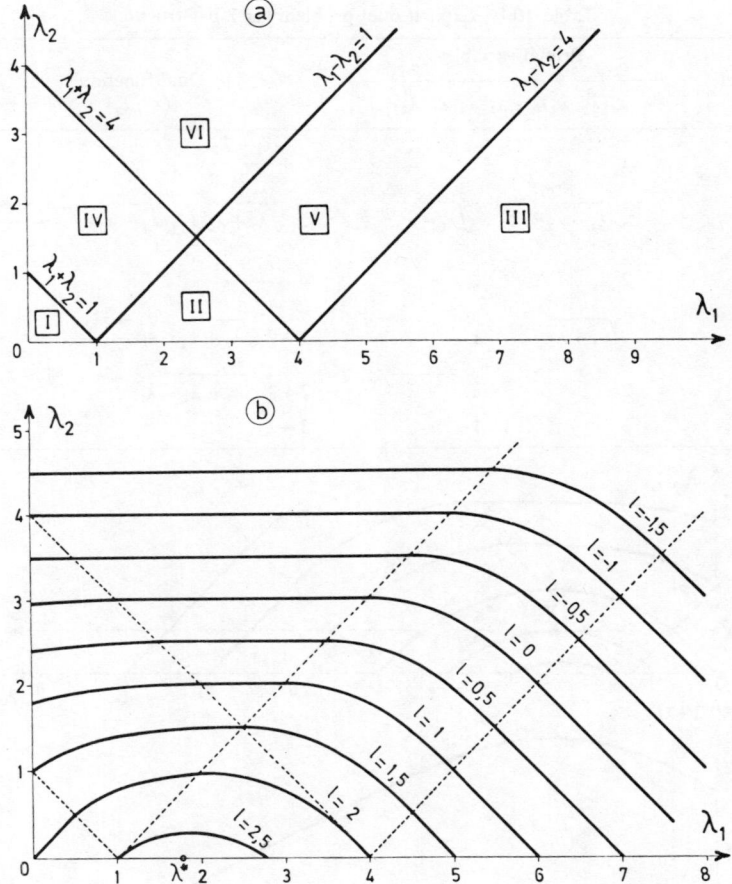

Fig. 10.2 Two-bar truss—dual problem (a) Discontinuity planes and subdomains (b) Contours of dual function

start with a projection method for nonlinear constraints, which permits generation of a sequence of steadily improving feasible designs (primal method). Subsequently, we show how the number of structural reanalyses can be reduced, by employing first-order Taylor series expansion of the behaviour constraints. The resulting approach can be viewed as a mixed primal/linearization method [13, 14].

As schematized in Fig. 10.4, the gradient projection method for nonlinear constraints proceeds as follows:

(i) Given a feasible starting point, evaluate the gradients of the objective function and the constraints; for a structural optimization problem, this step calls for a finite element analysis.

Table 10.1 Explicit dual problem for 2-bar truss

Definition domain	Primal variables		Dual function $l(\lambda_1, \lambda_2)$
	$a_1(\lambda_1, \lambda_2)$	$a_2(\lambda_1, \lambda_2)$	
I	1	1	$2 + \dfrac{\lambda_1 - \lambda_2}{2}$
II	$\sqrt{\lambda_1 + \lambda_2}$	$\sqrt{\lambda_1 - \lambda_2}$	$2(\sqrt{\lambda_1 + \lambda_2} + \sqrt{\lambda_1 - \lambda_2}) - \dfrac{3\lambda_1 + \lambda_2}{2}$
III	2	2	$4 - \dfrac{\lambda_1 + \lambda_2}{2}$
IV	$\sqrt{\lambda_1 + \lambda_2}$	1	$1 + 2\sqrt{\lambda_1 + \lambda_2} - \dfrac{\lambda_1 + 3\lambda_2}{2}$
V	2	$\sqrt{\lambda_1 - \lambda_2}$	$2 + 2\sqrt{\lambda_1 - \lambda_2} - \lambda_1$
VI	2	1	$3 - \lambda_2$

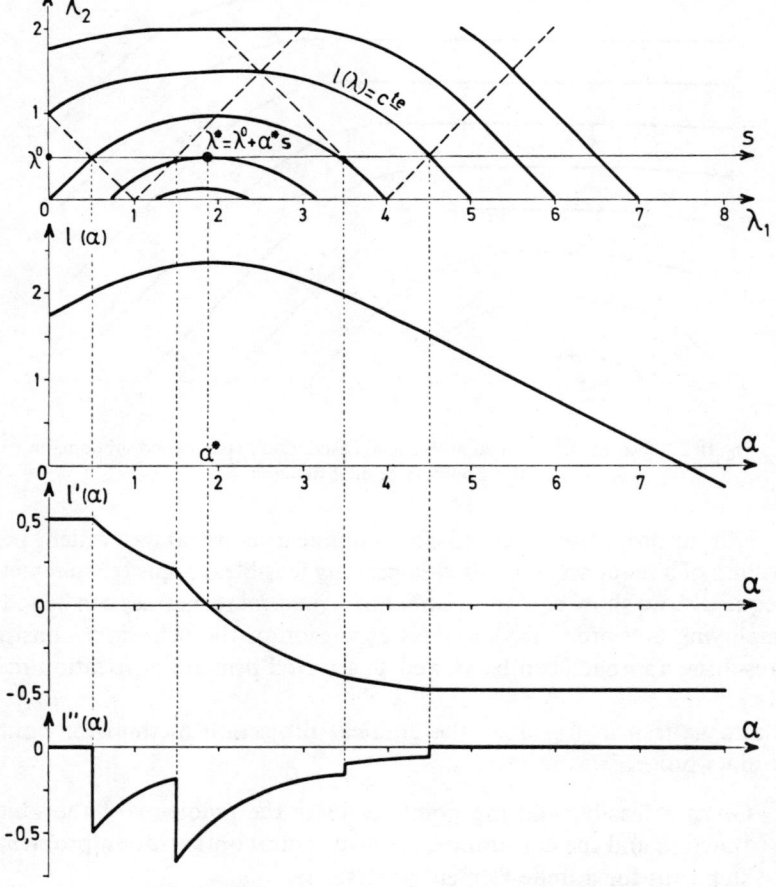

Fig. 10.3 Two-bar truss—line search in dual space

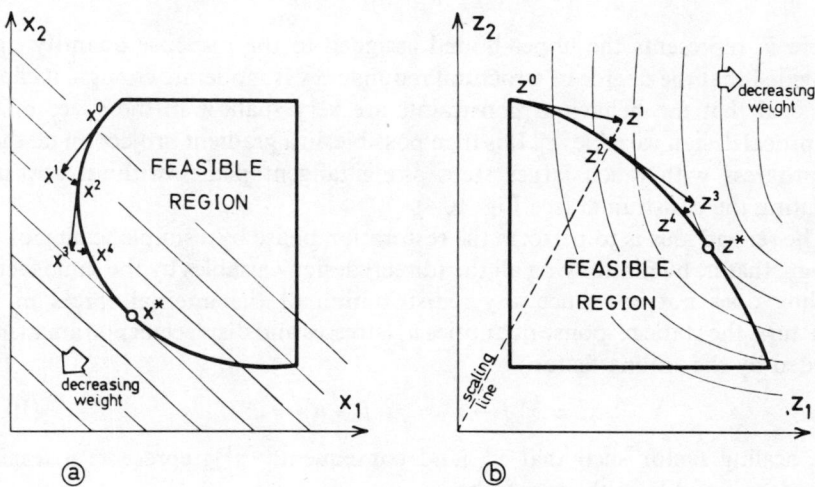

Fig. 10.4 Projection method (a) Direct variables (b) Reciprocal variables

(ii) Determine a search direction by projecting the objective function gradient on the plane tangent to the active constraints.
(iii) Compute the step length which minimizes the objective function along the search direction without violating any currently inactive constraint; this line search necessitates iteratively evaluating a maximal allowable step length, which implies a certain number of structural reanalyses.
(iv) Because the constraints are nonlinear, this minimization phase must be followed by a restoration phase that moves the design point back to the boundary of the feasible domain; this step also requires several function evaluations.
(v) After restoration, the design point is feasible and the whole process can be repeated by returning to step (i).

It is apparent that this projection method is very expensive, due to the need to evaluate the constraints and their gradients several times at each iteration. In addition, it is well known that in such an MP approach the number of iterations grows with the number of design variables. Therefore the computational cost may become prohibitively high for large-scale problems.

However, it is possible to construct an efficient projection method by taking three distinct actions. The first idea is to make the change of variables

$$z_i = \frac{1}{x_i} \tag{10.27}$$

whose effect is to lessen the nonlinear character of the static response quantities. The behaviour constraints will now be written (keeping for simplicity the same notation h and u as in Equation (10.4)):

$$h_j(z) \equiv \bar{u}_j - u_j(z) \geq 0 \qquad (10.28)$$

where $\bar{u}_j$ represents the upper bound assigned to the response quantity $u_j(z)$. Provided that the degree of structural redundancy is moderate enough, it can be expected that the behaviour constraints are very shallow in the space of the reciprocal design variables z_i. It is then possible, in a gradient projection method, to progress with much larger steps along tangent planes without seriously violating the constraints (see Fig. 10.4).

The second idea is to perform the restoration phase by a simple scaling of the design, that is, by multiplying all the (direct) design variables by the same factor. Scaling does not introduce any redistribution of the internal forces in the structure; the static response quantities u_j (stresses and displacements) are merely divided by the scaling factor f:

$$x_i^1 = x_i^0 f \rightarrow z_i^1 = z_i^0/f \rightarrow u_j^1 = u_j^0/f \qquad (10.29)$$

The scaling factor such that x^1 (and consequently z^1) represents a feasible boundary point is easily seen to be

$$f = \max\left\{\frac{u_j^0}{\bar{u}_j}\right\} \qquad (10.30)$$

In this way, the multiple structural reanalyses that were required in the restoration phase are avoided. Geometrically, scaling corresponds to a move along the straight line joining the origin to the current design point z^o, where the structural analysis is made (see Fig. 10.4).

The aim of the third idea is to avoid the structural reanalyses that are normally needed in the minimization phase (step (iii)). The line search will be made approximately, on the basis of the linearized forms of the behaviour constraints (10.28) (first-order Taylor series expansion):

$$\tilde{h}_j(z) \equiv \bar{u}_j - \left[u_j^0 + \sum_{i=1}^{n} \left(\frac{\partial u_j}{\partial z_i}\right)^0 (z_i - z_i^0)\right] \geq 0 \qquad (10.31)$$

It should be recalled that, in the space of the reciprocal variables z_i, the constraints h_j are close to planes and therefore the linearized form $\tilde{h}_j$ usually provide very good approximations.

In the modified projection method, only one structural analysis has to be accomplished at each iteration (step (ii)). However, with this method the number of iterations required to locate the optimum still increases with the number of design variables. We are, therefore, forced to adopt another idea. Since the linearized behaviour constraints (10.31) are in general accurate explicit approximations we can use them not only to evaluate the step length but also to generate a certain number of new search directions. A new structural analysis will then be needed only periodically, in order to update the linearized forms of the constraints. The resulting minimization phase now operates on the following explicit problem, obtained from the initial problem by replacing the behaviour constraints by approximate forms (10.31):

$$\text{minimize} \quad W(z) = \sum_{i=1}^{n} \frac{w_i}{z_i} \tag{10.32}$$

$$\text{subject to} \quad \tilde{h}_j(z) \geqslant 0 \tag{10.33}$$

$$\bar{z}_i \geqslant z_i \geqslant \underline{z}_i \tag{10.34}$$

We call this the *linearized problem*. If $\bar{k}$ denotes the number of search directions evaluated before reanalysing the structure, the minimization phase amounts to solving the linearized problem by the gradient projection algorithm for linear constraints (Section 9.5), but stopping this algorithm after $\bar{k}$ iterations. Only at this point is the structure reanalysed and the restoration phase is then performed through the scaling process (10.29). The linearized problem is reformed about this solution point and a new partial solution found.

Provided that the number of steps $\bar{k}$ is chosen sufficiently small, the mixed method should produce a sequence of feasible designs with decreasing values of the structural weight (primal philosophy). An alternative viewpoint is to recognize that the approximation made by linearizing the constraints with respect to the reciprocal design variables is of such high quality that the current explicit problem (10.32)–(10.34) can be solved exactly, rather than partially, after each structural reanalysis. This idea leads to abandoning the primal philosophy in favour of a pure linearization approach, which does not necessarily produce a sequence of steadily improving feasible designs. Therefore, depending upon the value of the number of steps $\bar{k}$, the mixed method behaves as a primal projection method ($\bar{k} = 1$), with high cost but guaranteed convergence, or as a linearization method ($\bar{k} = \infty$), with fast but uncertain convergence†. When $\bar{k}$ is limited to a given finite number, the method can be interpreted as a mix between primal and linearization approaches. This leads one to consider the number of steps $\bar{k}$ as a convergence control parameter that should be assigned high values for economy and reduced when divergence occurs.

10.3.1 Example We turn again to the 10-bar truss shown in Fig. 10.5 where the constraints impose lower bounds on the bar cross-sectional areas, maximal allowable stress limits, as well as upper bounds on the vertical displacements. The convergence curves represented in Fig. 10.5 give the variation of the structural

---

† Note that in conventional linearization methods the objective function is also linearized; this is not necessary in the present approach because the weight is a simple explicit function.

376   *Foundations of Structural Optimization: A Unified Approach*

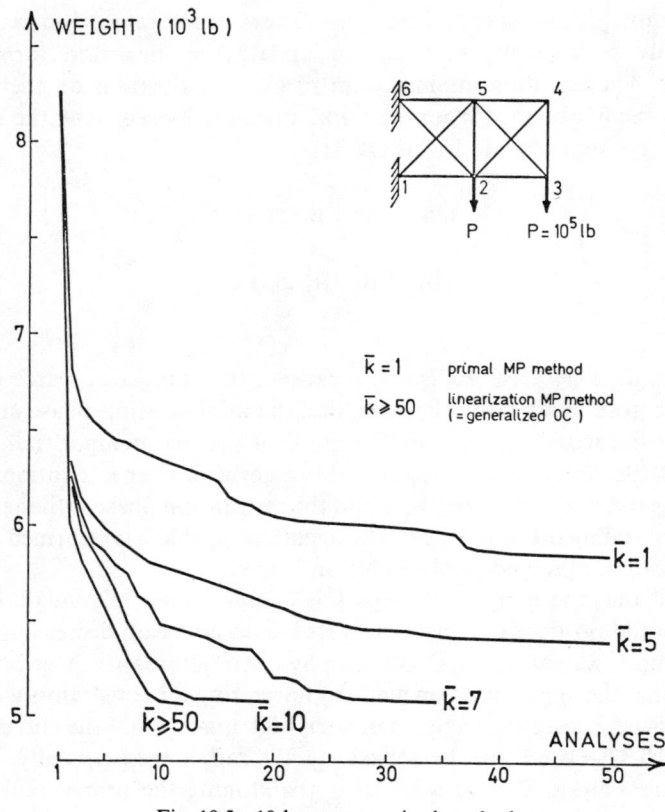

Fig. 10.5   10-bar truss—mixed method

weight as a function of the number of reanalyses for various values of $\bar{k}$ from 1 to 50. As expected, the choice of the parameter $\bar{k}$ largely affects the convergence properties of the mixed method. The case $\bar{k} = 1$ yields the pure projection method initially considered in this section. It requires a large number of structural reanalyses. When $\bar{k}$ is increased, the number of reanalyses is reduced. The limiting case $\bar{k} \geqslant 50$, which corresponds to solving exactly the linearized problem, produces the best results (pure linearization method).

10.4   RELATIONS BETWEEN OC AND MP APPROACHES

10.4.1   The Generalized OC as a Linearization Method

Continuing on the 10-bar truss example, Fig. 10.6 represents the iteration history produced by various OC techniques. Conventional OC approaches suffer from instability in the convergence process. However, the generalized OC (Section

10.2), as well as another recent OC technique [6], rapidly furnish the optimum design, with exactly the *same iteration history* as the mixed method with $\bar{k} \geq 50$ (Fig. 10.5), which is a MP linearization method.

In fact, it can be shown that the explicit approximations (10.5), derived in the OC approach by neglecting internal force redistribution, are identical to the linearized forms (10.31) of the behaviour constraints [13–15]. Indeed, the virtual strain energy densities Q_{ij} employed in OC techniques are nothing else than the gradients of the response quantities with respect to the reciprocal design variables:

$$Q_{ij} = \frac{\partial u_j}{\partial z_i} \tag{10.35}$$

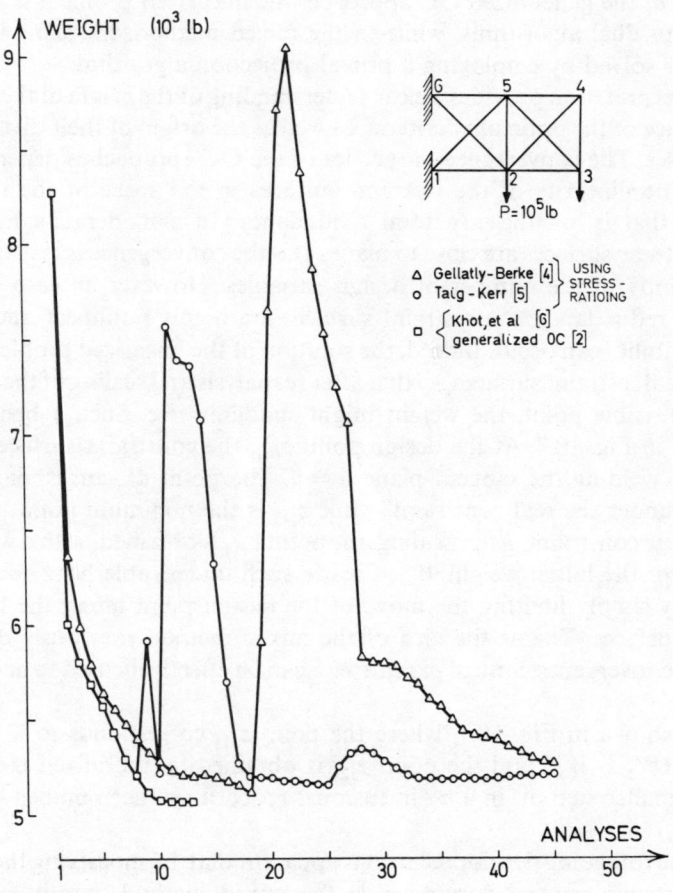

Fig. 10.6 10-bar truss—conventional and generalized optimality criteria

Furthermore, the definition of the Q_{ij} following from the virtual load technique clearly indicates that

$$u_j^0 = \sum_{i=1}^{n} Q_{ij}^0 z_i^0 \qquad (10.36)$$

Therefore (10.31) can be rewritten

$$\tilde{h}_j(z) \equiv \bar{u}_j - \sum_{i=1}^{n} Q_{ij}^0 z_i \geqslant 0 \qquad (10.37)$$

which is equivalent to (10.5), when restated in terms of the design variables x_i. In other words, the generalized OC approach can be defined in the MP terminology as a linearization method, because it proceeds by replacing the original problem (10.1)–(10.3) with a sequence of linearized problems of the form (10.32)–(10.34). Note that in the generalized OC approach, the linearized problem is solved by resorting to dual algorithms, while in the mixed method, the same linearized problem is solved by employing a primal projection algorithm.

This interpretation provides a clear understanding of the origin of the excellent performance of the optimality criteria, as well as the origin of their divergence in certain cases. The convergence properties of the OC approaches depend clearly upon the nonlinearity of the restraint surfaces in the space of the reciprocal variables, that is, on the structural redundancy. In a moderately hyperstatic structure, these surfaces are close to planes and the convergence is fast and stable, independently of the number of design variables. However, in case of strong structural redundancy, the restraint surfaces are highly nonlinear and convergence instability can occur. Indeed, the solution of the linearized problem lies far from the real restraint surfaces, so that after reanalysis and scaling of the design to obtain a feasible point, the weight might suddenly rise. Such a behaviour is illustrated in Fig. 10.7. At the design point $z_{(0)}$, the constraint surface $u = \bar{u}$ is linearized, yielding the tangent plane $\tilde{u} = \bar{u}$. The point z^* corresponds to the optimum under the real constraint, while $z_{(1)}$ is the minimum point under the approximate constraint. After scaling, the point $z'_{(1)}$ is obtained, with a weight W_1 greater than the initial weight W_0. Clearly such undesirable behaviour can be avoided by simply limiting the move of the design point along the linearized restraint surfaces. This is the idea of the mixed method previously described, where the convergence control parameter $\bar{k}$ can be effectively used to achieve this goal.

This is shown in Fig. 10.7, where the point $z_{(1)}$ corresponds to a pure OC approach ($W_1 > W_0$) and the point $z_{(2)}$ is obtained by the mixed method, by taking a smaller step or, in a n-dimensional space, a smaller number $\bar{k}$ of steps ($W_2 < W_0$).

From the foregoing developments it is apparent that, by modifying the value of the convergence control parameter $\bar{k}$, the mixed method permits a gradual transition between a primal MP approach ($\bar{k}$ small) and an OC approach ($\bar{k}$ large).

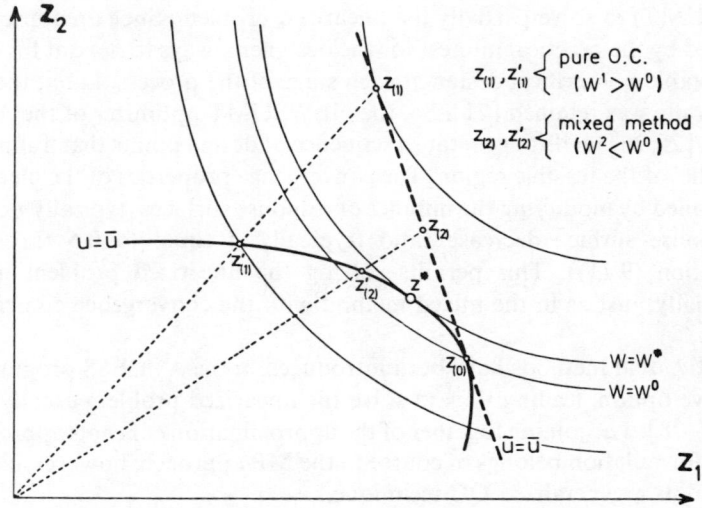

Fig. 10.7 Convergence of the optimality criteria and mixed methods

10.4.2 The Approximation Concepts Approach

In the context of the MP approach to structural optimization, the type of method previously described seems the only one to be efficient. The idea of linearizing the constraints with respect to the reciprocal variables was suggested by Reinschmidt et al. [16] as early as 1966. When looking at the history of the MP approach, initiated by Schmit in 1960 in a now famous paper [17], it appears that, after a long period of inefficiency, it has finally evolved into a powerful design procedure based on approximation concepts. Although developed independently of the mixed method, the approximation concepts approach of Schmit et al. [18–21] also relies on approximating the behaviour constraints by first-order Taylor series expansion with respect to the reciprocal design variables. Its origin can be traced back to the numerical investigation of Storaasli and Sobieszczanski [22] with regard to the accuracy of first-order Taylor series.

First, Schmit and Farshi [18] resorted to the method of inscribed hypersphere [23] in the space of the reciprocal variables, which led them to linearize all the functions involved in the problem statement, including the objective function. Subsequently, Schmit and Miura [19] recognized the importance of linearizing only the behaviour constraints and therefore they also considered the explicit problem (10.32)–(10.34), where the objective function keeps its nonlinear, but simple and explicit, form. In order to maintain a primal philosophy (a sequence of steadily improving feasible designs), Schmit and Miura [19, 20] employed either a feasible direction method (CONMIN [24]) or a barrier function method

(NEWSUMT) to solve partially the linearized problem. Such an approach was motivated by the practical interest for the designer always to have at his disposal an acceptable non-critical design at each stage of the process. Later, the barrier method only was retained [21]. So, the NEWSUMT optimizer of the ACCESS program [25, 26] tends to generate a sequence of design points that 'funnel down the middle' of the feasible region. The convergence properties of the method can be controlled by modifying the number of response surfaces (typically 1 or 2) and the response surface decrease ratio (typically, f equal to 0.5 through 0.1 in Equation (9.11)). This permits solving the linearized problem more or less partially, just as in the mixed method with the convergence control parameter K.

Recently, dual methods have been introduced in the ACCESS program as an alternative option, leading thus to solve the linearized problem exactly at each stage [27–28]. The joining together of the approximation concepts approach and the dual formulation belongs of course to the MP approach; however, it can also be viewed as a generalized OC technique.

10.4.3 The Constraint Gradients

In many MP approaches, the pseudo-loads technique is used to compute the gradients of the behaviour constraints (see Ref. [29], p. 242). This procedure requires a number of additional loading cases equal to the number of design variables times the number of applied loading cases. On the other hand, the virtual-load technique employed in OC approaches provides another way of evaluating the constraint gradients. The number of virtual-load cases to be introduced in the structural analysis phase is equal to the number of behaviour constraints retained in the linearized problem statement. As shown, for example, in Ref. [1], these two procedures can be analytically related to each other by taking account of the symmetry of the stiffness matrices. They also have been related to state-space methods [30].

It is interesting to note that again a primal versus dual opposition appears in the number of additional loading cases. For OC techniques it is equal to the number of dual variables, while, for MP methods, it is proportional to the number of primal variables.

10.4.4 Stress-ratioing

The generalized OC, as well as the mixed method and the approximation concepts approach, use a linearized expression for each (retained) behaviour constraint. However, in many conventional OC methods, only the displacement constraints are replaced with first-order explicit approximations while the stress constraints are dealt with stress-ratioing, which corresponds to adopting the 'fully stressed design' (FSD) philosophy [31]. In this approach, the implicit

Mathematical Programming and Optimality Criteria Methods 381

nonlinear stress constraints $\sigma_{il} \leqslant \bar{\sigma}_i$ are transformed into simple side constraints

$$x_i \geqslant \underline{\tilde{x}}_i \tag{10.38}$$

The minimum values $\underline{\tilde{x}}_i$ are given by the well-known stress-ratio formula

$$\underline{\tilde{x}}_i = x_i^0 \max_{l} \left\{ \frac{\sigma_{il}^0}{\bar{\sigma}_i} \right\} \tag{10.39}$$

where l is the load case index and $\bar{\sigma}_i$ denotes the allowable stress. Stress-ratioing can be interpreted as using zero-order approximation of the stresses on the scaling line, because it relies on explicit expressions that preserve the values of the stresses along that line, but not their derivatives [32]. Note that the explicit forms (10.5) of the constraints provided by the virtual-load procedure are first-order approximations on the scaling line, because they yield the exact values of the constraints and their gradients all along that line (the coefficients Q_{ij} are not affected by scaling).

A geometrical interpretation of first- and zero-order approximations is shown in Fig. 10.8 in the space of the reciprocal design variables. The analysis point is denoted $\mathbf{z}^0$ and the scaling line joins this point $\mathbf{z}^0$ to the origin. When using the first-order explicit approximations (10.31) or (10.37) (generalized OC), each real restraint surface is replaced with its tangent plane at its point of intersection with the scaling line (see Fig. 10.8(a)). When the zero-order explicit approximations (10.38) are used (FSD criterion), each real restraint surface is again approximated by a plane at its point of intersection with the scaling line. However, it is no longer the tangent plane, but the plane perpendicular to the ith axis of the design space (see Fig. 10.8(b)).

Example Stress-ratioing is obviously a computationally much less expensive procedure than linearization. However, it sometimes leads to divergence of the optimization process. In the 10-bar truss example previously discussed (Fig. 10.6), conventional OC techniques using stress-ratioing suffer from convergence instability even if the explicit problem is correctly solved at each stage. This is due to the fact that the stress constraint in one of the members is very critical, and its zero-order approximation is not accurate enough. Excellent results are nevertheless obtained in Ref. [6], by replacing the allowable stress constraint in each bar with the allowable relative displacement constraint between the two nodes connecting the bar. This procedure corresponds to linearizing the stress constraints with respect to the reciprocal variables (generalized OC).

It is clear that the idea of mixed method, i.e. solving partially the explicit problem, can still be employed when the stress constraints are dealt with by stress-ratioing. Figure 10.9 shows the results generated in this way for various values of the convergence control parameter $\bar{k}$. It may be observed that, during the eight first iterations, the convergence rate improves when $\bar{k}$ increases (just as in the case where the stress constraints are linearized; see Fig. 10.5). Subsequently, however,

382 *Foundations of Structural Optimization: A Unified Approach*

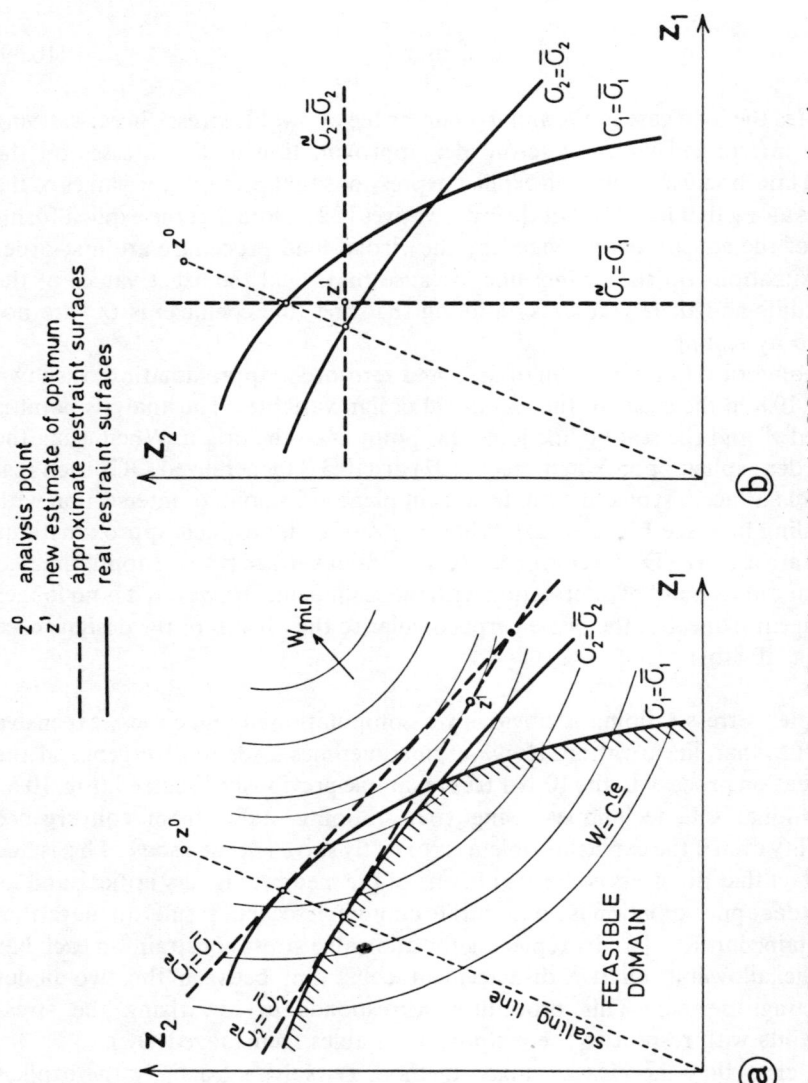

Fig. 10.8 Optimality criteria in the space of the reciprocal variables (a) GOC. First-order approximations by tangent planes (b) FSD. Zero-order approximations by normal planes

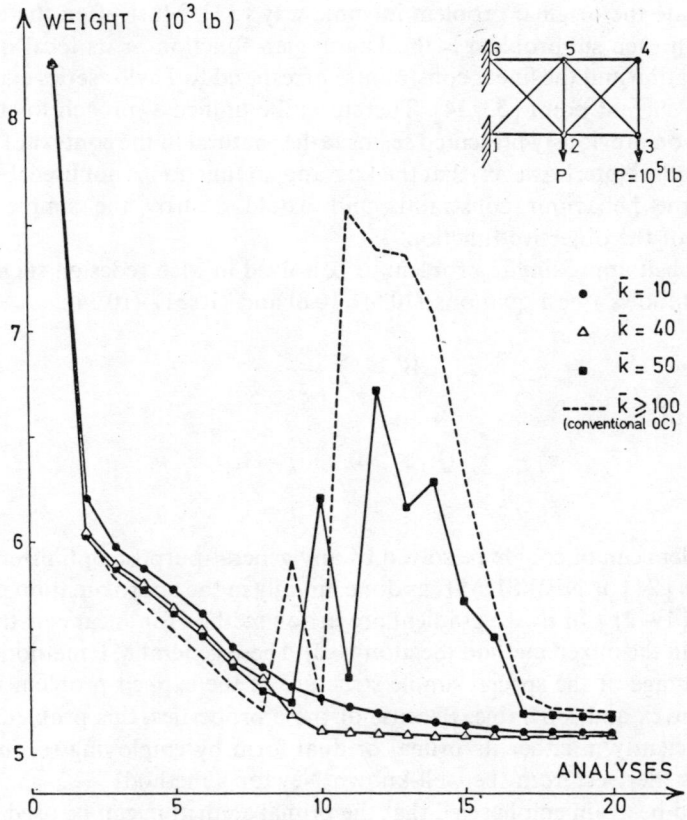

Fig. 10.9 10-bar truss—mixed method using stress-ratioing

the convergence properties of the optimization process are deteriorating for large values of $\bar{k}$, corresponding to an OC approach (the stress constraint in member 2–5 becomes too suddenly critical). By reducing $\bar{k}$, the convergence instability diminishes. The optimal choice is here $\bar{k} \simeq 40$, which yields fast and monotonic convergence. In this example, the interest of the mixed method is well demonstrated, as the conventional OC techniques, with stress-ratioing, do not normally lead to convergence. Note that even with $\bar{k} = 1$, the mixed method cannot be justified from an MP point of view, as the zero-order FSD approximations have been used. It works, however, quite well.

10.5 OPTIMIZATION ALGORITHMS

According to recent trends in the MP theory, a natural approach to a non-linear problem is to generate a sequence of linearly constrained subproblems which approximate the original problem in some way [33]. Most often the objective function in each subproblem is the Lagrangian function or its local quadratic approximation and the linear constraints correspond to Taylor series expansions about the current point [33, 34]. Therefore, the unified approach to structural optimization previously presented seems rather natural in the context of modern MP methods. Note, however, that the Lagrangian function is not used, because it involves the behaviour constraints and would destroy the simple explicit character of the objective function.

The explicit approximate problem to be solved in each redesign stage can be stated as follows (see Equations (10.6)–(10.8) and (10.31)–(10.34)):

minimize
$$W = \sum_{i=1}^{n} \frac{w_i}{z_i} \tag{10.40}$$

subject to

$$\bar{u}_j - \sum_{i=1}^{n} Q_{ij} z_i \geq 0 \qquad j = 1, \ldots, m \tag{10.41}$$

$$\bar{z}_i \geq z_i \geq \underline{z}_i \qquad i = 1, \ldots, n \tag{10.42}$$

This problem can of course be solved by any general-purpose optimizer such as CONMIN [24] or NEWSUMT, as done initially in the approximation concepts approach [19–21], or by the gradient projection method for linear constraints, as proposed in the mixed method (Section 10.3). These general MP methods do not take advantage of the special simple structure of the explicit problem which is strictly convex and separable. Because of these properties, this problem can be solved efficiently in either its primal or dual form by employing second-order algorithms (derived from the well-known Newton's method).

It should be again emphasized that the primal algorithm can be used to solve only partially the approximate problem (10.40)–(10.42), while the dual algorithm cannot, because intermediate points in the dual space usually correspond to highly infeasible points in the primal space. Thus, the capability of controlling the convergence of the overall optimization process is available in the SAMCEF [2, 35, 36] and ACCESS-3 [26, 27] programs only when a primal optimizer is selected. On the other hand, it is important to remember that using a dual algorithm yields results and convergence properties equivalent to those obtained using optimality criteria. The same is true for the primal algorithms if the approximate problems are solved completely.

10.5.1 Second-order Primal Algorithm—PRIMAL-2

PRIMAL 2 is a second-order projection algorithm especially well suited to the solution of problems with separable objective function and linear constraints. It uses a weighted projection operator of the form (8.78) to generate a sequence of Newton's search directions in the subspace formed by the intersection of the active constraint hyperplanes. Because the objective function (10.40) is separable the Hessian matrix is diagonal, which makes the second-order algorithm no more complicated than the well-known (first-order) gradient projection methods (see Section 8.5). It is worth mentioning that the algorithm can easily be modified so that only the main linear constraints (10.41) have to be introduced via the projection relations. The side constraints (10.42) can be treated separately just as explained in Section 8.8 for the first-order algorithm. The projection operator employed in PRIMAL-2 then takes the compacted form:

$$\tilde{\mathbf{P}} = \tilde{\mathbf{I}} - \tilde{\mathbf{H}}^{-1} \mathbf{A} (\mathbf{A}^T \tilde{\mathbf{H}}^{-1} \mathbf{A})^{-1} \mathbf{A}^T \qquad (10.43)$$

where $\mathbf{A}$ is defined in (8.90) and $\tilde{\mathbf{H}}$ is the compacted Hessian matrix:

$$\tilde{\mathbf{H}}_{ij} = 2 \frac{w_i}{z_i^3} \delta_{ij} \qquad i, j = 1, \ldots, \tilde{n} \qquad (10.44)$$

$\tilde{\mathbf{P}}$ is of dimension $\tilde{n}$, the number of free variables. The matrix $\tilde{\mathbf{H}}$ is diagonal, so that the only matrix to be inverted is explicitly given by

$$(\mathbf{A}^T \tilde{\mathbf{H}}^{-1} \mathbf{A})_{jk} = \frac{1}{2} \sum_{i=1}^{\tilde{n}} \frac{Q_{ij} Q_{ik} z_i^3}{w_i} \qquad (10.45)$$

The PRIMAL-2 alogrithm is described in detail in Section 6.2 of Ref. [2].

10.5.2 Second-order Dual Algorithm—DUAL-2

The dual function formulation, which exploits the separable form of the approximate problem, consists in maximizing the explicit dual function (10.16) subject to non-negativity constraints on the dual variables. DUAL-2 is a specially devised dual method which employs a second-order Newton-type of algorithm. It operates in a sequence of dual subspaces with gradually increasing dimensions, so that the operational dimensionality of the dual problem does not exceed the number of active behaviour constraints by more than one. Because this number is relatively low for many structural optimization problems of practical interest, the DUAL-2 optimizer is highly efficient.

The Newton search direction in a dual subspace can be written

$$\tilde{\mathbf{s}} = -[\tilde{\nabla}^2 l]^{-1} \tilde{\nabla} l \qquad (10.46)$$

where the tilde ($\sim$) means that the associated quantity has dimension $\tilde{m}$, the number of strictly positive dual variables. If the initial $\tilde{\nabla}^2 l$ is non-singular, and if additional non-zero dual variables are added one at a time, each subsequent $\tilde{\nabla}^2 l$ will be non-singular [2]. The iterative modification of the dual variables is then

$$\tilde{\lambda}^+ = \tilde{\lambda} + \alpha \tilde{\mathbf{s}} \qquad (10.47)$$

where α is the step length determined so that the dual function attains its maximum along the direction $\tilde{\mathbf{s}}$ in the current dual subspace. In a more recent version of the DUAL-2 algorithm, the line search is considerably simplified and most often a regular Newton unit step is taken [27, 28]. However, the existence of second-order discontinuity planes in the dual space must be taken into account (see Equations (10.22), (10.23)).

Because the primal problem is separable, the second derivatives of the dual function take on the explicit form (7.32) and are given by (10.45)[†]. This shows that in both primal and dual second-order algorithms, precisely the same matrix has to be inverted at each iteration. This matrix plays an important role in structural optimization, since it is also involved in many conventional OC techniques [5, 6] (see Equation (10.15)).

The DUAL-2 algorithm is described in more detail in Section 4.3 of Ref. [2], as well as in Section 3 of Ref. [27].

10.6 NUMERICAL EXAMPLES USING PRIMAL AND DUAL ALGORITHMS

In this section numerical results for several structural optimization problems are presented in brief summary form. Attention is focused on the comparison of various primal and dual optimizers available in the SAMCEF and ACCESS-3 computer programs, including the second-order algorithms previously described (PRIMAL-2 and DUAL-2). Detailed tabular input data and results can be found for all examples in Refs [2, 27, 36].

In addition, some advanced topics are introduced through specifically devised applications: the hybrid OC approach, the treatment of bending elements and the consideration of discrete design variables.

10.6.1 72-Bar Truss

The first example is the widely used 72-bar, four-level tower represented in Fig. 10.10. In addition to stress and minimum size constraints, displacement limits are imposed on the four uppermost nodes in the X and Y (horizontal) directions. By symmetry, the problem involves 16 independent design variables after linking. In a report describing computational experience Ref. [27], this problem was solved using the NEWSUMT option of the ACCESS-3 program, based on the approximation concepts approach. Three different pairs of values have been selected for the response factor decrease ratio and the number of response surfaces: (0.5×1), (0.3×2) and (0.1×3). Thus increasingly exact solutions are generated for each linearized problem and the approximation concepts approach changes from a pure primal solution scheme (with partial solution of each explicit problem), to a pure linearization technique (with complete solution of each explicit problem). The convergence curves of the weight

---

[†] The Hessian matrix of the Lagrangian function reduces to the Hessian matrix of the objective function because the constraints are linear.

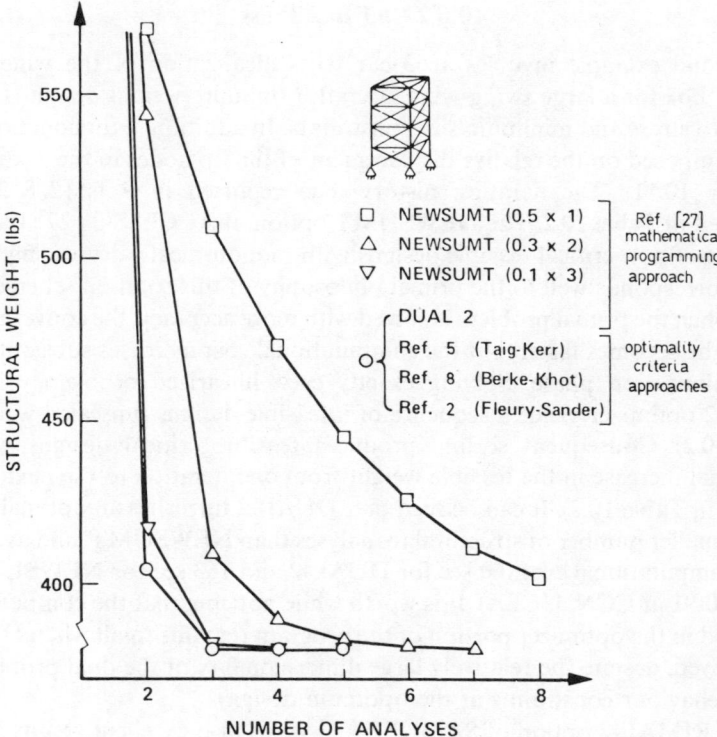

Fig. 10.10 72-bar truss example

with respect to the number of structural reanalyses are represented in Fig. 10.10. They clearly demonstrate that the more precise solutions of the linearized problems lead to faster convergence. In the limiting case where the explicit problem is solved exactly at each stage, the NEWSUMT optimizer would of course generate the same sequence of design points as the DUAL-2 optimizer.

In fact, the iteration history produced by SAMCEF (generalized OC), by ACCESS-3 (approximation concepts), and by the conventional OC techniques of Refs [5, 6, 8] are identical. These results numerically confirm that a unified approach to structural optimization has now emerged, which can be interpreted as an MP linearization method, or as a generalized OC approach.

It should be mentioned that the stress constraints are not very critical. Consequently they have been treated by stress-ratioing. It appears that this example is well adapted to the FSD approximation of the stress constraints. Any additional sophistication increases the computational cost without improving the global efficiency.

10.6.2 63-Bar Truss

The second example involves a 63-bar truss idealization of the wing carry-through box for a large swing-wing aircraft. Minimum-weight design is sought subject to stress and minimum size constraints. In addition, a torsional rotation limit is imposed on the relative displacement of the tip nodes in the X direction (see Fig. 10.11). The iteration history data reported in Refs [2, 8, 27] are compared in Table 10.2. The NEWSUMT option of ACCESS-3 [27] leads to a sequence of non-critical feasible designs with monotonically decreasing weight, which corresponds well to the primal philosophy of this solution scheme. Once again, when the primal problem is solved with more accuracy, the convergence of the weight becomes faster, but the computational cost increases substantially in the minimization phase. Solving exactly each linearized problem using the DUAL-2 optimizer yields a sequence of infeasible designs (unscaled weights in Table 10.2). Consequent scaling produces feasible critical designs with an occasional increase in the feasible weight from one iteration to the next (scaled weights in Table 10.2). It can be seen that DUAL-2 furnishes an optimal design after a smaller number of structural reanalyses than NEWSUMT, and at a much lower computational cost (60 sec for DUAL-2 and 163 sec for NEWSUMT on IBM 360-91 at CCN, UCLA). It is worth while noticing that the computer time expanded in the optimizer portion of the program remains small when DUAL-2 is employed, despite the relatively large dimensionality of the dual problem (25 active behaviour constraints at the optimum design).

The PRIMAL-2 option of SAMCEF [2] yields also excellent results for this problem. This is not surprising, since the mixed method used in SAMCEF consists here in solving almost exactly each linearized problem generated in sequence. As a result, the iteration histories produced by both computer programs are about the same. It should be noted that the good performance of the

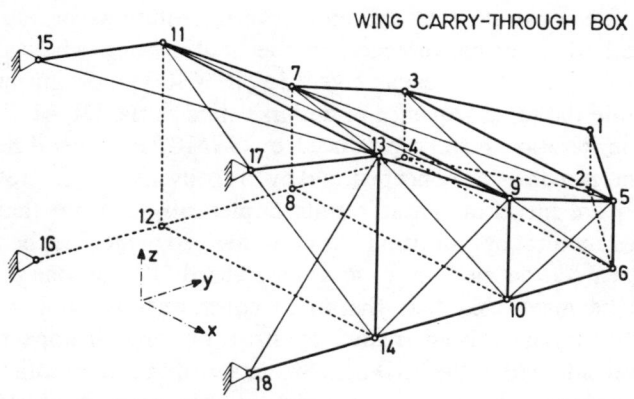

Fig. 10.11   63-bar truss

Table 10.2 Iteration history data for 63-bar truss

Analysis number	Weight (lb)					
	ACCESS-3 [27]				SAMCEF [2] PRIMAL-2	Conventional optimality criterion [8]
	NEWSUMT (0.5 × 1)	NEWSUMT (0.5 × 2)	DUAL 2			
			unscaled	scaled		
1	66628	66628	66628	30214	30214	30214
2	16914	12543	6706	7573	7680	7577
3	11137	8667	6316	6546	6591	6884
4	9338	7293	6195	6733	6398	6928
5	8305	6697	6157	6292	6270	6801
6	7620	6402	6138	6243	6246	6609
7	7154	6259	6129	6201	6199	6473
8	6836	6189	6124	6161	6159	6388
9	6620	6154	6121	6132	6126	6333
10	6467	6137	6120	6123	6123	6293
11	6362	6128	6119	6121	6121	6263
12	6289	6123	6118	6120	6120	6241
13	6238	6121	6118	6119	6119	6231
14	6203	6120			6118	6216
15	6178	6119			6118	6220
...	...					...
50						6159
CPU time (sec)						
total	108	163	60			
analysis	44	46	41			
optimization	59	113	14			

SAMCEF and ACCESS-3 programs is obtained at the expense of a large number of linearized stress constraints.

Looking at the results produced by the conventional OC technique of Ref. [8], it appears that this approach is not very good for this example, mainly because it employs stress-rationing. In contrast with the previous example, where the conventional OC behaved very well, zero-order stress constraint approximation leads to slow and unstable convergence: 50 structural reanalyses do not suffice to generate the optimum, while in methods based on first-order approximation of the stress constraints convergence is achieved within 15 stages.

10.6.2.1 Hybrid optimality criterion

Resorting to stress-ratioing offers two important advantages. First, when the virtual-load technique is used to compute the constraint gradients, the number of additional load cases is significantly reduced. Second, the number of behaviour constraints retained in each explicit approximate problem is also substantially reduced, because the stress constraints are now transformed into side constraints (see Equation (10.38)). This feature is especially beneficial with dual methods because the dimensionality of the dual problem corresponds to the number m of linearized constraints (10.41). On the other hand, it is well known that the FSD procedure does not always converge to the true optimum and sometimes is the source of instability or even divergence of the optimization process.

When dealing with large-scale problems, it is therefore necessary to adopt a compromise and to design the structure on the basis of a hybrid OC [32]. In this approach only a small number of stress constraints are linearized, while all the others are treated by stress-ratioing. The selection of constraints requiring first-order approximation can be made automatically on the basis of the following test:

$$\frac{\partial \sigma_i}{\partial z_i} \ll \frac{\sigma_i}{z_i} \qquad (10.48)$$

This condition arises from the fact that, in a statically determinate structure, zero- and first-order approximations of the stress constraints coalesce, since then:

$$\frac{\partial \sigma_i}{\partial z_k} = 0 \qquad \text{for } i \neq k \qquad (10.49)$$

Geometrically, the selection criterion (10.48) means that the stress constraint $\sigma_i = \bar{\sigma}_i$ must be approximated by its tangent plane rather than by a plane normal to the z_i axis, if these two planes have very different orientations (see Fig. 10.8).

Coming back to the 63-bar truss example, a numerical investigation has been done with the SAMCEF program [2] using the hybrid OC with automatic selection of the most critical stress constraints, as based on the criterion (10.48). The tolerance (denoted ε) for that selection has been made more and more severe, corresponding to 0, 6, 7, 10 and 27 linearized stress constraints. In this way, progressive transition from the conventional OC of Ref. [8] to the generalized OC is achieved. The results are illustrated in Fig. 10.12 as a function of the number of reanalyses and in Fig. 10.13 as a function of the total CPU time. It appears that there exists an economic optimum between conventional and generalized OC, obtained here when the 10 most critical stress constraints are linearized, the others being treated by stress-ratioing.

10.6.3 I-beam

Attention is now directed to the I-beam structure depicted in Fig. 10.14. The purpose of this example is to show that the generalized OC remains valid when natural frequency constraints are considered [36, 37]. The problem consists in minimizing the weight of the I-beam while controlling the frequencies of its three first eigenmodes: flange flexion, torsion, and web flexion. The analysis model involves 35 second-degree displacement elements, among which the 10 dia-

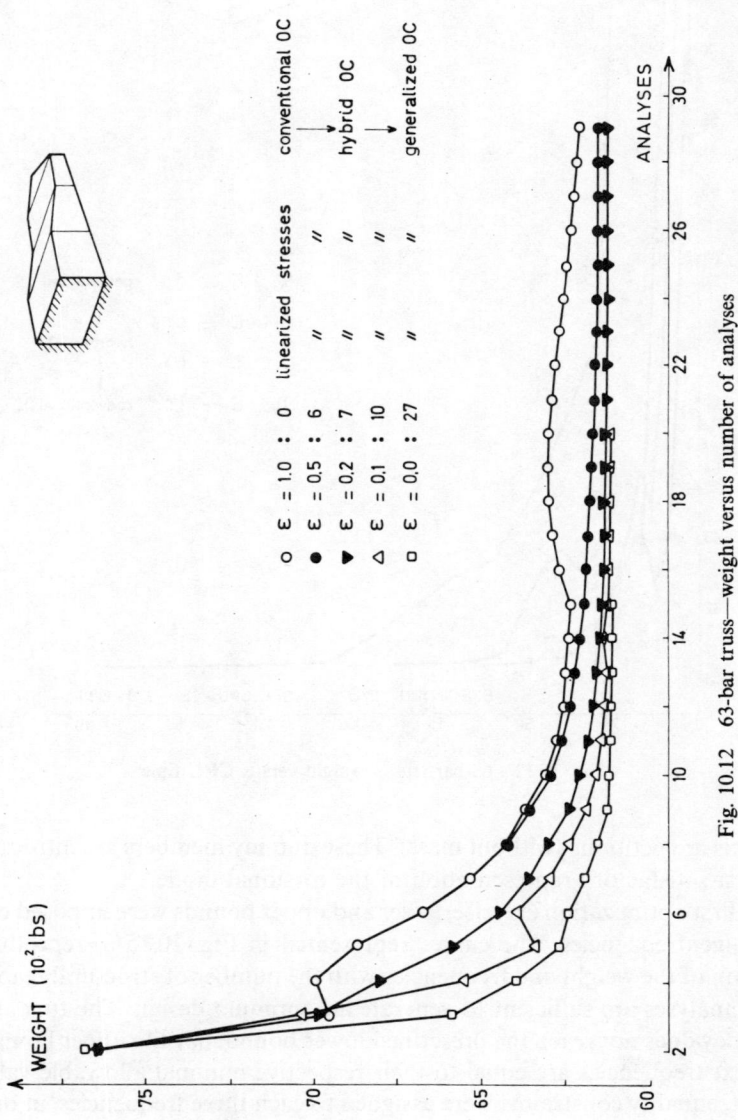

Fig. 10.12 63-bar truss—weight versus number of analyses

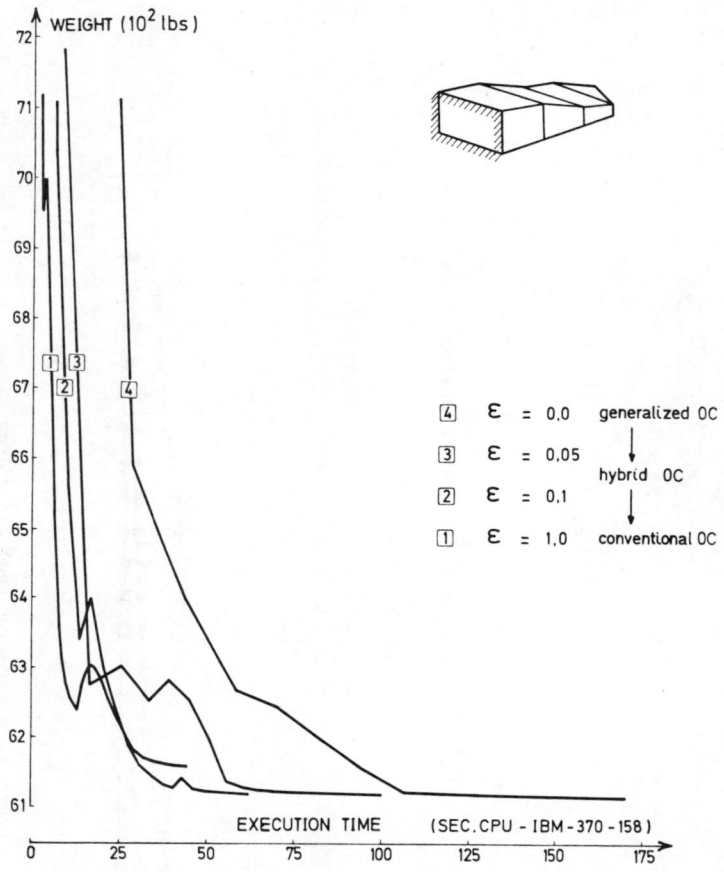

Fig. 10.13   63-bar truss—weight versus CPU time

phragms are fictitious (without mass). These dummy members are introduced to obtain a satisfactory representation of the torsional mode.

In a first optimization exercise, lower and upper bounds were imposed on each three eigenfrequencies. The curves represented in Fig. 10.15(a) reproduce the variation of the weight and frequencies with the number of structural reanalyses. Just 5 analyses are sufficient to generate an optimum design. The fundamental frequency does not reach the prescribed lower bound, nor the upper bound. The two next frequencies are equal to their respective minimal allowable value.

Next, equality constraints were assigned to each three frequencies, in order to reduce the fundamental frequency to 1 Hz, while keeping the two other frequencies at 1.2 Hz and 2.5 Hz, respectively. The iteration history presented in Fig. 10.15(b) again demonstrates the remarkable efficiency of the generalized OC approach. The increase in the weight after the first redesign stage is due to the

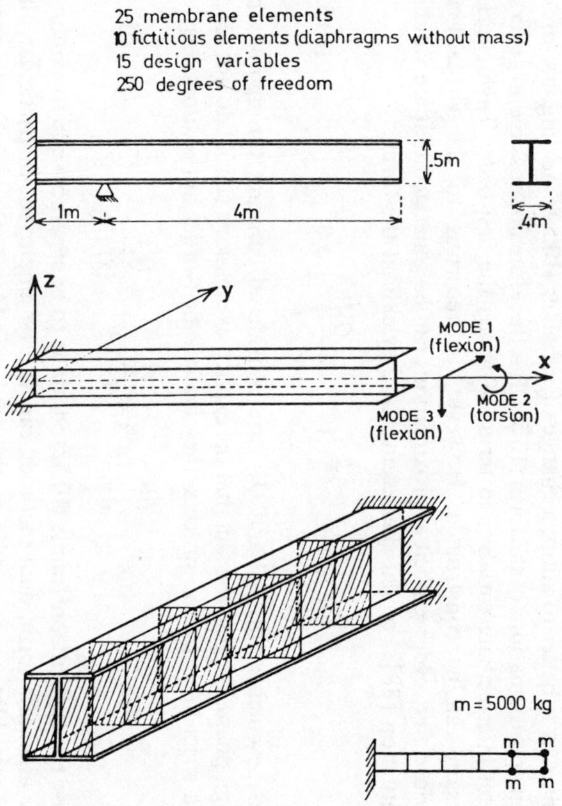

Mathematical Programming and Optimality Criteria Methods 393

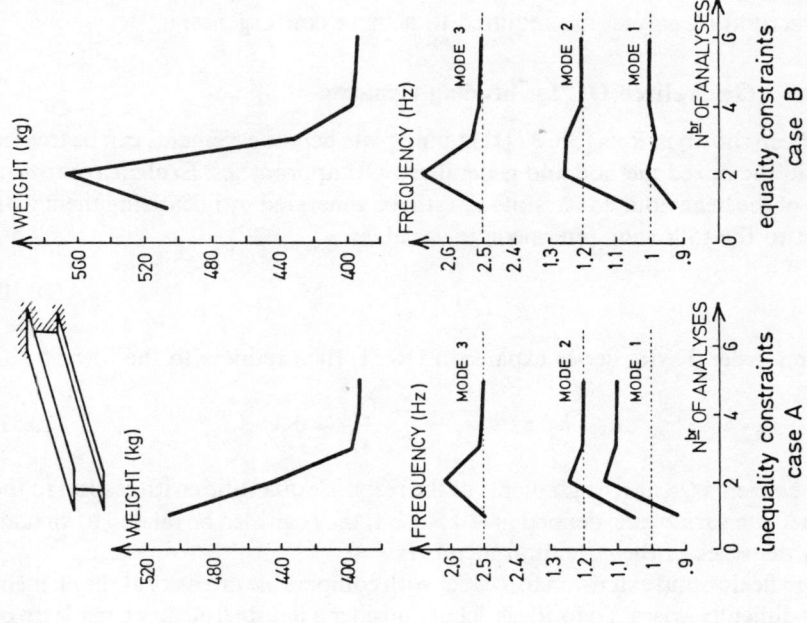

Fig. 10.15 Iteration history for I-beam (membrane model)

Fig. 10.14 I-beam model

equality constraint which is now imposed on the first frequency. Note that only six structural reanalyses are required to achieve convergence.

10.6.3.1 Generalized OC for bending elements

It has been shown in Refs [36, 37] that pure plate bending elements can be treated easily in the mixed method and generalized OC approaches. Explicit approximations of the behaviour constraints can still be generated by linearizing them with respect to the following intermediate variables:

$$z_i = \frac{1}{x_i^3} \tag{10.50}$$

The first-order Taylor series expansion (10.31) then reduces to the form:

$$\tilde{h}_j(x) \equiv \bar{u}_j - \sum_{i=1}^{n} \frac{Q_{ij}}{x_i^3} \geqslant 0 \tag{10.51}$$

The coefficients Q_{ij} are the gradients of the response quantities with respect to the intermediate variables z_i defined in (10.50), but they can also be related to virtual-energy densities in the structural members.

When flexion and extension forces act with comparable intensity at the element level, a difficulty arises. To fix ideas, let us consider a flat shell element, made up of a membrane and a plate stacked together. If the constraints are linearized with respect to the reciprocal design variables (10.27), their explicit approximations, given by expressions similar to (10.5), will be of high quality only if the structural members behave mainly in extension. Should the bending behaviour be dominant, it is better to adopt a change of variables (10.50), yielding first-order explicit approximations of the form (10.51). It is therefore not possible, in general, to select intermediate variables in terms of which the first-order Taylor series expansion can be constructed. However, by resorting to the virtual-load procedure, first-order explicit approximations can be generated that are valid in any situation [38]. For flat shell elements, they exhibit the form

$$\tilde{h}_j(x) \equiv \bar{u}_j - \sum_{i=1}^{n} \left(\frac{Q_{ij}^{(1)}}{x_i} + \frac{Q_{ij}^{(3)}}{x_i^3} \right) \geqslant 0 \tag{10.52}$$

where the coefficients $Q_{ij}^{(1)}$ and $Q_{ij}^{(3)}$ are considered constant throughout the redesign phase. Despite the fact that the explicit expressions (10.52) do not result from a strict linearization process, they are still first-order approximations:

$$\frac{\partial \tilde{h}_j}{\partial x_i}(x^0) = \frac{\partial h_j}{\partial x_i}(x^0) = \frac{Q_{ij}^{(1)}}{(x_i^0)^2} + 3\frac{Q_{ij}^{(3)}}{(x_i^0)^4} \tag{10.53}$$

Because the explicit constraints (10.52) continue to be separable and to exhibit a simple algebraic form, dual methods can still be employed. In particular, the primal variables x_i are related to the dual variables λ_j through fourth-order algebraic equations that can be solved in closed form.

The I-beam problem previously discussed was used to test the validity of these concepts. When the final design obtained with the membrane model is analysed by employing a more accurate model made up of flat shell elements, the torsional frequency constraint (mode 2) was seen to be violated by 10%. Therefore the problem was again solved with this new model, by resorting to the theory proposed in this section. Iteration history data are illustrated in Fig. 10.16 for both finite element models of the I-beam (inequality constraint case).

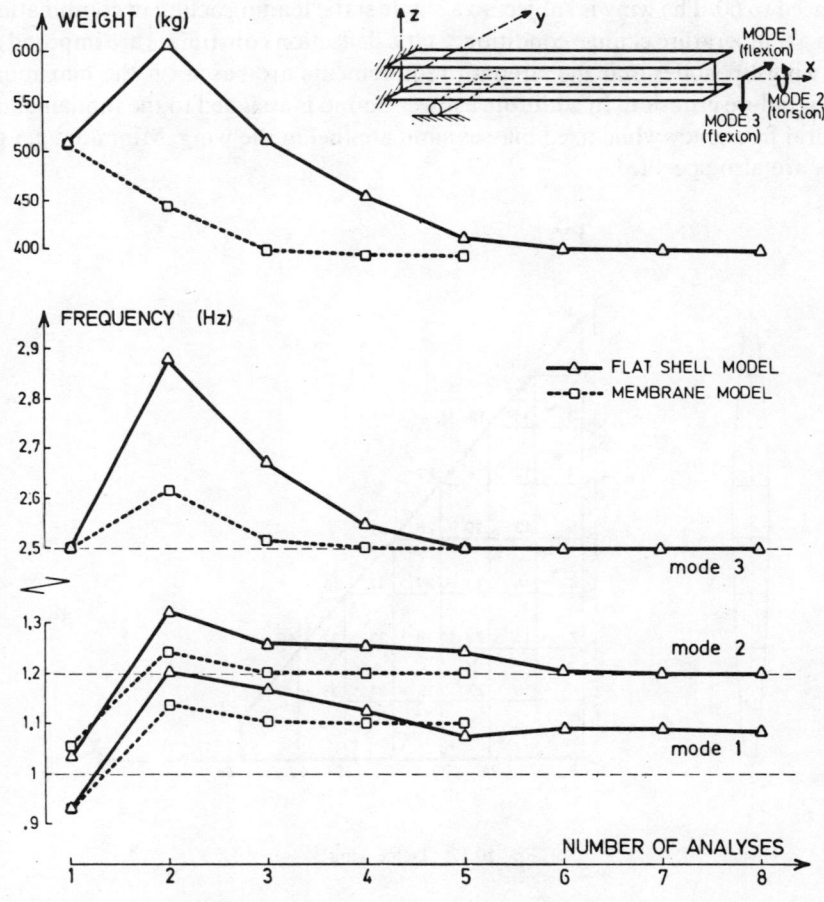

Fig. 10.16 Iteration history for I-beam (flat shell model)

10.6.4 Delta Wing

Consideration is now given to a thin delta wing structure with fibre composite skins and metallic webs (see Fig. 10.17). Several more and more realistic versions of the problem were studied in Refs [19–21, 27] using the ACCESS code. In the final version the skins are assumed to be made up of 0°, ±45° and 90° high-strength graphite epoxy laminates. The laminates are required to be balanced and symmetric and they are represented by stacking four constant-strain orthotropic elements in each triangular region shown in Fig. 10.17. Therefore the upper half of the delta wing is modelled using 252 orthotropic membrane elements to represent the skin and 70 symmetric shear panel elements for the vertical webs. However, after design variable linking, the number of independent variables is reduced to 60. The wing is subject to a single static loading acting in combination with a temperature change condition. Static deflection constraints are imposed at the wing-tip nodes and the strength requirements are based on the maximum strain failure criterion. In addition, a lower bound is assigned to the fundamental natural frequency while fixed masses simulate fuel in the wing. Minimum gauge sizes are also specified.

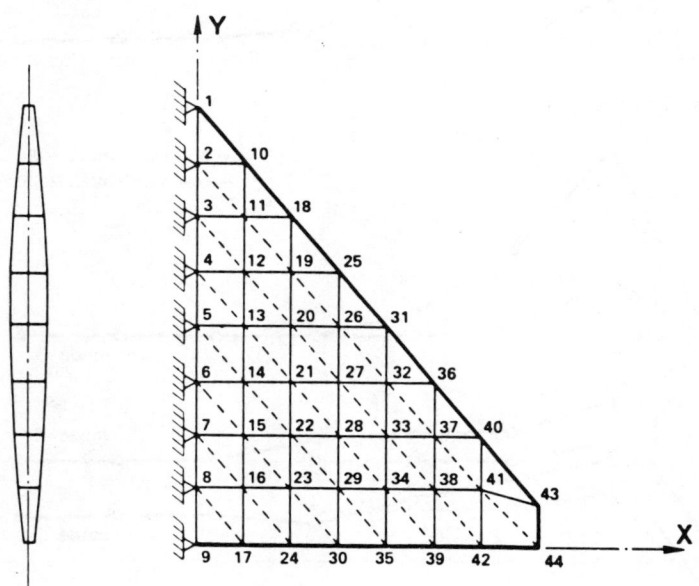

Fig. 10.17 Delta wing

Iteration history data presented in Table 10.3 were obtained using the NEWSUMT, DUAL-2 and DUAL-1 optimizers available in ACCESS-3 [27]. Since the fundamental natural frequency constraint is the main design driver in this example, its value as well as the weight for each design in the sequence is given. Initially the problem was treated using a continuous representation for all the design variables, although the fibre composite skins must in fact be described by

Table 10.3 Iteration history data for delta wing

Analysis number	NEWSUMT (continuous)		DUAL-2 (continuous)		DUAL-1 (mixed)	
	Weight ($\times 10^3$ lb)	Frequency (Hz)	Weight ($\times 10^3$ lb)	Frequency (Hz)	Weight ($\times 10^3$ lb)	Frequency (Hz)
1	86.82	2.829	86.82	2.829	86.82	2.829
2	70.26	2.650	21.39	2.016	20.83	2.009
3	58.11	2.516	16.75	1.961	17.17	2.000
4	49.16	2.396	14.34	1.937	15.40	1.974
5	42.64	2.293	14.85	2.007	14.99	2.000
6	38.04	2.209	13.98	1.987	14.40	1.994
7	33.93	2.127	13.81	2.003	14.08	1.996
8	29.86	2.042	13.62	2.007	13.92	2.003
9	26.78	2.010	13.41	2.005	13.52	1.998
10	24.56	2.009	13.24	2.004	13.41	2.003
11	22.64	2.010	13.10	2.002	13.37	2.009
12	20.95	2.010	12.99	2.001	13.29	2.003
13	19.48	2.010	12.91	2.001	<u>13.29</u>	2.000
14	18.21	2.010	12.85	2.000		
15	17.12	2.009	<u>12.81</u>	2.000		
...	...	...				
20	14.06	2.003				
25	13.56	2.002				
29	<u>13.47</u>	2.002				
CPU time (sec)						
total	719		261		253	
analysis	564		252		234	
optimization	145		2		12	

discrete variables (number of plies). The aim was simply to compare the efficiency of the NEWSUMT and DUAL-2 optimizers. It can be seen that the advantages of using the dual method approach are significant for the delta wing example. Not only is the computational effort expended in the optimizer portion of the program reduced dramatically (from 145 sec for NEWSUMT to 2 sec for DUAL-2), but the number of structural reanalyses required for convergence is also considerably decreased (29 and 15 stages for NEWSUMT and DUAL-2, respectively). Note that the NEWSUMT optimizer again produces feasible improved designs (primal solution scheme), while some of the designs in the DUAL-2 sequence are slightly infeasible with respect to the frequency constraint.

10.6.4.1 Discrete design variables

When some or all the design variables, instead of varying continuously, can only take on available discrete values, it can be shown that the OC equations (10.9)–(10.11), relating the primal variables x_i to the dual variables λ_j, must read as follows for each discrete variable [27, 39]:

$$x_i = x_i^k \quad \text{if } w_i x_i^k x_i^{k-1} < \sum_{j=1}^{m} Q_{ij}\lambda_j < w_i x_i^k x_i^{k+1} \qquad (10.54)$$

where it is understood that $\{x_i^k, k = 1, 2, \ldots\}$ denotes the set of available discrete values for the ith design variable. These expressions show that the dual space is subdivided into several regions, each of which corresponds to a distinct combination of available discrete values of the design variables. These regions are separated from each other by planes across which the first derivatives of the dual function are discontinuous. At the intersection of several first-order discontinuity planes (say p), there exist 2^p distinct possible gradients. However, the orthogonal projections of all the 2^p gradients onto the intersection of the p discontinuity planes are equal and define a unique ascent direction. Based on this observation, a specially devised gradient projection type of algorithm has been developed for maximizing the dual function in the mixed discrete–continuous design variable case. The resulting DUAL-1 optimizer is available in the ACCESS-3 program [26, 27].

Attention is now focused on the results obtained by DUAL-1 for the previously described delta wing example in the mixed continuous–discrete variable case. The metallic web thicknessess are still taken as continuous design variables, while the variables describing the laminated fibre composite skin are discrete (more precisely, integer variables representing the number of plies). Iteration histories are given in Table 10.3. It should be emphasized that DUAL-1 obtains a solution to the mixed variable problem in fewer stages (and less computer time) than DUAL-2 requires in the pure continuous case.

10.6.5 Aircraft Spoiler

The spoiler represented in Fig. 10.18 has been analysed in detail in an optimization exercise using the mixed method (Section 10.3). The structure is classically designed in light aluminium alloy sheet. The front spar and the secondary spar are joined by twelve ribs and covered by two skins reinforced by stringers. The spoiler is hinged at three points and actuated at one, in the midspan. The loads consist in pressure distribution on both faces, corresponding to two flight configurations. In one of them a flexibility constraint is imposed, which stipulates that the trailing edge has to remain straight within a tolerance $\varepsilon = 0.5$ mm, in order to eliminate contact with the flap. In the initial design this requirement was achieved by precambering the spoiler (see Fig. 10.18). This costly procedure has to be avoided in the final optimized design. So differential flexibility constraints are introduced which assign an upper limit $\varepsilon = 0.5$ mm to the absolute value of the difference between any two deflections along the trailing edge (see Fig. 10.20). In addition, maximum allowable stresses and minimum thicknesses are imposed, which differ from place to place depending on the material used and manufacturing considerations.

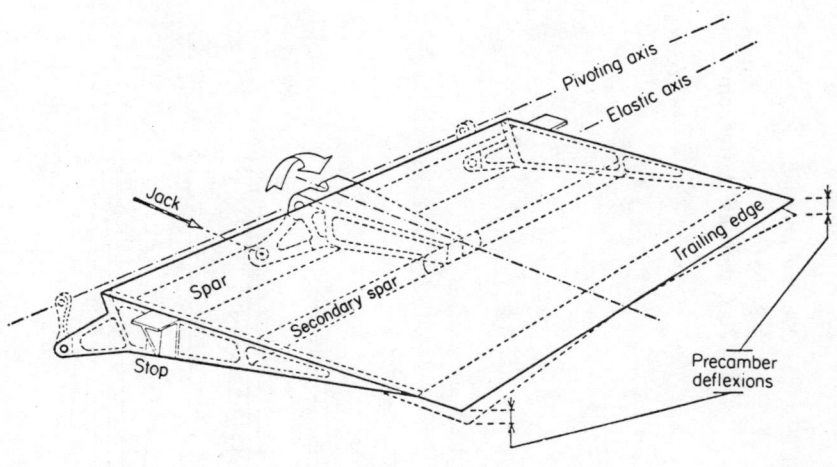

Fig. 10.18 Aircraft spoiler

Several finite element models of the structure were investigated, made up of 27, 64, 125 and 627 elements (see Ref. [36]). The final model is illustrated in Fig. 10.19. It involves 627 second-degree displacement elements and 2300 degrees of freedom. Based upon experience accumulated from the study of the simpler models, it was concluded that the mixed method had to be used for solving the spoiler problem. So a first-order projection algorithm was employed and it was necessary to limit the number of minimization steps $\bar{k}$ to avoid divergence of the process (note that the second-order primal algorithm previously described was not available at the time of the investigation). This means that any method based on optimality criteria (including dual algorithms) would not succeed in solving this problem.

Results are presented in Fig. 10.20. As expected from the experience gained with the simplified problems, a good convergence was obtained with the mixed method by setting $\bar{k} = 500$, that is, slightly below the number of design variables. Note that the initial scale-up of the weight is due to the fact that the original design of the spoiler did not satisfy the differential flexibility constraints when precambering was suppressed. Hence after scaling up the member sizes to obtain a feasible design, the weight jumps from 10 to 40 kg. After 13 structural reanalyses, the original weight of 10 kg is recovered, but it corresponds of course to a very different design. After each iteration, the deflection is increased; however, the trailing edge keeps about the same shape and remains straight within the specified tolerance.

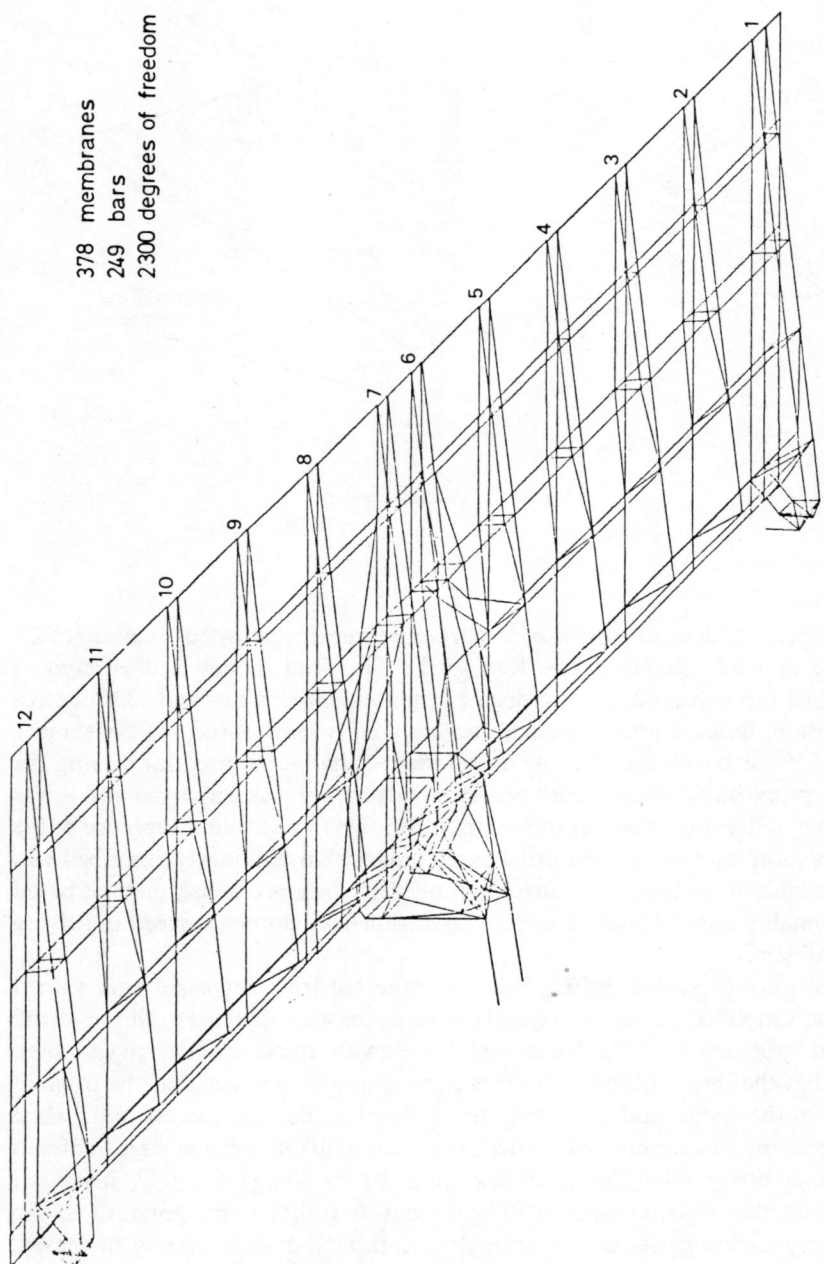

Fig. 10.19 Aircraft spoiler—finite element model

Mathematical Programming and Optimality Criteria Methods 401

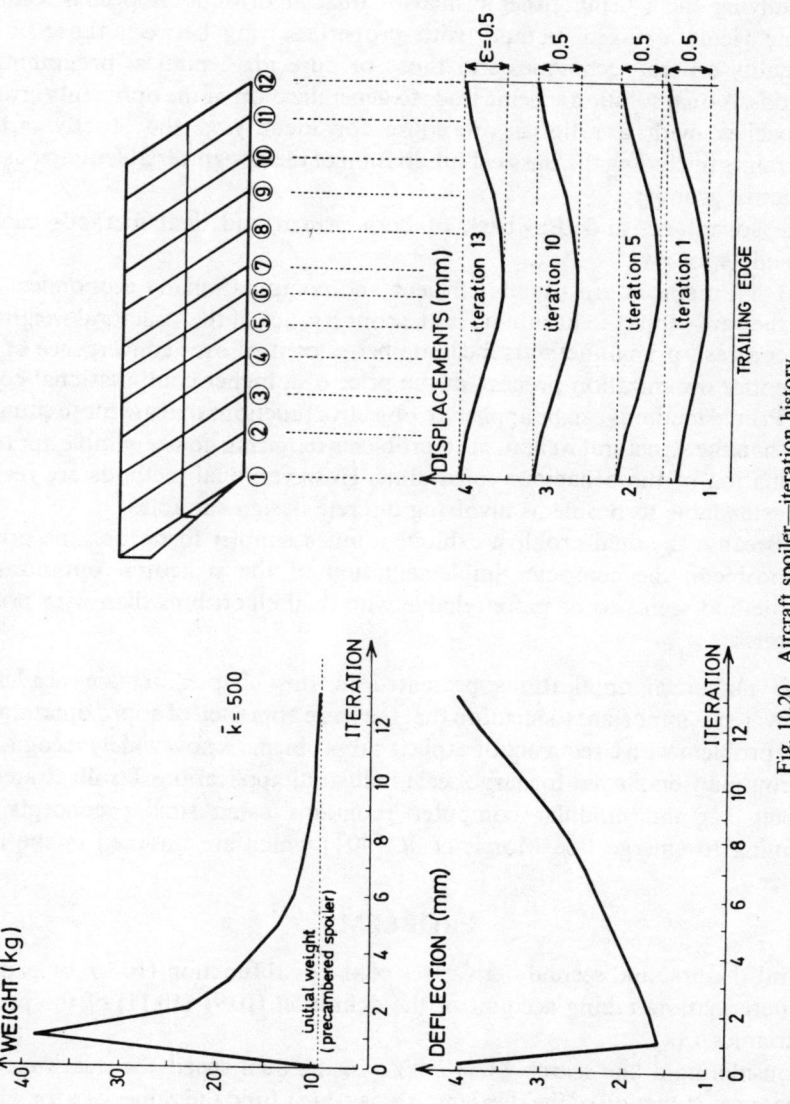

Fig. 10.20 Aircraft spoiler—iteration history

10.7 CONCLUDING REMARKS

A powerful and rather general approach to structural optimization is achieved by replacing the original problem with a sequence of explicit approximate problems and solving them using either primal or dual algorithms. A primal solution scheme yields a mixed method, with properties lying between those of the optimality criteria techniques and those of pure mathematical programming methods. A dual solution scheme leads to generalization of the optimality criteria approaches, with a rational procedure for identifying the strictly critical constraints (including the classical subdivision of the design variables into passive and active groups).

The advantages and drawback of both primal and dual methods can be outlined as follows:
1. Dual methods are usually efficient and computationally economical but they are subject to instability in the convergence of the structural weight. In contrast, primal methods facilitate better control over convergence of the entire optimization process, at the price of a higher computational cost.
2. Primal methods can be applied to objective functions that are more complex than the structural weight, or to problems requiring non-separable approximation of the behaviour constraints. However, dual methods are readily extendable to problems involving discrete design variables.
3. Because the dual problem exhibits a much simpler form than the primal problem, the computer implementation of the structural optimization method seems to be more reliable with dual algorithms than with primal ones.

The numerical applications presented in this chapter remain academic. However, it is important to mention that the basic approach of approximating the initial problem with a sequence of explicit subproblems is now widely recognized. It is routinely employed for large-scale industrial applications (as illustrated in Chapter 12) and modular computer programs using similar concepts are beginning to emerge (see Morris *et al.* [40]) which are outlined in the next chapter.

PROBLEMS

1. Find the first and second derivatives of the dual function (10.16) by explicit differentiation, taking account of the definition (10.9)–(10.11) of the primal variables $x(\lambda)$.
2. Considering a line search $\lambda^+ = \lambda + \alpha s$ in the dual space, the dual function becomes a function of the step length α, say $\phi(\alpha)$. Find the values of α for which $\phi''(\alpha)$ is discontinuous. Apply the result to the 2-bar truss example (Fig. 10.3).
3. Solve analytically the explicit problem (10.6)–(10.8) in the case $m = 1$ (one behaviour constraint). Apply the resulting formula to the 2-bar truss problem, by disregarding the non-critical constraint (10.25).

4. Give a geometrical interpretation of first- and zero-order approximations of the stress constraints in the space of the direct design variables (that is, Equations (10.5) and (10.38), respectively).
5. Considering the primal problem (10.40)–(10.42), devise a line-search procedure for minimizing the objective function along a descent direction (use an explicit form in term of the step length).
6. Consider the 2-bar truss problem (10.24), (10.25) in the pure discrete case, where the cross-sectional areas can only take on the values 1, 1.5 or 2. Draw the contours of the dual function in a two-dimensional dual space, by using Equation (10.16) with primal variables given by (10.54).

C.F.

REFERENCES

1. C. Fleury, A unified approach to structural weight minimization, *Comp. Meth. Appl. Mech. Eng.*, **20**(1), 17–38, 1979.
2. C. Fleury and G. Sander, Structural optimization by finite elements, *LTAS Report SA-58*, University of Liège, 1978.
3. C. Fleury and L. A. Schmit, Primal and dual methods in structural optimization, *J. Struct. Division*, ASCE, **106**(STS), 1117-133, 1980.
4. R. A. Gellatly and L. Berke, Optimal structural design, *AFFDL-TR-70-165*, 1971.
5. I. C. Taig and R. I. Kerr, Optimization of aircraft structures with multiple stiffness requirements, *AGARD CP-123*, paper 16, 1973.
6. N. S. Khot, L. Berke and V. B. .Venkayya, Comparison of optimality criteria algorithms for minimum weight design of structures, *AIAA J.*, **17**(2), 182–90, 1979.
7. J. Kiusalaas, Minimum weight design of structures via optimality criteria, *NASA TN-D-7115*, 1972.
8. L. Berke and N. S. Khot, Use of optimality criteria methods for large scale systems, *AGARD LS-70*, 1–29, 1974.
9. M. W. Dobbs and R. B. Nelson, Application of optimality criteria to automated structural design, *AIAA J.*, **14**(10), 1436–43.
10. D. Rizzi, Optimization of multi-constrained structures based on optimality criteria, *Proc. AIAA/ASME/SAE 17th Struct., Struct. Dynamics and Mat. Conf.*, King of Prussia, Pennsylvania, 448–62, 1976.
11. C. Fleury, Relation entre l'approche par critères d'optimalité et la programmation convexe en optimisation structurale, *Coll. Publ. Fac. Sc. Appl.*, University of Liège, **66**, 3–64, 1977.
12. C. Fleury, Structural weight optimization by dual methods of convex programming, *Int. J. Num. Meth. Engng.*, **14**(12), 1761–83, 1979.
13. C. Fleury, Optimization des structures par la méthode des éléments finis, *Coll. Publ. Fac. Sc. Appl.*, University of Liège, **59**, 63–102, 1976.
14. G. Sander and C. Fleury, A mixed method in structural optimization, *Int. J. Num. Meth. Engng.*, **13**(2), 385–404, 1978.
15. C. Fleury and G. Sander, Relations between optimality criteria and mathematical programming in structural optimization, *Proc. Symp. Applications of Computer Methods in Engineering*, University of Southern California, Los Angeles, **1**, 507–20, 1977.
16. K. F. Reinschmidt, A. C. Cornell and J. Brotchie, Iterative design and structural optimization, *J. Struct. Div.*, ASCE, **92**(ST6), 281–318, 1966.

17. L. A. Schmit, Structural design by systematic synthesis, *Proc. 2nd Conf. on Electronic Computation*, ASCE, New York, 105–22, 1960.
18. L. A. Schmit and B. Farshi, Some approximation concepts for structural synthesis, *AIAA J.*, **12**(5), 692–9, 1974.
19. L. A. Schmit and H. Miura, A new structural analysis/synthesis capability—ACCESS 1, *AIAA J.*, **14**(5), 661–71, 1976.
20. I. A. Schmit and H. Miura, Approximation concepts for efficient structural synthesis, *NASA CR-2552*, 1976.
21. L. A. Schmit and H. Miura, An advanced structural analysis/synthesis capability—ACCESS 2, *Int. J. Num. Meth. Engng.*, **12**(2), 353–77, 1978.
22. O. O. Storaasli and J. Sobieszczanski, On the accuracy of the Taylor approximation for structural resizing, *AIAA J.*, **12**(2), 231–3, 1974.
23. R. Baldur, Structural optimization by inscribed hyperspheres, *J. Eng. Mech. Div.*, ASCE, **98**(EM3), 503–8, 1972.
24. G. N. Vanderplaats, CONMIN—A Fortran program for constrained function minimization—User's manual, *NASA TMX-62-282*. 1973.
25. H. Miura and L. A. Schmit, ACCESS-2—Approximation concepts code for efficient structural synthesis—User's guide, *NASA CR-158949*, 1978.
26. C. Fleury and L. A. Schmit, ACCESS-3—Approximation concepts code for efficient structural synthesis—User's guide, *NASA-CR*, 1980 (to appear).
27. C. Fleury and L. A. Schmit, Dual methods and approximation concepts in structural synthesis, *NASA CR-3226*, 1980.
28. L. A. Schmit and C. Fleury, Structural synthesis by combining approximation concepts and dual methods, *AIAA J.*, 1980 (to appear).
29. R. L. Fox, *Optimization Methods for Engineering Design*, Addison-Wesley, 1971.
30. J. S. Arora and E. J. Haug, Methods of design sensitivity analysis in structural optimization, *AIAA J.*, **17**(9), 970–4, 1979.
31. R. H. Gallagher, Fully stressed design, in *Optimum Structural Design—Theory and Applications* (eds R. H. Gallagher and O. C. Zienkiewicz), John Wiley, London, 1973, pp. 143–77.
32. C. Fleury, An efficient optimality criteria approach to the minimum weight design of elastic structures, *Comput. Structures*, **11**(3), 163–73, 1980.
33. R. Fletcher, Methods related to Lagrangian functions, in *Numerical Methods for Constrained Optimization* (eds. P. E. Gill and W. Murray), Academic Press, London, 1974, pp. 219–39.
34. M. J. D. Powell, Algorithms for nonlinear constraints that use Lagrangian functions, *Math. Programming*, **14**, 224–48, 1978.
35. SAMCEF, *Système d'Analyse des Milieux Continus par Eléments Finis*, LTAS, University of Liège.
36. C. Fleury, Le dimensionnement automatique des structures élastiques, Doctoral dissertation, *LTAS Report SF-72*, University of Liège, 1978.
37. C. Fleury and G. Sander, Generalized optimality criteria for frequency constraints, buckling constraints and bending elements, *AFOSR-TR-80-0107*, 1979.
38. C. Fleury, Optimization of large flexural finite element systems, *Proc. NATO Advanced Study Institute on Optimization of Distributed Parameter Structures*, Iowa City, 1980 (to appear).
39. L. A. Schmit and C. Fleury, Discrete-continuous variable synthesis using dual methods, *AIAA J.*, 1980 (to appear).
40. A. J. Morris, P. Bartholomew and J. Dennis, A computer based system for structural design, analysis and optimization, *AGARD CP-280*, paper 20.

Foundations of Structural Optimization: A Unified Approach
Edited by A. J. Morris
© 1982 John Wiley & Sons Ltd.

Chapter 11
From a 'Black Box' to a Programming System

11.1 INTRODUCTION

This chapter is addressed to engineers who, having become acquainted with the theory of computer-based mathematical optimization in the preceding chapters, are contemplating implementation of these methods into the practice of their organization. To provide a rational basis for implementation decisions, a few alternative ways for such implementation shown in Fig. 11.1 are reviewed in the following section.

The review begins with identification of the principal generic components of an optimization procedure and proceeds to describe a 'black box' approach in which these components are packaged so that the inner workings of the whole procedure are shielded from the user's view. The modular concept of separation of the analysis and optimization parts of the code at a subroutine level is introduced next. The state-of-the-art in both approaches is illustrated by numerical examples from the references. The references, although numerous, are not intended to constitute an exhaustive bibliography of the subject. Instead, they are selected 'entry points' to the vast body of pertinent literature.

The review section concludes with a presentation of a 'programming system' which can be assembled of large, stand-alone programs to perform specialized tasks in a unifying framework of an optimization procedure. This order of presentation leads gradually to greater flexibility (adaptability) of the resulting optimization tool.

Since the 'programming system' occupies the highest level on that flexibility scale and is relatively more complex, the subsequent three sections are devoted to that concept. Sections 11.3 and 11.4 provide a description of the system components, their functions, and typical organizations, with references to existing systems reported in the literature. To streamline the discourse, corresponding mathematical details are contained in appendices at the end of the chapter which provide the reader with several useful techniques selectively

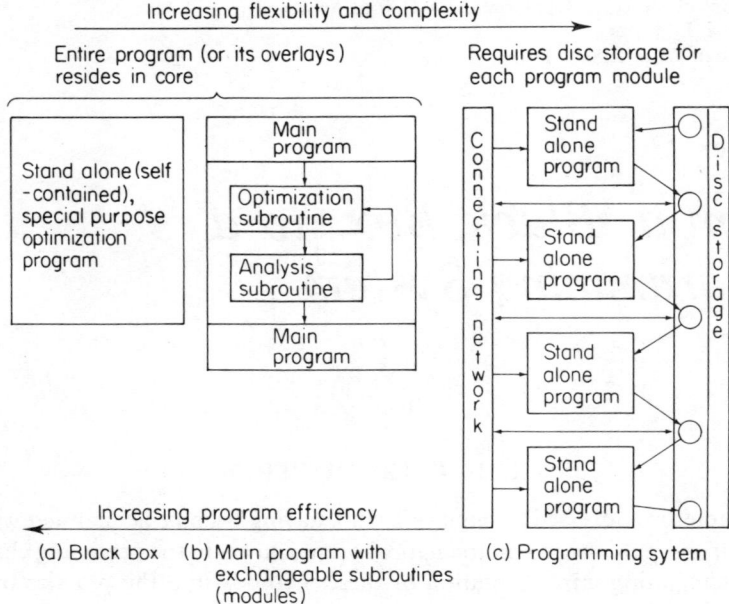

Fig. 11.1 Approaches for implementing optimization methods

extracted from many referenced sources. Section 11.5 illustrates the programming system adaptability by a few diversified numerical examples.

The paper concludes with a brief evaluation of advantages and disadvantages of the reviewed implementation approaches.

List of Symbols

Since we are drawing together many of the arguments and concepts introduced in the first ten chapters it is convenient to recall the definitions of the symbols used. This also allows the chapter to be read independently of the rest of the book if an outline only is required of modern systems.

E	Young's modulus
$F(\vec{x})$	objective function
$g(\vec{x})$	constraint function, overbar denotes a system level constraint
G	damping coefficient
$\mathbf{K}$	stiffness matrix
$\mathbf{M}$	mass matrix
$\mathbf{P}$	load vector
$\mathbf{u}$	displacement
$\mathbf{x}, \vec{\mathbf{x}}$	design variable and vector of design variables

ε	small constant
ϕ	a displacement function used as a Ritz function
v	Poisson's ratio
Ω	cumulative constraint
$\cong$	approximate equality
$\sim$	tilde denotes a condensed matrix or vector

Other symbols, used locally, are defined as needed.

11.2 BASIC IMPLEMENTATION SCHEMES

The function of an optimization procedure is to find a vector of design variables x_i that minimizes an objective function $F(x)$ while satisfying constraint equations $g(x)$. In a standard notation:

$$F(x) \to \min \tag{11.1}$$

subject to

$$g_j(x) \leq 0 \qquad 1 \leq j \leq m \tag{11.2}$$

It is worth observing that armed with the techniques described in Chapters 8, 9, and 10 we need, no longer, take weight as the objective function and, indeed, advantage is taken of this fact in this chapter and later in the book.

Regardless of the manner of implementation, the basic elements of an optimization procedure are: Optimizer (also referred to as 'search algorithm'), Analyser and Terminator organized in an iterative loop (optimization loop) as shown in Fig. 11.2. This procedure can be implemented on the computer in a number of ways which are reviewed in this section in order of increasing application flexibility.

The discussion begins with a 'black box' approach which is least flexible, but potentially most efficient in execution, simplest to use, and progresses to the concept of a programming system of potentially complete generality. Between these two extremes, the concept of using an optimizer in the form of a general-purpose subroutine is discussed.

11.2.1 Special-purpose 'Black Box'

Under a 'black box' approach, the inner workings of the program, such as might be illustrated in Fig. 11.2, are removed from the user's concern, leaving the input and output of data as the only means for user–program communication. An example of an optimization program packaged to appear as a 'black box' to its user is a procedure for aeroelastic optimization of wings whose application is discussed in Ref. [1]. This procedure illustrates well the advancement level in programs of the 'black box' category as it is capable of optimizing composite material wing covers subject to aeroelastic constraints for a choice of objective functions.

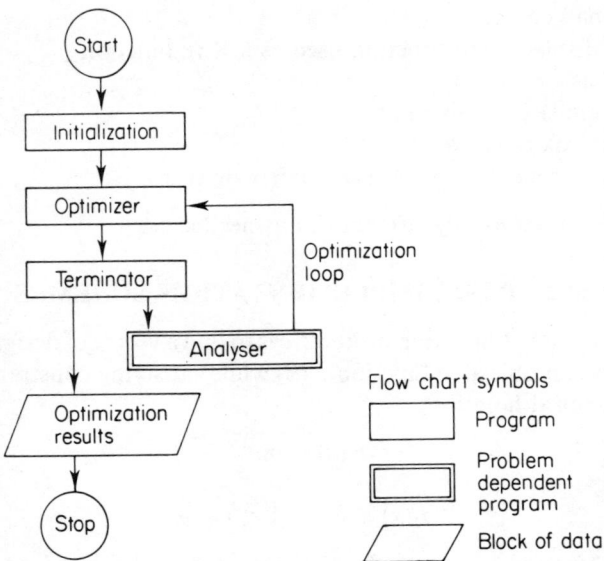

Fig. 11.2 Generic components and basic flow organization of an optimization procedure

The design variables are the ply thicknesses and fibre orientation angles in the wing covers, and the constraints include strength, flutter and divergence. For greater application flexibility, several physical quantities, important in aircraft design, constitute a pool from which the user may draw to form a list of constraints and to define an objective function. These quantities are: structural mass, first natural frequency, lift coefficient of a flexible wing, ratio of the aerodynamic load on a flexible wing to the corresponding load on a rigid wing, wing deflections (including relative deflections that determine camber and twist), roll control effectiveness and flutter speed. Any of these quantities can be added to the list of constraints, or be designated as the objective function. An objective function, which as we have seen above need not be structural weight, can also be formed as a weighted sum

$$F = \sum_i a_i Q_i$$

of these quantities Q_i with the weighting factors a_i.

One of the applications reported in Ref. [1] is for a fighter aircraft wing optimized for minimum structural mass at a design point defined by $M = 1.2$, altitude 4500 m, and a load factor of 8 g. Constraints are the wing flutter, divergence, and the laminate strength. The unidirectional fibre composite material layup is defined by three orientation angles constant over the entire wing and by spanwize and chordwize distributions of the total thickness of the material associated with each orientation angle as shown in Fig. 11.3. The design variables are the orientation angles and the coefficients of second-order polynomials in span and chord coordinates which govern the thickness distributions. Typical output information consists of a wing cover laminate definition illustrated by contour plots in Fig. 11.3. Separate contour plots are given for each orientation angle and for total thickness of the entire layup, showing a torsion-resisting material in layers 1 and 3, and a bending-resisting material in layer 2.

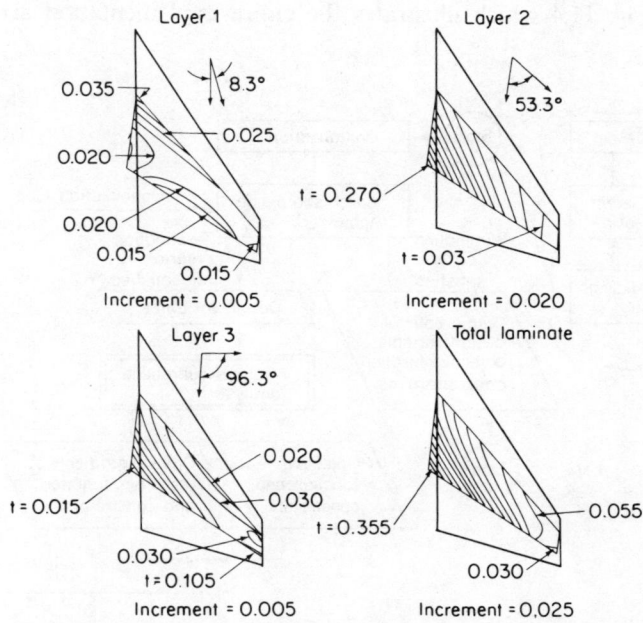

Fig. 11.3 Conceptual design aircraft wing—8 g weight-optimized composite laminate thickness distributions (Ref. [1])

Although the user is furnished with a considerable degree of the program adaptability for the problem at hand, he is not expected to adapt the program to accommodate problems of a type that had not been included among the program development functional objectives. This adaptability limitation is an inevitable consequence of the 'black box' approach.

11.2.2 Optimization Algorithm as a Subroutine

An implementation approach basically different from the 'black box' is embodied in an optimization program described in Refs [2] and [3]. Instead of dealing with

410 Foundations of Structural Optimization: A Unified Approach

all aspects of an optimization procedure, that is with input, analysis, search algorithm, termination and output, as 'black box' procedure would, it incorporates only a search algorithm that seeks a constrained minimum in n-dimensional design space. The search procedure is available in the form of a FORTRAN subroutine which, in one version of the code, calls an analysis subroutine. With this approach, the user must furnish an analysis subroutine suitable for the problem at hand and a main program whose functions are to accept input, to call the search subroutine, and to issue output. The search subroutine calls the analysis subroutine and passes to it the values of design variables, receiving in return the values of objective function and constraints as shown in Fig. 11.4 which illustrates the entire implementation arrangement,

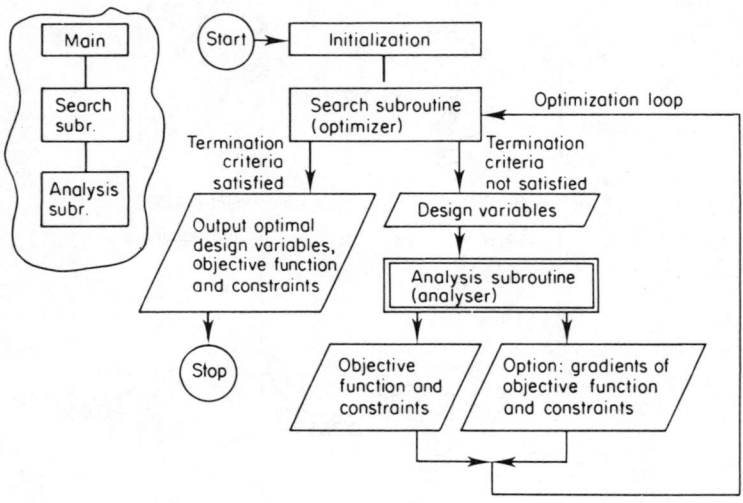

Fig. 11.4 Flow chart of optimization procedure with analysis subroutine called by search subroutine. *Inset*: corresponding block diagram module

including a block diagram in the inset to indicate the calling hierarchy of the modules. As seen in Fig. 11.4, the optimization iteration is accomplished entirely by the search and analysis subroutines placed in a loop similar to the one in Fig. 11.2, and return to the main program does not take place until the termination criteria built into the search procedure are satisfied.

Since this particular search algorithm (a usable-feasible directions method; Ref. [4]) uses derivatives of the objective function and constraints, an option is available to the user to code the analysis subroutine so that in addition to the objective function and constraints, their derivatives can also be computed. This option is exercised if the search subroutine signals a need for derivative values by setting a designated 'flag' integer to a specific value. This feature makes it possible to code the analysis subroutine so that the derivatives are not computed unless they are called for. By another 'flag' integer, the user can inform the search subroutine that no derivative information is to be expected from the analysis subroutine, and the search subroutine will carry out analyses for the variables systematically perturbed one at a time and compute finite difference approximations to the derivatives by a simple two-point incremental scheme. If a more accurate multipoint difference procedure for finite differences is desired, the analysis subroutine can be coded to include such capability. This manner of implementation gives the user freedom to form a complete optimization program by coupling an analysis subroutine pertinent to the problem at hand with a search subroutine.

In another version, the search subroutine does not contain calls to the analysis subroutine. Instead, it returns control to the main program after the new set of design variables is computed and the termination criteria checked. The analysis information is expected to come from the main program. This seemingly minor change in the search subroutine has an important impact on the organization of the optimization procedure. That organization is now up to the user, who must build into the main program the entire logic of optimization iteration. Typical logic for such a program is shown in the flow chart in Fig. 11.5, where a corresponding block diagram is also depicted. From comparison of Figs 11.4 and 11.5, it is evident that the organization shown in Fig. 11.5 gives the user more

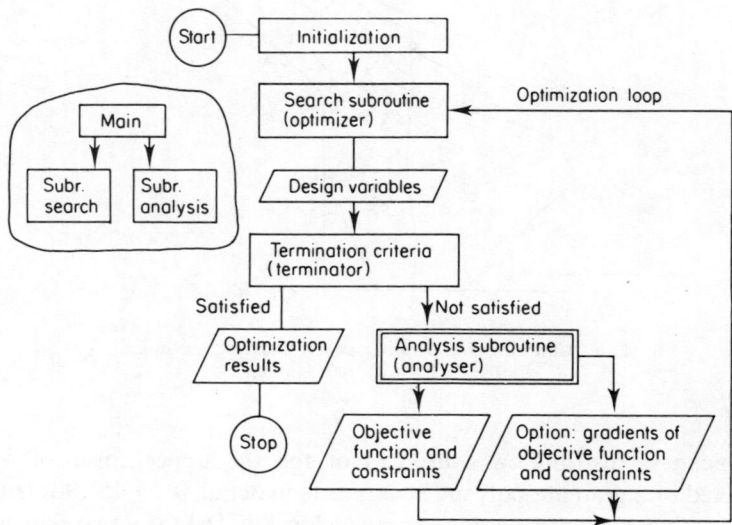

Fig. 11.5 Flow chart of optimization procedure with independent search and analysis subroutines. *Inset*: corresponding block diagram

flexibility. For example, the user may add to the optimization loop features that are known to be particularly effective, such as selected termination criteria, stop and restart capabilities, and additional print-outs of intermediate results.

In summary, the implementation approach described in this section has the advantages of modularity. For example, the user can isolate everything that depends on the physics of the problem in the analysis part of the program (analysis subroutine) and take advantage of the state-of-the-art in mathematics in selection of the search algorithm. This adaptability is illustrated by coupling the search algorithm from Ref. [4] with two different analysis algorithms to carry out a structural optimization and an aerodynamic optimization.

The structural optimization is reported in Ref. [5] for a supersonic transport aircraft wing. The aircraft is quite large, comparable to a Boeing 747 in weight, and has a wing area in excess of 800 m². The wing strength optimization is carried out for a 2.5 g load factor design point and consists of optimizing individual wing cover sandwich panels for minimum structural mass, subject to strength and local buckling constraints. Design variables, separate for each panel, are the sandwich depth and face sheet thickness. A typical result, illustrated in Fig. 11.6, is a face

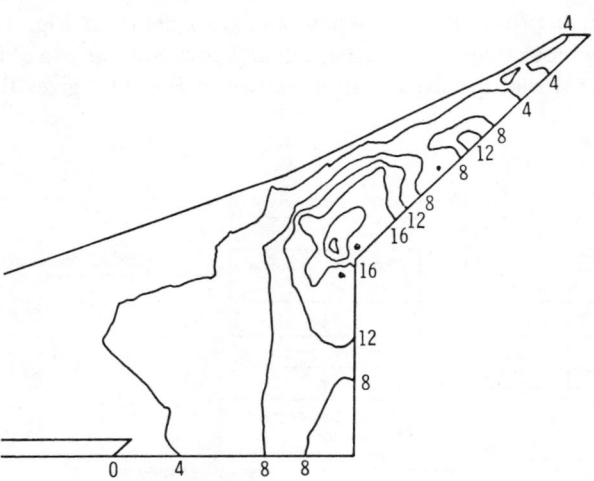

Fig. 11.6 Contour plot of wing cover thickness optimized for strength $0°/\pm 45°/90°$ laminate

sheet thickness distribution contour plot for the upper cover of a wing constructed of a graphite-polymide composite material, $0°/\pm 45°/90°$ laminate.

The aerodynamic optimization is reported in Ref. [6] for a two-dimensional airfoil. The objective is maximization of lift for $M = 0.75$ at a 0° angle of attack, subject to drag constraints. Transonic aerodynamics is included in the analysis.

From a 'Black Box' to a Programming System

The airfoil shape is defined as a linear combination of six basis shapes shown in Fig. 11.7(a) with the coefficients of the linear combinations being the design variables. The optimum shape and its characteristics shown in Fig. 11.7(b) are remarkable since they represent the so-called supercritical airfoil, although none of the basis shapes has a supercritical airfoil shape.

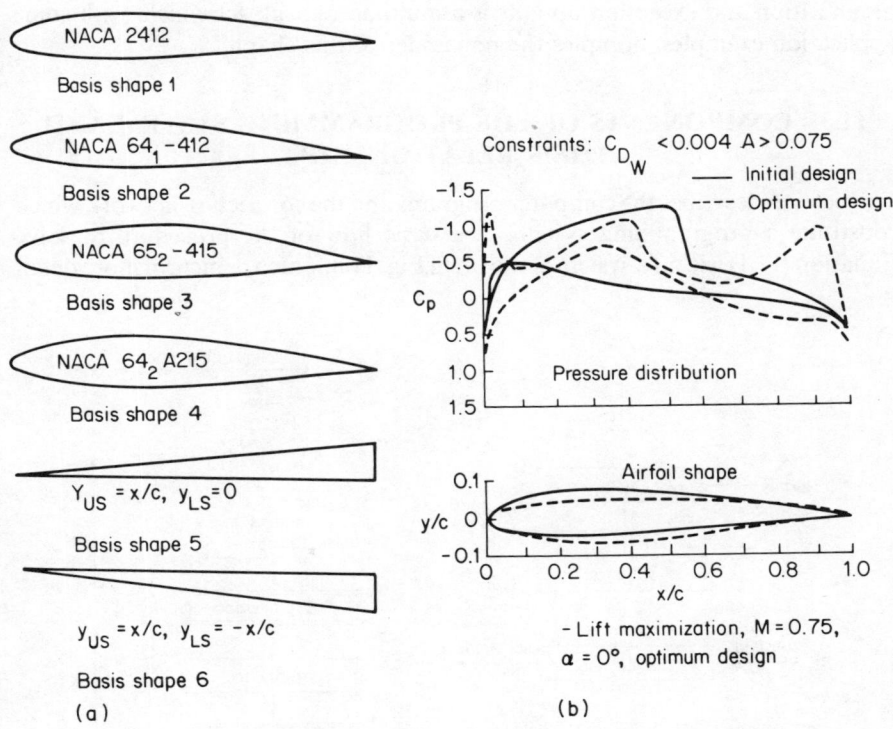

Fig. 11.7 Example of an airfoil optimization by means of a nonlinear mathematical programming method (Ref. [6]) (a) Basis shapes (b) Optimum shape and characteristics

11.2.3 Programming System

In comparison with the main program–subroutines arrangement, the concept of a programming system is the next logical step toward greater application flexibility. In a programming system (a term introduced in Ref. [7]), a user must furnish problem-dependent code modules in addition to input data, while only the input data are needed to execute an ordinary program or system of programs. A programming system removes the subroutine size restrictions imposed on the optimizer and analyser, and allows use of large stand-alone programs, or even systems of several large programs for the analysis and optimization functions. It also isolates the definitions of the design variables, objective function and constraints in separate problem-dependent user-supplied programs executed between the optimizer and the analyser. The system is operated by an executive command language and other software utilities that constitute a connecting network; thus, in principle, any optimization strategy can be implemented. The concept of a programming system with its components, functions, overall organization and execution options is a multifaceted subject which, with some application examples, occupies the remainder of this chapter.

11.3 COMPONENTS OF THE PROGRAMMING SYSTEM AND THEIR RELATIONSHIP

This section describes the computer programs and the connecting network which constitute a programming system. The basic flow of the procedure to solve Equation (11.1) within a system is shown in Fig. 11.8, which depicts in more detail

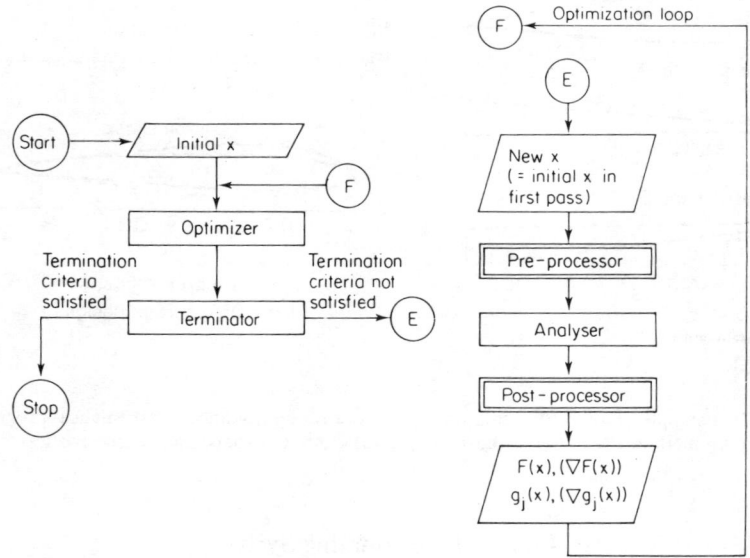

Fig. 11.8 Flow for a basic option (Option 1.1, Table 11.2) in a programming system

an organization identical to that in Fig. 11.2. The system's principal components are: Analyser, Pre- and Post-Processors, Optimizer, Terminator and the Connecting Network. Since a particular analyser is usually applicable to a very wide class of problems, it is shown as problem independent in the figures illustrating the programming system.

Description of the system's components and of the execution options for the entire optimization procedure is given in generic terms to emphasize that the system's organization does not depend on the internal details of its modules. Some of the basic functions of each component are specified briefly in the body of the paper; more detailed descriptions are given in the appendices.

11.3.1 Analyser

The function of the analyser is to compute values of the behaviour variables which carry the information on the physical object's response to the input quantities. In structural applications, the analyser may be a finite element program. Its input comprises structural cross-sectional dimensions, material properties, element connectivity data, nodal point coordinates, and loads. The behaviour variables consist of such quantities as displacements, internal forces, stresses, eigenvalues and eigenmodes for vibration and buckling, etc. The structural mass most commonly, but not exclusively, used as the objective function is also included in the output. Since the analyser is to be executed many times in the loop, it is obviously advantageous if the code is organized so that it may be split into a non-repeatable part and a repeatable part as shown in Fig. 11.9.

Non-repeatable and repeatable parts

The non-repeatable part is executed once outside of the optimization loop, and the repeatable part is included within the loop as shown in Fig. 11.9. Division between the two parts depends on the nature of the analysis and the character of the input variables and can be established by examination of a typical finite element program for structural analysis.

Most operations of such an analysis program are listed in the left-hand column of Table 11.1. The other four columns from left to right correspond to optimization with the following types of variables being free in the optimization procedure: cross-sectional dimensions, nodal coordinates, element-node connectivity, and type of displacement function used in a finite element. Table entries 'N' and 'R' signify allocation of the corresponding operation to non-repeatable or repeatable part, respectively, for each type of variable.

Computation of gradients

Most of the efficient mathematical optimization algorithms require not only the objective function and the constraint values but also their gradients, all evaluated for a given set of input values of the design variables x. Therefore the gradients are indicated as an optional output of the analyser in Figs 11.8 and 11.9. The gradients can be computed by a finite difference technique or by an analytical technique. Numerous literature sources (e.g., Refs [8], [9], [10], [11], and in

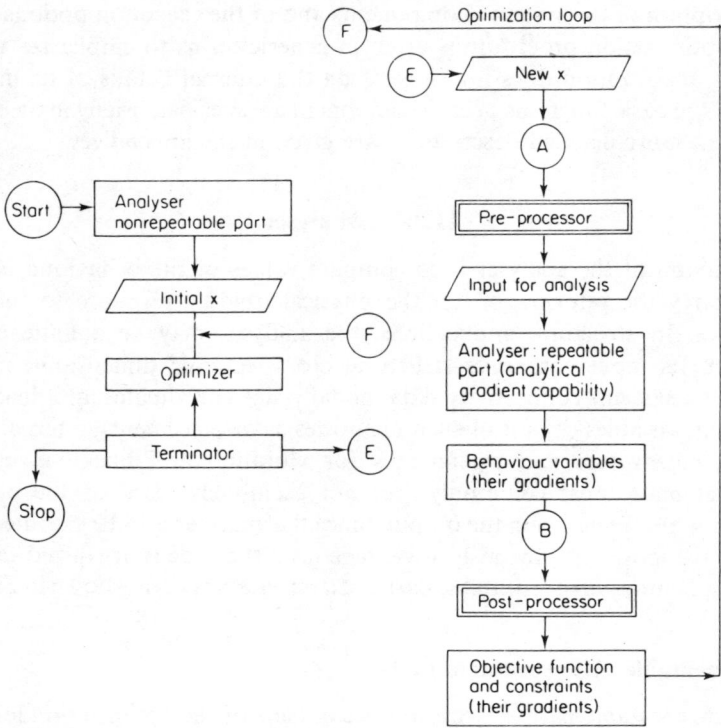

Fig. 11.9 Basic option flow (Fig. 11.8) modified to have the analyser split into non-repeatable and repeatable parts

Chapter 3 of this volume) describe the analytical techniques for computation of gradients (sensitivities) with respect to variables which affect the stiffness matrix linearly. Discussion of sensitivity in this chapter, therefore, is limited to showing in Appendix A how the higher-order derivatives can be computed by recursive formulas, and in Appendix B how analytical derivative techniques can be extended to include variables that affect the stiffness matrix in a nonlinear fashion.

Approximate analysis

The computer cost of large-scale structural optimization may be prohibitive unless an approximate analysis, significantly cheaper than the full analysis, is available as an option in the analyser. Appendix A contains a synopsis of several approximate analysis and reanalysis techniques, including the appropriate references, and provides an outline of a technique which reduces the problem dimensionality by condensation of the stiffness and mass matrices using orthogonal Ritz functions.

Table 11.1 Nonrepeatable (N) and repeatable (R) parts in finite element analysis based on displacement method

No.	Operation	Type of variable			
		Cross-sectional dimensions	Nodal coordinates	Connectivity	Elmt. displmnt. function type
1	Define material properties	N	N	N	N
2	Define coordinates of the nodes	N	R	N	N
3	Define each node's degrees of freedom	N	N	N	N
4	Define the loads	N	N	N	N
5	Define types of elements	N	N	N	R
6	Define cross-sectional dimensions	R	N	N	N
7	Define the element–node connectivity	N	N	R	N
8	Define supports	N	N	N	N
9	Define multipoint constraints	N	N	N	N
10	Compute elemental stiffness matrices	R	R	N	R
11	Compute elemental mass matrices	R	R	N	R
12	Assemble the structure stiffness matrix	R	R	R	R
13	Assemble the structure mass matrix	R	R	R	R
14	Decompose the stiffness matrix	R	R	R	R
15	Compute the displacements	R	R	R	R
16	Compute loads on elements	R	R	R	R
17	Assemble the structure geometrical stiffness matrix	R	R	R	R
18	Compute stresses	R	R	R	R
19	Compute eigenvalues and eigenmodes for vibration and/or buckling	R	R	R	R

11.3.2 Optimizer

The function of the optimizer is to search the design space by calculating a new vector of design variables x, on the basis of the values of the objective function and the constraints and, optionally, their gradients returned by the analyser in response to the previously issued vector x. Depending on the search algorithm employed, a record of changes in the design and behaviour variables over the past several passages through the optimization loop may be used to facilitate calculation for the new x. Abundant literature is available describing the inner workings of the optimization procedures (e.g. text references in [8] and [12]). In this discussion, therefore, the optimizer is viewed as a 'black box' and attention is focused on the type of data it requires from the rest of the system and on its execution options, since these influence organization of that system.

From this veiwpoint, the following execution options for the optimizer may be defined:

(a) Execution that requires current values of the objective function and constraints.
(b) Execution that requires current values of the objective function, constraints and their gradients.
(c) Execution accelerated because of linearity of either the objective function and/or constraints. That linearity can be detected by the optimizer itself or it can be determined by the optimizer's input.

As far as a record of past changes is concerned, it is assumed that such a record, if required, is set up and maintained within the optimizer. Also defined for the purpose of further discussion is 'the optimizer iteration', which is the sequence of operations the optimizer undertakes to generate a new x that 'improves' the object of the analysis. Depending on the execution option, the optimizer can send out requests for several executions of the analyser during one such iteration.

The type of algorithm (e.g., SUMT penalty technique, usable-feasible direction technique or even an optimality criteria algorithm) the optimizer employs to perform the search task is intentionally left unspecified to emphasize the freedom to use optimizers of various types.

11.3.3 Interface processors

The optimizer communicates with the analyser through a pre-processor and the analyser supplies the analysis information to the optimizer through a post-processor, as shown in the flow chart in Fig. 11.9. The pre-processor and the post-processor are user-supplied and problem-dependent. Capability of adding these two codes is the basis for the system's adaptability.

Pre-processor

In general, the function of the pre-processor is to convert the variables of the optimization process to a set of input parameters written in a format required by the analyser. For structural optimization, these parameters are structural member sizing and nodal point coordinate data which are actual physical design variables and which seldom are in a one-to-one direct equivalence relationship to the variables of the optimization process. Thus, in a typical application, the conversions within the pre-processor are not limited to format changes only but also include such commonly used techniques as variable linking, scaling, change from direct to reciprocal variables, etc.

Post-processor

The function of the post-processor is to compute the objective function and the constraints, and their gradients, if required, and to provide them in a format required by the optimizer. To do so the post-processor extracts the pertinent behaviour variables and, optionally, their gradients from the analysis output and combines them with the allowable values of these quantities in the equations specific to the design problem at hand. Frequently the allowable values, instead of being constants, are functions of x as, for example, in the case of local buckling constraints. Computation of such variable allowable values (and their derivatives) can be included among the post-processor's functions. The post-processor may also include a chain differentiation needed to calculate gradients of the structural behaviour variables with respect to those design variables which have a nonlinear influence on the stiffness matrix. Another example of a post-processor function is special handling of potentially discontinuous constraints. Appendix B provides details of these operations. For the sake of computational efficiency, the post-processor may also be equipped with logic to limit the set of constraints to only those whose probability of remaining or becoming active is high, as proposed in Ref. [10].

11.3.4 Terminator

The function of the terminator program is to monitor termination criteria and to issue appropriate commands to the connecting network to end the procedure when the criteria are satisfied. These criteria, discussed in Chapters 2 and 8, can be built into the optimizer as in the program described in Refs [2] and [3]. In such cases the role of the terminator is to provide an additional safeguard against indefinite continuation of the optimization iteration that may occur, if the termination criteria built into the optimizer fail. Another possible function of the terminator is to interrupt a partially completed optimization procedure to allow the user to evaluate the course it is taking and to decide on its continuation. This rôle can be adequately played by employing the duality principles discussed in Chapters 2 and 3 and exploited in other chapters, particularly Chapter 10. There are more complex cases, to be discussed later, in which the terminator incorporates problem-dependent termination criteria.

11.3.5 Connecting Network

The connecting network is frequently called 'executive software'. Its function is to carry out a computational process such as shown in Fig. 11.9, and its more complex forms to be considered subsequently. The connecting network that is shown in the block diagram in Fig. 11.1 (c) does not appear directly on the flow charts discussed herein, and it should be thought of as a medium through which the numerical processes illustrated in these flow charts are executed. Its role is completely analogous to that of the main program of Fig. 11.1(b) and Fig. 11.5, while other modules of the system are analogous to subroutines. The connecting network also enables the user to monitor progress of the optimization, to interrupt it at predetermined points and to restart without loss of information generated before the interruption. The latter capability opens the important possibility of an engineer overriding the optimizer decisions according to his judgment. Capability to perform the following functions will suffice to meet the operational requirements of a programing system:

1. Executing the programs in a computational sequence.
2. Performing logical functions required by the sequence, typically: branching on an if-test, looping, skipping to a labelled statement.
3. Storing, permanently or temporarily, and retrieving data generated by the programs or input externally.

A connecting network is made up of:

1. A repertoire of commands (job control language—JCL) for executing programs and for creating, changing, storing, retrieving and displaying the files.
2. The permanent or temporary information files that contain programs and numerical data.
3. The procedure files which contain JCL commands.

Several possibilities exist regarding the connecting network's relationship to the operating system of the computer on which it operates. Such a network, for example, can be coded to execute as a single program as in the program described in Ref. [13]. It may also be intertwined with the operating system to let the user invoke some of the operating system's JCL commands directly. An extreme case when the connecting network is a commercially available operating system with nothing added to it, is described in Ref. [14]. The same concept is used in the programming system reported in Ref. [15].

11.4 ORGANIZATION OF THE EXECUTION FLOW

A variety of execution flows can be set up using the components described previously. Organization of each flow option depends on the type of optimization procedure and on whether gradients are required by the optimizer and if so, whether these gradients can be generated analytically or by finite differences. The flow options discussed herein are the five shown in Table 11.2 and an additional option for the two-level optimization described in Appendix C.

Table 11.2 Optimization flow options

Method	No gradients supplied to optimizer	Gradients supplied to optimizer	
		Finite difference	analytical
NLP	1.1	1.2	1.3
PLA	Not applicable	2.2	2.3

11.4.1 Matrix of Options

The two optimization procedures defining rows of Table 11.2 are: nonlinear mathematical programming (NLP) and piecewise linear analysis (PLA). In the conventional NLP approach, the objective function and constraints are treated as non-linear functions of the design variables.

In the PLA procedure which was successfully used in a number of applications (Refs [10], [16], [17], and [18],) the optimization process progresses as a sequence of linear optimization subproblems (stages) within which the objective function and constraint functions are assumed to be linear. At the outset of each such stage, the functions and their gradients are computed once, and the gradients are assumed to be constant for the remainder of the stage. The objective function and the constraints are computed within the stage by using a linear approximation based on the Taylor series expansion:

$$f = f_0 + \nabla f_0 (\Delta x)$$

where f stands for the values of the objective function or constraint function at $x_0 + \Delta x$; ∇f is the gradient of the function with respect to the design variables; and subscript 0 indicates the values computed at the beginning of the stage.

The error of the linear approximation is controlled by limiting $x_0 + \Delta x$ to a neighbourhood of x_0 by side constraints on x or by an additional constraint limiting the change of the objective function allowed in one linear stage. The efficiency of PLA stems from replacing the full analysis of the physical problem with analysis by the linear extrapolation which in structural applications can be executed at least an order of magnitude faster than the full analysis. Additional time savings result when the optimizer is capable of executing faster if the problem is defined as linear (option (c) in Section 11.3.2). The number of consecutive linear stages required for overall convergence depends on the degree of nonlinearity.

The analyser computing capabilities corresponding to columns of Table 11.2 are: computation of the behaviour variables without gradients, calculation of gradients computed by finite differences, and calculation of gradients computed analytically. Each of the five resulting options defined by intersections of rows and columns in Table 11.2 calls for its own organization of the procedure flow.

11.4.2 Procedure Organization for Each Option

The organization of the flow for each procedure is described in this section using flow charts. Options 1.1 and 1.3 are covered by the flow chart in Fig. 11.9 in which the analyser is assumed to have a capability to produce gradients analytically (as indicated in parenthesis). When such capability does not exist, a finite difference procedure, shown in Fig. 11.10, is substituted for the portion of the flow chart

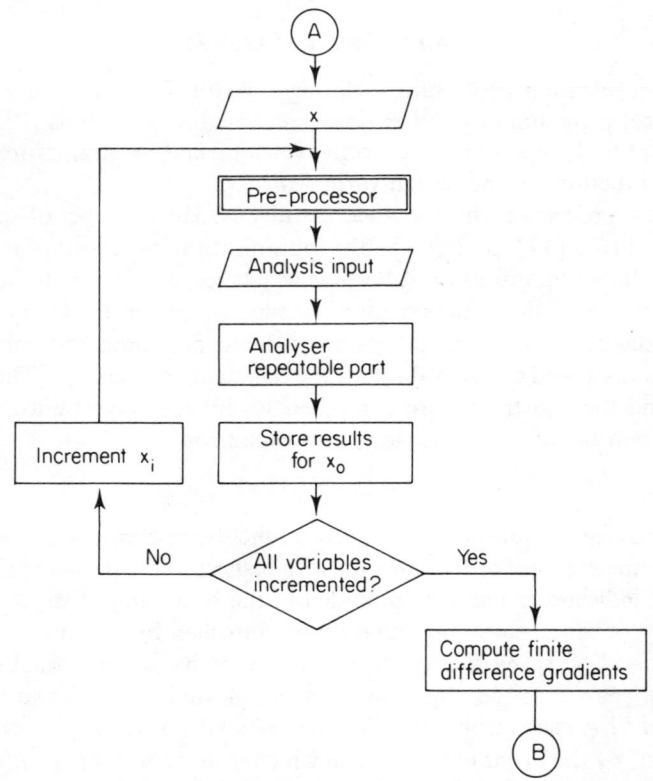

Fig. 11.10 Flow chart for finite difference gradient computation to be substituted for the portion of the flow chart between connectors A and B in Fig. 11.9 to form Option 1.2

shown in Fig. 11.9 contained between connectors A and B. The resulting flow chart corresponds to option 1.2.

Option 2.3 is organized as shown in Fig. 11.11. The inner loop comprises the optimizer and includes a very fast linear extrapolation in lieu of analyser

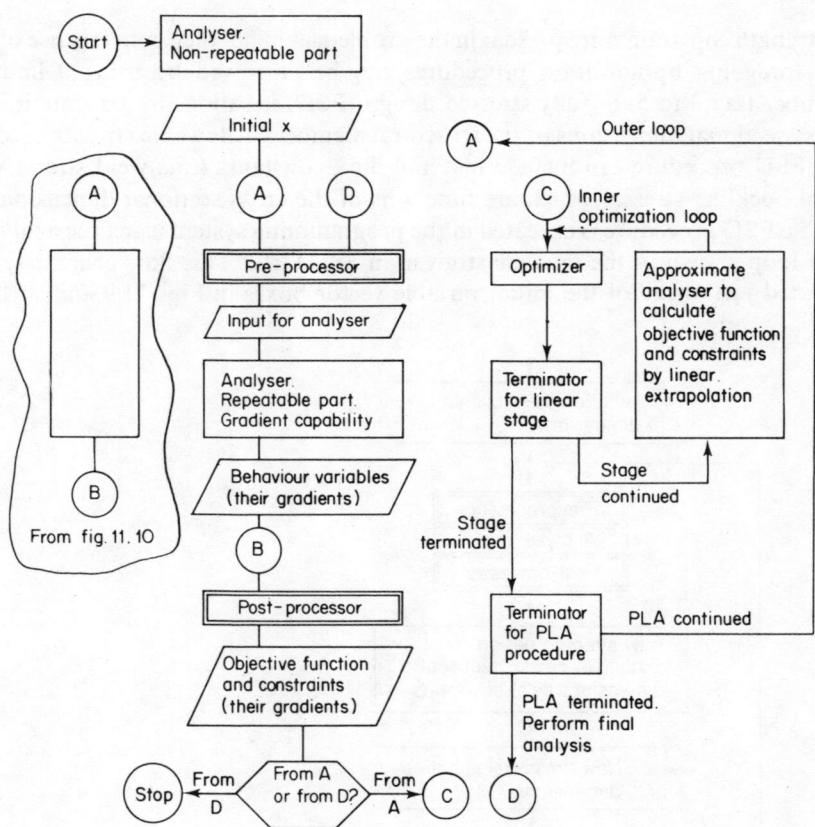

Fig. 11.11   Flow chart for Option 2.3 (Table 11.2)

execution so that the analyser with the associated processors can be moved outside of that loop. The outer loop represents consecutive linear stages when the analyser is used to update the behaviour variables and their gradients required for the next extrapolation. In this option, the analyser possesses an analytical gradient capability. If there is no such capability, option 2.2 is used. The difference in flow organization lies in substituting the flow chart from Fig. 11.10 between connectors A and B in the flow chart in Fig. 11.11 as shown in the inset.

The terminator shown in Fig. 11.11 is split in two parts to indicate the need for monitoring convergence of both inner and outer loops. This does not preclude implementing it as a single program.

There is one final execution of the analyser following termination of the PLA. It is necessary in order to obtain the final proof of the constraint satisfaction.

11.4.3 Generation of Initial Cross-sectional Dimensions by a Fully Stressed Design

If strength constraints are present in the problem at hand, then convergence of all the foregoing optimization procedures can be improved by using a limited number (say 3 to 5) of fully stressed design (FSD) iterations to generate initial cross-sectional dimensions of the structural members. Allowable stresses used in the FSD procedure can include material limit constants (e.g., yield stress) and local buckling stresses which are functions of the cross-sectional dimensions.

The FSD procedure is executed in the programming system using the analyser in a loop shown in the flow chart given in Fig. 11.12. This flow chart may be inserted just ahead of the initial variable vector boxes in Figs 11.9 and 11.11.

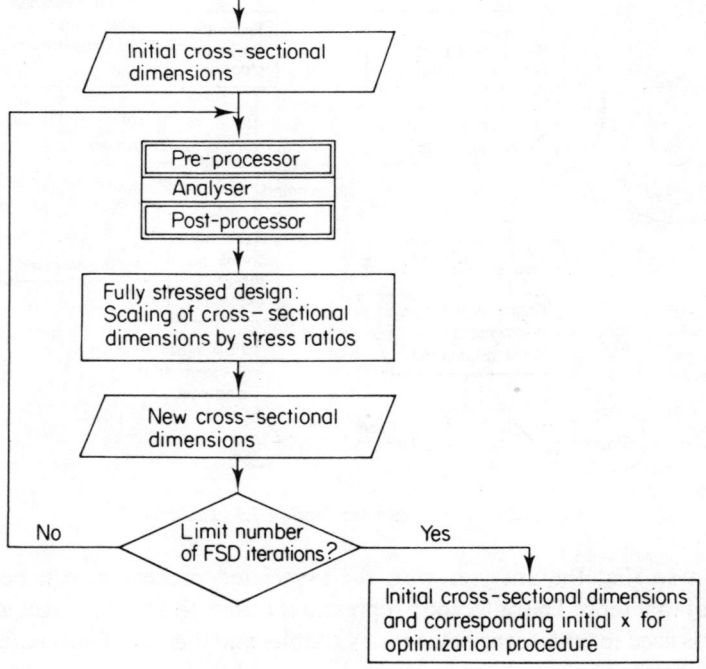

Fig. 11.12 Flow chart for generation of the initial cross-sections using fully stressed design (FSD) iterations

11.4.4 Cost of the data manipulations

To close the discussion of execution flow options (which include the two-level optimization procedure in Appendix C), one should point out that the capability of arranging the programming system components for execution in a vareity of ways is paid for in the increased cost of the data manipulation. Since each of the programs (system components) executes separately, the data manipulations for each program execution involve reading from (disc) storage the program code and input data and writing the output back on storage, as shown in Fig. 11.1. These data input and output operations are performed in addition to the other data manipulations the program may generate in the course of its execution and which represent an overhead cost. This is incurred repetitively in each passage of the optimization loop and, therefore, may be substantial. No comparable overhead cost exists in the main program–subroutine codes contained entirely in high-speed memory. In programs of a 'black box' type large enough to require use of storage (segmentation or overlaying), the overhead cost can usually be reduced by exploiting unique characteristics of the particular code and data.

There is one benefit of the intensive use of data storage in a programming system. It is an opportunity to access and review the intermediate results that flow from one system module to another. Such reviews enable an engineer to judge whether the optimization progresses in an acceptable manner, and to take appropriate action if it does not. Importance of such human participation in optimization processes, especially when a substantial computer cost is involved, cannot be overemphasized.

11.4.5 Adapting a Programming System to a Particular Application

It is important to distinguish between two basic forms in which a programming system exists. These are skeleton form and specialized form. The skeleton form, consists of the problem-independent parts of the system:

1. The optimizer program(s), the analyser program(s), and the programs controlling execution of the loops in Figs 11.10 (finite difference gradients) and 11.12 (FSD procedure).
2. Blocks of control statements to execute computational sequences for each execution option of the system.

For a specific application, the skeleton form must be turned into a specialized form. Problem-dependent interface processors and the input data must be created and stored as files. In addition, standard names in the JCL statement file corresponding to the option chosen must be replaced with names selected for the problem-dependent files. Thus, the specialized form ready to be executed consists of the skeleton form permanent files, an additional set of files containing the problem-dependent program and data, and a corresponding set of JCL statements.

Once the specialized form has been set up for a particular application, it can be protected from unauthorized alterations by using 'software locks' (passwords) on all its files except the input data files. Several such specialized forms can be created from the common skeleton form for a variety of applications. Each such 'frozen' specialized form can be used as a 'black box' for a given class of problems that differ only by input data.

In industrial applications preparation of specialized forms would fall naturally into the domain of a staff specialist while their application would be the task of production-oriented engineers. In research applications the system's modularity permits its major components, such as particular programs used for the analyser and the optimizer, to be replaced with other equivalent programs, and execution flows different from those described above can also be constructed. Thus, a programming system can be used as a test-bed for development of new optimization procedures. Although a programming system has been discussed here in the context of structural optimization, it could be adapted to synthesis problems in other engineering disciplines by an appropriate replacement of the analyser (e.g., replacing structural analysis by a computational aerodynamic program).

This concludes the description in generic terms of the components and organization options of a programming system linking analysis and optimization. The following section contains demonstration examples solved using two different systems described in Refs [13] and [15].

11.5 APPLICATION EXAMPLES FOR PROGRAMMING SYSTEMS

Implementations of the programming system approach are described in detail in Refs [13] and [15], which also provide the application examples to be discussed in this section. These are grouped by their sources.

The first is from Ref. [13] and illustrates resizing of a wing structure for flutter carried out by a system specialized for static strength and flutter optimization. This system is built on the basis of a finite element structural analysis program described in Refs [19] and [20]. The particular program has a modular organization of individual processors which communicate through a data base described in Ref. [21], and a command language (connecting network) to execute the processors and to manipulate the data. For optimization purposes, the program is augmented by processors for gradient computation, definition of design variables (including linking), and a general-purpose Newton–Raphson search algorithm. It also includes a processor to calculate aerodynamic loads. A single execution option, corresponding to option 1.1 in Table 11.2, is available to carry out strength and flutter optimizations sequentially by means of a penalty function technique with surface finite element thicknesses as design variables.

The subsequent four examples are from Ref. [15], which describes a system developed for applications whose precise nature and scope was left intentionally unspecified. For this purpose the definitions of variables, objective function and constraints are left for the user to define in the interface processors, and the operating system's command language is used as a connecting network to implement various execution options (e.g., Table 11.2 and Appendix C).

11.5.1 Example of Wing Flutter Resizing

Resizing of a delta wing for flutter is selected for the first example because of difficulty of the flutter optimization problem. The difficulty stems from complex relationships between the flutter mechanism and the wing vibratory and unsteady aerodynamic characteristics which depend upon the wing structure and mass variables. The particular example for a clipped delta wing is reported in Ref. [13] and shown in Fig. 11.13. The wing has a symmetric biconvex airfoil and is built up

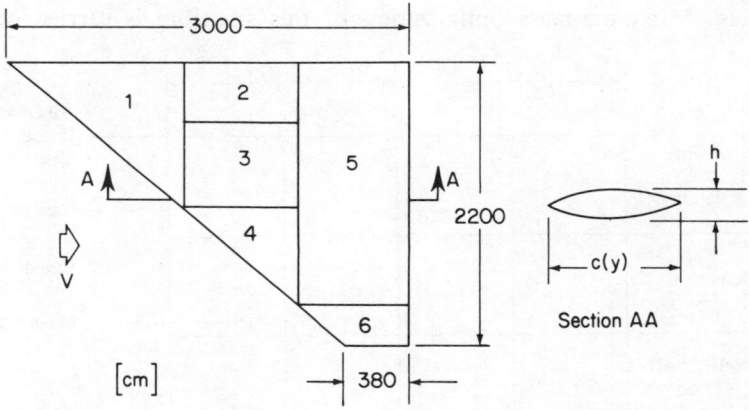

Fig. 11.13 Wing geometry (Ref. [13])

of honeycomb sandwich panels made of titanium. Thicknesses of the face sheets are used as design variables, and these are taken to be the same for upper and lower cover panels and constant in each of the six regions shown in Fig. 11.13. The flutter constraint is a velocity of Mach .6 at an altitude of 1500 m with a 20 per cent safety margin.

A Rayleigh–Ritz approximation (see Appendix A) is used to reduce problem dimensionality with the natural vibration modes used as the Ritz functions. A total of 42 analyses are needed in the course of the optimization; each analysis includes computation of the eigenfrequencies and eigenmodes of the condensed problem, and evaluation of the flutter speed and its derivatives. However, the natural vibration modes and frequencies of the full problem are calculated only five times.

The result is a structural mass of the cover panels of 3590 kg with most of the mass concentrated in panel 3, with panels 1, 2 and 6 remaining at minimum gauge (0.5 mm).

11.5.1 Miscellaneous Examples

The four examples from Ref. [15] presented in this subsection, when taken collectively, illustrate the variety of design variable formulations and the types of constraints that can be handled by a programming system approach. The design variables include cross-sectional dimensions with linking and in a reciprocal form (see Ref. [10]), and coordinates of the nodal points. The constraints include limits on stresses, displacements, natural frequencies, vibration modes, and buckling loads. All options of the optimization procedure shown in Table 11.2 were exercised in the course of generating the examples.

Example 1: Rod–panel structure The finite element model of a shear panel, stiffened in both directions with rods which have axial stiffness only, is shown in Fig. 11.14. Minimum-mass optimization of this structure is carried out for

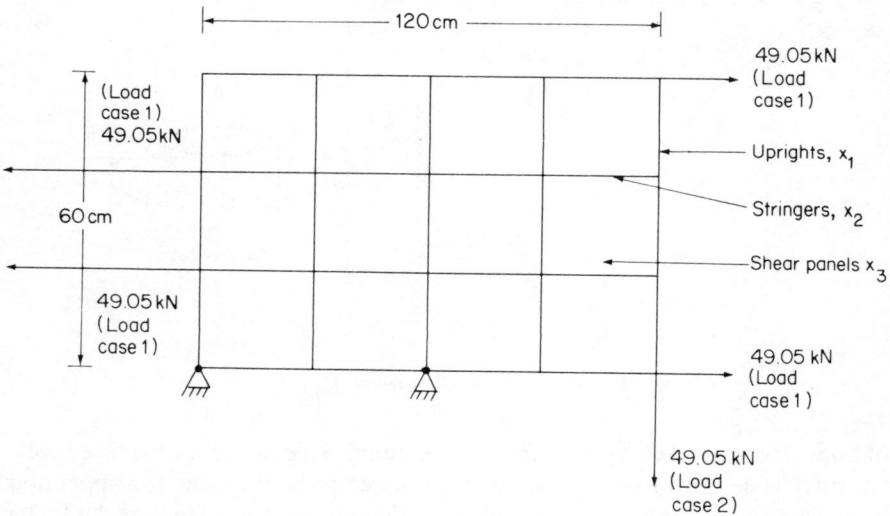

Fig. 11.14 Example 1, stiffened panel structure

different sets of design variables and constraints to illustrate adaptability of the programming system with respect to these quantities. Initially, three design variables were chosen, x_1 = cross-sectional area of transverse stringers, x_2 = cross-sections area of longitudinal stringers and x_3 = thickness of the panels. In this case and in all other examples, the cross-sectional dimension variables were handled by the optimizer in their reciprocal form to improve convergence (Ref. [10]). The structure was optimized, as a nonlinear optimization problem, with no gradients required from the analyser (option 1.1) and was started at minimum gauge. The set of constraints consisted of stresses corresponding to the two loading cases shown in Fig. 11.14. The starting values and the final optimized values of the design variables are given in Table 11.3. The history of iterations in the optimization process, shown in Fig. 11.15, indicates that the objective function converges smoothly and there is no significant variation after three iterations.

Table 11.3 Initial and final values of design variables for Example 1 (Fig. 11.14)

	Initial	Final
x_1	10.00 cm^2	8.46 cm^2
x_2	10.00 cm^2	4.30 cm^2
x_3	0.50 cm	0.17 cm

The rod-panel structure was next optimized considering ten design variables as shown in the inset in Fig. 11.16. A simple modification of only the pre-processor code implemented this change. The problem was treated by both NLP (options 1.1 and 1.2) and PLA (option 2.2). In this and all subsequent applications of the PLA, the objective function was allowed to change by 30 per cent in one linear stage, and there were no move limits on the design variables. Only three linear stages were needed for satisfactory convergence using the PLA option. The difference in the final objective function values obtained by the two procedures was only 1.8 per cent. The history of iterations of PLA (option 2.2) is shown in Fig. 11.16. Discontinuities of the objective functions shown in Fig. 11.15 at the beginning of each new stage reflect the result of new analyses performed at the outset of each stage. Figure 11.16 also illustrates the iterations performed by

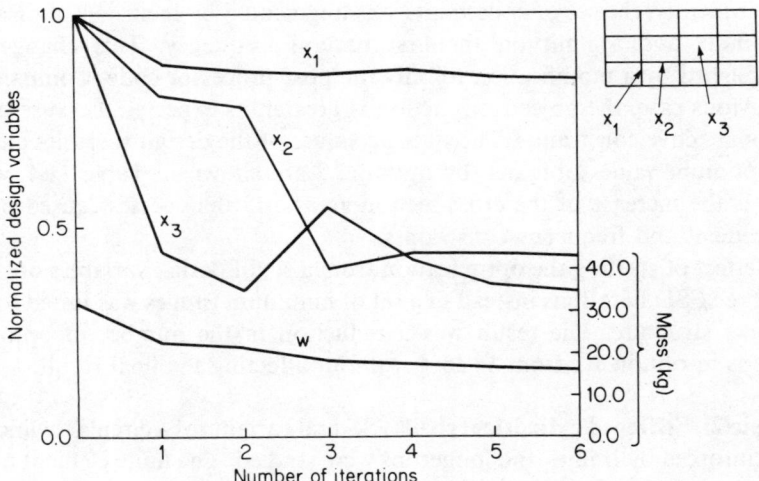

Fig. 11.15 History of iterations for stiffened panel structure—Option 1.1

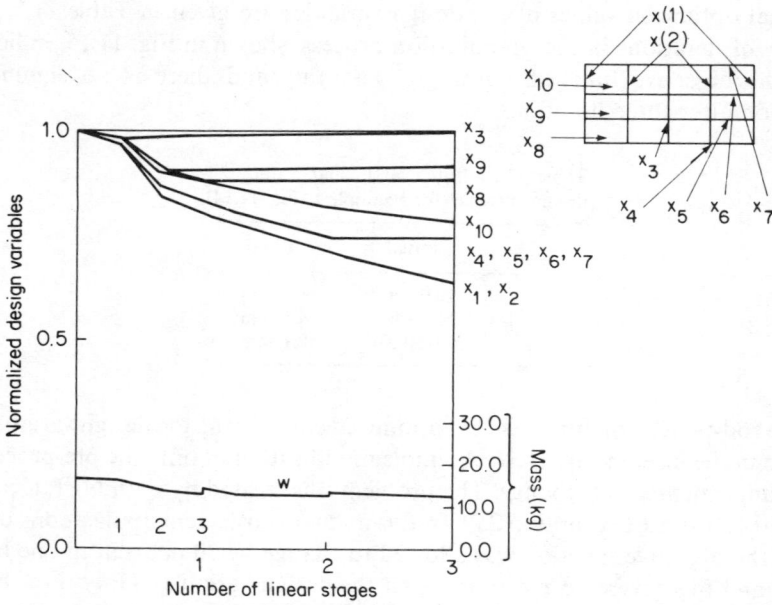

Fig. 11.16 History of iterations for stiffened panel structure—Option 2.2 *Inset*: the design variables

the optimizer within a linear stage, although for clarity this detail is shown only for the first linear stage. The increase of the number of design variables from 3 to 10 resulted in a 21 per cent lighter structure.

Subsequently, the set of constraints was augmented by inclusion of displacement limits and a limit on the first natural frequency. This change was accomplished by a modification to only the post-processor code. Compared to the previous case, the objective function is greater, as expected, because of the additional active constraints. The starting values of the design variables and the final optimum values obtained by option 2.2 are shown in Table 11.4, which indicates the increase of the cross-section areas and thicknesses caused by the displacement and frequency constraints.

The effect of starting the optimization from a set of design variables obtained from three FSD iterations instead of a set of minimum gauges was tested for the rod-panel structure. The result was a reduction in the number of optimizer iterations in option 1.1 from 14 to 4, without affecting the final result.

Example 2: Stiffened cylindrical shell Several variants of a circular cylindrical shell reinforced by frames and longerons were studied. The finite element model of the computationally largest variant (referred to as variant 1) is shown in Fig. 11.17. This variant is built up of membrane panels to represent the skin and beam

Table 11.4 Initial and final values of design variables for Example 1 (Figs 11.14, 11.16) for larger number of design variables

	Stress constraints only		Constraints on stress, displacement and frequency	
	Initial	Final	Initial	Final
x_1	6.51 cm^2	4.17 cm^2	6.51 cm^2	7.09 cm^2
x_2	6.52 cm^2	4.19 cm^2	6.52 cm^2	4.91 cm^2
x_3	8.11 cm^2	8.06 cm^2	8.11 cm^2	23.16 cm^2
x_4	5.61 cm^2	4.16 cm^2	5.61 cm^2	6.15 cm^2
x_5	5.63 cm^2	4.15 cm^2	5.63 cm^2	4.42 cm^2
x_6	5.62 cm^2	4.19 cm^2	5.62 cm^2	4.74 cm^2
x_7	5.66 cm^2	4.17 cm^2	5.66 cm^2	9.35 cm^2
x_8	1.91 cm	1.69 cm	1.91 cm	10.00 cm
x_9	1.86 cm	1.70 cm	1.86 cm	10.00 cm
x_{10}	0.18 cm	0.14 cm	0.18 cm	0.24 cm

elements (axial, bending and torsional stiffnesses) to simulate transverse frames and longerons. Each frame and longeron may be regarded as a lumped representation (Ref. [22]) of several real frames and longerons. One end of the shell is clamped around the circumference, and the other end is loaded by concentrated loads simulating distributed forces equivalent to a transverse force and a torque. This variant has a large cutout and a floor and represents a simplified model of a transport aircraft fuselage segment.

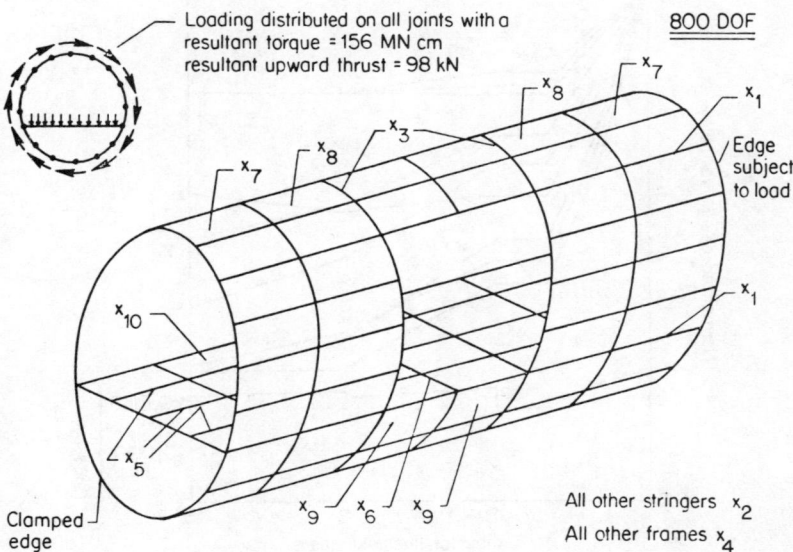

Fig. 11.17 Example 2, stiffened cylindrical shell, variant 1

This structure was expected to constitute a demanding test case, for the following two reasons. Firstly, the model contains 798 degrees of freedom, so it is a computationally large problem as far as optimization by mathematical programming is concerned. Secondly, the overall bending state of stress in a shell of this type depends on the in-plane stiffness of the frames; therefore, the design variables that govern the member cross-sectional dimensions become strongly coupled (e.g., Ref. [22]) and it is more difficult for the optimization process to converge.

Variant 1 is optimized by PLA using finite difference gradients (option 2.2) and the ten design variables shown in Fig. 11.17. As indicated in the figure, many structural parameters are linked to a single design variable. Variables x_1 through x_6 govern the cross-sectional areas of the beam elements which have a channel cross-section whose proportions remain constant as its area changes. Thus, the cross-sectional area becomes a single variable that governs all the beam's stiffness parameters. Variables x_7 through x_{10} govern the membrane panel thicknesses. Initial values of the variables are chosen arbitrarily within the range of values known to apply in typical aircraft fuselage structures. The history of iterations of optimization with stress constraints on beam elements and equivalent stress (Huber-von Mises stress) constraints on the panel elements is shown in Fig. 11.18. Convergence is quite good considering the problem size and the use of piecewise linear approximations. Table 11.5 shows the starting values for the design variables and the final optimum values. As might be expected, the elements

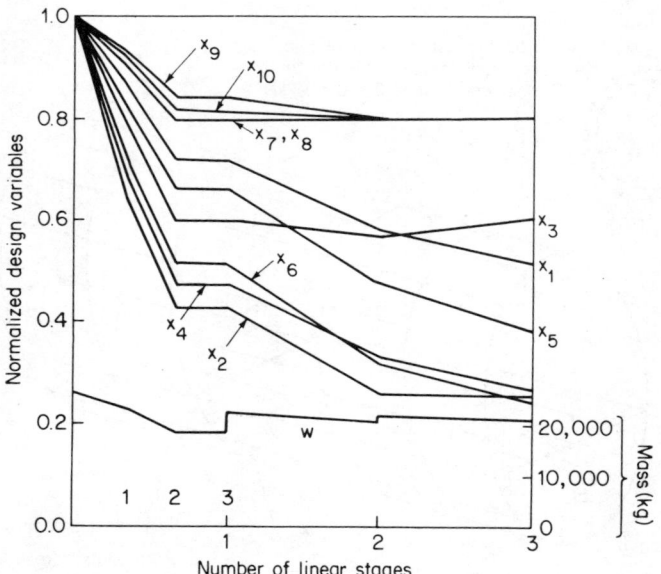

Fig. 11.18 History of iterations for Example 2, variant 1, using Option 2.2

Table 11.5 Initial and final values of design variables for Example 2, variant 1 (Fig. 11.17)

Design Variable	Initial	Final
$x_1{}^*$	0.749	0.397
x_2	0.400	0.100
x_3	0.625	0.376
x_4	0.435	0.115
x_5	0.667	0.256
$x_6{}^*$	0.500	0.122
x_7	0.125 cm	0.100 cm
x_8	0.125 cm	0.100 cm
x_9	0.125 cm	0.100 cm
x_{10}	0.125 cm	0.100 cm

* x_1 to x_6 have been normalized with respect to initial cross-sectional area 108.0 cm^2

flanking the cutout have 'grown' in the optimization process, as graphically illustrated in Fig. 11.19.

To further demonstrate the adaptability of the procedure, a simplified variant 2 of the shell structure is formed by eliminating the floor and two end bays and using rod elements to represent the longerons. Three design variables, one for

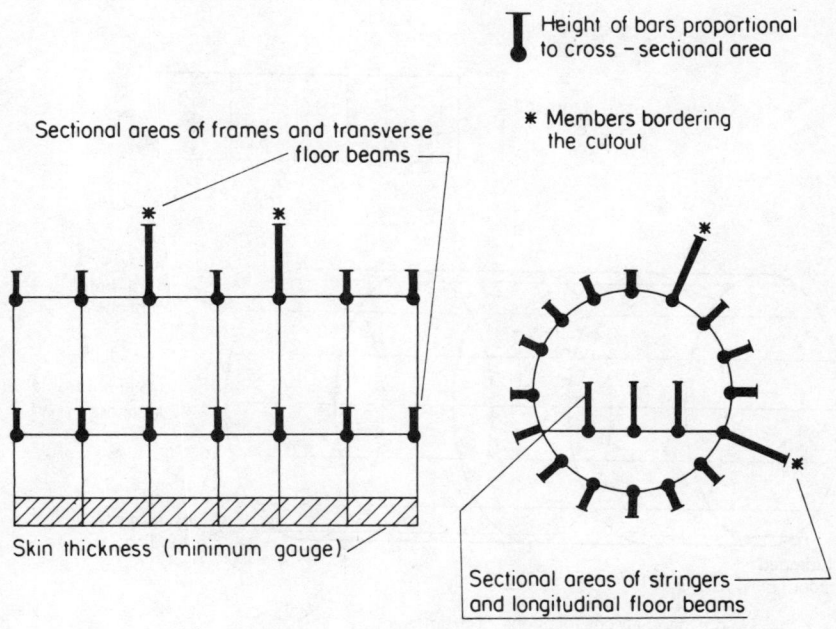

Fig. 11.19 Relative member sizes obtained by optimization for Example 2, variant 1

frames, one for longerons and one for the skin, are used in this case. Initially, the structure is optimized with stress constraints only and the resultant structure subsequently optimized with an additional overall shell buckling constraint. This requires a 21 per cent increase of the buckling load over and above the buckling load computed for the structure optimized with stress constraints only. Both optimizations are carried out using option 1.1. A comparison of these two results show that the structural mass increased by 9.6 per cent because of the additional buckling constraint. Additional optimization of this variant (with two loading cases), carried out with stress constraints only, using analytical gradients (option 2.3), reduces the execution time to approximately one-sixth of that required for option 1.1.

Locations of the node points in the finite element model represent design variables in a further simplified variant 3 of the shell structure. This variant has the cutout eliminated, is subject to only one loading case (transverse force), but has the longerons restored to the beam element representation. Previously defined cross-sectional variables are retained and three geometrical variables governed locations of the three intermediate frames. Optimization using option 1.1 with stress and overall shell buckling constraints results in an expected translation of the frames toward the loaded and unsupported end, where the buckling displacements are largest and an additional support is most needed. The initial and final frame locations are shown in Fig. 11.20.

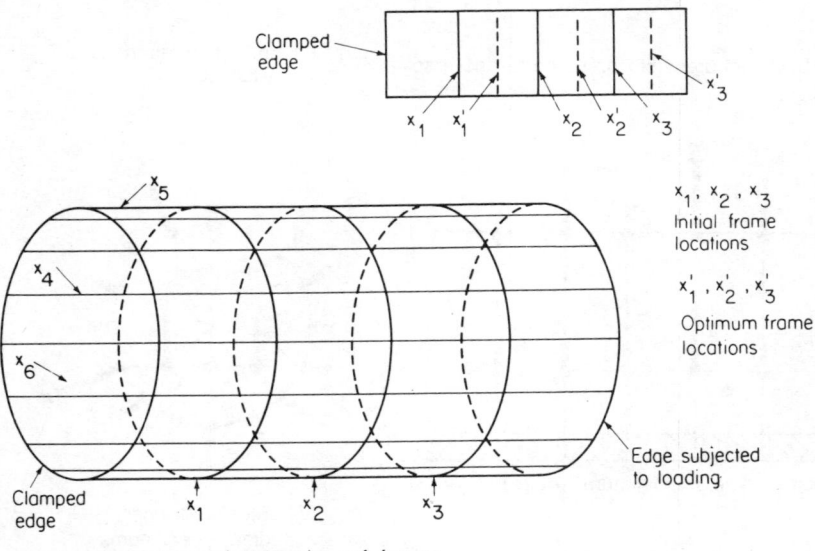

Fig. 11.20 Initial and optimized positions of transverse frames in Example 2, variant 3

Example 3: Portal framework The framework shown in Fig. 11.21 is optimized with geometrical variables only, to demonstrate the optimizer's ability to transform structural shapes. The variables defined in Fig. 11.21 are intended to allow the frame to transfrom itself into a truss. The constraints imposed on stress and the horizontal displacement are indicated in Fig. 11.21. Optimization, carried

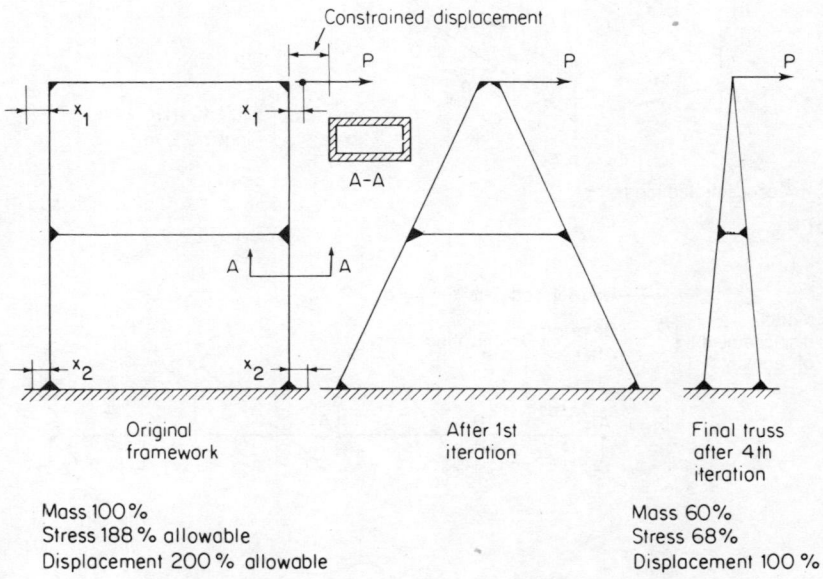

Fig. 11.21 Transformation of a framework (Example 3) to a truss by optimization with geometrical variables

out by means of option 1.1, has indeed produced the expected transformation of shape to an almost triangular truss. A side constraint on length of the top horizontal member, necessary to preserve that member's nonzero length to avoid a matrix singularity, has kept the top of the frame from shrinking to a point.

Example 4: Torsion box This example demonstrates 'tuning' of stiffness and mass distributions to achieve a prescribed change in an original structure's vibrational mode shapes. The structure is a torsion box shown in Fig. 11.22, built up of membrane panels, whose thicknesses are design variables as indicated in the figure. Because of concentrated nonstructural masses affixed to one side of the box as shown in Fig. 11.22(a) the vibration modes of the structure, initially uniformly sized to a minimum gauge, exhibit a distinct torsion-bending coupling in the first four modes. These modes are illustrated in Fig. 11.22(b) by the displacement of segment AB seen in view C (Fig. 11.22(a)). To reduce that

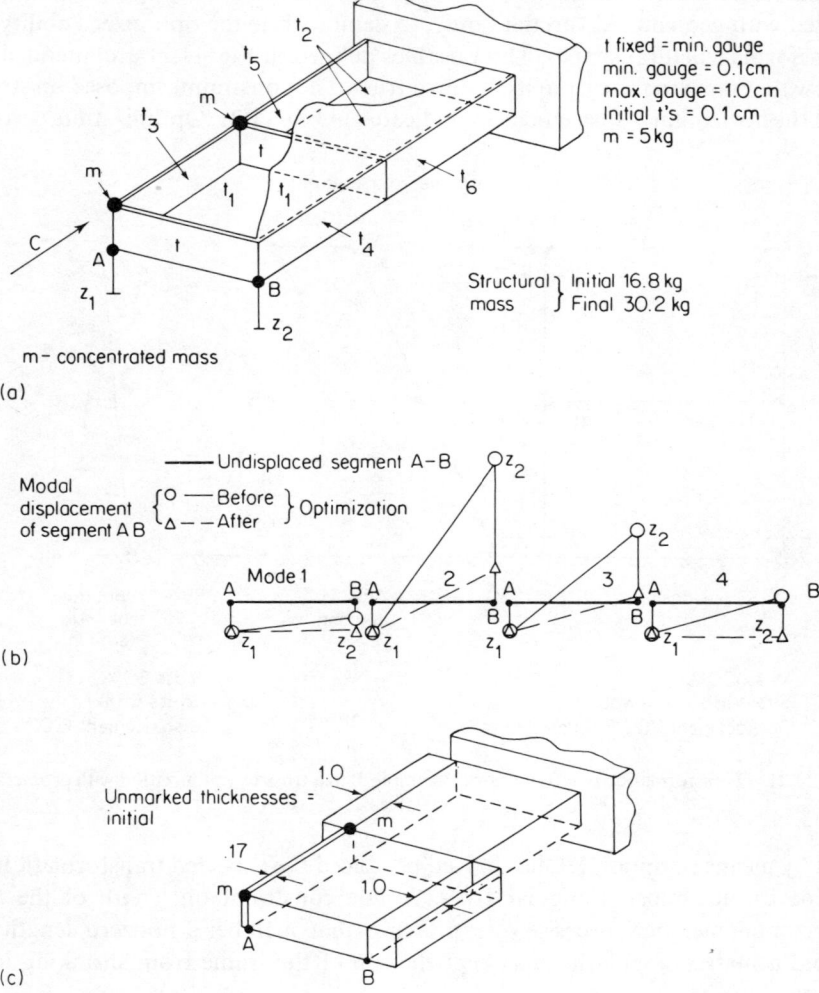

Fig. 11.22 Optimization of a torsion box (Example 4) to reduce torsion bending coupling in modes 1, 2, 3 and 4

coupling, constraints of

$$\left|\frac{z_2 - z_1}{z_1}\right| \leq 2$$

are imposed on the free-end vertical displacements z_1 (point A) and z_2 (point B) in modes 1 through 4. The result is a set of new modes shown in Fig. 11.22(b) which comply with the constraints. The thickness changes required to meet the constraints are indicated in Fig. 11.22(c). Note the increase of thickness t_3 and t_5 in the spar beam directly supporting the concentrated masses, and the increase of t_4 in a panel that forms a counterbalance to the fixed masses.

11.5.3 Summary of Applications

Summarizing the application examples, the following observations may be made. Transforming the system from one optimization option to another is simple to accomplish by changing the sequence in which components of the system were called for execution. Adaptation from one variable and constraint combination to another is carried out by changes in the pre- and post-processor codes. These adaptations, as well as changes from one structure to another, do not require any changes to the connecting network nor to the analyser and optimizer.

It is a routine matter to monitor the status of the optimization process by means of displaying the intermediate data files. Stopping and restarting are facilitated by storing intermediate data.

Among the options tested, option 2.3, the PLA with analytical gradients is by far the most cost-effective, especially for larger problems such as variant 2 in Example 2. Computer time required for that application was 40 sec of CPU time on a CDC-Cyber 175 for option 2.3, compared to 250 sec of CPU time consumed by the least efficient method, option 1.1. The bulk of the execution cost of a programming system execution lies in the file manipulation operations because the secondary (disc) storage and retrieval of data are required at interfaces of each two consecutive programs.

11.6 CONCLUDING REMARKS

Optimization implementation schemes are presented in order of increasing adaptability to diverse requirement which design practice impose on its supporting tools. The increased adaptability comes at a cost, in general, of greater tool development effort, greater amount of time needed to learn how to use the tool effectively, and greater computer overhead execution cost. The latter term refers to the cost of moving the data from one computational module of the procedure to another, as opposed to the cost of performing calculations within the modules.

The overhead cost is relatively large in the programming system scheme, where by definition the data flows from one module to another solely via the secondary (disc) storage route. That cost is the smallest in special-purpose 'black box' programs where every advantage can be taken of the problem's unique characteristics to use storage as little as possible.†

---

† This is true also for the so-called 'virtual memory' machines in which the core–disc duality of the memory still exists although it is hidden from the user by an operating system.

Consequently, one may offer the following comparative assessment:
1. A 'black box' is a good choice for problems:
   —which have a well-settled definition of objective function, constraints, and variables;
   —where the analysis is dimensionally small enough to require a minimum of the (disc) storage data manipulations;
   —which are likely to generate sufficient use to amortize the cost of development within a reasonable time.
2. A 'programming system' is a good choice for problems:
   —of such a broad scope that the possible definitions of their objective function, constraints, variables and solution strategies are too numerous or unknown at the time of development;
   —where the analysis is dimensionally so large that as to require substantial secondary (disc) storage data manipulations relative to the amount of such manipulations called for by the other components of the system.
   —where development effort can be lessened by using existing, stand-alone programs;
   —which are of interest to a group of engineers who are likely to be divided into developers and users.
3. The main program–subroutines concept is a choice that fits naturally between the above two extremes.

APPENDIX A SELECTED TECHNIQUES FOR THE ANALYSER APPROXIMATE ANALYSIS

The goal of a design-oriented approximate analysis is to obtain a solution for a modified structure at a sharply reduced cost, compared to the full analysis, and with accuracy adequate for the purposes of deciding how to modify the structure next. Several techniques exist for such analysis. They are categorized in Fig. A1 and briefly reviewed. Subsequently, a particular technique for dimensionality reduction is described with some mathematical detail.

Techniques for Approximate Analysis

The starting point of the review is a relatively highly refined finite element model (left-hand column in Fig. A1). The objective is to reduce the number of times that model is fully analysed in a repetitive optimization loop. One way to proceed is to extrapolate from the initially fully analysed structure using a Taylor series (Refs [9], [23], [24]) or a perturbation method (Ref. [25]). The latter expresses the structural changes in the form of a power series of a small parameter and ultimately leads to an algorithm identical to that of the former, although it has an advantage of not requiring analytical expressions for stiffness matrix derivatives.

The other category of methods is problem dimensionality reduction, which results in a significantly smaller number of unknowns to be repetitively reeval-

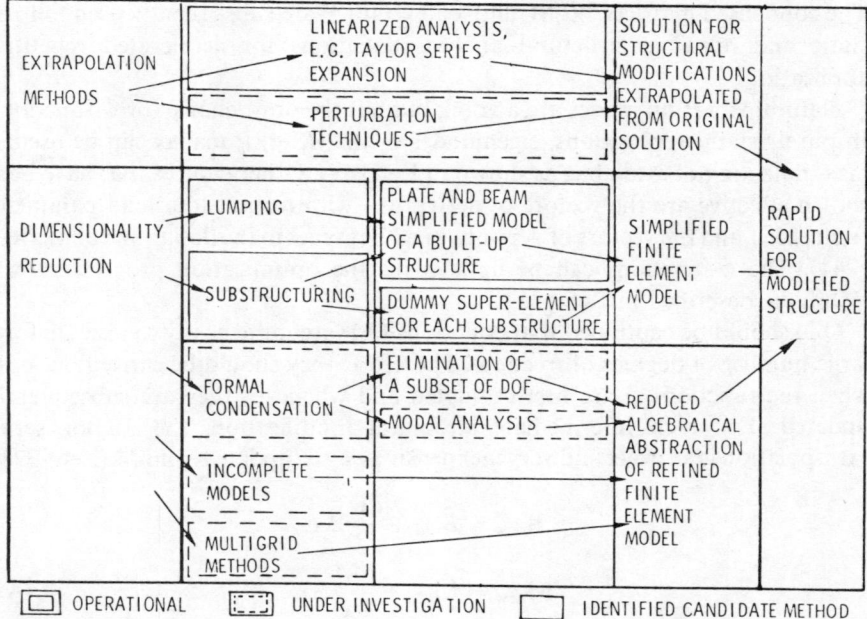

Fig. A1 Candidate methods to accelerate structural analysis

uated in the optimization loop. The reduction can be achieved by lumping, that is by creating a simplified finite element model correlated with the refined one, as shown in Refs [26] and [27]. It can also be obtained by substructuring and by use of 'super-elements' for geometrically regular substructures (Refs [28] and [29]).

Formal condensation by either an algebraic elimination of a subset of degrees of freedom (Ref. [30]), or by a Rayleigh–Ritz approach has also been used for the dimensionality reduction with considerable success (Refs [31], [32], [33]). Two other recently proposed techniques in this class are the incomplete models method (Ref. [34]), and a multigrid approach which systematically alternates between the refined and simplified grid meshes in geometrically identical finite element models (Ref. [35]).

The references given are a sample of a large body of literature that deals with the subject. A recent comprehensive survey is available in Ref. [36].

Condensation Technique using Ritz Functions

This technique uses a set of orthogonal displacement functions ϕ (Ritz functions) to condense the stiffness, $\mathbf{K}$, and mass, $\mathbf{M}$, matrices, and the load vector, $\mathbf{P}$ to:

$$\begin{aligned} \tilde{\mathbf{K}} &= \phi^T \mathbf{K} \phi \\ \tilde{\mathbf{M}} &= \phi^T \mathbf{M} \phi \\ \tilde{\mathbf{P}} &= \phi^T \mathbf{P} \end{aligned} \tag{A1}$$

The condensed matrices, $\tilde{\mathbf{K}}$, $\tilde{\mathbf{M}}$, and load vector $\tilde{\mathbf{P}}$ are subsequently used to form static and dynamic structural analysis equations for accelerated repetitive application.

Natural vibration modes are a good, but not the only, choice for ϕ-functions. In purely static applications, eigenmodes of the $(K - \lambda I)$ matrix can be used, if mass data are not available, as shown in Ref. [37]. Other choices that have been shown effective are the vectors of derivatives with respect to a load parameter (Ref. [33]), and the vectors of derivatives with respect to the design variables (Ref. [24]). The ϕ-functions can be updated as the optimization progresses by a technique described in Ref. [31]).

One should be cautioned that equations (A1) are quite costly to execute for a large number of degrees of freedom. Therefore, they should be carried out only when the functions ϕ are first evaluated and whenever they are subsequently updated. For each intermediate structural modification, the Taylor series extrapolation is an exact and very inexpensive way to update $\tilde{\mathbf{K}}$ and $\tilde{\mathbf{M}}$ (Ref. [37]):

$$\tilde{\mathbf{K}}_{\text{new}} = \tilde{\mathbf{K}}_{\text{old}} + \frac{\partial \tilde{\mathbf{K}}}{\partial x} \Delta x$$

$$\tilde{\mathbf{M}}_{\text{new}} = \tilde{\mathbf{M}}_{\text{old}} + \frac{\partial \tilde{\mathbf{M}}}{\partial x} \Delta x \tag{A2}$$

where

$$\frac{\partial \tilde{\mathbf{K}}}{\partial x} = \phi^T \frac{\partial \mathbf{K}}{\partial x} \phi; \quad \frac{\partial \tilde{\mathbf{M}}}{\partial x} = \phi \frac{\partial \mathbf{M}}{\partial x} \phi \tag{A3}$$

and x is a design variable which linearly affects $\mathbf{K}$ and $\mathbf{M}$.

This updating technique can easily be extended to include variables that influence $\mathbf{K}$ and $\mathbf{M}$ matrices nonlinearly by the technique described in Appendix B, section on 'Post-processing for sensitivity analysis'.

Higher-order Derivatives

An analytical technique for efficient calculation of the higher-order derivatives in static structural analysis using a finite element method is given in Ref. [23]. Assuming a design variable independent right-hand side in the standard load deflection equations $\mathbf{K}\mathbf{u} = \mathbf{P}$, and differentiating n times with respect to a design variable, one obtains recursive equations for $\partial^n \mathbf{u}/\partial x^n$:

$$\mathbf{K}\frac{\partial^n \mathbf{u}}{\partial x_i^n} = -\frac{\partial^n \mathbf{K}}{\partial x_i^n}\mathbf{u} - n\left(\frac{\partial \mathbf{K}}{\partial x_i}\frac{\partial^{n-1}\mathbf{u}}{\partial x_i^{n-1}} + \frac{\partial^2 \mathbf{K}}{\partial x_i^2}\frac{\partial^{n-2}\mathbf{u}}{\partial x_i^{n-2}} + \ldots + \frac{\partial^{n-1}\mathbf{K}}{\partial x_i^{n-1}}\frac{\partial \mathbf{u}}{\partial x_i}\right) \tag{A4}$$

which can be solved for $\partial^n \mathbf{u}/\partial x_i^n$, having been solved already for all derivatives of order less than n.

If, for design variables x_i, such quantities as thickness of membrane panels or cross-sectional moments of inertia for beam elements are selected, then $\partial^n \mathbf{K}/\partial x^n = 0$, for $n \geqslant 2$, and Equation (A4) can be simplified to:

$$\mathbf{K} \frac{\partial^n \mathbf{u}}{\partial x_i^n} = -n \frac{\partial \mathbf{K}}{\partial x_i} \frac{\partial^{n-1} \mathbf{u}}{\partial x_i^{n-1}} \tag{A5}$$

One application in which the derivatives of order 2 and higher are useful is a Taylor expansion with many terms to predict approximately the structural behaviour in a unidirectional search. In such a search, which occurs in many search algorithms, all design variables are in effect linked through the search-direction vector to a single variable which is the step length in the search direction. Therefore, the higher-order derivatives need to be evaluated with respect to that single variable only, the mixed derivatives do not exist, and the multi-term Taylor expansion approximation becomes an efficient one.

APPENDIX B SELECTED TECHNIQUES FOR A POST-PROCESSOR

This appendix provides details for some techniques particularly useful in post-processor functions.

Post-processing for Sensitivity Analysis

This kind of post-processing is needed when, as is true for most of the current finite element analysis programs, the analyser is capable only of computing derivatives with respect to variables that influence the stiffness matrix linearly, e.g. cross-sectional area A and moment of inertia I, thickness of a membrane panel t, etc. Then, if derivatives are needed with respect to variables that affect the stiffness matrix nonlinearly, e.g., depth h of an I-beam, they can be obtained by a chain differentiation, i.e.:

$$\frac{\partial u}{\partial z} = \frac{\partial u}{\partial x} \frac{\partial x}{\partial z} \tag{B1}$$

where

u = displacement
z = variable of which the stiffness matrix is a nonlinear function
x = variable of which the stiffness matrix is a linear function

An I-beam constitutes a good example. Suppose that the depth h of the I-beam shown in Fig. B1 is a design variable, then

$$I_x \cong \frac{(h - 2t_1)^3 t_2}{12} + \frac{bt_1(h - t_1)^2}{2} \tag{B2}$$

and

$$\frac{\partial I_x}{\partial h} = \frac{(h - 2t_1)^2 t_2}{4} + bt_1(h - t_1) \tag{B3}$$

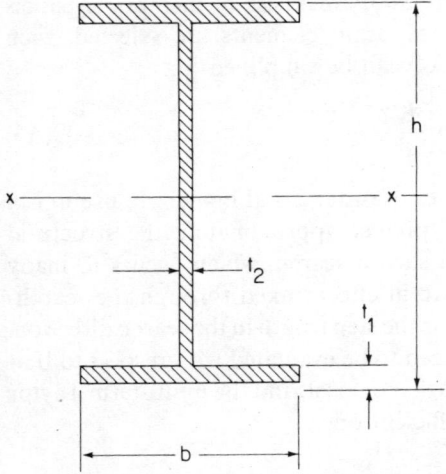

Fig. B1 Cross-section of an I-beam

For this example, h is a variable of category z, and I_x is a variable of category x and the derivative of displacement **u** with respect to h according to equation (B1) is:

$$\frac{\partial u}{\partial h} = \frac{\partial u}{\partial I_x}\frac{\partial I_x}{\partial h} \tag{B4}$$

Equation (B4) is coded in the post-processor, which also includes a code to compute Equation (B3); the values of the $\partial u / \partial I_x$ term is available from the analyser.

Continuing the example, assume now that the derivative of stress due to bending moment M_x is needed for the I-beam with respect to h. Applying engineering beam theory,

$$\sigma = \frac{M_x c}{I_x} \tag{B5}$$

(where $c = \tfrac{1}{2}h$ for I-beam)
and

$$\frac{\partial \sigma}{\partial h} = I_x^{-2}\left[\left(\frac{\partial M_x}{\partial h}c + M_x\frac{\partial c}{\partial h}\right)I_x - M_x c\frac{\partial I_x}{\partial h}\right] \tag{B6}$$

where

$$\frac{\partial M_x}{\partial h} = \frac{\partial M_x}{\partial I_x}\frac{\partial I_x}{\partial h}$$

Again, the value for term $\partial M_x/\partial I_x$ is obtained from the analyser, while the other terms and Equation (B6) are coded in the post-processor. In this manner, which

From a 'Black Box' to a Programming System

may be generalized without difficulty to any problem at hand, the analytical derivatives can be made available to the optimizer for variables of any type. The higher-order derivatives can be included by combining this approach with the technique described in Appendix A in the section on the higher-order derivatives.

Potentially Discontinuous Constraints

In cases where the analyser's output is used in the post-processor to calculate constraint values that may, potentially, be discontinuous functions of design variables, it is necessary to prepare measures in the post-processor for this eventuality, as illustrated by example of a wing flutter analysis.

Suppose that a flutter speed is evaluated, using the standard V–g method, as corresponding to a crossing point A for mode 1, while mode 2 produces no crossing, as shown in Fig. B2 by continuous curves. Now assume that the gradient

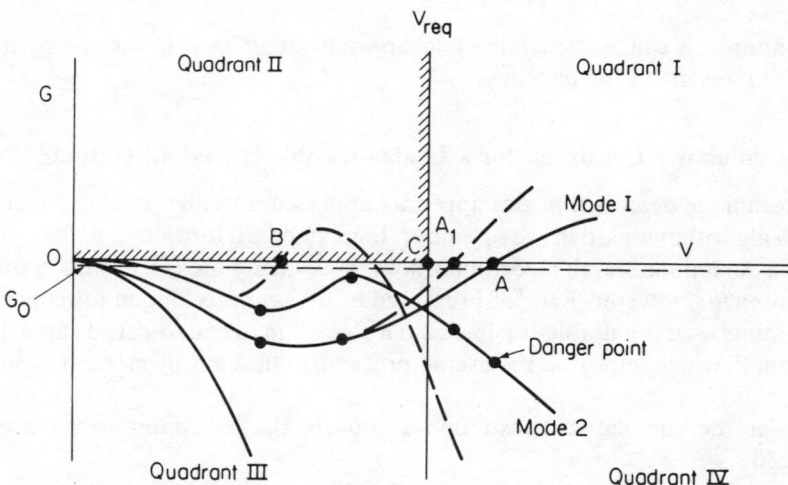

Fig. B2 A typical V–g diagram and the definition of the continuous flutter constraint

analysis of mode 1 flutter indicates that the mode 1 crossing will move to point A_1 (dashed curve), toward satisfaction of the required flutter speed value represented by C, for a certain change in the design variables. If the optimizer carries out that change, it may turn out in the next analysis that while the desired improvement was achieved in mode 1, mode 2, heretofore inactive and therefore ignored, has produced a new crossing at B (dashed curve), resulting in a drop in the flutter velocity. Since this drop would contradict the previous gradient information, the search algorithm may break down.

The remedy is to restore continuity by guarding against *all potential* failure modes rather than only against those found to be active at one stage of analysis. In the flutter example, the remedy may be implemented in the form of a continuous flutter constraint, described in Ref. [13]. The constraint is constructed to keep the judiciously selected 'danger points' shown in Fig. B2 from moving into quadrant II. For each danger point, a separate constraint equation is written, and depending on the quadrant in which a given 'danger point' is located, that equation reads:

in quadrant I $\quad g = \dfrac{V_{req}}{V} - 1$

in quadrant II $\quad g = -\dfrac{G}{G_0}\left(\dfrac{V_{req}}{V} - 1\right)$

in quadrant III $\quad g = -\dfrac{G}{G_0}$

in quadrant IV $\quad g = \dfrac{V_{req}}{V} - \dfrac{G}{G_0} - 1$

This approach can be generalized to apply in other similar cases, e.g., those involving resonance or buckling.

Cumulative Constraint for a Usable-feasible Directions Optimizer

The technique described in this appendix applies if a usable-feasible directions search algorithm is used in the optimizer. In its standard formulation, the usable-feasible directions search algorithm tracks individually the constraints active in the problem. However, Ref. [38] reported a successful technique to reduce the constraints used in a usable-feasible search algorithm to one so-called cumulative constraint, which improved the overall procedure efficiency by more than 30 per cent.

Under the cumulative constraints approach, the constraint values are all summed:

$$\Omega = -\varepsilon + \sum_{i=1}^{m} (\langle g_i \rangle)^2$$

where

$$\langle g_i \rangle = \begin{cases} g_i, & \text{if } g_i > 0 \\ 0 & \text{if } g_i \leqslant 0 \end{cases}$$

assuming the standard formulation:

$$g_i = \dfrac{\text{actual behaviour value}}{\text{allowable value}} - 1$$

An alternative equivalent formulation is

$$\Omega = -\varepsilon + \sum_{i=1}^{m} (\tfrac{1}{2}(g_i + |g_i|))^2$$

The quanity Ω is a value of a single (cumulative) constraint which the usable-feasible direction algorithm is given to handle. The purpose of the small arbitrary parameter ε is to assure satisfaction of the Ω constraint when each $g_i = 0$, and $\langle g_i \rangle^2$ is used instead of $\langle g_i \rangle$ in Equation (B1) to assure continuity of $\Omega(\vec{x})$ and $\nabla\Omega(\vec{x})$ with respect to design variables x in the design space. Geometrically, $\Omega(\vec{x})$ is an envelope of all the participating constraint boundaries $g_i(\vec{x})$.

APPENDIX C TWO-LEVEL OPTIMIZATION METHOD

A two-level structural optimization procedure, which is another flow option, was introduced in Ref. [39]. It is a decomposition scheme for breaking a large optimization problem with many design variables into several subproblems, each with a relatively smaller number of design variables. The decomposition is achieved by partitioning the set of design variables into a subset of so-called system variables which directly enter the equations that govern the whole structure behaviour in response to the applied loads, and several subsets of so-called local variables which define detailed cross-sectional dimensions of the individual structural components, or groups of components, as shown in Figs C1 and C2. Similarly, the constraints are partitioned into the system constraints which are functions of the system variables and external loads applied to the structure, and local constraints which depend on the local variables and internal forces acting on each component.

A panel, shown in Fig. C1, stiffened by two sets of orthogonal stringers and divided for analysis purposes into several quadrilateral, orthotropic, membrane finite elments is an instructive example for the approach. For analysis purposes, the panel can be represented by three layers of three materials:

1. An isotropic sheet of thickness t, Young's modulus E, Poisson ratio v to represent the panel's upper cover.

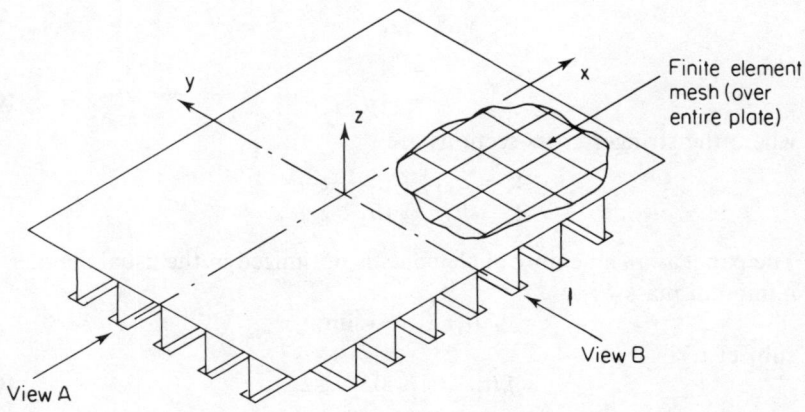

Fig. C1 Plate stiffened by orthogonal stringers

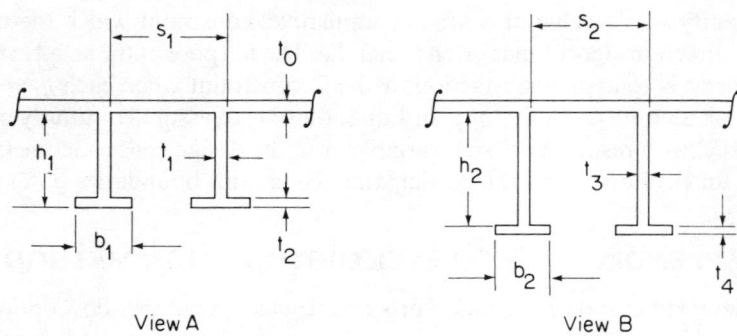

Fig. C2 Detailed cross-sectional dimensions of stiffened plate

2. An orthotropic sheet of thickness t_x, moduli $E_x = E$, $E_y = 0$, and constants $v_{12} = v_{21} = 0$, to represent smeared stringers parallel to axis x.
3. An orthotropic sheet of thickness t_y, $E_x = 0$, $E_y = E$, $v_{12} = v_{21} = 0$, to represent the other stringers.

Thicknesses t, t_x, t_y for each finite element or groups of elements (linking) are system variables and are used for solution of the panel internal forces N_x, N_y, N_{xy} acting on each finite quadrilateral membrane element, and for the panel displacements.

The displacements may be restricted, thus becoming system constraints. The other system constraints are stresses, because in this example they also depend directly on system variables t, t_x, t_y. With forces N_x, N_y, N_{xy} known for each component, the local constraints such as local buckling can be evaluated as functions of local variables defined as cross-sectional dimensions $t_0, t_1, t_2, b_1, h_1, t_3, t_4, b_2, h_2$, shown in Fig. C2.

The optimization procedure progresses as follows:
1. Local variables are initialized and system variables are computed as
$$t = t_o$$
$$\begin{aligned} t_x &= A_x/S_1 \\ t_y &= A_y/S_2 \end{aligned} \tag{C1}$$
where the stringer cross-sections are
$$\begin{aligned} A_x &= b_1 t_2 + (h_1 - t_2)t \\ A_y &= b_2 t_4 + (h_2 - t_4)t_3 \end{aligned} \tag{C2}$$

2. The panel as an assembly of elements is optimized in the usual manner for minimum mass F:
$$F(t, t_x, t_y) \to \min$$
subject to
$$\bar{g}_i(t, t_x, t_y) \leqslant 0, \quad i \in I \tag{C3}$$
where g_i constraints represent restricted displacements and stresses.

3. Local optimization is carried out for each group of finite elements with common local variables. Forces N_x, N_y, N_{xy} acting on each element of a group are known from the last analysis in step 2. In the optimization, the objective function is a sum of squares of residuals of Equation (C1), that is,

$$F_j = (t_x - A_x/S_1)^2 + (t_y - A_y/S_2)^2 \tag{C4}$$

where t_x, t_y are determined in step 2. Thus, the optimization is:

$$F_j \text{ (local variables)} \to \min$$

subject to local constraints g_i (local variables) $\leqslant 0$.

4. Iteration is repeated from step 2 until both a constrained minimum is achieved in step 2 and $F_j \leqslant F_{\min} \simeq 0$ for each group of finite elements in step 3.

The procedure flow chart is shown in Fig. C3. The separate optimization operators indicated in the flow chart for the system and for each component can be carried out by any of the options specified in Table 11.2. Typically, two different analysers are used in a two-level scheme. For example: a finite element analysis in the system optimization, and a program specialized to analyse local constraints in the component optimization. Regardless of the types of analysers used, the key idea of the procedure is to enforce equality of the elemental

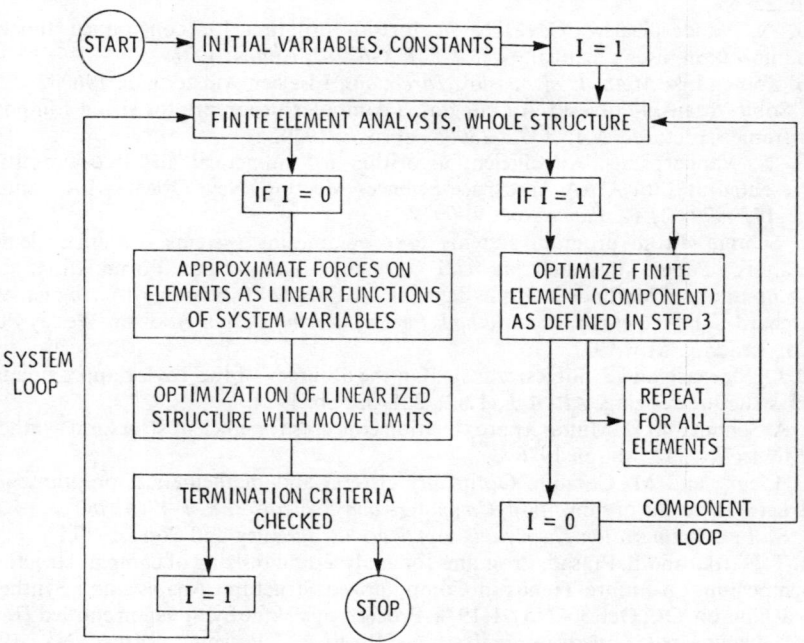

Fig. C3 Flow chart for two-level optimization procedure

stiffnesses defined from the system side and from the component side by reducing to near zero the residuals in the objective function in local optimization.

If the panel in Fig. C1 was divided in four zones each having a common set of nine local design variables, the total number of variables would be 36. Using the two-level approach, this total breaks down into one system problem of $4 \times 3 = 12$ variables, and four local subproblems of nine variables each. As reported in Ref. [39], the computer cost of solution of a decomposed problem is reduced compared to the cost of solution of the original problem, while the final results are practically the same and the convergence behaviour of the procedure is satisfactory.

J.S.

REFERENCES

1. W. E. Tripplett, Aeroelastic tailoring studies in fighter aircraft design, *AIAA Paper No. 79–0725*, AIAA/ASME/ASCE/AHS 20th Structures, Structural Dynamics and Materials Conference. A Collection of Technical Papers on Structures, St Louis, MO, April 4–6, 1979, pp. 72–8.
2. G. N. Vanderplaats, The computer design and optimization, in *Computing in Applied Mechanics* (R. F. Hartung, ed.), AMD-vol. 18, American Society of Mech. Eng., v. 1976, pp. 25–48.
3. G. N. Vanderplaats, CONMIN—A fortran program for constrained function minimization, user's manual, *NASA TM X-62282*, August 1973.
4. G. Zoutendijk, *Methods of Feasible Directions*, Elsevier, Amsterdam, 1960.
5. J. Sobieszczanski-Sobieski, An integrated computer procedure for sizing composite airframe structures. *NASA TP-1300*, February 1979.
6. G. N. Vanderplaats, An efficient algorithm for numerical airfoil optimization. Presented at 17th AIAA Aerospace Sciences Meeting, New Orleans, LA, January 15–17, 1979. *AIAA Paper No. 79–0079*.
7. E. Schrem, From program systems to programming systems for finite element analysis. Paper presented at U.S.-Germany Symposium: Formulations and Computational Methods in Finite Element Analysis. MIT, Boston, MA, August 1976.
8. Richard L. Fox, *Optimization Methods for Engineering Design*, Addison-Wesley Publ. Co., Reading, MA, 1971.
9. O. O. Storaasli and J. Sobieszczanski, On the accuracy of the Taylor approximation for structure resizing, *AIAA J.*, **12**(2), 231–3, Feb. 1974.
10. L. A. Schmit and H. Miura, Approximation concepts for efficient structural synthesis, *NASA CR-2552*, March 1976.
11. C. Fleury and M. Geradin, Optimality criteria and mathematical programing in structural weight optimization, *Computers and Structures*, **8**, 7–17, 1978.
12. S. S. Rao, *Optimization Theory and Applications*, J. Wiley and Sons, 1979.
13. R. T. Haftka and B. Prasad, Programs for analysis and resizing of complex structures. Symposium on Future Trends in Computerized Structural Analysis and Synthesis, Washington, DC, Oct. 30–Nov. 1, 1978. Proceedings of the Symposium entitled *Trends in Computerized Structural Analysis and Synthesis*, Pergamon Press, NY, 1978, pp. 323–32.

14. J. Sobieszczanski, Building a computer aided design capability using a standard time share operating system. *Proceedings of the ASME Winter Annual Meeting, Integrated Design and Analysis of Aerospace Structures*, Houston, TX, Nov. 30–Dec. 5, 1975, pp. 93–112.
15. J. Sobieszczanski-Sobieski and R. B. Bhat, Adaptable structural synthesis using advanced analysis and optimization coupled by a computer operating system, *AIAA Paper No. 79-0723*, AIAA/ASME/ASCE/AHS 20th Structures, Structural Dynamics and Materials Conference. A Collection of Technical Papers on Structures, St Louis, MO, April 4–6, 1979.
16. L. A. Schmit and R. K. Ramanathan, A multilevel approach for minimum weight design including local and system buckling constraints, *Proceedings of the AIAA/ASME 18th Structures, Structural Dynamics and Materials Conference*, San Diego, CA, March 21–23, 1977, pp. 58–70.
17. J. H. Starnes and R. T. Haftka, Preliminary design of composite wings for buckling, strength and displacement constraints, *AIAA Paper No. 78-466*, A Collection of Technical Papers, AIAA/ASME 19th Structures, Structural Dynamics and Materials Conference, Bethesda, MD, April 3–5, 1978.
18. M. S. Anderson and W. J. Stroud, A general panel sizing computer code and its application to composite structural panels, *AIAA Paper No. 78-467*, A collection of Technical Papers, AIAA/ASME 19th Structures, Structural Dynamics and Materials Conference, Bethesda, MD, April 3–5, 1978.
19. W. D. Whetstone, Computer analysis of large linear frames, *J. of Structural Division, ASCE*, November 1969, pp. 2401–17.
20. G. L. Giles, Computer aided methods for analysis and synthesis of supersonic cruise aircraft structures, *Proceedings of the SCAR Conference, NASA CP-001*, Part 2, November 1976, pp. 637–58.
21. G. Giles and R. T. Haftka, SPAR data handling utilities, *NASA TM 78701*, September 1978.
22. J. Sobieszczanski, Sizing of complex structure by the integration of several different optimal design algorithms, AGARD Lecture Series No. 70 on structural Optimization, *AGARD-LS-70*, September 1974.
23. O. O. Storaasli and J. Sobieszczanski, Design oriented structural analysis, *AIAA Paper No. 73-338*, Presented at AIAA/ASME/SAE 14th Structures, Dynamics and Materials Conference, Williamsburg, VA, March 1973.
24. A. K. Noor and H. E. Lowder, Approximate techniques of structural reanalysis, *Computers and Structures*, **4**, 801–12, 1974.
25. J. C. Chen and B. K. Wada, Matrix perturbation for structural dynamic analysis, *AIAA Journal*, **15**(8), August 1977.
26. R. H. Ricketts and J. Sobieszczanski, Simplified and refined structural modeling for economical flutter analysis and design, *AIAA Paper No. 77-421*, AIAA 18th Structures, Structural Dynamics and Materials Conference, San Diego, CA, March 1977.
27. J. Sobieszczanski-Sobieski, D. Gross, W. Kurtze, J. Newsom, G. Wrenn and W. Greene, Supersonic cruise research aircraft structural studies: methods and results, NASA CP pending, November 1979.
28. A. K. Noor, H. A. Kamel and R. E. Fulton, Substructuring techniques—status and projections, *Computers and Structures*, **9**(1), 1978.
29. O. Egeland and P. O. Aaraldsen, SESAM-69—A general purpose finite element program, *Computers and Structures*, **4**, 41–68, 1974.
30. J. S. Przemieniecki, *Theory of Matrix Structural Analysis*, ch. 6, McGraw-Hill Book Co., 1968.

31. B. O. Almroth, P. Stern and F. A. Brogan, Automatic choice of global shape functions in structural analysis, *AIAA J.*, May 15, 1978.
32. A. K. Noor, C. M. Andersen and J. M. Peters, Global-local approach for nonlinear shell analysis, *Proceedings of Seventh Conference on Electronic Computation*, ASCE, St. Louis, MO, August 1979, pp. 634–657.
33. A. K. Noor and J. M. Peters, Reduced basis technique for nonlinear analysis of structures, *AIAA Paper No. 79-0747*, 116–126, A Collection of Technical Papers on Structures, AIAA/ASME/ASCE/AHS, 20th Structures, Dynamics and Materials Conference, St. Louis, MO, April 1979.
34. A. Berman and W. G. Flannelly, Theory of incomplete models of dynamic structures, *AIAA Journal*, **9**(8), 1481–7, August 1971.
35. A. Brandt, Multi-level adaptive solutions to boundary-value problems, *Mathematics of Computation*, **31**, 333–90, 1977.
36. J. S. Arora, Survey of structural reanalysis techniques, *J. of Structural Division, ASCE*, **ST4**, 783–802, April 1976.
37. J. Sobieszczanski-Sobieski and P. Hajela, Accuracy of an approximate static structural analysis technique based on stiffness matrix eigenmodes, *AIAA Paper No. 79-0748*, 127–36, A Collection of Technical Papers on Structures, AIAA/ASME/ASCE/AHS, 20th Structures, Dynamics and Materials Conference, St. Louis, MO, April 1979.
38. D. W. Gross and J. Sobieszczanski-Sobieski, Application to aircraft design of nonlinear optimization methods which include probabilistic constraints. *AIAA Paper No. 80-0153*, Presented at AIAA 18th Aerospace Sciences Meeting, Pasadena, CA, January 14–16, 1980.
39. L. A. Schmit and R. K. Ramanathan, Multilevel approach to minimum weight design including Buckling constraints, *AIAA Journal*, **16**(2), 97–104, February 1973.

Foundations of Structural Optimization: A Unified Approach
Edited by A. J. Morris
© 1982 John Wiley & Sons Ltd.

Chapter 12
Optimization of Aircraft Structures

12.1 ANALYSIS METHODS

We now want to turn our attention to the application of one of the types of system described in Chapter 11 to the solution of actual aircraft structural design problems. However, before detailing the optimization program, we present the different analysis methods occurring in the optimization process.

Indeed, as we have seen earlier, a useful optimization method needs an analysis method efficient enough to iterate without excessive costs. In the present case the analysis modules are based upon the Dassault–Breguet ELFINI code which links, around the finite element method, the main branches of an aircraft structures analysis (Fig. 12.1). That is:
— calculation of linear and nonlinear static stresses;
— static aeroelasticity, calculation and management of load cases;
— dynamics, computed with a reduction method, gives natural frequencies, flutter speed, transient response (Figs 12.4 and 12.5)

All Dassault–Breguet aircraft have been computed with this tool since 1970 (Figs 12.2 and 12.3).

The most significant aspects of these optimization codes are:
— the generation and management of meshes in an interactive or batch mode using the 'topological method' which describes nodes and elements by blocks of 'constant properties' in an index space;
— the solution of the equilibrium equation by an efficient improvement of the Frontal Gauss Method providing the factorized stiffness matrix and allowing a cheap computation of displacements under numerous cases of loads or 'dummy' loads;
— a monitor manages the calls to the different modules with a dynamic allocation of memory and allows an easy introduction of iterations between the modules and the branches of the program.

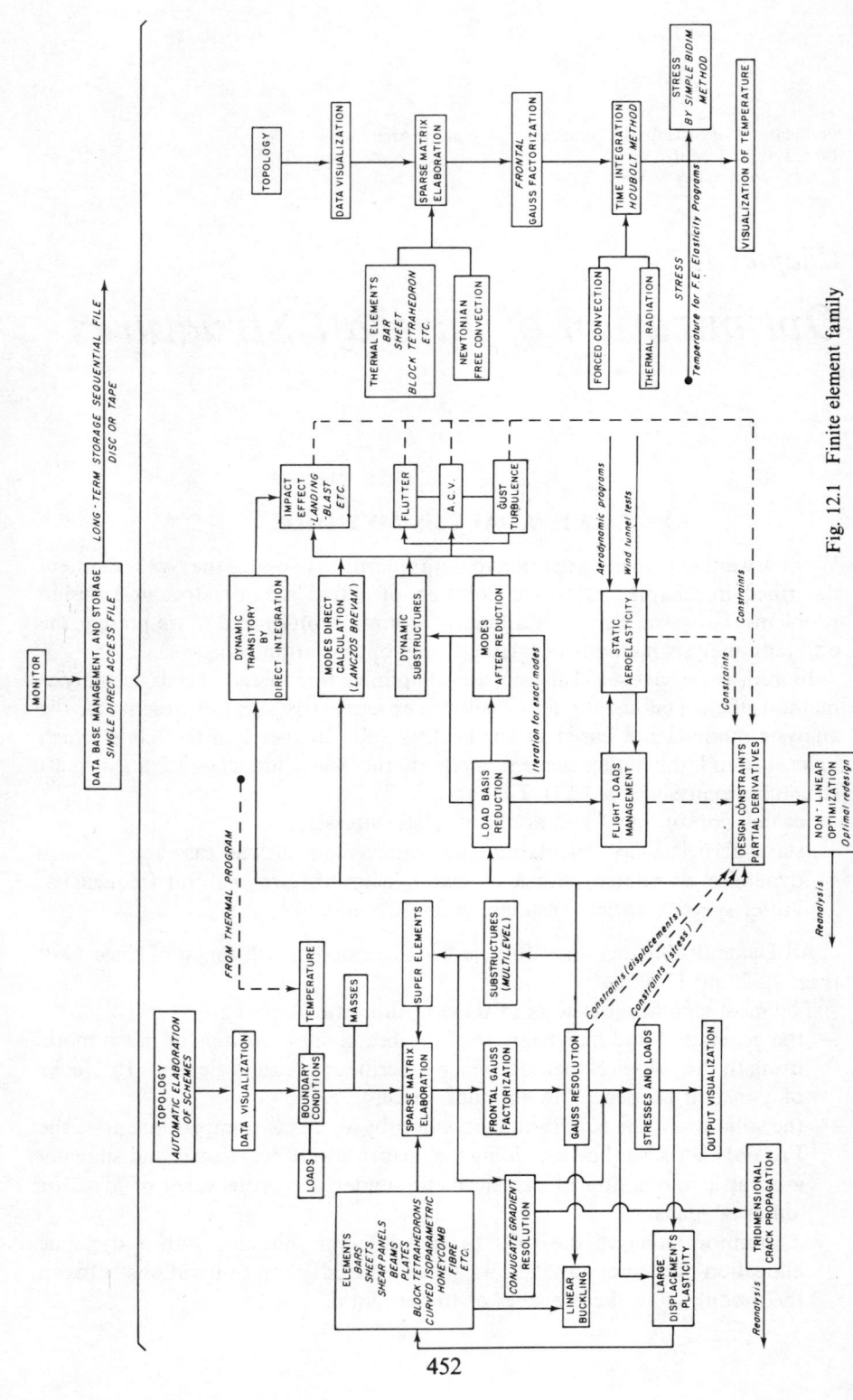

Fig. 12.1 Finite element family

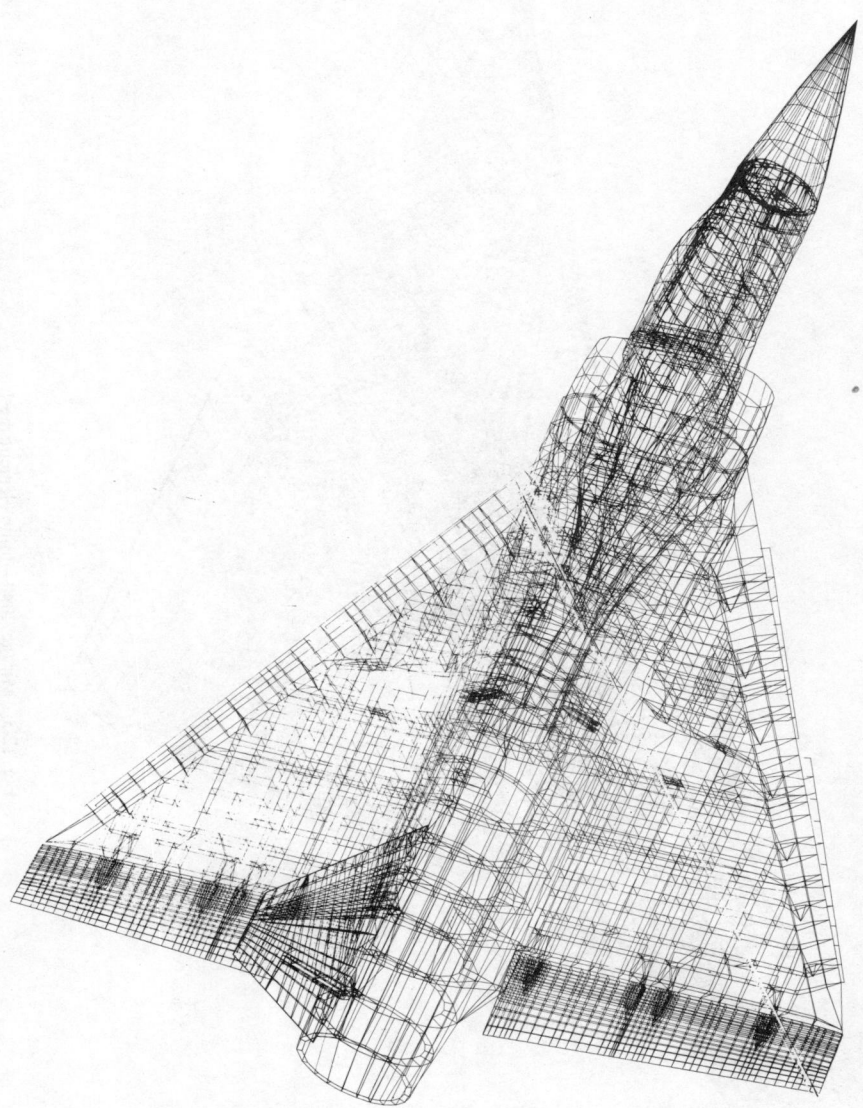

Fig. 12.2 Mirage 2000—finite element mesh

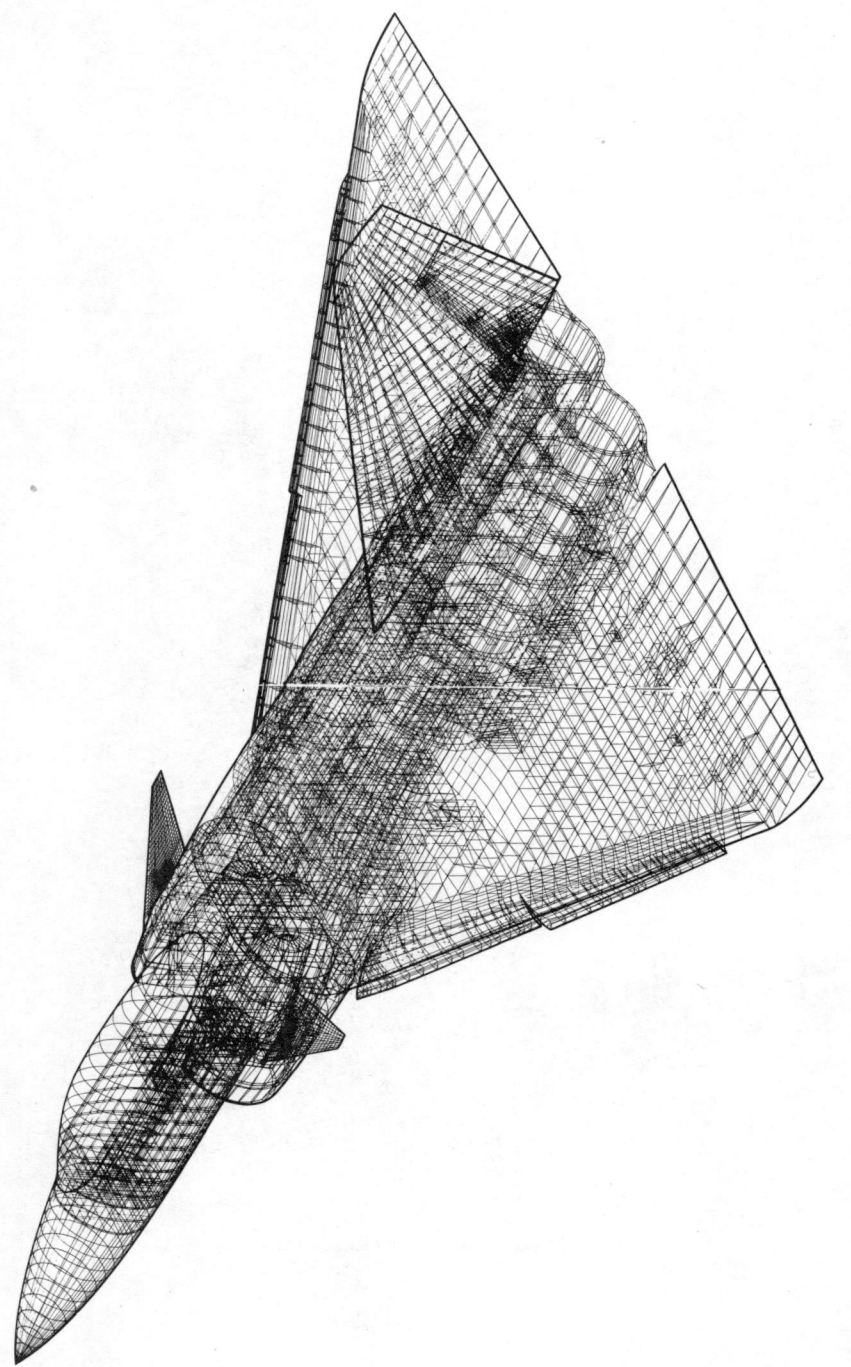

Fig. 12.3 Mirage 4000—finite element mesh

Optimization of Aircraft Structures

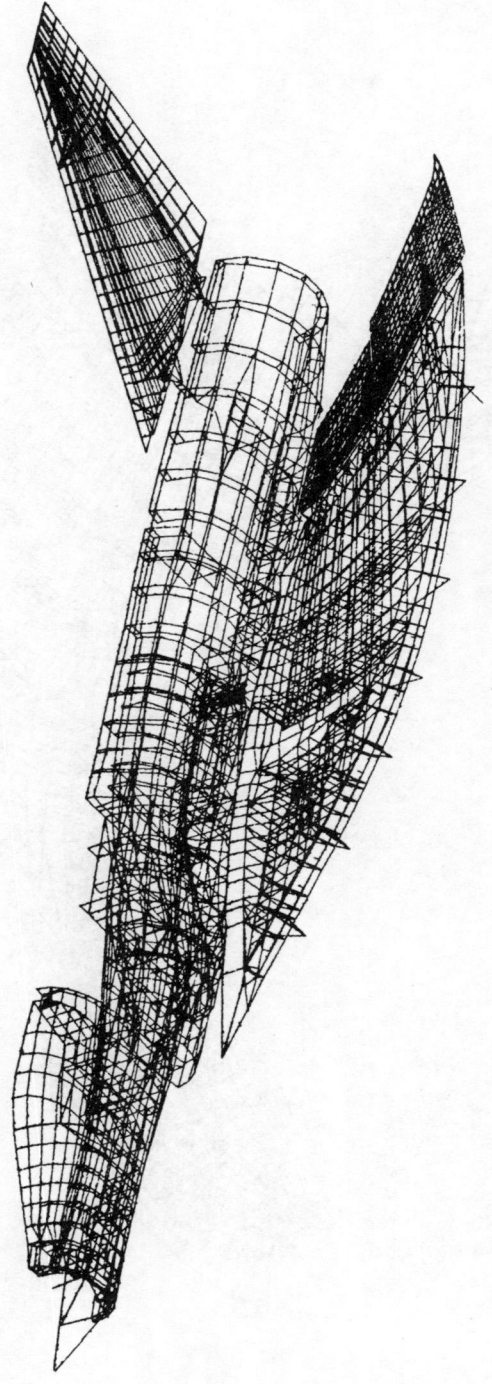

Fig. 12.4 Mirage 2000—Mode No. 1. $F = 11.38\,\text{Hz}$; $MG = 47.63\,\text{kg m}^2$

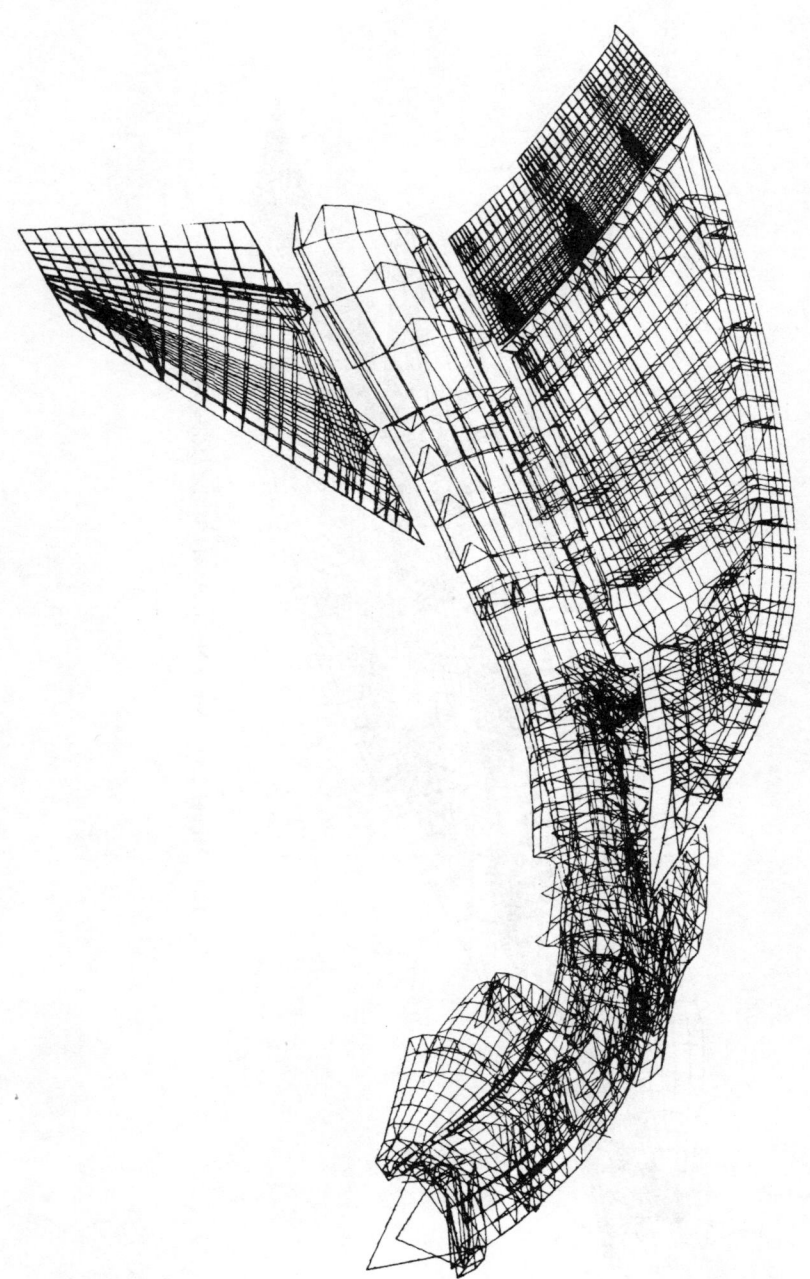

Fig. 12.5 Mirage 2000—Mode No. 2. $F = 12.53$ Hz; $MG = 458$ kg m^2

12.2 STRUCTURAL OPTIMIZATION METHOD

The general layout of the optimization system (Fig. 12.6) developed by Dassault–Breguet is similar to that described in Chapter 11 but is specific to aircraft structures. We describe the system in some detail and to this end repeat some of the theory given earlier in order to provide a complete picture.

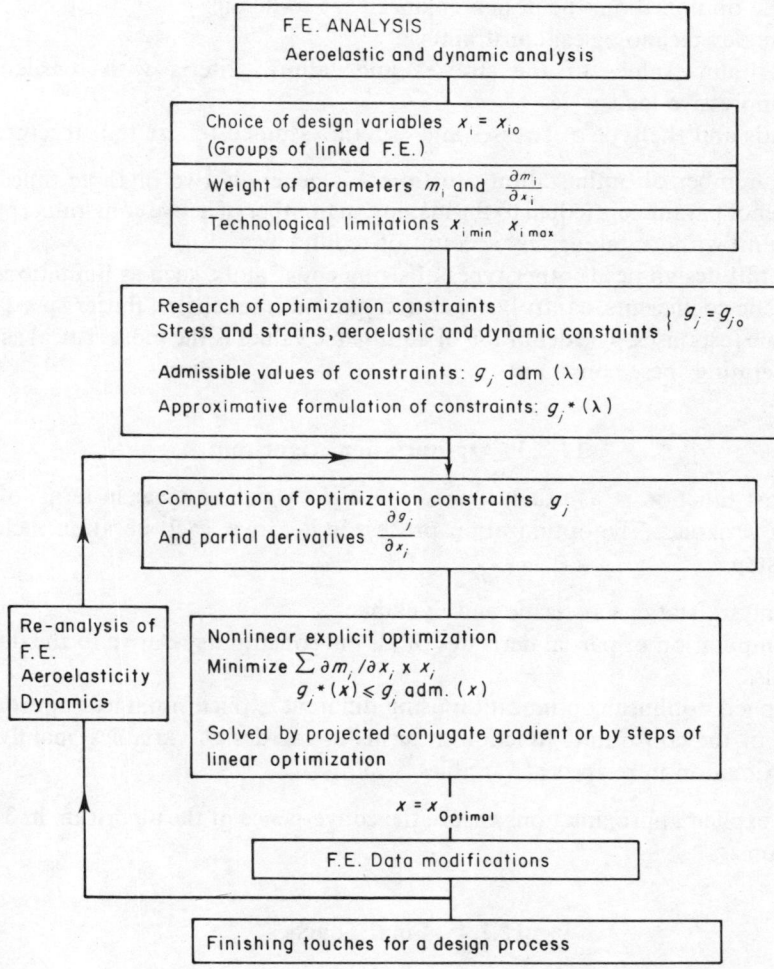

Fig. 12.6 General organization of optimization

12.2.1 Design Variables

The optimization design variables are characterized by a group of linked finite elements (Fig. 12.7). The choice of these variables depends on the geometric gradient of the stresses and takes into account the tooling rules.

12.2.2 Optimization Constraints

In static optimization, the design engineer has to define:
— complex technological constraints;
— maximum values of the stresses and failure criteria with tensile and compressive loads;
— loads and the type of stresses and criteria assumed to size the structure.

The number of optimization constraints is generally two or three times the number of parameters (equal to the maximum number of active constraints at the optimum) without taking any account of redundancy.

Aircraft design needs other types of sizing constraints, such as limitations on aerolastic coefficients, control efficiencies, natural frequencies, flutter speed and dynamic responses. The definition of admissible values is the most critical aspect in generating these constraints.

12.2.3 Optimization Algorithm

The cost function is, as usual, structural weight which is linear in terms of the design variables. The optimization process is iterative, each iteration includes three steps:

(1) analysis, static, aeroelastic and dynamic;
(2) computation of partial derivates of all the constraints relative to the design variables;
(3) explicit nonlinear optimization using different explicit nonlinear approximations of the constraints function in terms of the design variables, mainly the approximation in reciprocal variables.

These explicit approximations ensure the convergence of the algorithm in 3 to 5 iterations.

12.2.4 Final Touches

Generally, the numerical optimum obtained from the optimization algorithm needs some modification since it will not normally represent a realizable design. The final touch consists in making small modifications to the design parameters iteratively to give a prediction of the final values of the constraints, cost function and their derivatives.

Optimization of Aircraft Structures 459

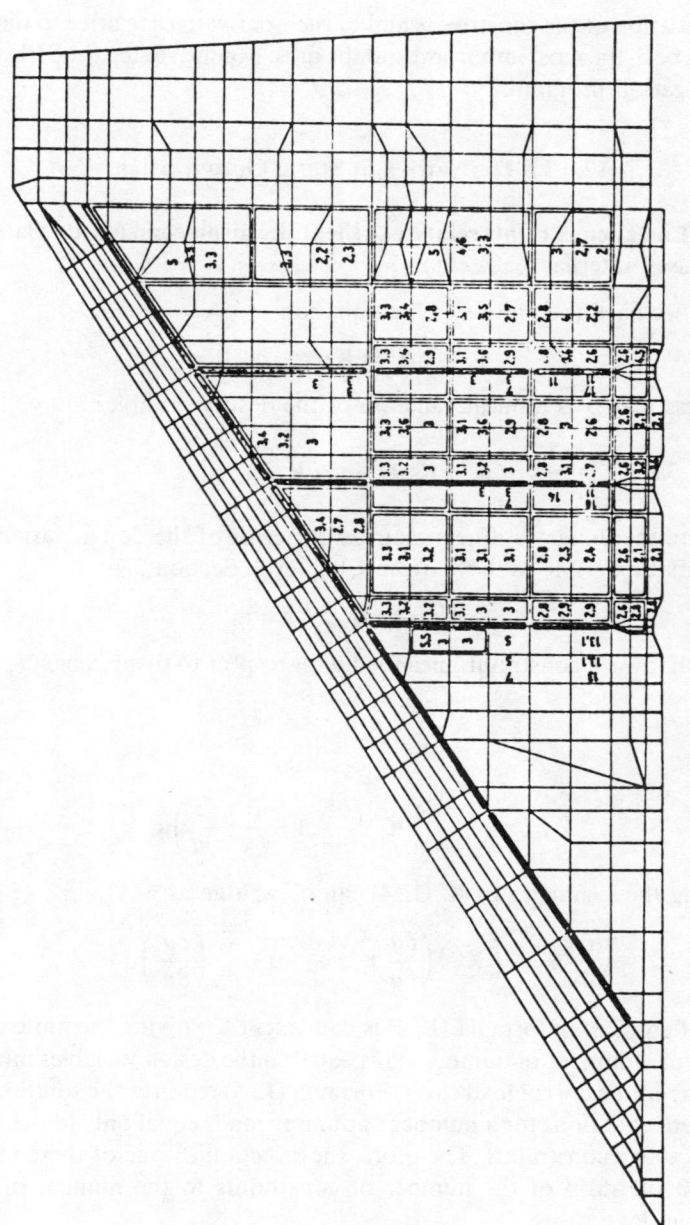

Fig. 12.7 Mirage 4000—optimization parameters on the delta wing

12.3 COMPUTATION OF CONSTRAINTS AND CONSTRAINT DERIVATIVES

The computation of the constraints and of their derivatives relative to the design variables is both the most important and the most expensive step (in CPU time) of our optimization program.

12.3.1 Derivatives in Static Optimization

12.3.1.1 Linear constraints relative to the static displacements (displacements, elastic stresses, internal loads etc.)

As before, we have the finite element equation

$$\mathbf{P} = \mathbf{K}\mathbf{u} \tag{12.1}$$

The stiffness matrix is a linear function of the design variables x_i:

$$\mathbf{K} = \mathbf{K}_0 + \sum_i x_i (\partial \mathbf{K}/\partial x_i) \tag{12.2}$$

$(\partial \mathbf{K}/\partial x_i)$ are the unitary stiffness matrices in terms of the design variables.

Differentiating (12.1) we have, from Chapter 3, Section 3.4.1

$$\Delta \mathbf{u} = -\mathbf{K}^{-1} \Delta \mathbf{K} \mathbf{u} \tag{12.3}$$

Knowing that for a constraint linearized with respect to displacements, such as stress

$$g = (\partial g/\partial u)u$$

then

$$\partial g/\partial x = -\frac{\partial g}{\partial u}\mathbf{K}^{-1}\frac{\partial \mathbf{K}}{\partial x}\mathbf{u} + \frac{\partial}{\partial x}\left(\frac{\partial g}{\partial u}\right)\mathbf{u} \tag{12.4}$$

Considering the symmetry of $\mathbf{K}$, (12.4) can be written as:

$$\frac{\partial g}{\partial x} = -\left(\mathbf{K}^{-1}\left(\frac{\partial g}{\partial u}\right)^{\mathrm{T}}\right)^{\mathrm{T}}\frac{\partial \mathbf{K}}{\partial u}\mathbf{u} + \frac{\partial}{\partial x}\left(\frac{\partial g}{\partial u}\right)\mathbf{u} \tag{12.5}$$

The cost of employing formula (12.4) is equivalent to solving the finite element equations for a number of dummy loads equal to the design variables multiplied by the number of external load cases. Formula (12.5) requires the solution of the finite element equations for a number of dummy loads equal only to the number of selected active constraints. Therefore, the selection of one of these formulas depends on the ratio of the number of constraints to the number of design variables and load cases.

It may be observed that formula (12.4) is, perhaps, the most useful since the addition or modification of constraints does not require a complete re-solving of the finite element equations. Nevertheless, the selection of the most efficient formula is important as the calculation of derivatives requires typically about 60 per cent of the total CPU time for a complete solution of a structural optimization problem in aircraft design.

12.3.1.2 Analysis and derivatives of aeroelastic coefficients

The computation and the derivation of aeroelastic coefficients are relatively complex. They are detailed in Ref. [1] and summed up in Appendix 1. As before, the relative cost is equivalent to a dummy load for each constraint on an aeroelastic coefficient.

12.3.2 Derivatives in Dynamic Optimization

For this problem the vibration modes computed in a condensed basis give a good approximation to the modes in the finite element basis. Therefore we are employing derivatives in dynamic optimization which are not exact (as in static optimization) but approximate. The main hypothesis is, therefore, that dummy loads occurring in the computation of these derivatives could be expressed in the reduced basis. Because of this condensed basis (requiring about fifty degrees of freedom), the cost of dynamic derivatives is lower than that of obtaining static derivatives in the finite element basis (requiring many thousands degrees of freedom).

12.3.2.1 Derivatives of natural frequencies

The dynamic equation used in our analysis is given by:

$$\mathbf{M}\mathbf{u}'' + \mathbf{K}\mathbf{u} = 0$$

The Rayleigh quotient $\omega^2(u) = \mathbf{u}^T\mathbf{K}\mathbf{u}/\mathbf{u}^T\mathbf{M}\mathbf{u}$ is stationary and equal to ω^2 when

$$\mathbf{u} = \mathbf{V}, \text{ the natural deformation}$$

Therefore:

$$\Delta\omega^2 = \mathbf{V}^T\Delta\mathbf{k}\mathbf{V}/\mathbf{V}^T\mathbf{M}\mathbf{V} - \omega^2\mathbf{V}^T\Delta\mathbf{M}\mathbf{V}/\mathbf{V}^T\mathbf{M}\mathbf{V}$$

The cost of these derivations is relatively low.

12.3.2.2 Derivatives of extrema in transient response

The general form for the equations of dynamic motion in the finite element base is

$$\mathbf{M}\ddot{\mathbf{U}} + \mathbf{K}\mathbf{U} = \mathbf{F}(t) \qquad (12.6)$$

The solution in the condensed basis $\mathbf{U} = \mathbf{V}\mathbf{u}$ is obtained by the integration of

$$\mathbf{m}\ddot{\mathbf{u}} + \mathbf{k}\mathbf{u} = \mathbf{f}(t) \qquad (12.7)$$

with $\quad\quad\quad \mathbf{m} = \mathbf{V}^T \mathbf{M} \mathbf{V}, \mathbf{k} = \mathbf{V}^T \mathbf{K} \mathbf{V}, \mathbf{f}(t) = \mathbf{V}^T \mathbf{F}(t)$

The dynamic stress σ can be written: $\sigma = (\partial\sigma/\partial \mathbf{U})\mathbf{U} = (\partial\sigma/\partial \mathbf{U})\mathbf{V}\mathbf{u} = (\partial\sigma/\partial u)\mathbf{u}$
Derivation of Equation (12.6) gives

$$\mathbf{M}\Delta\mathbf{U}'' + \mathbf{K}\Delta\mathbf{U} = -\Delta\mathbf{M}\mathbf{U}'' - \Delta\mathbf{K}\mathbf{U} \tag{12.8}$$

which is similar to Equation (12.6) except we now have a second term representing the excitation. If this dummy excitation can be expressed in the condensed basis, we obtain

$$\mathbf{m}\Delta\mathbf{u}'' + \mathbf{k}\Delta\mathbf{u} = -\Delta\mathbf{m}\mathbf{u}'' - \Delta\mathbf{k}\mathbf{u} \tag{12.9}$$

$$\Delta\mathbf{m} = \mathbf{V}^T \Delta\mathbf{M}\mathbf{V} \quad\quad \Delta\mathbf{k} = \mathbf{V}^T \Delta\mathbf{K}\mathbf{V}$$

At the extremum the variation of $\sigma(t)$ is obtained by the following procedure:

$$\Delta\sigma(t) = (\partial\sigma/\partial u)\Delta u(t)$$

$$d\sigma(x,t) = (\Delta\sigma/\Delta x)dx \tag{12.10}$$

in which the second term is identically zero.

12.3.2.3 Analysis and derivatives of flutter speed

Turning to the analysis and derivatives of the flutter speed these computations are complex. They are detailed in Ref. [1] and summed up in Appendix 2. The cost of analysis and derivatives is low because they are also computed in the condensed basis.

12.4 EXPLICIT OPTIMIZATION AND SUB-ITERATION

12.4.1 Explicit Optimization

The main idea in this section is to replace the exact formulation $\sigma(x)$ of the constraints, which is, in any case, only implicitly known by a FE analysis, by an explicit approximation $\sigma^*(x)$

$\sigma^*(x)$ is selected so that:

— at $x = x_{\text{ANALYSIS}} \sigma^*(x) = \sigma_{\text{FE}}$ and $\partial\sigma^*(x)/\partial x = \partial\sigma_{\text{FE}}/\partial x$
— $\sigma^*(x)$ must be exact for statically determinate structures
— $\sigma^*(x)$ must be well behaved when $x \to 0$ and $x \to \infty$

Following the arguments of Chapter 3, there is reason to believe that the most efficient explicit approximation is to use reciprocal variables,

with $\quad\quad z_i = 1/x_i \quad$ and $\quad a_{ij} = -\dfrac{1}{z_i^2}\partial\sigma_{j\text{FE}}/\partial x_i$

together with $\quad\quad \sigma_j^*(z) = a_{0j} + \sum_i a_{ij} z_i \quad$ and $\quad a_{0j} = \sigma_{\text{FE}} \mp \sum_i a_{ij}(z_i - z_{i0})$

The optimization problem becomes:

$$\text{minimize} \quad \sum_i m_i/z_i$$

$$\text{subject to} \quad \begin{cases} a_{0j}\sum_i a_{ij}z_i \leq \sigma_{j\,\max} \\ \phantom{a_{0j}\sum_i}z_i \leq 1/x_{i\min} \end{cases}$$

With constraints on local buckling criteria, the admissible values are not constant but are taken as functions of a single parameter obtained from the classical local buckling problem.

The explicit optimization problem is solved using a conjugate projected gradient method with an efficient normalization of the tangent Hessian. The basis of this type of algorithm is explained, in detail, in Chapter 9.

12.4.2 Sub-iteration Process

The good results and excellent convergence of this type of algorithm in static optimization are mainly due to the explicit approximation of constraints in $1/x_i$ as described in Chapters 3 and 10. But for other types of constraints such as natural frequencies, flutter speed, dynamic responses, this explicit form in $1/x_i$ has no theoretical base and in certain cases has given rise to poor convergence.

Since the cost of calculating dynamic constraints is relatively low compared to the cost of an analysis it is possible to carry out a sub-iteration (Fig. 12.8) in order to improve convergence.

The convergence of the sub-iteration process is ensured by applying more limits to the design variables and by relaxing the constraints to permit the detection of infeasible directions.

12.5 EXAMPLES AND COMMENTS

12.5.1 Optimization of a Metallic Delta Wing

This example is one of the first applications of our algorithm to a real problem. Unlike most demonstrations of structural optimization programs the algorithm starts from an initial design which represents a reasonable engineering solution to the design problem.

The FE mesh (Fig. 12.9) has about 4000 degrees of freedom and the stiffness of the fuselage is obtained using a super-element technique.

We have 88 design variables on the wing (panel thicknesses and areas of spars and rib flanges) and 198 constraints representing:

(1) stresses under two load cases (stabilized load factor and roll) which consists of:
(a) principal stresses at the intrados panel;

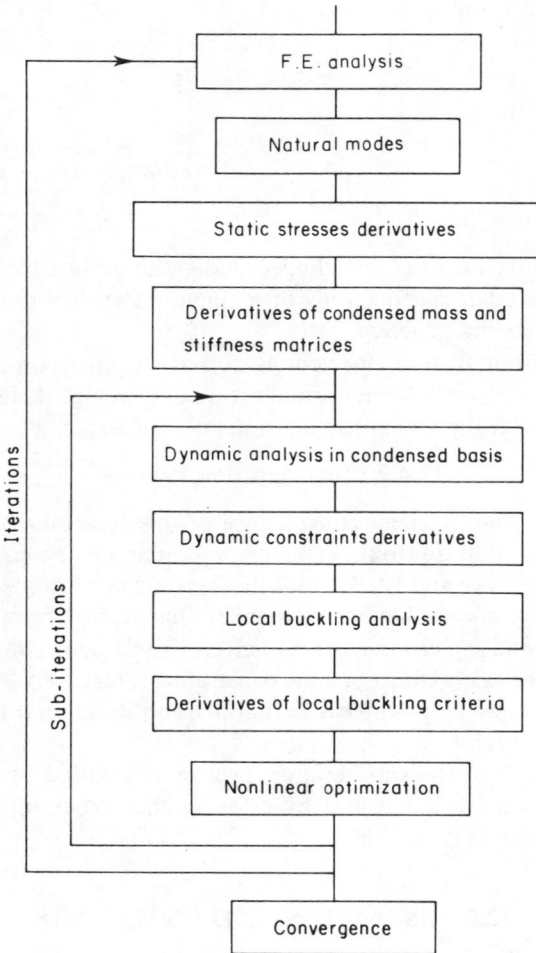

Fig. 12.8 Sub-iteration algorithms

(b) normal stresses in spars and rib flanges;
(c) shear stresses in spars and rib webs;
(d) plastic local buckling on extrados;
(2) bending moments limited at each 'spar fuselage' attachment (technological constraints, cost saving);
(3) static aeroelasticity constraints:
(a) limited variation on pitch moment;
(b) limited variation of control efficiencies.

Convergence is obtained after three iterations and the predicted value of the constraints is very good (Fig. 12.10). The algorithm gives a saving of 41 kg for the half-wing structure compared with initial design.

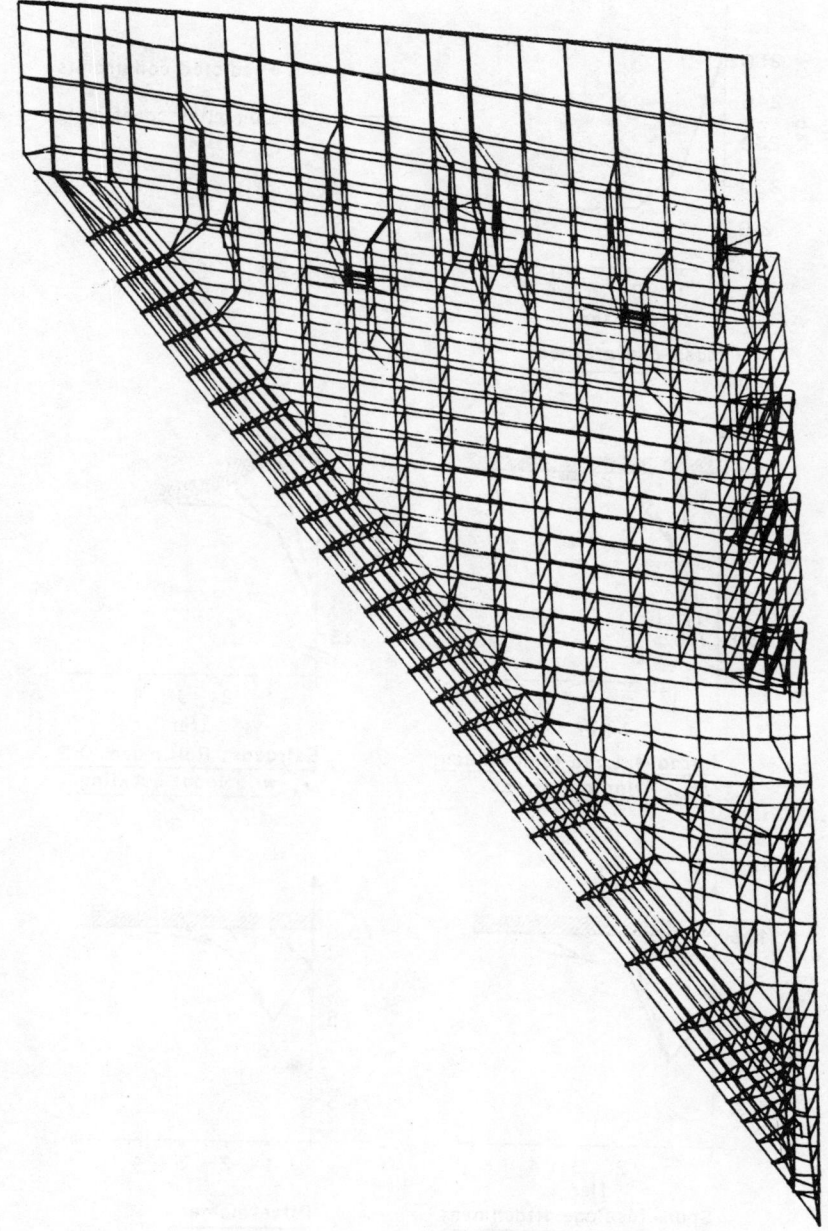

Fig. 12.9 Three-spar delta wing on Mirage 4000—finite element mesh

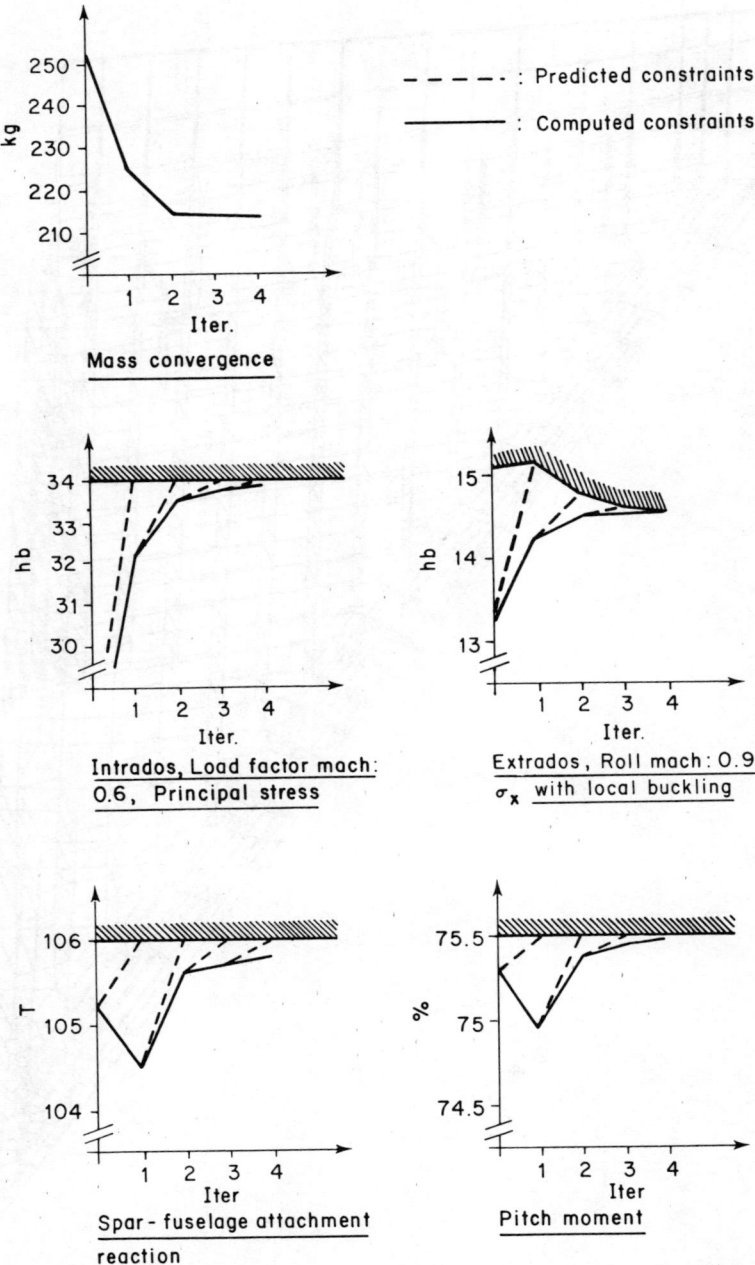

Fig. 12.10 Three-spar delta wing on Mirage 4000—convergence

12.5.2 Optimization of a Carbon Epoxy Empennage under Stress and Flutter Speed Constraints

In this case the finite element mesh is relatively dense near to the root of the empennage (Fig. 12.11). There are altogether more than 5000 degrees of freedom. The structure is optimized with 135 design variables representing the thickness of each ply (four design variables in the same mesh) and areas of ribs and spar flanges.

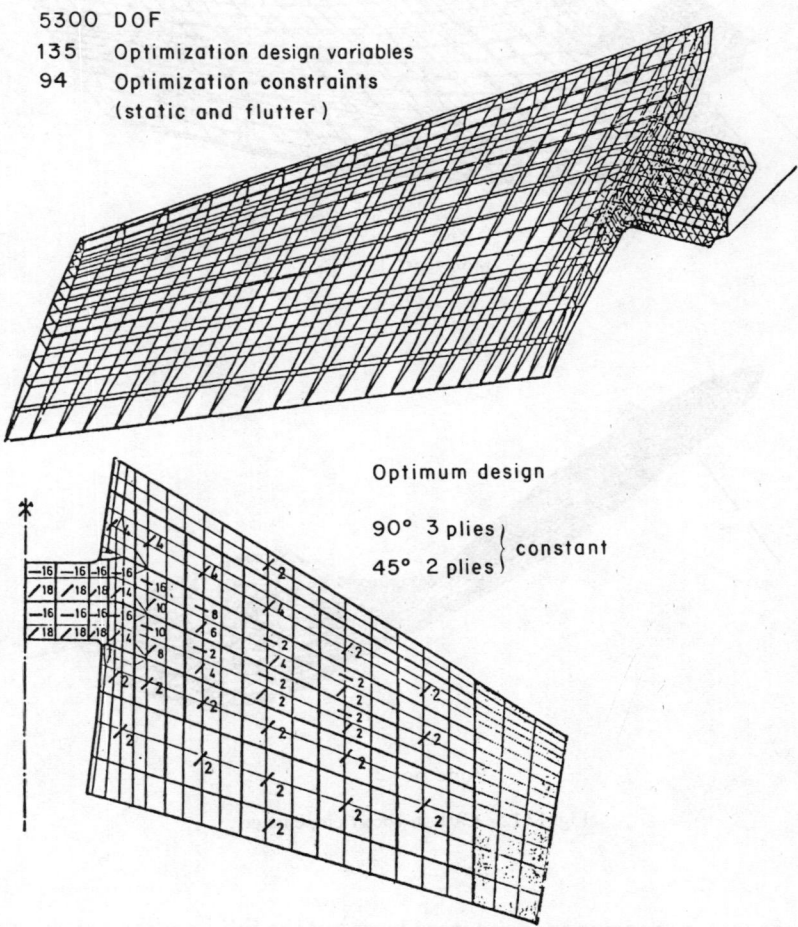

Fig. 12.11 Carbon epoxy emphennage

The 93 stress constraints empoly the Hill failure criteria in the carbon panel but the most important constraint is requirement that the flutter speed be greater than 360 m/s at Mach 0.9 (Fig. 12.12).

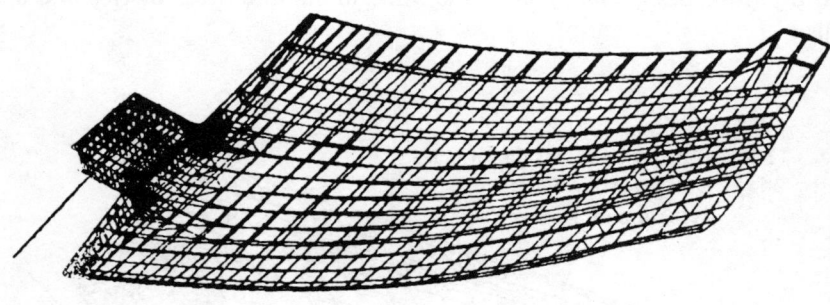

Mode no. 1

Mode no. 2

Fig. 12.12   Carbon epoxy empennage

There are ten modes in the condensed basis but the sub-iteration process is not employed. Convergence is obtained after 5 to 6 iteration with a relaxation of the admissible value of the flutter speed because of its great difference with respect to the initial value. The prediction of flutter speed is rather good using a linearization in reciprocal variables ($z_i = 1/x_i$) (Fig. 12.13).

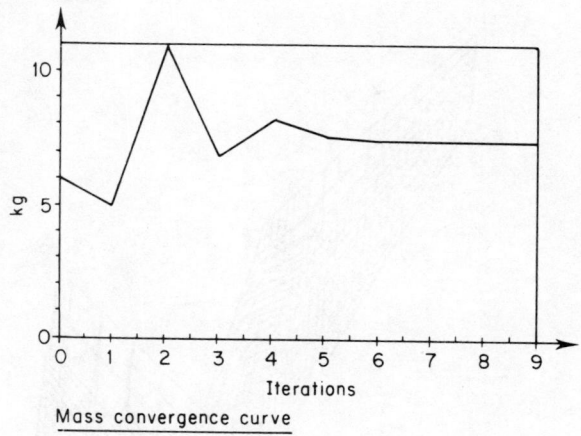

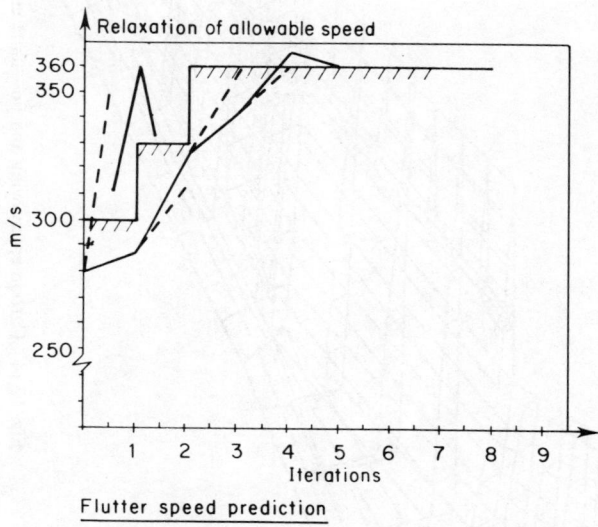

Fig. 12.13 Flutter speed prediction

12.5.3 Optimization of a Carbon Epoxy Wing with a Vertical Fin on a Fighter Aircraft

This particular problem represents one of our most significant applications in dynamic optimization. Two vertical fins are fixed on masts at the trailing edge of the wing. The finite element mesh is not very dense (Fig. 12.14) with 3000 degrees of freedom.

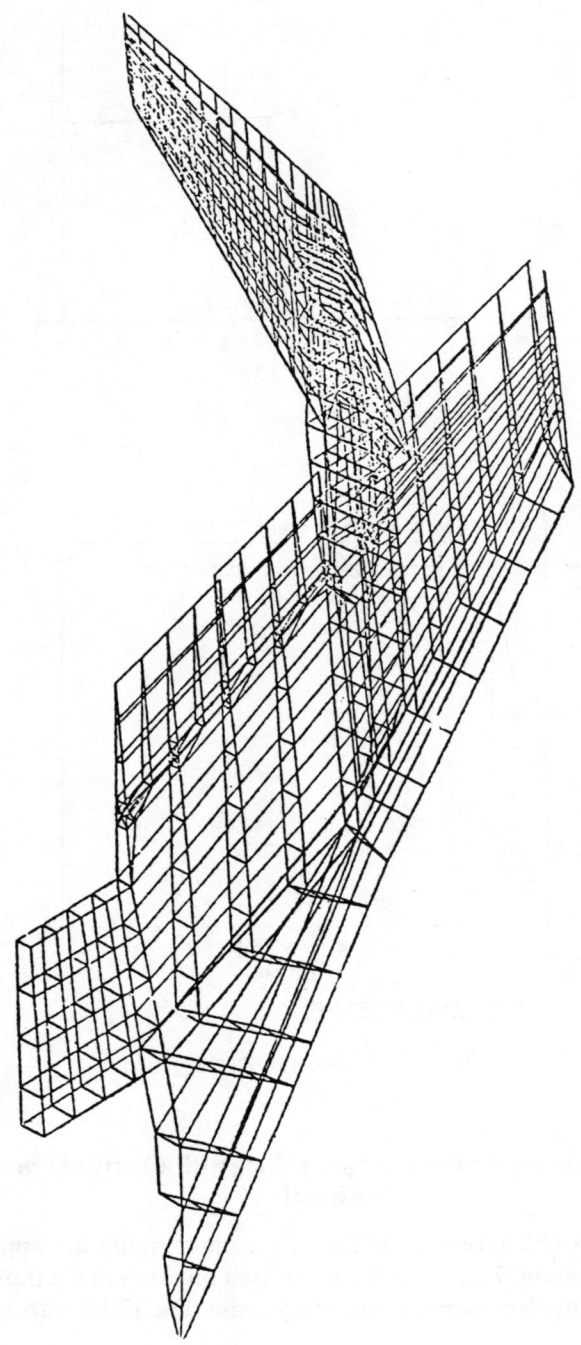

Fig. 12.14 Carbon epoxy wing and fin—finite element mesh

Optimization of Aircraft Structures

The main problem raised by this type of structure is the control of flutter occurring on the wing (due to the position of the fin on the wing) and on the fin which is completely manoeuvrable.

It is possible to separate the optimization of the wing from the optimization of the fin because there is little interaction between the two lifting surfaces.

12.5.3.1 Optimization of the wing

The characteristics of this optimization are:
— 149 design variables for the thickness of each ply, the areas of spars and rib flanges, and so on;
— 143 optimization constraints using Hill failure criteria in carbon fibre for two load cases;
— flutter speed to be greater than 450 m/s (Fig. 12.15);
— 20 modes in the condensed basis;
— the sub-iteration process has not been used because the quality of the condensed basis is not good enough.

Convergence is achieved in 5 to 6 iterations and the prediction of flutter speed is good (Fig. 12.16).

Two optimizations have been carried out, one with a fin and one without, and the two compared in (Fig. 12.17) with respect to fibre layout.

12.5.3.2 Optimization of the fin

The fin represents the most complex problem that we have had to optimize and it gave rise to great difficulties in convergency.

The fin, fixed at the end of a mast, pivots around its axis as a complete structure and has two natural flutter modes (Fig. 12.18): one very rough and one rather smooth.

The characteristics of this optimization are:
— 150 design variables representing the thickness of each ply. We also introduce inertia block parameters on the leading edge of the fin;
— 106 optimization constraints using the Hill failure criteria for carbon fibre in one load case;
— optimization constraints on flutter speeds.

In controlling flutter with only the flutter speeds, there is an oscillation on the convergence curve, the second natural flutter mode appearing and disappearing (Fig. 12.19).

We avoid the oscillations on the convergence by the introduction, as optimization constraints of damping coefficients for several speeds.

The optimization constraints for flutter control are (Fig. 12.20):
— flutter speeds for $g = 0$;
— damping coefficients for several speeds.

472   *Foundations of Structural Optimization: A Unified Approach*

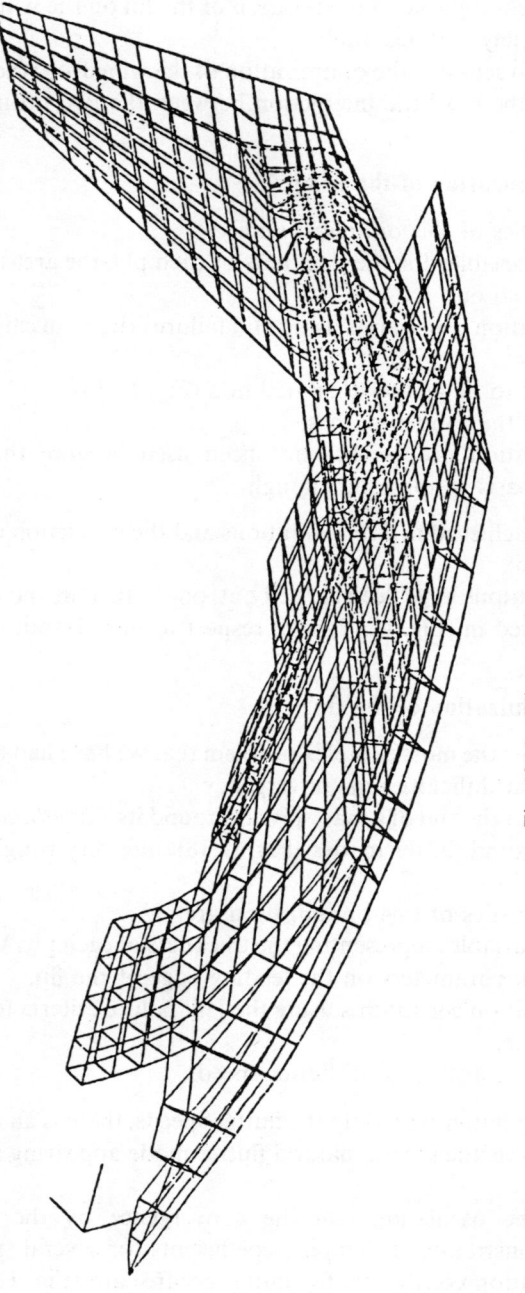

Fig. 12.15   Wing and fin—mode No. 1

Optimization of Aircraft Structures 473

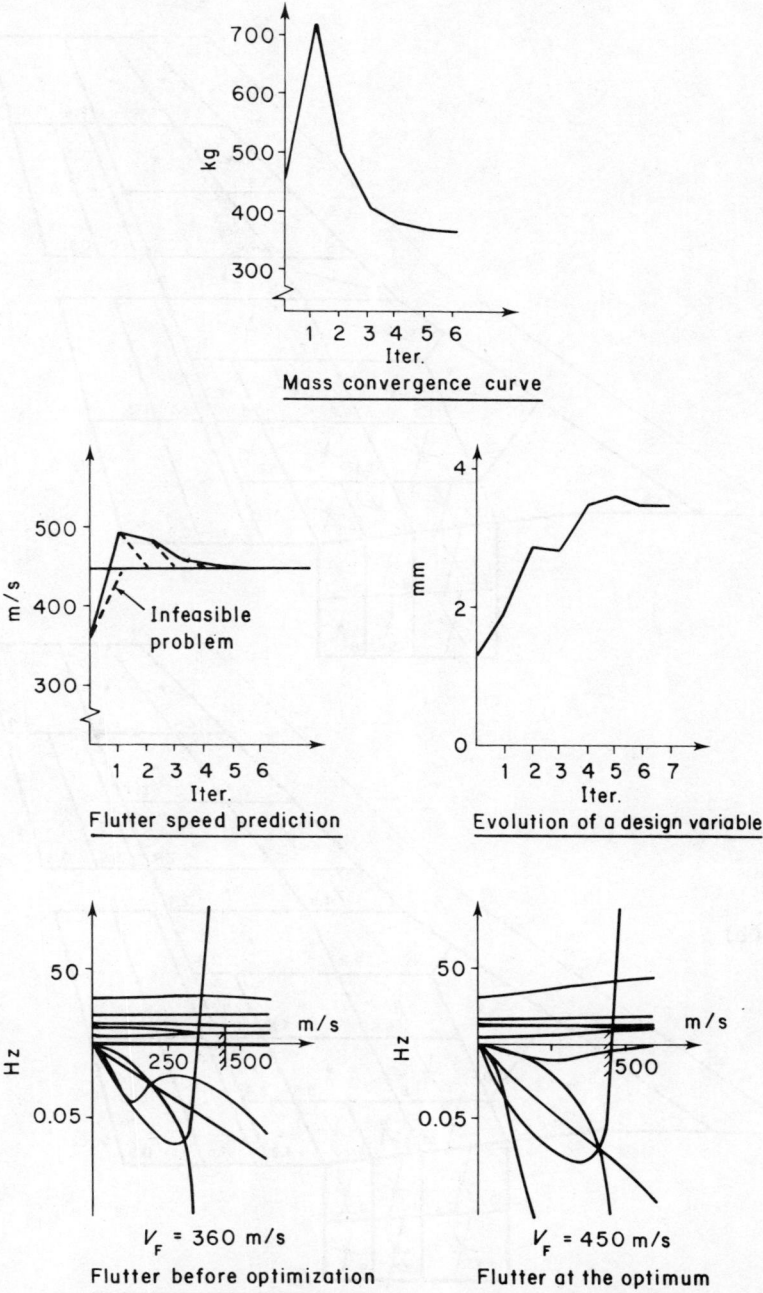

Fig. 12.16 Carbon epoxy wing

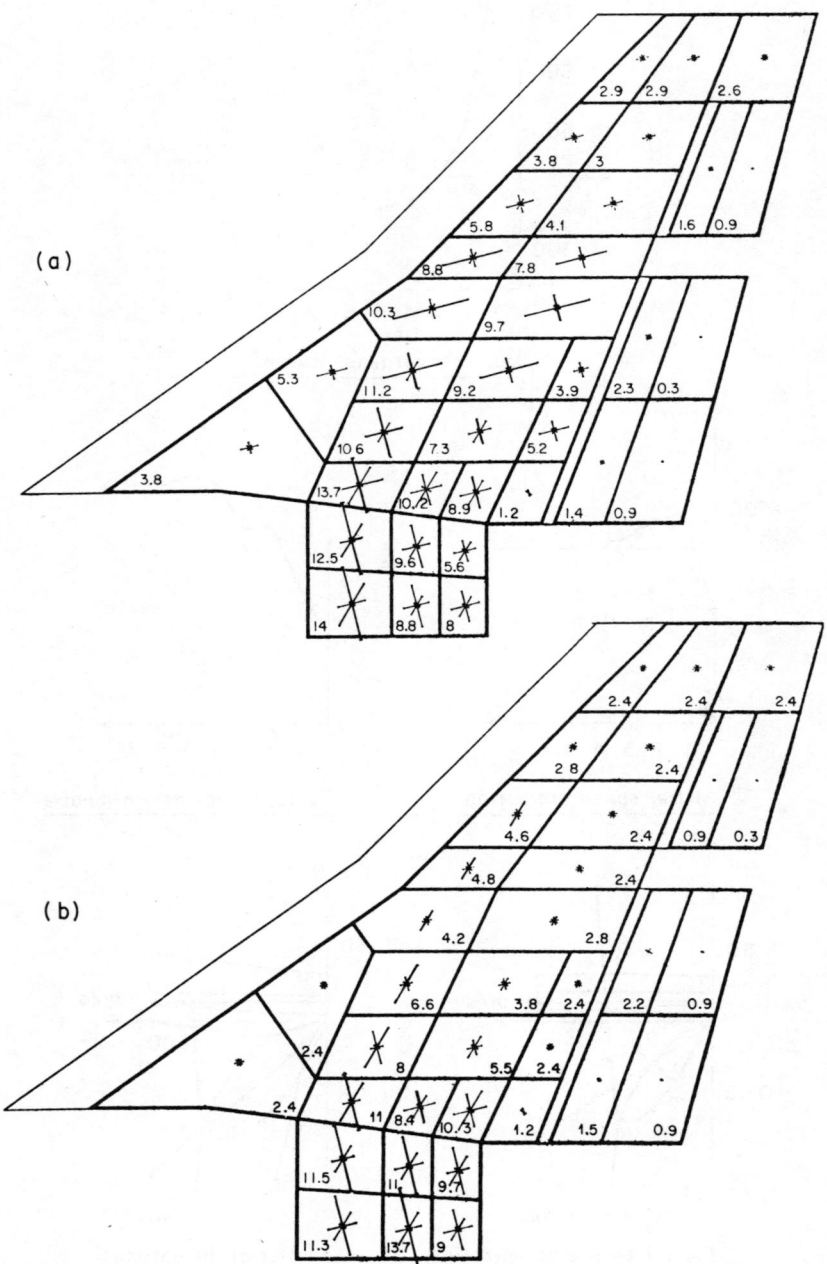

Fig. 12.17 Carbon epoxy wing—optimum design (a) with fin (b) without fin

Optimization of Aircraft Structures 475

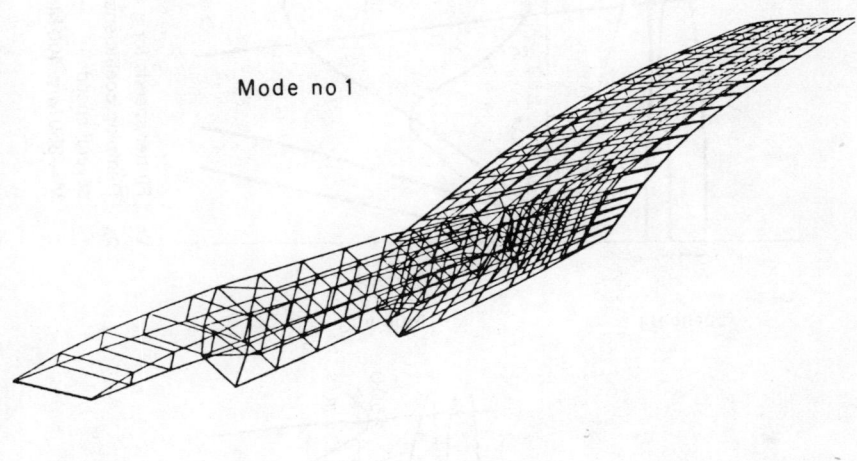

Mode no 1

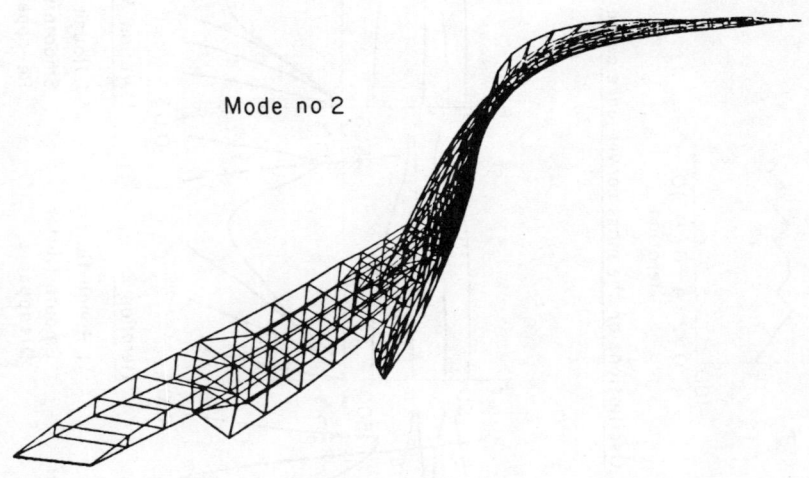

Mode no 2

Fig. 12.18 Carbon epoxy wing

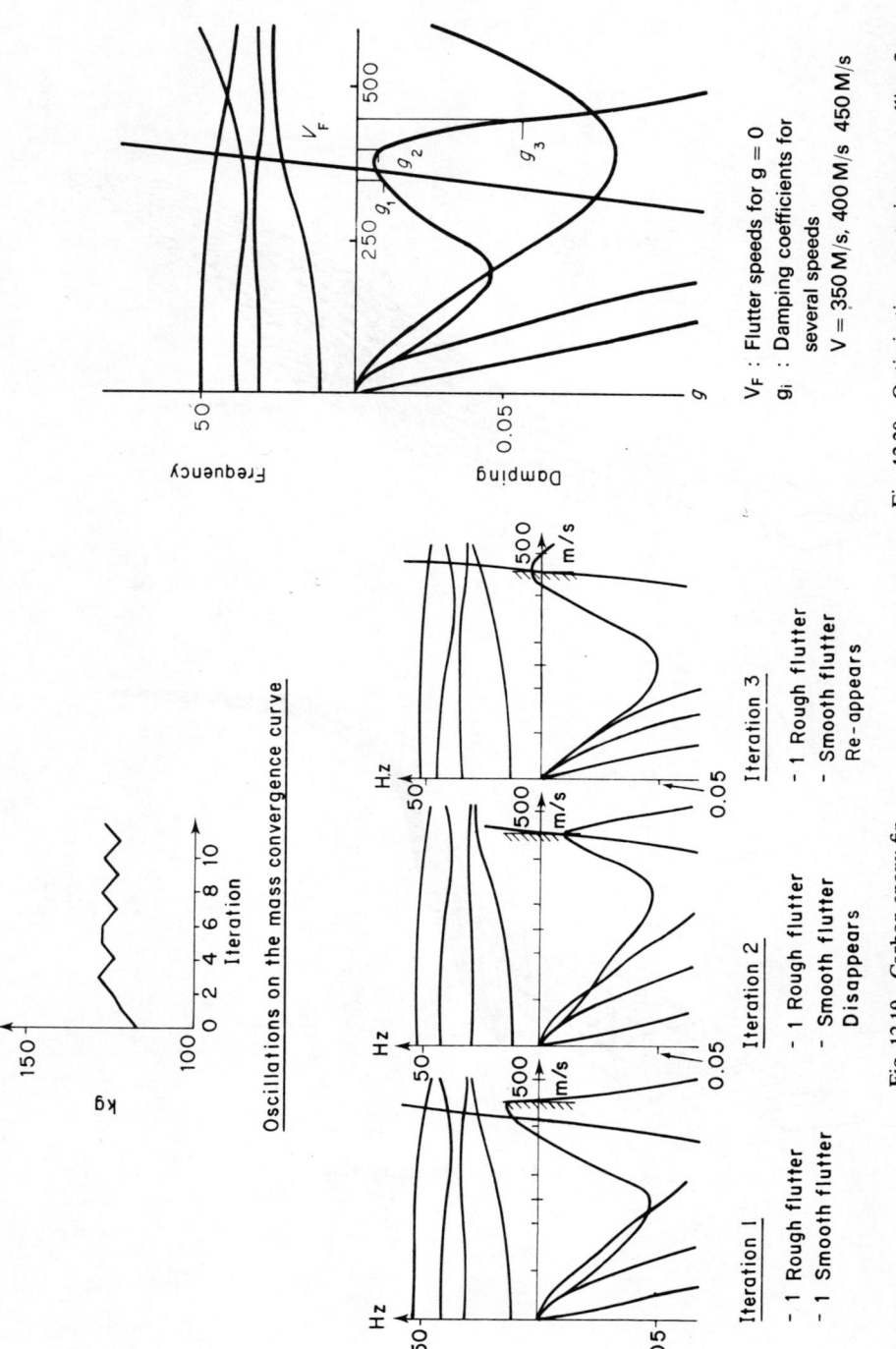

Fig. 12.19 Carbon epoxy fin

Fig. 12.20 Optimization constraints controlling flutter

Optimization of Aircraft Structures

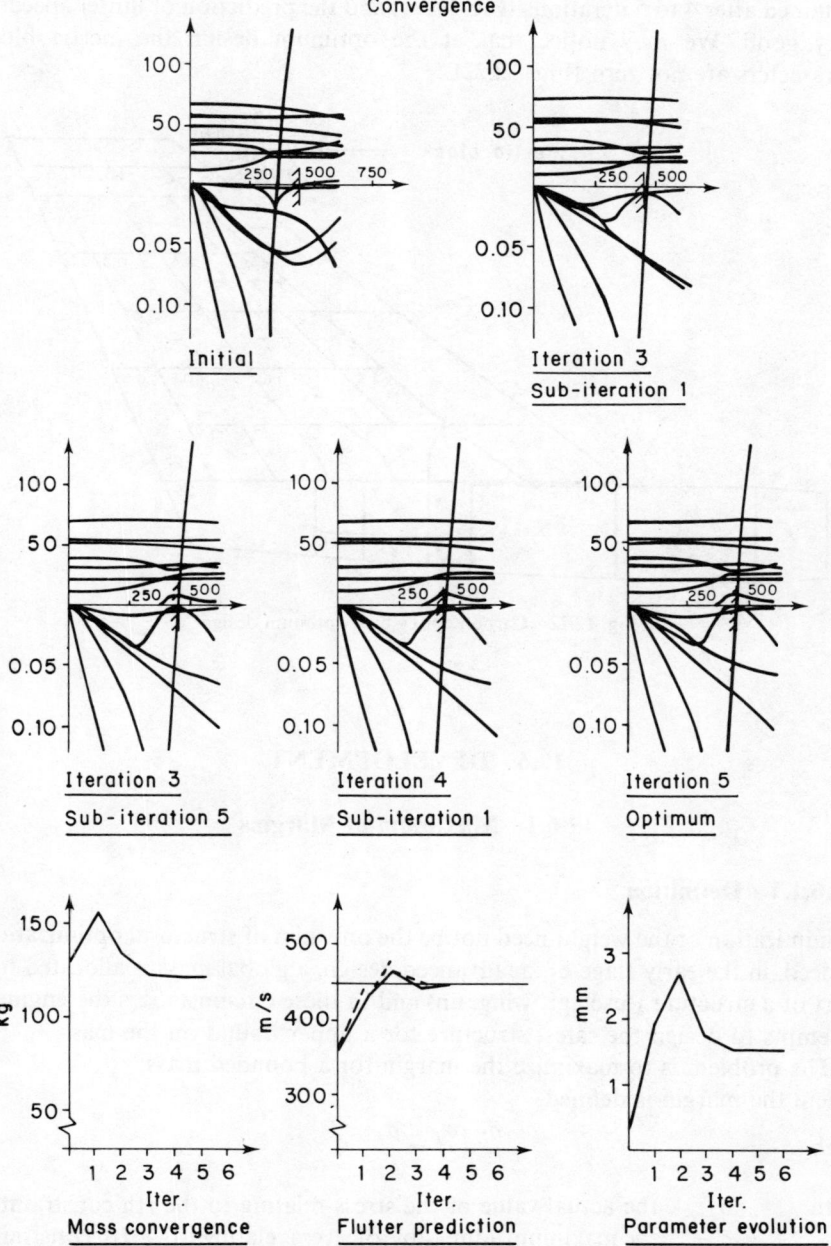

Fig. 12.21 Carbon epoxy fin

With these constraints and using the sub-iteration process, convergence is obtained after 4 to 5 iterations (Fig. 12.21) and the prediction of flutter speeds is very good. We may notice that at the optimum design the inertia block parameters are not zero (Fig. 12.22).

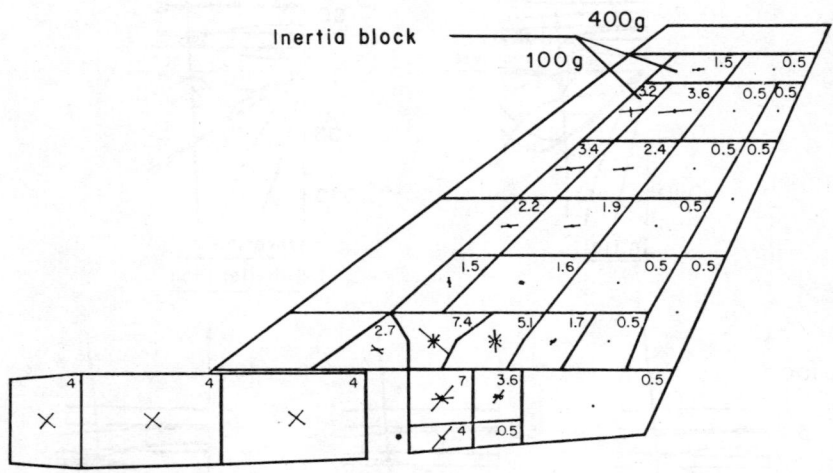

Fig. 12.22 Carbon epoxy fin—optimum design

12.6 DEVELOPMENT

12.6.1 Maximum of Margins

12.6.1.1 Definition

Minimization of the weight need not be the only aim of structural optimization. Indeed, in the early stage of an advanced design, a global mass is allocated to a part of a structure (fuselage, wing, fin) and in these circumstances the engineer attempts to design the safest structure for a upper bound on the mass.

The problem is to maximize the margin for a bounded mass; where the margin is defined

with
$$\frac{\bar{\sigma} - \sigma_j}{\sigma_j} = \frac{\bar{\sigma}_j}{\sigma_j} - 1$$

with σ_j the actual value of the stress relating to the j th constraint
          $\bar{\sigma}_j$ the maximum admissible of stress relating to the j th constraint.

The algorithm has to minimize the function $\mu_j = \sigma_j / \bar{\sigma}_j$ subject to the constraint on the mass: $\sum_{m_i} x_j \leqslant M_0$ where M_0 represents the maximum mass allowed.

To minimize the μ_j we define $\mu = \max_j (\mu_j)$ and the problem has the following form:

$$\min \mu$$
$$\sigma_j \leqslant \mu\bar{\sigma}_j \quad \text{for constraints concerned with the defined margin}$$
$$\sigma_k \leqslant \bar{\sigma}_k \quad \text{for other constraints}$$

In reciprocal variables $z_i = 1/x_i$ we have:
minimize μ

$$\sigma_{0j} + \sum_i \frac{\partial \sigma_j}{\partial z_i}(z_i - z_{i0}) \leqslant \mu\bar{\sigma}_j$$

$$\sigma_{0k} + \sum_i \frac{\partial \sigma_k}{\partial z_i}(z_i - z_{i0}) \leqslant \bar{\sigma}_k$$

$$\sum \frac{a_i}{z_i} \leqslant M_0$$
$$0 \leqslant \mu \leqslant 1$$

This optimization problem, with a linear objective function and nonlinear constraints (mainly mass constraint), is solved by the sequence of linear programs using move limits.

12.6.1.2 Minimization of stresses at the edge of a hole

The case of reinforcing a hole in a plate is an example where the reinforcing material matters less than decreasing the maximum stress around the hole.

When the problem is solved (Fig. 12.23) using the method described in this chapter, the convergence to a global solution is rapid. Nevertheless, the values of the design parameters do not completely converge since the optimum is rather flat. Thus for a given margin there exist several possible distributions of material.

12.6.2 'Sparsity' of the Partial Derivatives matrix

Increase in the size of problems which can be handled is amongst other things limited by the size of the partial derivatives matrix. But many of these partial derivatives are near zero, and thus we artificially annul partial derivatives which satisfy the following criteria.

The derivative of constraint j relative to parameter i is annulled if:

$$\sigma_j + \left(\frac{\partial \sigma_j}{\partial z_i}\right) \Delta z_{i\max} \times N_{\text{para}} \times R \leqslant \sigma_{j\text{adm}}$$

where R is an arbitrary reduction factor and N_{para} the number of design variables.

We are able to show on real applications that we can annul 80% of the coefficients without creating convergence problems (Fig. 12.24).

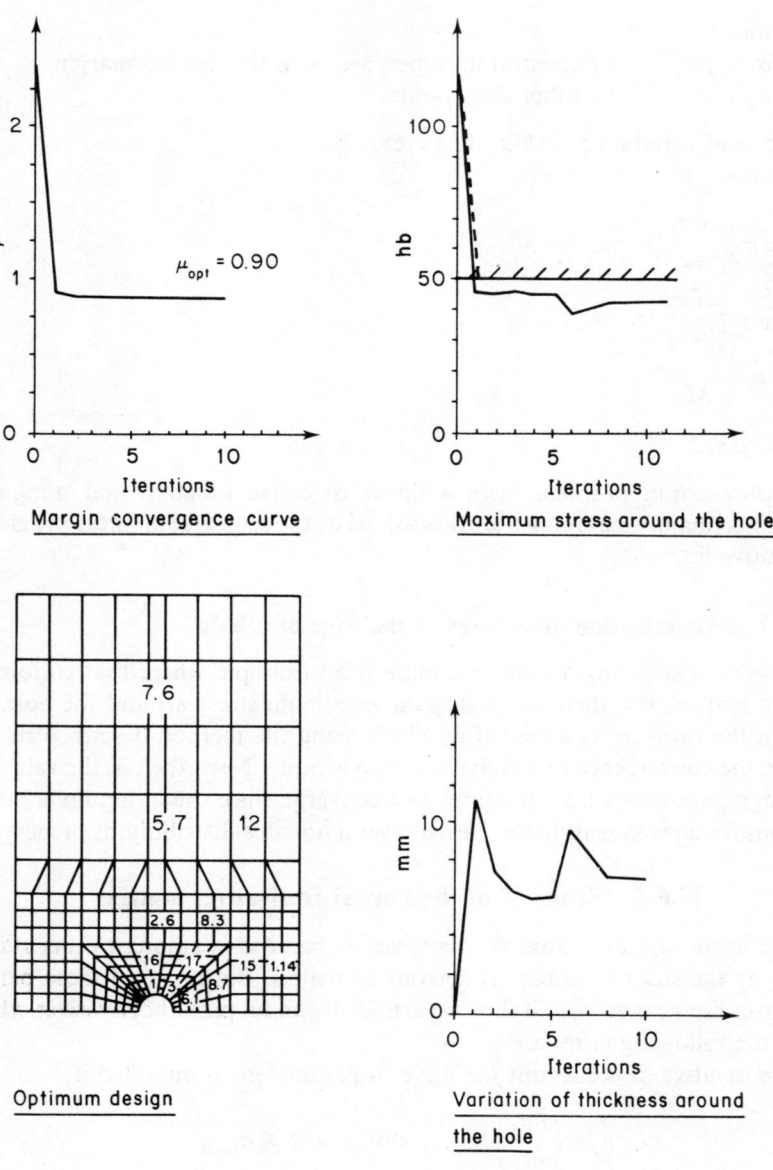

Fig. 12.23 Margin maximization

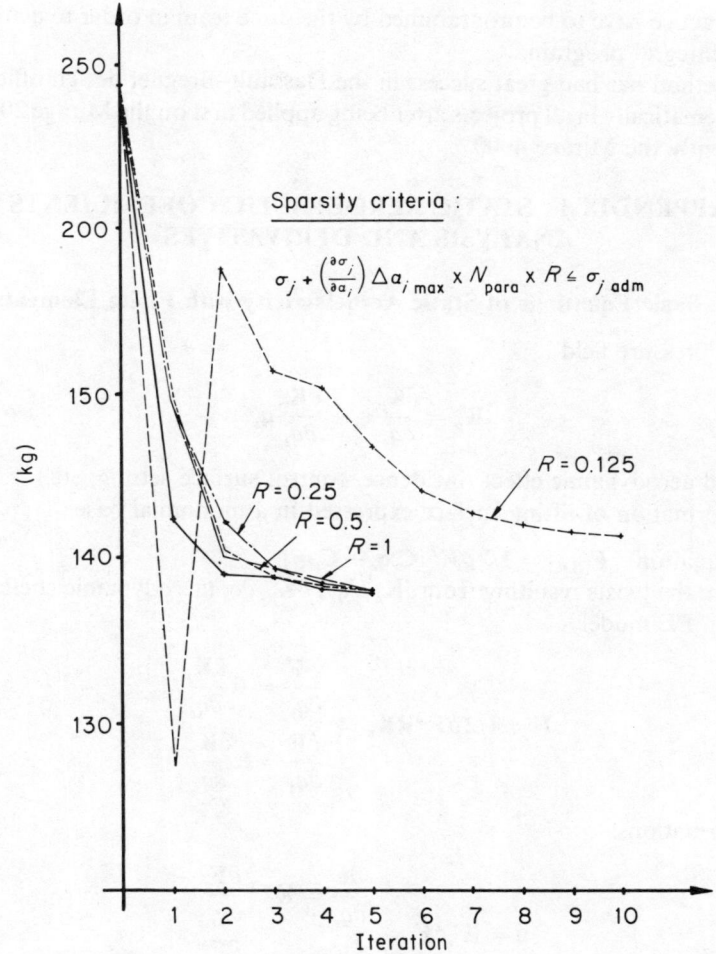

Fig. 12.24 Sparsity of the partial derivatives matrix

12.7 CONCLUSIONS

Our approach to the optimization problem presents three important qualities:
— easy introduction of new constraints without any need to be concerned with the redundancy of the structure;
— optimality conditions are satisfied at the convergence point;
— calculation times are acceptable for usual practice.

Another advantage of the process is the apparent theoretical simplicity which makes the main concepts of the algorithm understandable to non-specialists. In addition, the analysis, derivatives and optimization processes are closely related and in practice have to be programmed by the same team in order to achieve an efficient integral program.

The method has had great success in the Dassault–Breguet design office. It is used systematically in all projects after being applied first on the Mirage 2000 and subsequently the Mirage 4000.

APPENDIX 1 STATIC AEROELASTIC COEFFICIENTS ANALYSIS AND DERIVATIVES

1 Basic Equations of Static Aeroelasticity with Finite Elements

Discrete pressure field

$$\mathbf{K}_p = \frac{\partial \mathbf{K}_p}{\partial q_r}\mathbf{q}_r + \frac{\partial \mathbf{K}_p}{\partial q_s}\mathbf{q}_s$$

q_r = rigid aerodynamic effect (incidence, control surface setting, etc.)
q_s = deformation of lifting surface expressed in a monomial base.

Flight equation $\quad F_{CDG} = 1/2\rho V^2 (\mathbf{C}_r \mathbf{q}_r + \mathbf{C}_s \mathbf{q}_s)$
$\mathbf{C}_r, \mathbf{C}_s$ are the twists resulting from $\partial \mathbf{K}_p/\partial q_r$, $\partial \mathbf{K}_p/\partial q_s$ (aerodynamic coefficients)
Loads on FE model

$$\mathbf{F} = 1/2\rho V^2 \mathbf{R}\mathbf{K}_p \Rightarrow \begin{aligned} \frac{\partial \mathbf{F}}{\partial q_r} &= \mathbf{R}\frac{\partial \mathbf{K}_p}{\partial q_r} \\ \frac{\partial \mathbf{F}}{\partial q_s} &= \mathbf{R}\frac{\partial \mathbf{K}_p}{\partial q_s} \end{aligned}$$

FE deformations:

$$\mathbf{u} = \mathbf{K}^{-1}\mathbf{F} \Rightarrow \begin{aligned} \frac{\partial \mathbf{u}}{\partial q_r} &= \mathbf{K}^{-1}\frac{\partial \mathbf{F}}{\partial q_r} \\ \frac{\partial \mathbf{u}}{\partial q_s} &= \mathbf{K}^{-1}\frac{\partial \mathbf{F}}{\partial q_s} \end{aligned}$$

Smoothed transition FE $\mathbf{q}_s = \mathbf{L}\mathbf{X}$

$$\mathbf{q}_s = 1/2\rho V^2 (\mathbf{A}_1 \mathbf{q}_r + \mathbf{A}_2 \mathbf{q}_s) \qquad \begin{aligned} \mathbf{A}_1 &= \mathbf{L}\frac{\partial \mathbf{u}}{\partial q_r} \\ \mathbf{A}_2 &= \mathbf{L}\frac{\partial \mathbf{u}}{\partial q_s} \end{aligned} \qquad (1)$$

Flexible effect elimination

$$\mathbf{q}_s = 1/2\rho V^2 \mu \mathbf{q}_r, \; \mu = \mathbf{D}^{-1}\mathbf{A}_1, \; \mathbf{D} = [\mathbf{I} - 1/2\rho V^2 \mathbf{A}_2] \qquad (2)$$

Flexible aerodynamic coefficients

$$\mathbf{C} = \mathbf{C}_r + 1/2\rho V^2 \mathbf{C}_s \mu \qquad (3)$$

2 Derivative Relative to Structural Parameters

Differentiation of (1) knowing that $\Delta(\mathbf{K}^{-1}\mathbf{F}) = \mathbf{K}^{-1}\Delta\mathbf{K}\mathbf{X}$:

$$\Delta\mathbf{q}_s = -1/2\rho V^2 \mathbf{L}\mathbf{K}^{-1}\Delta\mathbf{K}(\partial\mathbf{u}/\partial q_r\,\mathbf{q}_r + \partial\mathbf{u}/\partial q_s\mathbf{q}_s) + 1/2\rho V^2 \mathbf{A}_2 \Delta\mathbf{q}_s$$

By eliminating $\mathbf{q}_s$ from (2)

$$\Delta\mathbf{q}_s = -1/2\rho V^2 \mathbf{D}^{-1}\mathbf{L}\mathbf{K}^{-1}\Delta\mathbf{K}(\partial\mathbf{u}/\partial q_r + \partial\mathbf{u}/\partial q_s\mu)\mathbf{q}_r = \Delta\mu\mathbf{q}_r$$

Differentiation of (3):

$$\Delta\mathbf{C} = 1/2\rho V^2 \mathbf{C}_s \Delta\mu$$
$$\Delta\mathbf{C} = 1/2\rho V^2 \mathbf{C}_s \mathbf{D}^{-1}\mathbf{L}\mathbf{K}^{-1} - \Delta\mathbf{K}(\partial\mathbf{u}/\partial q_r + \partial\mathbf{u}/\partial q_s\mu)$$

Preferable in the form

$$\Delta\mathbf{C} = 1/2\rho V^2 (\mathbf{K}^{-1}(\mathbf{C}_s \mathbf{D}^{-1}\mathbf{L})^T)^T - \Delta\mathbf{K}[(\partial\mathbf{u}/\partial q_r) + (\partial\mathbf{u}/\partial q_s)\mu]$$

which means solving the equilibrium equation for a single load for each aerodynamic coefficient at a given Mach and dynamic pressure

$$\Delta\mathbf{C} = 1/2\rho V^2 \mathbf{C}_s \mathbf{D}^{-1}(\mathbf{K}^{-1}\mathbf{L}^T)^T - \Delta\mathbf{K}[\partial\mathbf{u}/\partial q_r) + (\partial\mathbf{u}/\partial q_s)\mu]$$

which requires q_s solutions (several tens), which is absorbed for all Mach numbers and dynamic pressures.

APPENDIX 2 DERIVATION OF FLUTTER SPEED (J. P. BREVAN'S METHOD)

1 Analysis

We attempt to solve the flutter equation at a given Mach number

$$\left[\mathbf{K}(\lambda) - \omega^2(1+ig)^2\mathbf{M}(\lambda) - \rho V^2 \mathbf{A}\left(\frac{\omega}{V}\right)\right]\mathbf{q} = 0$$

$\mathbf{K}$ — reduced stiffness matrix
$\mathbf{M}$ — reduced mass matrix
ω — frequency of the solution
g — damping ($g = 0 \Rightarrow$ flutter)
V — velocity
$\mathbf{A}$ — matrix of aerodynamic forces
$\mathbf{P}$ — left solution
$\mathbf{q}$ — right solution

In the simplified form

$$[\mathbf{D}(\lambda, \omega, g, V)]\mathbf{q} = 0 \qquad (1)$$
$$\mathbf{P}_t[\mathbf{D}(\lambda, \omega, g, V)] = 0 \qquad (2)$$

with $g = 0$ (or given).

2 Differentiation

$$(1) \Rightarrow \Delta(\mathbf{Dq}) = \Delta \mathbf{Dq} + \mathbf{D}\Delta \mathbf{q} = 0$$

by multiplying with $\mathbf{p}_t$,

$$\mathbf{p}_t \Delta \mathbf{Dq} + \mathbf{p}_t \mathbf{D}\Delta \mathbf{q} = 0$$

noting that the second term is identically zero we obtain

$$\mathbf{p}_t \Delta \mathbf{Dq} = 0$$

If we fix the damping g,

$$\mathbf{p}_t[\partial \mathbf{D}/\partial \lambda]\mathbf{q} + \mathbf{p}_t[\partial \mathbf{D}/\partial \omega]\mathbf{q} + \mathbf{p}_t[\partial \mathbf{D}/\partial V]\mathbf{q} = 0 \qquad (3)$$

This complex equation gives the derivatives of the frequency of the solution and the flutter speed.

C. P.
G. L.

REFERENCES

1. Optimisation des structures sur modéle d'éléments finis, Marché DRET n°76.34039.
2. C. Fleury, Le dimensionnement automatique des structures élastiques, Thèse Université de Liège, 1977.
3. Polak, *Computational method in optimization—a unified approach*, Academic Press, New York, 1971.
4. C. Petiau, N. S n° 1300 (AMD-BA), Resolution des grands systèmes linéaries creux.
5. C. Petiau, 14ème congrès International Aéronautique PARIS juin 1979. Maillage par la méthode topologique conversationnelle et optimisation dans les calculs de structure.
6. Rapport DRET 76/452, Optimisation d'une pièce plane élastique pour minimiser les constraintes, (1978).
7. AGARD, Lectures series n°70 on structural optimisation, (1974).

Special Applications and Techniques

Foundations of Structural Optimization: A Unified Approach
Edited by A. J. Morris
© 1982 John Wiley & Sons Ltd.

Chapter 13

Structural Optimization with Material Selection

13.1 INTRODUCTION

Rational design of structures requires the optimal determination of the three kinds of design parameters: (a) sizing, (b) configuration, and (c) material selection parameters. The sizing parameters are the ones that can specify the gross cross-sections of structural members. For example, in this book the cross-sectional areas of truss members and/or the thicknesses of panels have been taken as the representative sizing parameters. The configuration parameters are the quantities that can define the shape of a whole structure. Therefore, when referred to truss structures, the coordinates of nodal points can be considered as the configuration parameters as done in Chapters 14 and 15.

Finally, material selection parameters are the indicators that can distinguish which material must be used in a particular structural member.

In an absolutely rational structural design, all these parameters should be integrally employed. However, such a general approach simultaneously employing these three kinds of design parameters is extremely difficult. At the present time, only a partial solution has been given to this general structural optimization problem. As we have seen in earlier chapters, efficient solution methods were first established for the pure sizing problems, which preassign configuration as well as material constitution. A practically meaningful approximate optimum for the problem can now be obtained for a reasonable number of structural analyses by using either the mathematical programming based methods (Chapters 8 and 9) or the optimality criteria methods (Chapters 5 and 6).

As compared with the success in the pure sizing problems, it might be said that the present state of the art on the combined configuration-sizing problems is still in its infancy. The reason why the configuration optimization problems are difficult to solve resides in the fact that the displacement quantities are highly nonlinear functions of the configuration parameters. An explicit and high-quality approximation for displacements has not been found for the configuration parameters. Therefore, more structural analyses are generally required to attain an optimum. In spite of this inherent difficulty, when the configuration parameters are included in structural optimization, it is observed that the optimum weight can be considerably reduced from the optimum achieved through pure sizing [1] and this is discussed in Chapters 14 and 15.

How material selection influences the structural optimum is a completely unanswered question. Recently, various new materials have been produced for a variety of industrial applications so that designers can select, in principle, the most suitable materials for structural members corresponding to the required use. However, it might be argued that there is no rational and efficient way of determining the optimal material distribution of structures. Therefore, the present investigation aims to develop an efficient algorithm for the material-selection problems and at the same time obtain an insight into the optimal material constitution.

A new solution method is proposed based on the use the mini-max dual method of Chapters 3 and 10 and the approximation concept [2, 3]. The approximation concept is used to replace an original nonlinear and implicit structural optimization problem with a sequence of explicit and high-quality approximate problems. Each approximate problem is solved in the dual space with respect to the Lagrange multipliers and the material properties. This is achieved through the use of the mini-max dual method. Finally, it is shown that the tacit use of the mini-max dual method can avoid the inherent combinatorial nature associated with the material-selection problems.

13.2 PROBLEM FORMULATION

Although the applicability of the procedure presented herein is not necessarily restricted to truss structures, we consider here only the minimum-weight design of trusses in order to present the arguments in the simplest mathematical form.

The simplest sizing optimization problem of a truss structure takes the following nonlinear mathematical programming form:

$$\min \sum_{i=1}^{n} \rho_i x_i l_i \qquad (13.1)$$

subject to

$$|\sigma_{ij}| \leq \sigma_i^{(a)} \qquad i = 1, \ldots, q; j = 1, \ldots, l \qquad (13.2)$$

(PS) $\qquad |U_{ij}| \leq U_i^{(a)} \qquad i = 1, \ldots h; \qquad j = 1, \ldots, l \qquad (13.3)$

$$x_i \geq x_i^{(L)} \qquad i = 1, \ldots n \qquad (13.4)$$

The problem posed by Equations (13.1) to (13.4) is a well-known min-gauge, stress and displacement constrained optimization problem. The design parameters are simply the cross-sectional areas of the independent truss members.

Note that the allowable compressive and tensile stresses are assumed to be the same in order to simplify the formulation. The problem has multiple loading conditions and the use of the design variable linking concept has also been assumed.

It is widely recognized that regardless of the optimization method employed the use of the approximation concept [2, 3] is absolutely necessary in order to solve the (PS) problem (Equations (13.1) to (13.4)) efficiently. The approximation concept consists of the following fundamental procedures:

(a) Construct a first-order Taylor series expansion of the displacement response quantities around a certain design point with respect to the intermediate design variables.
(b) Create an approximate optimization problem by substituting the expansion in (a) into the behavioural constraints.
(c) Select potentially critical constraints by the posture table concept in order to throw out many redundant constraints.
(d) Solve the approximate problem of small size by some optimization method.
(e) Repeat from steps (a) to (d) until convergence is attained.

There are two points now available to produce an optimum for the approximate problem: to employ either primal space methods or dual space methods. The primal space methods directly solve the approximate problems by using one of the standard mathematical programming methods. The unknowns in the design space are the independent cross-sectional areas. The alternative dual space methods seek for the optimum Lagrange multipliers for the potentially critical constraints. The corresponding primal optimum is recovered after the optimum Lagrange multipliers have been obtained. The so-called discretized optimality criteria methods can be shown to be a special application of the dual space methods.

The comparison between the two different space methods reveals the following facts:

(a) The primal space methods treat the original design variables as unknowns so that the efficiency is easily influenced by the number of design variables.
(b) The dual space methods, on the other hand, seek for the optimum Lagrange multipliers for potentially critical constraints so that the efficiency depends on the number of these constraints.

Since there are relatively few critical constraints at an optimum, the problem size in the dual space (i.e., the dual space dimension) generally tends to be smaller than the primal space dimension. In addition, when dual space methods are employed, the actual computer implementation becomes easier, because the dual problem is essentially an unconstrained convex program. The most primitive steepest-ascent method is sufficient for locating the optimum in the dual space.

However, in the case of simple pure sizing problems, the choice between the two methods will be a matter of preference, because the efficiencies of both methods are essentially the same when measured with respect to the required number of structural analyses.

The advantages associated with the dual space methods accrue when these are applied to structural optimization problems other than the simple sizing problem (PS). One example is a material-selection problem considered in this chapter. This problem can be stated as follows:

'Allocate to each independent design group a suitable material from a set of candidate materials and simultaneously determine the cross-sectional areas such that the minimum weight is achieved.'

Such a combined material-selection and sizing problem can be described by

$$\min \sum_{i=1}^{n} \bar{\rho}_i x_i l_i \qquad (13.5)$$

subject to

$$|\bar{\sigma}_{ij}| \leqslant \bar{\sigma}_i^{(a)} \qquad i = 1, \ldots, q; \qquad j = 1, \ldots, l \qquad (13.6)$$

(MS)
$$|\bar{U}_{ij}| \leqslant \bar{U}_i^{(a)} \qquad i = 1, \ldots, h; \qquad j = 1, \ldots, l \qquad (13.7)$$

$$x_i \geqslant x_i^{(L)} \qquad i = 1, \ldots, n \qquad (13.8)$$

$$(\rho_k, E_k, \sigma_k^{(a)}) \in S_i \qquad i = 1, \ldots, n; \qquad k = 1, \ldots, p \qquad (13.9)$$

This problem is termed the (MS) problem. New side constraints Equation (13.9) are formally added to the original (PS) problem. Equation (13.9) indicates that the members in the ith independent group have to take one of the materials listed in the set S_i. The material set S_i is specified by the material properties: density ρ, Young's modulus E and allowable stress $\sigma^{(a)}$. The upper-bar notations have been used to indicate the quantities that are influenced by the material selection. For example, the allowable stress for the ith member depends on what material is used for that member.

The characteristics of the (MS) problem can be described as follows:
(a) The value of the objective function Equation (13.5) is dependent on not only the cross-sectional areas but also the material selection because the mass density ρ_i is no longer constant. The value of ρ_i must be selected from the set S_i. Thus, the objective function (13.5) is now a continuous function of the cross-sectional area and a discrete function with respect to the material selection.
(b) The response quantities σ_{ij} and U_{ij} are influenced by the material selection because they are functions of the Young's modulus.
(c) The allowable stresses can not remain and are also dependent on the selected material. Together with the observation in (b) this implies that the behavioural constraints for the (MS) problem change their shape and move abruptly in the design space, according to which material is considered.

These particular characteristics make the (MS) problem quite difficult to solve. Unless some special and efficient method is provided, very many sizing problems have to be solved by the fixing material distribution on a one by one basis. For example, in the present formulation, it has been assumed that there are p different materials for each of the n independent design groups (see Equation (13.9)). This yields p^n distinct material combinations in a structure. Therefore, in order to obtain a global minimum, these p^n pure sizing problems have to be solved. However, the number p^n is quite large. Take, for example, $p = 2$ and $n = 10$ (this corresponds to a relatively small material-selection problem). Then more than one thousand pure sizing problems have to be dealt with. The existing efficient methods for the pure sizing problems may require ten structural analyses for each sizing problem necessitating ten thousand structural analyses to obtain an optimum.

The extent of this combinatorial problem has prevented the progress of research for the material-selection optimization problems. However, the proposed method based on the mini-max dual method can tacitly handle the combinatorial nature of the problem, without increasing the required number of structural analyses.

13.3 APPROXIMATE PROBLEM

Following the approximation concept described in Section 13.2, the behavioural constraints (13.6) and (13.7) are approximated by introducing the first-order Taylor series expansion of the displacement response quantities.

The expansion takes the form of

$$U_{ij} \cong U_{ij}^0 + \sum_{k=1}^{n} \frac{\partial U_{ij}^0}{\partial \beta_k} (\beta_k - \beta_k^0) \tag{13.10}$$

at a specified design point 0. The variable β_k is a so-called intermediate variable. For the material-selection problems, the following intermediate variable is recommended to assure the quality of the approximation defined by Equation (13.10):

$$\beta_k = 1/(x_k \overline{E}_k) \tag{13.11}$$

That is, the variable β_k is a reciprocal of axial stiffness.

Since the identity

$$U_{ij}^0 - \sum_{k=1}^{n} \frac{\partial U_{ij}^0}{\partial \beta_k} \beta_k^0 = 0 \tag{13.12}$$

can be easily proved by assuming that the external loads are independent of the design parameters, Equation (13.10) can be simplified to

$$U_{ij} \cong \sum_{k=1}^{n} \frac{\partial U_{ij}^0}{\partial \beta_k} \beta_k \tag{13.13}$$

Equation (13.13) is a high-quality approximation for the material-selection problem. Indeed, it should be noted that when the material is fixed (i.e., $\overline{E}_k =$

const.), Equation (13.13) collapses to the usual expansion for the pure sizing problems (see Refs [2] and [3]).

It is now obvious that the displacement constraints (13.7) can be approximated by

$$-U_i^{(a)} \leq \sum_{k=1}^{n} \frac{\partial U_{ij}^0}{\partial \beta_k} \beta_k \leq U_i^{(a)} \qquad (13.14)$$

On the other hand, the approximation for the stress constraints (13.6) needs some manipulation. From (13.7) the ith member stress due to the jth loading condition is described by

$$\sigma_{ij} = \bar{E}_i B_i \hat{U}_{ij} \qquad (13.15)$$

where the symbol B_i is the design parameter-independent stress matrix and the $\hat{U}_{ij}$ is a displacement vector of the ith member. Substitution of Equation (13.13) into Equation (13.15) creates the approximation:

$$\sigma_{ij} \cong \bar{E}_i \sum_{k=1}^{n} a_{ijk}^0 \beta_k \qquad (13.16)$$

where

$$a_{ijk}^0 = B_i \frac{\partial U_{ij}^0}{\partial \beta_k} \qquad (13.17)$$

Thus, the stress constraints (13.6) can be written as

$$-\frac{\bar{\sigma}_i^{(a)}}{\bar{E}_i} \leq \sum_{k=1}^{n} a_{ijk}^0 \beta_k \leq \frac{\bar{\sigma}_i^{(a)}}{\bar{E}_i} \qquad (13.18)$$

and this leads to a new optimization problem created about a specified design point 0:

$$\min \sum_{i=1}^{n} \bar{\rho}_i x_i l_i \qquad (13.19)$$

subject to

$$\sum_{i=1}^{n} \frac{b_{ij}^0}{x_i \bar{E}_i} \leq \bar{d}_j \qquad j = 1, \ldots, m \qquad (13.20)$$

(AMS) $\qquad\qquad x_i \geq x_i^{(l)} \qquad i = 1, \ldots, n \qquad (13.21)$

$$(\rho_k, E_k, \sigma_k^{(a)}) \in S_i \qquad i = 1, \ldots, n; k = 1, \ldots, p \qquad (13.22)$$

The approximate constraints (13.20) represent either stress or displacement constraints and at this stage, we have selected the m potentially critical constraints. The right-hand side in (13.20), $\bar{d}_j$, generally depends on the material selection so that the upper-bar notation has been used. If the jth constraint is a stress constraint in the ith independent group, then it is given by

$$\bar{d}_j = \bar{\sigma}_i^{(a)}/\bar{E}_i \qquad (13.23)$$

and if it is the ith displacement,

$$\bar{d}_j = U_i^{(a)} \tag{13.24}$$

The approximate material selection problem (AMS) represent by Equations (13.19) to (13.22) is now solved instead of the original (MS) problem. The (AMS) problem becomes explicit and separable with respect to the cross-sectional areas and the material selection. However, its efficient solution is still difficult in the primal space because of the discrete side requirements, Equation (13.22). Therefore, the (AMS) problem is solved by recourse to the dual space, considered in the next section.

13.4 THE DUAL METHOD

The Lagrangian function for the (AMS) problem becomes

$$L = \sum_{i=1}^{n} \left(\bar{\rho}_i x_i l_i + \frac{e_i}{x_i \bar{E}_i} \right) - \sum_{j=1}^{m} \lambda_j \bar{d}_j \tag{13.25}$$

where

$$e_i = \sum_{j=1}^{m} \lambda_j b_{ij}^0 \tag{13.26}$$

We now divide the potentially critical constraints into two groups: stress and displacement constraints. Then the second term in the right-hand side of Equation (13.25) can be rewritten as

$$\sum_{j=1}^{m} \lambda_j \bar{d}_j = \sum_{j \in I_1} \lambda_j \bar{d}_j + \sum_{j \in I_2} \lambda_j \bar{d}_j \tag{13.27}$$

where I_1 and I_2 represent the stress and displacement constraint sets, respectively.

However, noting Equation (13.23), the stress summation term is transformed to

$$\sum_{j \in I_1} \lambda_j \bar{d}_j = \sum_{i=1}^{n} \lambda_1' \frac{\bar{\sigma}_i^{(a)}}{\bar{E}_i} \tag{13.28}$$

where

$$\lambda_1' = \sum_{j \in G_i} \lambda_j \tag{13.29}$$

The new quantity $\lambda_{i'}$ defined by Equation (13.29) is the summation of the Lagrange multipliers for the stress constraints in the ith independent design group G_i. On the other hand, the second term in Equation (13.27) is simply

$$\sum_{j \in I_2} \lambda_j \bar{d}_i = \sum_{j \in I_2} \lambda_j U_j^{(a)} \tag{13.30}$$

Thus the following separable Lagrangian function is obtained:

$$L = \sum_{i=1}^{n} \left(\bar{\rho}_i x_i l_i + \frac{e_i}{x_i \bar{E}_i} - \lambda_i' \frac{\bar{\sigma}_i^{(a)}}{\bar{E}_i} \right) - \sum_{j \in I_2} \lambda_j U_j^{(a)} \qquad (13.31)$$

which is separable with respect to the design variable group.

The corresponding dual objective function is given from Chapters 2 and 3 by

$$L(\lambda) = \min_{\substack{x_i^{(L)} \leq x_i \\ S_i}} \sum_{i=1}^{n} \left(\bar{\rho}_i x_i l_i + \frac{e_i}{x_i \bar{E}_i} - \lambda_i' \frac{\bar{\sigma}_i^{(a)}}{\bar{E}_i} \right) - \sum_{j \in I_2} \lambda_j U_j^{(a)} \qquad (13.32)$$

Since the side requirements on the cross-sectional area (13.21) and the material selection by (13.22) are separable with respect to the design variable group, the minimization and the summation operations in (13.32) can be exchanged. Thus,

$$L(\lambda) = \sum_{i=1}^{n} \min_{\substack{x_i^{(L)} \leq x_i \\ S_i}} \left(\bar{\rho}_i x_i l_i + \frac{e_i}{x_i \bar{E}_i} - \lambda_i' \frac{\bar{\sigma}_i^{(a)}}{\bar{E}_i} \right) - \sum_{j \in I_2} \lambda_j U_j^{(a)} \qquad (13.33)$$

In order to simplify the representation we introduce a dual objective function for each independent group as

$$L_i(\lambda) = \min_{\substack{x_i^{(L)} \leq x_i \\ S_i}} \left(\bar{\rho}_i x_i l_i + \frac{e_i}{x_i \bar{E}_i} - \lambda_i' \frac{\bar{\sigma}_i^{(a)}}{\bar{E}_i} \right) \qquad (13.34)$$

Then the final form of the dual objective function can be given by

$$L(\lambda) = \sum_{i=1}^{n} L_i(\lambda) - \sum_{j \in I_2} \lambda_j U_j^{(a)} \qquad (13.35)$$

It is now obvious that the minimization operation shown in the right-hand side of Equation (13.34) must be performed for the evaluation of the dual objective function. The required operation is performed for both the cross-sectional area and the material selection. However, the two minimization operations are independent of each other so that the separate treatment is possible. First of all, perform the minimization with respect to x_i by fixing the material; the following expression is obtained:

$$L_i(\lambda) = \min_{S_i} \left(\bar{\rho}_i x_i^* l_i + \frac{e_i}{x_i^* \bar{E}_i} - \lambda_i' \frac{\bar{\sigma}_i^{(a)}}{\bar{E}_i} \right) \qquad (13.36)$$

where

$$x_i^* = \begin{cases} \sqrt{e_i/(\bar{\rho}_i \bar{E}_i l_i)} & \text{if } e_i \geq (\bar{\rho}_i \bar{E}_i l_i x_i^{(L)})^2 \\ x_i^{(L)} & \text{otherwise} \end{cases} \qquad (13.37)$$

The minimization with respect to the material selection remains and is carried out numerically. That is, we pick up materials one by one from the set S_i and evaluate the function in the bracket of Equation (13.36). The material that gives the minimum value is a solution. This procedure can determine the material selection and the dual objective function value simultaneously.

The relation (13.37) is an important formula in the sense that it connects the dual design variables (the Lagrange multipliers) to the primal design variables (the cross-sectional areas). Once the optimal Lagrange multipliers and the corresponding materials have been determined, it can be used to obtain the optimal cross-sectional areas. It is interesting to note in Equation (13.37) that the cross-sectional area is a function of the material selection.

Drawing together the main theme of our discussion, the following dual algorithm is recommended as an implementable structural optimization tool.

Step 1   Set a starting point λ^0 in the dual space.

Step 2   Determine the material distribution and the cross-sectional areas for all independent design groups by numerically solving the sub-optimization problem:

$$\min_{S_i} \left(\bar{\rho}_i x_i^* l_i + \frac{e_i}{x_i^* \bar{\bar{E}}_i} - \lambda_i' \frac{\bar{\sigma}_i^{(a)}}{\bar{\bar{E}}_i} \right) \qquad (13.38)$$

Denote the solution of this problem as

$$(\bar{\rho}_i, \bar{\bar{E}}_i, \bar{\sigma}_i^{(a)}) \quad \text{and} \quad \bar{\bar{x}}_i \quad i = 1, \ldots, n$$

Step 3   Compute the jth component of the gradient vector in the dual space by

$$S_j = \sum_{i=1}^{n} \frac{b_{ij}^0}{\bar{x}_i \bar{E}_i} - \bar{\bar{d}}_j \qquad j = 1, \ldots, m \qquad (13.39)$$

where

$$\bar{\bar{d}} = \frac{\bar{\sigma}_i^{(a)}}{\bar{E}_i} \quad \text{or} \quad U_j^{(a)} \qquad (13.40)$$

depending on whether the jth constraint is for stress or displacement.

Step 4   Perform a unidimensional search by

$$\lambda_j = \lambda_j^0 + \alpha S_j \qquad j = 1, \ldots, m \qquad (13.41)$$

with the restriction $\lambda_j \geq 0$. Denote the optimal step size as α^*.

Step 5   Set $\lambda_j^0 = \lambda_j^0 + \alpha^* S_j \qquad j = 1, \ldots, m \qquad (13.42)$

Step 6   Repeat step 2 to step 5 until convergence is attained.

Step 7   When convergence is attained, solve Equation (13.38) to determine the optimal material selection and cross-sectional areas.

The above algorithm is constructed on the basis of employing the steepest-ascent method in the dual space. Note that the present dual algorithm reduces to the usual dual method employed for the pure sizing problems when the minimization operation with respect to the material selection is disregarded.

Since the approximate primal problem is handled instead of the original problem a rigorous search in the dual space is not necessary. A rather rough search in the dual space being somewhat better. The dual method usually yields a sequence of a slightly infeasible designs so that a uniform scaling procedure for the cross-sectional areas is necessary to obtain a sequence of feasible designs. Since the present method seeks for the optimum Lagrange multipliers, the redundant constraints can be easily identified by using the fact that the Lagrange multiplier for a redundant constraint is zero. Therefore, when some of the Lagrange multipliers become almost zero after one dual solution the corresponding constraints are deleted from the potentially critical constraint set in the next iteration. Such dynamic selection of the potentially critical constraints can further enhance the efficiency of the dual method.

13.5 NUMERICAL EXAMPLES

A computer program (MATOP) has been written for the combined material-selection and sizing truss optimization problems, and the approximation concept and the mini-max dual method implemented. The computations were carried out with the M-180 computer at Kajima Corporation.

The standard 10- and 72-bar truss problems are solved as the combined material-selection and sizing problems. Two different materials are provided for all numerical examples; one is a high-tension steel and the other is an aluminium. The material constants (density, Young's modulus and allowable stress) are summarized in Table 13.1. The aluminium properties in Table 13.1 are the same as

Table 13.1 Material properties

Materials	Density (pci)	Young's modulus (psi)	Allowable stress (psi)
High-tension steel	0.284	0.298×10^8	$\pm 0.45 \times 10^5$
Aluminium	0.100	0.100×10^8	$\pm 0.25 \times 10^5$

those used in the previous studies for the pure sizing problems (see for example, Refs [2] and [3]). The results reported in Ref. [2] will be conveniently taken for the numerical comparison purpose to the present combined results.

(a) Ten Bar Truss

This familiar problem (see Fig. 13.1) is solved by allowing each structural number to take either steel or an aluminium material. The problem specification is shown in Fig. 13.1. The following four different material-selection problems were solved:

Case 1A: Load case 1, stress constrained.
Case 1B: Load case 1, stress and displacement constrained.
Case 2A: Load case 2, stress constrained.
Case 2B: Load case 2, stress and displacement constrained.

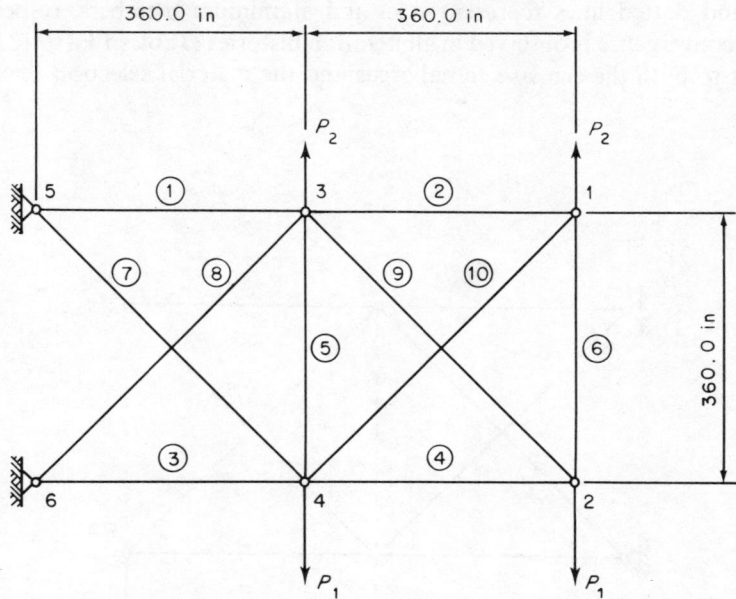

Loading Case 1 : Single load $P_1 = 100$ K, $P_2 = 0$
Loading Case 2 : Single load $P_1 = 150$ K, $P_2 = 50$ K
Lower limits : 0.1 in² (All members)
Upper limits : None

Fig. 13.1 Ten-bar truss

The displacements at all nodes are restricted within ± 2.0 in. in cases 1B and 2B.

All computations are initiated from the same starting design, i.e., the starting materials for all members are high-tension steels and all initial cross-sectional areas are $x_i^0 = 10.0$ in² $(i = 1, \ldots, 10)$.

The iteration histories for all four problems are summarized in Tables 13.2 to 13.5. In the tables, 's' and 'a' are used to indicate that the material used is a steel or an aluminium. The comparison between the present results and the known all-aluminium designs is shown in Table 13.6, referring the results reported in Ref. [2].

The optimum in Case 1A is the known all-aluminium design and no weight saving is achieved by the material selection in this case. The best result is obtained after 16 structural analyses. In Case 1B, the combined steel–aluminium design is lighter by about 4% compared to the all-aluminium optimal design and required 19 analyses. The optimum for Case 2A is an all-aluminium design obtained after 20 analyses. Case 2B gave a combined steel–aluminium design after 17 structural analyses which is 3 per cent lighter than the all-aluminium design.

The material distributions in Cases 1B and 2B are shown in Fig. 13.2, in which solid and dotted lines represent steel and aluminium members, respectively. Stable convergence is observed in all iteration histories (Tables 13.2 to 13.5) with respect to both the cross-sectional areas and the material selection.

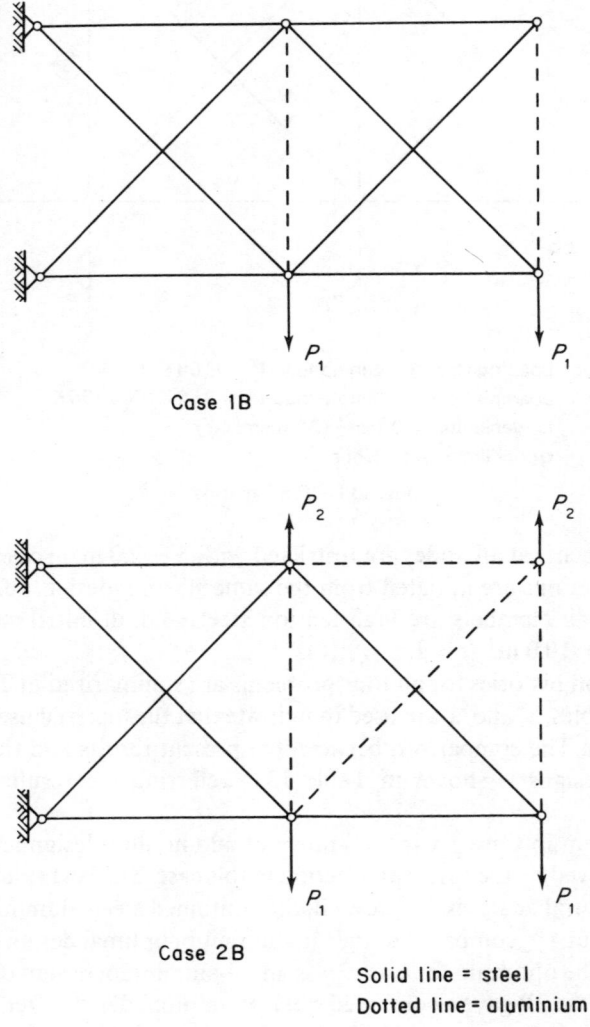

Fig. 13.2 Material distribution in Cases 1B and 2B

Table 13.2 Results of Case 1A problem

Stage	x_1	x_2	x_3	x_4	x_5	x_6	x_7	x_8	x_9	x_{10} (in^2)	W (lb)
0	10.0(s)	10.0(s)	10.0(s)	10.0(s)	10.0(s)	10.0(s)	10.0(s)	10.0(s)	10.0(s)	10.0(s)	11918.0
1	8.884(a)	2.048(a)	9.794(a)	3.220(a)	0.730(s)	2.048(a)	6.886(a)	6.246(a)	4.284(a)	0.952(s)	2034.8
2	8.504(a)	0.524(s)	10.401(a)	3.827(a)	0.406(s)	0.524(s)	7.761(a)	5.546(a)	5.066(a)	0.732(s)	2008.1
3	8.791(a)	1.220(a)	11.681(a)	4.539(a)	0.138(s)	0.419(s)	8.998(a)	5.408(a)	6.051(a)	0.592(s)	2128.3
4	9.572(a)	0.352(s)	12.438(a)	4.868(a)	0.118(a)	0.355(s)	9.715(a)	5.983(a)	6.650(a)	1.487(a)	2257.6
5	9.598(a)	0.288(s)	12.051(a)	4.940(a)	0.119(a)	0.846(a)	9.299(a)	6.139(a)	6.763(a)	1.210(a)	2213.3
6	8.256(a)	0.584(a)	10.064(a)	4.304(a)	0.100(a)	0.582(a)	7.682(a)	5.382(a)	5.900(a)	0.834(a)	1868.0
7	9.800(a)	0.189(s)	11.653(a)	5.141(a)	0.128(a)	0.553(a)	8.811(a)	6.485(a)	7.074(a)	0.265(s)	2178.5
8	8.780(a)	0.391(a)	10.218(a)	4.648(a)	0.100(a)	0.391(a)	7.661(a)	5.885(a)	6.399(a)	0.557(a)	1926.8
9	10.134(a)	0.129(s)	11.582(a)	5.375(a)	0.129(a)	0.129(s)	8.622(a)	6.859(a)	7.427(a)	0.173(s)	2197.7
10	10.059(a)	0.282(a)	11.349(a)	5.356(a)	0.112(a)	0.112(s)	8.405(a)	6.856(a)	7.402(a)	0.141(s)	2163.3

Table 13.2 (Continued)

Stage	x_1	x_2	x_3	x_4	x_5	x_6	x_7	x_8	x_9	$x_{10}(\text{in}^2)$	$W(\text{lb})$
11	9.371(a)	0.208(a)	10.471(a)	5.006(a)	0.112(a)	0.210(a)	7.723(a)	6.421(a)	6.922(a)	0.112(s)	2002.2
12	8.424(a)	0.148(a)	9.331(a)	4.518(a)	0.100(a)	0.148(a)	6.855(a)	5.800(a)	6.248(a)	0.222(a)	1789.8
13	8.006(a)	0.110(a)	8.800(a)	4.303(a)	0.100(a)	0.100(a)	6.443(a)	5.535(a)	5.956(a)	0.156(a)	1692.3
14	7.935(a)	0.100(a)	8.093(a)	3.941(a)	0.100(a)	0.100(a)	5.782(a)	5.558(a)	5.570(a)	0.119(a)	1596.7
15	7.956(a)	0.100(a)	8.069(a)	3.938(a)	0.100(a)	0.100(a)	5.746(a)	5.585(a)	5.580(a)	0.100(a)	1595.5
16	7.944(a)	0.100(a)	8.072(a)	3.950(a)	0.100(a)	0.100(a)	5.742(a)	5.604(a)	5.597(a)	0.100(a)	1598.4
17	7.976(a)	0.100(a)	8.072(a)	3.951(a)	0.100(a)	0.100(a)	5.742(a)	5.606(a)	5.598(a)	0.100(a)	1598.6
18	7.976(a)	0.100(a)	8.072(a)	3.951(a)	0.100(a)	0.100(a)	5.742(a)	5.606(a)	5.597(a)	0.100(a)	1598.6
19	7.974(a)	0.100(a)	8.072(a)	3.950(a)	0.100(a)	0.100(a)	5.742(a)	5.604(a)	5.595(a)	0.100(a)	1598.3
20	7.973(a)	0.100(a)	8.071(a)	3.950(a)	0.100(a)	0.100(a)	5.742(a)	5.602(a)	5.595(a)	0.100(a)	1598.1
21	7.969(a)	0.100(a)	8.070(a)	3.949(a)	0.100(a)	0.100(a)	5.742(a)	5.600(a)	5.592(a)	1.100(a)	1597.6

Table 13.3 Results of Case 1B problem

Stage	x_1	x_2	x_3	x_4	x_5	x_6	x_7	x_8	x_9	$x_{10}(\text{in}^2)$	$W(\text{lb})$
0	10.0(s)	10.0(s)	10.0(s)	10.0(s)	10.0(s)	10.0(s)	10.0(s)	10.0(s)	10.0(s)	10.0(s)	11918.0
1	28.182(a)	6.894(a)	29.850(a)	10.435(a)	4.348(a)	5.042(a)	19.910(a)	18.570(a)	13.060(a)	9.960(a)	6182.1
2	27.011(a)	1.770(s)	30.627(a)	11.699(a)	1.510(a)	3.443(a)	21.176(a)	16.965(a)	14.984(a)	7.698(a)	5952.1
3	26.735(a)	1.399(s)	31.048(a)	12.772(a)	0.189(a)	2.312(a)	21.671(a)	16.527(a)	16.519(a)	6.146(a)	5871.6
4	27.529(a)	1.186(s)	29.749(a)	13.071(a)	0.163(a)	1.845(a)	19.834(a)	17.780(a)	17.149(a)	5.065(a)	5772.0
5	28.114(a)	0.985(s)	29.116(a)	13.440(a)	0.172(a)	1.184(a)	18.915(a)	18.587(a)	17.930(a)	1.385(s)	5715.9
6	28.553(a)	0.770(s)	28.370(a)	13.906(a)	0.166(a)	0.698(a)	17.942(a)	19.313(a)	18.667(a)	1.089(s)	5664.1
7	10.012(s)	0.473(s)	25.502(a)	4.640(s)	0.147(a)	1.061(a)	14.873(a)	20.083(a)	19.520(a)	0.669(s)	5378.2
8	10.174(s)	0.375(s)	8.524(s)	4.744(s)	0.1136(a)	0.385(a)	12.438(a)	6.821(s)	6.777(s)	0.531(s)	5129.9
9	10.224(s)	0.230(s)	8.347(s)	4.819(s)	0.140(a)	0.140(a)	11.815(a)	6.944(s)	6.882(s)	0.326(s)	5072.8
10	10.316(s)	0.259(s)	9.420(s)	4.653(s)	0.139(a)	0.139(a)	15.105(a)	6.677(s)	6.581(s)	0.367(s)	5269.1

Table 13.3 (*Continued*)

Stage	x_1	x_2	x_3	x_4	x_5	x_6	x_7	x_8	x_9	$x_{10}(\text{in}^2)$	$W(\text{lb})$
11	10.317(s)	0.361(s)	9.605(s)	4.632(s)	0.141(a)	0.141(a)	15.673(a)	6.602(s)	6.550(s)	0.510(s)	5330.8
12	10.328(s)	0.417(s)	9.801(s)	4.614(s)	0.139(a)	0.139(a)	5.426(s)	6.535(s)	6.526(s)	0.590(s)	5340.8
13	10.341(s)	0.445(s)	9.900(s)	4.600(s)	0.140(a)	0.140(a)	5.541(s)	6.509(s)	6.503(s)	0.630(s)	5368.7
14	10.394(s)	0.452(s)	9.544(s)	4.641(s)	0.126(a)	0.126(a)	15.808(a)	6.559(s)	6.563(s)	0.640(s)	5362.9
15	10.438(s)	0.417(s)	9.341(s)	4.677(s)	0.120(a)	0.120(a)	4.724(s)	6.605(s)	6.614(s)	0.590(s)	5231.4
16	11.548(s)	0.387(s)	10.058(s)	5.256(s)	0.100(a)	0.100(a)	4.776(s)	7.430(s)	7.433(s)	0.548(s)	5712.0
17	10.316(s)	0.220(s)	8.401(s)	4.873(s)	0.105(a)	0.105(a)	3.399(s)	6.886(s)	6.891(s)	0.311(s)	4970.5
18	10.313(s)	0.136(s)	8.201(s)	4.910(s)	0.101(a)	0.101(a)	3.122(s)	6.964(s)	6.964(s)	0.192(s)	4906.3
19	11.105(s)	0.251(a)	8.659(s)	5.331(s)	0.110(a)	0.110(a)	3.077(s)	7.567(s)	7.539(s)	0.122(s)	5229.4
20	12.579(s)	0.119(a)	9.576(s)	6.096(s)	0.119(a)	0.119(a)	3.063(s)	8.663(s)	8.621(s)	0.119(a)	5849.3
21	17.172(s)	0.154(a)	12.554(s)	8.450(s)	0.154(a)	0.318(a)	3.018(s)	12.021(s)	11.951(s)	0.154(a)	7835.7

Table 13.4 Results of Case 2A problem

Stage	x_1	x_2	x_3	x_4	x_5	x_6	x_7	x_8	x_9	$x_{10}(\text{in}^2)$	$W(\text{lb})$
0	10.0(s)	10.0(s)	10.0(s)	10.0(s)	10.0(s)	10.0(s)	10.0(s)	10.0(s)	10.0(s)	10.0(s)	11918.0
1	8.313(a)	0.544(s)	9.592(a)	3.605(a)	1.229(s)	3.208(a)	6.889(a)	5.728(a)	4.950(a)	0.835(s)	2091.3
2	7.536(a)	0.384(s)	9.927(a)	4.075(a)	0.812(s)	2.565(a)	7.535(a)	4.770(a)	5.616(a)	0.543(s)	1980.9
3	7.385(a)	0.249(s)	10.867(a)	4.690(a)	0.495(s)	2.259(a)	8.577(a)	4.277(a)	6.487(a)	0.352(s)	2018.9
4	6.858(a)	0.100(s)	11.188(a)	4.503(a)	0.335(s)	2.356(a)	9.362(a)	3.327(a)	6.387(a)	0.141(s)	1932.6
5	6.038(a)	0.135(a)	10.922(a)	4.196(a)	0.127(s)	2.089(a)	9.481(a)	2.627(a)	5.967(a)	0.191(a)	1784.6
6	6.320(a)	0.103(a)	10.650(a)	4.149(a)	0.139(a)	2.272(a)	9.209(a)	3.029(a)	5.861(a)	0.146(a)	1779.7
7	5.973(a)	0.100(a)	10.051(a)	3.954(a)	0.100(a)	2.057(a)	8.560(a)	2.763(a)	5.609(a)	0.100(a)	1667.6
8	5.962(a)	0.100(a)	10.061(a)	3.948(a)	0.100(a)	2.064(a)	8.563(a)	2.759(a)	5.604(a)	0.100(a)	1667.3
9	5.959(a)	0.100(a)	10.060(a)	3.948(a)	0.100(a)	2.062(a)	8.562(a)	2.757(a)	5.600(a)	0.100(a)	1666.7
10	5.956(a)	0.100(a)	10.056(a)	3.948(a)	0.100(a)	2.058(a)	8.560(a)	2.756(a)	5.597(a)	0.100(a)	1666.0

Table 13.4 (*Continued*)

Stage	x_1	x_2	x_3	x_4	x_5	x_6	x_7	x_8	x_9	x_{10} (in^2)	W (lb)
11	5.955(a)	0.100(a)	10.055(a)	3.948(a)	0.100(a)	2.056(a)	8.559(a)	2.756(a)	5.595(a)	0.100(a)	1665.7
12	5.954(a)	0.100(a)	10.055(a)	3.949(a)	0.100(a)	2.055(a)	8.559(a)	2.756(a)	5.593(a)	0.100(a)	1665.6
13	5.953(a)	0.100(a)	10.055(a)	3.949(a)	0.100(a)	2.054(a)	8.559(a)	2.756(a)	5.592(a)	0.100(a)	1665.5
14	5.952(a)	0.100(a)	10.055(a)	3.949(a)	0.100(a)	2.054(a)	8.559(a)	2.756(a)	5.590(a)	0.100(a)	1665.3
15	5.951(a)	0.100(a)	10.054(a)	3.949(a)	0.100(a)	2.053(a)	8.559(a)	2.756(a)	5.588(a)	0.100(a)	1665.1
16	5.949(a)	0.100(a)	10.053(a)	3.949(a)	0.100(a)	2.053(a)	8.559(a)	2.755(a)	5.585(a)	0.100(a)	1664.8
17	5.948(a)	0.100(a)	10.053(a)	3.948(a)	0.100(a)	2.052(a)	8.559(a)	2.755(a)	5.584(a)	0.100(a)	1664.7
18	5.948(a)	0.100(a)	10.053(a)	3.948(a)	0.100(a)	2.053(a)	8.560(a)	2.755(a)	5.583(a)	0.100(a)	1664.6
19	5.948(a)	0.100(a)	10.052(a)	3.948(a)	0.100(a)	2.052(a)	8.559(a)	2.755(a)	5.583(a)	0.100(a)	1664.5
20	5.949(a)	0.100(a)	10.054(a)	3.949(a)	0.100(a)	2.052(a)	8.561(a)	2.755(a)	5.584(a)	0.100(a)	1664.8
21	5.949(a)	0.100(a)	10.055(a)	3.949(a)	0.100(a)	2.053(a)	8.561(a)	2.755(a)	5.584(a)	0.100(a)	1664.9

Table 13.5 Results of Case 2B problem

Stage	x_1	x_2	x_3	x_4	x_5	x_6	x_7	x_8	x_9	x_{10} (in^2)	W(lb)
0	10.0(s)	10.0(s)	10.0(s)	10.0(s)	10.0(s)	10.0(s)	10.0(s)	10.0(s)	10.0(s)	10.0(s)	11918.0
1	27.294(a)	1.886(s)	30.132(a)	11.360(a)	8.370(a)	9.559(a)	20.449(a)	17.654(a)	14.596(a)	8.075(a)	6408.7
2	24.942(a)	1.219(s)	30.854(a)	12.806(a)	4.655(a)	7.390(a)	22.070(a)	14.933(a)	16.706(a)	5.280(a)	6031.2
3	23.202(a)	0.752(s)	31.383(a)	13.633(a)	1.886(a)	5.782(a)	23.291(a)	12.826(a)	18.141(a)	1.048(s)	5722.7
4	21.860(a)	0.413(s)	31.479(a)	14.207(a)	0.161(a)	4.631(a)	23.840(a)	11.371(a)	18.988(a)	0.567(a)	5487.8
5	21.874(a)	0.193(s)	31.066(a)	14.565(a)	0.169(a)	4.196(a)	23.140(a)	11.370(a)	19.680(a)	0.273(a)	5405.4
6	7.800(s)	0.214(a)	27.084(a)	4.549(s)	0.136(s)	6.204(a)	18.004(a)	4.338(s)	19.437(a)	0.302(a)	5022.3
7	7.815(s)	0.149(a)	26.367(a)	4.594(s)	0.149(a)	5.631(a)	17.266(a)	4.422(s)	19.523(a)	0.149(a)	4951.4
8	7.647(s)	0.136(a)	9.360(s)	4.780(s)	0.136(s)	3.416(a)	19.457(a)	4.117(s)	6.620(s)	0.136(a)	4910.1
9	7.647(s)	0.142(a)	9.344(s)	4.791(s)	0.142(a)	3.439(a)	19.403(a)	4.106(s)	6.645(s)	0.142(a)	4910.6
10	7.727(s)	0.132(a)	27.016(a)	4.747(s)	0.132(a)	3.677(a)	17.791(a)	4.188(s)	6.665(s)	0.132(a)	4871.5

Table 13.5 (*Continued*)

Stage	x_1	x_2	x_3	x_4	x_5	x_6	x_7	x_8	x_9	$x_{10}(\text{in}^2)$	$W(\text{lb})$
11	7.736(s)	0.135(a)	9.098(s)	4.715(s)	0.135(a)	3.463(a)	17.439(a)	4.203(s)	6.668(s)	0.135(a)	4804.1
12	7.738(s)	0.112(a)	8.963(s)	4.778(s)	0.112(a)	3.640(a)	5.515(s)	4.131(s)	6.757(s)	0.112(a)	4712.7
13	7.796(s)	0.114(a)	8.816(s)	4.855(s)	0.114(a)	2.817(a)	14.492(a)	4.112(s)	6.866(s)	0.114(a)	4635.3
14	7.803(s)	0.113(a)	8.791(s)	4.859(s)	0.113(a)	2.541(a)	4.762(s)	4.127(s)	6.871(s)	0.113(a)	4577.4
15	7.821(s)	0.111(a)	8.707(s)	4.847(s)	0.111(a)	2.242(a)	14.406(a)	4.170(s)	6.854(s)	0.111(a)	4607.1
16	7.817(s)	0.111(a)	8.712(s)	4.848(s)	0.111(a)	2.530(a)	4.733(s)	4.165(s)	6.856(s)	0.111(a)	4568.4
17	7.818(s)	0.111(a)	8.712(s)	4.847(s)	0.111(a)	2.534(a)	4.733(a)	4.166(s)	6.855(s)	0.111(a)	4568.4
18	7.823(s)	0.111(a)	8.711(s)	4.849(a)	0.111(a)	2.405(a)	13.942(a)	4.171(a)	6.857(s)	0.111(a)	4590.7
19	7.820(s)	0.111(a)	8.703(a)	4.846(a)	0.111(a)	2.401(a)	4.785(s)	4.171(a)	6.853(s)	0.113(a)	4570.6
20	7.819(s)	0.111(a)	8.707(s)	4.847(s)	0.111(a)	2.402(a)	4.789(s)	4.168(s)	6.854(s)	0.111(a)	4571.4
21	7.819(s)	0.111(a)	8.707(s)	4.847(s)	0.111(a)	2.403(a)	4.790(s)	4.168(s)	6.854(s)	0.111(a)	4571.4

Table 13.6 Comparison of ten-bar truss results

	Case 1A		Case 1B		Case 2A		Case 2B	
	Present study	Schmit, Farshi[2]	Present study	Schmit, Farshi	Present study	Schmit, Farshi	Present study	Schmit, Farshi
x_1	7.956(a)	7.938(a)	10.313 (s)	33.432(a)	5.948(a)	5.948(a)	7.817(s)	24.290(a)
x_2	0.100(a)	0.1 (a)	0.136(s)	0.1 (a)	0.100(a)	0.1 (a)	0.111(a)	0.1 (a)
x_3	8.069(a)	8.062(a)	8.201(s)	24.260(a)	10.052(a)	10.052(a)	8.712(s)	23.346(a)
x_4	3.938(a)	3.938(a)	4.910(s)	14.260(a)	3.948(a)	3.948(a)	4.848(s)	13.654(a)
x_5	0.100(a)	0.1 (a)	0.101(a)	0.1 (a)	0.100(a)	0.1 (a)	0.111(a)	0.1 (a)
x_6	0.100(a)	0.1 (a)	0.101(a)	0.1 (a)	2.052(a)	2.052(a)	2.530(a)	1.970(a)
x_7	5.746(a)	5.745(a)	3.122(s)	8.338(a)	8.559(a)	8.559(a)	4.733(s)	12.670(a)
x_8	5.585(a)	5.569(a)	6.964(s)	20.740(a)	2.755(a)	2.754(a)	4.165(s)	12.544(a)
x_9	5.580(a)	5.569(a)	6.964(s)	19.690(a)	5.583(a)	5.583(a)	6.856(s)	21.971(a)
x_{10}(in^2)	0.100(a)	0.1 (a)	0.192(s)	0.1 (a)	0.100(a)	0.1 (a)	0.111(a)	0.1 (a)
W(lb)	1595.5	1593.2	4906.3	5089.0	1664.5	1664.5	4568.4	4691.8

Table 13.7 Results of seventy-two bar truss

Stage	x_1	x_2	x_3	x_4	x_5	x_6	x_7	x_8	x_9
0	5.000(s)	5.000(s)	5.000(s)	5.000(s)	5.000(s)	5.000(s)	5.000(s)	5.000(s)	5.000(s)
1	0.719(a)	0.208(s)	0.208(a)	0.208(a)	0.945(a)	0.208(s)	0.208(a)	0.208(a)	1.511(a)
2	0.835(a)	0.225(s)	0.156(a)	0.156(a)	1.048(a)	0.224(s)	0.156(a)	0.156(a)	1.701(a)
3	0.723(a)	0.221(s)	0.171(a)	0.171(a)	0.911(a)	0.219(s)	9.171(a)	0.171(a)	0.543(s)
4	0.701(a)	0.224(s)	0.164(a)	0.164(a)	0.899(a)	0.221(s)	0.164(a)	0.164(a)	0.547(s)
5	0.439(a)	0.224(s)	0.151(a)	0.188(a)	0.650(a)	0.220(s)	0.151(a)	0.151(a)	0.543(s)
6	0.146(a)	0.270(s)	0.213(a)	0.122(s)	0.244(s)	0.259(s)	0.118(a)	0.134(a)	0.640(s)
7	0.199(a)	0.198(s)	0.100(s)	0.146(s)	0.177(s)	0.183(s)	0.100(a)	0.112(a)	0.452(s)
8	0.167(a)	0.187(s)	0.109(s)	0.158(s)	0.177(s)	0.180(s)	0.109(a)	0.109(a)	0.442(s)
9	0.195(a)	0.189(s)	0.112(s)	0.163(s)	0.180(s)	0.180(s)	0.105(a)	0.105(a)	0.442(s)
10	0.168(a)	0.186(s)	0.113(s)	0.165(s)	0.177(s)	0.179(s)	0.106(a)	0.106(a)	0.439(s)
11	0.191(a)	0.188(s)	0.116(s)	0.167(s)	0.179(s)	0.179(s)	0.106(a)	0.106(a)	0.440(s)
12	0.169(a)	0.186(s)	0.116(s)	0.168(s)	0.177(s)	0.178(s)	0.106(a)	0.106(a)	0.438(s)
13	0.193(a)	0.187(s)	0.118(s)	0.170(s)	0.179(s)	0.179(s)	0.106(a)	0.106(a)	0.439(s)
14	0.171(a)	0.185(s)	0.119(s)	0.171(s)	0.177(s)	0.178(s)	0.105(a)	0.105(a)	0.436(s)
15	0.191(a)	0.187(s)	0.121(s)	0.173(s)	0.179(s)	0.178(s)	0.106(a)	0.106(a)	0.437(s)
16	0.1693(a)	0.1850(s)	0.1211(s)	0.1741(s)	0.1768(s)	0.1770(s)	0.1053(a)	0.1053(a)	0.4348(s)
Ref. [1]	0.158(a)	0.594(a)	0.341(a)	0.608(a)	0.264(a)	0.548(a)	0.1(a)	0.151(a)	1.107(a)

x_{10}	x_{11}	x_{12}	x_{13}	x_{14}	x_{15}	$x_{16}^{(in^2)}$	$W^{(lb)}$	Dual dimension
5.000(s)	5.000(s)	5.000(s)	5.000(s)	5.000(s)	5.000(s)	5.000(s)	12113.8	27
0.208(s)	0.208(a)	0.208(a)	2.142(a)	0.217(s)	0.244(a)	0.208(a)	453.9	23
0.222(s)	0.156(a)	0.156(a)	0.835(s)	0.223(s)	0.156(a)	0.156(a)	466.8	22
0.217(s)	0.171(a)	0.171(a)	0.812(s)	0.218(s)	0.171(a)	0.171(a)	454.0	18
0.220(s)	0.164(a)	0.164(a)	0.821(s)	0.220(s)	0.164(a)	0.164(a)	455.2	17
0.218(s)	0.151(a)	0.151(a)	0.840(s)	0.219(s)	0.151(a)	0.151(a)	436.8	16
0.255(s)	0.118(a)	0.118(a)	0.941(s)	0.256(s)	0.118(a)	0.118(a)	496.5	12
0.180(s)	0.100(a)	0.100(a)	0.659(s)	0.180(s)	0.100(a)	0.100(a)	371.3	10
0.178(s)	0.109(a)	0.109(a)	0.654(s)	0.178(s)	0.109(a)	0.109(a)	367.8	10
0.178(s)	0.105(a)	0.105(a)	0.657(s)	0.178(s)	0.105(a)	0.105(a)	369.7	10
0.177(s)	0.106(a)	0.106(a)	0.652(s)	0.177(s)	0.106(a)	0.106(a)	367.0	10
0.177(s)	0.106(a)	0.106(a)	0.653(s)	0.178(s)	0.106(a)	0.106(a)	369.1	10
0.176(s)	0.106(a)	0.106(a)	0.649(s)	0.177(s)	0.106(a)	0.106(a)	366.6	10
0.177(s)	0.106(a)	0.106(a)	0.652(s)	0.177(s)	0.106(a)	0.106(a)	369.2	10
0.176(s)	0.105(a)	0.105(a)	0.648(s)	0.176(s)	0.105(a)	0.105(a)	366.6	10
0.176(s)	0.106(a)	0.106(a)	0.649(s)	0.176(s)	0.106(a)	0.106(a)	368.7	10
0.1752(s)	0.1053(a)	0.1053(a)	0.6458(s)	0.1755(s)	0.1053(a)	0.1053(a)	366.2	
0.579(a)	0.1(a)	0.1(a)	2.078(a)	0.503(a)	0.1(a)	0.1(a)	388.6	

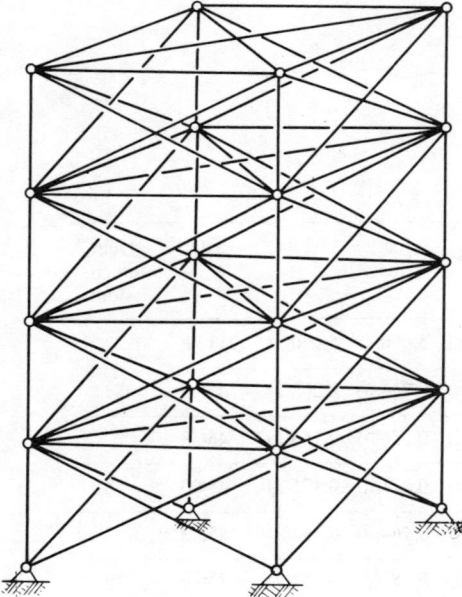

Fig. 13.3 Seventy-two bar truss

(b) Seventy-two Bar Truss

The seventy-two bar truss shown in Fig. 13.3 is solved as a combined material-selection and sizing problem with the same materials as in the ten-bar problem. The problem specifications (loading conditions, minimum gauge, numbering of nodes and members, and design variable linking) are the same as in Ref. [2].

The iteration history is summarized in Table 13.7, together with the all-aluminium design of Ref. [2]. A 6 per cent lighter combined steel–aluminium design is obtained after 17 analyses with stable convergence. Figure 13.4 shows the optimal material distribution for the present problem.

13.6 CONCLUSIONS

An efficient optimization method could be given for the combined material-selection and sizing structural optimization problems. The combinatorial nature of the problem was handled as a sub-optimization problem in the dual space.

As can be seen from the numerical results, the method possesses stable convergence properties with respect to both the cross-sectional areas and the material selection. The computational burden measured by the required structural analysis number is observed to be of the same order as in the most efficient solution for the pure sizing problems. In the present research, unnecessarily many iterations were intentionally carried out to investigate the nature of the

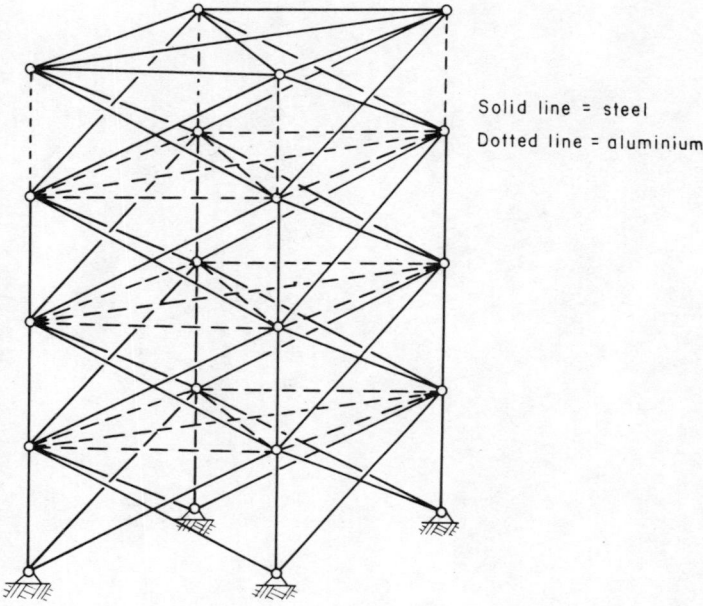

Fig. 13.4 Material distribution of seventy-two bar truss

convergence. However, as can be seen from all the iteration histories ten structural analyses are usually enough to obtain practically meaningful approximate solutions. This fact implies that the simultaneous determination of the material constitution and the sizing variables is practicable even for a large structure.

The weight saving in all examples was not as significant as it might have been because only two materials (steel and aluminium) were considered. However, if other materials with high performance (i.e., small density and large Young's modulus and allowable stress) were used, a much more significant weight saving would be possible.

K. I.

REFERENCES

1. K. Imai, Configuration optimization of trusses by the multiplier method, PhD dissertation, UCLA, 1978.
2. L. A. Schmit, Jr. and B. Farshi, Some approximation concepts for structural synthesis, *AIAAJ.*, **12**(5), 692–9, May 1974.
3. L. A. Schmit, Jr. and H. Miura, Approximation concepts for efficient structural synthesis, *NASA CR-2552*, March 1976.

Foundations of Structural Optimization: A Unified Approach
Edited by A. J. Morris
© 1982 John Wiley & Sons Ltd.

Chapter 14
Optimal Geometry in Truss Design

14.1 INTRODUCTION

Optimum structural design is typically concerned with the problem of finding in some sense 'best' structures for given purposes (e.g. to carry some given loads at a given height above the ground). To attack this formidable problem with mathematical methods one must restrict the search to some proper subset of all possible structures.

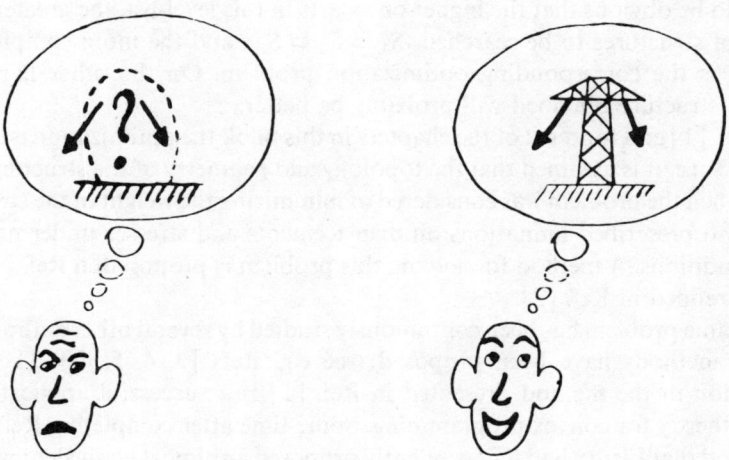

Fig. 14.1 Conceptual design

A common assumption (to which we adhere) to make the problem a bit more quantitative, is that the structure can be divided into 'elements', connected in a discrete number of 'nodes'. This may be done either in a natural way (like bars in a truss system) or according to the finite element method (FEM).

The optimization may now be started on essentially three different levels, dependent on what kind of subset of structures (S_1, S_2 and S_3) is to be considered.

Level 1 $S_1 = \{$'all' possible structures$\}$
By this we mean all structures which satisfy the original functional requirements.

Level 2 $S_2 = \{$'all' possible structures with a given topology and a given material$\}$
On this level the topology of the structure is held fixed, i.e. questions like the following have already been answered: Should the structure be composed of only bars, or should some other elements be considered? Which elements should be connected to each other, and how many elements should there be?
The choice of material is also excluded from the optimization on this level.

Level 3 $S_3 = \{$'all possible structures with a given topology, a given material and a given geometry$\}$
On this level the geometry of the structure is also held fixed, i.e. the coordinates of the nodes are given. What remain to be optimized on level 3 are the sizes of the transverse dimensions of the elements, e.g. cross section areas of bars and thicknesses of membrane plates.

It should be obvious that the higher one starts in this level-list, the greater is the subset of structures to be searched ($S_1 \supset S_2 \supset S_3$), and the more complex and difficult is the corresponding optimization problem. On the other hand, the optimal structure obtained will probably be better.

In Ref. [1] and for most of the chapters in this book the optimization is started at level 3', i.e. it is assumed that the topology and geometry of the structure were given. Then the problem was considered of minimizing the weight of the structure subject to prescribed limitations on displacements and stresses under multiple load conditions. A method for solving this problem is proposed in Ref. [1] and further refined in Ref. [2].

The same problem has been continuously studied by several other authors, and various methods have been proposed (see e.g. Refs [3, 4, 5, 6]). The main innovation in the method presented in Ref. [1] is a successful application of duality theory for convex programming. Some time after completing Ref. [2] it was found that Fleury had independently proposed an almost equivalent method (Ref. [7]; and discussed in Chapter 10).

In this paper we 'climb up' one level and start the optimization at level 2, i.e. it is *not* assumed that the geometry of the structure is given.

We will only consider linearly elastic structures, and to simply the presentation we assume that the structure is a truss system, composed of pin-jointed bars. However, we believe that the method proposed could be used also on more general structures.

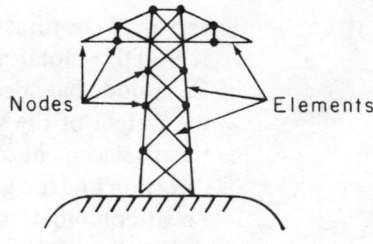

Fig. 14.2 Notation

According to the FEM vocabulary, the bars in the truss system are called 'elements' and the joints 'nodes' (see Fig. 14.2).

Shape optimization is also treated in Ref. [8].

14.2 FORMULATION OF THE OPTIMIZATION PROBLEM

The problem we consider is to minimize the weight of a 3-dimensional truss system subject to:
— displacement constraints;
— stress constraints; and
— element buckling constraints
under multiple load conditions.

There are two kinds of variables in the optimization:
— cross-sectional areas of the elements and
— coordinates of the nodes.

It is a natural requirement that the structure should be kept symmetric in some sense. This implies that the variables cannot, in general, vary independently of each other. Some variables (e.g. some coordinates) may be fixed from the beginning. These restrictions can be expressed as linear relations between the variables.

The optimization problem can mathematically be formulated as follows:

P: $\quad$ min $\quad w(x,c)$

subject to:
$$d_i(x,c) \leqslant d_i^{\max} \quad i = 1, \ldots, m_d$$
$$\sigma_i(x,c) \leqslant \sigma_i^{\max} \quad i = 1, \ldots, m_\sigma$$
$$b_i(x,c) \leqslant 0 \quad i = 1, \ldots, m_b$$
$$A\binom{x}{c} = h$$
$$x_j^{\min} \leqslant x_j \leqslant a_j^{\max} \quad j = 1, \ldots, \text{KELEM}$$
$$c_j^{\min} \leqslant c_j \leqslant c_j^{\max} \quad j = 1, \ldots, 3 \cdot \text{KNODE}$$

where:

$\mathbf{x} = (x_1, \ldots, x_{\text{KELEM}})^T \quad$ = vector of area variables.

$\quad\quad\quad\quad\quad\quad\quad\quad\quad\quad$ KELEM = total number of elements.

$\quad\quad\quad\quad\quad\quad\quad\quad\quad\quad x_j$ = cross-sectional area of the jth element.

$\mathbf{c} = (c_1, \ldots, c_{3 \cdot \text{KNODE}})^T$ = vector of coordinate variables.
$\quad$ KNODE = total number of nodes.
$\quad$ Each node has 3 coordinates.

$w(x, c)$ = the weight of the structure.

$d_i(x, c)$ = the displacement of a given node in a given direction under a given load condition, or a linear combination of such displacements.

$\sigma_i(x, c)$ = some relevant stress (tensile or compressive) in a given element under a given load condition.

$b_i(x, c) \leqslant 0$ = a buckling constraint in a given element under a given load condition.

$\mathbf{A} \begin{pmatrix} x \\ c \end{pmatrix} = \mathbf{h}$ = linear constraints to ensure that the structure is kept symmetric, or that some nodes have prescribed coordinates etc. (A is a given matrix, h is a given vector).

$x_j^{\min}, x_j^{\max}, c_j^{\min}, c_j^{\max}$ = explicit lower and upper bounds on the variables (due to, e.g., manufacturing and geometrical factors).

$d_i^{\max}, \sigma_i^{\max}$ = largest permitted values of $d_i(x, c)$ and $\sigma_i(x, c)$.

P is a nonlinearly constrained optimization problem. The functions d_i, σ_i and b_i are not explicitly given. Instead, for given values on the variables **x** and **c**, we shall assume that the function values are obtained by an FEM calculation. Since such calculations may be comparatively expensive, it is necessary to find methods which require as few function evaluations as possible for solving P.

It is, for example, hardly realistic to solve P by some 'standard' method such as the augmented Lagrangian method or the generalized reduced gradient method. The number of function evaluations required by these methods is empirically known to grow rapidly when the size of the problem grows, and even for a modest problem could require several hundred. This implies several hundred FEM calculations which would be unacceptable.

The philosophy when developing methods for solving P must therefore be to 'make as much use of each function evaluation as possible'.

14.3 GRADIENTS OF THE CONSTRAINT FUNCTIONS

A convenient property of the constraint functions d_i, σ_i and b_i is that their gradients ∇d_i, $\nabla \sigma_i$ and ∇b_i can be evaluated without too much additional effort, once the functions themselves have been evaluated by an FEM calculation. The subject of this chapter is to show how these evaluations can be accomplished and supplements the discussion in Chapters 3 and 4.

Optimal Geometry in Truss Design

Before turning to the constraint functions we note that the objective function may be written

$$w(x, c) = \rho \sum_{e=1}^{\text{KELEM}} x_e \|r_e\|$$

where

ρ = density (= a given constant)

x_e = cross-sectional area of the eth element

r_e = a vector connecting the two nodes of the eth element

($\|r_e\|$ = the length of the eth element; Fig. 14.3)

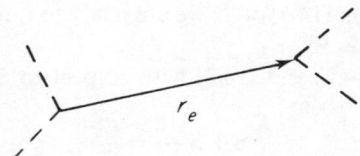

Fig. 14.3 Definition of r_e

It is obvious that the gradient of w, with respect to x and c, can easily be analytically expressed.

14.3.1 Displacement Constraints

Consider a given displacement constraint $d_i(x, c) \leq d_i^{\max}$ at a given point $(\mathbf{x}, \mathbf{c})$. We recall that $d_i(x, c)$ is the displacement of a given node in a given direction under a given load case.

We now need some definitions. Let:

p = a vector composed of the external node forces corresponding to the given loadcase; $p \in R^{3 \cdot \text{KNODE}}$.

u = a vector composed of the node displacements under the given load case, i.e. the displacements caused by p; $u \in R^{3 \cdot \text{KNODE}}$.

K = the 'stiffness matrix' for the structure which from Chapter 3 is symmetric and positive definite.

We then have the following relation defined in Chapter 3:

$$\mathbf{K}\mathbf{u} = \mathbf{p} \qquad (14.1)$$

q: the displacement $d_i(x, c)$ may be written $d_i(x, c) = \mathbf{q}^T \mathbf{u}$, where **q** is a conveniently chosen vector (e.g., if $d_i(x, c)$ is simply the displacement of the j_0th node in the $\hat{\mathbf{n}}$-direction, $\hat{\mathbf{n}}$ a unit vector, then **q** should be composed of $\hat{\mathbf{n}}$ at the j_0th node and zero vectors at all other nodes).

v: it is customary to consider **q** as a vector of external node forces. The corresponding vector **v** of node displacements satisfies the relation

$$\mathbf{Kv} = \mathbf{q} \tag{14.2}$$

v may be obtained by letting the 'forces' **q** appear as additional loadcase in the FEM calculations.

We now get the following expression for d_i which we employed in earlier chapters:

$$d_i(x, c) = \mathbf{q}^T \mathbf{u} = \mathbf{u}^T \mathbf{Kv} \ (= \mathbf{p}^T \mathbf{v}) \tag{14.3}$$

(**q** and **v** are employed in Chapters 3, 4, 10 and referred to as 'virtual load case' and 'virtual displacements', and then $d_i(x, c)$ may be thought of as 'virtual work'.)

After this reformulation (14.3) of d_i, we are ready to differentiate it. Let ξ be an arbitrary variable, i.e. $\xi = x_j$ or c_j.

Differentiate (14.1), (14.2) and (14.3) with respect to ξ:

$$\frac{\partial \mathbf{K}}{\partial \xi} \mathbf{u} + \mathbf{K} \frac{\partial \mathbf{u}}{\partial \xi} = \frac{\partial \mathbf{p}}{\partial \xi} \tag{14.4}$$

$$\frac{\partial \mathbf{K}}{\partial \xi} \mathbf{v} + \mathbf{K} \frac{\partial \mathbf{v}}{\partial \xi} = \frac{\partial \mathbf{q}}{\partial \xi} \tag{14.5}$$

$$\frac{\partial d_i}{\partial \xi} = \frac{\partial \mathbf{u}^T}{\partial \xi} \mathbf{Kv} + \mathbf{u}^T \frac{\partial \mathbf{K}}{\partial \xi} \mathbf{v} + \mathbf{u}^T \mathbf{K} \frac{\partial \mathbf{v}}{\partial \xi} \tag{14.6}$$

Combining (14.4), (14.5) and (14.6) gives:

$$\frac{\partial d_i}{\partial \xi} = \mathbf{v}^T \frac{\partial \mathbf{p}}{\partial \xi} + \mathbf{u}^T \frac{\partial \mathbf{q}}{\partial \xi} - \mathbf{u}^T \frac{\partial \mathbf{K}}{\partial \xi} \mathbf{v} \tag{14.7}$$

For pure displacement constraints **q** is independent of both x_j and c_j and hence $\partial \mathbf{q}/\partial \xi = 0$.

We will also assume that **p** is independent of x_j and c_j so that $\partial \mathbf{p}/\partial \xi = 0$. (A case where $\partial \mathbf{p}/\partial \xi \neq 0$ is discussed in Section 14.3.4.)

Equation (14.7) then reduces to:

$$\frac{\partial d_i}{\partial \xi} = -\mathbf{u}^T \frac{\partial K}{\partial \xi} \mathbf{v} = - \sum_{e=1}^{\text{KELEM}} \mathbf{u}_e^T \frac{\partial \mathbf{K}_e}{\partial \xi} \mathbf{v}_e \tag{14.8}$$

where $\mathbf{K}_e$ is the element stiffness matrix and $\mathbf{u}_e$ and $\mathbf{v}_e$ are the vectors of node displacements for the eth element.

Although we have shown how $\partial \mathbf{K}_e/\partial \xi$ can be calculated in Chapter 3 for the sake of continuity we outline the main points in computing these derivatives.

Recalling the definition of the stiffness matrix for a bar structure we have:

$$\mathbf{K}_e = \begin{pmatrix} \tilde{\mathbf{K}}_e & -\tilde{\mathbf{K}}_e \\ -\tilde{\mathbf{K}}_e & \tilde{\mathbf{K}}_e \end{pmatrix} \quad \text{where } \tilde{\mathbf{K}}_e \text{ is the matrix}$$

$$\tilde{\mathbf{K}}_e = \frac{E x_e}{\|\mathbf{r}_e\|^3} \mathbf{r}_e \mathbf{r}_e^T \tag{14.9}$$

with

x_e = cross-sectional area of the element
$\mathbf{r}_e$ = a vector connecting the two nodes of the element
E = Young's modulus = a given constant.

From (14.9) it is obvious that $\partial \mathbf{K}_e/\partial x_j$ and $\partial \mathbf{K}_e/\partial c_j$ can both be expressed analytically. For $\partial \mathbf{K}_e/\partial x_j$ it is very easy:

$$\frac{\partial \mathbf{K}_e}{\partial x_j} = \begin{cases} \dfrac{1}{x_e} \mathbf{K}_e & \text{if } j = e \\ \\ \mathbf{0} & \text{if } j \neq e \end{cases}$$

($\mathbf{0}$ = the zero matrix)

For $\partial \mathbf{K}_e/\partial c_j$ the formula is a bit more complicated but still straight forward:

$$\frac{\partial \mathbf{K}_e}{\partial c_j} = E x_e \left(\frac{-3}{\|\mathbf{r}_e\|^4} \frac{\partial \|\mathbf{r}_e\|}{\partial c_j} \mathbf{r}_e \mathbf{r}_e^T + \frac{1}{\|\mathbf{r}_e\|^3} \frac{\partial \mathbf{r}_e}{\partial c_j} \mathbf{r}_e^T + \frac{1}{\|\mathbf{r}_e\|^3} \mathbf{r}_e \frac{\partial \mathbf{r}_e^T}{\partial c_j} \right).$$

It should be noted that $\partial \mathbf{K}_e/\partial c_j \neq 0$ *only* if c_j stands for a coordinate for one of the two nodes of the eth element. This implies that $\partial \mathbf{K}_e/\partial c_j = 0$ for the vast majority of indices j.

14.3.2 Stress Constraints

Let σ_e be the tensile stress in a given element (the eth element) under a given load case p.

Introduce the virtual load case q composed of forces $\mathbf{f}_q$ and $-\mathbf{f}_q$ according to Fig. 14.4; $\mathbf{f}_q$ should have the magnitude $1/\|\mathbf{r}_e\| \Rightarrow \mathbf{f}_q = \mathbf{r}_e/\|\mathbf{r}_e\|^2$.

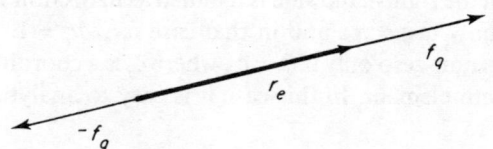

Fig. 14.4 Definition of forces and r_e

Let E be Young's modulus and let $\Delta \|\mathbf{r}_e\|$ be the increase in length of the given element, caused by the given load case p. We then have:

$$\frac{\sigma_e}{E} = \frac{\Delta \|\mathbf{r}_e\|}{\|\mathbf{r}_e\|} = \mathbf{q}^T \mathbf{u} = \mathbf{v}^T \mathbf{K} \mathbf{u} \qquad (14.10)$$

($\mathbf{u}$ and $\mathbf{v}$ are, as before, the node displacements associated with p and q respectively).

σ_e may thus be considered as a 'generalized' displacement. The only difference from a 'pure' displacement is that **q** is no longer a constant vector. We then get (compare Equation (14.7)):

$$\frac{1}{E}\frac{\partial \sigma_e}{\partial \xi} = -\mathbf{v}^T \frac{\partial \mathbf{K}}{\partial \xi}\mathbf{u} + \mathbf{u}^T \frac{\partial \mathbf{q}}{\partial \xi} \qquad (14.11)$$

The first term on the right-hand side is handled as in Section 14.3.1. The second term is even easier:

If $\xi = x_j$ then $\partial \mathbf{q}/\partial \xi = 0$.

If $\xi = c_j$ then $\partial \mathbf{q}/\partial \xi \neq 0$ only if c_j is a coordinate for one of the two nodes of the eth element. In that case it is easy to analytically differentiate $\mathbf{f}_q$, and thus also **q**, with respect to ξ.

If σ_e is the *compressive* stress in a given element, then the only difference from above is that $\mathbf{f}_q$ and $-\mathbf{f}_q$ should change directions, i.e. $\mathbf{f}_q = -\mathbf{r}_e/\|\mathbf{r}_e\|^2$.

14.3.3 Buckling Constraints

Let σ_e be the compressive stress in the eth element under a given loadcase. The Euler buckling constraint for the element may be written:

$$b_e \equiv \frac{\sigma_e}{E} - \frac{kx_e}{\|\mathbf{r}_e\|^2} \leq 0 \qquad (14.12)$$

where k is a constant which depends on the cross-sectional shape. (If the cross section is circular then $k = \pi/4 \approx 0.7854$.)

We then get:

$$\frac{\partial b_e}{\partial \xi} = \frac{1}{E}\frac{\partial \sigma_e}{\partial \xi} - \frac{k}{\|\mathbf{r}_e\|^2}\frac{\partial x_e}{\partial \xi} + \frac{2kx_e}{\|\mathbf{r}_e\|^3}\frac{\partial \|\mathbf{r}_e\|}{\partial \xi} \qquad (14.13)$$

The first term on the right-hand side is handled as in Section 14.3.2. The second term is non-zero only if $\xi = x_e$, and in that case $\partial x_e/\partial \xi = 1$.

The third term is non-zero only if $\xi = c_j$, where c_j is a coordinate for one of the two nodes of the eth element. In this case it is easy to analytically differentiate $\|\mathbf{r}_e\|$ with respect to ξ.

14.3.4 Consideration of the Gravity Loads

In the previous sections we assumed that the load vector **p** is a constant vector, i.e. $\partial \mathbf{p}/\partial \xi = 0$. This assumption is obviously not correct if the element's own weight is a substantial part of the loads.

Let us consider a displacement constraint where from (14.7) and (14.8) we have:

$$\frac{\partial d_i}{\partial \xi} = -\sum_e \mathbf{u}_e^T \frac{\partial \mathbf{K}_e}{\partial \xi}\mathbf{v}_e + \mathbf{v}^T \frac{\partial \mathbf{p}}{\partial \xi}$$

As the first term on the right-hand side has already been discussed in Section 14.3.1, we consider only the second term.

We assume that the only varying part in **p** is the element's own weight:

$$\mathbf{p} = \mathbf{p}_0 + \sum_{e=1}^{\text{KELEM}} \mathbf{p}_e, \text{ where } \frac{\partial \mathbf{p}_0}{\partial \xi} = 0$$

$$= \mathbf{v}^T \frac{\partial \mathbf{p}}{\partial \xi} = \sum_e \mathbf{v}_e^T \frac{\partial \mathbf{p}_e}{\partial \xi}$$

The distributed gravity load $\mathbf{p}_e$ is lumped into the two equal vertical forces $\mathbf{f}_q$ shown in Fig. 14.5.

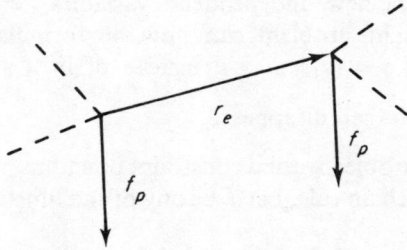

Fig. 14.5 Definition of forces

$||\mathbf{f}_p|| = \dfrac{\rho g x_e}{2}||\mathbf{r}_e||$, where g = acceleration of gravity ≈ 9.81 m/s^2

It is now obvious that $\mathbf{f}_p$, and thus also $\mathbf{p}_e$, can easily be analytically differentiated.

14.4 INDEPENDENT VARIABLES

In this section we focus our attention on the linear constraints $A \begin{pmatrix} x \\ c \end{pmatrix} = \mathbf{h}$.

We do not consider completely general constraints of this form. In practice such constraints imply that the variables are divided into disjunctive groups, so that all variables within a given group depend linearly on a single variable. We assume that this is the case in problems discussed here. Let:

$J_x \equiv \{1, \ldots, \text{KELEM}\}$ = index set for the area variables
$J_c \equiv \{1, \ldots, 3 \cdot \text{KNODE}\}$ = index set for the coordinate variables
$J_x(j) \equiv$ index set for the jth group of area variables, $j = 0, 1, \ldots, n_x$.
$J_c(j) \equiv$ index set for the jth group of coordinate variables, $j = 0, 1, \ldots, n_c$.

These index sets satisfy:

$$J_x(0) \cup J_x(1) \cup \ldots \cup J_x(n_x) = J_x$$
$$J_c(0) \cup J_c(1) \cup \ldots \cup J_c(n_x) = J_c$$

$$J_x(i) \cap J_x(j) = \emptyset \text{ if } i \neq j$$
$$J_c(i) \cap J_c(j) = \emptyset \text{ if } i \neq j$$

Now, under our assumptions, there are known constants $\bar{x}_{e0}, \bar{x}_{e1}, \bar{c}_{v0}, \bar{c}_{v1}$ for all $e \in J_x$ and $v \in J_c$, such that:

$$x_e = \bar{x}_{e0} \qquad \text{for all } e \in J_x(0) \qquad (14.14)$$
$$x_e = \bar{x}_{e0} + \bar{x}_{e1}\alpha_j \qquad \text{for all } e \in J_x(j), j = 1, \ldots, n_x \qquad (14.15)$$
$$c_v = \bar{c}_{v0} \qquad \text{for all } v \in J_c(0) \qquad (14.16)$$
$$c_v = \bar{c}_{v0} + \bar{c}_{v1}\gamma_j \qquad \text{for all } v \in J_c(j), j = 1, \ldots, n_c \qquad (14.17)$$

where α_j and γ_j are the new, 'independent', variables.

The minimum-weight problem can now be formulated in the variables $\boldsymbol{\alpha} = (\alpha_1, \ldots \alpha_{n_x})^T$ and $\boldsymbol{\gamma} = (\gamma_1, \ldots \gamma_n)^T$ instead of in $\mathbf{x}$ and $\mathbf{c}$, thus the linear constraints $\mathbf{A}\begin{pmatrix} x \\ c \end{pmatrix} = \mathbf{h}$ then disappear.

The gradients of the objective and constraint functions, with respect to $\boldsymbol{\alpha}$ and $\boldsymbol{\gamma}$, are calculated by the chain rule. Let f be any of the functions w, d_i, σ_i or b_i. We then have:

$$\frac{\partial f}{\partial \alpha_j} = \sum_{e \in J_x(j)} \frac{\partial f}{\partial x_e} \frac{\partial x_e}{\partial \alpha_j} = \sum_{e \in J_x(j)} \frac{\partial f}{\partial x_e} \bar{x}_{e1} \qquad j = 1, \ldots, n_x \qquad (14.18)$$

$$\frac{\partial f}{\partial \gamma_j} = \sum_{v \in J_c(j)} \frac{\partial f}{\partial c_v} \frac{\partial c_v}{\partial \gamma_j} = \sum_{v \in J_c(j)} \frac{\partial f}{\partial c_v} \bar{c}_{v1} \qquad j = 1, \ldots, n_c \qquad (14.19)$$

and we already know, from Section 14.3, how to calculate $\dfrac{\partial f}{\partial x_e}$ and $\dfrac{\partial f}{\partial c_v}$.

From now on, therefore, we consider P in the following form:

P: min $w(\alpha, \gamma)$

     subject to $d_i(\alpha, \gamma) \leqslant d_i^{\max}$ $i = 1, \ldots, m_d$
                $\sigma_i(\alpha, \gamma) \leqslant \sigma_i^{\max}$ $i = 1, \ldots, m$
                $b_i(\alpha, \gamma) \leqslant 0$ $i = 1, \ldots, m_b$
                $\alpha_j^{\min} \leqslant \alpha_j \leqslant \alpha_j^{\max}$ $j = 1, \ldots, n_x$
                $\gamma_j^{\min} \leqslant \gamma_j \leqslant \gamma_j^{\max}$ $j = 1, \ldots, n_c$

where the constants $\{\alpha_j^{\min}, \alpha_j^{\max}\}_{j=1}^{n_x}$ and $\{\gamma_j^{\min}, \gamma_j^{\max}\}_{j=1}^{n_c}$ are easily calculated from the known constants

$$\{x_j^{\min}, x_j^{\max}, \bar{x}_{j0}, \bar{x}_{j1}\}_{j=1}^{\text{KELEM}} \text{ and } \{c_j^{\min}, c_j^{\max}, \bar{c}_{j0}, \bar{c}_{j1}\}_{j=1}^{3\text{KNODE}}.$$

It is not a serious limitation to assume that $\bar{x}_{e0} = 0$ and $\bar{x}_{e1} = 1$ for all $e \in J_x(1) \cup \ldots \cup J_x(n_x)$, i.e. that:

$$x_e = \alpha_j \text{ for all } e \in J_x(j), j = 1, \ldots, n_x \qquad (14.20)$$

This assumption is made because it greatly simplifies the forthcoming presentation. Equation (14.20) then replaces (14.15).

14.5 METHOD FOR SOLVING THE OPTIMIZATION PROBLEM

It is shown in Sections 14.3 and 14.4 that the gradients of the objective and constraint functions, with respect to α and γ, are comparatively easy to calculate, once the functions themselves have been evaluated by an FEM calculation. We also mentioned, in Section 14.2, that we would like to do 'as much as possible' between each FEM calculation. A natural approach is then to solve a 'subproblem' after each FEM calculation. Each subproblem should be an approximation of the original problem, based on the gradient information, and it must of course be comparatively easy to solve.

The immediate suggestion would be to linearize both the objective function and the constraint functions and to solve a linear programming problem. However, *if the geometry is held fixed*, i.e. if γ is fixed, then we get a *much* better approximation of the original problem if we linearize the constraint functions in the *inverse* variables $1/\alpha_j$ introduced in Chapter 3 while the objective function is linearized in the original variables α_j (Ref. [9]). Although the corresponding subproblem is not a linear programming problem, it can be solved in an efficient way by dual methods of convex programming as shown in Refs [7] and [2].

It is not trivial to generalize the above observations to the case when the geometry is *not* fixed. However, we would like to use an approach which becomes equivalent to the method in Ref. [2] *if* the geometry is held fixed.

14.5.1 General Approach, Flowchart

Let P be written:

P: $\min w(\alpha, \gamma)$
subject to $g_i(\alpha, \gamma) \leq 0$ $i = 1, \ldots, m = m_d + m_\sigma + m_b$
$\alpha_j^{\min} \leq \alpha_j \leq \alpha_j^{\max}$ $j = 1, \ldots, n_x$
$\gamma_j^{\min} \leq \gamma_j \leq \gamma_j^{\max}$ $j = 1, \ldots, n_c$

The general approach to solve P is shown in the flowchart of Fig. 14.6.

The purpose of introducing parameters $h_j^{(k)}$ is to stabilize the algorithm by ensuring that $\gamma^{(k+1)}$ lies within a prescribed 'box' around $\gamma^{(k)}$. This is also the purpose of the constraints $\frac{1}{2}\alpha_j^{(k)} \leq \alpha_j \leq 2\alpha_j^{(k)}$.

$h_j^{(k)}$ should normally be chosen such that $h_j^{(k)} \to 0$ when $k \to \infty$, but $h_j^{(k)}$ must not approach zero 'to fast', since this may imply that the sequence $\{\alpha^{(k)}, \gamma^{(k)}\}_{k=1}^{\infty}$ converges towards a non-optimal solution of P. In particular, it is necessary that

$$\sum_{k=1}^{\infty} h_j^{(k)} = \infty$$

A simple choice is $h_j^{(k)} = C_j/(k + D_j)$, where C_j and D_j are constants.

The most crucial step in the general approach above is to choose suitable functions $\bar{w}^{(k)}$ and $\bar{g}_i^{(k)}$. This is the subject of the next section.

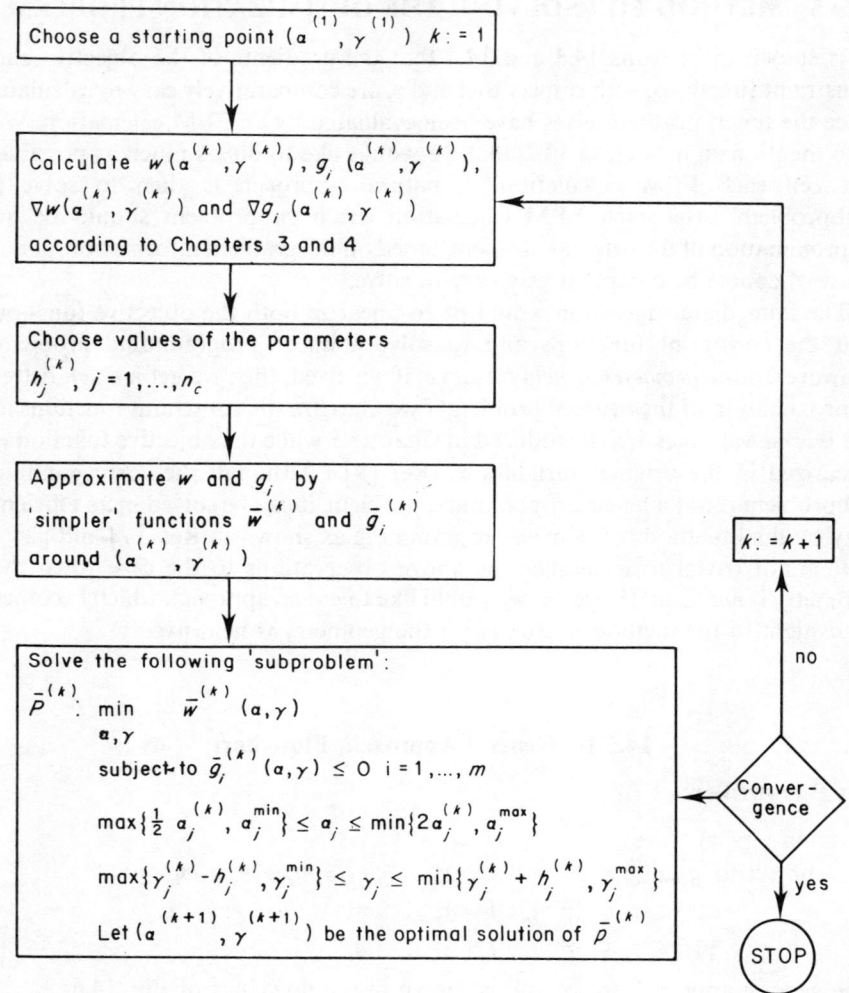

Fig. 14.6 Flowchart for optimization problem

14.5.2 How to Choose $\bar{w}^{(k)}$ and $\bar{g}_i^{(k)}$

We linearize w by a first-degree Taylor expression at the given iteration point $(\alpha^{(k)}, \gamma^{(k)})$:

$$w(\alpha, \gamma) \approx \bar{w}^{(k)}(\alpha, \gamma) \equiv w(\alpha^{(k)}, \gamma^{(k)})$$
$$+ \sum_{j=1}^{n_c} \frac{\partial w}{\partial \alpha_j}(\alpha_j - \alpha_j^{(k)}) + \sum_{j=1}^{n_c} \frac{\partial w}{\partial \gamma_j}(\gamma_j - \gamma_j^{(k)}) \qquad (14.21)$$

where all derivatives are evaluated in $(\alpha^{(k)}, \gamma^{(k)})$.

w is a linear function in α, i.e.:

$$w(\alpha, \gamma) = w_0(\gamma) + \sum_{j=1}^{n_x} v_j(\gamma)\alpha_j \tag{14.22}$$

($w_0(\gamma)$ = the weight of the elements with fixed cross-sectional areas.)

From (14.22) it is obvious that $v_j(\gamma^{(k)}) = \partial w/\partial \alpha_j$, evaluated in $(\alpha^{(k)}, \gamma^{(k)})$, and thus:

$$w(\alpha^{(k)}, \gamma^{(k)}) = w_0^{(k)} + \sum_{j=1}^{n_x} \frac{\partial w}{\partial \alpha_j} \alpha_j^{(k)}$$

where $w_0^{(k)} \equiv w_0(\gamma^{(k)})$.

Equation (14.21) may then be written:

$$\bar{w}^{(k)}(\alpha, \gamma) = w_0^{(k)} + \sum_{j=1}^{n_x} \frac{\partial w}{\partial \alpha_j} \alpha_j + \sum_{j=1}^{n_c} \frac{\partial w}{\partial \gamma_j}(\gamma_j - \gamma_j^{(k)}) \tag{14.23}$$

Let
$$s_j^{(k)} \equiv \begin{cases} \theta h_j^{(k)} & \text{if } \frac{\partial w}{\partial \gamma_j} \geq 0 \text{ at } (\alpha^{(k)}, \gamma^{(k)}) \\ -\theta h_j^{(k)} & \text{if } \frac{\partial w}{\partial \gamma_j} < 0 \text{ at } (\alpha^{(k)}, \gamma^{(k)}) \end{cases}$$

where θ is a constant > 1. We have mostly used $\theta = 3$.

We now introduce new variables **y** and **z** by the transformations:

$$y_j = \frac{\alpha_j^{(k)}}{\alpha_j} \Leftrightarrow \alpha_j = \frac{\alpha_j^{(k)}}{y_j} \tag{14.24}$$

$$z_j = \frac{s_j^{(k)}}{\gamma_j - \gamma_j^{(k)} + s_j^{(k)}} \Leftrightarrow \gamma_j = \gamma_j^{(k)} - s_j^{(k)} + \frac{s_j^{(k)}}{z_j} \tag{14.25}$$

We then express $\bar{w}^{(k)}$ in terms of these new variables:

$$\bar{w}^{(k)}(y, z) = w_0^{(k)} + \sum_{j=1}^{n_x} \frac{\partial w}{\partial \alpha_j} \frac{\alpha_j^{(k)}}{y_j} + \sum_{j=1}^{n_c} \frac{\partial w}{\partial \gamma_j} s_j^{(k)} \left(\frac{1}{z_j} - 1\right) \tag{14.26}$$

There are explicit bounds on the variables α_j:

$$\frac{1}{2}\alpha_j^{(k)} \leq \alpha_j \leq 2\alpha_j^{(k)}$$

and
$$\alpha_j^{\min} \leq \alpha_j \leq \alpha_j^{\max}$$

When translating these bounds to y_j, by (14.24), they become:

$$\frac{1}{2} \leq y_j \leq 2$$

and
$$\frac{\alpha_j^{(k)}}{\alpha_j^{\max}} \leq y_j \leq \frac{\alpha_j^{(k)}}{\alpha_j^{\min}}$$

which may equivalently be written:

$$\max\left\{\frac{1}{2}, \frac{\alpha_j^{(k)}}{\alpha_j^{\max}}\right\} \leqslant y_j \leqslant \min\left\{2, \frac{\alpha_j^{(k)}}{\alpha_j^{\min}}\right\} \qquad (14.27)$$

The bounds on the variables γ_j are:

$$\max\{\gamma_j^{\min}, \gamma_j^{(k)} - h_j^{(k)}\} \leqslant \gamma_j \leqslant \min\{\gamma_j^{\max}, \gamma_j^{(k)} + h_j^{(k)}\}$$

When translating this to z_j, by (14.25), it becomes:

$$\frac{\Theta}{\Theta + \delta_j^+} \leqslant z_j \leqslant \frac{\Theta}{\Theta + \delta_j^-} \qquad (14.28)$$

where $\delta_j^+ = \min\{1, \max\{\delta_j^{\min}, \delta_j^{\max}\}\}$
$\delta_j^- = \max\{-1, \min\{\delta_j^{\min}, \delta_j^{\max}\}\}$
and $\delta_j^{\min} = \dfrac{\Theta(\gamma_j^{\min} - \gamma_j^{(k)})}{s_j^{(k)}}$

$$\delta_j^{\max} = \frac{\Theta(\gamma_j^{\max} - \gamma_j^{(k)})}{s_j^{(k)}}$$

Note that $\delta_j^{\min} \cdot \delta_j^{\max} \leqslant 0$ so that $\delta_j^+ \geqslant 0$ and $\delta_j^- \leqslant 0$.

In ordinary cases we have $\delta_j^+ = 1$ and $\delta_j^- = -1$, and then (14.28) may simply be written:

$$\frac{\Theta}{\Theta + 1} \leqslant z_j \leqslant \frac{\Theta}{\Theta - 1} \quad \text{or, with } \Theta = 3: \frac{3}{4} \leqslant z_j \leqslant \frac{3}{2}$$

Since $y_j > 0$ and $z_j > 0$ for all feasible **y** and **z** and since $\partial w/\partial \alpha_j \geqslant 0$ and $(\partial w/\partial \gamma_j) s_j^{(k)} \geqslant 0$, we see from (14.26) that $\bar{w}^{(k)}$ is a *convex* function in the variables **y** and **z**.

We now turn to the constraint functions $g_i (= d_i, \sigma_i \text{ or } b_i)$, which we linearize in the (**y**, **z**)-space, in contrast to w which is linearized in the (α, γ)-space.

$$\bar{g}^{(k)}(y, z) = g_i(y^{(k)}, z^{(k)}) + \sum \frac{\partial g_i}{\partial y_j}(y_j - y_j^{(k)}) + \sum \frac{\partial g_i}{\partial z_j}(z_j - z_j^{(k)})$$

Let $\quad g_i^{(k)} \equiv g_i(y^{(k)}, z^{(k)}) = g_i(\alpha^{(k)}, \gamma^{(k)})$

From (14.24) and (14.25) we see that $\alpha_j = \alpha_j^{(k)}$ corresponds to $y_j = 1$, and $\gamma_j = \gamma_j^{(k)}$ corresponds to $z_j = 1$. Thus $y_j^{(k)} = 1$ and $z_j^{(k)} = 1$.

The derivatives $\partial g_i/\partial y_j$ and $\partial g_i/\partial z_j$ are obtained by the chain rule:

$$\frac{\partial g_i}{\partial y_j} = \frac{\partial g_i}{\partial \alpha_j} \frac{d\alpha_j}{dy_i} = \frac{\partial g_i}{\partial \alpha_j}(-\alpha_j^{(k)}) \qquad (14.29)$$

$$\frac{\partial g_i}{\partial z_j} = \frac{\partial g_i}{\partial \gamma_j} \frac{d\gamma_j}{dz_j} = \frac{\partial g_i}{\partial \gamma_j}(-s_j^{(k)}) \qquad (14.30)$$

The subproblem $\overline{P}^{(k)}$ may thus be written:

$$\overline{P}^{(k)}: \min_{y,z} \quad w_0^{(k)} + \sum_j \frac{\partial w}{\partial \alpha_j} \frac{\alpha_j^{(k)}}{y_j} + \sum_j \frac{\partial w}{\partial \gamma_j} s_j^{(k)} \left(\frac{1}{z_j} - 1\right)$$

$$\text{subject to} \quad g_i^{(k)} - \sum_j \frac{\partial g_i}{\partial \alpha_j} \alpha_j^{(k)}(y_j - 1) - \sum_j \frac{\partial g_i}{\partial \gamma_j} s_j^{(k)}(z_j - 1) \leq 0$$

$$\max \left\{\frac{1}{2}, \frac{\alpha_j^{(k)}}{\alpha_j^{\max}}\right\} \leq y_j \leq \min \left\{2, \frac{\alpha_j^{(k)}}{\alpha_j^{\min}}\right\}$$

$$\frac{\Theta}{\Theta + \delta_j^+} \leq z_j \leq \frac{\Theta}{\Theta + \delta_j^-}$$

where all derivatives are evaluated in $(\alpha^{(k)}, \gamma^{(k)})$.

The introduction of the variables y_j and z_j gives several advantages. Firstly, the subproblem $\overline{P}^{(k)}$ is now in a form which is perfectly suited for application of convex programming duality (see Section 14.6). Also, if $n_c = 0$, i.e. if all the coordinates are fixed, then the method is equivalent to that in Ref. [2], which has proved to work very well.

Secondly, if we choose a form for $\overline{P}^{(k)}$ where *both* the objective function and the constraint functions are linear, then the optimal solution of $\overline{P}^{(k)}$ will normally be located on the boundary of the 'box', and it will thus be difficult to decide *if* and *when* convergence has occurred. As it is now, with a strictly convex objective function, the boxes will in most cases be superfluous in the final iterations, and the convergence may be established by the following lemma:

Lemma 1 *If the subproblem $\overline{P}^{(k)}$ has the optimal solution $y_j = 1$ for $j = 1, \ldots, n_x$ and $z_j = 1$ for $j = 1, \ldots, n_c$, then $(\alpha^{(k)}, \gamma^{(k)})$ is a Kuhn–Tucker point (i.e. most likely a local optimum) of the original problem* P.

The straightforward proof is left to the Appendix at the end of the chapter.

14.5.3 How to Obtain Perfectly Feasible Solutions

For two reasons, at least, it is advantageous if, whenever needed, we could easily obtain a feasible solution of P, say $(\tilde{\alpha}^{(k)}, \tilde{\gamma}^{(k)})$, which is also in some sense 'close' to $(\alpha^{(k)}, \gamma^{(k)})$ (the optimal solution of subproblem $\overline{P}^{(k-1)}$).

The first reason is that if $(\alpha^{(k)}, \gamma^{(k)})$ is 'very infeasible' then it might happen that the next subproblem $\overline{P}^{(k)}$ has no feasible solutions, i.e. we can not solve $\overline{P}^{(k)}$. If *in that case* we approximate w and g_i (by $\overline{w}^{(k)}$ and $\overline{g}_i^{(k)}$) around $(\tilde{\alpha}^{(k)}, \tilde{\gamma}^{(k)})$ instead of around $(\alpha^{(k)}, \gamma^{(k)})$, then $\overline{P}^{(k)}$ always has feasible solutions.

The second reason is that, in practice, we always stop the iterations before the sequence $\{(\alpha^{(k)}, \gamma^{(k)})\}_k$ has converged 'exactly'. Thus we can not be sure that the last $(\alpha^{(k)}, \gamma^{(k)})$ is a feasible solution of P, although it is often very nearly so.

Under the assumptions that:
— the load cases do not depend on (α, γ);
— $J_x(0) = \phi$ (see Section 14.4), i.e. no cross-sectional areas are fixed; and
— the explicit bounds on $\alpha_j, j = 1, \ldots, n_x$, do not become active;

we can obtain an explicit feasible solution by increasing the α-variables. To do this, without undue deterioration of the objective function value, we proceed as follows: Set $(\alpha, \gamma) = (\mu\alpha^{(k)}, \gamma^{(k)})$ where $\mu > 0$, and consider problem P under the restriction that the only variable is μ:

$$\min_{\mu} \quad w(\mu\alpha^{(k)}, \gamma^{(k)})$$

$$\text{subject to} \quad d_i(\mu\alpha^{(k)}, \gamma^{(k)}) \leq d_i^{\max}$$

$$\sigma_i(\mu\alpha^{(k)}, \gamma^{(k)}) \leq \sigma_i^{\max}$$

$$\frac{\sigma_i(\mu\alpha^{(k)}, \gamma^{(k)})}{E} \leq \frac{kx_i(\mu)}{\|r_i^{(k)}\|^2}$$

Under the assumptions above, we have:

$$w(\mu\alpha^{(k)}, \gamma^{(k)}) = \mu w(\alpha^{(k)}, \gamma^{(k)}) \equiv \mu w^{(k)}$$

$$d_i(\mu\alpha^{(k)}, \gamma^{(k)}) = \frac{1}{\mu} d_i(\alpha^{(k)}, \gamma^{(k)}) \equiv \frac{1}{\mu} d_i^{(k)}$$

$$\sigma_i(\mu\alpha^{(k)}, \gamma^{(k)}) = \frac{1}{\mu} \sigma_i(\alpha^{(k)}, \gamma^{(k)}) \equiv \frac{1}{\mu} \sigma_i^{(k)}$$

$$x_i(\mu) = \mu x_i^{(k)}$$

and so we get the problem:

$$\min_{\mu} \quad \mu w^{(k)}$$

$$\text{subject to} \quad \frac{1}{\mu} d_i^{(k)} \leq d_i^{\max}$$

$$\frac{1}{\mu} \sigma_i^{(k)} \leq \sigma_i^{\max}$$

$$\frac{1}{\mu} \frac{\sigma_i^{(k)}}{E} \leq \mu \frac{kx_i^{(k)}}{\|r_i^{(k)}\|^2}$$

with the explicit optimal solution $\mu = \mu^{(k)}$, where

$$\mu^{(k)} \equiv \max \left\{ \max_i \left(\frac{d_i^{(k)}}{d_i^{\max}} \right), \max_i \left(\frac{\sigma_i^{(k)}}{\sigma_i^{\max}} \right), \sqrt{\max_i \left(\frac{\sigma_i^{(k)} \|r_i^{(k)}\|^2}{x_i^{(k)} kE} \right)} \right] \right\}$$

Now $(\tilde{\alpha}^{(k)}, \tilde{\gamma}^{(k)}) \equiv (\mu^{(k)}\alpha^{(k)}, \gamma^{(k)})$ is a feasible solution of P.

14.6 USING DUALITY TO SOLVE THE SUBPROBLEMS

The subproblem $\overline{P}^{(k)}$ is a convex programming problem. If we neglect the (rare) case that $\partial w/\partial \gamma_j = 0$ for some j, we can further state that $\overline{P}^{(k)}$ is a *strictly* convex programming problem, since the objective function is strictly convex in y and z and the constraint functions are linear. (If $\partial w/\partial \gamma = 0$ for some j, say $j \in J_0$, we may avoid possible difficulties by adding the term $\varepsilon \sum_{j \in J_0} (z_j - 1)^2$ to the objective function; ε should be a 'small' positive constant.)

To solve $\overline{P}^{(k)}$ we use the method proposed in Refs [1] and [2]. However, to make this chapter self-contained we repeat the arguments below.

First a short comment on the notation: since there are a limited number of characters in the alphabet and since we have already used the majority of them, some of the characters used in this section, e.g. b_i and c_j, will *not* have the same meaning as in the previous and forthcoming sections. The notation in this section should thus be considered as 'locally defined'.

After removing the constant

$$w_0^{(k)} - \sum \frac{\partial w}{\partial \gamma_j} s_j^{(k)}$$

from the objective function, the subproblem may be written:

$\overline{P}$: $\min_x \sum_{j=1}^{n} \frac{c_j}{x_j}$

subject to $\sum_{j=1}^{n} a_{ij} x_j \leqslant b_i \quad i = 1, \ldots, m$

$x \in X \equiv \{x \in R^n | x_j^{\min} \leqslant x_j \leqslant x_j^{\max}\}$

where $n = n_a + n_c$ and $m = m_d + m_\sigma + m_b$.

$c_j, a_{ij}, b_i, x_j^{\min}$ and $x_j^{\max}$ are known constants.

We shall now apply the duality theory for convex programming, which for completeness is summarized below and essentially repeats the arguments of Chapters 2 and 3 (for an exhaustive description see Ref. [10]):

Consider the problem:

PC: $\min \; f(x)$

subject to $g_i(x) \leqslant 0 \quad i = 1, \ldots, m$

$x \in X$

where X is a convex, compact subset of R^n and f and g_i are convex, continuous functions $X \to R$.

The 'Lagrangian' for problem PC is defined, for $x \in X$ and $\lambda = (\lambda_1, \ldots, \lambda_m)^T$ with all $\lambda_i \geqslant 0$, as:

$$L(x, \lambda) \equiv f(x) + \sum_{i=1}^{m} \lambda_i g_i(x)$$

The 'dual objective function' for problem PC is defined, for $\lambda \geq 0$ (i.e. $\lambda_i \geq 0$ for all i), as:
$$\Phi(\lambda) \equiv \min_{x \in X} L(x, \lambda) = \min_{x \in X} \{ f(x) + \sum \lambda_i g_i(x) \}$$

Since X is compact, and f and g_i are continuous, $\Phi(\lambda)$ is well defined.

Φ is a *concave* function in λ. This follows from the fact that $\Phi(\lambda)$ is the pointwise minimum of a collection of functions which are linear in λ.

If PC is called the 'primal problem' then the corresponding 'dual problem' is:

$\max_{\lambda \geq 0} \{ \min_{x \in X} L(x, \lambda) \}$, which may be written:

DC: $\qquad\qquad\max\ \Phi(\lambda)$
$\qquad\qquad$subject to $\quad \lambda_i \geq 0 \quad i = 1, \ldots, m$

Under very weak assumptions, which are justified in our cases, for example there is a point $\mathbf{x}_0 \in X$ such that $g_i(x_0) < 0$ for all i, the following holds:
(i) Problem DC has an optimal solution $\bar{\lambda}$.

(ii) If $\bar{\mathbf{x}}$ is an unique optimal solution of the problem: $\min_{x \in X} L(x, \bar{\lambda})$, then $\bar{\mathbf{x}}$ is also the unique optimal solution of problem PC.

We now return to our subproblem $\bar{P}$ and apply these results.

The Lagrangian for problem $\bar{P}$ is, by definition,
$$L(x, \lambda) \equiv \sum_j \frac{c_j}{x_j} + \sum_i \lambda_i \left(\sum_j a_{ij} x_j - b_i \right)$$
$$= \sum_j \left(\frac{c_j}{x_j} + x_j \lambda^T a_j \right) - \lambda^T b$$

where $\quad \lambda = \begin{pmatrix} \lambda_1 \\ \vdots \\ \lambda_m \end{pmatrix}, \quad a_j = \begin{pmatrix} a_{1j} \\ \vdots \\ a_{mj} \end{pmatrix}$ and $b = \begin{pmatrix} b_1 \\ \vdots \\ b_m \end{pmatrix}$

$L(x, \lambda)$ is defined for $\mathbf{x} \in X$ and $\lambda \geq \mathbf{0}$ (i.e. $\lambda_i \geq 0$ for all i).

To obtain $\Phi(\lambda)$, $L(x, \lambda)$ must be minimized with respect to $\mathbf{x} \in X$. This is easily accomplished, since the minimization of $L(x, \lambda)$ is divided into n simple one-dimensional minimizations.

Straightforward calculations show that, for a given $\lambda \geq \mathbf{0}$, $L(x, \lambda)$ is uniquely minimized by $\mathbf{x} = \mathbf{x}(\lambda)$, where (for $j = 1, \ldots, n$):

$$x_j(\lambda) = \begin{cases} x_j^{\min} & \text{if } \lambda^T a_j \geq \dfrac{c_j}{(x_j^{\min})^2} \\[2mm] \sqrt{\dfrac{c_j}{\lambda^T a_j}} & \text{if } \dfrac{c_j}{(x_j^{\max})^2} \leq \lambda^T a_j \leq \dfrac{c_j}{(x_j^{\min})^2} \\[2mm] x_j^{\max} & \text{if } \lambda^T a_j \leq \dfrac{c_j}{(x_j^{\max})^2} \end{cases} \qquad (*)$$

We then get the following dual objective function:

$$\Phi(\lambda) = L(x(\lambda), \lambda) = \sum_j \left(\frac{c_j}{x_j(\lambda)} + x_j(\lambda) \lambda^T a_j \right) - \lambda^T b$$

with $x_j(\lambda)$ as in (∗).

We thus get an *explicit* expression for Φ as a function of λ.

It is also very easy to calculate the derivatives of Φ:

$$\frac{\partial \Phi}{\partial \lambda_i} = \sum_j a_{ij} x_j(\lambda) - b_i$$

with $x_j(\lambda)$ as in (∗).

Instead of solving $\overline{P}$ we may then solve the dual problem $\overline{D}$:

$\overline{D}$:   max $\Phi(\lambda)$

                   subject to $\lambda_i \geq 0, \quad i = 1, \ldots, m$

If $\overline{\lambda}$ is the optimal solution of $\overline{D}$, then we get the optimal solution $\overline{x}$ of $\overline{P}$ by simply letting $\overline{x}_j = x_j(\overline{\lambda})$, with $x_j(\lambda)$ defined by (∗).

$\overline{D}$ is considerably easier to solve than $\overline{P}$ because of two properties:
— simple constraints (only non-negatively restrictions on the variables);
— few variables (often $m << n$).

($\overline{P}$ has n variables, m linear constraints and $2n$ simple bounds on the variables, while $\overline{D}$ has m variables, no linear constraints and m simple bounds on the variables.)

Since Φ is a continuous, concave function with a gradient which is continuous and easy to calculate, problem $\overline{D}$ may be solved by an arbitrary gradient method (e.g. a conjugate gradient method or a quasi-Newton method), slightly modified to take care of the non-negativity restrictions. We have successfully used a very simple steepest-ascent method.

14.7 THE INFLUENCE OF THE LOADING MAGNITUDE ON THE OPTIMAL SOLUTION

In this section we return to our original notation and study the following questions:
— Assume that all loads are multiplied by a parameter $t > 0$. (The same t for all load cases.) How does the optimal solution of the minimum-weight problem P depend on t?
— Assume that the value of Young's modulus E is changed (due to a change in material), and/or that the cross-sectional shape of the elements is changed (i.e. the constant k in the definition of $b_i(\alpha, \gamma)$ in Section 14.3.3). What effect does this have on the optimal solution of P?

The answers are rather pleasing: in many cases the optimal geometry and the optimal ratio between the different cross-sectional areas are unchanged.

Throughout this section we assume that the load cases do not depend on the variables (α and γ) and that $J_a(0) = \emptyset$ see Section 14.4), i.e. that there are no fixed cross-sectional areas.

14.7.1 Displacement Constraints

The displacement of a given node in a given direction under a given load case, may be written:

$$d_i(\alpha, \gamma) = \frac{t}{E}\overline{d}_i(\alpha, \gamma)$$

where $\overline{d}_i(\alpha, \gamma)$ is the size of the displacement if $t = 1$ and $E = 1$.

Let $\tau \equiv E/t$ and consider the minimum-weight problem under displacement constraints only:

P_d:
$$\begin{aligned}\min \quad & w(\alpha, \gamma) \\ \text{subject to} \quad & d_i(\alpha, \gamma) \leqslant d_i^{\max} \quad i = 1, \ldots, m_d\end{aligned}$$

which may equivalently be written:

$P_d(\tau)$:
$$\begin{aligned}\min \quad & w(\alpha, \gamma) \\ \text{subject to} \quad & \overline{d}_i(\alpha, \gamma) \leqslant \tau d_i^{\max} \quad i = 1, \ldots, m_d\end{aligned}$$

w and $\overline{d}_i$ have the following properties:

$$w(\mu\alpha, \gamma) = \mu w(\alpha, \gamma) \tag{14.31}$$

$$\overline{d}_i(\mu\alpha, \gamma) = \frac{1}{\mu}\overline{d}_i(\alpha, \gamma) \tag{14.32}$$

Lemma 2 *Let $\tau_0 > 0$ and $\tau_1 > 0$ be given. Assume that (α_0, γ_0) is an optimal solution of $P_d(\tau_0)$.*
Let $\alpha_1 = \frac{\tau_0}{\tau_1}\alpha_0$ and $\gamma_1 = \gamma_0$.
Then (α_1, γ_1) is an optimal solution of $P_d(\tau_1)$.
This lemma is proved in the Appendix.

The minimum-weight problem under displacement constraints has the following property:

The optimal geometry and the optimal ratio between the different cross-sectional areas are independent of t and E.

14.7.2 Stress Constraints

The stress in a given element under a given load case, may be written:

$$\sigma_i(\alpha, \gamma) = t\overline{\sigma}_i(\alpha, \gamma)$$

where $\bar{\sigma}_i(\alpha, \gamma)$ is the size of the stress if $t = 1$.
In contrast to d_i, σ_i does not depend on E.

Let $\tau = 1/t$ and consider the minimum-weight problem under stress constraints only:

$P_\sigma(\tau)$:
$$\min \; w(\alpha, \gamma)$$
$$\text{subject to} \quad \bar{\sigma}_i(\alpha, \gamma) \leqslant \tau \sigma_i^{\max} \quad i = 1, \ldots, m$$

$\bar{\sigma}_i$ has the property:
$$\bar{\sigma}_i(\mu\alpha, \gamma) = \frac{1}{\mu} \bar{\sigma}_i(\alpha, \gamma) \tag{14.33}$$

Lemma 3 *Let $\tau_0 > 0$ and $\tau_1 > 0$ be given.*
Assume that (α_0, γ_0) is an optimal solution of $P_\sigma(\tau_0)$.
Let $\alpha_1 = \dfrac{\tau_0}{\tau_1} \alpha_0$ and $\gamma_1 = \gamma_0$
Then (α_1, γ_1) is an optimal solution of $P_\sigma(\tau_1)$.
The proof of Lemma 3 is identical to the proof of Lemma 2.

The minimum-weight problem under stress constraints thus has the following property:

The optimal geometry and the optimal ratio between the different cross-sectional areas are independent of t (and of E).

14.7.3 Buckling Constraints

The buckling constraints may be written (see Sections 14.3.3 and 14.7.2):

$$\frac{t}{E} \frac{\bar{\sigma}_i(\alpha, \gamma)}{x_i} \leqslant \frac{k}{\|r_i\|^2} \quad i = 1, \ldots, m_b$$

Let
$$h_i(\alpha, \gamma) \equiv \frac{\bar{\sigma}_i(\alpha, \gamma)}{x_i} \quad i = 1, \ldots, m_b$$

and
$$g_i(\gamma) \equiv \frac{1}{\|r_i\|^2} \quad i = 1, \ldots, m_b$$

and
$$\tau \equiv \frac{kE}{t}$$

Then the buckling constraints become:

$$h_i(\alpha, \gamma) \leqslant \tau g_i(\gamma) \quad i = 1, \ldots, m_b \tag{14.34}$$

We now consider the minimum-weight problem under buckling constraints only:

$P_b(\tau)$:
$$\min \; w(\alpha, \gamma)$$
$$\text{subject to} \quad h_i(\alpha, \gamma) \leqslant \tau g_i(\gamma) \quad i = 1, \ldots, m_b$$

h_i has the following property:

$$h_i(\mu\alpha, \gamma) = \frac{\bar{\sigma}_i(\mu\alpha, \gamma)}{\mu x_i} = \frac{1}{\mu^2} h_i(\alpha, \gamma) \qquad (14.35)$$

Lemma 4 Let $\tau_0 > 0$ and $\tau_1 > 0$ be given.
Assume that (α_0, γ_0) is an optimal solution of $P_b(\tau_0)$.
Let $\alpha_1 = \sqrt{\left(\dfrac{\tau_0}{\tau_1}\right)} \alpha_0$ and $\gamma_1 = \gamma_0$.
Then (α_1, γ_1) is an optimal solution of $P_b(\tau_1)$.
The proof may be found in the Appendix.

The minimum-weight problem under element buckling constraints thus has the following property:

The optimal geometry and the optimal ratio between the different cross-sectional areas are independent of t, E and k.

14.7.4 Mixed Constraints

As we have seen, the situation is very nice if there are displacement constraints *only*, or stress constraints *only*, or buckling constraints *only*.

It is easy to show that the combination of displacement/stress constraints has similar properties:

$P_{d\sigma}$:
$$\begin{aligned}
\min \quad & w(\alpha, \gamma) \\
\text{subject to} \quad & d_i(\alpha, \gamma) \leq d_i^{\max} & i = 1, \ldots, m_d \\
& \sigma_i(\alpha, \gamma) \leq \sigma_i & i = 1, \ldots, m_\sigma
\end{aligned}$$

The optimal geometry and the optimal ratio between the different cross-sectional areas, in problem $P_{d\sigma}$, are independent of t (but not necessarily of E). This statement is proved exactly as the corresponding statements in Sections 14.7.1 and 14.7.2.

If, however, there are *both* buckling constraints *and* displacement constraints or stress constraints, the situation is more complicated. A two-bar truss exemplifies this.

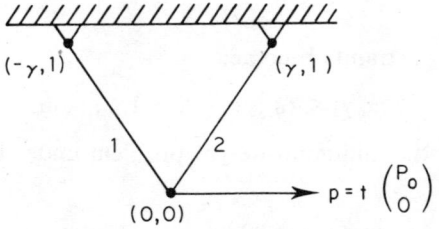

Fig. 14.7 2-bar truss

Two elements, numbered 1 and 2 in Fig. 14.7, have the same cross-sectional area $= \alpha$ and length $= \sqrt{(1+\gamma^2)}$.
The weight $= w(\alpha, \gamma) = 2\rho\alpha\sqrt{(1+\gamma^2)}$.
There is a tensile stress constraint on element 1 and a buckling constraint on element 2.
Elementary calculations show that:

the tensile stress in element $1 = \sigma_1$
$\quad=$ the compressive stress in element $2 = \sigma_2$
$= \dfrac{tP_0}{2\alpha} \dfrac{\sqrt{(1+\gamma^2)}}{\gamma}$.

Thus we have the constraints:

$$\frac{tP_0}{2\alpha} \frac{\sqrt{(1+\gamma^2)}}{\gamma} \leqslant \sigma^{\max} \qquad \text{(CON1)}$$

$$\frac{tP_0}{2E\alpha^2} \frac{\sqrt{(1+\gamma^2)}}{\gamma} \leqslant \frac{k}{1+\gamma^2} \qquad \text{(CON2)}$$

which may be written:

$$\alpha \geqslant \frac{tP_0}{2\sigma^{\max}} \frac{\sqrt{(1+\gamma^2)}}{\gamma} \qquad \text{(CON1)}$$

$$\alpha^2 \geqslant \frac{tP_0}{2kE} \frac{(1+\gamma^2)^{3/2}}{\gamma} \qquad \text{(CON2)}$$

or, with $C_1 = \dfrac{P_0}{2\sigma^{\max}}$ and $C_2 = \left(\dfrac{P_0}{2kE}\right)^{1/2}$

$$\alpha \geqslant tC_1 \frac{(1+\gamma^2)^{1/2}}{\gamma} \qquad \text{(CON1)}$$

$$\alpha \geqslant \sqrt{(t)}\, C_2 \frac{(1+\gamma^2)^{3/4}}{\gamma^{1/2}} \qquad \text{(CON2)}$$

It is obvious that, for a given γ, α should be chosen as small as possible, i.e.

$$\alpha = \alpha(\gamma) = \max\left\{tC_1 \frac{(1+\gamma^2)^{1/2}}{\gamma}, \sqrt{(t)}C_2 \frac{(1+\gamma^2)^{3/4}}{\gamma^{1/2}}\right\}$$

Thus the minimum-weight problem may be written in terms of γ only:

$$\min 2\rho\alpha(\gamma)\sqrt{1+\gamma^2}$$

$$\Leftrightarrow \min_{\gamma} \left\{2\rho \frac{(1+\gamma^2)}{\gamma} \max\left\{tC_1, \sqrt{(t)}C_2\gamma^{1/2}(1+\gamma^2)^{1/4}\right\}\right\}$$

It can be shown that this problem has the following solution:

(I) If $t > \sqrt{2}\left(\dfrac{C_2}{C_1}\right)^2$ then

$\gamma_{opt} = 1$ and $\alpha_{opt} = t\sqrt{2}C_1$

The tensile stress constraint (CON1) is active while the buckling constraint (CON2) is not.

(II) If $t < \dfrac{\sqrt{5}}{4}\left(\dfrac{C_2}{C_1}\right)^2$ then

$\gamma_{opt} = \dfrac{1}{2}$ and $\alpha_{opt} = \sqrt{t}\dfrac{5^{3/4}}{2}C_2$

Now CON 2 is active while CON 1 is not.

(III) If $\dfrac{\sqrt{5}}{4}\left(\dfrac{C_2}{C_1}\right) \leqslant t \leqslant \sqrt{2}\left(\dfrac{C_2}{C_1}\right)$ then

$$\gamma_{opt} = \dfrac{1}{\sqrt{2}}\left[\left\{1 + 4\left(\dfrac{C_1}{C_2}\right)^4 t^2\right\}^{1/2} - 1\right]^{1/2}$$

$$\alpha_{opt} = tC_1 \cdot \dfrac{(1 + \gamma_{opt}^2)^{1/2}}{\gamma_{opt}}$$

$$= tC_1\left(\dfrac{\{1 + 4(C_1/C_2)^4 t^2\}^{1/2} + 1}{\{1 + 4(C_1/C_2)^4 t^2\}^{1/2} - 1}\right)$$

Now both CON 1 and CON 2 are active.

14.8 TEST PROBLEMS

The method described in Sections 14.3–14.6 has been coded in FORTRAN and tested on some simple problems.

First of all the validity of the algorithm is checked on some very simple three-bar problems, where the optimal solutions could be analytically derived.

14.8.1 3-bar Pyramid

The coordinates of the nodes are shown in Fig. 14.8 and each element has a cross-sectional area $= \alpha$. A horizontal force $p = (0, P_0, 0)$ is acting in the top node. The length of each element is $\sqrt{(1 + \gamma^2)}$.

The weight of the pyramid is:

$$w(\alpha, \gamma) = C\alpha\sqrt{(1 + \gamma^2)}$$

where $C = 3\rho$, $\rho =$ the density.

Optimal Geometry in Truss Design

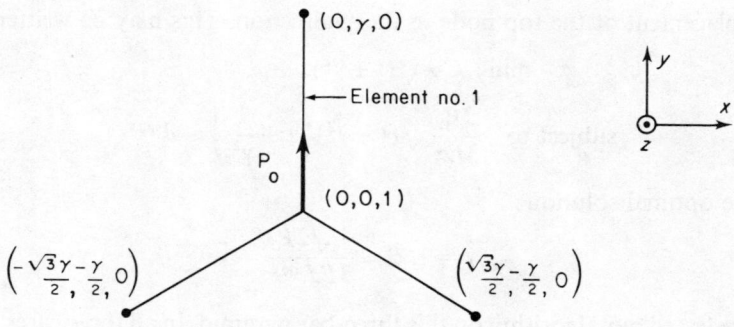

Fig. 14.8 3-bar pyramid observed from above

The compressive stress in element no. 1 is found to be:

$$\sigma_1 = \frac{2P_0}{3\alpha}\sqrt{\left(1+\frac{1}{\gamma^2}\right)}$$

Thus the minimum-weight problem subject to a compressive stress constraint (in element no. 1) may be written:

$$\min_{\alpha,\gamma} \quad C\alpha\sqrt{(1+\gamma^2)}$$
$$\text{subject to} \quad \frac{2P_0}{3\alpha}\sqrt{\left(1+\frac{1}{\gamma^2}\right)} \leqslant \sigma_1^{\max}$$

with the optimal solution:

$$\hat{\gamma}=1, \alpha = \frac{2\sqrt{2}P_0}{3\sigma_1^{\max}}$$

(It is in accordance with Section 14.7 that $\hat{\gamma}$ is independent of P_0.)

If we consider instead the minimum-weight problem subject to an element buckling constraint (in element no. 1) we get:

$$\min_{\alpha,\gamma} \quad C\alpha\sqrt{(1+\gamma^2)}$$
$$\text{subject to} \quad \frac{2P_0}{3E\alpha}\sqrt{\left(1+\frac{1}{\gamma^2}\right)} - \frac{k\alpha}{1+\gamma^2} \leqslant 0$$

with the optimal solution:

$$\hat{\gamma}=\frac{1}{2}, \hat{\alpha} = \left(\frac{5\sqrt{5}P_0}{6kE}\right)^{1/2}$$

(Again, this is in accordance with Section 14.7.)

We consider finally the minimum-weight problem subject to a constraint on the displacement of the top node in the *y*-direction. This may be written:

$$\min_{\alpha,\gamma} \quad C\alpha \sqrt{(1+\gamma^2)}$$

$$\text{subject to} \quad \frac{2P_0}{3E\alpha} \sqrt{(1+\gamma^2)} \left(1+\frac{1}{\gamma^2}\right) \leq d^{\max}$$

with the optimal solution

$$\hat{\gamma} = 1, \quad \hat{\alpha} = \frac{4\sqrt{2}P_0}{3Ed^{\max}}$$

We have tested our algorithm on this three-bar pyramid. In all three cases (stress constraint, buckling constraint and displacement constraint) the algorithm arrived at the same solution as obtained above analytically. Independently of starting point the algorithms required approximately 10 iterations and achieved an optimum, with an error in the variables less than 1%.

14.8.2 39-bar Tower

We now consider a more complex problem involving a symmetric 39-bar tower with topology shown in Fig. 14.9. There are 39 elements and 15 nodes, i.e. 36 degrees of freedom since the bottom nodes are fixed.

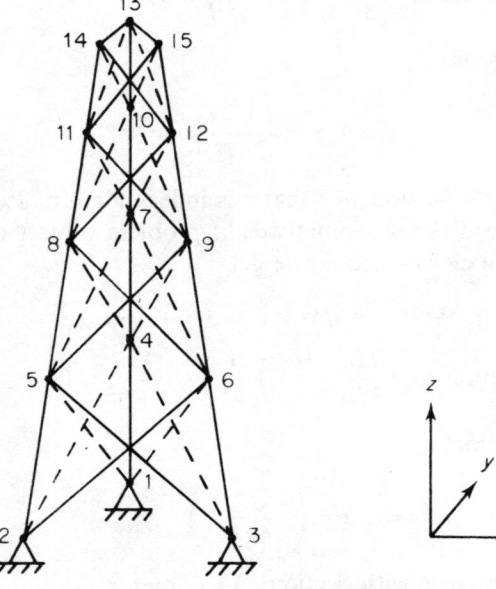

Fig. 14.9 39-bar tower

The three bottom nodes (1, 2 and 3) have the coordinates:

$$(0, 1, 0), \left(-\frac{\sqrt{3}}{2}, -\frac{1}{2}, 0\right) \text{ and } \left(\frac{\sqrt{3}}{2}, -\frac{1}{2}, 0\right)$$

while the three top nodes (13, 14 and 15) have the coordinates:

$$(0, 0.28, 4), (-0.14\sqrt{3}, -0.14, 4) \text{ and } (0.14\sqrt{3}, -0.14, 4).$$

The other node coordinates are permitted to vary during the optimization.

The tower must be kept symmetric which implies that there are 6 coordinate variables γ_j:

γ_1 = the z-coordinate of the nodes 4, 5 and 6.
γ_2 = the z-coordinate of the nodes 7, 8 and 9.
γ_3 = the z-coordinate of the nodes 10, 11 and 12.
γ_4 = the 'radius' of the tower at the level of the nodes 4, 5 and 6.
γ_5 = the 'radius' of the tower at the level of the nodes 7, 8 and 9.
γ_6 = the 'radius' of the tower at the level of the nodes 10, 11 and 12.
(γ_4 defines uniquely the x- and y-coordinates of the nodes 4, 5 and 6, etc.)

There are 5 area variables α_j:

α_1 = the cross-sectional area of the elements defined by the node-pairs (1, 4), (2, 5) and (3, 6).
α_2 = the cross-sectional area of the elements defined by the node-pairs (4, 7), (5, 8) and (6, 9).
α_3 = the cross-sectional area of the elements defined by the node-pairs (7, 10), (8, 11) and (9, 12).
α_4 = the cross-sectional area of the elements defined by the node-pairs (10, 13), (11, 14) and (12, 15).
α_5 = the cross-sectional area of the rest of the elements.

Thus we have $n_x = 5$ and $n_c = 6$ in P.

Displacement constraint

First we consider the minimum-weight problem subject to a single displacement constraint under a single load.

The load consists of three identical horizontal forces, acting on the three top nodes in the positive y-direction as seen in Fig. 14.10. The displacement constraint claims that the sum of the displacements in the positive y-direction of the three top nodes must not exceed a given quantity. The algorithm arrived at the geometry and proportions of the cross-sectional areas shown in Fig. 14.11.

We are convinced that this is at least a local optimum of P, since all the y_j and $z_j \approx 1$ in the final iterations (see Lemma l). We also believe that it is a global optimum, since we have tried different starting points and always obtained the same results.

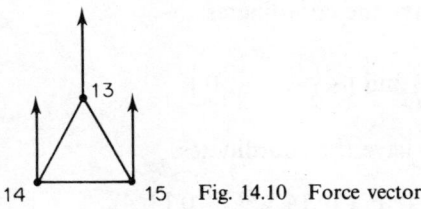

Fig. 14.10 Force vectors

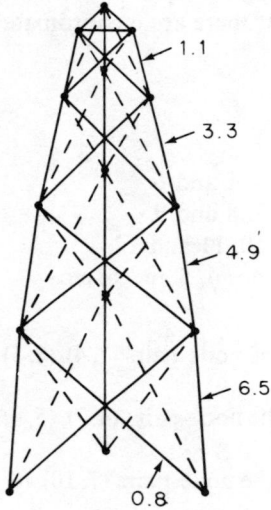

Fig. 14.11 Geometry and cross-sectional areas

The number of iterations needed was as high as 30–40. However, there is still much to be done to speed up the algorithm. One crucial point is to find better box-sizes (see Section 14.5.1) than those that we have used.

Element buckling constraints

We now consider the minimum-weight problem subject to buckling constraints in *all* the elements, under a single load. The load consists of three identical forces, acting in the three top nodes. Each force is of the type $(0, P_0, -2P_0)$, i.e. the vertical component is twice as big as the horizontal one.

It is, of course, not necessary to know in advance which of the 39 buckling constraints are active in the optimal solution, but a qualified guess could reduce the required solution time. A possible approach is to pick out, say, 10 elements and then to solve the problem subject to these 10 constraints. Then a check can be made to see if the guess was correct, i.e. if all the 39 constraints became satisfied in the solution obtained. If not, more constraints are included and the problem is solved again, etc. (On this problem we made a correct guess in our second trial.)

This kind of procedure can be particularly useful on problems with a large number of constraints of which only a fraction are active in the optimal solution.

The algorithm arrived at the geometry and (proportions of the) cross-sectional areas of Fig. 14.12. The number of iterations needed is of the same size as for the single displacement constraint problem. It is encouraging that the number of iterations needed does not seem to be very sensitive to the number of constraints.

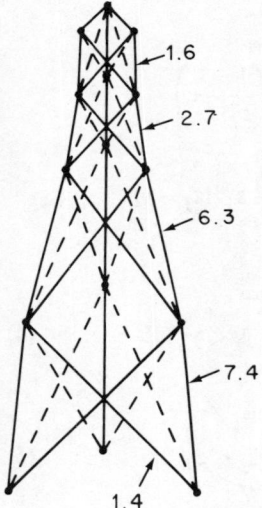

Fig. 14.12 Geometry and cross-sectional areas

APPENDIX PROOF OF SOME LEMMAS

Proof of Lemma 1

If $(\bar{y}, \bar{z})$ is an optimal solution of $\bar{P}^{(k)}$, then $(\bar{y}, \bar{z})$ is a Kuhn–Tucker point of $\bar{P}^{(k)}$, i.e. there are multipliers $\lambda_i \geq 0$ such that, for all j:

$$\frac{\partial \bar{w}^{(k)}}{\partial y_j} + \sum_i \lambda_i \frac{\partial \bar{g}_i^{(k)}}{\partial y_j} \quad \begin{matrix} \geq 0 \\ = 0 \\ \leq 0 \end{matrix} \quad \begin{matrix} \text{if} \\ \text{if} \\ \text{if} \end{matrix} \quad \begin{matrix} \bar{y}_j = Y_j^{\min} \\ y_j^{\min} < \bar{y} < y_j^{\max} \\ \bar{Y}_j = Y_j^{\max} \end{matrix} \tag{1}$$

$$\frac{\partial \bar{w}^{(k)}}{\partial z_j} + \sum_i \lambda_i \frac{\partial \bar{g}_i^{(k)}}{\partial z_j} \quad \begin{matrix} \geq 0 \\ = 0 \\ \leq 0 \end{matrix} \quad \begin{matrix} \text{if} \\ \text{if} \\ \text{if} \end{matrix} \quad \begin{matrix} \bar{z}_j = z_j^{\min} \\ z_j^{\min} < \bar{z}_j < z_j^{\max} \\ \bar{z}_j = z_j^{\max} \end{matrix} \tag{2}$$

$$\lambda_i \bar{g}_i^{(k)}(\bar{y}, \bar{z}) = 0 \quad \text{for all } i \tag{3}$$

where all derivatives are calculated in $(\bar{y}, \bar{z})$, and where:

$$y_j^{\min} = \max\left\{\frac{1}{2}, \frac{\alpha_j^{(k)}}{\alpha_j^{\max}}\right\}, \quad y_j^{\max} = \min\left\{2, \frac{\alpha_j^{(k)}}{\alpha_j^{\min}}\right\},$$

$$z_j^{\min} = \frac{\theta}{\theta + \delta_j^+} \quad \text{and} \quad z_j^{\max} = \frac{\theta}{\theta} + \delta_j^-.$$

δ_j^+ and δ_j^- are defined in Section 14.5.2.

Using the expressions:

$$\bar{w}^{(k)}(y, z) = w_0^{(k)} + \sum_j \frac{\partial w}{\partial \alpha_j} \frac{\alpha_j^{(k)}}{y_j} + \sum_j \frac{\partial w}{\partial \gamma_j} s_j^{(k)}\left(\frac{1}{z_j} - 1\right)$$

and

$$\bar{g}_i^{(k)}(y, z) = g_i^{(k)} - \sum_j \frac{\partial g_i}{\partial \alpha_j} \alpha_j^{(k)}(y_j - 1) - \sum_j \frac{\partial g_i}{\partial \gamma_j} s_j^{(k)}(z_j - 1)$$

Equations (1), (2) and (3) become:

$$-\frac{\partial w}{\partial \alpha_j} \frac{\alpha_j^{(k)}}{\bar{y}_j^2} - \sum_i \lambda_i \frac{\partial g_i}{\partial \alpha_j} \alpha_j^{(k)} \begin{cases} \geq 0 & \text{if } \bar{y}_j = Y_j^{\min} \\ = 0 & \text{if } y_j^{\min} < \bar{y}_j < y_j^{\max} \\ \leq 0 & \text{if } \bar{y}_j = y_j^{\max} \end{cases} \quad (4)$$

$$-\frac{\partial w}{\partial \gamma_j} \frac{s_j^{(k)}}{\bar{z}_j^2} - \sum_i \lambda_i \frac{\partial g_i}{\partial \gamma_j} s_j^{(k)} \begin{cases} \geq 0 & \text{if } \bar{z}_j = z_j^{\min} \\ = 0 & \text{if } z_j^{\min} < \bar{z}_j < z_j^{\max} \\ \leq 0 & \text{if } \bar{z}_j = z_j^{\max} \end{cases} \quad (5)$$

$$\lambda_i\left(g_i^{(k)} - \sum_j \frac{\partial g_i}{\partial \alpha_j} \alpha_j^{(k)}(\bar{y}_j - 1) - \sum_j \frac{\partial g_i}{\partial \gamma_j} s_j^{(k)}(\bar{z}_j - 1)\right) = 0 \quad (6)$$

Putting $\bar{y}_j = 1$ and $\bar{z}_j = 1$ in (4), (5) and (6), they become:

$$\frac{\partial w}{\partial \alpha_j} + \sum_i \lambda_i \frac{\partial g_i}{\partial \alpha_j} \begin{cases} \leq 0 & \text{if } \alpha_j^{(k)} = \alpha_j^{\max} \\ = 0 & \text{if } \alpha_j^{\min} < \alpha_j^{(k)} < \alpha_j^{\max} \\ \geq 0 & \text{if } \alpha_j^{(k)} = \alpha_j^{\min} \end{cases} \quad (7)$$

(e.g. $\bar{y}_j = 1 = y_j^{\min} \Leftrightarrow 1 = \max\left\{\frac{1}{2}, \frac{\alpha_j^{(k)}}{\alpha_j^{\max}}\right\} \Leftrightarrow \alpha_j^{(k)} = \alpha_j^{\max}$)

$$\frac{\partial w}{\partial \gamma_j} + \sum_i \lambda_i \frac{\partial g_i}{\partial \gamma_j} \begin{cases} \leq 0 & \text{if } \gamma_j^{(k)} = \gamma_j^{\max} \\ = 0 & \text{if } \gamma_j^{\min} < \gamma_j^{(k)} < \gamma_j^{\max} \\ \geq 0 & \text{if } \gamma_j^{(k)} = \gamma_j^{\min} \end{cases} \quad (8)$$

$$\lambda_i g_i^{(k)} = 0, \text{ i.e. } \lambda_i g_i(\alpha^{(k)}, \gamma^{(k)}) = 0 \quad (9)$$

We also know that $\bar{y}_j = 1$, $\bar{z}_j = 1$ is a feasible solution of

$$\bar{P}^{(k)} = \bar{g}_i^{(k)}(\bar{y}, \bar{z}) \leq 0 \Rightarrow g_i^{(k)} \leq 0$$

$$\Leftrightarrow g_i(\alpha^{(k)}, \gamma^{(k)}) \leq 0 \quad (10)$$

Now, (7), (8), (9) and (10) are precisely the Kuhn–Tucker conditions for the point $(\alpha^{(k)}, \gamma^{(k)})$ to be a local optimum of the original problem P.

Proof of Lemma 2

(i) By Equation (14.32),

$$\bar{d}_i(\alpha_1, \gamma_1) = \frac{\tau_1}{\tau_0} \bar{d}_i(\alpha_0, \gamma_0)$$

$$\leq \frac{\tau_1}{\tau_0} \tau_0 d_i^{\max}$$

(since $\bar{d}_i(\alpha_0, \gamma_0) \leq \tau_0 d_i^{\max}$)

$$= \tau_1 d_i^{\max}$$

i.e. (α_1, γ_1) is a *feasible* solution of $P_d(\tau_1)$.

(ii) Let (α, γ) be an arbitrary feasible solution of $P_d(\tau_1)$, i.e. $\bar{d}_i(\alpha, \gamma) \leq \tau_1 d_i^{\max}$. Then $\left(\frac{\tau_1}{\tau_0}\alpha, \gamma\right)$ is a feasible solution of $P_d(\alpha_0)$.

Since (α_0, γ_0) is an optimal solution of $P_d(\tau_0)$ we then know that

$$w(\alpha_0, \gamma_0) \leq w\left(\frac{\tau_1}{\tau_0}\alpha, \gamma\right) = \frac{\tau_1}{\tau_0} w(\alpha, \gamma)$$

We then get:

$$w(\alpha_1, \gamma_1) = w\left(\frac{\tau_0}{\tau_1}\alpha_0, \gamma_0\right) = \frac{\tau_0}{\tau_1} w(\alpha_0, \gamma_0)$$

$$\leq \frac{\tau_0}{\tau_1} \frac{\tau_1}{\tau_0} w(\alpha, \gamma) = w(\alpha, \gamma).$$

Thus (α_1, γ_1) is an optimal solution of $P_d(\tau_1)$.

Proof of Lemma 4

(i) By Equation (14.35),

$$h_i(\alpha_1, \gamma_1) = \frac{\tau_1}{\tau_0} h_i(\alpha_0, \gamma_0)$$

$$\leq \frac{\tau_1}{\tau_0} \tau_0 g_i(\gamma_0) = \tau_1 g_i(\gamma_0) = \tau_1 g(\gamma_1)$$

i.e. (α_1, γ_1) is a feasible solution of $P_b(\tau_1)$.

(ii) Let (α, γ) be an arbitrary feasible solution of $P_b(\tau_1)$, i.e. $h_i(\alpha, \gamma) \leq \tau_1 g_i(\gamma)$.

Then $\left(\sqrt{\dfrac{\tau_1}{\tau_0}}\alpha, \gamma\right)$ is a feasible solution of $P_b(\tau_0)$ (due to (14.35)).

Since (α_0, γ_0) is an optimal solution of $P_b(\tau_0)$ we then know that:

$$w(\alpha_0, \gamma_0) \leq w\left(\sqrt{\dfrac{\tau_1}{\tau_0}}\alpha, \gamma\right) = \sqrt{\dfrac{\tau_1}{\tau_0}}w(\alpha, \gamma)$$

We then get:

$$w(\alpha_1, \gamma_1) = w\left(\sqrt{\dfrac{\tau_0}{\tau_1}}\alpha_0, \gamma_0\right) = \sqrt{\dfrac{\tau_0}{\tau_1}}w(\alpha_0, \gamma_0)$$

$$\leq \sqrt{\dfrac{\tau_0}{\tau_1}}\sqrt{\dfrac{\tau_1}{\tau_0}}w(\alpha, \gamma) = w(\alpha, \gamma)$$

Thus (α_1, γ_1) is an optimal solution of $P_b(\tau_1)$.

K.S.

REFERENCES

1. K. Svanberg, Optimum Structural design using duality, *TRITA-MAT-1979-6*, Royal Institute of Technology, Stockholm, 1959.
2. K. Svanberg, An algorithm for optimum structural design using duality (paper presented at the Tenth International Symposium on Mathematical Programming, Montreal, August 1979), UTM-Report no. 43, Royal Institute of Technology, Stockholm, 1979.
3. C. Fleury, A unified approach to structural weight minimization, *Comp. Meth. in Appl. Mech. and Eng.*, **30**, 17–38, 1979.
4. C. Fleury and M. Geradin, Optimality criteria and mathematical programming in structural weight optimization, *J. Computer and Structures*, **8**, 7–17, 1978.
5. R. A. Gellatly and L. Berke, Optimality-criterion-based algorithms, in Gallagher and Zienkiewicz (Eds), *Optimum Structural Design, Theory and Applications*, John Wiley & Sons, 1973.
6. A. B. Templeman, A dual approach to optimum truss design, *J. Struct. Mech.* **4**, 235–55, 1976.
7. C. Fleury, Structural weight optimization by dual methods of convex programming, *J. Num. Meth. in Engineering*, **14**, 1761–83, 1979.
8. O. C. Zienkiewicz and J. S. Campbell, 'Shape optimization and sequential linear programming', in Gallagher and Zienkiewicz (Eds), *Optimum Structural Design, Theory and Applications*, John Wiley & Sons, 1973.
9. G. Sander and C. Fleury, A mixed method in structural optimization, *J. Num. Meth. in Engineering*, **13**, 385–404, 1978.
10. A. M. Geoffrion, Duality in nonlinear programming: a simplified application-oriented development, *SIAN Review*, **13**(1), 1971.

Foundations of Structural Optimization: A Unified Approach
Edited by A. J. Morris
© 1982 John Wiley & Sons Ltd.

Chapter 15
Shape Optimal Design of Elastic Structural Elements†

15.1 THE SHAPE OPTIMIZATION PROBLEM

Most optimal structural design problems involve design variables that are either of the form of a vector of design parameters or a function defining a dimension of a two- or three-dimensional body. The design variable then appears in the state equations, cost functional, and constraints. Most optimization methods are thus oriented toward determining design variables that define the structure under consideration.

In many design problems, particularly machine and structural elements, the shape of a two- or three-dimensional solid plays the role of design. For structures in which the deformation field is defined by equations of elasticity over the domain, the effect of a domain shape variation on the displacement field is not easily determined. Even if the domain is defined by a function that locates its boundary, the influence of a shape variation on the displacement field must be determined.

The idea of material derivative of continuum mechanics is used in Section 15.3 to develop an explicit formula for variation of functionals in terms of domain shape change. This result is used in Sections 15.4 and 15.5 for solution of a problem of maximum torsional stiffness of a shaft.

15.2 SHAPE OPTIMAL DESIGN

15.2.1 Numerical Methods for Shape Optimization

One of the first treatments of the general problem of selection of shape of a structure as the design variable is presented by Zienkiewicz and Campbell [1]. They formulate the shape optimal design problem using a finite element model

---

† Research supported by US National Science Foundation Project ENG 77–19967.

and treat the location of the nodal points of the finite element model as design variables. They then calculate derivatives of stiffness and load matrices with respect to design parameters, obtain derivatives of structural response measures, and employ sequential linear programming for numerical solution. They present examples associated with dams and rotating turbine machinery. Ramakrishnan and Francavilla [2] employ a similar finite element formulation, but they use a penalty function method for numerical optimization. Francavilla, Ramakrishnan, and Zienkiewicz [3] employ the finite element method of Refs [1, 2] for fillet optimization in order to minimize stress concentration.

More basic approaches for surface contouring to minimize stress concentration are presented by Tvergaard, in selecting the optimum shape of a fillet [4]. He uses a stress field model of the fillet with a finite-dimensional family of perturbations allowed in the boundary shape, defined in terms of coordinate parameters. He also employs a variational analysis of the stress field equations to obtain derivatives of stress with respect to the design variables and uses sequential linear programming to iteratively construct an optimum design. Kristensen and Madsen [5] formulate a class of shape optimal design problems for planar solids which generalizes the approach of Tvergaard [4]. They use orthogonal polynomials to locate the boundary of the body and treat the coefficients in these polynomials as design parameters. A finite element approach is employed to model structural response and obtain derivatives of stress with respect to design variables, and employs sequential linear programming to solve the optimization problem. The elementary problem of the optimum shape of a hole in a biaxial stress field is solved analytically and the method is illustrated numerically on more complex problems. Bhavikatti and Ramakrishnan [6] present a refinement of the formulation of Refs [1, 2, and 3] for optimum design of fillets in flat and round tension bars. A polynomial is used with design variables as coefficients to characterize the shape of the fillet and a finite element model is then employed to calculate stress in the body. Minimization of the stress concentration factor is investigated, as well as, minimum volume design, and design for uniform stress distribution along the fillet boundary as optimality criterion. Derivatives of response measures with respect to design parameters are calculated using a finite element model, and sequential linear programming is employed for numerical solution.

A function-space gradient projection method of optimal design of the shape of two-dimensional elastic bodies has recently been presented by Chun and Haug [7, 8], using design sensitivity analysis methods similar to those presented by Rousselet and Haug [9] and a gradient projection method for iterative optimization. The design objective in this work is weight minimization, with constraints on Von Mises yield stress and shear stress distribution on the boundary.

15.2.2 Shape of Cross-section of Shafts in Torsion

The problem of optimization of cross-sectional shape of torsion members was first addressed by Henry [10]. He develops an analytical method for location of the boundary in terms of a small number of parameters and iteratively selects these parameters to minimize weight, subject to constraints on cross-sectional geometry and boundary stress. He treats the problem of a shaft with grooves required for keyways.

Banichuk [11, 12] formulates a general problem of selecting the optimum shape of cross-section for a non-homogeneous shaft to maximize torsional stiffness with a given amount of material available. He uses the fact that the functional minimized by the warping potential in a variational formulation of the boundary-value problem is proportional to the torsional stiffness of the shaft. He then takes variations of this functional with respect to both the warping potential and boundary variation, using the material-derivative concept of continuum mechanics, and obtains a necessary condition for optimum location of the boundary. He treats both simply connected and multiply connected cross-sections. Kurshin and Onoprienko [13] treat the same problem of maximum torsional stiffness of a shaft with a doubly connected cross-section, using a complex-variable method to determine the optimum boundary.

Banichuk [14] subsequently presents an extension of the torsional stiffness maximization problem for rods, using optimal distribution of a given amount of stiffening material around the boundary. The method he employs is a direct extension of that used in Refs [11 and 12]. Turvitch [15] presents an alternate analytical technique for optimizing the shape of an interior boundary that is associated with inhomogeneity in the material, using a coordinate system associated with the warping function and obtaining necessary and sufficient conditions of optimality.

15.2.3 Shapes of Holes in Planar Solids

Neuber [16, 17] and Cherpanov [18] discuss the problem of finding a hole shape in a planar solid to make tangential normal stress acting on the boundary constant, under the assumption that this is a condition of optimality. Cherpanov [18] cites considerable earlier Soviet literature addressing the same design objective. Wheeler [19] investigates conditions under which constant tangential normal stress along the boundary of a hole, fillet, or notch is a valid optimality criterion for minimum peak stress. He develops criteria for axisymmetric torsion problems and plane problems involving holes, notches, and fillets.

Banichuk [20] formulates the problem of selecting hole shape in an infinite plane body that is in biaxial tension of infinity to minimize peak stress in the vicinity of the hole. Using the maximum principle for harmonic functions, he proves that the optimum hole shape leads to constant tangential normal stress around the boundary of the hole, hence proving the optimality criterion employed in Refs [16–18] for this class of problems. In a related paper [21], Banichuk treats the problem of finding the optimum hole shape to minimize the maximum value of the second invariant of the stress tensor deviator over an entire plate, which is subjected to uniform bending at infinity. Again using the maximum principle for harmonic functions, and an intricate argument, he shows that the maximum stress occurs at the hole boundary and that for certain classes of problems the hole boundary is uniformly stressed.

15.2.4 Miscellaneous Shape Optimal Design Problems

Banichuk and Karihaloo [22] seek the shape of the cross-section of a cylindrical bar to minimize weight, subject to constraints on torsional stiffness and bending stiffness. A variational formulation of the torsion problem and a Lagrange multiplier technique are used to adjoin constraints to the cost function. Using a material-derivative calculation, the first variation of the augmented cost function is taken with respect to shape, and optimality criteria are derived. Parbery and Karihaloo [23] use the same method to optimize hollow cylinders, with constraints on torsional and bending stiffness.

Cherkaev [24] presents a theoretical treatment of the problem of boundary shape selection to minimize the volume of a structure, subject to a lower bound constraint on natural frequency. He develops a general necessary condition and shows that a variational formulation of Prager can be applied to obtain the same result. As a final note, Durelli and Rajaiah [25] present an experimental method using photo-elasticity to find the optimum shape of a hole in a flat plate, under uni-axial load, to minimize stress concentration.

15.2.5 Related Literature of Domain Optimization

Cea, Zolesio, and Rousselet [26–33] present techniques and applications, from fields other than structural optimization, for selecting optimum domain. They cite substantial literature in this general field and give examples, from diverse engineering disciplines, that hold potential for optimality criteria and direct numerical methods for optimization of the shape of structures.

15.3 ANALYSIS OF THE EFFECT OF SHAPE VARIATION

Consider a domain Ω in two or three dimensions, as shown schematically in Fig. 15.1. It is presumed that the boundary Γ of Ω is smooth enough so that an outward unit normal vector **n** is defined. One method of defining a variation in the domain Ω is to let $\mathbf{V}(X)$ be a vector field of deformation, for $X \in \Omega$. A one-parameter family of perturbed domains may then be defined by the mapping

$$\mathbf{x} = \mathbf{X} + t\mathbf{V}(X) \qquad X \in \Omega, \, t \in R^1 \tag{15.1}$$

Shape Optional Design of Elastic Structural Elements

As shown in Fig. 15.1, for t small, the deformed domain Ω_t has its boundary near Γ; in fact, $\Omega = \Omega_0$.

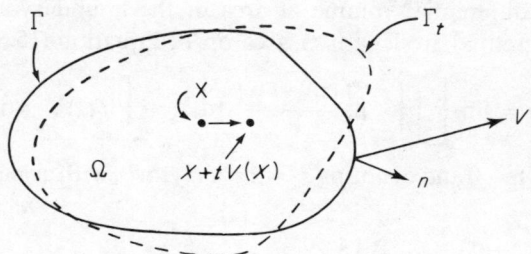

Fig. 15.1 Variation of domain

The process of deforming Ω to Ω_t may be viewed [26, 27, 28] as a dynamic process of deforming a continuum, with t playing the role of time and $\mathbf{V}(X)$ playing the role of a 'velocity field'. With this viewpoint, the idea of the material derivative of continuum mechanics may be used to advantage.

Let z_t be the solution of the boundary-value problem

$$Az_t = q \quad \text{in } \Omega_t \tag{15.2}$$

$$Bz_t = 0 \quad \text{on } \Gamma_t \tag{15.3}$$

where q is defined everywhere in a neighbourhood of Ω. It is presumed that the operator A with boundary conditions of Equation (15.3) is symmetric, i.e.

$$\iint_\Omega \eta A v \, d\Omega = \iint_\Omega v A \eta \, d\Omega$$

for all η and v satisfying Equation (15.3).

Consider now a functional

$$\psi = \iint_\Omega f(z) \, d\Omega \tag{15.4}$$

which depends on Ω in two ways. First, there is an obvious dependence on Ω through the process of integration. Second, and more subtle, the function z depends on Ω through the boundary-value problem of Equations (15.2) and (15.3). If Ω is perturbed, as in Equation (15.1),

$$\psi(t) = \iint_{\Omega_t} f(z_t) \, d\Omega_t \tag{15.5}$$

The variation of ψ is thus

$$\delta\psi = \left.\frac{d\psi(t)}{dt}\right|_{t=0} = \lim_{t \to 0} \frac{1}{t}\left[\iint_{\Omega_t} f(z_t) \, d\Omega_t - \iint_\Omega f(z) \, d\Omega\right] \tag{15.6}$$

As in the usual derivation of the material derivative of continuum mechanics, let Ω_* be the intersection of Ω and Ω_t. For t small, one can write two integrals in Equation (15.6) as sums of integral over Ω_* and over a boundary strip. Approximating differential volume or area of the boundary strip as $t\mathbf{V}\cdot\mathbf{n}\,ds$, where ds is differential arc length or area on Γ, Equation (15.6) becomes

$$\delta\psi = \lim_{t\to 0}\left[\iint_{\Omega_*}\left(\frac{f(z_t)-f(z)}{t}\right)d\Omega_* + \int_\Gamma f(z)\mathbf{V}\cdot\mathbf{n}\,ds\right] \quad (15.7)$$

Since $\Omega_* \to \Omega$ as $t \to 0$ and denoting $z' = \partial z_t/\partial t$, (15.7), with arguments of z_t and z the same in Ω_*,

$$\delta\psi = \iint_\Omega \frac{\partial f}{\partial z}z'\,d\Omega + \int_\Gamma f\mathbf{v}\cdot\mathbf{n}\,ds \quad (15.8)$$

denoting

$$\dot{d} = \frac{d}{dt}(t\mathbf{v}\cdot\mathbf{n}) = \mathbf{v}\cdot\mathbf{n} \quad (15.9)$$

as the rate of normal movement of the boundary Γ, (15.8) can finally be written as

$$\delta\psi = \iint_\Omega \left[\frac{\partial f}{\partial t}\dot{z} - \frac{\partial f}{\partial z}(\nabla z\cdot\mathbf{v})\right]d\Omega + \int_\Gamma f\dot{d}\,ds \quad (15.10)$$

where $\dot{z} = z' + \nabla z \cdot \mathbf{v}$ is the variation in the solution of (15.2) and (15.3), due to the normal variation $\dot{d}$ in the boundary Γ.

It is important to note that even though the 'velocity field' $\mathbf{V}(X)$ was used in (15.1) to define perturbation of the domain, it does not arise explicitly in the result of (15.10). That is, any field $\mathbf{V}$ that gives the same normal movement $\dot{d}$ at the boundary Γ has the same effect on ψ.

Recall that $\dot{z}$ is determined by the boundary-value problem of (15.2) and (15.3) for the perturbed and unperturbed domains. Thus, $\dot{z}$ is determined by $\dot{d}$, but in a complicated way. One can, however, use an adjoint variable method [34] to write $\delta\psi$ of (15.10) explicitly in terms of $\dot{d}$. To do so, first write the variational form of Equations (15.2) and (15.3) using symmetry of A, as

$$\iint_{\Omega_t} c(z_t,\lambda)\,d\Omega_t = \iint_{\Omega_t} \lambda q\,d\Omega_t \quad (15.11)$$

for all λ satisfying (15.3), where $c(z_t,\lambda)$ is a symmetric, lowest order bilinear form in z_t and λ such that

$$\iint_{\Omega_t}(Az_t)\lambda\,d\Omega_t = \iint_{\Omega_t} c(z_t,\lambda)\,d\Omega_t \quad (15.12)$$

It is shown in Ref. [9] that the right side of (15.12) is differentiable, so one can apply the result of (15.10) to both sides of (15.11), to obtain the identity

$$\iint_\Omega [c(z',\lambda)+c(z,\lambda')]\,d\Omega + \int_\Gamma c(z,\lambda)\dot{d}\,ds = \iint_\Omega \lambda'q\,d\Omega + \int_\Gamma \lambda q\dot{d}\,ds \quad (15.13)$$

Since $Bz_t = 0$ on Γ_t, $B\dot{z} = 0$ on Γ_t, and $\dot{\lambda} = \lambda' + \nabla\lambda \cdot \mathbf{v} = 0$ [29], (15.12) at $t = 0$ allows one to rewrite the first term of (15.13), yielding

$$\iint_\Omega \dot{z} A\lambda \, d\Omega = \int_\Gamma [\lambda q - c(z, \lambda)] \dot{d} \, ds + \iint_\Omega [c(\nabla z \cdot \mathbf{v}, \lambda) + \lambda' q + c(z, \nabla\lambda \cdot \mathbf{v})] \, d\Omega \tag{15.14}$$

Choosing the adjoint variable λ to satisfy

$$A\lambda = \frac{df}{dz} \quad \text{in } \Omega \tag{15.15}$$

$$B\lambda = 0 \quad \text{on } \Gamma$$

(15.10) may be written, using (15.14) and (15.15), as

$$\delta\psi = \int_\Gamma [f + \lambda q - c(z, \lambda)] \dot{d} \, ds + \iint_\Omega \bigg[c(\nabla z \cdot \mathbf{v}, \lambda) + c(z, \nabla\lambda \cdot \mathbf{v}) \\ - q\nabla\lambda \cdot \mathbf{v} - \frac{\partial f}{\partial z}(\nabla z \cdot \mathbf{v}) \bigg] d\Omega \tag{15.16}$$

where the last term can be transformed to a boundary integral in specific cases [29].

Thus, by solving the original boundary-value problem of (15.2) and (15.3) for z and the adjoint boundary-value problem of (15.15) for λ, (15.16) gives an explicit formula for the variation of the functional ψ of (15.4), in terms of normal boundary variation $\dot{d}$. This result can now be used in optimizing the domain Ω.

15.4 MAXIMUM TORSIONAL STIFFNESS OF A SHAFT

Consider a cylindrical shaft with cross-sectional shape shown in Fig. 15.2. The objective is to find the shape of Ω of given area A to maximize torsional stiffness. From the St. Venant theory of torsion, the torsional stiffness is

$$\psi_0 = 2 \iint_\Omega z \, d\Omega \tag{15.17}$$

where

$$-\nabla^2 z = 2 \quad \text{in } \Omega \\ z = 0 \quad \text{on } \Gamma \tag{15.18}$$

The constraint of given area is just

$$\psi_1 = \iint_\Omega d\Omega = A \tag{15.19}$$

Carrying out the integration by parts called for in (15.12), one has

$$\iint_\Omega (-\nabla^2 z)\lambda \, d\Omega = \iint_\Omega \nabla z \cdot \nabla\lambda \, d\Omega \tag{15.20}$$

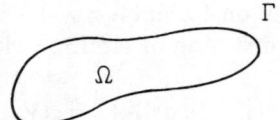

Fig. 15.2 Cross-section of shaft

so
$$c(z, \lambda) = \nabla z \cdot \nabla \lambda = z_{x_1} \lambda_{x_1} + z_{x_2} \lambda_{x_2} \tag{15.21}$$

Thus, since $z = \lambda = 0$ on Γ, (15.16) gives

$$\delta \psi_0 = \int_\Gamma (-\nabla z \cdot \nabla \lambda) d \, ds + \iint_\Omega [\nabla(\nabla z \cdot \mathbf{v}) \cdot \nabla \lambda + \nabla z \cdot \nabla(\nabla \lambda \cdot \mathbf{v}) \\ - z \nabla \lambda \cdot \mathbf{v} - z \nabla z \cdot \mathbf{v}] d\Omega \tag{15.22}$$

where, from (15.15), λ satisfies

$$\begin{aligned} -\nabla^2 \lambda &= 2 & \text{in } \Omega \\ \lambda &= 0 & \text{on } \Gamma \end{aligned} \tag{15.23}$$

But (15.18) and (15.23) imply $\lambda = z$, so (15.22) becomes simply

$$\delta \psi_0 = -\int_\Gamma (\nabla z \cdot \nabla z) d \, ds + \iint_\Omega [2\nabla(\nabla z \cdot \mathbf{v}) \cdot \nabla z - 4\nabla z \cdot \mathbf{v}] d\Omega$$

Integrating by parts and noting $-\nabla^2 z = 2$, this is

$$\delta \psi_0 = \int_\Gamma [-\nabla z \cdot \nabla z \, d + 2(\nabla z \cdot \mathbf{v})(\nabla z \cdot \mathbf{n})] ds$$

Finally, since $z = 0$ on Γ, $\nabla z = \dfrac{\partial z}{\partial n} n$ on Γ and one has

$$\begin{aligned} \delta \psi_0 &= \int_\Gamma \left(\frac{\partial z}{\partial n} \right)^2 d \, ds \\ &= \int_\Gamma \nabla z \cdot \nabla z \, d \, ds \end{aligned} \tag{15.24}$$

Since the integrand of ψ_1 in (15.19) does not depend on z, one has simply

$$\delta \psi_1 = \int_\Gamma d \, ds \tag{15.25}$$

A necessary condition of optimality for this problem is that $\delta \psi_0 = 0$ for all d such that $\delta \psi_1 = 0$. Thus, there exists a Lagrange multiplier v such that

$$\int_\Gamma [(\nabla z \cdot \nabla z) + v] d \, ds = 0 \tag{15.26}$$

for arbitrary $\dot{d}$ on Γ. This requires that

$$\nabla z \cdot \nabla z + v = 0 \tag{15.27}$$

on the optimum boundary.

One can now attempt to find a curve Γ enclosing area A (15.19), for which the solution z of (15.18) satisfies (15.27). In general, such free boundary problems are not easy to solve. A direct optimization method is thus used in the following section to solve this problem.

15.5 ITERATIVE SHAPE OPTIMAL DESIGN OF A SHAFT

The shaft shape optimal design problem of Section 15.4 is now solved, using a gradient projection numerical method. While the problem is specific, the method is quite general and may be used to solve rather broad classes of optimal design problems.

For the shaft, one wishes to find a direction $\dot{d}$ that increases ψ_0 as much as possible; i.e. maximizes $\delta\psi_0$ of Equation (15.24), subject to the condition $\delta\psi_1 = A - \psi_1$. This latter condition corrects any error that may be present in the constraint of Equation (15.19). To find the desired direction $\dot{d}$, first normalize $\dot{d}$ by

$$\int_\Gamma (\dot{d})^2 \, ds = 1 \tag{15.28}$$

A necessary condition satisfied by $\dot{d}$ is then that $dL/d\dot{d} = 0$, where the constraints of Equations (15.25) and (15.28) are incorporated in the Lagrangian.

$$L = -\nabla z \cdot \nabla z \dot{d} + \mu \dot{d} + \gamma \dot{d}^2 \tag{15.29}$$

That is,

$$-\nabla z \cdot \nabla z + \mu + 2\gamma \dot{d} = 0$$

so

$$\dot{d} = \frac{1}{2\gamma} \nabla z \cdot \nabla z - \frac{\mu}{2\gamma} \tag{15.30}$$

But, to satisfy the condition $\delta\psi_1 = A - \psi_1$,

$$\frac{1}{2\gamma}\left[\int_\Gamma \nabla z \cdot \nabla z \, ds - \mu l \right] = A - \psi_1$$

where

$$l = \int_\Gamma ds \tag{15.31}$$

Thus

$$\mu = \frac{1}{l}\left\{ -2\gamma(A - \psi_1) + \int_\Gamma \nabla z \cdot \nabla z \, ds \right\} \tag{15.32}$$

From Equation (15.30),

$$\dot{d} = \frac{1}{2\gamma}\left[\nabla z \cdot \nabla z - \frac{1}{l}\int_{\Gamma} \nabla z \cdot \nabla z \, ds\right] + \frac{1}{l}(A - \psi_1) \qquad (15.33)$$

The first term in Equation (15.33) is the negative projected gradient of ψ_0 onto the constraint boundary of Equation (15.19) [34]. In order to increase ψ_0, $\gamma < 0$ must be chosen [34] as a step size.

The second term in Equation (15.33) provides the needed constraint error correction. One may now do a one-dimensional search or use some other method to choose the step size γ in Equation (15.33) to determine the desired normal movement $\dot{d}$ of the boundary. This calculation is repeated until the optimum design is reached.

At each iteration of this optimization algorithm, the boundary-value problem of Equation (15.18) must be solved on the current estimate Ω of the optimum domain. A finite element model of the domain with triangular elements, shown in Fig. 15.3, is used for this purpose, with regridding to account for shape modification carried out as follows:

(i) the movement of each boundary node is approximated to be the average of those of adjacent boundary elements;
(ii) the direction of movement of each boundary node coincides with the direction of a diagonal line that connects nodal point 1 and the boundary node;
(iii) new locations of other nodal points are calculated simply by dividing the diagonal lines into equal length segments.

For iterative optimization, the initial design of shaft (square cross-section), its finite element model, and a schematic drawing of boundary nodal point movements are as shown in Fig. 15.3. A sequence of changed boundaries of the shaft as iterations proceed is shown in Fig. 15.4. A plot of cost function vs. design iteration is shown in Fig. 15.5. The total CPU time on an IBM 370-168 computer for all calculations done in executing ten iterations was 5.6 sec.

E.J.H.

Shape Optional Design of Elastic Structural Elements

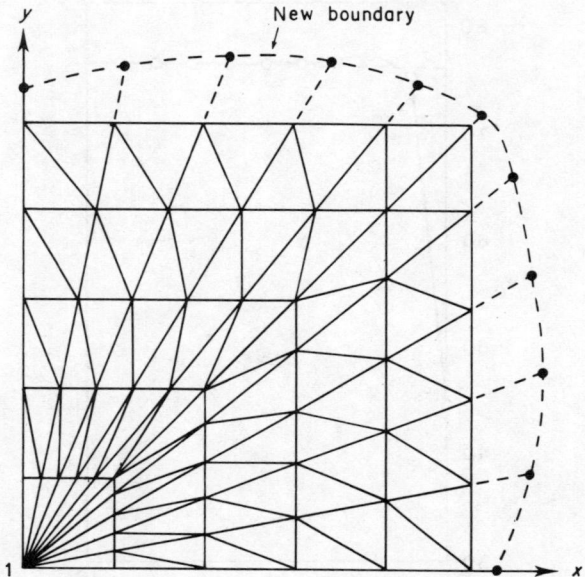

Fig. 15.3 Finite element model of initial design of shaft

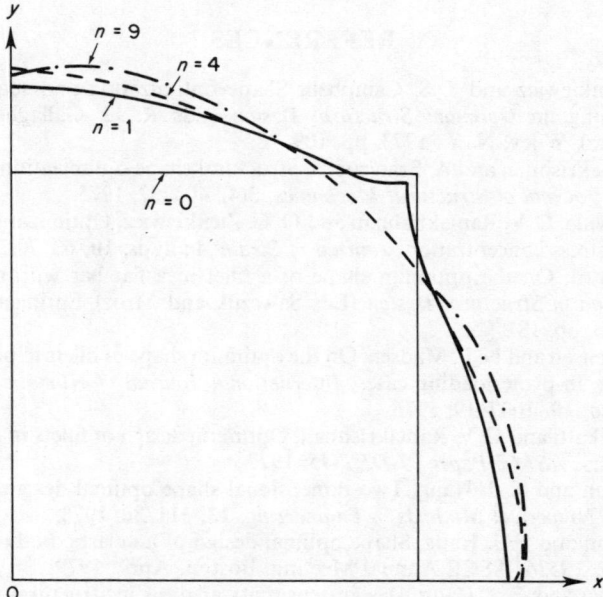

Fig. 15.4 Change of boundary after iterations

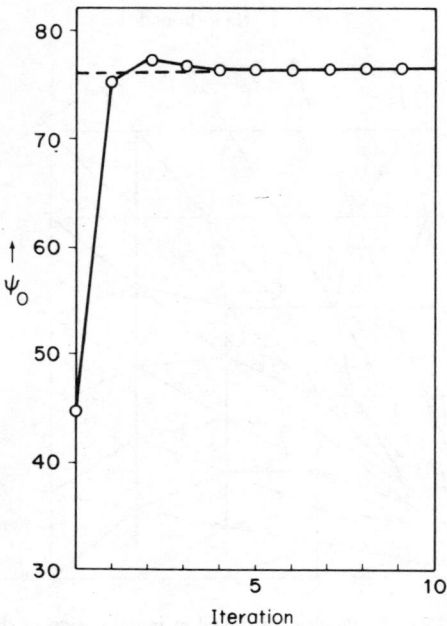

Fig. 15.5 Iteration versus torsional constraint of the shaft

REFERENCES

1. O. C. Zienkiewicz and J. S. Campbell, Shape optimization and sequential linear programming, in *Optimum Structural Design* (Eds R. H. Gallagher and O. C. Zienkiewicz), Wiley, N. Y., 1973, pp. 109–26.
2. C. V. Ramakrishnan and A. Francavilla, Structural shape optimization using penalty functions, *Journal of Structural Mechanics*, **3**(4), 403–32, 1975.
3. A. Francavilla, C. V. Ramakrishnan and O. C. Zienkiewicz, Optimization of shape to minimize stress concentration, *Journal of Strain Analysis*, **10**, 63–70, 1975.
4. V. Tvergaard, On the optimum shape of a fillet in a flat bar with restrictions, in *Optimization in Structural Design* (Eds Sawczuk and Mroz), Springer-Verlag, New York, 1975, pp. 181–95.
5. E. S. Kristensen and N. F. Madsen, On the optimum shape of fillets in plates subjected to multiple in-plane loading cases, *International Journal of Numerical Methods in Engineering*, **10**, 1007–19, 1976.
6. S. S. Bhavikatti and C. V. Ramakrishnan, Optimum design of fillets in flat and round tension bars, *ASME Paper 77-DET-45*, 1977
7. Y. W. Chun and E. J. Haug, Two dimensional shape optimal design, *International Journal of Numerical Methods in Engineering*, **13**, 311–36, 1978.
8. Y. W. Chun and E. J. Haug, Shape optimal design of an elastic body of revolution, *Preprint No. 3516*, ASCE Annual Meeting, Boston, April, 1979.
9. B. Rousselet and E. J. Haug, Design sensitivity analysis in structural mechanics III: Shape variation, in *Optimization of Distributed Parameter Structures* (Eds E. J. Haug and J. Cea), Sijthoff-Noordhoff, Alphen aan den Rijn, to appear 1981.

10. A. S. Henry, The analytic design of torsion members, PhD Thesis, University of Iowa, 1971.
11. N. V. Banichuk, Optimization of elastic bars in torsion, *International Journal of Solids and Structures*, **12**, 275–86, 1976.
12. N. V. Banichuk, On a variational problem with unknown boundaries and the determination of optimal shapes of elastic bodies, *PMM*, **39**(6), 1037–47, 1975.
13. L. M. Kurshin and P. M. Onoprienko, Determination of the shapes of doubly-connected bar sections of maximum torsional stiffness, *PMM*, **40**(6), 1020–6, 1976.
14. N. V. Banichuk, On a two dimensional optimization problem in elastic bar torsion theory, *Soviet Applied Mechanics*, **11**(5), 38–44, 1976.
15. E. L. Turvitch, On isoparimetric problems for domains with partly known boundaries, *Journal of Optimization Theory and Applications*, **20**(1), 65–79, 1976.
16. H. Neuber, Der zugbeanspruchte flachstabl mit optimalem querschnittubergang, *Forsch Ingenieurwesen*, **35**, 29–30, 1969.
17. H. Neuber, Zur optimierung der spannungskonzentration, *Continuum Mechanics and Related Problems of Analysis*, Nauka, Moscow, 1972, pp. 375–80.
18. G. P. Cherpanov, Inverse problems of the plane theory of elasticity, *PMM*, **38**(6), 963–79, 1974.
19. L. Wheeler, On the role of constant-stress surfaces in the problem of minimizing elastic stress concentration, *International Journal of Solids and Structures*, **12**, 779–89, 1976.
20. N. V. Banichuk, Optimality conditions in the problem of seeking the hole shapes in elastic bodies, *PMM*, **41**(5), pp. 920–5, 1977.
21. N. V. Banichuk, Optimizing mole shape in plates working in bending, *Soviet Applied Mechanics*, **12**(3), 72–8, 1977.
22. N. V. Banichuk and B. L. Karihaloo, Minimum-weight design of multi-purpose cylindrical bars, *International Journal of Solids and Structures*, **12**, 267–73, 1976.
23. R. D. Parbery and B. L. Karihaloo, Minimum-weight design of hollow cylinders for given lower bounds on torsional and flexural rigidities, *International Journal of Solids and Structures*, **13**, 1271–80, 1977.
24. A. V. Cherkaev, On the question of formulating the problem of optimal design of freely oscillating structures, *PMM*, **42**(1), 194–7, 1978.
25. A. J. Durelli and K. Rajaiah, Optimum hole shapes in finite plates under uniaxial load, *Journal of Applied Mechanics*, **46**, 691–5, 1979.
26. J. Cea, Solution of a model problem by variational methods and examples of problems of shape optimal design, in *Optimization of Distributed Parameter Structures* (Eds E. J. Haug and J. Cea), Sijthoff-Noordhoff, Alphen aan den Rijn, to appear 1981.
27. J. Cea, Definition of boundaries for shape design, in *Optimization of Distributed Parameter Structures* (Eds E. J. Haug and J. Cea), Sijthoff-Noordhoff, Alphen aan den Rijn, to appear 1981.
28. J. Cea, continuous steepest descent in Nilbert space and 'romain spaces', in *Optimization of Distributed Parameter Structures* (eds E. J. Haug and J. Cea), Sijthoff-Noordhoff, Alphen aan den Rijn, to appear 1981.
29. J. P. Zolesio, The material derivative (or speed) method for shape optimization, in *Optimization of Distributed Parameter Structures* (Eds E. J. Haug and J. Cea), Sijthoff-Noordhoff, Alphen aan den Rijn, to appear 1981.
30. J. P. Zolesio, Speed method in several examples, in *Optimization of Distributed Parameter Structures* (Eds E. J. Haug and J. Cea), Sijthoff-Noordhoff, Alphen aan den Rijn, to appear 1981.
31. B. Rousselet, Implementation of shape optimal design algorithms, in *Optimization of Distributed Parameter Structures* (Eds E. J. Haug and J. Cea), Sijthoff-Noordhoff, Alphen aan den Rijn, to appear 1981.

32. J. Cea, Other methods in shape optimal design, in *Optimization of Distributed Parameter Structures* (Eds E. J. Haug and J. Cea), Sijthoff-Noordhoff, Alphen aan den Rijn, to appear 1981.
33. B. Rousselet, Dependence of eigenvalues on shape, in *Optimization of Distributed Parameter Structures* (Eds E. J. Haug and J. Cea), Sijthoff-Noordhoff, Alphen aan den Rijn, to appear 1981.
34. E. J. Haug and J. S. Arora, *Applied Optimal Design*, Wiley–Interscience, New York, 1979.

Foundations of Standard Optimization: A Unified Approach
Edited by A. J. Morris
© 1982 John Wiley & Sons Ltd.

Chapter 16

Optimization of Structures in which Repeated Eigenvalues Occur[†]

16.1 INTRODUCTION

The purpose of this chapter is to formulate and analyse structural optimization problems in which multiple eigenvalues may occur. It has recently been shown by Olhoff and Rasmussen [1] and Masur and Mroz [2] that in a certain clamped column optimized for maximum buckling load, a repeated eigenvalue may occur. In a recent paper [3], Prager and Prager have shown that repeated eigenvalues can be predicted for an optimal column, using a finite dimensional mode. In this chapter, elementary examples are presented to show that for some values of defining parameters for problems, repeated eigenvalues occur at an optimum design, and for other values only simple eigenvalues occur. It is therefore not clear *a priori* whether repeated eigenvalues will occur in a specific problem.

A Lagrange multiplier method is applied to an example to show that the formal Lagrange multiplier method is not valid in general, when repeated eigenvalues occur. Recent results concerning differentiability of eigenvalues of structures presented in Refs [4] and [5] and vector space optimization theory [6] are used to develop valid necessary conditions of optimality.

16.2 EXAMPLES OF STRUCTURAL OPTIMIZATION PROBLEMS WITH MULTIPLE EIGENVALUES

The purpose of this section is to formulate and analyse elementary examples of optimization problems that exhibit multiple eigenvalues. Some of the examples are simple enough that they can be solved in closed form. These examples are used in later sections to test the methods and theories developed.

[†] Research supported by US National Science Foundation Project ENG 77–19967.

16.2.1 A Simple Spring–Mass Optimal Design Problem with Repeated Natural Frequencies

Consider the spring–mass system shown in Fig. 16.1. The eigenvalue problem for small-amplitude vibration of the rigid body is simply derived as

$$\begin{bmatrix} 4k_1 + k_2 & k_2 \\ k_2 & 4k_1 + k_2 \end{bmatrix} \mathbf{y} = \zeta \begin{bmatrix} 2 & 1 \\ 1 & 2 \end{bmatrix} \mathbf{y} \tag{16.1}$$

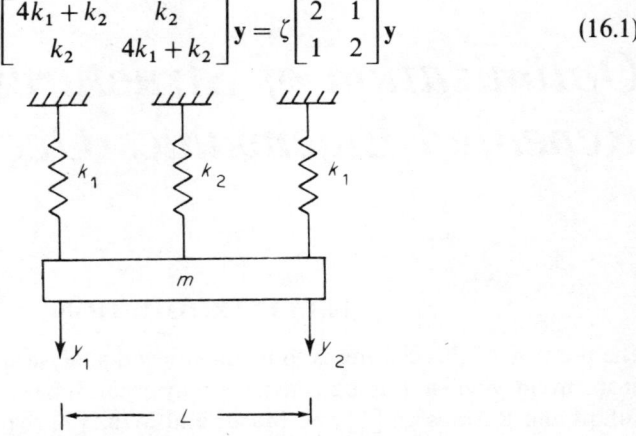

Fig. 16.1 Two degree of freedom spring–mass system

where $\zeta = 2\omega^2 m/3$, m is the mass of the bar, $I = mL^2/12$ is the polar moment of inertia, and horizontal motion of the bar is ignored.

The optimal design objective is to find design variables k_1 and k_2 to minimize weight of the spring supports, which is presumed to be of the form

$$\psi_0 = c_1 k_1 + c_2 k_2 \tag{16.2}$$

where c_1 and c_2 are known constants, subject to constraints that the eigenvalues be no lower that $\zeta_0 > 0$ and the spring constants are non-negative. In inequality constraint form, this is

$$\left. \begin{aligned} \psi_1 &= \zeta_0 - \zeta_1 \leqslant 0 \\ \psi_2 &= \zeta_0 - \zeta_2 \leqslant 0 \\ \psi_3 &= -k_1 \leqslant 0 \\ \psi_4 &= -k_2 \leqslant 0 \end{aligned} \right\} \tag{16.3}$$

The eigenvalues of Equation (16.1) are $\zeta_1 = (4k_1 + 2k_2)/3$ and $\zeta_2 = 4k_1$, which gives $\omega_1^2 = (2k_1 + k_2)/m$ and $\omega_2^2 = 6k_1/m$. Thus, the constraints of Equations (16.3) become

$$\left. \begin{aligned} \psi_1 &= \zeta_0 - \frac{4k_1 + 2k_2}{3} \leqslant 0 \\ \psi_2 &= \zeta_0 - 4k_1 \leqslant 0 \\ \psi_3 &= -k_1 \leqslant 0 \\ \psi_4 &= -k_2 \leqslant 0 \end{aligned} \right\} \tag{16.4}$$

Equations (16.2) and (16.4) form a linear programming problem. The feasible set is shown graphically in Fig. 16.2. Note that the slope of the line connecting points A and B in Fig. 16.2 is -2. The level lines of the cost function of Equation (16.2) are straight, with slope equal to $-c_1/c_2$. The cost function decreases as level lines of the cost function move to the lower left. It is clear that point A (repeated eigenvalue) is the optimum design if $c_1/c_2 \geqslant 2$ and point B (simple eigenvalue) is the optimum design if $c_1/c_2 \leqslant 2$.

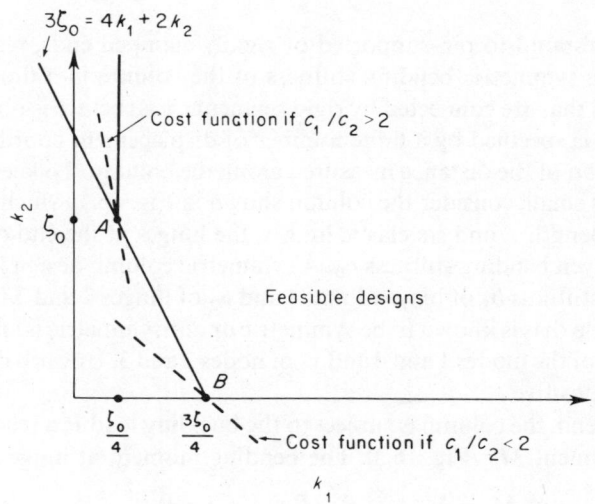

Fig. 16.2 Feasible region in design space for 2 DOF system

This result is quite interesting, since for certain values of parameters in the problem a repeated eigenvalue (A) occurs at the optimum design and for other values of the design parameters, only a simple eigenvalue (B) occurs at the optimum design. It is expected that this uncertainty as to whether repeated roots arise will also occur in more complex optimal design problems.

16.2.2 A Column Problem with Repeated Eigenvalues

In Ref. [3], Prager and Prager present an elegant analysis of a simple column buckling problem that exhibits repeated eigenvalues at an optimum point. The analysis and results for a column with elastically clamped ends are summarized here.

Rotation of the end sections of Fig. 16.3 by an angle θ_0 is opposed by a clamping moment $M_0 = b_0 \theta_0$, where b_0 is a given constant. The cases $b_0 = 0$ and

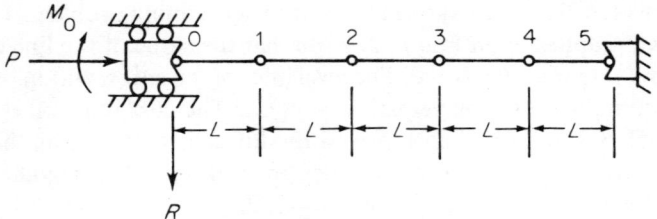

Fig. 16.3 Elastically supported column

$b_0 = \infty$ correspond to pin-supported or rigidly clamped ends, respectively. By localizing the symmetric bending stiffness of the column in a finite number of elastic hinges that are connected by rigid segments, a structure is obtained whose deformation is specified by a finite number of displacement coordinates, rather than a function of the distance measured along the column. To keep the number of unknowns small, consider the column shown in Fig. 16.3, which has five rigid segments of length L and six elastic hinges, the hinges at the end of the column having the given bending stiffness b_0. A symmetric column design is specified by the bending stiffness b_1 of hinges 1 and 4 and b_2 of hinges 2 and 3 in Fig. 16.3. A buckling mode that is known to be symmetric or antisymmetric is specified by the deflection y_1 of the modes 1 and 4 and y_2 of nodes 2 and 3. Upward deflections are regarded as positive.

At the left end, the column is subject to the buckling load P, a reaction R, and a clamping moment M_0 (Fig. 16.3). The bending moment at hinge i is

$$M_i = M_0 - iLR - Py_i \qquad i = 0, \ldots, 4 \tag{16.5}$$

where $y_0 = 0$. If θ_i denotes the relative rotation of the segments meeting at hinges i, considered positive if counterclockwise rotation of the segment to the right of i exceeds that of the segment to the left of i, then

$$M_i = b_i\theta_i = b_i(y_{i+1} - 2y_i + y_{i-1})/L \qquad i = 0, \ldots, 4 \tag{16.6}$$

where $y_{-1} = y_0 = 0$. It is convenient to introduce a reference stiffness b^* and define the dimensionless variables

$$\bar{P} = PL/b^*, \; \bar{R} = RL/b^*, \; \bar{M}_i = M_i/b^*, \; \bar{y}_i = y_i/L \tag{16.7}$$

Note that with these dimensionless variables, Equations (16.5) and (16.6) yield

$$M_0 - iR - Py_i = b_i(y_{i+1} - 2y_i + y_{i-1}) \qquad i = 0, 1, 2 \tag{16.8}$$

For a symmetric buckling mode, $y_3 = y_2$ and $R = 0$. For $i = 0, 1, 2$, Equation (16.8) yields

$$\left. \begin{aligned} M_0 &= b_0 y_1 \\ M_0 - P_s y_1 &= b_1(y_2 - 2y_1) \\ M_0 - P_s y_2 &= b_2(-y_2 + y_1) \end{aligned} \right\} \tag{16.9}$$

Optimization of Structures in which Repeated Eigenvalues Occur

where P_s is the buckling load of the symmetric mode. When M_0 from the first of these equations is substituted into the other two, linear homogenous equations for y_1 and y_2 are obtained that admit a nontrivial solution only if

$$P_s^2 - (b_0 + 2b_1 + b_2)P_s + b_0(b_1 + b_2) + b_1 b_2 = 0 \qquad (16.10)$$

The smaller root of this equation is the symmetric buckling load.

Assume that the cost of the design b_1, b_2 is fixed as

$$b_1 + b_2 = 1 \qquad (16.11)$$

and find a design that has the greatest buckling load. In view of Equation (16.11), Equation (16.10) reduces to

$$P_s^2 - (1 + b_0 + b_1)P_s + b_0 + b_1(1 - b_1) = 0 \qquad (16.12)$$

For an antisymmetric buckling mode, $y_3 = -y_2$ and $R = 2M_0/(\Omega)$, because bending moment and deflection vanish at the centre of the column. Proceeding as above, one obtains the quadratic equation for the buckling load P_a of the antisymmetric mode as

$$P_a^2 - (3 + 0.6b_0 - b_1)P_a + b_0(1.8 - 1.6b_1) + 5b_1(1 - b_1) = 0 \qquad (16.13)$$

The smaller root of this equation is the antisymmetric buckling load.

In Fig. 16.4, the smaller of the two buckling loads P_s and P_a is shown as a function of b_1, for fixed values of b_0. To indicate important features, consider the

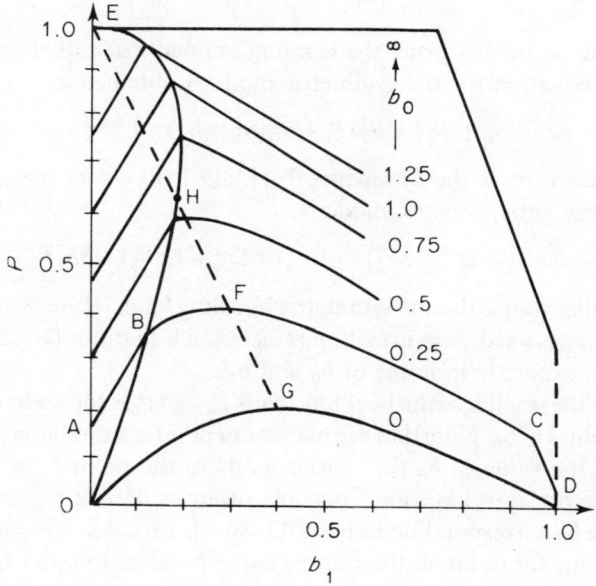

Fig. 16.4 Buckling loads for optimum columns

case $b_0 = 0.25$, for which the variation of the buckling load is shown by the line ABCD. The arcs AB and CD correspond to antisymmetric buckling, while the arc BC corresponds to symmetric buckling. At point B in Fig. 16.4, that is for $b_1 = 0.122$, both symmetric and antisymmetric buckling are possible and the buckling load is a double eigenvalue. A similar observation applies to point C, that is to $b_1 = 0.953$. The arc BC, however, has its greatest ordinate at point F, so the optimum design for $b_0 = 0.25$ corresponds to $b_1 = 0.300$, which buckles in a symmetric mode, the buckling load being a simple eigenvalue.

Along the curve OBE in Fig. 16.4, the loads P_s and P_a for symmetric and antisymmetric buckling have the same value. The maxima of the arcs for symmetric buckling lie on the straight line EFG, with the equation $P = 1 - 2b_1$, which intersects the curve OBE at the point H, corresponding to $b_0 = 0.526$, $b_1 = 0.190$. For $b_0 < 0.526$, the optimum design buckles symmetrically, the buckling load being a simple eigenvalue. For $b_0 > 0.526$, however, the buckling load is a double eigenvalue and the optimum design may buckle in a symmetric or an antisymmetric mode or in any linear combination of the two. Note in Fig. 16.4 that the smallest of the buckling loads is or is not differentiable at the optimum design, depending on whether the buckling load is a single or double eigenvalue. In a slightly different context, this has already been pointed out by Masur and Mroz [2].

In many structures, it is natural to express the bending stiffness as the square of the design variable, i.e.,

$$M_i = b_i^2 \theta_i \qquad i = 0, \ldots, 4 \tag{16.14}$$

If one proceeds as before, using the bending moments of Equation (16.14), the characteristic equation for the symmetric mode is obtained as

$$P_s^2 - (b_0^2 + 2b_1^2 + b_2^2)P_s + b_0^2(b_1^2 + b_2^2) + b_1^2 b_2^2 = 0 \tag{16.15}$$

and the smaller root is the symmetric buckling load. Also, the characteristic equation for the antisymmetric mode is

$$P_a^2 - (0.6b_0^2 + 2b_1^2 + 3b_2^2)P_a + b_0^2(0.2b_1^2 + 1.8b_2^2) + 5b_1^2 b_2^2 = 0 \tag{16.16}$$

where the smaller root is the antisymmetric buckling load. If one assumes that the cost of the design is fixed, as in Equation (16.11), the Equations (16.15) and (16.16) can be written explicitly in terms of b_0 and b_1.

In Fig. 16.5, the smaller of the buckling loads P_s and P_a is shown as a function of b_1, for fixed values of b_0. Note that the maximum point is a double eigenvalue if b_0 is larger than the value of b_0 that corresponds to the point C in Fig. 16.5. In contrast to the previous case, local maxima occur as double eigenvalues for b_0 larger than the b_0 corresponding to point D. Another local maximum occurs as a simple eigenvalue for b_0 larger than the b_0 corresponding to point E. At the left-hand ordinate of Fig. 16.5, if $b_0 > 1.29099$, then $P = P_s = 1$, which is a simple

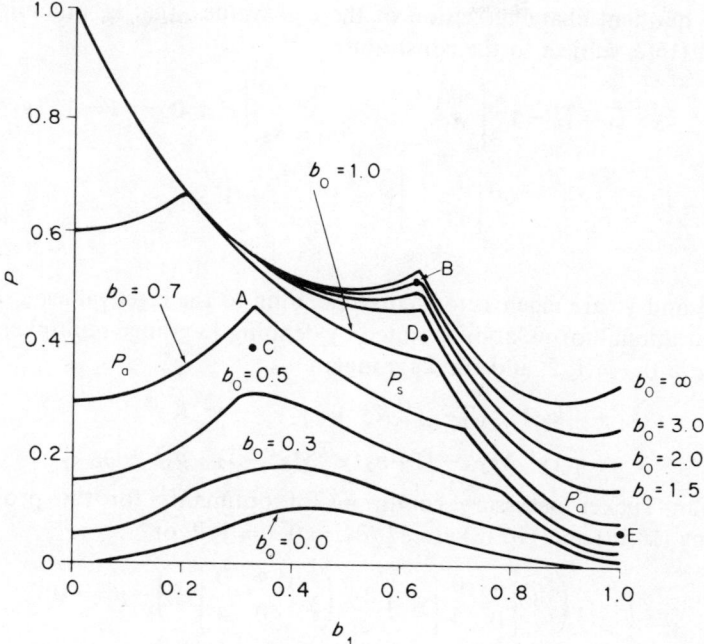

Fig. 16.5 Buckling loads for optimum columns

eigenvalue. If $b_0 = 1.29099$, then $P = P_s = P_a = 1$, which is a double eigenvalue. For $b_0 < 1.29099$, $P = P_a = 0.6 b_0^2$, which is a simple eigenvalue.

Comparing the plots of smallest eigenvalue vs. b_1 for the two column problems considered, one notes that: (1) the curve of Fig. 16.4 is concave, so a relative maximum is a global maximum; (2) the curve of Fig. 16.5 is not concave and indeed several relative maxima occur; and (3) when a repeated eigenvalue occurs at an optimum, the eigenvalue is not differentiable. Thus, if one used $dP/db_1 = 0$ as an 'optimality criterion', this would be in error.

16.3 FORMAL LAGRANGE MULTIPLIER ANALYSIS OF THE TWO DEGREE OF FREEDOM SPRING–MASS EXAMPLE

It is interesting to apply formally the well-known optimization results obtained by Olhoff and Rasmussen [1] and Prager and Taylor [7], using variational formulations of the eigenvalue problem with Lagrange multipliers. To evaluate the Lagrange multiplier method in the spirit of this foregoing work, consider the elementary two degree of freedom spring–mass vibration problem of Section 16.2.1. The idea of the formal Lagrange multiplier method is to use the

Rayleigh quotient characterization of the eigenvalues; that is, minimize ψ_0 of Equation (16.2) subject to the constraints

$$\zeta_0 - \zeta_i = \zeta_0 - \mathbf{y}^{i\mathrm{T}} \begin{bmatrix} 4k_1 + k_2 & k_2 \\ k_2 & 4k_1 + k_2 \end{bmatrix} \mathbf{y}^i \leqslant 0 \qquad i = 1, 2 \quad (16.17)$$

$$\mathbf{y}^{i\mathrm{T}} \begin{bmatrix} 2 & 1 \\ 1 & 2 \end{bmatrix} \mathbf{y}^i = 1 \qquad i = 1, 2$$
$$-\mathbf{k}_i \leqslant 0 \qquad i = 1, 2 \quad (16.18)$$

where $\mathbf{y}^1$ and $\mathbf{y}^2$ are eigenvectors corresponding to the eigenvalues ζ_1 and ζ_2.

The variational formulation is stated by defining Lagrange multipliers $\gamma_i \geqslant 0$, η_i, and $\mu_i \geqslant 0$, $i = 1, 2$, and the Lagrangian

$$L = c_1 k_1 + c_2 k_2 + \gamma_1(\zeta_0 - \mathbf{y}^{1\mathrm{T}}\mathbf{K}\mathbf{y}^1) + \gamma_2(\zeta_0 - \mathbf{y}^{2\mathrm{T}}\mathbf{K}\mathbf{y}^2)$$
$$+ \eta_1(\mathbf{y}^{1\mathrm{T}}\mathbf{M}\mathbf{y}^1 - 1) + \eta_2(\mathbf{y}^{2\mathrm{T}}\mathbf{M}\mathbf{y}^2 - 1) - \mu_1 k_1 - \mu_2 k_2 \quad (16.19)$$

The Kuhn–Tucker necessary conditions of optimality for the problem of Equations (16.17) and (16.18) are $\partial L/\partial k_i = 0$, $i = 1, 2$, or

$$c_1 - \gamma_1 \left(\mathbf{y}^{1\mathrm{T}} \begin{bmatrix} 4 & 0 \\ 0 & 4 \end{bmatrix} \mathbf{y}^1 \right) - \gamma_2 \left(\mathbf{y}^{2\mathrm{T}} \begin{bmatrix} 4 & 0 \\ 0 & 4 \end{bmatrix} \mathbf{y}^2 \right) - \mu_1 = 0 \quad (16.20)$$

$$c_2 - \gamma_1 \left(\mathbf{y}^{1\mathrm{T}} \begin{bmatrix} 1 & 1 \\ 1 & 1 \end{bmatrix} \mathbf{y}^1 \right) - \gamma_2 \left(\mathbf{y}^{2\mathrm{T}} \begin{bmatrix} 1 & 1 \\ 1 & 1 \end{bmatrix} \mathbf{y}^2 \right) - \mu_2 = 0 \quad (16.21)$$

and

$$\gamma_1(\zeta_0 - \mathbf{y}^{1\mathrm{T}}\mathbf{K}\mathbf{y}^1) = 0, \quad \gamma_2(\zeta_0 - \mathbf{y}^{2\mathrm{T}}\mathbf{K}\mathbf{y}^2) = 0 \quad (16.22)$$

$$\mu_1 k_1 = 0, \quad \mu_2 k_2 = 0 \quad (16.23)$$

Consider the known optimum design $k_0 = (\zeta_0/4, \zeta_0)$, where a double eigenvalue occurs. From Equation (16.23), $\mu_1 = \mu_2 = 0$ and the eigenvectors $\mathbf{y}^i$ are solutions of the equation

$$(\mathbf{K} - \zeta_0 \mathbf{M})\bigg|_{k=k_0} \mathbf{y}^i = \begin{bmatrix} 0 & 0 \\ 0 & 0 \end{bmatrix} \mathbf{y}^i = 0 \qquad i = 1, 2 \quad (16.24)$$

Hence any nonzero vector $\mathbf{y}$ is an eigenvector. Eigenvectors may now be selected and substituted into Equations (16.20) and (16.21), which must, of course, be satisfied.

With the M-normal eigenvectors

$$y^1 = \frac{1}{\sqrt{2}} \begin{bmatrix} 1 \\ 0 \end{bmatrix}, \quad y^2 = \frac{1}{\sqrt{2}} \begin{bmatrix} 0 \\ 1 \end{bmatrix} \quad (16.25)$$

Equations (16.20) and (16.21), with $\mu_i = 0$, $i = 1, 2$, yield

$$\left. \begin{array}{l} c_1 = 2\gamma_1 + 2\gamma_2 \geqslant 0 \\ c_2 = \tfrac{1}{2}\gamma_1 + \tfrac{1}{2}\gamma_2 \geqslant 0 \end{array} \right\} \quad (16.26)$$

Thus by observation, the necessary condition is only satisfied if $c_1 = 4c_2 \geq 0$. But from Section 16.2.1, it is known that $k_0 = (\zeta_0/4, \zeta_0)$ is an optimum design as long as $c_1 \geq 2c_2 > 0$. Thus the supposed necessary conditions of Equations (16.20) and (16.21) are in fact not necessary at all.

Consider another set of eigenvectors, which are M-orthonormal,

$$\mathbf{y}^1 = \frac{1}{\sqrt{2}}\begin{bmatrix}1\\0\end{bmatrix}, \quad \mathbf{y}^2 = \frac{1}{\sqrt{6}}\begin{bmatrix}1\\-2\end{bmatrix} \tag{16.27}$$

With these eigenvectors, Equations (16.20) and (16.21) become

$$\left.\begin{array}{l} c_1 = 2\gamma_1 + \tfrac{10}{3}\gamma_2 \geq 0 \\ c_2 = \tfrac{1}{2}\gamma_1 + \tfrac{1}{6}\gamma_2 \geq 0 \end{array}\right\} \tag{16.28}$$

From Equation (16.28), one has

$$\gamma_1 = \frac{1}{8}(-c_1 + 20c_2) \geq 0 \quad \text{and} \quad \gamma_2 = \frac{1}{8}(3c_1 - 12c_2) \geq 0$$

which give

$$20c_2 \geq c_1 \geq 4c_2 \geq 0 \tag{16.29}$$

These are also not valid necessary conditions. It is clear at this point that the formal Lagrange multiplier method is not valid for the repeated eigenvalue case. This might have been suspected, based on the non-differentiability of the eigenvalue noted in Section 16.2.2.

Try now the set of eigenvectors

$$\mathbf{y}^1 = \frac{1}{\sqrt{6}}\begin{bmatrix}1\\1\end{bmatrix}, \quad \mathbf{y}^2 = \frac{1}{\sqrt{2}}\begin{bmatrix}1\\-1\end{bmatrix} \tag{16.30}$$

Substituting $\mathbf{y}^1$ and $\mathbf{y}^2$ of Equation (16.30) into Equations (16.20) and (16.21), one obtains

$$\left.\begin{array}{l} c_1 = \tfrac{4}{3}\gamma_1 + 4\gamma_2 \geq 0 \\ c_2 = \tfrac{2}{3}\gamma_1 \geq 0 \end{array}\right\} \tag{16.31}$$

or

$$\begin{array}{l} \gamma_1 = \tfrac{3}{2}c_1 \geq 0 \\ \gamma_2 = \tfrac{1}{2}(c_1 - 2c_2) \geq 0 \end{array} \tag{16.32}$$

This gives the correct result $c_1 \geq 2c_2 \geq 0$. Hence, for this problem there is a set of eigenvectors for which the variational formulation may be valid.

To find out 'what went wrong' in the formal Lagrange multiplier method, note that with $\mathbf{y}^i$ of Equation (16.25), the values of the Rayleigh quotients are

$$\left.\begin{array}{l} f_1(k) \equiv \mathbf{y}^{1\mathrm{T}}\mathbf{K}(k)\mathbf{y}^1 = \tfrac{1}{2}(4k_1 + k_2) \\ f_2(k) \equiv \mathbf{y}^{2\mathrm{T}}\mathbf{K}(k)\mathbf{y}^2 = \tfrac{1}{2}(4k_1 + k_2) \end{array}\right\} \tag{16.33}$$

For $k_0 = (\zeta_0/4, \zeta_0)$, $f_1 = f_2 = \zeta_0$ and the gradients of the Rayleigh quotients are, from Equation (16.33),

$$\nabla f_1 = \nabla f_2 = \begin{bmatrix} 2 \\ \frac{1}{2} \end{bmatrix} \qquad (16.34)$$

From the algebraic solution of ζ_i, one has $\zeta_1 = \frac{2}{3}(2k_1 + k_2)$ and $\zeta_2 = 4k_1$. Thus the gradients of the actual eigenvalues are

$$\nabla \zeta_1 = \begin{bmatrix} \frac{4}{3} \\ \frac{2}{3} \end{bmatrix}, \quad \nabla \zeta_2 = \begin{bmatrix} 4 \\ 0 \end{bmatrix} \qquad (16.35)$$

Thus, $\nabla \zeta_i \neq \nabla f_i$, so the wrong derivatives were used in the Kuhn–Tucker optimality criteria, leading to erroneous results. It is clear that one must be very careful in the analysis of derivatives of repeated eigenvalues.

16.4 DIRECTIONAL DERIVATIVES OF EIGENVALUES

To see why the foregoing difficulty arises, it is helpful to review the engineering perturbation analysis of the eigenvalue problem, which starts with

$$\mathbf{A}(k)\mathbf{y}^i = \zeta_i \mathbf{B}(k)\mathbf{y}^i, \quad \mathbf{y}^{iT}\mathbf{B}(k)\mathbf{y}^i = 1 \qquad (16.36)$$

Premultiplying Equation (16.36) by a constant vector $\mathbf{v}^T$,

$$\mathbf{v}^T \mathbf{A}(k)\mathbf{y}^i = \zeta_i \mathbf{v}^T \mathbf{B}(k)\mathbf{y}^i \qquad (16.37)$$

Denoting variation with an over-dot and treating $\mathbf{k}$, $\mathbf{y}^i$, and ζ_i as independent, we obtain

$$\dot{\mathbf{v}}^T \mathbf{A}\mathbf{y}^i + \mathbf{v}^T \dot{\mathbf{A}}\mathbf{y}^i = \dot{\zeta}_i \mathbf{v}^T \mathbf{B}\mathbf{y}^i + \zeta_i \mathbf{v}^T \dot{\mathbf{B}}\mathbf{y}^i + \zeta_i \mathbf{v}^T \mathbf{B}\dot{\mathbf{y}}^i \qquad (16.38)$$

Rewriting, using symmetry of $\mathbf{A}$ and $\mathbf{B}$, and setting $\mathbf{v} = \mathbf{y}^i$,

$$\mathbf{y}^{iT}\dot{\mathbf{A}}\mathbf{y}^i - \zeta_i \mathbf{y}^{iT}\dot{\mathbf{B}}\mathbf{y}^i + \dot{\mathbf{y}}^{iT}(\mathbf{A}\mathbf{y}^i - \zeta_i \mathbf{B}\mathbf{y}^i) = \dot{\zeta}_i \mathbf{y}^{iT}\mathbf{B}\mathbf{y}^i \qquad (16.39)$$

Putting $i = j$ in Equation (16.39) and using Equation (16.36), we obtain the formula

$$\dot{\zeta}_i = \mathbf{y}^{iT}\dot{\mathbf{A}}\mathbf{y}^i - \zeta_i \mathbf{y}^{iT}\dot{\mathbf{B}}\mathbf{y}^i \qquad i = 1, 2 \qquad (16.40)$$

But, the example treated in Section 16.3 shows that this is 'wrong.'

To see what goes wrong at repeated eigenvalues, consider the two degree of freedom example of Section 16.21. The eigenvalue ζ_0 is repeated if $k_1 = \zeta_0/4$ and $k_2 = \zeta_0$. If k is perturbed to $k_1 = \zeta_0/4 + h_1$, $k_2 = \zeta_0 + h_2$, the eigenvalue equation is

$$\begin{bmatrix} 2\zeta_0 + 4h_1 + h_2 & \zeta_0 + h_2 \\ \zeta_0 + h_2 & 2\zeta_0 + 4h_1 + h_2 \end{bmatrix} \mathbf{y}_h = \zeta_h \begin{bmatrix} 2 & 1 \\ 1 & 2 \end{bmatrix} \mathbf{y}_h \qquad (16.41)$$

where the subscript h denotes dependence on h. The solution is $\zeta_h^1 = \zeta_0 + (4h_1 + 2h_2)/3$, $\zeta_h^2 = \zeta_0 + 4h_1$. If $4h_1 \neq h_2$, $\zeta_h^1 \neq \zeta_h^2$ and the M-orthonormal

Optimization of Structures in which Repeated Eigenvalues Occur

eigenvectors are

$$\mathbf{y}_h^1 = \frac{1}{\sqrt{6}} \begin{bmatrix} 1 \\ 1 \end{bmatrix}, \quad \mathbf{y}_h^2 = \frac{1}{\sqrt{2}} \begin{bmatrix} 1 \\ -1 \end{bmatrix} \qquad (16.42)$$

which happen to be independent of h. If, on the other hand, $4h_1 = h_2$, anything is an eigenvector, e.g.,

$$\bar{\mathbf{y}}^1 = \frac{1}{\sqrt{2}} \begin{bmatrix} 1 \\ 0 \end{bmatrix}, \quad \bar{\mathbf{y}}^2 = \frac{1}{\sqrt{2}} \begin{bmatrix} 0 \\ 1 \end{bmatrix} \qquad (16.43)$$

To see what has happened geometrically, Fig. 16.6 shows the eigenvectors just calculated. The perturbation of design from $\mathbf{k}_0$, which gives a repeated eigenvalue, transforms $\bar{\mathbf{y}}^i$ of Equation (16.43) to $\mathbf{y}_h^i$ of Equation (16.42). Clearly, this transformation is not continuous, so the perturbation analysis of Equation (16.38) is meaningless at the repeated eigenvalue.

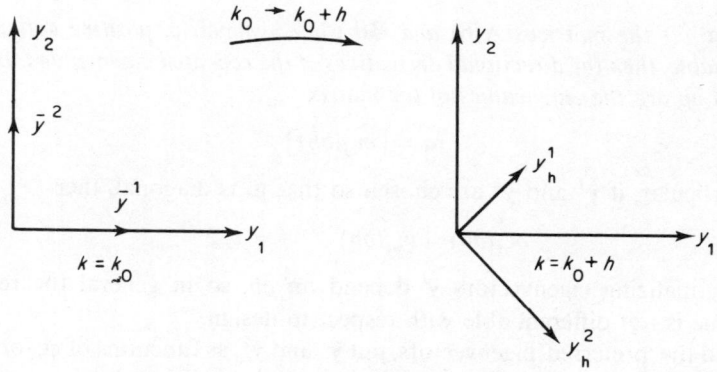

Fig. 16.6 Transformation of eigenspace

To be more precise, let $h_1 = \varepsilon$ and $h_2 = \varepsilon$. Then

$$\|\dot{\mathbf{y}}^1\| = \|\bar{\mathbf{y}}^1 - \mathbf{y}_h^1\| = \left\| \frac{1}{\sqrt{6}} \begin{bmatrix} \sqrt{3} - 1 \\ -1 \end{bmatrix} \right\| = \left(\frac{5 - 2\sqrt{3}}{6} \right)^{1/2}$$

$$\|\dot{\mathbf{y}}^2\| = \|\bar{\mathbf{y}}^2 - \mathbf{y}_h^2\| = \left\| \frac{1}{\sqrt{2}} \begin{bmatrix} -1 \\ 2 \end{bmatrix} \right\| = \left(\frac{5}{2} \right)^{1/2}$$

Since this is true for any $\varepsilon > 0$, arbitrarily small, $\|\dot{\mathbf{y}}^i\|$ do not approach zero as $\|h\| = \sqrt{2}\varepsilon \to 0$. Thus the linearization of Equation (16.37) is meaningless, since $\mathbf{y}(h)$ is not even continuous, much less differentiable.

Note that if Equation (16.40) is to have meaning for $\mathbf{y}^{iT}\mathbf{B}\mathbf{y}^2 = 0$, a special set of eigenvectors must be chosen. If $\mathbf{y}^1$ and $\mathbf{y}^2$ are B-orthonormal eigenvectors corresponding to a repeated eigenvalue ζ, so are

$$\left.\begin{aligned}\bar{\mathbf{y}}^1 &= \mathbf{y}^1 \cos\phi + \mathbf{y}^2 \sin\phi \\ \bar{\mathbf{y}}^2 &= -\mathbf{y}^1 \sin\phi + \mathbf{y}^2 \cos\phi\end{aligned}\right\} \quad (16.44)$$

for any rotation angle ϕ.

If Equation (16.40) is to be valid, ϕ must be chosen so that the left-hand side of Equation (16.39), from which Equation (16.40) was derived, is 0 for $i \neq j$ and $\bar{\mathbf{y}}^i$. An easy calculation shows that the eigenvectors of Equation (16.30) satisfy this condition for the spring–mass problem of Section 16.2.1.

We are now in a position to develop the argument. Let

$$m_{ij} \equiv \frac{\partial}{\partial b}(\mathbf{y}^{iT}\mathbf{A}(b)\mathbf{y}^i)\delta b - \zeta\frac{\partial}{\partial b}(\mathbf{y}^{iT}\mathbf{B}(b)\mathbf{y}^i)\delta b \quad (16.45)$$

which depends on the direction δb of design change; i.e. $m_{ij} = m_{ij}(\delta b)$. The following theorem is proved in Ref. [5]:

Theorem *If the matrices $\mathbf{A}(b)$ and $\mathbf{B}(b)$ are symmetric, positive definite, and differentiable, then the directional derivatives of the repeated eigenvalue $\zeta(b)$ in the direction δb are the eigenvalues of the matrix*

$$\mathbf{m} = [m_{ij}(\delta b)] \quad (16.46)$$

In particular, if $\bar{\mathbf{y}}^1$ and $\bar{\mathbf{y}}^2$ are chosen so that $\mathbf{m}$ is diagonal, then

$$\delta\zeta_i(\delta b) = \bar{m}_{ii}(\delta b) \quad i = 1, 2$$

The diagonalizing eigenvectors $\bar{\mathbf{y}}^i$ depend on δb, so in general the repeated eigenvalue is not differentiable with respect to design.

To find the 'preferred' eigenvectors, put $\bar{\mathbf{y}}^1$ and $\bar{\mathbf{y}}^2$, as functions of ϕ, for any β-orthonormal eigenvectors $\mathbf{y}^1$ and $\mathbf{y}^2$, into $\bar{m}_{12}(\bar{\mathbf{y}}^1, \bar{\mathbf{y}}^2, \delta b)$, and put

$$\begin{aligned}0 = \bar{m}_{12}(\bar{\mathbf{y}}^1, \bar{\mathbf{y}}^2, \delta b) &\equiv \bar{m}_{12} \\ &= -\cos\phi\sin\phi\, m_{11} + (\cos^2\phi - \sin^2\phi)m_{12} \\ &\quad + \sin\phi\cos\phi\, m_{22} \\ &= \tfrac{1}{2}\sin 2\phi\,(m_{22} - m_{11}) + \cos 2\phi\, m_{12}\end{aligned} \quad (16.47)$$

where $m_{ij} = m_{ij}(\mathbf{y}^1, \mathbf{y}^2, \delta b)$. Thus

$$\phi = \phi(\delta b) = \tfrac{1}{2}\arctan\left[\frac{2m_{12}(\mathbf{y}^1, \mathbf{y}^2, \delta b)}{m_{11}(\mathbf{y}^1, \mathbf{y}^2, \delta b) - m_{22}(\mathbf{y}^1, \mathbf{y}^2, \delta b)}\right] \quad (16.48)$$

With these eigenvectors, $\bar{m}_{12} = \bar{m}_{21} = 0$ and

$$\begin{aligned}\delta\zeta_1(\delta b) = \bar{m}_{11} &= \tfrac{1}{2}[m_{11}(\delta b) + m_{22}(\delta b)] \\ &\quad + \sin 2\phi(\delta b) m_{12}(\delta b) \\ &\quad + \tfrac{1}{2}\cos 2\phi(\delta b)[m_{11}(\delta b) - m_{22}(\delta b)]\end{aligned} \quad (16.49)$$

$$\delta\zeta_2(\delta b) = \bar{m}_{22} = \tfrac{1}{2}[m_{11}(\delta b) + m_{22}(\delta b)]$$
$$- \sin 2\phi(\delta b) m_{12}(\delta b)$$
$$- \tfrac{1}{2}\cos 2\phi(\delta b)[m_{11}(\delta b) - m_{22}(\delta b)] \tag{16.50}$$

where dependence of m_{ij} on the fixed $\mathbf{y}^1$ and $\mathbf{y}^2$ is suppressed.

Since $\phi(\delta b)$ depends on δb, these equations show that $\delta\zeta_i(\delta b)$ are not linear in δb (even though $m_{ij}(\delta b)$ are linear in δb). Thus the repeated eigenvalue is not, in general, differentiable.

16.5 RIGOROUS OPTIMALITY CRITERIA

A generally valid and rigorous optimality criteria for the problem of minimizing $\psi_0(b)$, subject to the condition $\zeta_0 - \zeta_i \leq 0$, $i = 1, 2, \ldots$, is as follows [8]: There exist multipliers λ_0 and $\lambda_1 \geq 0$, not both zero, such that

$$\lambda_0 \frac{\partial \psi_0}{\partial b} \delta b - \lambda_1 \inf_{\xi^T \xi = 1} [\xi^T m(y^1, y^2, \delta b)\xi] \geq 0 \tag{16.51}$$

for all δb consistent with constraints and $\lambda_1(\zeta_0 - \zeta(b_0)) = 0$.

This condition cannot be reduced to equality form, in general. Special cases where it does are known [8]. Otherwise, it is difficult to use in finding solutions.

An alternative and equivalent formulation of the two degree of freedom spring–mass example of Section 16.2.1 is to choose $k_i \equiv b_i$, $i = 1, 2$, to maximize the fundamental eigenvalue, subject to the condition that cost is fixed. That is, maximize

$$\bar{\psi}_0 = \min_{i=1,2} \zeta_i \tag{16.52}$$

subject to the conditions

$$\bar{\psi}_1 = c_1 b_1 + c_2 b_2 = 0$$
$$b_i \geq 0, \, i = 1, 2 \tag{16.53}$$

A necessary condition for this problem is obtained from results presented by Masur and Mroz in Ref. [2]. They state the necessary condition

$$\delta\zeta_1(\delta b) \times \delta\zeta_2(\delta b) \leq 0 \tag{16.54}$$

for all design variations δb that are consistent with constraints. At the known optimum design $b_1 = \zeta_0/4$ and $b_2 = \zeta_0$, both k_i are positive, so δb need only satisfy

$$c_1 \delta b_1 + c_2 \delta b_2 = 0 \tag{16.55}$$

To use the necessary condition of Equation (16.54), $\delta\zeta_i(\delta b)$ must be calculated. But $\delta\zeta_i(\delta b)$ are eigenvalues of the matrix $\mathbf{m}(b_0, h)$, which were calculated from Equation (16.41) as $4\delta b_1/3 + 2\delta b_2/3$ and $4\delta b_1$. Using this result, one has the necessary condition

$$(4\delta b_1/3 + 2\delta b_2/3)4\delta b_1 \leq 0 \tag{16.56}$$

for all δb_i satisfying Equation (16.55). Solving Equation (16.55) for δb_2 and substituting into Equation (16.56), one has

$$\left(\frac{4}{3} - \frac{2}{3}\frac{c_1}{c_2}\right) 4(\delta b_1)^2 \leq 0$$

which is just

$$\frac{c_1}{c_2} \geq 2$$

and is the correct result, as shown in Section 16.2.1.

E.J.H.

REFERENCES

1. N. Olhoff and S. H. Rasmussen, On single and biomodal optimum buckling loads of clamped columns, *Int. J. Solids and Structures*, **13**, 605–14, 1977.
2. E. F. Masur and Z. Mroz, Singular solutions in structural optimization problems, in I.U.T.A.M. Conference on *Variational Methods in the Mechanics of Solids* (Ed. S. Nemat-Nasser), Springer-Verlag, to appear.
3. S. Prager and W. Prager, A note on optimal design of columns, *Int. J. Mech. Sci.*, **21**, 249–51, 1979.
4. E. J. Haug and B. Rousselet, Design sensitivity analysis in structural mechanics I: Static response variations, *J. Str. Mech.* **8**(1), 17–41, 1980.
5. E. J. Haug and B. Rousselet, Design sensitivity analysis in structural mechanics II: Eigenvalue variations, *J. Str. Mech.*, **8**(2), 1980.
6. B. M. Pschenichnyi, *Necessary Conditions for an Extremum*, Marcel Dekker, New York, 1971.
7. W. Prager and J. E. Taylor, Problems of optimal structural design, *J. Applied Mechanics*, **35**(1), 102–6, 1968.
8. E. J. Haug and K. K. Choi, Optimization of structures with repeated eigenvalues, in *Optimization of Distributed Parameter Structures* (Eds E. J. Haug and J. Cea), Sijthoff-Noordhoff, Amsterdam, to appear 1981.

Foundations of Structural Optimization: A Unified Approach
Edited by A. J. Morris
© 1982 John Wiley & Sons Ltd.

Chapter 17

Structural Optimization by Geometric Programming

17.1 INTRODUCTION TO GEOMETRIC PROGRAMMING

There are two ways of introducing geometric programming; one through the inequality relating the arithmetic and geometric mean, the other via the classical Lagrangian duality relationship which has formed one of the central themes of this book. The concepts of the theory are more easily introduced by the former approach and the present section follows this course. We introduce the basic inequality and then proceed to develop the optimization method itself with the aid of a trivial structural example. In the next section the analysis is repeated but the alternative Lagrangian viewpoint is adopted.

We begin by noting that a squared term cannot be negative and hence for two positive numbers a_1, a_2,

$$\left(\frac{a_1+a_2}{2}\right)^2 \geq \left(\frac{a_1+a_2}{2}\right)^2 - \left(\frac{a_1-a_2}{2}\right)^2 = a_1 a_2$$

thus

$$a_1^{1/2} a_2^{1/2} \leq \frac{a_1}{2} + \frac{a_2}{2}$$

This inequality can be developed [1, 2] into the more general form

$$\prod_{i=1}^{n} x_i^{q_i} \leq \sum_{i=1}^{n} q_i x_i$$

where the terms q_1, $i = 1, \ldots, n$ are a set of arbitrary non-negative weights satisfying the equality

$$\sum_{i=1}^{n} q_i = 1$$

The left-hand side of the above inequality is a weighted geometric mean of the variables x_i which are again arbitrary, non-negative weights, and the right-hand side is the corresponding weighted arithmetic mean. The inequality itself is known as the arithmetic–geometric inequality or simply the geometric inequality.

We now introduce $2n$ new positive variables y_i, δ_i ($i = 1, 2, \ldots, n$), defined by

$$y_i = q_i x_i$$
$$\delta_i = q_i \Delta$$
$$\text{and} \quad \Delta = \sum_{i=1}^{n} \delta_i$$

Thus the inequality becomes

$$\left\{ \prod_{i=1}^{n} (y_i/\delta_i)^{\delta_i/\Delta} \right\} \Delta \leqslant \sum_{i=1}^{n} y_i$$

or alternatively,

$$\left\{ \prod_{i=1}^{n} (y_i/\delta_i)^{\delta_i} \right\} \Delta^{\Delta} \leqslant \left\{ \sum_{i=1}^{n} y_i \right\}^{\Delta}$$

which we call the generalized arithmetic–geometric inequality. There are now two observations which can be made with regard to this inequality. Because of the arbitrary nature of the variables, y_i can represent the terms of a generalized positive polynomial or, briefly 'posynomial'. An example of such a posynomial containing two terms and three variables would be

$$\sum_{i=1}^{n} y_i = y_1 + y_2 = a x_1^{0.7} x_2^{-1} + b x_2^{0.5} x_3$$

A posynomial may thus be defined as a function consisting of a sum of terms which comprise a positive coefficient multiplied by products of variables, with each variable raised to an arbitrary power. The second observation stems from the first in that the simple inequality given above can form the basis of an effective optimization technique providing the function being minimized is written in terms of posynomials. This technique, known as geometric programming, was originally proposed by Duffin [3] and Zener [4] who together with Peterson [5] developed the subject into a usable theory. Because a rigorous discussion is already available, the remaining part of the section will avoid a detailed mathematical treatment and will concentrate on drawing the basic equations of the theory from a simple structural optimization problem.

Consider then the problem of finding a minimum-weight design for the two-bars truss, shown in Fig. 17.1, with specific weights ρ which must carry a vertical load of $2P$. The design variables consist of the two bar cross-sectional areas which, because of symmetry, can be denoted by the single variable x_i, and the overall height x_2. The constraints on the structure are simple stress constraints which require that the bar stresses must not exceed a prescribed value σ_0. Using

Structural Optimization by Geometric Programming

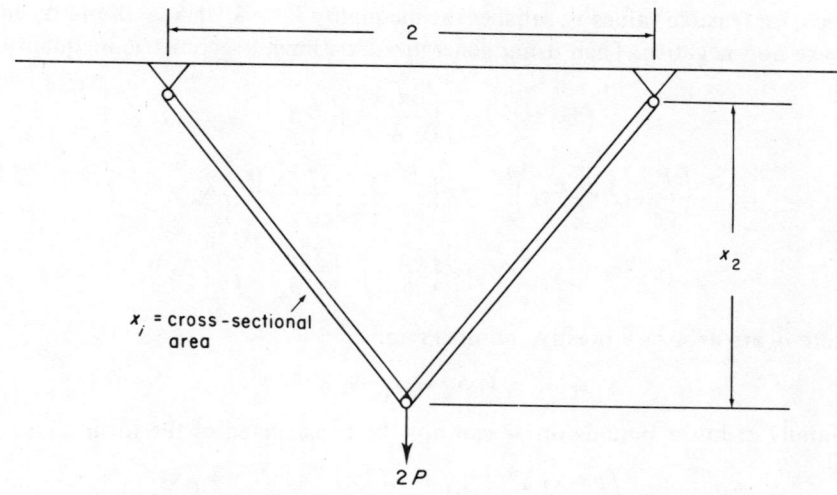

Fig. 17.1 Two-bar truss

this notation, the problem can be written as

$$\text{minimize} \quad W(x) = \rho x_1 (1 + x_2^2)^{1/2}$$

$$\text{subject to} \quad \frac{P}{\sigma_0} \frac{(1 + x_2^2)^{1/2}}{x_1 x_2} \leqslant 1$$

This simple problem can now be used as a vehicle to illustrate the approach of geometric programming, which is to use the generalized arithmetic–geometric inequality to construct a dual problem having a particularly simple form. However, in order to use this inequality all of the various expressions must be in posynomial form. By recalling the definition of a posynomial, it is immediately seen that the term $(1 + x_2^2)^{1/2}$ does not fit into this category. A new variable x_3 is, therefore, introduced through the inequality

$$x_3 \geqslant 1 + x_2^2$$

The original optimization problem has now become

$$\text{minimize} \quad W(x) = \rho x_1 x_3$$

$$\text{subject to} \quad 1 \geqslant g_1(x) = \frac{P}{\sigma_0} x_1^{-1} x_2^{-1} x_3$$

$$1 \geqslant g_2(x) = x_3^{-2} + x_2^2 x_3^{-2}$$

Consider now a new function

$$\overline{W}(x, \Delta) = (\rho x_1 x_3)^{\Delta_1} \left(\frac{P}{\sigma_0} x_1^{-1} x_2^{-2} x_3 \right)^{\Delta_2} (x_3^{-2} + x_2^2 x_3^{-2})^{\Delta_3}$$

which, for feasible values x_i, satisfies the inequality $W \geq \overline{W}$ if $\Delta_1 = 1$ and Δ_2 and Δ_3 are non-negative. Then using generalized arithmetic–geometric inequality,

$$(\rho x_1 x_3)^{\Delta_1} \geq \left\{\left(\frac{\rho x_1 x_3}{\delta_1}\right)^{\delta_1}\right\} \Delta_1^{\Delta_1}$$

$$\left(\frac{P}{\sigma_0} x_1^{-1} x_2^{-2} x_3\right)^{\Delta_2} \geq \left\{\left(\frac{P}{\sigma_0} \frac{x_1^{-1} x_2^{-2} x_3}{\delta_2}\right)^{\delta_2}\right\} \Delta_2^{\Delta_2}$$

$$(x_3^{-2} + x_2^2 x_3^{-2})^{\Delta_3} \geq \left\{\left(\frac{x_3^{-2}}{\delta_3}\right)^{\delta_3} \left(\frac{x_2^2 x_3^{-2}}{\delta_4}\right)^{\delta_4}\right\} \Delta_3^{\Delta_3}$$

where δ_i are arbitrary positive numbers and

$$\Delta_1 = \delta_1 = 1, \quad \Delta_2 = \delta_2, \quad \Delta_3 = \delta_3 + \delta_4$$

A family of lower bounds on W can now be constructed of the form

$$W \geq \overline{W} \geq \left(\frac{\rho x_1 x_2}{\delta_1}\right)^{\delta_1} \delta_1^{\delta_1} \left(\frac{P}{\sigma_0} \frac{x_1^{-1} x_2^{-1} x_3}{\delta_2}\right)^{\delta_2} \delta_2^{\delta_2} \left(\frac{x_3^{-2}}{\delta_3}\right)^{\delta_3}$$

$$\left(\frac{x_2^2 x_3^{-2}}{\delta_4}\right)^{\delta_4} (\delta_3 + \delta_4)^{(\delta_3 + \delta_4)}$$

which reduces to

$$W \geq V(x, \delta) = (\rho)^{\delta_1} \left(\frac{P}{\sigma_0}\right)^{\delta_2} \left(\frac{1}{\delta_3}\right)^{\delta_3} \left(\frac{1}{\delta_4}\right)^{\delta_4} (\delta_3 + \delta_4)^{(\delta_3 + \delta_4)}$$

$$x_1^{(\delta_1 - \delta_2)} x_2^{(-\delta_2 + 2\delta_4)} x_3^{(\delta_1 + \delta_2 - 2\delta_2 - 2\delta_4)}$$

with $\delta_1 = 1$.

This final inequality puts us in a position whereby we can move to the central idea of geometric programming which lies in the creation of a dual function. Following the principles laid down in Chapter 2, the dual is given by

$$V(\delta) = \min_{x \in X} V(x, \delta)$$

where $X \in R^n$ and the variables δ_i are called the dual variables. Thus we must set $\partial V(x, \delta)/\partial x_i = 0$ for $i = 1, 2, 3$, and so obtain the three *orthogonality conditions*,

$$\begin{aligned} \delta_1 - \delta_2 &= 0 \\ -\delta_2 + 2\delta_4 &= 0 \\ \delta_1 + \delta_2 - 2\delta_3 - 2\delta_4 &= 0 \end{aligned}$$

It now follows that since

$$W \geq V(x, \delta)$$

then

$$W \geq V(\delta)$$

where, because of the orthogonality conditions

$$V(\delta) = (\rho)^{\delta_1} \left(\frac{P}{\sigma_0}\right)^{\delta_2} \left(\frac{1}{\delta_3}\right)^{\delta_3} \left(\frac{1}{\delta_4}\right)^{\delta_4} (\delta_3 + \delta_4)^{(\delta_3 + \delta_4)}$$
$$\delta_1 = 1$$

This latter condition $\delta_1 = 1$ is a particular form of a general requirement that the δ variables associated with the objective function should sum to unity; this is known as the normality condition.

The final form for the dual is defined by the function $V(\delta)$ together with the normality and orthogonality conditions, and if the constrained minimum for the weight is given by $W(x^*)$, then $W(x^*) \geq V(\delta)$. Furthermore, if $\delta^* \in D$, $D \in R^n$, is the dual vector which maximizes $V(\delta)$ and satisfies the orthogonality and normality conditions, then $V(\delta^*) = W(x^*)$. As indicated earlier, the dual formulation is the key to geometric programming since we shall always seek the solution $W(x^*)$ by maximizing the dual function.

In applying this procedure to our problem an unusual situation arises in that the dual variables $\delta_i (i = 1, \ldots, 4)$ are completely specified by the three orthogonality conditions and the normality condition. Thus, the dual problem has a single feasible point, which is also the maximum; hence

$$\delta_1^* = 1, \quad \delta_2^* = 1, \quad \delta_3^* = \tfrac{1}{2}, \quad \delta_4^* = \tfrac{1}{2}$$

and

$$V(\delta^*) = \rho \frac{P}{\sigma_0} \left(\frac{1}{1/2}\right)^{1/2} \left(\frac{1}{1/2}\right)^{1/2} (1/2 + 1/2)^{(1/2 + 1/2)}$$
$$= 2\rho P/\sigma_0$$

Thus

$$W(x^*) = \frac{2\rho P}{\sigma_0}$$

The example illustrates a powerful feature of geometric programming where the optimum has been found by solving a set of linear equations. Such a solution is only possible when the number of equations is equal to the number of dual variables. Unfortunately, for most problems there are more variables than equations and the difference between these is known as the *degree of difficulty*. A simple rule-of-thumb for calculating this is to note the number of primal variables n and the number of posynomial terms m; then the degree of difficulty is given by $m - n - 1$. For the above problem there are three primal variables (x_1, x_2, x_3); four terms, one in the objective function W, one and two respectively in constraints g_1 and g_2, giving a zero degree of difficulty.

Zero degree of difficulty problems have two main advantages. In addition to obtaining optima by a simple solution procedures, the optimizing dual vector δ^* is independent of the coefficients of the objective function and the constraints.

Thus in the above problem changes in the optimum value $W(x^*)$ for variations in the parameters ρ, P, σ_0 can be computed directly from the expression for $V(\delta^*)$ since the dual variables δ_i^* ($i = 1, \ldots, 4$) do not vary.

In moving to non-zero degree of difficulty optimization problems, life is no longer so simple. Here the solution must be found by maximizing $V(\delta)$ subject to the orthogonality and normality conditions. At first sight there would appear to be little advantage in substituting a constrained maximization for the original constrained design problems. However, it should be noted that the orthogonality and normality conditions are always linear irrespective of the degree of nonlinearity of the original design constraints. Furthermore, the number of free variables in $V(\delta)$ is equal to the degree of difficulty and not to the number of design variables. In certain problems this leads to a reduction in the dimensionality of the optimization problem. We shall return to solution procedures at a later stage in this chapter.

To sum up, having expressed our primal problem in the form

$$\text{minimize} \quad W(x)$$
$$\text{subject to} \quad 1 \geqslant g_i(x) \quad i = 1, \ldots, k$$

where $W(x)$ and the constraint functions $g_i(x)$ are posynomials, we introduce the function

$$\overline{W}(x, \Delta) = W(x)^{\Delta_1} g_1(x)^{\Delta_2} g_3(x)^{\Delta_3} \cdots g_k(x)^{\Delta_{k+1}}$$

which satisfies

$$W(x) \geqslant \overline{W}(x, \Delta)$$

when the constraints $g_i(n)$ are satisfied, $\Delta_1 = 1$ (normality condition) and $\Delta_2, \Delta_3, \ldots, \Delta_{k+1}$ are non-negative. The generalized arithmetic–geometric inequality gives

$$\overline{W}(x, \Delta) \geqslant V(x, \delta)$$

where $V(x, \delta)$ can be written in the form

$$V(x, \delta) = V(\delta) x_1^{h_1(\delta)} x_2^{h_2(\delta)} \cdots x_n^{h_n(\delta)}$$

Thus, if the dual variables satisfy the orthogonality conditions

$$h_i(\delta) = 0 \quad i = 1, 2, \ldots, n$$

we have

$$W(x) \geqslant \overline{W}(x, \Delta) \geqslant V(x, \delta) = V(\delta)$$

and if $\delta = \delta^*$ maximizes $V(\delta)$ subject to the normality and orthogonality conditions, $V(\delta^*)$ is a lower bound for $W(x)$ subject to the primal constraints. In fact, as we shall see in the next section, the problem

$$\text{maximize} \quad V(\delta)$$
$$\text{subject to} \quad \Delta_1(\delta) = 1$$

and

$$h_i(\delta) = 0 \quad i = 1, 2, \ldots, n$$

is a dual to the primal problem, and $W(x^*) = V(\delta^*)$ where x^* minimizes $W(x)$ subject to the primal constraints.

It should now be clear that we are not dealing with a mere optimization trick based upon the arithmetic–geometric mean, but with a firmly based general optimization technique. Although the authoritative work, is that of Duffin *et al.*[5], the reader may be interested in an alternative, and more engineering based, approach which can be found in a subsequent book due to Zener[6]. This latter book contains several examples of engineering design including a discussion of some structural optimization problems.

17.2 LAGRANGIAN INTERPRETATION

The power and attraction of geometric programming lies in the primal–dual relationship derived at the close of Section 17.2 by an appeal to the generalized geometric inequality. Although this provides a satisfactory approach for establishing the basic equations and functions, it does not reveal the full significance of the dual formulation. An illuminating and rigorous procedure is to re-create the dual problem by setting up the associated Lagrangian and employing the methods outlined in Chapter 2. By following this path we gain a clearer insight into the significance of the dual variables and can demonstrate the concavity of the dual objective function. Viewed in this light, geometric programming is simply a special case of convex programming. This alternative interpretation will not be pursued here since the interested reader may find a very adequate treatment in the book by R. Tyrell Rockafellar[7].

The particular method adopted in this section for creating the dual geometric problem is due to Zangwill[8], who attributes it in turn to M. Cannon and P. Wolfe. Again, we develop the arguments by employing the example discussed in the previous section; for convenience we present the analysis in a series of steps.

Step 1   We recall that the problem to be solved takes the form

$$\text{minimize} \quad W(x) = \rho x_1 x_2$$

$$\text{subject to} \quad 1 \geqslant g_1(x) = \left(\frac{P}{\sigma_0}\right) x_1^{-1} x_2^{-1} x_3$$

and

$$1 \geqslant g_2(x) = x_3^{-2} + x_2^2 x_3^{-2}$$

In order to cast the problem in a more manageable form we follow Duffin, Peterson and Zener [5] in employing the transformation

$$x_i = e^{y_i} \quad i = 1, 2, 3$$

and writing the coefficients of the polynomial terms in the objective and constraint functions as e^{b_k} ($k = 1, 2, 3, 4$):

$$\rho = e^{b_1}$$
$$P/\sigma_0 = e^{b_2}$$
$$1 = e^{b_3} = e^{b_4}$$

The problem becomes

$$\text{minimize} \quad e^{z_1}$$
$$\text{subject to} \quad 1 \geq e^{z_2}$$
$$\text{and} \quad 1 \geq e^{z_3} + e^{z_4}$$

where the new variables z_k ($k = 1, \ldots, 4$) are defined by

$$z_1 = b_1 + y_1 + y_3$$
$$z_2 = b_2 - y_1 - y_2 + y_3$$
$$z_3 = b_3 - 2y_3$$
$$z_4 = b_4 + 2y_2 - 2y_3$$

More compactly, we write

$$z_k - \sum_{j=1}^{3} a_{kj} y_i = b_k \qquad k = 1, 2, 3, 4$$

where each row $(a_{1j}, a_{2j}, a_{3j}, a_{4j})$ of the matrix **a**, called the exponent matrix, contains the exponents of the design variables x_i ($i = 1, 2, 3$) in the 4 posynomial terms of the problem. In this case,

$$\mathbf{a} = \begin{pmatrix} 1 & 0 & 1 \\ -1 & -1 & 1 \\ 0 & 0 & -2 \\ 0 & 2 & -2 \end{pmatrix}$$

Step 1 is completed by taking natural logarithms of the objective and constraint functions to obtain the transformed problem

$$\text{minimize} \quad \ln(e^{z_1})$$
$$\text{subject to} \quad 0 \geq \ln(e^{z_2})$$
$$0 \geq \ln(e^{z_3} + e^{z_4})$$
$$\text{and} \quad \mathbf{z} - \mathbf{a}\mathbf{y} = \mathbf{b}$$

where we have retained the terms $\ln(e^{z_1})$, $\ln(e^{z_2})$ in place of z_1, z_2 in order to preserve generality, and where $\mathbf{z} = (z_1, z_2, z_3, z_4)^T$, etc.

Step 2 Following the method described in Chapter 2, the Lagrangian associated with transformed problem can be written as

$$L(z, y, \lambda, \delta) = \ln e^{z_1} + \lambda_1 \ln e^{z_2} + \lambda_2 \ln(e^{z_3} + e^{z_4})$$
$$+ \sum_{k=1}^{4} \delta_k \left(b_k - z_k + \sum_{j=1}^{3} a_{kj} y_j \right)$$

and the corresponding dual problem is

$$\text{maximize} \quad L(z, y, \lambda, \delta)$$

Structural Optimization by Geometric Programming

subject to

$$\frac{\partial L}{\partial z_1} = \frac{e^{z_1}}{e^{z_2}} - \delta_1 = 1 - \delta_1 = 0$$

$$\frac{\partial L}{\partial z_2} = \lambda_1 \frac{e^{z_2}}{e^{z_2}} - \delta_2 = \lambda_1 - \delta_2 = 0$$

$$\frac{\partial L}{\partial z_3} = \frac{\lambda_2 e^{z_3}}{(e^{z_3} + e^{z_4})} - \delta_3 = 0$$

$$\frac{\partial L}{\partial z_4} = \frac{\lambda_2 e^{z_4}}{(e^{z_3} + e^{z_4})} - \delta_4 = 0$$

$$\frac{\partial L}{\partial y_j} = \sum_{k=1}^{4} \delta_k a_{kj} = 0 \qquad j = 1, 2, 3$$

or, alternatively $\mathbf{a}\boldsymbol{\delta} = \mathbf{0}$

together with $\lambda_i \geqslant 0, \delta_k \geqslant 0 \qquad i = 1, 2;\ k = 1, 2, 3, 4$

Although both sets of variables λ and δ act as Lagrangian multipliers, we shall follow the usual custom, in geometric programming, of referring to the set δ as the dual variables and reserve the term 'Lagrangian multiplier' for the set λ.

To conclude step 2 we introduce the term $\lambda_0 = \delta_1$ and observe that

$$y_j \left(\sum_{k=1}^{4} \delta_k a_{kj} \right) = 0 \qquad j = 1, 2, 3$$

and hence the dual problem reduces to

maximize $\quad L(z, \lambda, \delta) = \lambda_0 \ln e^{z_1} + \lambda_1 \ln e^{z_2} + \lambda_2 \ln (e^{z_3} + e^{z_4})$
$\qquad\qquad\qquad\qquad + \delta_1 (b_1 - z_1) + \delta_2 (b_2 - z_2) + \delta_3 (b_3 - z_3)$
$\qquad\qquad\qquad\qquad + \delta_4 (b_4 - z_4)$

subject to $\qquad \mathbf{a}^T \boldsymbol{\delta} = \mathbf{0}$

and $\quad \lambda_0 = \delta_1$

$\qquad \lambda_2 = \delta_2$

$$\frac{\lambda_2 e^{z_3}}{(e^{z_3} + e^{z_4})} = \delta_3$$

$$\frac{\lambda_2 e^{z_4}}{(e^{z_3} + e^{z_4})} = \delta_4 \qquad \delta_i \geqslant 0, i = 1, 2, 3, 4$$

Notice that combining the final two constraints gives

$$\lambda_2 = \delta_3 + \delta_4$$

indicating that the Lagrangian multiplier for the second primal constraint is composed of the sum of the associated dual variables. This is general and applies to any multi-termed constraint which is cast in the appropriate posynomial form.

Step 3 Although the final form of the dual problem given above is fairly compact, it is still possible to perform a further contraction. To do this we note that the constraint
$$\lambda_0 = \delta_1$$
is, in fact,
$$\frac{\lambda_0 e^{z_1}}{e^{z_1}} = \delta_1$$
Thus, by taking the natural logarithm of this expression we have
$$\ln(\lambda_0) + \ln(e^{z_1}) = \ln(\delta_1) + \ln(e^{z_1})$$
If we now multiply throughout by the dual variable δ_1 and selectively employ the expression $\delta_1 = \lambda_0$, we find that
$$\lambda_0 \ln(\lambda_0) + \delta_1 z_1 = \delta_1 \ln(\delta_1) + \lambda_0 \ln(e^{z_1})$$
Adding the term $\delta_1 b_1$ to each side and rearranging leads to the expression
$$\lambda_0 \ln(e^{z_1}) + \delta_1(b_1 - z_1) = \ln \lambda_0^{\lambda_0} + \ln\left(\frac{e^{b_1}}{\delta_1}\right)^{\delta_1}$$
$$= \ln\left\{\left(\frac{\rho}{\delta_1}\right)^{\delta_1} \lambda_0^{\lambda_0}\right\}$$

Taking the two, more complex constraints,
$$\frac{\lambda_2 e^{z_3}}{(e^{z_3} + e^{z_4})} = \delta_3$$
$$\frac{\lambda_2 e^{z_4}}{(e^{z_3} + e^{z_4})} = \delta_4$$
and proceeding in the same way, we obtain
$$\delta_3 \ln(e^{z_3} + e^{z_4}) - \delta_3 z_3 + \delta_3 b_3 = \delta_3 \ln \lambda_2 - \ln[\delta_3^{\delta_3}] + \delta_3 \ln(e^{b_3})$$
$$\delta_4 \ln(e^{z_3} + e^{z_4}) - \delta_4 z_4 + \delta_4 b_4 = \delta_4 \ln \lambda_2 - \ln[\delta_4^{\delta_4}] + \delta_4 \ln(e^{b_4})$$
by recalling $b_3 = b_4 = 1$, these may be combined to give
$$\lambda_2 \ln(e^{z_3} + e^{z_4}) + \delta_3(b_3 - z_3) + \delta_4(b_4 - z_4) = \ln \lambda_2^{\lambda_2} + \ln\left(\frac{1}{\delta_3}\right)^{\delta_3} + \ln\left(\frac{1}{\delta_4}\right)^{\delta_4}$$
$$= \ln\left\{\left(\frac{1}{\delta_3}\right)^{\delta_3}\left(\frac{1}{\delta_4}\right)^{\delta_4} \lambda_2^{\lambda_2}\right\}$$

The final dual constraint $\lambda_1 = \delta_2$ can be similarly manipulated and the dual problem becomes
$$\text{maximize} \quad L(\delta) = \ln\left\{\left(\frac{\rho}{\delta_1}\right)^{\delta_1}\left(\frac{p}{\sigma_0 \delta_2}\right)^{\delta_2}\left(\frac{1}{\delta_3}\right)^{\delta_3}\left(\frac{1}{\delta_4}\right)^{\delta_4} \lambda_0^{\lambda_0} \lambda_1^{\lambda_1} \lambda_2^{\lambda_2}\right\}$$

subject to
$$\mathbf{a}^T \delta = 0$$
where
$$\lambda_0 = 1, \lambda_1 = \delta_2, \lambda_2 = \delta_3 + \delta_4$$

Writing in $(v(\delta))$ in place of $L(\delta)$ then the objective function can take the alternative form

$$v(\delta) = \left(\frac{\rho}{\delta_1}\right)^{\delta_1} \left(\frac{p}{\sigma_0 \delta_2}\right)^{\delta_2} \left(\frac{1}{\delta_3}\right)^{\delta_3} \left(\frac{1}{\delta_4}\right)^{\delta_4} \lambda_0^{\lambda_0} \lambda_1^{\lambda_1} \lambda_2^{\lambda_2}$$

given in Section 17.1.

As indicated earlier, the Lagrangian multiplier associated with a multi-termed primal constraint is split into component dual variables. The values taken by such variables indicate the contribution to the numerical value of a constraint by a specific component within the posynomial. This provides a technique for deriving optimal values for the primal variables from a known solution to the dual problem. For example, denoting optimal values by asterisks, we have for the second primal constraint,

$$\frac{\lambda_2^* e^{z_3^*}}{(e^{z_3^*} + e^{z_4^*})} = \delta_3^*$$

$$\frac{\lambda_2^* e^{z_4^*}}{(e^{z_3^*} + e^{z_4^*})} = \delta_4^*$$

Assuming the constraint to be active,

$$e^{z_3^*} + e^{z_4^*} = 1$$

and transforming back to the original primal variables, we have,

$$\lambda_2^* (x_3^*)^{-2} = \delta_3^*$$
$$\lambda_2^* (x_2^*)^{-2} (x_3^*)^{-2} = \delta_4^*$$

or

$$(x_3^*)^{-2} = \delta_3^*/(\delta_3^* + \delta_4^*)$$
$$(x_2^*)^2 (x_3^*)^{-2} = \delta_4^*/(\delta_3^* + \delta_4^*)$$

which can be solved for x_2^* and x_3^*. Either the remaining constraint or the objective function can now be used to solve for the optimal value of the remaining design variable x_1^*. We note, in passing, that this type of consideration indicates that the optimum solution requires that the second primal constraint be active.

A direct corollary from the above arguments arises in the case of inactive constraints. Here the associated Lagrangian multiplier takes a zero value and the non-negative condition requires that all the constituent dual variables must also be zero. In the case of a numerical solution technique the passing of multi-term constraint from the active to the passive set implies that the constituent dual variables become zero *together*.

As may be guessed, the primal–dual relationship established for this simple problem can be generalized to form the general primal and dual problems of geometric programming. Consider an objective function g_0 which is the sum of m_0 terms dependent upon a set of variables $(x_1, x_2, \ldots, x_n)$,

$$g_0(x) = \sum_{i=1}^{m_0} c_i x_1^{a_{i1}} x_2^{a_{i2}}, \ldots, x_n^{a_{in}}$$

where, as in the case of the demonstration example, the c_i's are positive constants and the powers a_{ij} are real numbers. We can rewrite this expression in a more compact notation as

$$g_0(x) = \sum_{i=1}^{m_0} P_i(x)$$

where

$$P_i(x) = c_i \prod_{j=1}^{n} x_j^{a_{ij}}$$

which constitutes a posynomial formulation for the objective function. The general form for the primal program requires that we find a vector $\mathbf{x}$ which minimizes g_0 subject to the constraints

$$x_i \geq 0 \qquad i = 1, 2, \ldots, n$$

and

$$g_1(x) \leq 1, \quad g_2(x) \leq 1, \ldots, g_p(x) \leq 1$$

where these additional g's are given by

$$g_k(x) = \sum_{i=l_k}^{m_k} P_i(x) \qquad k = 1, \ldots, p$$

with

$$l_1 = m_0 + 1, \ldots, l_k = m_{(k-1)} + 1 \qquad k = 1, \ldots, p$$

and

$$m_0 = \text{the number of terms in } g_0$$
$$m_1 - m_0 = \text{the number of terms in } g_1$$
$$\vdots$$
$$m_p - m_{p-1} = \text{the number of terms in } g_p$$

noting that $m = m_p = $ the total number of terms, i.e. the summation of all the terms in all posynomials.

The dual programme associated with this primal requires finding a vector $\boldsymbol{\delta}$ that maximizes the function

$$\ln[v(\delta)] = \sum_{i=1}^{m} \delta_i \{\ln c_i - \ln \delta_i\} + \sum_{k=1}^{p} \lambda_k(\delta) \ln \lambda_k(\delta)$$

where

$$\lambda_k(\delta) = \sum_{i=l_k}^{m_k} \delta_i \qquad k = 1, \ldots, p$$

subject to the constraints

$$\delta_i \geq 0 \quad i = 1, \ldots, m$$

the normality condition,

$$\sum_{i=1}^{m_0} \delta_i = 1$$

and the orthogonality condition,

$$\sum_{i=1}^{m} a_{ij}\delta_i = 0 \quad j = 1, 2, \ldots, n$$

The function $\ln[v(\delta)]$ is termed the dual function, and the variables δ_i are called dual variables.

As in the case of the simple example, each dual variable δ_i is associated with the ith term $c_i x_1^{a_{i1}} \ldots x_n^{a_{in}}$ of the primal program, thus each term in each constant g_k is associated with a specific dual variable δ_i. Similarly, each factor $\lambda_k^{\lambda_k}$ is associated with a primal constraint $g_k(x) \leq 1$.

Finally, we observe that if the vector $\boldsymbol{\delta}^*$ is a maximizing point for the dual program then, analogous to the earlier demonstration, the minimizing point $\mathbf{x}^*$ for the primal problem satisfies the system of equations

$$\delta_i^* = \begin{cases} P_i(x^*)/g_0(x^*) & i = 1, 2, \ldots, m_0 \\ \lambda_k(\delta^*)P_i(x^*) & i = m_0 + 1, \ldots, m;\ k = 1, 2, \ldots, p \end{cases}$$

where each P_i in the expression $\lambda_k(\delta^*)P_i(x^*)$ is a term contained in the posynomial constraint equation $g_k(x^*)$. Furthermore, the usual Kuhn–Tucker conditions require

$$\lambda_k(\delta^*)[1 - g_k(x^*)] = 0 \quad i = 1, 2, \ldots, p$$

and, thus, the dual variables which correspond to tight constraints ($g_k(x^*) = 1$) are all positive, whilst those associated with loose constraints ($g_k(x^*) < 1$) are *all* zero.

17.3 REDUCTION TO INDEPENDENT VARIABLES

In the previous two sections we observed that the dual geometric programming problem contains a set of equality constraints known as the normality and orthogonality conditions. The existence of these constraints clearly implies that not all of the dual variables can be regarded as independent. Indeed, if the problem contains m dual variables and n orthogonality conditions, the number of independent variables is equal to $m - n - 1$ which was previously defined as the 'degree of difficulty' and which we denote henceforth by the letter d.

As with the case of linear programming, this property can be exploited to reduce both the number of variables employed in the problem and, potentially, the number of constraints. In order to realize this property we return to the general form for the dual given at the close of Section 17.2 where the problem is to find a vector δ^* which maximizes

$$\sum_{i=1}^{m} \delta_i (\ln c_i - \ln \delta_i) + \sum_{k=1}^{p} \lambda_k(\delta) \ln \lambda_k(\delta)$$

subject to the dual equality constraints

$$\sum_{i=1}^{m} \delta_i = 1 \qquad \text{(normality)}$$

$$\sum_{i=1}^{m} a_{ij} \delta_i = 0 \qquad j = 1, 2, \ldots, n \qquad \text{(orthogonality)}$$

and the inequality constraints $\delta_i \geq 0 \qquad i = 1, 2, \ldots, n$.

The process of reduction begins by dividing the dual vector δ into two groups with δ_I containing n variables and a set δ_{II} which contains the remainder. The orthogonality conditions can now be written as

$$\mathbf{P}\delta_I + \mathbf{Q}\delta_{II} = 0$$

and these equations solved in terms of the vector δ_{II} to give

$$\delta = \left\{ \frac{-\mathbf{Q}^T \mathbf{P}^{-1}}{\mathbf{I}} \right\} \delta_{II}$$

where $\mathbf{I}$ is a $(d+1) \times (d+1)$ identity matrix. The final step in the reduction is commenced by introducing the matrix $\mathbf{b}$, defined by

$$\tilde{\mathbf{b}} = \{\tilde{b}_{ij}\} = \left\{ \frac{-\mathbf{Q}^T \mathbf{P}}{\mathbf{I}} \right\} \qquad i = 1, 2, \ldots, d+1; j = 1, 2, \ldots, m$$

which gives the expressions

$$\delta = \tilde{\mathbf{b}} \delta_{II}$$

The normality condition is now satisfied by the two final moves in the analysis. Firstly, the number of independent variables is reduced by one through the introduction of a new vector of dual variables $\{1, r_1, r_2, \ldots, r_d\}$ and, secondly, by replacing the terms $\tilde{b}_{ij}$ by b_{ij}, where

$$b_{ij} = \frac{\tilde{b}_{ij}}{\sum_{k=1}^{m_0} \tilde{b}_{ik}} \qquad j = 1, 2, \ldots, m$$

$$b_{ij} = \frac{\tilde{b}_{ij}}{\sum_{k=1}^{m_0} \tilde{b}_{ij}} - b_{ij} \qquad \begin{array}{l} i = 2, 3, \ldots, d+1; \\ j = 1, 2, \ldots, m \end{array}$$

The vector of old dual variables δ can now be written in terms of the variables $\mathbf{r}$ through the expressions

$$\delta = \mathbf{b}\left\{\dfrac{1}{\mathbf{r}}\right\}$$

which give a set δ satisfying the orthogonality and normality conditions.

Using these new terms, the dual geometric programming problem can be defined as finding vector $\mathbf{r}^*$ which maximizes the function

$$\sum_{i=1}^{m} \delta_i(r)(\ln c_i - \ln \delta_i(r)) + \sum_{k=1}^{p} \lambda_k(r)\ln \lambda_k(r)$$

subject to the inequality constraints

$$\delta(r) = \mathbf{b}\left\{\dfrac{1}{\mathbf{r}}\right\} \geqslant 0$$

where

$$\lambda_k(r) = \lambda_k^{(0)} + \sum_{i=1}^{d} r_i \lambda_k^{(i)} \qquad k = 1, \ldots, p$$

with

$$\lambda_k^{(i)} = \sum_{j=l_k}^{m_k} b_{ij} \qquad i = 1, 2, \ldots, d+1$$

and where the terms l_k, m_k, p are defined in Section 17.2. The problem has, therefore, been reduced to that of maximizing a nonlinear objective function subject only to simple linear constraints. As seen in Chapter 8, this type of problem can be efficiently solved by employing a projected gradient technique. In addition, the objective function can be shown to be concave and the Hessian matrix easily evaluated, which indicates that a Newton direction be projected onto the constraints. Although the Newton-projected gradient method is not the only solution procedure available, it has proved to be fast and reliable when applied to a range of problems.

17.4 THE GENERATION OF MINIMUM-WEIGHT DESIGNS FOR STATISTICALLY DETERMINATE PIN-JOINTED FRAMEWORKS

In the earlier part of the chapter a simple problem involving the minimum-weight design of a stress-constrained pin-jointed framework is employed as a vehicle for developing the basic forms of geometric programming. We now return to this class of structure and examine a more complex problem with the aid of the technique developed in the previous section.

The particular framework considered is shown in Fig. 17.2 (page 590) and supports two alternative loading systems of magnitudes P_1 and P_2. The design variables are the cross-sectional areas x_i ($i = 1, 2, \ldots, 6$) and we seek a minimum-weight design subject to constraints on the vertical displacements of the free nodes and the bar stress levels. This problem can be defined as one of finding a vector $\{x^*\}$ which

$$\text{minimizes} \quad \frac{W}{\rho l} = 2(x_1 + x_2) + 2.236(x_3 + x_4 + x_5) + x_6$$

subject to the stress constraints

$$1 \geq \frac{T_i}{\sigma^*} x_i^{-1} \quad i = 1, 2, \ldots, 6$$

and the displacement constraints

$$1 \geq 8Ax_1^{-1} + 8Ax_2^{-1} + 5\sqrt{5}Ax_3^{-1} + 10\sqrt{5}Ax_4^{-1}$$

$$1 \geq 5\sqrt{5}Ax_4^{-1} + \frac{5\sqrt{5}}{2}Ax_5^{-1} + 2Ax_6^{-1}$$

$$1 \geq 4\sqrt{5}Bx_1^{-1} + 4\sqrt{5}Bx_2^{-1} + 10Bx_3^{-1} + 20Bx_4^{-1}$$

$$1 \geq \frac{35}{4}Bx_4^{-1} + \frac{25}{4}Bx_5^{-1}$$

where W is the structural weight and T_i the load occurring in the ith bar under either the influence of system P_1 or P_2, whichever gives rise to the numerically largest value (either tensile or compressive). The specific weight of the material of the structure is denoted by ρ, which is taken as constant for all bars. The terms A and B are given by

$$A = \frac{P_1 l}{\Delta^* E}, \quad B = \frac{P_2 l}{\Delta^* E}$$

where E denotes Young's modulus and l a fixed length shown in Fig. 17.2. Finally, Δ^* represents the maximum allowable vertical displacements at the nodes 1, 2, and $\pm \sigma^*$ is the maximum range for the individual bar stresses. This particular structural problem has been previously considered by Chern and Prager from a conventional optimality criterion approach [9].

The above formulation constitutes a primal geometric problem and the associated dual requires that a vector of dual variables $\{\delta^*\}$ be found which

$$\text{maximizes} \quad \ln(v(\delta)) = \ln\left\{\left(\frac{2}{\delta}\right)^{\delta_i}\left(\frac{2}{\delta_2}\right)^{\delta_2}\cdots\left(\frac{25B}{4\delta_{25}}\right)^{\delta_{25}} \lambda_1^{\lambda_1} \lambda_2^{\lambda_2} \lambda_3^{\lambda_3} \lambda_4^{\lambda_4}\right\}$$

subject to the normality condition $\sum_{i=1}^{6} \delta_i = 1$

the orthogonality conditions $\sum_{k=1}^{25} a_{jk} \delta_k = 0 \quad j = 1, 2, \ldots, 6$

and the positivity conditions $\delta_i \geq 0 \quad i = 1, \ldots$

where the Lagrangian multipliers associated with the displacements are given by

$$\lambda_1 = \delta_{13} + \delta_{14} + \delta_{15} + \delta_{16}$$
$$\lambda_2 = \delta_{17} + \delta_{18} + \delta_{19}$$
$$\lambda_3 = \delta_{20} + \delta_{21} + \delta_{22} + \delta_{23}$$
$$\lambda_4 = \delta_{24} + \delta_{25}$$

In order to proceed with the solution method of Section 17.3 we write the orthogonality conditions in a form where the exponent matrix is given explicitly, i.e.,

$$\begin{bmatrix} 1 & 0 & 0 & 0 & 0 & 0 \\ 0 & 1 & 0 & 0 & 0 & 0 \\ 0 & 0 & 1 & 0 & 0 & 0 \\ 0 & 0 & 0 & 1 & 0 & 0 \\ 0 & 0 & 0 & 0 & 1 & 0 \\ 0 & 0 & 0 & 0 & 0 & 1 \\ -1 & 0 & 0 & 0 & 0 & 0 \\ 0 & -1 & 0 & 0 & 0 & 0 \\ 0 & 0 & -1 & 0 & 0 & 0 \\ 0 & 0 & 0 & -1 & 0 & 0 \\ 0 & 0 & 0 & 0 & -1 & 0 \\ 0 & 0 & 0 & 0 & 0 & -1 \\ -1 & 0 & 0 & 0 & 0 & 0 \\ 0 & -1 & 0 & 0 & 0 & 0 \\ 0 & 0 & -1 & 0 & 0 & 0 \\ 0 & 0 & 0 & -1 & 0 & 0 \\ 0 & 0 & 0 & -1 & 0 & 0 \\ 0 & 0 & 0 & 0 & -1 & 0 \\ 0 & 0 & 0 & 0 & 0 & -1 \\ -1 & 0 & 0 & 0 & 0 & 0 \\ 0 & -1 & 0 & 0 & 0 & 0 \\ 0 & 0 & -1 & 0 & 0 & 0 \\ 0 & 0 & 0 & -1 & 0 & 0 \\ 0 & 0 & 0 & -1 & 0 & 0 \\ 0 & 0 & 0 & 0 & -1 & 0 \end{bmatrix}^T \begin{bmatrix} \delta_1 \\ \delta_2 \\ \delta_3 \\ \delta_4 \\ \delta_5 \\ \delta_6 \\ \delta_7 \\ \delta_8 \\ \delta_9 \\ \delta_{10} \\ \delta_{11} \\ \delta_{12} \\ \delta_{13} \\ \delta_{14} \\ \delta_{15} \\ \delta_{16} \\ \delta_{17} \\ \delta_{18} \\ \delta_{19} \\ \delta_{20} \\ \delta_{21} \\ \delta_{22} \\ \delta_{23} \\ \delta_{24} \\ \delta_{25} \end{bmatrix} = \mathbf{0}$$

The exponent matrix can be partioned into the two matrices **P** and **Q**, where, from Section 17.3,

$$\mathbf{P} = \begin{bmatrix} 1 & 0 & 0 & 0 & 0 & 0 \\ 0 & 1 & 0 & 0 & 0 & 0 \\ 0 & 0 & 1 & 0 & 0 & 0 \\ 0 & 0 & 0 & 1 & 0 & 0 \\ 0 & 0 & 0 & 0 & 1 & 0 \\ 0 & 0 & 0 & 0 & 0 & 1 \end{bmatrix}$$

$$\mathbf{Q} = \begin{bmatrix} -1 & 0 & 0 & 0 & 0 & 0 \\ 0 & -1 & 0 & 0 & 0 & 0 \\ 0 & 0 & -1 & 0 & 0 & 0 \\ 0 & 0 & 0 & -1 & 0 & 0 \\ 0 & 0 & 0 & 0 & -1 & 0 \\ 0 & 0 & 0 & 0 & 0 & -1 \\ -1 & 0 & 0 & 0 & 0 & 0 \\ 0 & -1 & 0 & 0 & 0 & 0 \\ 0 & 0 & -1 & 0 & 0 & 0 \\ 0 & 0 & 0 & -1 & 0 & 0 \\ 0 & 0 & 0 & -1 & 0 & 0 \\ 0 & 0 & 0 & 0 & -1 & 0 \\ 0 & 0 & 0 & 0 & 0 & -1 \\ -1 & 0 & 0 & 0 & 0 & 0 \\ 0 & -1 & 0 & 0 & 0 & 0 \\ 0 & 0 & -1 & 0 & 0 & 0 \\ 0 & 0 & 0 & -1 & 0 & 0 \\ 0 & 0 & 0 & -1 & 0 & 0 \\ 0 & 0 & 0 & 0 & -1 & 0 \end{bmatrix}$$

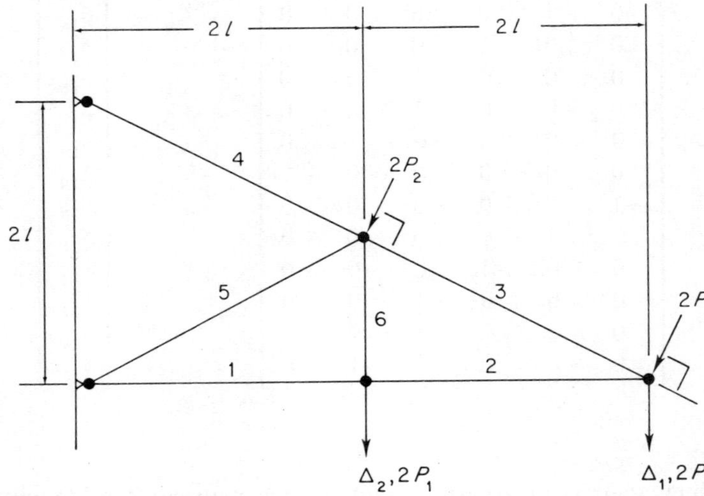

Fig. 17.2 Framework subject to multiple loads

Because the matrices **P** and **Q** are respectively diagonal and sparse, the form for $-\mathbf{Q}\mathbf{P}^{-1}$ is particularly simple; thus dual problem becomes:

maximize
$$\ln(v(\delta))$$
subject to

$$\begin{bmatrix} 1 & -1 & -1 & -1 & -1 & -1 \\ 0 & 1 & 0 & 0 & 0 & 0 \\ 0 & 0 & 1 & 0 & 0 & 0 \\ 0 & 0 & 0 & 1 & 0 & 0 \\ 0 & 0 & 0 & 0 & 1 & 0 \\ 0 & 0 & 0 & 0 & 0 & 1 \\ 1 & -1 & -1 & -1 & -1 & -1 \\ 0 & 1 & 0 & 0 & 0 & 0 \\ 0 & 0 & 1 & 0 & 0 & 0 \\ 0 & 0 & 0 & 1 & 0 & 0 \\ 0 & 0 & 0 & 0 & 1 & 0 \\ 0 & 0 & 0 & 0 & 0 & 1 \\ 1 & -1 & -1 & -1 & -1 & -1 \\ 0 & 1 & 0 & 0 & 0 & 0 \\ 0 & 0 & 1 & 0 & 0 & 0 \\ 0 & 0 & 0 & 1 & 0 & 0 \\ 0 & 0 & 0 & 1 & 0 & 0 \\ 0 & 0 & 0 & 0 & 1 & 0 \\ 1 & -1 & -1 & -1 & -1 & -1 \\ 0 & 1 & 0 & 0 & 0 & 0 \\ 0 & 0 & 1 & 0 & 0 & 0 \\ 0 & 0 & 0 & 1 & 0 & 0 \\ 0 & 0 & 0 & 0 & 1 & 0 \\ 0 & 0 & 0 & 0 & 0 & 1 \end{bmatrix} \begin{bmatrix} 1 \\ r_1 \\ r_2 \\ r_3 \\ r_4 \\ r_5 \end{bmatrix} \geqslant \mathbf{0}$$

This description of the dual linear constraints is achieved by exploiting the full process described in Section 17.3.

In order to get a feel for the kind of numbers involved, we can take a particular case with $P_1 = P_2 = P$, $P/\sigma^* = 4\,\mathrm{m}^2$ and $A = B = 1\,\mathrm{m}^2$. These values give a specific form to the dual objective function $\ln(v(\delta))$ and by applying a Newton-based projected gradient technique, the maximizing vector of dual variables $(\delta_1^*, \delta_2^*, \ldots, \delta_{25}^*)$ is found to be (0.18587, 0.18508, 0.23086, 0.32704, 0.052409, 0.01866, 0.18411, 0.18331, 0.22915, 0.32443, 0.00000, 0.00000, 0.00176, 0.00177, 0.00171, 0.00261, 0.00000, 0.00000, 0.00000, 0.00000, 0.00000, 0.0000, 0.05249, 0.01866) and $V(\delta^*) = 433.13$. Using the procedure of Section 17.3 this solution can be converted into the minimum for the primal structural optimization problem with cross-sectional areas (40.253, 40.081, 44.718, 63.348, 10.168, 8.082) and a minimum structural weight of $433.13\,\rho l$. Since the problem has been solved from the dual side we would expect the numerical procedure to give values which slightly violate the primal constraints; and this is indeed the case.

Having obtained a solution, we can use the information inherent in the dual form to examine the mathematical structure of the optimum. We begin by evaluating the optimizing values for the Lagrangian multipliers λ_i ($i = 1, 2, 3, 4$) which is achieved by employing the optimizing dual variables δ^* in the definitions of multipliers given above. This gives $\lambda_1^* = 0.923$, $\lambda_2^* = 0.000$, $\lambda_3^* = 0.009$, $\lambda_4^* = 0.000$, indicating that the two displacement constraints containing x_5 and x_6 are passive. Intuitively, we expect that the design variables x_5 and x_6 are contained within the set of stress constraints, which is confirmed by observing that $\delta_7 = \delta_8 = \delta_9 = \delta_{10} = 0.000$, whilst $\delta_{11} = 0.0062$ and $\delta_{12} = 0.019$. This separation of variables into stress or displacement constraints is a general rule which can be seen through the simple relationship which exists between the dual variables δ and the Lagrangian multipliers λ.

Once an optimum value has been found, the influence of a particular constraint on this value is readily calculated by setting to zero the associated dual variables δ and observing the effect on the dual function $\ln[v(\delta)]$. Recalling that we are dealing with a dual problem, this procedure leads to a lower bound on the true optimum and opens up the possibility of bounding the minimum by solving a sub-problem containing a reduced constraint set.

So far, it has been assumed that the problems being solved are constructed from posynomial forms, and the statically determinate framework employed above came within this category. However, with multiple loads applied to such a structure it is quite possible for negative signs to appear in the displacement constraint equations. Fortunately, in the case of statically determinate frameworks a transformation exists which restores the desired posynomial formulation. For more general structural optimization such an expedient may not exist and recourse must be made to a variant of geometric programming known as complementary geometric programming.

17.5 GEOMETRIC PROGRAMMING AND GENERAL PROBLEMS

Although the purpose of this section is to introduce the technique of complementary geometric programming, we begin by generalizing the ideas introduced in Section 17.1 where an auxiliary variable is required in order to create the required posynomial form.

Suppose that a given minimization problem contains a positive term of the form

$$F(x) = \frac{x_1 x_3 + x_1 x_2 + x_2 x_3}{x_1^2 - x_2^2 - x_1 x_3}$$

which may be part of the objective function or a constraint but is clearly a non-posynomial function. A posynomial form can now be generated by introducing two new variables x_4, x_5, through the relations

$$x_4 \geq x_1 x_3 + x_1 x_2 + x_2 x_3$$
$$x_5 \leq x_1^2 - x_2^2 - x_1 x_3$$

This reduces the original function to the very compact form

$$F(x) = x_4 x_5^{-1}$$

but also introduces two additional constraints

$$1 \geqslant x_1 x_3 x_4^{-1} + x_1 x_2 x_4^{-1} + x_2 x_3 x_4^{-1}$$
$$1 \geqslant x_1^{-2} x_2^2 + x_1^{-1} x_3 + x_1^{-2} x_5$$

The concept of introducing new variables can be applied for more general functions of the form

$$f(x) - \left(\sum_{i=1}^{k} C_i x_1^{a_{i1}} x_2^{a_{i2}} \ldots x_m^{Q_{im}} \right)^n$$

where n is an arbitrary real-valued number and, for $f(x)$ in the numerator all the coefficients are positive whilst for $f(x)$ in the denominator *only* C_1 is positive. Following the above arguments, we introduce new variables z_1 and z_2 where $z_1 \geqslant f(x)$ for $f(x)$ a numerator and $f(x) \geqslant z_2$ for $f(x)$ a denominator. These inequalities are important since they ensure that $f(x)$ and z_1 have the same infimum whilst $f(x)$ and z_2 share the same supremum. Thus the resulting geometric programming problems created by these transformations are compact.

Unfortunately, life is not usually so convenient as to supply problems which can be transformed into posynomial form by this technique. In these situations we must have recourse to an alternative method hinted at in the previous section whereby posynomial functions are created by an approximation procedure. At this stage two points emerge: first, any procedure must create functions which satisfy the inequalities of the last paragraph; second, it is clearly convenient to approximate each function by a single-term posynomial and in this way we can contract the size of the problem. However, the price paid is that we create a primal geometric programming problem which does not map the entire range of the original problem. The solution can now only be achieved by solving a sequence of geometric programming problems.

Before discussing the solution procedure in detail, let us look at the actual approximating formula to be employed. Consider a minimization problem which again contains a non-posynomial term $f(x)$ in either the objective function or a constraint. If this term is written as a function of n variables x_i, $i = 1, 2, \ldots, n$, then we seek an approximating single term posynomial $f(x, \tilde{x})$ defined by

$$f(x, \tilde{x}) = f(\tilde{x}) \prod_{i=1}^{n} \left(\frac{x_i}{\tilde{x}_i} \right)^{a_i}$$

with

$$a_i = \left[\frac{x_i}{f(x)} \frac{\partial f}{\partial x_i} \right]_{x = \tilde{x}}$$

and where $\tilde{\mathbf{x}} = (\tilde{x}_1, \tilde{x}_2, \ldots, \tilde{x}_n)$ represents a feasible point about which this approximation is taken. With the problem now cast in posynomial form, it can be solved by the method of Section 17.3 to give a point $\mathbf{x}^*$ which optimizes the new geometric programming problem. However, this represents only an estimate of the actual optimizing vector which may not be sufficiently close to constitute a satisfactory solution to the real problem. In such a case a new approximation is required giving rise to a second geometric programming problem which is again solved. Clearly, we are dealing with an algorithm which solves a sequence of geometric programs. This algorithm can be computerized in the following outlined form:

Step 1 Set $i = 0$ and select values for tolerance $\varepsilon_1, \varepsilon_2$

Step 2 Select a feasible approximating point

$$\tilde{x} \in X = \{x \mid x \in R^n, \quad g(x) \leqslant 1, \quad x \geqslant 0\}$$

Step 3 For non-posynomial functions form a single term approximately posynomial

Step 4 Solve the resulting geometric programming problem and obtain an estimate of the minimum value of the cost function $f(x^{(i)})$ and an estimate of the minimizing vector $x^{(i)}$

Step 5 If $f(x^{(i)}) - f(x^{(i-1)}) \leqslant \varepsilon_i$ and max $\left|x_k^{(i)} - x_k^{(i-1)}\right| \leqslant \varepsilon_2$ ($k = 1, \ldots, n$), go to Step 7

Step 6 Set $\tilde{x}_k = x_k^{(i)}$ ($k = 1, 2, \ldots, n$), and return to Step 3

Step 7 Accept $x^{(i)}$ as the optimizing vector and print results.

This whole algorithmic process has been termed 'complementary geometric programming' by Avriel and Williams [10], who noted that the process is guaranteed to converge to a stationary value providing the above inequalities are satisfied. For polynomial functions this represents no problem since the removal of negative terms in any numerator, or positive terms in a denominator, by the approximating scheme automatically satisfies the inequalities. This is, of course, providing the actual functions generate positive values. For non-polynomial functions there is no similar simple rule of thumb and each function must then be examined individually.

As an illustration of the method, consider the following simple problem:

$$\text{minimize} \quad g_o(x) = \left(x_2^2 - \frac{x_1}{2}\right)^2$$

subject to the constraints

$$1 \geqslant g_1(x) = x_1$$

$$1 \geqslant g_2(x) = \frac{2x_1^{-1}}{(1 + 2x_2^2)}$$

$$x_1 \geqslant 0, \, x_2 \geqslant 0$$

The algorithm is started by setting the tolerances $\varepsilon_1, \varepsilon_2$ at 0.1 and selecting $x_1 = 1$, $x_2 = 1$ as a suitable approximating point $\tilde{x}$. Consider the objective function $g_0(x)$; then

$$g_0(\tilde{x}) = \tfrac{1}{4}$$

If the approximating formulae are then applied it is found that

$$a_1 = \frac{x_1}{g_0} \frac{2g_0}{2x_1}\bigg|_{x=\tilde{x}} = -2$$

and, similarly, $a_2 = 8$, giving approximate form for $g_0(x)$,

$$g_0(x,\tilde{x}) = \tfrac{1}{4} x_1^{-2} x_2^8$$

The other non-posynomial $g_2(x)$ is approximated by the function

$$g_2(x,\tilde{x}) = \tfrac{2}{3} x_1^{-1} x_2^{-4/3}$$

This completes Step 3 of the algorithm and the problem is now one of finding the minimum of

$$g_0(x,\tilde{x}) = \tfrac{1}{4} x_1^{-2} x_2^8$$

subject to

$$1 \geqslant g_1(x) = x_1$$
$$1 \geqslant g_2(x,\tilde{x}) = \tfrac{2}{3} x_1^{-1} x_2^{-4/3}$$
$$x_1 \geqslant 0, \quad x_2 \geqslant 0$$

The problem has a zero degree of difficulty which can be easily solved to yield a solution $x_1^{(1)} = 1, x_2^{(1)} = 0.738, f(x^{(1)}) = g_0(x,\tilde{x}) = 0.0219$. Taking these values as the new approximating point, the cycle is repeated to give the solution $x_1^{(2)} = 1, x_2^{(2)} = 0.708, f(x^{(2)}) = g_0(x,\tilde{x}) = 0.0001$. We now check these numbers against the previously set tolerances and observe that $f(x^{(2)}) - f(x^{(1)}) = 0.0128 < \varepsilon_1$, and $\max |x_1^{(2)} - x_1^{(1)}| = 0.03 < \varepsilon_2$. Hence the second iteration gives values which satisfy the previously set tolerances and thus it provides a satisfactory solution to the main problem.

17.5.1 Application to a Design Problem

Having developed an elaborate piece of machinery for solving complex problems by casting them into geometric programming form, we now apply this procedure to an actual design problem. The example is concerned with designing a ship bulkhead subject to a set of design code constraints. It is typical of many problems where structures are constructed according to a set of design or cost codes and where the objective function depends upon the structural dimensions. In such cases the relevant functions can often be expressed in terms of algebraic forms which can be transformed into posynomials.

The particular problem being described here has been previously solved by Kavlie, Kowalik and Moe [11] and concerns the design of a corrugated bulkhead for a liquid cargo tanker shown in Fig. 17.3. The requirement is to obtain a minimum-weight design by permitting variations in bulkhead thickness and shape whilst conforming to the design limitations imposed by the Norwegian shipping code *Norske Veritas*. The code lays down rules which control the

596 Foundations of Structural Optimization: A Unified Approach

buckling strength of the panel, the bending stresses, the corrosion vulnerability and the geometric shape.

Several preliminaries are necessary and the first task requires the calculation of the maximum bending stress σ_b according to the rules of *Norske Veritas*. It is assumed that each stiffener shown in Fig. 17.3(a) is clamped at each end and subject to a constant load $\gamma h x_2$, where γ is the specific gravity of water (0.001 kg/cm^3), h is the mean height of pressure at the mid-span, and x_2 is the dimension shown in Fig. 17.3(b). Considering the stiffener as a beam, the

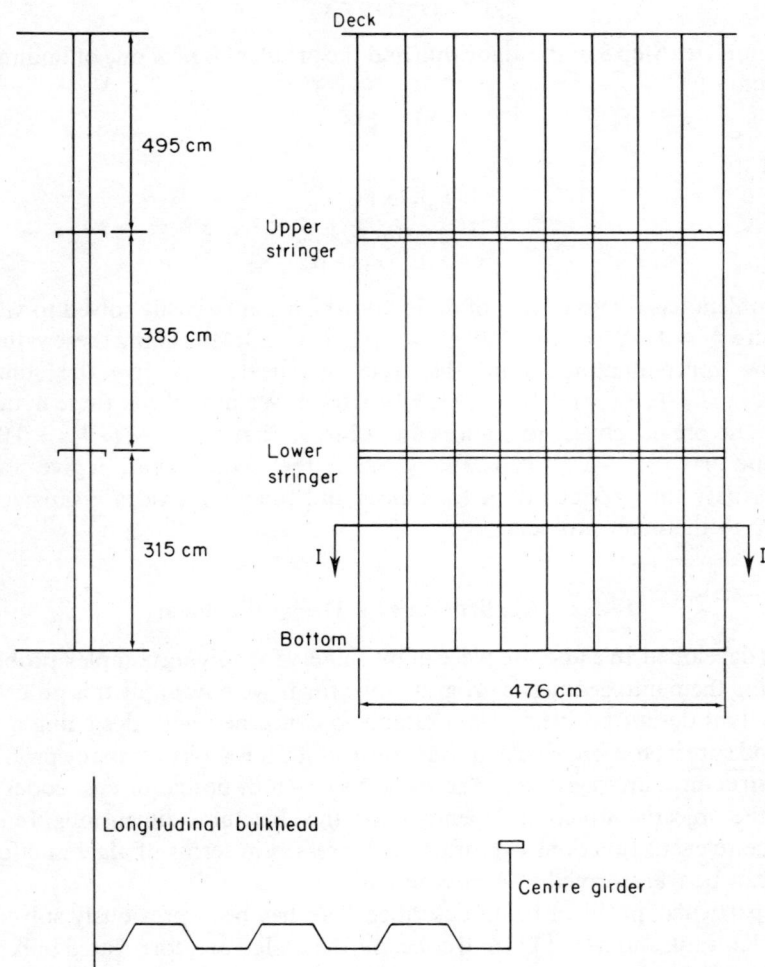

Fig. 17.3(a) Bulkhead layout

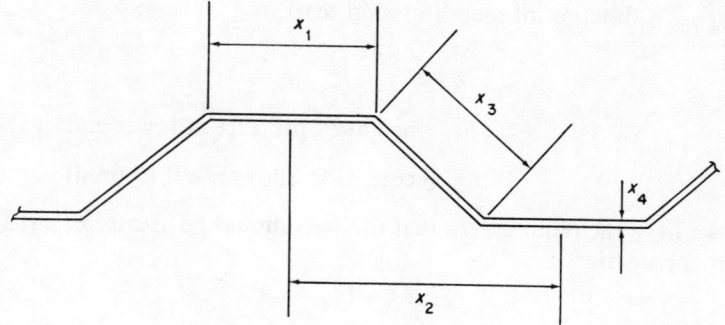

Fig. 17.3(b) Design variables

maximum bending moment is given by

$$M = \frac{\gamma h x_2 l^2}{12} \quad (\text{kg}-\text{cm})$$

where l is the length of the stiffener measured from top to bottom. Thus the maximum bending stress is given by

$$\sigma_b = \frac{M}{Z}$$

where z is the section modulus, which is obtained by Kavlie, Kowalik and Moe in the form

$$Z = \frac{x_4}{2}\left(\frac{x_3}{3} + x_1 e\right)(x_3 - (x_2 - x_1)^2)^{1/2}$$

where e is a term representing the effectiveness of the flange and is dimensionless. The code then indicates that the maximum bending stress σ_b must not exceed $1200\,\text{kg/cm}^2$ which, together with the last two equations, gives rise to the constraint.

$$1 \geq \frac{6\gamma h x_2 l^2 x_2 x_4^{-1}}{(3x_1 e + x_3)(x_3^2 - (x_2 - x_1)^2)^{1/2}}$$

The second constraint is derived from a requirement that the section modulus $J = 0.5 Z (x_3^2 - (x_2 - x_1)^2)^{1/2}$ be restricted, which leads to

$$1 \geq \frac{12(\gamma h x_2 l^2)^{4/3} x_4^{-1}}{2.2(3x_1 e + x_3)(x_3^2 - (x_2 - x_1)^2)^{1/2}}$$

The thickness of each of the panels is subject to the constraint

$$x_4 \geq \begin{cases} x_{4\,\text{min}} \\ 3.9b\sqrt{h_1} + K_2 \end{cases}$$

where $x_{4\,\text{min}}$ = function of length of ship (cm)

$$b = \begin{cases} 1.05x_1 \\ 1.05x_3 \end{cases} (m)$$

h_1 = height of pressure at lower end of the panel

K_2 = corrosion allowance 0.15 (mm)

Finally we have the requirement that the web should be greater or equal to the depth of corrugation, i.e.:

$$1 \geqslant x_4^{-1}(x_3^2 - (x_2 - x_1)^2)^{1/2}$$

In discussing this problem in detail it is more convenient not to deal with the entire problem but to examine the optimizing of the bottom panel alone. For this panel the weight, which is the objective function, may be defined as

$$W(x) = 476\rho\, 315 x_4 (x_1 + x_3)/x_2$$

where ρ is the density of the material in ton/cm^3, and following Kavlie et al. is set at 7.850×10^{-6}. Grouping together these various terms, the problem becomes:

minimize $\quad W(x) = g_0 = 1.177(x_1 x_2^{-2} x_4 + x_2^{-2} x_3 x_4)$

$$1 \geqslant g_1 = \frac{53.64 x_2 x_4^{-1}}{(2.4x_1 + x_3)(x_3^2 - (x_2 - x_1)^2)^{1/2}}$$

$$1 \geqslant g_2 = \frac{26.4(8.94 x_2)^{4/3} x_4^{-1}}{(2.4x_1 + x_3)(x_3^2 - (x_2 - x_1)^2)^{1/2}}$$

$$1 \geqslant g_3 = 0.0156 x_1 x_4^{-1} + 0.15 x_4^{-1}$$

$$1 \geqslant g_4 = 0.0156 x_3 x_4^{-1} + 0.15 x_4^{-1}$$

$$1 \geqslant g_5 = 1.05 x_4^{-1}$$

The two constraints g_1 and g_2 contain non-posynomial terms which require approximation, but before doing this greater accuracy can be achieved by introducing the new variable x_5 by the relationship

$$1 \geqslant x_5^2 x_3^{-2} + (x_2 - x_1)^2 x_3^{-2}$$

The term $(x_2 - x_1)^2$ now requires approximation in addition to $(2.4 x_1 + x_3)$.

Following the algorithm of Section 17.5, the problem should be solved iteratively, but it turns out that a single step is sufficient. We form the approximating geometric program by using the approximating point $x_1 = x_3 = 50$, $x_2 = 90$, and the primal problem is

minimize $\quad g_0 = 1.177(x_1 x_2^{-1} x_4 + x_2^{-1} x_3 x_4)$

subject to

$$1 \geqslant g_1 = 15.77 x_1^{-0.706} x_2 x_3^{-0.294} x_4^{-1} x_5^{-1}$$

$$1 \geqslant g_2 = 143.65 x_1^{-0.706} x_2^{1.333} x_3^{-0.294} x_4^{-1} x_5^{-1}$$

Structural Optimization by Geometric Programming

$$1 \geqslant g_3 = x_5^2 x_3^{-2} + 0.045 x_1^{-2.5} x_2^{4.5} x_3^{-2}$$
$$1 \geqslant g_4 = 0.0156 x_1 x_4^{-1} + 0.15 x_4^{-1}$$
$$1 \geqslant g_5 = 0.0156 x_3 x_4^{-1} + 0.15 x_4^{-1}$$
$$1 \geqslant g_6 = 1.05 x_4^{-1}$$

which has a degree of difficulty of 5.

The exponent matrix associated with this problem is therefore

$$\begin{pmatrix} 1 & -1 & 0 & 1 & 0 \\ 0 & -1 & 1 & 1 & 0 \\ -0.706 & 1 & -0.294 & -1 & -1 \\ -0.706 & 1.333 & -0.294 & -1 & -2 \\ 0 & 0 & -2 & 0 & 2 \\ -2.5 & 4.5 & -2 & 0 & 0 \\ 1 & 0 & 0 & -1 & 0 \\ 0 & 0 & 0 & -1 & 0 \\ 0 & 0 & 1 & -1 & 0 \\ 0 & 0 & 0 & -1 & 0 \\ 0 & 0 & 0 & -1 & 0 \end{pmatrix}$$

In order to obtain the transformed variables **r** according to the method of Section 17.3, the matrix comprising the first five columns of the above must be inverted. Unfortunately, this matrix is singular but it can be made non-singular by the simple expedient of interchanging various columns. In the present case we place the third row beneath the eleventh, thus effectively moving all but the first and second rows up one place. Performing the algebraic operations gives a set of transformed dual variables:

$$\begin{bmatrix} \delta_1(r) \\ \delta_2(r) \\ \cdot \\ \cdot \\ \cdot \\ = \\ \cdot \\ \cdot \\ \cdot \\ \delta_{11}(r) \end{bmatrix} = \begin{bmatrix} -0.07795 & 0.64250 & 0.64250 & 0.64250 & 0.64250 & 1.06580 \\ 1.07795 & -0.64250 & -0.64250 & -0.64250 & -0.64250 & -1.06580 \\ 0.37495 & 0.37495 & 0 & -0.37495 & -0.37495 & -1.87420 \\ 0.37495 & -0.37495 & 0 & -0.37495 & -0.37495 & -0.62458 \\ 0.10890 & 0.10882 & 0 & 0.10882 & 0.10882 & -0.00002 \\ 0.62505 & +\ -0.62505 & -0.62505 & -0.62505 & -0.62505 & -0.62505 \\ 0 & 1 & 0 & 0 & 0 & 0 \\ 0 & 0 & 0.62505 & 0 & 0 & 0 \\ 0 & 0 & 0 & 1 & 0 & 0 \\ 0 & 0 & 0 & 0 & 1 & 0 \\ 0 & 0 & 0 & 0 & 0 & 2.49925 \end{bmatrix} \begin{bmatrix} r_1 \\ r_2 \\ r_3 \\ r_4 \\ r_5 \end{bmatrix}$$

The transformed dual problem is then one of finding a vector r_i^* ($i = 1, 2, \ldots, 5$) which maximizes

$$\ln[V(r)] = \delta_1(r) \ln(1.177) - \delta_1(r) \ln(\delta_1(r)) + \delta_2(r) \ln(1.177) - \delta_2(r) \ln(\delta_2(r))$$
$$+ \delta_3(r) \ln(15.77) + \delta_4(r) \ln(143.65) + \delta_6(r) \ln(0.045) - \delta_6(r) \ln(\delta_6(r))$$
$$+ \delta_7(r) \ln(0.015) - \delta_7(r) \ln(\delta_7(r)) + \delta_8(r) \ln(0.15)$$
$$- \delta_8(r) \ln(\delta_8(r)) + \delta_9(r) \ln(0.015) - \delta_9(r) \ln(\delta_9(r))$$

$$+ \delta_{10}(r)\ln(0.15) - \delta_{10}(r)\ln(\delta_{10}(r)) + \delta_{11}(r)\ln(1.05)$$
$$+ (\delta_5(r) + \delta_6(r))\ln(\delta_5(r) + \delta_6(r))$$
$$+ (\delta_7(r) + \delta_8(r))\ln(\delta_7(r) + \delta_8(r))$$
$$+ (\delta_9(r) + \delta_{10}(r))\ln(\delta_9(r) + \delta_{10}(r))$$

subject to the positivity conditions

$$\delta_i(r) \geqslant 0 \qquad i = 1, 2, \ldots, 11$$

Although this problem requires a computer algorithm in order to achieve a complete solution, it is possible to employ the bound procedures on this sub-problem. In fact, the solution to the current sub-problem is close enough to that of the actual design to suggest these bounds bracket the desired optimum. A wide range of possibilities exist for bounding the solution but we shall consider only the simplest option by setting $r_i = 0$ ($i = 2, \ldots, 5$). This leaves r_1 as the only variable and by taking account of the positivity of the variables $\delta(r)$ we find that this variable must lie within the range

$$1 \geqslant r_1 \geqslant 0.121\,318$$

Any number within this range can be selected and substituted into the expressions for $\delta_j(r)$. In particular, $r_1 = 0.5$ gives a vector $(\delta_1, \delta_2, \ldots, \delta_{11})$ with values (0.243, 0.757, 0.187, 0.187, 0.163, 0.312, 0.500, 0, 0, 0), which gives $V(r) = 1.11$. However, the original starting design given by Kavlie *et al.* has a weight 1.79 tons for the bottom panel. Thus the minimum weight for this panel is 1.45 ± 0.34 tons and a trivial calculation has yielded an answer which constitutes a meaningful bound on the optimum.

This ability of geometric programming to yield lower bounds can be exploited in a much more adventurous way. We could seek a bound by removing the dual variables δ_5 and δ_6 which, in turn, would remove the influence of the constraint $1 \geqslant g_3$. In addition to deleting constraints, this procedure also reduces the degree of difficulty of the structural optimization problem. By removing sufficient constraints the problem can be reduced to one having a zero degree of difficulty with the ensuing simplicity in achieving a solution. For many problems the structural engineer can exploit this technique to ascertain the influence of the constraints by solving a group of low-order, or zero, degree of difficulty each using a different selection of the total number of constraint equations. In addition to giving the designer a feel for the problem, it can also be used to suggest sensible starting values for the solution of the main problem.

17.6 CONDENSED GEOMETRIC PROGRAMMING

In the previous section we approximated only those terms which did not present themselves in posynomial form. However, it has been suggested that, because single-term posynomials or monomials are easy to handle, all constraints should be reduced to this form even where they occur in multi-term posynomial form.

Although this form of condensed programming is not guaranteed to create compact problems, it has proved extremely effective in the solution of certain structural optimization problems [12].

As a vehicle for discussion we employ the problem of finding the minimum-weight design of a structure of given geometry subject to specific constraints. Weight is only one example of a design parameter, and other linear objective functions can be employed without affecting the arguments; however, a nonlinear function would require a modified theory. Following the normal custom, it is assumed that the structural responses are modelled by the finite element method and that the components of the resulting stiffness matrix are linear in terms of the design variables. Although this normally implies that the finite elements employed in the formulation are simple, it is by no means necessary that this should be the case. For simplicity a single loading case is assumed, and the design constraints are taken to be limits on element stress levels and nodal displacements. Nevertheless, the arguments advanced below apply when satisfaction of more general constraints is necessitated and also in the case of alternative loading systems.

Taking into account these considerations, the design problem may be posed in the following form. Choose a set of design variables $x_1, x_2, \ldots, x_n$ which minimize the objective function

$$W(x) = \sum_{i=1}^{n} \omega_i x_i, \quad \mathbf{x} = [x_1, x_2, \ldots, x_n]^T$$

subject to the constraints   MWP

$$\bar{u}_j \geq u_j(x) \geq -\bar{u}_j \quad j = 1, \ldots, m$$
$$\bar{\phi}_i \geq \phi_i(x) \quad i = 1, \ldots, n$$

where ω_i represents the specific weight of element i, u_j is the maximum allowable displacement of certain prescribed nodes of which there are m, and ϕ_i is a yield function for the ith element and may be a direct stress in the case of bar elements or a more general criterion for complex elements. The term $\bar{\phi}_i$ denotes the maximum permitted value which the yield function may attain in the ith element. This now constitutes the primal structural optimization problem but does not necessarily take the form appropriate to a standard primal geometric programming problem. At this point it is now possible to employ the device mentioned above where all the constraints are reduced to single-term posynomials to yield a condensed problem. The basic philosophy of complementary programming can then be utilized and the solution to the design problem sought by means of a sequence of geometric programming sub-problems. Although recourse to a sequence of optimization problems may appear to be a time-consuming artifice, the simplicity of the dual formulation which results from a condensed geometric programming interpretation leads to an efficient algorithm for the class of problems being considered.

If the log-linear approximation of Section 17.5 is taken for the constraints of the structural optimization problem, the resulting forms are given by

$$\bar{u}_j \geq u_j(\tilde{x}) \prod_{i=1}^{n} (x_i/\tilde{x}_i)^{e_{ji}} \quad j = 1, 2, \ldots, m; \quad \hat{\mathbf{x}} = [\tilde{x}_i, \tilde{x}_2, \ldots, \tilde{x}_n]^T$$

$$\bar{\phi}_p \geq \phi_p(\tilde{x}) \prod_{i=1}^{n} (x_i/\tilde{x}_i)^{e_{ki}} \quad p = 1, \ldots, n; \, k = m+1, \ldots, m+n$$

where x denotes a feasible point about which the approximation is taken. The exponents are given by

$$e_{ji} = \left[\frac{x_i}{u_j(x)} \frac{\partial u_j}{\partial x_i}\right]_{x=\tilde{x}} \quad j = 1, 2, \ldots, m; \quad i = 1, 2, \ldots, n$$

$$e_{ki} = \left[\frac{x_i}{\phi_p(x)} \frac{\partial \phi_p}{\partial x_i}\right]_{x=\tilde{x}} \quad p = 1, 2, \ldots, n; \quad k = m+1, \ldots, m+n;$$
$$i = 1, 2, \ldots, n$$

where the derivatives are calculated using the assumed linearity of the stiffness matrix as follows:

$$\frac{\partial \mathbf{u}}{\partial x_i} = \mathbf{K}^{-1}(\mathbf{k}_i/x_i)\mathbf{u}, \quad \mathbf{u} = [u_1(x), u_2(x), \ldots u_m(x)]^T$$

$$\frac{\partial \phi_p}{\partial x_i} = \frac{\partial \phi_p}{\partial \sigma_{kj}} \frac{\partial \sigma_{kj}}{\partial x_i}$$

where $\mathbf{K}$ is the global stiffness matrix and k_i the stiffness matrix of the element associated with the design variable x_i. The components of the stress tensor σ_{kj} for each specific point are obtained from the vector of modal displacements $\mathbf{u}$ by employing the usual matrix of material constants and direction cosines.

Applying the approximating forms and the above expressions, we can obtain the primal geometric sub-problem which forms an approximation to the full primal structural optimization problem. Thus the new problem is:

$$\text{minimize} \quad W(x) = \sum_{i=1}^{n} c_i x_i$$

subject to

$$1 \geq g_j(x) = c_q \prod_{i=1}^{n} (x_i/\tilde{x}_i)^{e_{ji}} \quad j = 1, 2, \ldots, n+m; \quad q = n+1, \ldots, 2n+m \quad (17.1)$$

$$x_i \geq 0 \quad i = 1, 2, \ldots, n$$

with
$$c_i = \omega_i \quad i = 1, 2, \ldots, n$$
$$= u_j(\tilde{x})/\bar{u}_j \quad i = n+1, \ldots, n+m; \quad j = 1, \ldots, m$$
$$= \phi_k(\tilde{x})/\bar{\phi}_k \quad i = n+m+1, \ldots, 2n+m; \quad k = 1, \ldots, m$$

which is a primal geometric programming problem with a degree of difficulty $d = m + n - 1$.

Following the usual procedures, we can generate the associated dual geometric programming problem in terms of a set of independent dual variables $\mathbf{r} = [r_1, r_2, \ldots, r_d]^r$. The problem becomes

maximize
$$\ln[v(r)] = \sum_{i=1}^{2n+m} \delta_i(r) \ln c_i - \sum_{i=1}^{n} \delta_i(r) \ln \delta_i(r)$$

subject to
$$\delta_i(r) = b_{0i} + \sum_{j=1}^{d} r_j b_{ji} \geq 0 \qquad i = 1, 2, \ldots, 2n+m$$

$$r_j \geq 0 \qquad j = 1, 2, \ldots, d \tag{17.2}$$

where the components b_{ij} satisfy the subsidiary conditions

$$\sum_{i=1}^{n} b_{0i} = 1$$

$$\sum_{i=1}^{n} b_{ji} = 0 \qquad j = 1, 2, \ldots, d$$

This dual problem can be solved to yield a maximizing vector $\mathbf{r}^*$ and, consequentially, a minimizing vector $\mathbf{x}^*$ for the associated primal problem. However, these values represent only an estimate of the actual optimization vectors which may not be sufficiently close to constitute a satisfactory solution to the real structural design problem. In this situation the solution vector $\mathbf{x}^*$ may be taken as the approximating point and a new geometric programming problem can be constructed about this point. It is tempting to demand that $\mathbf{x}^*$ be a feasible point for the original structural optimization problem, but this requirement is superfluous as demonstrated in the next section.

We may observe that the dual problem has a large linear programming component; indeed, if the term $\sum_{i=1}^{n} \delta_i(r) \ln \delta_i(r)$ is deleted from the objective function, the problem is entirely linear. This linear programming problem is important and is used to establish an initial set of active constraints when numerical solutions are sought.

Although we indicated in the previous section that the use of condensed geometric programming, in the manner described, does not necessarily create a compact set, procedures are available to circumvent this difficulty. In particular, the introduction of move limits on the primal design variable is sufficient and is also easily accommodated. Move limits are used frequently in this book to provide artificial constraints on noncompact sub-problems in the solution of general nonlinear optimization problems where some form of linear approximation is employed.

The dual problem (17.2) with the possible addition of move limits represents the actual problem whose properties we shall now examine. It should be noted that it is quite proper to solve the primal problem (17.1) in place of (17.2), and indeed, it is possible to argue that this forms the most effective solution process.

17.6.1 Some Properties of Condensed Structural Optimization Problems

17.6.1.1 Concavity

An important advantage of the geometric programming method is the ready availability of the Hessian matrix for the dual objective function, and this opens the way for Newton-type methods of solution. However, for the particular form of the condensed geometric programming dual used in the present analysis the concavity properties of the objective function are restricted, and this requires a controlled use of methods relying on inverting the Hessian.

For the dual problem (17.2), twice differentiating the objective function gives a compact Hessian in the form

$$H(l,j) = - \sum_{i=1}^{n} \frac{b_{il} b_{ij}}{\delta_i} \qquad j, l = 1, 2, \ldots, d$$

which is a symmetric $d \times d$ matrix. Concavity may now be proved by showing that it is negative definite. However, the relationship between the n primal design variables and the d dual variables make it clear that this Hessian can only be nonsingular if $d \leq n$. In addition, there exists conditions on the components of H, namely

$$\sum_{i=1}^{n} b_{ij} = 0 \qquad j = 1, 2, \ldots, d$$

which lead to a singular matrix if $n = d$.

These restrictions on the nonsingular form of the Hessian matrix require that the original theorem of Duffin et al. [5] concerning the concavity of the dual function $\ln[v(r)]$ must be slightly modified for the present form of condensed geometric programming:

Theorem *Suppose that a condensed geometric program of the form* (17.1) *has a degree of difficulty d with $n - 1 \geq d > 0$, and let the column of the matrix b_{ij} ($i = 1, 2, \ldots, 2n + m$; $j = 1, 2, \ldots, d$) satisfy the conditions $\sum_{i=1}^{n} b_{ij} = 0$ ($j = 1, 2, \ldots, d$) if the submatrix with elements b_{ij} for $i = 1, 2, \ldots, n$ and $j = 1, 2, \ldots, d$ has rank d, then the Hessian matrix H is nonsingular for strictly positive values.*

Consideration of the expression for the degree of difficulty d, coupled with the observation that each primal constraint is a single term posynomial, establishes the following corollary to this theorem:

Corollary *The solution to the primal structural optimization problem with n design variables has at most n active constraints.*

Although this limit on the number of potentially active constraints is advantageous in controlling the problem size, it does require that any solution technique taking advantage of the availability of the Hessian must contain an effective active set strategy.

17.6.1.2 Optimality criteria for condensed geometric programs

In order to gain a clearer insight into the properties of condensed geometric programming, we need to focus attention on the optimality criteria associated with the method. For convenience we take the displacement constraints only for the original primal problem (MWP) and an alternative form for the dual problem which requires finding vectors δ, λ which

$$\text{maximize} \quad V(\delta, \lambda) = \prod_{i=1}^{n} \left(\frac{c_i}{\delta_i}\right)^{\delta_i} \prod_{j=1}^{m} \left(\frac{u_j}{\bar{u}_j \tilde{x}^{e_{ji}} \ldots \tilde{x}^{e_{jn}}}\right)^{\lambda_j}$$

subject to a normality condition

$$\sum_{i=1}^{n} \delta_i = 1$$

and orthogonality conditions

$$\delta_1 + \sum_{j=1}^{m} \lambda_j e_{ji} = 0$$

where the vectors δ, λ are now the dual variables. This new description is the more common form for the dual problem but can be transformed easily into the formulation (17.1). In order to reduce the number of constraint equations, a set of values for $\delta_i (i = 1, 2, \ldots, n)$ may be chosen which satisfy the normality condition exactly, namely

$$\delta_i = \frac{c_i x_i}{W(x)}$$

and the dual condensed problem becomes:

maximize

$$V(\lambda) \qquad\qquad (17.3)$$

subject to

$$\frac{c_i x_i}{W(x)} + \sum_{j=1}^{m} \lambda_j e_{ji} = 0$$

This formulation represents a particular form for the optimality conditions of the condensed geometric programming problem associated with a displacement-constrained structural optimization problem.

At first sight this does not provide any new insight into the nature of the formulation, but the significance of the method can be illuminated by considering the optimality-criterion approach used in some of the more successful structural optimization computer programs as shown in Chapter 5. The iterative formulas for the optimality criterion method are constructed by satisfying the differential forms of the associated Lagrangian function. In the normal derivation the Lagrangian is constructed for the standard displacement-constrained structural optimization problem. For the present case it is more convenient to employ a logarithmic version of the structural optimization problem but at the same time following the logic associated with the derivation of optimality criterion methods. Thus, the new primal problem becomes one of finding a vector **x** which

minimizes

$$\ln[W(x)] = \ln\left[\sum_{i=1}^{n} c_i x_i\right]$$

subject to

$$\ln \bar{u}_j \geqslant \ln u_j \quad j = 1, \ldots, m$$

The Lagrangian associated with this problem is given by

$$L(x, \mu) = \ln\left[\sum_{i=1}^{n} c_i x_i\right] + \sum_{j=1}^{m} \mu_j \{\ln u_j - \ln \bar{u}_j\}$$

with the optimality criterion obtained by differentiating this function with respect to the primal design variables:

$$\frac{\partial L}{\partial x_i} = \frac{c_i}{W(x)} + \sum_{j=1}^{m} \mu_j \frac{1}{u_j} \frac{\partial u_j}{\partial x_i} - 0$$

or

$$\frac{c_i x_i}{W(x)} + \sum_{j=1}^{m} \mu_j e_{ji} = 0$$

By setting $\mu_j = \lambda_j$ ($j = 1, 2, \ldots, m$) we can now recover the constraint equations of the dual condensed geometric programming problem. The two approaches are thus attempting to satisfy the same optimality equations and in this sense are equivalent.

Although both techniques are members of the same family and are similar in requiring a structural analysis to set up the basic equation, they differ in the actual method of solution. The geometric programming approach establishes a sub-problem which is optimized by maximizing (17.3) subject to the dual constraints. The sub-optimization problem is simple in structure, and a variety of linear or quadratic programming methods are available for its solution. Candidate solutions for the minimizing design variables x_i^* ($i = 1, \ldots, n$) are therefore provided by the solutions to those sub-problems. By contrast, optimality criterion methods perform no sub-optimization and estimate the minimizing values for the design variable $\mathbf{x}^* = [x_1^*, x_2^*, \ldots, x_n^*]^T$ from up-date formulas based on a constraint satisfaction philosophy.

17.6.1.3 Scalar invariance

Because the approximating procedure used in the generation of condensed geometric programs is not conservative, in the case of structural optimization there is no guarantee that the optimum for a given sub-problem is a feasible point for the original problem. However, as we indicated in Section 17.6, the approximating procedure requires a feasible point $\mathbf{x} = [x_1, x_2, \ldots, x_n]^T$ for the generation of a new condensed geometric programming sub-problem. Normally, with structural design problems feasible points are generated by scaling the design with respect to the most violated constraint. This procedure could be applied in the present case but is rendered unnecessary by the fact that the condensed geometric programming formulation is invariant to scaling. Thus, the requirement that the approximating point is feasible is no longer necessary.

Since the constraint functions are all homogeneous functions, the scalar invariance conjecture can best be proved by considering a general positively homogeneous polynomial function $f(x)$ of degree t. Applying the approximation scheme to the function about a point $\mathbf{x}$, we obtain single-term posynomial

$$f(\tilde{x}, x) = f(\tilde{x}) \prod_{i=1}^{n} \left(\frac{x_i}{\tilde{x}_i}\right)^{e_i(\tilde{x})}, \qquad e_i(\tilde{x}) = \frac{x_i}{f(\tilde{x})} \frac{\partial f}{\partial x_i}\bigg|_{x=\tilde{x}}$$

If a uniform scaling s is taken, then the function becomes

$$f(s\tilde{x}, x) = f(s\tilde{x}) \prod_{i=1}^{n} \left(\frac{x_i}{s\tilde{x}_i}\right)^{e_i(s\tilde{x})}$$

$$= s^t f(\tilde{x}) \prod_{i=1}^{n} \left(\frac{x_i}{\tilde{x}_i}\right)^{e_i(s\tilde{x})} \prod_{i=1}^{n} \left(\frac{1}{s}\right)^{e_i(s\tilde{x})}$$

$$= \frac{s^t f(\tilde{x})}{s^\gamma} \prod_{i=1}^{n} \left(\frac{x_i}{\tilde{x}_i}\right)^{e_i(s\tilde{x})}$$

where

$$\gamma = \sum_{i=1}^{n} e_i(s\tilde{x})$$

Now,

$$e_i(s\tilde{x}) = \frac{s\tilde{x}}{s^t f(\tilde{x})} s^{t-1} \frac{\partial f}{\partial x_i}\bigg|_{x=\tilde{x}} = e_i(\tilde{x})$$

and further, by Euler's theorem,

$$\sum_{i=1}^{n} e_i(\tilde{x}) = \frac{t f(\tilde{x})}{f(\tilde{x})} = t$$

hence,

$$f(s\tilde{x}, x) = \frac{s^t f(\tilde{x})}{s^t} \prod_{i=1}^{n} \left(\frac{x_i}{\tilde{x}_i}\right)^{e_i(\tilde{x})} = f(\tilde{x}, x)$$

It may be observed that a similar analysis is available for negatively homogeneous polynomials, and thus the conjecture is proved.

Turning to the actual structural optimization problem, the constraint functions $u_j(x)$ and $\phi_i(x)$ ($j = 1, 2, \ldots, m; i = 1, 2, \ldots, n$) can be described in terms of homogeneous polynomials. Thus, our earlier remark that feasible values for the strucutral design variables are not required is seen to be justified. Any algorithm attempting to solve the structural problem by condensing the constraints to single-term polynomials does not therefore theoretically require any constraint satisfaction steps. In addition, scalar invariance implies that a problem which is dependent upon a single dominant design variable is solved exactly by optimizing a single condensed geometric programming problem.

17.6.1.4 Linearity

We conclude this section by briefly noting that the most obvious property of the dual problem (17.2) is the strong linearity of the formulation and that this can be used advantageously in any computerized algorithm. Such linearities are to be expected since the essence of geometric programming, and a central point of the theory, is the exploitation of inherent linearities. In the case of (17.2), the only nonlinear term occurs in the dual objective function and is associated with the linear form for the structural weight used in the objective function for the primal problem. If the primal objective function is also reduced to a single-term posynomial, the associated dual geometric programming problem is entirely linear.

A total linearization of the dual problem is not usual, and solutions are normally sought with the aid of a standard mathematical programming technique. In view of the existence of the Hessian and the linearity of the constraint, a strong candidate method for solving the dual problem is a Newton-based projected gradient technique. The application of such a technique then brings into play the restriction on the number of constraints described in Section 7.6.1.1. Under this restriction an effective active-set strategy is required with the additional necessity of selecting an initial active set. The strong linearity of the dual problem can now be exploited for making such an initial selection.

This may be achieved by deleting the logarithmic term from the objective function of (17.1), leaving a function of the form

$$\sum_{i=1}^{n} \delta_i(y) \ln c_i$$

which is to be maximized subject to the dual linear constraints. The resulting linear program is easily solved to provide an initial set of active constraints which satisfy the corollary of Section 17.6.1. These may be noted and the main dual problem solved by re-introducing the nonlinear term into the objective function. The principle of this approach is similar to those introduced in the study of convex separable analysis [13].

17.6.2 Conclusions

The use of an approximation formula for collapsing the constraints of optimum design problems is known to produce a powerful solution algorithm which employs a sequence of geometric programs. The present chapter indicates that the power of the technique is based on a formulation which exhibits some important properties. Two of these are of particular significance and are concerned with the dual condensed geometric programming/optimality criterion correspondence and the strong linear nature of the dual method. The first clearly shows that the geometric programming approach and the optimality criterion method are in reality members of the same family of structural optimization methods. The second indicates that there is a strong similarity between a solution technique based a sequence of condensed geometric programs and a sequence of linear programs. The corollary is that there exists a stronger familiar correspondence between the sequential linear programming and the optimality criterion methods of solving structural optimization problems than is generally recognized.

The concavity properties of the dual problem clearly impose limitations on the number of active constraints which would cause numerical problems when solutions to the sub-problems are being sought. However, the inherent linearity may be exploited to provide a procedure for selecting a satisfactory number of active constraints. An additional numerical advantage is the invariance property which alleviates the usual necessity for scaling when the optimized sub-problem generates an infeasible point for the main structural optimization problem.

Although many algorithms exist for the solution of geometric programming problems, the special properties of the condensed geometric programming problem associated with optimized structural design suggest that improvements in numerical performance are available.

A.J.M.

REFERENCES

1. G. H. Hardy, J. E. Littlewood and G. Ploya, *Inequalities*, Cambridge University Press, 1959.
2. N. D. Kazarinoff, *Geometric Inequalities*, Random House, 1961.
3. R. J. Duffin, Cost minimization problems treated by geometric means, *Operations Research*, **10**, 1962.
4. C. M. Zener, A mathematical aid in optimising engineering designs, *Proc. Natn. Acad. Sci.*, **47**, 537–9, 1961.
5. R. J. Duffin, E. L. Peterson and C. M. Zener, *Geometric Programming*, John Wiley, 1967.
6. C. M. Zener, *Engineering Design by Geometric Programming*, Wiley–Interscience, 1971.
7. R. Tyrell Rockerfeller, *Convex Analysis*, Princeton University Press, 1970.
8. W. I. Zangwill, *Non-Linear Programming: A Unified Approach*, Prentice-Hall, 1969.

9. J-M. Chern and W. Prager, Minimum weight design of statically determinate trusses subject to multiple constraints, *Int. J. Solids Struct.*, **7**(8), 931–40, 1971.
10. M. Avriel and A. C. Williams, Complementary geometric programming, *SIAM J. Appl. Math.*, **19,** 125–41, 1970.
11. D. Kavlie, J. Kowalik and J. Moe, Structural optimisation by means of non-linear programming, *International Symposium on the Use of Digital Computers in Structural Engineering,* University of Newcastle upon Tyne, 1967.
12. A. B. Templeman and S. K. Winterbottom, Structural design applications of geometric programming, *AGARD 21 Symposium on Structural Optimisation*, Milan, Italy, 1973.
13. G. B. Dantzig, *Linear Programming and Extensions*, Princeton University Press, 1963.

Index

acceleration procedures, 340
A-conjugate directions, 309–310
adjoint variables, 482–483
aerodynamic
   coefficients, 482–483
   forces, 483
aerodynamics, 6
aerodynamics unsteady, 427
aeroelasticity
   coefficients, 458, 461, 482
   static, 451, 482
aircraft
   industry, 13
   spoiler, 398
   technology, 1
airfoil, 427
algorithm, 2, 8, 36–8, 40
   augmented Lagrangian, 516
   conjugate direction, 307
   conjugate gradient, 308–309, 332
   convergence, 39, 113, 115, 158, 179, 185, 193, 201, 209, 249, 259, 263, 267, 270, 283, 307, 311, 325, 328, 337–338, 340, 345, 352, 369, 375, 378, 380–381, 387–388, 394, 397, 399, 422–3, 428–429, 448, 456, 463, 478, 495
   domain of convergence, 39
   linear convergence, 40–41
   order of convergence, 39
   quadratically convergent, 40
   speed of convergence, 40
   superlinear convergence, 40
   convergence ratio, 352
   descent, 280, 302
   dual, 369, 378, 384, 386, 499, 599
   efficiency, 41
   first order projection, 326
   generalized reduced gradient, 516
   gradient projection, 322, 330, 332, 375
   implicit, 97
   iteration, 41
   primal, 384, 386, 399
   quasi-Newton, 313, 315
   second order, 384
   second order dual, 385
   second order primal, 384
   second order projection, 329, 384
   starting piont, 39
   unconstrained minimization, 339
   usable feasible directions, 445
aluminium, 152
aluminium alloy, 6, 250
A-orthogonal, 317
applied load vector, 133
approximate
   analysis, 416, 432
   problem, 491
   relations, 132
approximations
   explicit, 462, 488
   first order, 381, 389, 390, 394, 489, 491, 524
   log linear, 601
   Rayleigh–Ritz, 427
   zero order, 381, 390
arches, 6
axial, 431
axis
   principal 283

bars, 514
bar cross-sectional areas, 3, 7, 45
   forces, 45, 47, 68, 72, 166
   stresses, 45

barrier function transformation, 360
  basic principles, 43
  basis, 328–329
  condensed, 461–462, 471
beam, 4
  element, 87
  of uniform strength, 5
bending, 4, 431
  moment, 442, 564
bilinear form, 550
binomial theorem, 107, 125
bisection
  iteration, 290
black box, 83
boundary point, 373
bounding the optimum weight, 54
bounds, 57, 72, 345, 525, 592, 600
  upper, 33, 60, 274, 277, 330–331, 392
  lower, 33, 60, 63, 274, 277, 290, 330, 392
boron epoxy, 146
box beam, 133
buckling, 7, 78, 126, 147, 415
  antisymmetric, 596
  constraints, 7, 80, 433, 515, 520, 533, 535
  load, 78, 104, 158, 274, 364, 428, 563, 564
  mode, 7, 128, 562
  resonance, 444
  symmetric, 564
  strength, 596

canonical convergence rate, 352
cantilever
  beam, 88
  box beam, 145
Carnot cycle, 6
Cartesian coordinate system, 15
Cauchy's steepest descent, 37–38, 41
Cauchy
  Schwarz inequality, 316
CDC
  Cyber-173, 97
Choleski decomposition, 303
circular cylindrical shell, 7
coalescing buckling mode, 7
column
  elastically supported, 562
compatibility
  conditions, 239, 244, 249, 252, 254
  evaluations, 250

composite
  elements, 152
  material, 133, 395–396, 408, 412, 467, 469, 471
  wing structure, 133
compression, 6
computer time, 437
conditions for constrained minima, 26
condition number, 283
configuration parameters, 482
conjugate
  directions, 276, 319
  gradient method, 310–311
  mass matrix, 91
constrained
  non-fully, 260
  optimization problem, 24
constraints
  active, 29–30, 35, 115, 128–130, 143, 158, 162, 166, 192, 240, 246, 249, 250, 257, 263, 271, 279, 320–321, 325, 327, 337, 350, 353–354, 373, 430, 458, 489, 492, 585, 604
  active behavior, 369
  active side, 320
  adjoint, 548
  approximate, 375, 389, 492
  aeroelastic, 464
  behaviour, 369, 370, 373, 389
  compatibility, 244, 246, 252, 259
  concave, 30, 32
  critical point, 94, 496
  cumulative, 444
  deflection, 390, 396, 430
  design, 595
  detection, 246
  discard, 258
  discontinuous, 443
  displacement, 54, 75, 105, 115, 130–133, 143, 162, 239, 241, 263–264, 365, 369, 430, 488, 492–493, 513, 601
  dual, 57
  element stress, 245
  equality, 24, 26, 109, 119, 274, 344, 357, 394
  flexibility, 396
  flutter, 444, 467
  frequency, 390, 396, 430
  gauge, 4, 240

Index

gradient of, 73, 106, 160, 280, 320–322, 330–331, 368, 380, 468, 526
inactive, 29, 35, 240, 249, 322, 348, 373, 585
inequality, 24, 27, 29, 110, 347–348, 357, 369
linear, 52, 320, 325, 327–328, 330, 332, 351, 375, 384, 531
linear equality, 326, 328
linear inequality, 326
linearized, 49, 350, 390, 460, 523, 526
linearized stress, 390
local, 447
local buckling, 145, 412, 419, 420
multiple displacements, 57, 71, 109, 110, 121, 123–124, 128
non-linear, 51, 257, 350, 352, 371, 373
passive, 4, 99, 115, 130–131, 157, 161
population, 259
primal, 368, 579
qualification, 30
redundant, 496
regular linear, 320
side, 274, 276, 285, 330–331, 336, 360, 369, 381, 390, 421, 435
single, 58
stress, 47–48, 66–67, 75, 104, 115, 118–119, 123–124, 130–133, 137, 143, 149, 157–168, 227, 234, 241, 249, 364–365, 381, 389, 390, 428, 432, 458, 467, 488, 490, 492–493, 515, 519, 532, 535, 537, 574, 588, 601
twist, 149
violated, 342, 355, 374
continuity
functions, 11
control
efficiency, 464
parameter-stepsize, 117
controlling parameter, 342
convergence, 2
acceleration, 345
control parameter, 375, 378, 380
critieria, 281, 296
finite step, 316
global, 296, 329, 351, 356
monotonic, 383
order of, 301
order two, 304
rate, 345
ratio, 283, 297
convexity, 20, 49, 52–53
convex programming problem, 30–31, 275
corrosion vulnerability
allowance, 598
vulnerability, 596
cost function, 3
cost of manufacture, 3
CP time, 97, 145, 162
Crank–Nicholson (control difference), 96
cubic interpolation, 291, 293, 297
curve fitting, 290

damping coefficients, 406, 471, 483
data base, 83, 426
dead weight, 6
degree of difficulty, 577, 585, 600, 602, 604
degree of difficulty zero, 577, 595
derivatives, 73, 77, 84, 96, 370, 440
aeroelastic coefficients, 461, 483
approximate, 262
constrained, 27, 29, 277, 419, 420, 531
displacement, 74
dynamic optimization, 461
flutter speed, 462, 483
higher order, 440, 443
mass matrix, 92
material, 547, 549
modules, 83
natural frequencies, 461, 483
Rayleigh quotient, 568
second, 293, 386
stiffness matrix, 92, 438
stress, 74–75
transient response, 461
descent
condition, 321
methods, 279
design
cycle times, 1
fully constrained optimal, 258
fully stressed, 390, 430
fully vertex, 270
group, 491
sensitivity method, 546
ship bulkhead, 595
variable linking, 386, 432
variables, 3, 7, 8, 185, 274
vertex, 267

DFP Davidon–Fletcher–Powell, 315, 327, 332, 340
differentiability, 11
differentiation chain, 419–420
dimension of the space, 14
Dirac delta function, 94
direct operating cost
    operating cost, 3
    stress, 7
directional derivative, 318
directions
    conjugate, 316–317
    descent, 352
    Newton, 326
    orthogonal, 280
    steepest descent, 325
    usable feasible, 335
displacements, 3, 7, 45, 46, 239, 274, 428
    buckling, 434
    constrained, 49, 52, 53
    finite element, 237
    generalized, 520
    limits, 430
    method, 101
    vector, 102, 517
discontinuities, 278, 368, 429
discontinuity planes, 368–369, 385, 398
divergence, 302, 408
dome structure, 157
downhill direction, 280, 283, 318–320
dual, 36
    form, 384
    formation, 67
    formulation 367, 369
    mini-max, 488, 496
    sub-space, 385
duality, 31, 33, 54, 73, 514
    gap, 263
    principle, 72
    theory, 72
dummy load method, 52, 65, 75, 86, 451, 460
dynamic
    loads, 99
    pressure, 483
    stiffness, 100

eigenfrequencies, 392
eigenmodes, 390, 415, 440
eigenspace, 569
eigenvalue, 78, 91, 283, 303, 338, 340, 344

derivatives, 79
diagaonalizing, 570
double, 564, 566
fundamental, 571
lowest, 78
multiple, 559
non-differential, 567
repeated, 561, 565, 567–569
repeated directional derivative, 570
unit, 315, 317
eigenvector, 78, 283 303, 315, 317, 566–567
    B-normal, 569–570
    M-normal, 566, 568
efficiency of algorithms, 39
elements, 12
    active, 130, 192
    bar, 145, 364, 514
    beam, 364, 432, 441
    bending, 386, 394
    membrane, 396, 444, 446, 514
    membrane plate, 364
    panel, 436
    passive, 108, 115, 130, 131
    plate, 105, 246
    plate bending, 394
    rod, 433
    shear panel, 396
    shell, 394, 395
    skin beam, 430
    stiffness matrices, 74–75, 77, 87, 102
    super, 439
    thickness, 73, 75
engineering beam theory, 442
equations of equilibrium, 85
equations non-linear, 350
equilibrium,
    conditions, 45, 68, 243
    equations, 69, 103, 165
equivalent interior constraints, 93
errors
    round off, 330
estimates
    first and second order multiplier, 330
Euclidean
    metric, 319, 325
    norm, 318
Euler
    forward difference, 96
    theorem, 607
extrapolation schemes, 340

Index

extension
    linear, 341
    quadratic, 341

factorization, 304
failure criteria (Hill), 468, 471
feasible
    condition, 352
    design, 60, 103, 371
    directions, 352–354
    direction methods, 335, 379
    direction usable, 352
    domain, 320, 350, 353, 373
    dual space, 278
    function, 527
    point, 24, 320, 322, 339, 593, 603
    region, 4, 24, 50, 52, 93
    solution, 543
    starting point, 371
flange, 464
flexion, 390
flexibility coefficient, 102, 106–107, 115, 118, 129–130, 132–133, 161, 220, 227, 365, 383
flutter, 408, 426, 471
    speed, 408, 451, 458, 463, 471, 484
finite
    difference, 83, 415, 421–422, 425
    approximations, 41
    procedures, 96, 422
finite dimensional space, 3, 4
finite element, 2, 3, 8, 104
    analysis, 73, 364, 371, 426, 441, 564
    bar, 145, 364, 514
    beam, 364, 432, 441
    bending, 386, 394
    force method, 238, 241, 244, 247, 263
    membrane, 73, 145, 240, 364, 396, 441. 444, 446, 514
    method, 78, 99, 101, 428, 460, 513, 546, 601
    model, 75, 86, 96, 149, 273, 391, 399, 438
    package, 83
    program, 83, 415
    plate bending, 394
    shear panel, 396
    shell, 394–395
first-order Taylor series, 374, 379, 394
formula recursion, 350
forced linear-vibration problem, 78

forces, 140
    element, 243
    internal, 374
    internal membrane, 242
    redundant, 237, 243–244, 252
    self-equilibrating, 237, 242
frequency, 91, 274, 483
    fundamental, 90–91
fully-stressed design, 7, 119–120, 140, 162, 215, 227, 250, 252, 255, 267, 380, 423, 425
function, 16
    augmented Lagrangian, 344, 346, 348
    auxiliary, 338–339
    barrier, 336
    concave, 22, 30, 33, 276, 530–531
    convex, 22, 26, 30, 33, 66, 276, 526
    differentiable, 17, 26
    dual, 277, 278, 297, 346–347, 368–369, 398
    dual objective, 55, 59, 63, 68, 69, 77, 494, 530–531
    extended barrier, 340
    global minimum, 26
    gradient, 17
    gradient objective, 352, 371
    graph, 17
    hyperplane, 17
    ill-conditioned, 338
    inverse, 336
    inverse barrier, 340
    linear, 22, 318, 525
    local minimum, 26
    logarithmic, 336
    logarithmic barrier, 337, 360
    mapping, 17
    monotonic decreasing, 337
    non-quadratic objective, 310
    objective, 3, 24, 63, 274, 280, 283, 297, 311, 322, 326, 327, 352, 364, 373, 490
    positive definite quadratic, 327
    quadratic, 40, 283, 288, 290, 307, 316
    quadratic extended barrier, 341
    quadratic loss, 342
    Ritz, 416
    second derivatives (dual), 368
    separable objective, 384
    shape, 73
    tangent, 18
    zangwill loss, 342

generalized
  Newton method, 303
  norm, 319
  optimality criterion, 363–365
geometric stiffness matrix, 78–79, 85, 87, 90
global
  efficiency, 387
  mass matrix, 91
  minimum, 25
  solution, 25, 332, 479
  stiffness matrix, 74
gradient, 303
  computation, 426
  constraints function, 421, 522
  direction, 38
  finite difference, 432
  methods, 282
  negative, 283, 321
  negative projected, 554
  objective function, 331, 350, 373, 421, 522
  primal constraint, 277
  projection, 321, 331, 351, 398, 546
graphite epoxy, 146, 152

H-conjugate, 327
hinges
  elastic, 562
hole shape, 547
Hyperplane, 320, 368
Hyperplane
  active constraint, 325

I-beam, 390, 395
ill-conditioned, 303, 342, 345, 347
indeterminate
  frameworks, 66
  pin-jointed framework, 72
inertia
  polar moment, 560
inequality
  generalized geometric arithmetic, 575, 576
infeasible primal points, 346
inner product, 313
instabilities, 3
integrated backwards, 95
interior penalty function method, 93
internal bar forces, 45, 64
internal bar distribution, 255

interval of uncertainty, 287, 288
invariances
  scalar, 607
inverse
  Hessian, 328
  thickness, 76
  variables, 77
isolines, 353
iteration, 36
  index, 345
  sub, 471

Kuhn–Tucker
  conditions, 30, 32–33, 53, 58, 66–67, 71, 73, 77, 100, 239, 241, 274–275, 321, 337, 366, 543, 566, 568
  optimality, 76–77, 79
  point, 527, 541
  sufficient conditions, 62

Lagrangian, 76, 105, 119, 121, 124, 127, 239, 241, 263, 325, 529, 553, 579, 580
  function, 32–33, 35, 53, 55, 58, 61–62, 66–67, 69, 71, 277, 319, 344, 351, 356–358, 384, 386, 493, 606
  multipliers, 27, 53, 54, 57–59, 61, 63, 68, 70, 77, 99, 107–109, 111, 113, 115–116, 121–124, 126, 128–131, 134, 139, 143, 160, 166, 173, 179, 185, 192–193, 209, 210, 220, 227, 238–241, 246, 248–249, 252, 259, 262, 267, 270, 274, 276, 319, 320, 322, 327–329, 331, 337, 342, 344–346, 348, 357, 366–369, 489, 495–496, 548, 552
  multiplier analysis, 565
  multipliers negative, 115, 134
  seperable, 364
laminated fibre composite skin, 398
language command
  command, 426–427
  job control, 420
lateral heat transfer, 95
line
  search, 281, 283, 285–286, 294, 303, 311, 315, 320, 322, 332, 352, 354, 373, 385
  segment, 21

linear
  approximations, 325, 340, 350, 603
  extrapolation, 423
  interpolation, 291
  programming, 35, 238, 247, 249, 265, 267, 270, 274, 328, 353, 479, 523, 546, 585, 602
  stability, 78
  temperature element, 96
  variety, 308
linearly
  constrained minimization problem, 275, 299, 319
  dependent set, 15
  design, 249
  indpendent set, 15, 315, 317
linearized
  dual, 72
  forms, 374
  problem, 375, 378, 386, 388
linearization
  methods, 354
loading
  multiple static, 364
local
  buckling, 7, 463, 464
  element buckling, 100, 537, 540
  minimum, 24, 30, 353
  solution, 24
load factor, 127
loads
  gravity, 520
looping, 420
lower bounds, 4

Mach number, 483
margins, 478
mass density, 123
mathematical analysis, 12
mathematical
  programming, 2, 3, 11, 100, 263, 273, 489
  programming problem, 274, 340
matrix, 16
  derivative, 263, 479
  exponent, 580, 589, 599
  flexibility, 244, 254
  Hessian, 17, 23, 26, 117, 276–277, 303, 311–312, 317, 319, 325–326, 329, 338–339, 342, 344, 346, 352, 358, 369, 385–386, 587, 604

Jacobian, 17
linearly independent, 306
mass, 89–90, 406, 416, 483
orthogonal projection, 350
positive-definite, 23
positive semi-definite, 26, 53, 302, 304, 306, 314–316, 326, 332, 570
projection, 321–322, 328, 335
rank, 319, 321, 330
sparse, 590
singular, 304
stiffness, 244, 380, 406, 416, 441, 451, 460, 483, 517–518, 601–602
symmetric positive-definite, 319
maximum allowable stress, 143
maximum strain failure criteria, 396
mean
  geometric, 573
  arithmetic, 573
members
  passive, 367
members of a set, 12
method
  adjoint variable, 550
  barrier function, 336, 340
  BFGS, 317, 358
  conjugate gradient, 531
  cutting plane, 355
  dual, 337, 345, 380, 390, 394, 489, 523, 578
  envelope, 267
  first order projection, 335, 367
  gradient projection, 319, 329, 350, 374, 383, 385
  mixed, 370, 375, 394, 398–399
  multiplier, 344, 347, 360
  newton-projected gradient, 587, 591, 608
  penalty, 341, 346
  primal, 345, 350, 353, 399, 489
  projection, 371, 373
  rank one, 317
  secant, 291
  simplex, 328
  state space, 370, 375, 384, 394, 398–399
  steepest ascent, 495
  transformation, 336, 340
  variable metric, 319, 351
minimization
  one-dimensional, 335, 367
  sequential, 280

minimum
  gauge, 427, 430, 435
  member sizes, 239
  unconstrained, 29
  weight design, 3
modularity, 426
modulus section, 597
moments of inertia, 441
mononomial, 600
monotonicity, 388
move limits, 357, 479, 603
multiple-loading conditions, 78

N-dimensional, 13
natural
  frequency, 89, 364, 396, 428, 430, 451, 458, 463
  frequency constraint, 548
  vibration modes, 427, 440
necessary condition for minimum, 25
neighbourhood, 30, 330, 549
neutral axis, 5
Newton
  algorithm, 303
  iteration, 351
  method, 39, 301, 303, 319, 329, 332, 340, 358, 368, 384
  type, 325, 604
  search, 271
Newton–Raphson
  algorithm, 123
  generalized, 258
  method, 118, 126, 131, 137, 161, 238, 263, 270, 277, 286, 367
  search, 239, 426
  technique, 238
nodal
  coordinates, 415
  displacements, 65, 73, 166
  forces, 74
non-linear equations, 358
non-linear programming problem, 275
normal modes, 89
normality conditions, 577, 585, 589, 605
normalized buckling mode, 80
number
  condition, 283, 339, 344

oblique projection operator, 321
optima
  identification, 36
  location, 36

optimal geometry, 513, 531
optimality
  conditions, 345
  conditions of second order, 352
  criterion, 106, 108, 110–111, 115–116, 121, 123–124, 126–128, 130, 160, 162, 239, 364, 370, 376, 378, 380, 384, 386–387, 394, 399, 487
  criterion (discrete), 489
  criterion (hybrid), 389
  criterion method, 99, 365, 605
optimization
  aerodynamic, 412
  aeroelastic, 407
  dynamic, 469
  flutter, 426
  method (dual), 346
  method (primal), 346
  non-linearly constrained, 516
  of fin, 471
  procedure, 129
  strength, 426
optimizer
  usable–feasible directions, 444
  usable–feasible directions, 444
optimizing point, 37
optimum design, 7
orthogonal
  operator, 350
  projection operator, 321
orthogonality, 15, 16, 303, 306, 308
  conditions, 576–577, 585, 589, 605
orthotropic sheet, 445

panels
  honeycomb sandwich, 427
  membrane, 441
pattern
  one point, 286
penalty
  function transformation, 344
  two-point, 287
  three-point, 293, 296
photo-elasticity, 548
pin-jointed framework, 3, 6, 43–44, 46, 57
pitch moment, 464
plate flat, 548
ply thickness, 149
point
  dual, 370
  interior, 336

polynomial
  approximation, 290
  cubic, 291
  fit, 287
  Hermite interpolating, 291
  homogeneous, 607
  Lagrange interpolating, 293
  orthogonal, 564
  quadratic, 293
  second order, 408
portal framework, 435
positive definite, 300
positivity conditions, 589
post-processor, 419, 420, 441–443
posynomial, 574–575, 580, 583, 592, 600, 604, 608
pre- and post-processors, 414
pre-processors, 419
principal minor, 90
problem
  adjoint boundary-value, 551
  approximate, 384, 389
  3-bar truss, 259, 264
  10-bar truss, 250, 269
  22-bar truss, 259, 264
  boundary value, 549
  column buckling, 561
  convex, 53, 275, 332, 355, 356
  dual, 31, 33, 35, 55, 57, 59–60, 80, 276–277, 279, 368–369, 530–531, 575, 577, 580–581, 591
  dual geometric, 579, 581, 587, 603, 605, 608
  dual relationship, 584
  linear, 306
  mixed variable, 398
  non-linear optimization, 428
  non-linear programming, 354, 429
  optimization, 25, 56, 275, 280, 299, 318, 325
  primal, 31, 33, 56–57, 276, 367, 530, 598
  quadratic, 306, 358
  quadratic programming, 332
  quasi-unconstrained, 279, 285
  separable, 386
  sequence of linear programming, 354
  spring–mass optimal design, 560
  unconstrained, 336
program
  convex, 489, 514, 523, 529
  primal, 583, 589, 601, 602, 605–606

programming
  approximation, 356
  complementary geometric, 592, 601
  condensed geometric, 604–605
  geometric, 573
  mathematical, 365, 370, 375, 384, 387
  non-linear mathematical, 365, 421
  recursive linear, 354–356
  recursive quadratic, 357
  separable, 384
  system, 425
projected
  Newton method, 326
  operator, 325, 327
pseudo-
  load, 77
  load system, 54
push-off factors, 353
pyramid
  3-bar, 535

Q-conjugate, 332
quadratic, 309
  approximation, 300, 302, 310, 358, 384
  convergence, 313
  function, 310, 313
  interpolation, 293, 296–297
  problem, 328
  programming problem, 275, 335
  termination, 276
quasi-
  Newton, 315, 319, 325, 327, 330, 338
  Newton method, 288, 311
  Newton updates, 359
  unconstrained problem, 275, 277, 319
  unconstrained sub-problem, 335

Rayleigh quotient, 566–577
Rayleigh–Ritz approach
  function of n variables, 17
  line, 13
  line origin, 14
  numbers, 11, 13, 14
  n-vectors, 14
  plane, 14
  real, 439
  vector space, 14
reciprocal
  areas, 51
  design variables, 102

recurrence
  design variables, 102
  relation, 106, 111, 121, 124–126, 128, 130, 132, 136, 140, 161, 166, 173, 179, 185, 192, 201, 209, 220, 227
  relation for multiple constraints, 111
reduction factor, 479
region
  convex, 50, 63, 384
  feasible, 35
  infeasible, 24
  multiply, 547
  non-convex, 50
  simply-connected, 547
regular, 30
reinforcing material, 479
response
  dynamic, 458, 463
  surfaces, 380
  transient, 451
restoration
  algorithm, 351
  phase, 351, 373, 375
  step, 350
rib, 467, 471
rigid-jointed framework, 3
roll control effectiveness, 408
round-off error, 270

saddle
  point, 32
  point conditions, 33, 346
safety margin, 427
sandwich panels, 7, 412
scaled weight, 113
scaling, 419
scaling
  parameter, 103, 104
scheme
  dual solution, 364
  primal solution, 370
search
  algorithm, 409, 441
  direction, 37, 280, 283, 350, 353, 385
  function, 36–37
  one-dimensional, 277, 554
  procedure, 409–410
  technique, 11
second moments of area, 3
second order term, 311
sensitivities, 85

sensitivity
  analysis, 441
  process, 84
separable programming problem, 276–277
sequences, 19, 20, 39
  accumulation points, 19
  cluster points, 19, 20
  converges, 19
  divergent, 19
  limit, 19
  limit of, 20
  limiting points, 19
  subsequence, 19
series
  first order Taylor, 371
set
  active, 285, 322, 327, 329
  belongs, 12
  boundary point, 13
  bounded, 20
  cartesian product, 13
  closed, 18
  closed interval, 18
  compact, 20, 529, 603
  contains, 12
  convex, 21–22, 275, 529
  convex combination, 21
  convex envelope, 22
  convex hull, 22
  direct product, 13
  disjoint, 13
  domain of definition, 17
  elements of, 12
  greatest lower bound, 20
  half-open interval, 18, 20
  includes, 12
  infimum, 20
  interior point, 18
  intersection, 12
  least upper bound, 20
  lower bound, 20
  neighbourhood, 18, 24
  null, 12
  open, 18
  open interval, 18
  point, 14
  supremum, 20, 593
  union, 13, 22
set theory, 11
shape optimization, 545
shear panels, 145, 428

single
  constrained problem, 106
  displacement constrained problem, 52
singular, 303
skeleton form, 425
slender bars, 78
software
  executive, 420
solution
  errors, 263
  sensitivity, 57
  vertex, 249
space
  design, 369
  dual, 346, 348, 368, 384, 386, 398, 495
  Euclidean n-space, 16
  feasible, 370
  primal, 384
  real n-space, 14
  spanned, 15
  subspace, 15
  three-dimensional, 16
span chord coordinates, 409
spar, 7, 464, 467, 471
stable, 8
stability, 2, 6
  constraint problem, 126–127, 258
  equation, 104
starting point, 36, 283
static equilibrium, 64
statical determinate pin-jointed framework, 72
statically
  determinate structures, 6, 43, 45, 57, 106, 390
  indeterminate structures, 64, 366
  indeterminate pin-jointed framework, 64, 72
stiffened
  cylindrical panels, 7
  cylindrical shells, 430
stiffner bending, 562, 564
stiffness
  constrained problems, 49
  matrix, 46, 65, 73, 78, 102–104
step length, 280, 285, 290, 294, 302, 320, 322, 326, 333, 353, 373, 385, 441
step size, 131, 142, 144
strain, 73
  displacement matrix, 102
  displacement relation, 104

energy, 102, 123
energy density, 58, 67, 79, 123
strength, 412
  critical framework, 46, 49, 66
  requirements, 396
stress, 548
  bending, 596, 597
  concentration, 73, 546
  constrained problem, 119, 121, 123
  constraint design, 152
  constraints, 6
  distribution, 5
  distribution uniform, 546
  Huber-von Mises, 432
  limits, 46
  maximum, 73
  ratio method, 252, 389
  resultants, 73
  strain matrix, 102
  strain relation, 104
  tangential normal, 547
  tensor deviator, 543
  vector, 76
stress ratioing, 8, 47, 380, 381, 383
stresses, 3, 4, 83, 239, 274, 428
  bending, 432
  local buckling, 424
stringer
  cross-sections, 446
structural analyser, 414
structural analysis
  analysis, 2, 4, 64, 380, 409, 419, 421
  elements, 440, 490
  geometry, 514
  mass, 3, 408, 427
  optimization, 4
  parameters, 3
  statically determinate, 237
  symmetry, 128
  weight, 3
structure
  determinate, 119
  hyperstatic, 378
  rod panel, 428, 429
subdomain, 370
subproblem, 592, 600
  geometric, 601, 606
subset, 12
subspace, 332
sufficient condition, 25
supercritical airfoil, 412

superlinear, 245
  rate, 346
symmetric rank-one formulae, 313
symmetric conditions, 129
system stability, 100

tangent, 281
  plane, 350, 351, 373–374, 378, 381, 390
Taylor expansion, 330, 421, 441, 524
Taylor series extrapolation, 440, 489, 491
technique
  condensation, 439
  curve fitting, 286
  linearization, 335
  penalty function, 418, 426
  usable-feasible direction, 418
tension, 6
termination criteria, 37–38, 412, 419, 420
terminator, 414, 419, 420
theorem
  expanding subspace, 308
thin-walled components, 78
tolerance, 330
topology of the structure, 514
torsion, 390
  bending coupling, 435
  box, 435
torsional
  frequency, 395
  mode, 390
  stiffness, 431, 547
tower
  39-bar, 538
transient response, 92, 96
Tresca, 76
trusses, 6, 7
  2-bar, 534
  3-bar, 164
  10-bar, 134, 137, 142, 375, 381, 510
  63-bar, 388
  72-bar, 386, 510
  200-bar, 142
  statically determinate, 365
  structures, 487
twist, 155
  constraint design, 153

uniform strength, 5
unimodal search, 296

unit
  dummy load, 46
  loads, 72, 75, 77
unstable, 6
update
  inverse DFP, 317
  rank one, 313
  rank two, 315
upper bound, 4

values
  starting, 430
variables
  active design, 366
  adjoint, 551
  complex, 547
  configuration design, 349
  decision, 26
  dependent, 26
  discrete design, 386, 397
  duals, 57, 59, 68, 70, 276–277, 380, 385, 394, 398, 495, 578, 581, 583, 585–586, 589, 603
  geometry, 434
  gradient, 30
  independent, 26, 491, 521
  integer, 398
  inverse, 59, 373, 523
  linking, 419, 489, 515
  local, 446–447
  maximum size, 115
  minimum, 115
  mixed discrete continuous design, 398
  passive design, 336, 368, 369
  primal, 279, 495
  reciprocals, 48, 57, 71, 374–375, 381, 419, 428, 458, 468, 479, 491
  slack, 347
  state, 26
  transformed dual, 599
variation boundary, 547
variational formulation, 548
vertex
  non-optimal, 355
  solutions, 47, 355
vectors, 15, 25
  base, 320
  basis, 16
  column, 16
  distance, 16

dual gradient, 495
field of deformation, 548
function, 17
gradient, 308
length, 16, 318
linear combinations, 15, 16
linear dependent, 15
linear independent, 30
linear space, 14
load, 85, 101, 406, 517
negative gradient, 308
orthogonal, 15, 16
projected, 321
row, 16
row gradient, 328
space, 15
spanning, 15
subspace, 15
unit normal, 548
velocity, 483
field, 549, 550
Venn diagram, 12
vibration
frequencies, 83
mode, 89, 91–92, 427–428
small amplitudes, 560
vibrational mode, 435
virtual
displacement vector, 102, 106, 121
load, 103, 106, 122–123, 166, 365, 378, 380, 389, 394
load vector, 111, 113, 115, 121, 124, 518
memory, 437
nodal displacements, 70, 518
strain energy, 57, 77, 106, 113, 122, 124–125
strain energy density, 106–107, 111, 365, 375, 394
strains, 70, 546
work, 518
Von Mise, 76

webs, 7, 464, 598
web flexion, 390
weights
non-negative, 573
wing
carry-through box, 388
deflections, 408
delta, 395, 398
flutter, 408
flutter analysis, 443
flutter resizing, 427
structure, 149
optimization, 471
supersonic transport aircraft, 412
vibratory, 427

yield criterion, 76

zigzag, 354
zigzagging, 329, 330

RAYMOND H. FOGLER LIBRARY
DATE DUE

**BOOKS ARE SUBJECT TO
RECALL AFTER TWO WEEKS**